A PEARSON AUSTRALIA CUSTOM BOOK

ELEN1000
Electrical Systems

This custom book is compiled from:

ELECTRICAL ENGINEERING
CONCEPTS AND APPLICATIONS
ZEKAVAT

CURTIN UNIVERSITY

Suite 1001 Level 10
151 Castlereagh St
Sydney NSW 2000
Australia
Ph: 02 9454 2200
www.pearson.com.au

Project Management Team Leader:	Jill Gillies
Custom Specialist:	Lucie Bartonek
Project Manager:	Linda Chryssavgis
Production Controller:	Karen Young

ISBN: 978 1 4886 1050 9

TABLE OF CONTENTS

ABOUT THIS CUSTOM BOOK

Welcome to **ELEN1000 Electrical Systems**.

The material included in this custom book has been specifically chosen to meet your course requirements.

Please be aware that chapter, section and page numbers from the original source text *Electrical Engineering: Concepts and Applications* by Zekavat still appear in this book. The reference for this source text is:

Zekavat, S.A.R. (2013). *Electrical Engineering: Concepts and applications*. Upper Saddle River, NJ: Pearson Higher Education, Inc.

The Table of Contents refers to the page numbers of this book, not the source text. These page numbers also appear in the Navigation Bar at the top of each page.

This custom book also includes a customised index. The page references in the index refer to the page number of the custom book found in the Navigation Bar, not the source text page numbers.

Navigation Bar – Use this information to navigate through this custom book. The page numbers here run continuously from beginning to end.

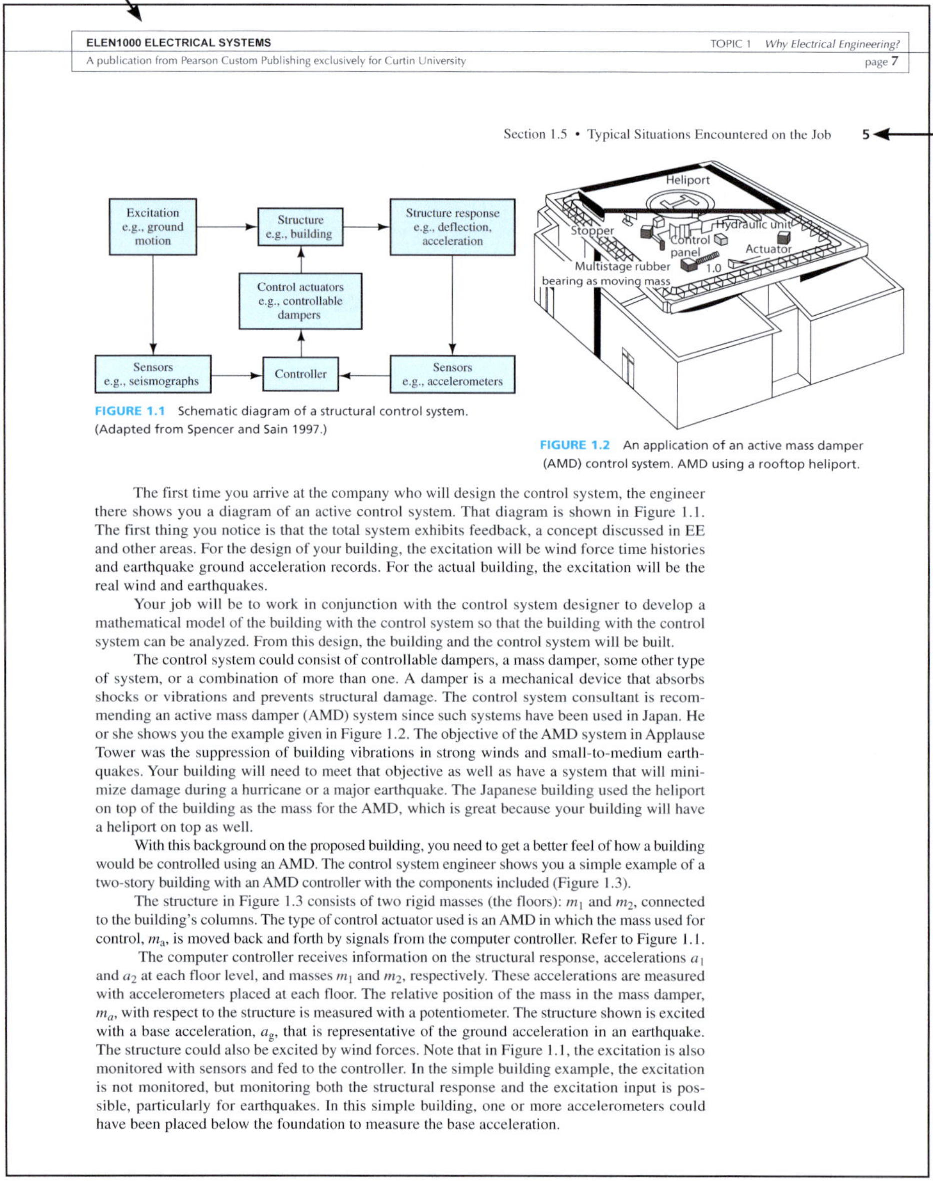

Source Book Page Number – This is the original page number.

ELEN1000 ELECTRICAL SYSTEMS — TOPIC 1 *Why Electrical Engineering?*

A publication from Pearson Custom Publishing exclusively for Curtin University — page 7

Section 1.5 • Typical Situations Encountered on the Job 5

FIGURE 1.1 Schematic diagram of a structural control system. (Adapted from Spencer and Sain 1997.)

FIGURE 1.2 An application of an active mass damper (AMD) control system. AMD using a rooftop heliport.

The first time you arrive at the company who will design the control system, the engineer there shows you a diagram of an active control system. That diagram is shown in Figure 1.1. The first thing you notice is that the total system exhibits feedback, a concept discussed in EE and other areas. For the design of your building, the excitation will be wind force time histories and earthquake ground acceleration records. For the actual building, the excitation will be the real wind and earthquakes.

Your job will be to work in conjunction with the control system designer to develop a mathematical model of the building with the control system so that the building with the control system can be analyzed. From this design, the building and the control system will be built.

The control system could consist of controllable dampers, a mass damper, some other type of system, or a combination of more than one. A damper is a mechanical device that absorbs shocks or vibrations and prevents structural damage. The control system consultant is recommending an active mass damper (AMD) system since such systems have been used in Japan. He or she shows you the example given in Figure 1.2. The objective of the AMD system in Applause Tower was the suppression of building vibrations in strong winds and small-to-medium earthquakes. Your building will need to meet that objective as well as have a system that will minimize damage during a hurricane or a major earthquake. The Japanese building used the heliport on top of the building as the mass for the AMD, which is great because your building will have a heliport on top as well.

With this background on the proposed building, you need to get a better feel of how a building would be controlled using an AMD. The control system engineer shows you a simple example of a two-story building with an AMD controller with the components included (Figure 1.3).

The structure in Figure 1.3 consists of two rigid masses (the floors): m_1 and m_2, connected to the building's columns. The type of control actuator used is an AMD in which the mass used for control, m_a, is moved back and forth by signals from the computer controller. Refer to Figure 1.1.

The computer controller receives information on the structural response, accelerations a_1 and a_2 at each floor level, and masses m_1 and m_2, respectively. These accelerations are measured with accelerometers placed at each floor. The relative position of the mass in the mass damper, m_a, with respect to the structure is measured with a potentiometer. The structure shown is excited with a base acceleration, a_g, that is representative of the ground acceleration in an earthquake. The structure could also be excited by wind forces. Note that in Figure 1.1, the excitation is also monitored with sensors and fed to the controller. In the simple building example, the excitation is not monitored, but monitoring both the structural response and the excitation input is possible, particularly for earthquakes. In this simple building, one or more accelerometers could have been placed below the foundation to measure the base acceleration.

TOPIC 1

Why Electrical Engineering?

The content of this topic is compiled from:

Chapter 1 (pp. 1–12), *Electrical Engineering: Concepts and Applications*

By S.A. Reza Zekavat

CHAPTER 1

Why Electrical Engineering?

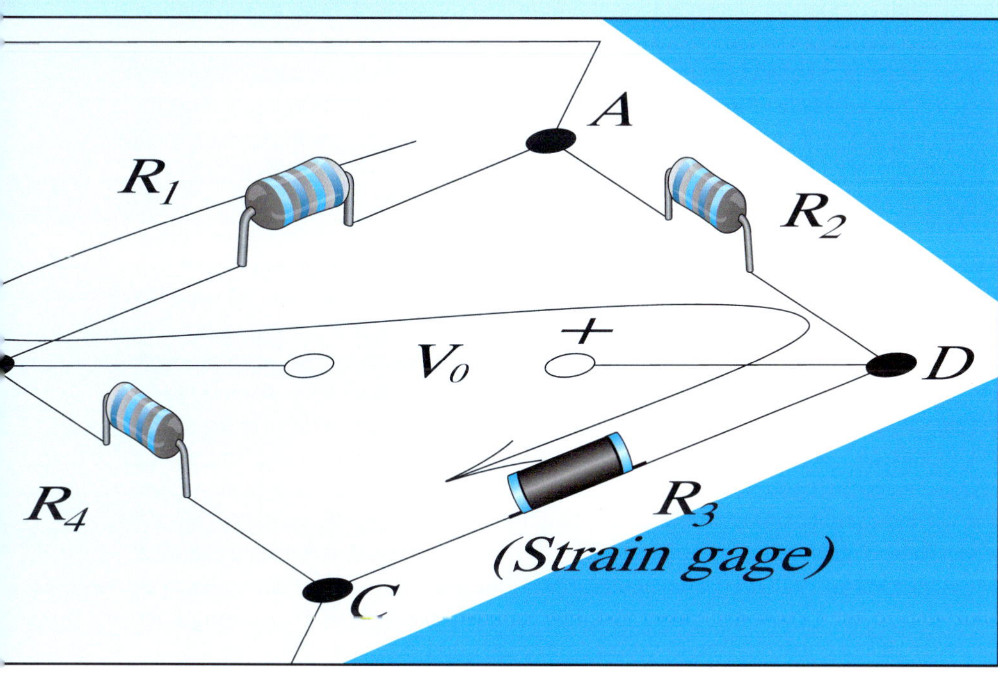

1.1 INTRODUCTION

If you are reading these words, then you are probably an engineering student who is about to take a course in electrical engineering (EE), or possibly an engineer who wants to learn about EE. In either case, it is safe to say that you are *not* majoring in EE nor are you already an electrical engineer. So, why are you doing this to yourself?

As an engineering student, there are two likely possible reasons: (1) You are being forced to because it is required for your major, and/or (2) you believe that it will help you pass the Fundamentals of Engineering (FE) examination that you will take before you graduate or shortly thereafter. If you are already an engineer, then you are likely reading this book because you need to learn EE for the FE exam that you put off until after graduation, or you need to learn EE to perform your job better. Studying EE because it is required for you to graduate or because you want all the help you can get to pass the FE exam are both laudable reasons. But, the second possible reason mentioned earlier for our hypothetical practicing engineer needs to be considered further. Can learning EE help you in your engineering career? The short answer is, yes. The long answer follows.

1.2 ELECTRICAL ENGINEERING AND A SUCCESSFUL CAREER

As a practicing engineer, you will work on projects that require a wide range of different engineers and engineering disciplines. Communication among those engineers will be vital to the successful completion of the project. You will be in a better position to communicate with the engineers working on electrical systems of all sorts if you have a basic background in EE. Certainly—through this course alone—you will not be able to design complicated electrical systems, but you will be able to get a feel of how the system works and be better able to discuss the implications of areas where the non-EE system you are designing and the electrical system overlap. For example, mechanical engineers often design packages for electronic systems where heat dissipation due to electronic components can be a major problem. In this instance, the non-EE engineer should be able to help the EE with component placement for optimum heat dissipation. In short, no engineer works in isolation and the more you can communicate with other engineers the better.

The company that hires you out of engineering school understands how important communication is. Thus, they will most likely have training programs that help their engineers learn more about the specific engineering that they will perform as well as other engineering disciplines with which they will be associated. If you have taken EE as an engineering student, then you will have a good foundation for learning EE topics specific to that company, which will make your on-the-job training easier and, consequently, less expensive for your employer. Saving money for your employer is always a good thing. So, by having taken an EE course, you will be a more promising hire for many companies.

In addition, there will be instances in your engineering career where you will be working directly with electrical or electronic components that you need to understand in some depth. For example, many engineers work in manufacturing processing and will need to work with products that have electrical/electronic content. Likewise, engineers often work with systems that used to be mechanical, but are now electronic (e.g., electronic fuel injection, electronic gas pedals). In the course of your work, you may also need to perform tests in which the test apparatus uses a Wheatstone bridge. If that is the case, then you need to know how a Wheatstone bridge, which is an electric circuit, works to use the equipment adequately. In addition, most mechanical measurements involve converting the mechanical quantity to an electrical signal. Finally, if you need to purchase electrical components and equipment you will need a fundamental background in EE to talk to the vendor in an intelligent manner and get the type of equipment that your company needs. Thus, by knowing some EE, you will be better able to obtain and use electrical components and electrical equipment.

Another reason for learning the principles and practices of EE is that you may be able to make connections between your engineering discipline and EE that lead to creative problem solutions or even inventions. For instance, maybe your job will require you to monitor a system on a regular basis that requires you to perform a significant number of tedious by-hand techniques. Your familiarity with the monitoring process, combined with your background in EE, might allow you to teach yourself enough in-depth EE to design and build a prototype monitoring system that is faster and less hands-on. This type of invention could lead to a patent or could lead to a significant savings in monitoring costs for your company. In this scenario, you would have been able to do the work yourself and would thus gain ownership of your work and ideas, that is, the design and fabrication of a prototype monitoring system. Learning EE (as well as other engineering fundamentals outside your specific discipline) may allow you to make connections that could lead to creative solutions to certain types of engineering problems.

In conclusion, studying EE will not only help you pass the FE exam, but it will make you more marketable, give you capabilities that will enhance your engineering career, and increase your self-confidence, all of which may allow you to solve problems in ways you cannot now imagine.

1.3 WHAT DO YOU NEED TO KNOW ABOUT EE?

Electric circuits are an integral part of nearly every product on the market. In any engineering career, you will need a working knowledge of circuits and the various elements that make up a circuit, including resistors, capacitors, transistors, power supplies, switches, and others. You need to know circuit analysis techniques and by learning these gain an understanding of how voltage, current, and power interact. You will likely be required to purchase equipment during your career, so you will need to learn how to determine technical specifications for that equipment. Working with electrical equipment exposes you to certain hazards with which you must be aware. Thus, you will need to learn to respect electrical systems and work with them safely.

Although engineers in all disciplines are expected to understand and use electrical systems, power sources, and circuits in many job assignments, expert knowledge is not required. For example, many non-EEs are plant managers and are called upon to manage heating and air conditioning (HVAC) systems. While the engineer will not be asked to design the system or its components, a basic EE knowledge is useful for day-to-day management. The practicing engineer must be able to design and analyze simple circuits and be able to convey technical requirements to vendors, electricians, and electrical and computer engineers.

A typical on-the-job application is data acquisition from temperature, pressure, and flow sensors that monitor process or experimental equipment. Process control and monitoring situations in plants and refineries require working knowledge of data acquisition and logging, signal processing, analog-to-digital (A/D) conversion, and interfacing valves and other control devices with controllers. In-line, real-time chemical analysis is sometimes necessary, as well as monitoring process temperatures, pressures, and flow rates with in-line sensors.

Familiarity with power generation and general knowledge of generators, electric motors, and power grids is also beneficial to engineers. A major job objective is frequently to reduce utility expenses, primarily electricity costs. For example, EE knowledge is necessary to design and operate cogeneration systems for simultaneous production of heat and electricity. In such situations, high-pressure steam can be throttled through a turbine-generator system for power production, and the lower-pressure exhaust steam is available for plant use. Alternatively, natural gas can be combusted in a gas turbine to generate electricity, and the hot gas exhaust can make steam in a boiler. One issue is how to operate the system to match electricity use patterns in the plant.

Process engineers also need to recapture energy (as electricity) from process streams possessing high thermodynamic availability, that is, streams at high pressure and/or temperature. Often, this can be done by putting the process stream through an isentropic expander (turbine) and using the resulting shaft work to operate an electric generator.

Electrochemistry involves knowledge and use of potentiostats, battery testing equipment, cyclic voltammetry measurements, electrode selection, and electrochemical cells. Electroplating operations are also of interest to chemical engineers. A background in EE will help you understand these and other related processes.

1.4 REAL CAREER SUCCESS STORIES

The bottom line for engineers is frequently the economic consequence of operating a process. The profit motive is paramount, with safety and environmental considerations providing constraints in operation. Electrical engineering principles often directly affect a process's profitability and operability. Learning and applying the concepts in this textbook may help you get noticed (favorably) in a future job by saving money for your company.

Consider the case of a chemical engineering graduate who began work a few years ago in a major refinery that had recently implemented a cogeneration system that produced steam and electric power simultaneously. Generation of high-pressure steam in a natural gas fired boiler, followed by expansion of the steam through a turbine, produced shaft work that was used to operate a generator. For internal plant use, this generated electricity was valued at the retail electricity price. (External sale of excess electricity is regulated by the Public Utility Resource Power Act (PUPA), and the price is the cost the utility company incurs to make incremental electricity, i.e., the utility company's "avoided cost.") The new employee did an economic analysis, looking at the trade-off between the equipment capital investments and operating costs versus the anticipated electricity savings. The bottom line is that the employee's recommendation was to "Turn it off!" since the cogeneration scheme was losing money. This result was not popular with the refinery's management (at first), but the employee was right on target with the analysis and recommendation. Saving money for the refinery provided a jump-start to a very successful career. Working knowledge of power generation cycles and equipment was necessary to do this critical analysis.

1.5 TYPICAL SITUATIONS ENCOUNTERED ON THE JOB

The case studies in this section are intended to illustrate, with discipline-relevant projects, how the general principles presented in this textbook may be applied in a real-life job or research setting. After reading the descriptions, you should better understand where, how, and why electrical engineering fits into an engineer's job. Until you have completed this course, some of the terms in these case studies may be unfamiliar. This illustrates the importance of completing this course prior to encountering these situations in real, on-the-job situations. For more detailed information on the equipment and processes described in the case studies, the interested reader can view the PowerPoint® presentations that accompany this textbook.

1.5.1 On-the-Job Situation 1: Active Structural Control

Imagine that you are a young civil engineer in Charleston, SC, who is working on the design and construction of a 15-story building that will have motion-sensitive equipment in it and must be able to withstand both hurricane winds and earthquakes. (Charleston has a history of strong earthquake events.) Since the building motion must be controlled during both moderate wind and earthquake events as well as during hurricanes and a major earthquake, active structural control is required. Passive structural control systems may be used in conjunction with active control systems, but the scenario described here pertains to an active control system.

You have been chosen by your boss to be the liaison between the company that will design and install the active control system and your company. You will need to give them information about the building that will allow them to design the control system, and you will need to feed information from them back to engineers in your company who are designing the building. If the building is being designed in the manner described, then the design process is a simple feedback loop: the initial building design affects the initial design of the control system, which in turn affects the building design, which in turn affects the control system design, and so on until both designs are compatible. So it is clear that you must have a reasonable understanding of the building design (your field) and a reasonable understanding of the active control system (not your field—the EE concepts are found here).

In general, structural control is the control of dynamic behavior of structures such as building and bridges. This type of control becomes important when the structure is relatively flexible, such as tall buildings and long-span bridges, or is sensitive to damage, such as historic buildings in earthquake regions. Engineers want to control structures: (1) to prevent damage and/or occupant discomfort during typical events, relatively high wind, and small earthquakes, and (2) to prevent collapse of the structure during large events, for example, major earthquakes. In your case, the building contents are vibration sensitive and the building is susceptible to damage from hurricanes and earthquakes.

Structural control is performed in one of three ways: passive control, active control, or hybrid control. *Passive control* is accomplished using the mass, stiffness, and damping built into the system. A passive control system cannot be readily altered. *Active control*, which is the method we will examine in this case study, is accomplished using control actuators, which require external energy, to modify the dynamic behavior of the system. An active control system can be adaptive. *Hybrid control* is simply a combination of passive and active control.

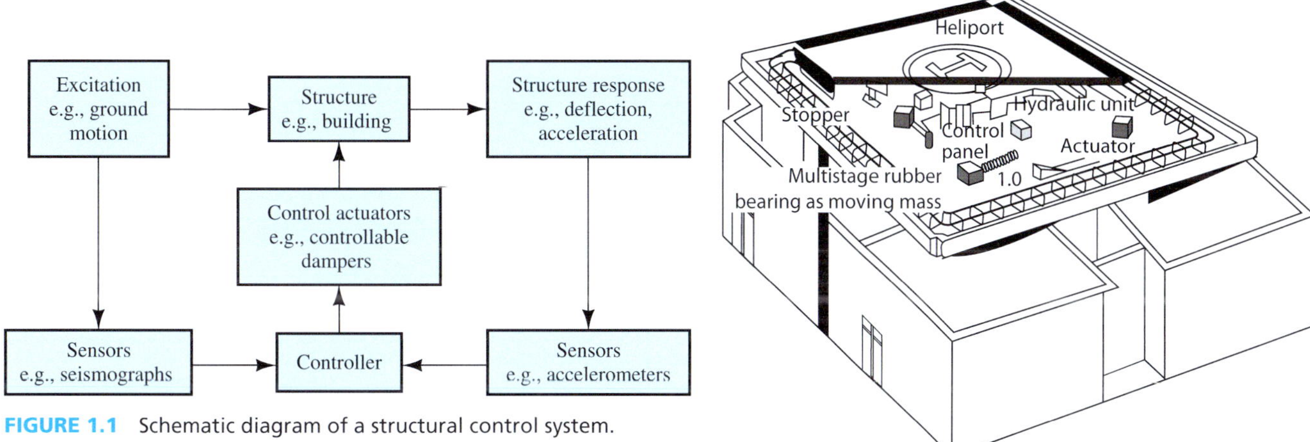

FIGURE 1.1 Schematic diagram of a structural control system. (Adapted from Spencer and Sain 1997.)

FIGURE 1.2 An application of an active mass damper (AMD) control system. AMD using a rooftop heliport.

The first time you arrive at the company who will design the control system, the engineer there shows you a diagram of an active control system. That diagram is shown in Figure 1.1. The first thing you notice is that the total system exhibits feedback, a concept discussed in EE and other areas. For the design of your building, the excitation will be wind force time histories and earthquake ground acceleration records. For the actual building, the excitation will be the real wind and earthquakes.

Your job will be to work in conjunction with the control system designer to develop a mathematical model of the building with the control system so that the building with the control system can be analyzed. From this design, the building and the control system will be built.

The control system could consist of controllable dampers, a mass damper, some other type of system, or a combination of more than one. A damper is a mechanical device that absorbs shocks or vibrations and prevents structural damage. The control system consultant is recommending an active mass damper (AMD) system since such systems have been used in Japan. He or she shows you the example given in Figure 1.2. The objective of the AMD system in Applause Tower was the suppression of building vibrations in strong winds and small-to-medium earthquakes. Your building will need to meet that objective as well as have a system that will minimize damage during a hurricane or a major earthquake. The Japanese building used the heliport on top of the building as the mass for the AMD, which is great because your building will have a heliport on top as well.

With this background on the proposed building, you need to get a better feel of how a building would be controlled using an AMD. The control system engineer shows you a simple example of a two-story building with an AMD controller with the components included (Figure 1.3).

The structure in Figure 1.3 consists of two rigid masses (the floors): m_1 and m_2, connected to the building's columns. The type of control actuator used is an AMD in which the mass used for control, m_a, is moved back and forth by signals from the computer controller. Refer to Figure 1.1.

The computer controller receives information on the structural response, accelerations a_1 and a_2 at each floor level, and masses m_1 and m_2, respectively. These accelerations are measured with accelerometers placed at each floor. The relative position of the mass in the mass damper, m_a, with respect to the structure is measured with a potentiometer. The structure shown is excited with a base acceleration, a_g, that is representative of the ground acceleration in an earthquake. The structure could also be excited by wind forces. Note that in Figure 1.1, the excitation is also monitored with sensors and fed to the controller. In the simple building example, the excitation is not monitored, but monitoring both the structural response and the excitation input is possible, particularly for earthquakes. In this simple building, one or more accelerometers could have been placed below the foundation to measure the base acceleration.

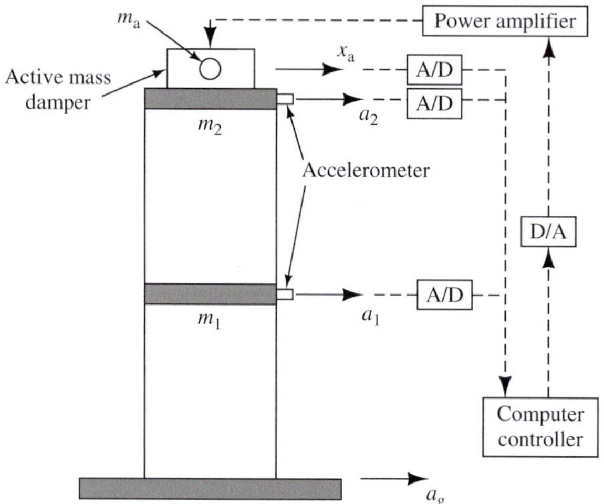

FIGURE 1.3 Simple building structure and AMD control system.

As you will note from Figure 1.3 and its description, there are a number of components that you will not understand without a basic EE knowledge. For example: What do the terms D/A and A/D mean? How does an accelerometer work? Why do you need an amplifier? How does the computer controller do its job? What is a potentiometer? Without an introduction to EE—like the one provided in this text—you will have difficulty in a job setting like this, and your learning curve will be much steeper, making your life more difficult. Note: If you would like to know the explanation now to the topics introduced in this example, see Chapter 11 of this book.

1.5.2 On-the-Job Situation 2: Chemical Process Control

In this scenario, imagine that you are a chemical engineering graduate who has just taken a job with a petroleum company in Baton Rouge, LA. Dwindling reserves of easy-to-access petroleum, as well as government subsidies for alternative fuel ventures, has led your company to diversify into fuel-grade ethanol production.

The process of interest at your plant (see Figure 1.4) is the application of corn (grain) fermentation to produce a dilute aqueous solution of ethyl alcohol that is further purified by distillation. Corn, sugar, yeast, water, and nutrients are continuously fed into a fermenter, which produces CO_2, spent yeast and grain, and a dilute (~8%) ethanol–water solution. This slurry is filtered and clarified to remove the yeast and grain before being sent to a preheater (see Figure 1.5). The heated ethanol–water solution is then fed to a distillation column that produces a 95% to 96% pure ethanol product as its overheads. Due to the large amount of energy required to separate the ethanol and water, it is desired to keep the temperature of the dilute feed elevated. Plant data, however, reveal that the feed is currently well below its boiling temperature.

Your plant manager intends to make the process profitable, so your first major assignment is a blanket imperative to "Reduce costs!" wherever possible. Since this is your first week on the job, you want to do a superlative job without panicking (or damaging any equipment). Thinking about the problem, you remember from school that it is more economical to preheat the distillation column feed than to introduce it into the column cold. You therefore decide to further preheat the temperature of the feed stream to its boiling point.

Preheating of the dilute ethanol–water feed in a heat exchanger (using Dowtherm®, an industrial heat transfer fluid produced by Dow Chemical) would significantly reduce the amount of energy needed for the separation, reducing the cost of operation (i.e., reducing the reboiler steam heating duty). The easiest approach would be to heat the feed at a constant rate. However, several

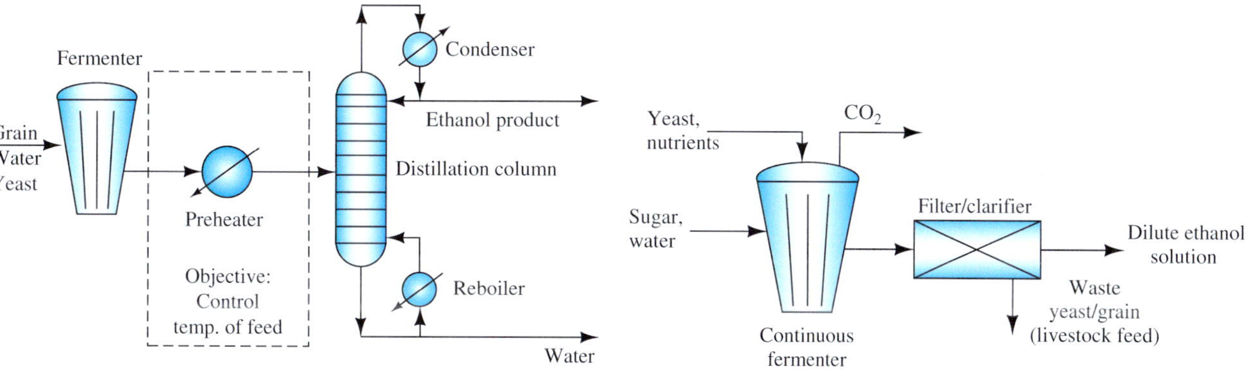

FIGURE 1.4 Ethanol distillation flow diagram.

FIGURE 1.5 Filtering of fermenter products.

key variables of the feed stream, including temperature, flow rate, and ethanol concentration are subject to change, requiring a variable preheating rate. The rate at which the feed is preheated is also affected by the temperature of the Dowtherm®, which may also vary with time. All of these factors make the heat load on the preheater vary with time.

Knowing these considerations, you decide to carefully control the rate at which the feed is preheated to lower operating costs (i.e., lower the reboiler duty and Dowtherm® consumption). Applying a feedback control strategy is your choice of an efficient way to regulate the column's feed temperature.

A basic feedback system is seen in Figure 1.6. A variable of a stream exiting a process is measured and sent to a controller. The controller compares the signal with a predetermined set point and determines the appropriate action to take. A signal is then sent from the controller to a piece of equipment that changes a variable affecting the process, resulting in a measured variable that is closer to the set point. For our case study, the temperature exiting the preheater is monitored and compared with a predetermined set point. A computer then determines what action must be taken, and adjusts the flow of Dowtherm® to the preheater, raising the temperature of the column feed as needed. Figure 1.7 shows the basic setup for the preheater.

The control scheme you wind up choosing for the process consists of five major components:

1. A thermocouple that measures the temperature of the feed exiting the preheater and produces an analog signal (TC) (Figure 1.8)
2. An analog-to-digital signal converter (A/D)
3. A proportional controller (CPU) that compares the feed temperature with a predetermined set point

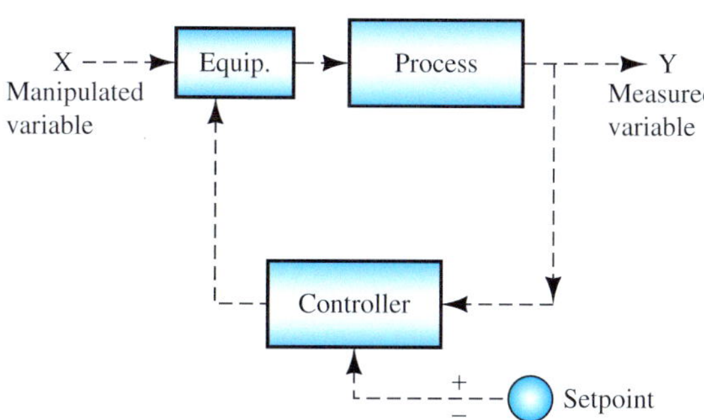

FIGURE 1.6 Basic feedback control strategy.

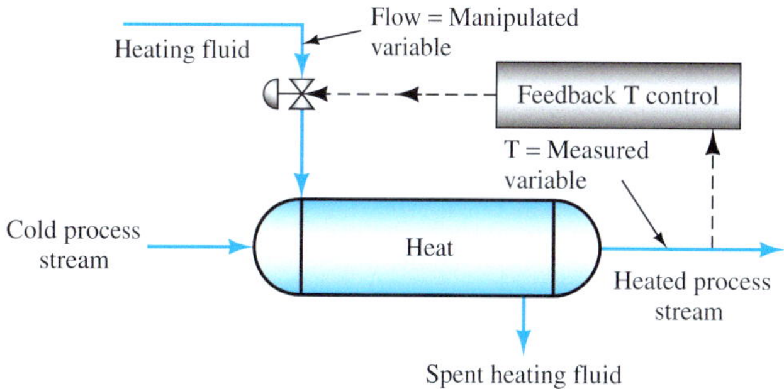

FIGURE 1.7 Schematic diagram of a generalized preheater.

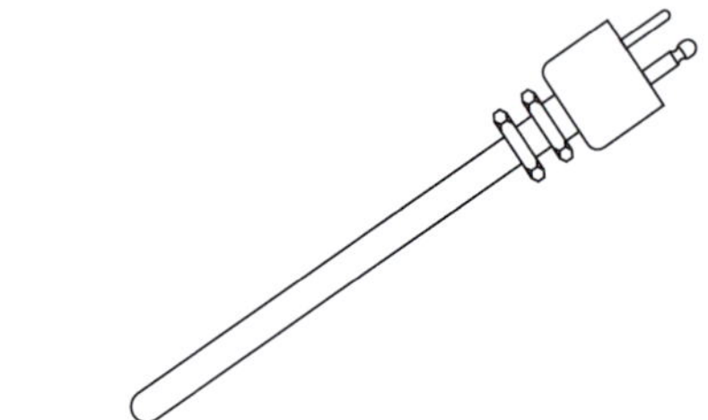

FIGURE 1.8 A thermocouple.

FIGURE 1.9 Pneumatic control valve (© Alexander Malyshev/Alamy).

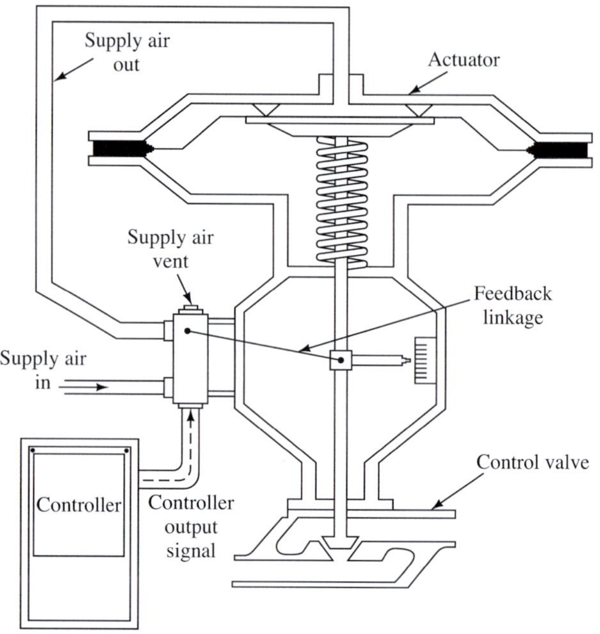

FIGURE 1.10 Diagram of a pneumatic control valve.

4. A digital-to-analog signal converter (D/A)
5. A pneumatic diaphragm control valve (Figures 1.9 and 1.10)

The dilute column feed to the ethanol separator is fed to the preheater where its temperature is increased by thermal contact with the Dowtherm®. As the feed exits the preheater, the thermocouple (TC) measures its temperature and emits an analog voltage signal (see Figure 1.11). The A/D converts the analog signal to a digital signal, and sends it to the CPU where the temperature is compared to a predetermined set-point temperature.

The CPU produces a digital signal proportional to the error (difference in temperatures), which is converted back to an analog signal by the D/A. This analog signal is interpreted as an analog air signal, typically scaled to 3 to 15 psig. The air pressure signal adjusts the valve stem position on the pneumatic diaphragm control valve, manipulating the amount of Dowtherm® flowing in the preheater. By keeping the measured temperature error below a specified tolerance, the temperature of the feed to the column is kept at its optimum and costs are kept down.

Net profit goes up as a result, and you get a big pat on the back—and maybe a big raise!

Clearly, to do this on-the-job assignment, you must be able to apply EE skills and knowledge in a plant environment. As this scenario illustrates, you must often define the problem, select a solution strategy, and pick equipment to implement it. The interested reader is encouraged to examine this textbook's related topics (see Table 1.1). Chapter 11 explains many fundamentals required for understanding the basics of a PC-controlled system.

1.5.3 On-the-Job Situation 3: Performance of an Off-Road vehicle prototype

A mechanical engineer working on the design and development of an off-road vehicle, SUV, or snowmobile is likely to be assigned the task of monitoring the field performance of a prototype during prescribed maneuvers, either to confirm that it is operating within design specifications, or to identify and troubleshoot malfunctions. A sample of vehicle performance features is given below:

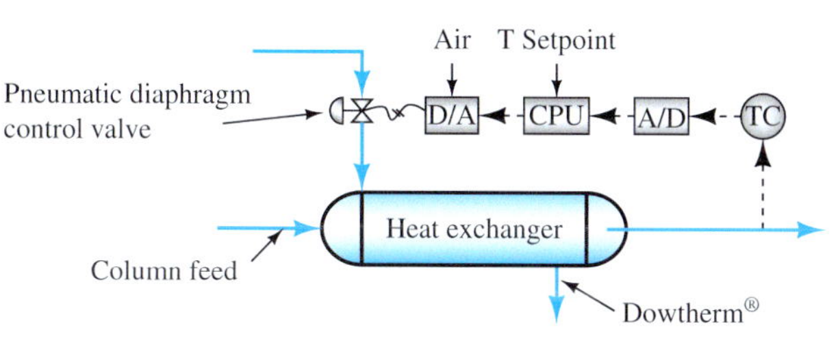

FIGURE 1.11 Detailed control scheme for situation 2.

TABLE 1.1 EE Connections to Process Control

Case Study	Textbook Topic
Pneumatic Valve[a]	Analog Actuator
Thermocouple[a]	Sensors Data Acquisition
CPU	Microprocessor Control Feedback Control
Digital and Analog Signal	A/D and D/A Conversion

[a] PowerPoint presentations for these topics accompany this textbook.

Engine

- Dynamic temperature in piston, connecting rod, or cylinder wall
- Dynamic pressure in combustion chamber
- Dynamic stress in piston, connecting rod, or cylinder wall
- Temperature fluctuations in water or oil
- Output torque fluctuations
- Bearing clearance (related to lateral load)

Suspension and Drive Train

- Stress in suspension elements
- Deflections and clearances in suspension elements
- Windup and backlash in gear train
- Temperature during braking

Vehicle Interior

- Vibration of steering wheel, dash, mirrors, and floor
- Sound pressure level at driver's ear or passenger's ear
- Vibration at passenger's seat
- Driver's head–neck vibration

Even though your contribution to the design itself may be "purely" mechanical (a rarity!), as a mechanical engineer you are expected to interface strongly with other disciplines to *evaluate the actual performance of that design*. The mechanical engineer would be expected to specify the appropriate measurements, and even to select the sensors that will perform reliably in this demanding environment. Because each sensor has its own particular electrical requirements, you must work closely with a technical staff of instrumentation, computer, and electronic professionals capable of advising you, describing your options, and implementing your decisions on these options. This will require that you have a reasonable understanding not only of the vehicle's design but also the terminology and electrical features of the instrumentation used by your staff.

In this situation, you are part of the engineering team engaged in testing an SUV prototype during field tests, to ensure that the interactions between the shock tower and the shock absorber are within specifications.

The shock tower provides the attachment between the shock absorber (as seen at its threaded end in Figure 1.12) and the frame supporting the shock tower. In your particular assignment, it might be necessary to measure: (a) the stress history in the shock tower (fatigue life); (b) the history of the load transmitted through the shock absorber (attachment life); or (c) the history of the relative displacement between the shock absorber shaft and the shock tower (blowout of the rubber washer).

In each case, you are trying to create a voltage analog—a voltage signal proportional to the measured phenomenon—of the stress/load/displacement history. The voltage form is usually most desirable because it can be easily sampled and stored in files for computer-aided analysis.

FIGURE 1.12 Front shock tower. (Photo courtesy of Rob Robinette.)

STRESS IN THE SHOCK TOWER

The force exerted on the shock tower by the shock absorber causes the shock tower to deform slightly, in proportion to that load. If, at a key location on the tower, the strain associated with that deformation can be sensed, it could be used to produce a voltage proportional to that strain, or to the stress at that location. This is accomplished using a device called a strain gage, a fine metal wire, or foil, which is glued to the chosen point on the surface of the shock tower (see Figure 1.13).

As the surface beneath the strain gage stretches or contracts, the resistance of the wire increases or decreases in proportion to that strain. This very small change of resistance is measured using a Wheatstone bridge that is especially adapted to measure extremely small *changes* in resistance while ignoring the relatively large basic resistance of the gage. Figure 1.14 represents the Wheatstone bridge. A Wheatstone bridge (discussed in Chapter 11) works like a balance scale in that an unknown quantity is measured by comparing it with a known quantity.

The output of the Wheatstone bridge is a voltage proportional to the change of resistance of the deforming strain gage. This strain is, in turn, proportional to the stress at the attachment point of the gage.

LOAD ON THE SHOCK TOWER

This can be measured through a specially designed load cell attached to the shock tower, with one side fastened to the shock absorber shaft and the other fastened to the tower structure. Any load exerted by the shock absorber on the shock tower must pass through the load cell, which produces a voltage proportional to that load. This can be accomplished using a transducer containing a piezoelectric crystal, a crystal that develops a charge proportional to its deformation under load (see Figure 1.15). Chapter 11 discusses sensors that use different properties including piezoelectric.

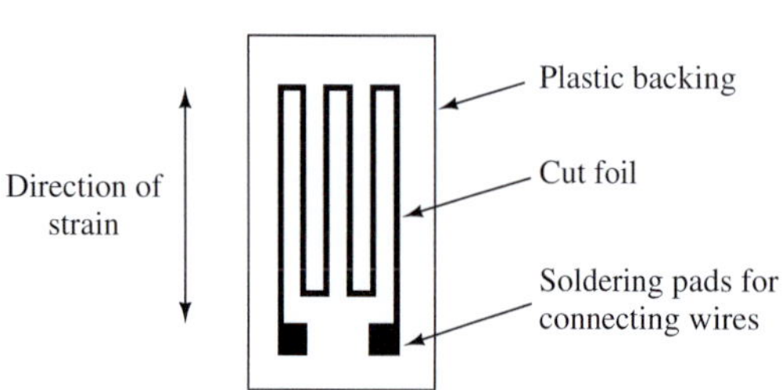

FIGURE 1.13 Electrical resistance strain gage.

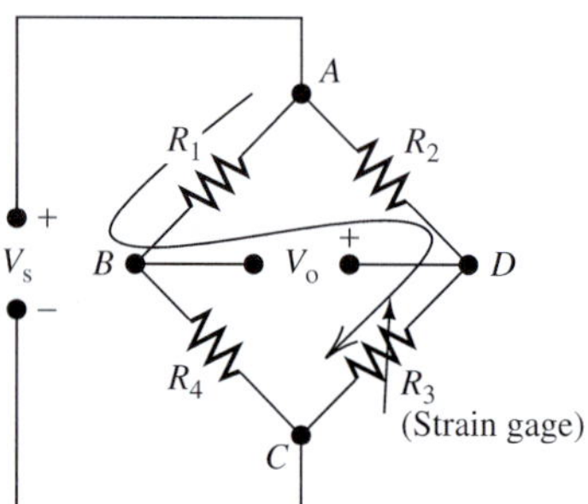

FIGURE 1.14 Wheatstone bridge circuit incorporating a single strain gage.

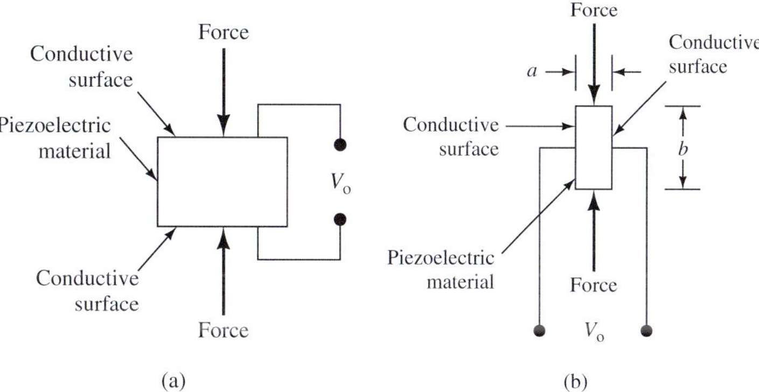

FIGURE 1.15 Piezoelectric crystals: (a) longitudinal and (b) transverse effect.

Because the charge developed during deformation is small, it is difficult to measure this charge without dissipating the accumulated electrons through the measurement device. A device with very high "input impedance" is necessary to accomplish this measurement without degrading the signal itself. This small charge is measured with the aid of an operational amplifier—configured with extremely high input impedance to prevent drainage of the signal, and extremely high output gain to produce a strong output voltage proportional to that charge.

The operational amplifier (Figure 1.16) is configured in a so-called charge amplifier circuit to produce a voltage proportional to the small charge signal, which is, in turn, proportional to the deformation of the piezoelectric crystal under load. The principles of the operational amplifier and the charge amplifier will be explained in Chapter 8.

RELATIVE DISPLACEMENT AT AN ATTACHMENT POINT

This can be measured through a capacitive load cell, one plate attached to the end of the shock absorber, the other plate attached to the shock tower. As the shock tower and the shock absorber move relative to each other, the gap between the capacitors changes and the net capacitance of the capacitor changes in inverse proportion to the gap distance (see Figure 1.17).

This capacitor is incorporated into a capacitive load cell, consisting of the capacitor and a stiff nonconducting elastic medium between the plates. The load applied by the shock absorber is passed through the elastic medium into the shock tower.

The capacitive load cell is connected into the feedback arm of a simple operational amplifier circuit, which produces a voltage proportional to the dynamic component of the displacement signal, while tending to ignore the static component (see Figure 1.18). Chapter 8 explains the fundamentals of operational amplifiers. This separation can be enhanced by low-pass filtering. Chapter 7 explains the principles of low-pass filters.

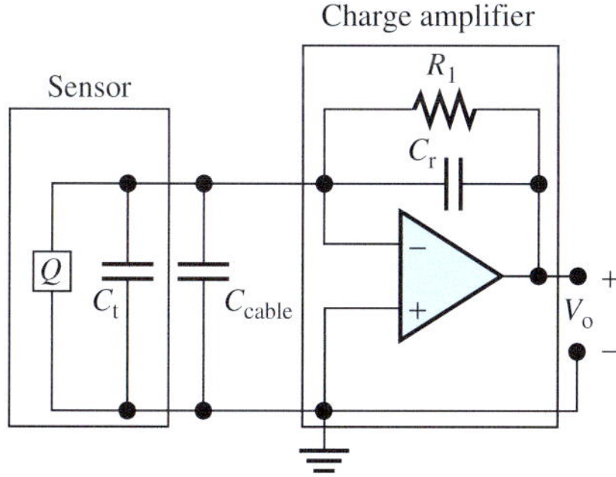

FIGURE 1.16 Charge amplifier circuit, incorporating a piezoelectric sensor and operational amplifier.

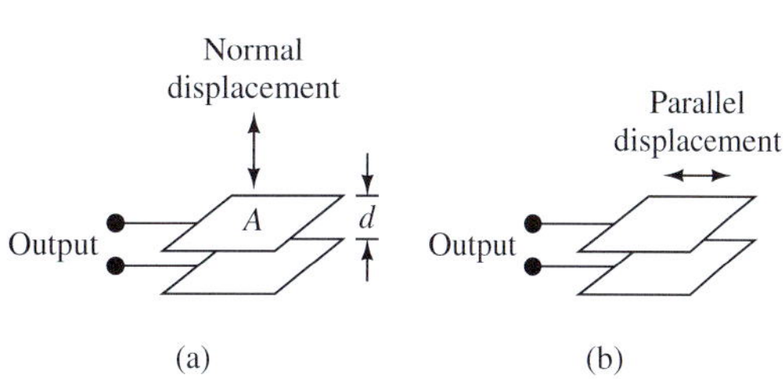

FIGURE 1.17 Capacitive transducers, sensitive to (a) normal and (b) parallel displacements.

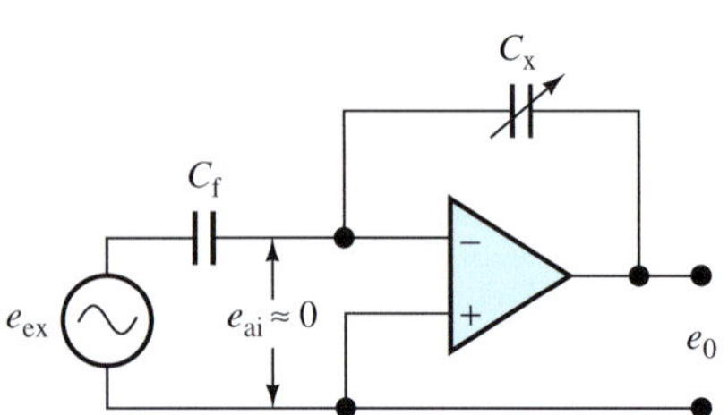

FIGURE 1.18 Operational amplifier with capacitive transducer in feedback arm.

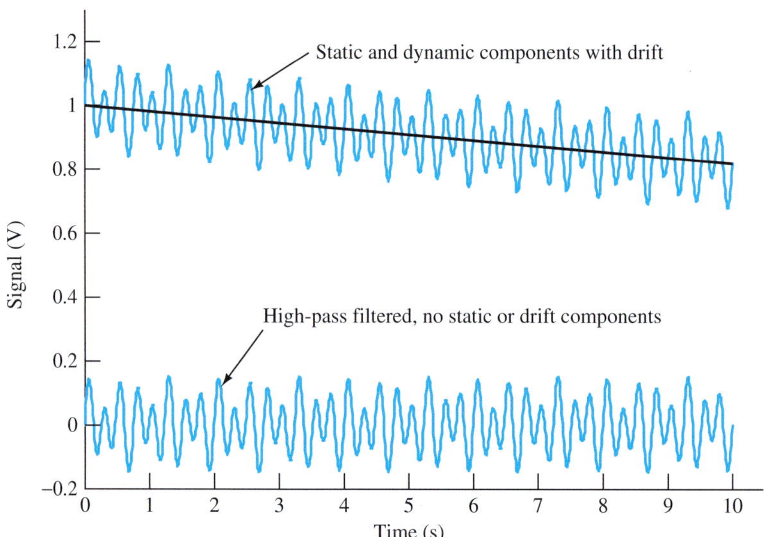

FIGURE 1.19 Separation of dynamic and static components.

The separation of the static and dynamic components allows the engineer to observe the actual dynamic component, which is usually superimposed on a fairly large, but already known, static component (see Figure 1.19).

It should be clear that a mechanical engineer with no understanding of electrical transducers will have to place his or her success in the hands of the staff, who will be obliged to select the appropriate transducers, assure that they respond to only the desired phenomenon, produce signals in the desired format, and are correctly interpreted. Nevertheless, it is the engineer, not the staff, who will be held responsible for the acquisition and interpretation of that data.

Further Reading

Battaini, M., Yang, G., and Spencer, B. F., Jr. (2000) "Bench-Scale Experiment for Structural Control," *Journal of Engineering Mechanics*, ASCE, 126(2), pp. 140–148. (Available at www.uiuc.edu/sstl/default.html)

Spencer, B. F., Jr. and Sain, M. K. (1997) "Controlling Buildings: A New Frontier in Feedback," *IEEE Control Systems*, vol. 17, no. 6, pp. 19–35.

TOPIC 2
Fundamentals of Electric Circuits

The content of this topic is compiled from:
Chapter 2 (pp. 13–60), *Electrical Engineering: Concepts and Applications*
By S.A. Reza Zekavat

CHAPTER 2

Fundamentals of Electric Circuits

2.1 INTRODUCTION

This chapter introduces the main variables and tools needed to analyze an electric circuit. You may wonder: How can an understanding of circuit theory help me better understand and solve real-world problems in my field of engineering?

As you learned through the case studies in Chapter 1, there is a clear link between many engineering disciplines and electrical engineering. In Chapter 2, we will examine more examples that will clarify the applications of electric circuits in your engineering field. After completing this chapter, you should better understand the link between engineering fields and circuit theory. In addition, through many application-based examples, the chapter will clarify how circuit theory can be applied to solve problems in your field of interest. Please note that the application-based examples you will encounter in this chapter are not the only application-based examples. Many more application-based examples and problems are provided in other chapters.

Circuits consist of individual elements that together form a model structure that can be used to simplify the process of analyzing and interpreting the behavior of complex engineering structures. The two main variables needed to analyze electric circuits are (1) charge flow or current, and (2) voltage. Current is measured by placing an Ammeter (ampere meter) in series, and voltage is measured by placing a voltmeter across elements in the circuit. The three main tools needed to analyze circuits are (1) Kirchhoff's current law (KCL), (2) Kirchhoff's voltage law (KVL), and (3) Ohm's law. Analysis of an electric circuit includes computation of the voltage across an element or the current going through that element. The following sections discuss the aforementioned laws, and detail how each is used to compute current and voltage.

Circuit models enable engineers to analyze the impact of different individual elements of interest. For instance, this chapter discusses problem solving and the generation of waveforms using the advanced computer software package PSpice. You will learn about and run PSpice applications to study the effect of each variable within a circuit. Please note that we give the PSpice tutorial in a distributive format. Thus, a basic tutorial is provided in this chapter, and new concepts in PSpice are introduced in Chapters 3–13.

Circuits enable engineers to investigate the impact of different elements of a system via software, that is, without actually having to build the engineering structure. This often reduces the time and the cost of the analysis. However, circuit models can be used to analyze systems in more than just electrical system applications. Circuit modeling can also be used in other engineering areas. For example, in a hydraulic machine, hydraulic fluid is pumped to gain a high pressure and transmitted throughout the machine to various actuators and then returned to the reservoirs. The hydraulic fluid circulation path and the associated elements can be modeled using a circuit model. Figure 2.1(a) represents a hydraulic structure.

Figure 2.1(b) represents a circuit model schematic structure for the hydraulic system, consisting of the following elements: hydraulic cylinder, pump, filter, reservoir, control valve, and retract/extend. Here, pipes play the role of interconnectors. Pipe friction in mechanics measures the resistance. Therefore, the pipes should ideally be frictionless to represent high conductivity.

In electrochemical cells (e.g., batteries), oxidation and reaction processes generate electrical energy. The energy is transferred through an electrical conducting path from zinc to copper. As shown in Figure 2.2(a), classical electrochemical circuit elements may include zinc and copper metals and a lamp to show the power. The zinc and copper sulfides and the copper wire are the interconnecting parts or conductors. The equivalent circuit of Figure 2.2(a) is shown in Figure 2.2(b). This circuit simplifies the structure of Figure 2.2(a) to include only a resistor, voltage source, and connectors (e.g., copper wires).

In the case of electric circuits, a circuit model includes various types of circuit elements connected in a path by conductors. Therefore, in an electric circuit, the interconnecting materials are conductors. Copper wires are usually used as conductors. These interconnectors may be

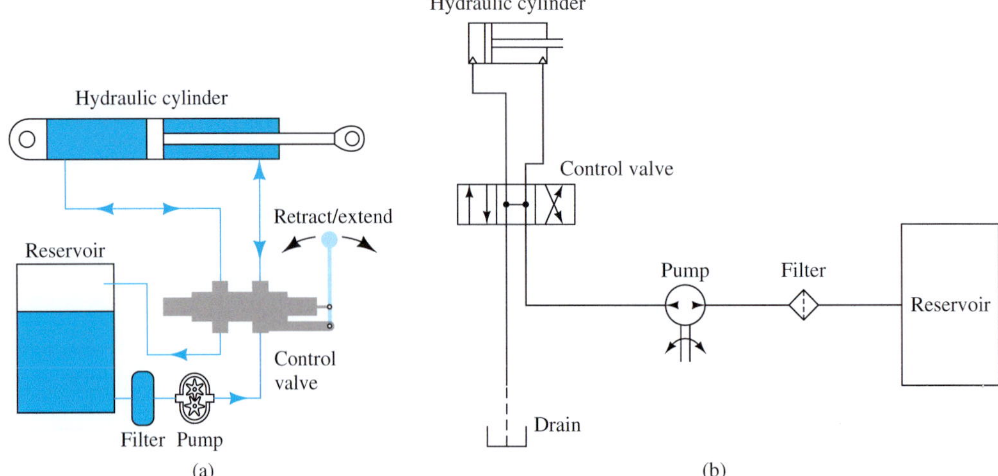

FIGURE 2.1
(a) Hydraulic circuit structure and (b) its equivalent schematic structure.

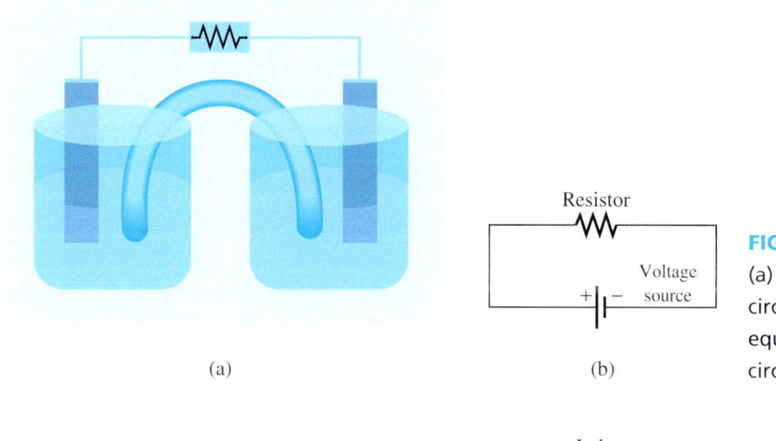

FIGURE 2.2 (a) Electrochemical circuit and (b) its equivalent schematic circuit.

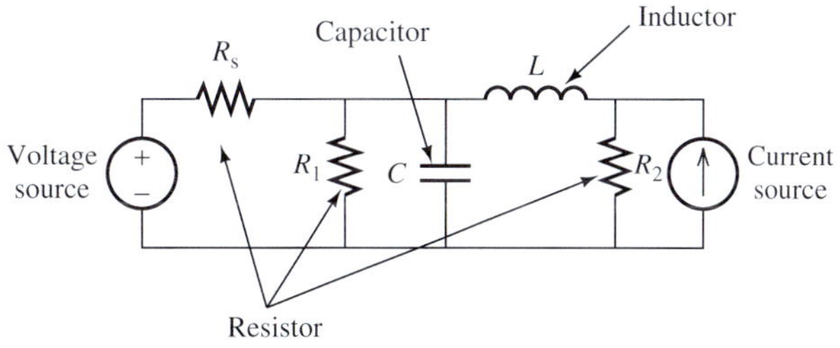

FIGURE 2.3 An electric circuit.

as small as one micrometer (e.g., in electronic circuits) or as large as hundreds of kilometers (e.g., in power transmission lines that may connect generated hydroelectric power to cities; see Chapter 9). Voltage source, resistance, capacitance, and inductance are examples of circuit elements. An example of an electric circuit is shown in Figure 2.3.

2.2 CHARGE AND CURRENT

Fundamental particles are composed of positive and negative electric charges. Niels Bohr (1885–1962) introduced an atom model [Figure 2.4(a)] in which electrons move in orbits around a nucleus containing neutrons and protons. Protons have a positive charge, electrons have a negative charge, and neutrons are particles without any charge. Particles with the same charge repel each other, while opposite charges attract each other [Figure 2.4(b)]. In normal circumstances, the atom is neutral, that is, the number of protons and electrons are equivalent.

Charge is measured in *coulombs*. Electrons are the smallest particles and each possesses a negative charge equal to $e = -1.6 \times 10^{-19}$ C. Equivalently, the charge of each proton is $+1.6 \times 10^{-19}$ C.

Electric charge (q), measured by the discrete number of electrons, corresponds to:

$$q = ne \tag{2.1}$$

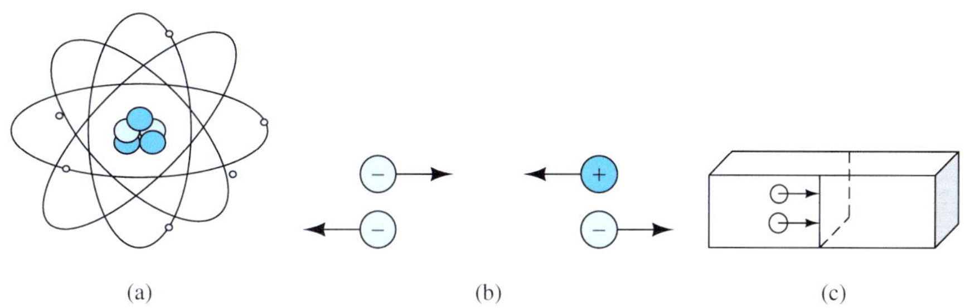

FIGURE 2.4 (a) Bohr atom model, (b) interaction between the same and the opposite charges and (c) charges are passing through a cross section at a constant rate.

where n is a positive or negative integer. *Electric current* is the rate at which charges flow through a conductor; specifically, the amount of charges moving from a point per unit of time.

In Figure 2.4(c), charges are moving through a cross section of a conductor with a constant current, i. During time t, the total amount of charges passing through the cross section is:

$$q = i \times t \tag{2.2a}$$

Then the constant current, i, can be expressed as:

$$i = \frac{q}{t} \tag{2.2b}$$

When the current is time-varying, denoted as a function of time t, Equation (2.2a) can be expressed in integration form:

$$q(t) = \int_{t_0}^{t} i(\tau)\mathrm{d}\tau \tag{2.3a}$$

In Equation (2.3a), $q(t)$ is the total amount of charge flowing through the cross section within the time t_0 to t $(t–t_0)$, t_0 is the initial time, and t is the time at which we intend to measure the charge. Equation (2.3a) states that the total charge flowing through a cross section at any given time period is the integration of the current passing through that cross section from the initial time t_0 up to the time t.

Equation (2.2b) shows that the current is the rate of variation of charge within a specific time period. Therefore, the relationship between the time-varying current, the charge, and the time corresponds to:

$$i(t) = \frac{\mathrm{d}q(t)}{\mathrm{d}t} \tag{2.3b}$$

where $\mathrm{d}q(t)$ represents the variation of charges and i shows the current measured in amperes. In fact, 1 *ampere* (A) = 1 *coulomb/second* (C/s). Microamperes (μA $= 10^{-6}$ A), milliamperes (μA $= 10^{-3}$ A), and kiloamperes (kA $= 10^{+3}$ A) are also widely used for measuring current. For instance, kA is the unit commonly used in power distribution systems, while μA is used regularly in digital microelectronic systems.

Current has both value and direction. In general, the direction of the movement of *positive* charges is defined as the conventional direction of the current (Figure 2.5). However, in an electric circuit with various elements, for the purpose of analysis of a circuit, we may assign arbitrary directions to the current that flows through an element. These arbitrarily selected directions for current determine the direction of voltage (see details in Section 2.3). Figure 2.6 shows an example of how current direction can be assigned. Here, the current variables i_A, i_B, i_C, i_E are assigned to different circuit elements. Note that we do not need to know the real direction of current; rather, arbitrary directions can be assigned to each branch. Real directions are determined after performing the calculations: if the calculated voltage or current is positive, we can deduce that we have selected the actual direction; if the calculated voltage or current is negative, the actual direction would be the inverse of the selected one.

If the magnitude and the direction of the current are constant over time, it is referred to as *direct current* and it is usually denoted by I; if the amount of the current or its direction changes

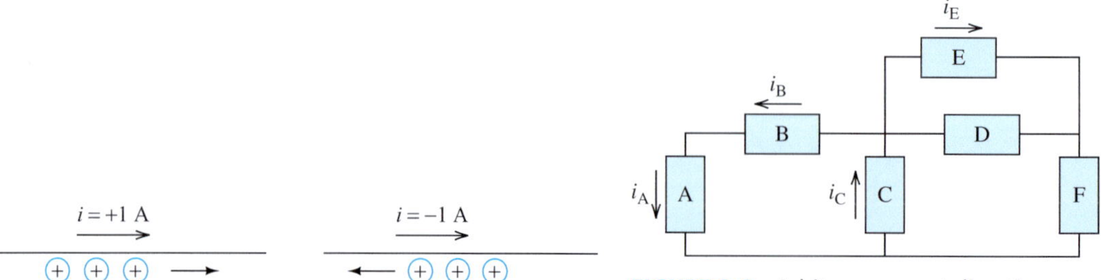

FIGURE 2.5 Current and the movement of positive charges.

FIGURE 2.6 Arbitrary current direction assignment in an electric circuit.

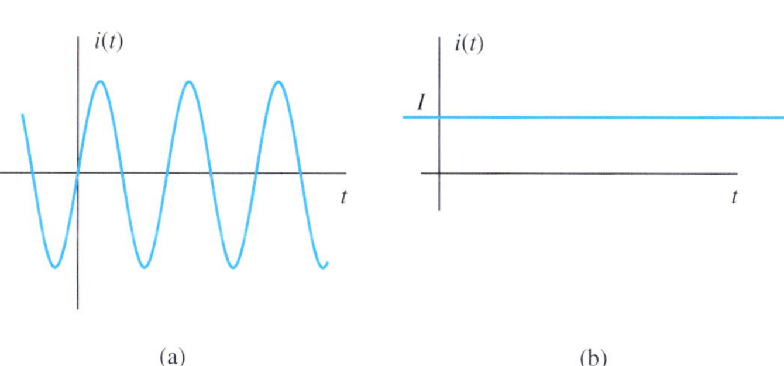

FIGURE 2.7 (a) Alternating and (b) direct currents.

FIGURE 2.8 Water gravitational potential energy at the top of the waterfall is converted to kinetic energy.

over time, it is called *time-varying current* and it is usually denoted by $i(t)$. Alternating currents are an example of time-varying currents. In alternating currents, both the current amplitude and current direction periodically change with time. Direct and alternating currents are abbreviated as DC and AC, respectively. Figure 2.7 represents examples of DC and AC.

2.3 VOLTAGE

To better explain the phenomenon of voltage, let us examine an analogy based on potential energy. Potential energy is the energy that is able to do work if it is converted to another type of energy. For example, the water at the top of a waterfall has a potential energy with respect to the bottom of the waterfall due to gravity forces. As water moves down, this potential energy is gradually converted to kinetic energy. As a result, at the bottom of the waterfall, the stored energy that we observed at the top of the waterfall shows itself as full kinetic energy (Figure 2.8). Therefore, the potential energy *defined between the two points* (the top and the bottom of the waterfall) is converted to kinetic energy (resulting in the current of the water flow). This represents a real-life example: dams, in fact, use this technique to create electricity (hydroelectric power). In hydroelectric dams, potential energy is converted to kinetic energy that is applied to turbines and electric generators to create electrical energy.

Similarly, the difference in voltage between two points in a circuit forces or *motivates electrons* (charges) to travel. This is why voltage is referred to as *electromotive force* (EMF). In an electric circuit, differences in potential energy at different locations (called voltage) force electrons to move. Therefore, electric current is the result of a difference in the amount of potential energy at two points in a circuit. Like the example of the dammed water outlined earlier, where potential energy was converted to the movement of water, here, the voltage makes the charges move. Voltage is measured in volts (V).

One joule (1 J) of energy is needed to move one coulomb (1 C) of charge through one volt (1 V) of potential difference. Therefore:

$$1 \text{ Volt (V)} = 1 \text{ Joule/Coulomb (J/C)}.$$

A DC voltage is a constant voltage, which is always either negative or positive. In contrast, an AC voltage alternates its size and sign with time. In Figure 2.9, v_A represents the voltage across element A; v_B represents the voltage across element B; and v_F represents the voltage across

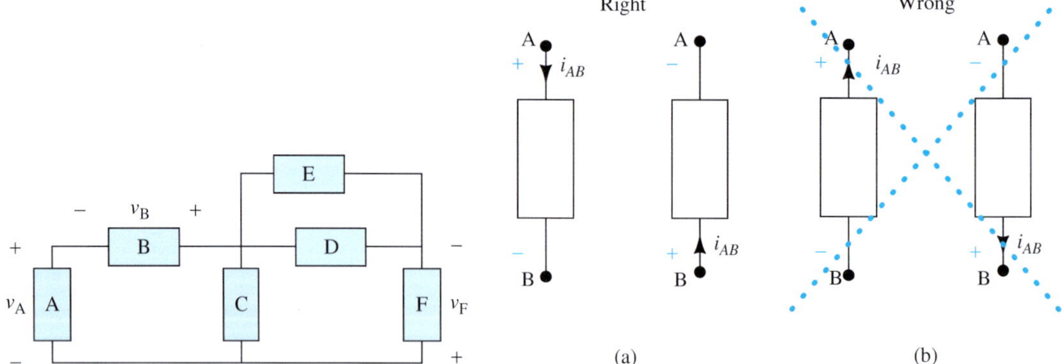

FIGURE 2.9 Arbitrary voltage polarity assignment in an electric circuit.

FIGURE 2.10 Current and voltage respective direction: (a) correct; (b) incorrect.

element F. The same indexing approach can be used to represent the voltage across any given element. In Section 2.2, we explained that the current flow direction within an element can be assigned arbitrarily. However, if we select the current direction, the voltage direction is determined based on the selected current direction. Similarly, the voltage polarity across an element can be determined arbitrarily as shown in Figure 2.9. However, if we select the voltage direction across an element, the corresponding current is determined relative to the voltage. This point is discussed further in the next section.

2.4 RESPECTIVE DIRECTION OF VOLTAGE AND CURRENT

In the prior two sections, we explained that the current flow direction within an element and the voltage polarity across an element might be determined arbitrarily. Here, we will clarify this concept: for an element that consumes power, the current flows into the side of the element that has the greatest amount of positive voltage and out of the side with the greatest amount of negative voltage. Once the direction of the current has been assigned, the assignment of voltage polarity must follow. The alternative case is also true: once the polarity of an element has been assigned, the direction of the current flow must follow from the known polarity.

Please note that the selected current flow and the voltage polarity may not be consistent with real (actual) current flow and voltage direction. If, upon calculation, the current flow is computed to be a positive number, then the directions assigned were correct. On the other hand, if the current flow is calculated to be a negative number, then the directions assigned were the reverse of the actual current flow and voltage direction. Figure 2.10 summarizes the rule for the selection of voltage polarity as it relates to current direction in an element that consumes power. Figure 2.10(a) represents correct selection. Here, the current enters from the positive sign of the voltage across the element and leaves from the negative sign of the voltage across the element. Figure 2.10(b) shows an incorrect selection.

2.5 KIRCHHOFF'S CURRENT LAW

Circuit analysis refers to characterizing the current flowing through and voltage across every circuit element within a given circuit. Some general rules apply when analyzing any circuit with any number of elements. However, before discussing these rules, we need to define other terms that are commonly used in circuit analysis literature: a *node* and a *branch*.

A **node** is the connecting point of two (or more) elements of a circuit.

A **branch** represents a circuit element that is located between any two nodes in a circuit.

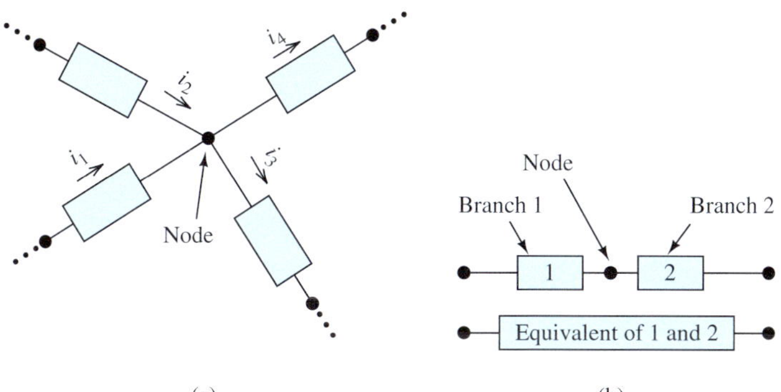

FIGURE 2.11 Definition of a node and branch in an electric circuit.

Therefore, nodes are the starting and ending points of branches. Several branches may meet at a single node. In Figure 2.11(a), four circuit elements in four branches are connected to each other at a common node. In some special cases, two elements might be connected back to back as shown in Figure 2.11(b). In that case, we may consider each one as a branch, and their connecting point as a node. We may also replace them with an equivalent element. In that case, we consider the element that is the equivalent of both as one branch.

Kirchhoff's current and voltage laws are the two fundamental laws in electric circuits. Russian scientist Gustav Robert Kirchhoff (1824–1887) introduced the two laws that now bear his name. These laws allow the calculation of currents and voltages in electric circuits with multiple loops using simple algebraic equations.

(Used with permission from © INTERFOTO/ Alamy.)

Kirchhoff's current law (KCL) states that the net current entering a node in a circuit is zero. Some currents enter into a node and some leave the node. Thus, based on this law, the sum of the currents entering a node is equal to the sum of the currents leaving that node. KCL results from the law of the conservation of charge:

> **Law of the conservation of charge:** Charges are always moving in a circuit and cannot be stored at a point; charge can neither be destroyed nor created.

EXAMPLE 2.1 Series Elements: Currents in the "Same Direction"

In Figure 2.12, Node A is between two elements 1 and 2, and the current entering Node A (i_1) equals the current leaving Node A (i_2). In the scenario shown in Figure 2.12, Node A connects only two branches. In this case, we say element 1 is in series with element 2. Therefore, currents passing through the two elements in series are always the same in size and in direction.

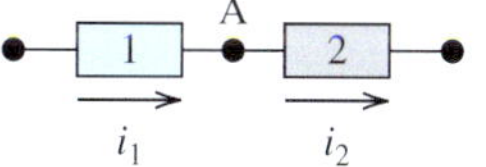

FIGURE 2.12 Kirchhoff's current law for two series elements with similar current direction assignment.

EXAMPLE 2.2 Series Elements: Currents in "Opposite Directions"

As discussed earlier, we can assign arbitrary directions for the branch currents. For example, in the simple circuit of Figure 2.13, we have assigned the current directions different from Figure 2.12. In this case, the algebraic sum of the currents entering the node equals $i_1 + i_2$. Based on KCL, if we set this sum to zero, then $i_2 = -i_1$. This confirms the same notion that the currents through two series elements are equal in size and direction.

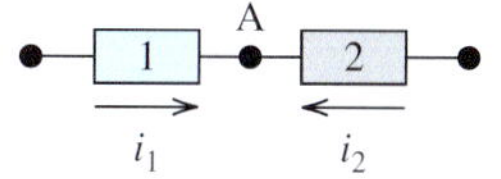

FIGURE 2.13 Kirchhoff's current law for two series elements with opposite current direction assignment.

EXAMPLE 2.3 KCL

KCL can also be applied to a node with more than two branches. Consider the scenario of applying KCL equations to the node represented in Figure 2.11. In this case:

$$i_1 + i_2 = i_3 + i_4 \tag{2.4}$$

or

$$i_1 + i_2 - i_3 - i_4 = 0 \tag{2.5}$$

Knowing this, we can resketch Figure 2.11 as the one shown in Figure 2.14. Applying Kirchhoff's law to the node shown in this figure leads to:

$$i_1 + i_2 + (-i_3) + (-i_4) = 0 \tag{2.6}$$

This equation is equivalent to Equation (2.5).

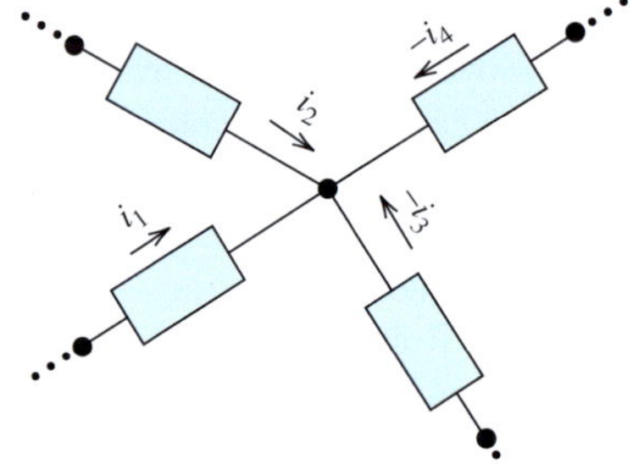

FIGURE 2.14 Kirchhoff's current law for arbitrary current assignment.

In general, KCL can be applied to every closed surface (e.g., closed sphere) within any circuit.

> Based on the Kirchhoff's current law:
>
> *The net current entering any closed surface is zero.*

A node is indeed a special case of a closed surface when the area of that closed surface tends to be zero. Using Figure 2.15, a generalized form of KCL for a node containing N branches can be shown as:

$$\sum_{k=1}^{n} i_k = \sum_{k=n+1}^{N} i_k \tag{2.7}$$

In Equation (2.7), $i_k, k \in \{1,2, ..., n\}$ are currents entering Node A, while $i_k, k \in \{n+1, ..., N\}$ are currents leaving Node A.

For instance, if we apply KCL to the closed surface shown by dotted line in Figure 2.16, the equation will be:

$$i_1 + i_2 + i_3 + i_4 + i_5 = 0 \tag{2.8}$$

The closed surface can be considered as a black box. The black box covers a portion of circuit that includes multiple branches and circuit elements.

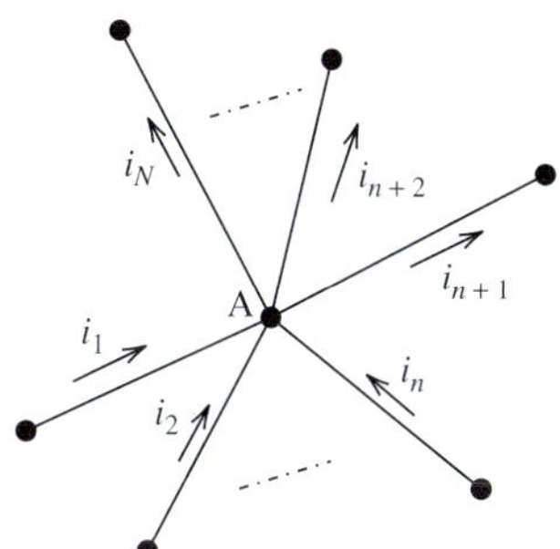

FIGURE 2.15 The generalized Kirchhoff's current law.

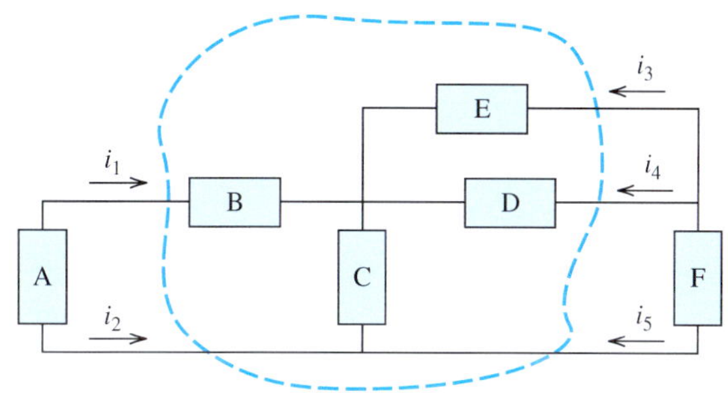

FIGURE 2.16 Generalized KCL for a closed surface.

In conclusion, by applying KCL, only four steps are needed to find an unknown current entering or leaving a node or a closed surface. The steps include the following:

1. Determine currents entering the node (or closed surface).
2. Determine currents leaving the node (or closed surface).
3. Apply KCL to the nodes.
4. Use the created set of equations to find the unknown current.

EXAMPLE 2.4 KCL

Find i_C in Figure 2.17.

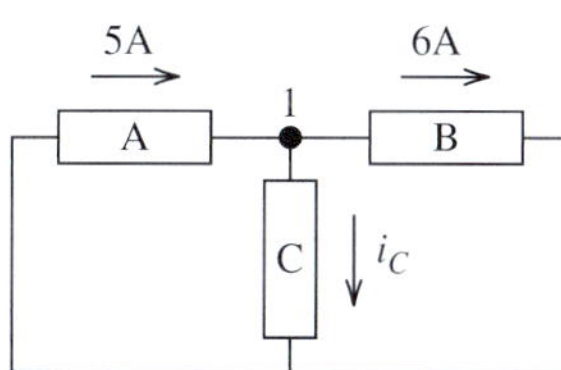

FIGURE 2.17 Circuit diagram for Example 2.4.

SOLUTION

- The current from only one branch enters Node 1—the one that passes through element A. Therefore, if we show the current entering Node 1 by i_A, we can express this as:

$$i_A = 5\text{A}$$

- The currents passing through elements B and C leave Node 1. Therefore, if we indicate the current leaving Node 1 by i_B, we see that:

$$i_A = 5 = i_B + i_C = 6\text{A} + i_C$$

Using this equation, we find that:

$$i_C = -1\text{ A}$$

Therefore, the i_C current size is 1 A with a direction opposite to the one shown in Figure 2.17.

EXAMPLE 2.5 KCL

Find the currents i_B, i_C, and i_D in Figure 2.18.

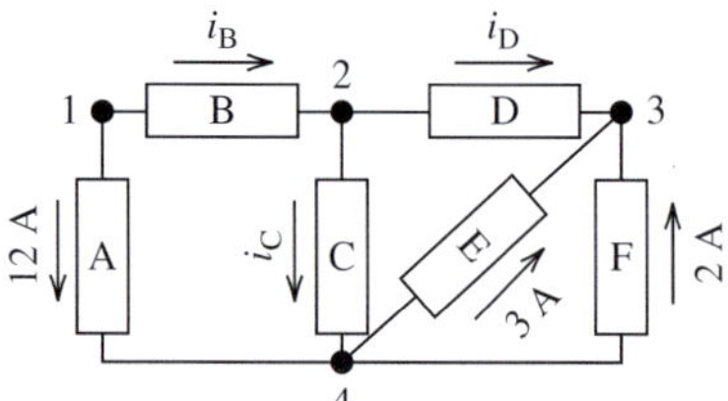

FIGURE 2.18 Circuit diagram for Example 2.5.

SOLUTION

To find the desired currents, apply KCL to Nodes 1, 2, and 3.

1. Currents entering Nodes 1, 2, and 3 are, respectively:

$$i_{i1} = 0 \text{ A}$$
$$i_{i2} = i_B$$
$$i_{i3} = i_D + 3 \text{ A} + 2 \text{ A} = i_D + 5 \text{ A}$$

2. Currents leaving Nodes 1, 2, and 3 are, respectively:

$$i_{o1} = i_B + 12 \text{ A}$$
$$i_{o2} = i_C + i_D$$
$$i_{o3} = 0 \text{ A}$$

3. Applying KCL, we show that $i_{i1} = i_{o1}$, $i_{i2} = i_{o2}$, and $i_{i3} = i_{o3}$, therefore,

$$i_B = -12 \text{ A}$$
$$i_D = -5 \text{ A}$$
$$i_C = -7 \text{ A}$$

The signs of the currents we calculated (negative) show that the direction of each is opposite to the initially assigned direction.

EXERCISE 2.1

Can all unknown currents in Figure 2.18 be found if we assume that element D is shorted? If yes, find all currents for this scenario. Note that in this case we say elements E and C are in parallel.

2.6 KIRCHHOFF'S VOLTAGE LAW

Before discussing the second fundamental law used to analyze electric circuits, we must first introduce another variable in circuits, called a **loop**. A loop is a closed path that starts at a node, proceeds through some circuit elements, and returns to the starting node. In other words, a loop is formed by tracing a closed path in a circuit without passing through any node more than once.

For instance, in Figure 2.19, Path 1 is a loop, because its starting and ending points are the same and it passes by each node in its path only once. However, Path 2 is not a loop. Although Path 2 starts and ends at the same node, it is not a loop because it passes by Node 1 twice.

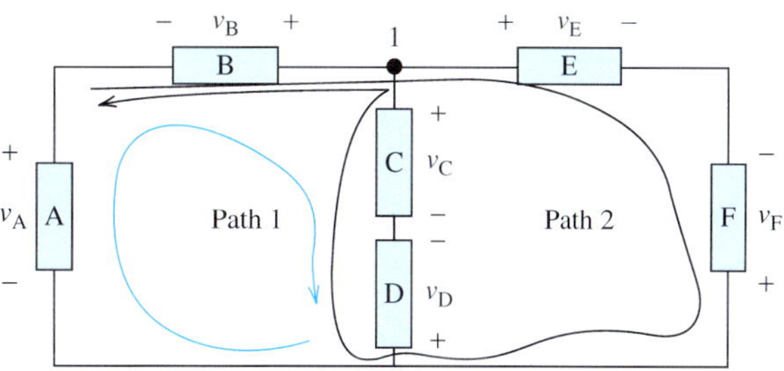

FIGURE 2.19 Loop definition—Path 1 is a loop but Path 2 is not.

Kirchhoff's voltage law states that:

The algebraic sum of the voltages equals zero for any loop in an electrical circuit.

Each defined loop in a circuit can have an arbitrary direction. Using the KVL equation, we can assign positive and negative signs, respectively, to the rising and dropping voltages. For simplicity, from now on, we will consider the polarity of voltages across each element to follow the direction of the currents through that element, as explained in detail in Figure 2.10. When voltages rise, the polarity changes from negative to positive; when voltages drop, the polarity changes from positive to negative. To apply KVL, the algebraic sum of the raised and dropped voltages in the direction of the loop is set equal to zero. Then, we can determine the unknown voltage in the loop.

Figure 2.20 outlines the procedure for writing KVL equations. When following the loop, we consider the change of the voltage polarity of an element and assign the relevant sign for the voltage in the KVL equation. For Figure 2.20, the equation is:

$$-v_A + v_B - v_C = 0 \tag{2.9}$$

We can also move the negative voltages to the other side of the equation; therefore, Equation (2.9) can also be written as:

$$v_B = v_A + v_C$$

This is really another way of writing the KVL: it equates the sum of the absolute values of the dropping voltages to that of the rising voltages.

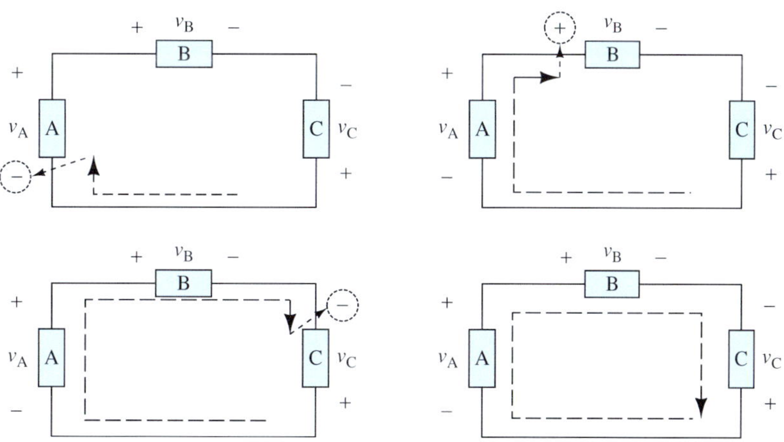

FIGURE 2.20 KVL procedure.

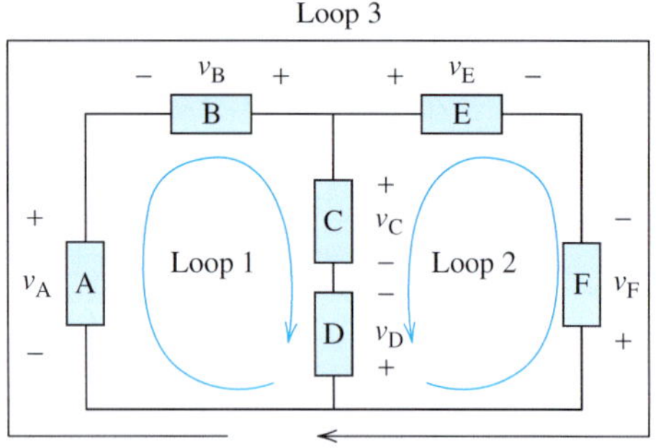

FIGURE 2.21 KVL example.

Figure 2.21 shows three loops; KVL can be applied to each. If we do this, the generated set of KVL equations applied to loops 1, 2, and 3, respectively, corresponds to:

$$\begin{aligned} \text{Loop 1:} \quad & -v_A - v_B + v_C - v_D = 0 \\ \text{Loop 2:} \quad & v_F - v_E + v_C - v_D = 0 \\ \text{Loop 3:} \quad & -v_A - v_B + v_E - v_F = 0 \end{aligned}$$

Assuming one of these voltages (e.g., v_D) is known, other unknown voltages can be easily extracted via this set of algebraic equations.

EXAMPLE 2.6 KVL

Find v_B in Figure 2.22.

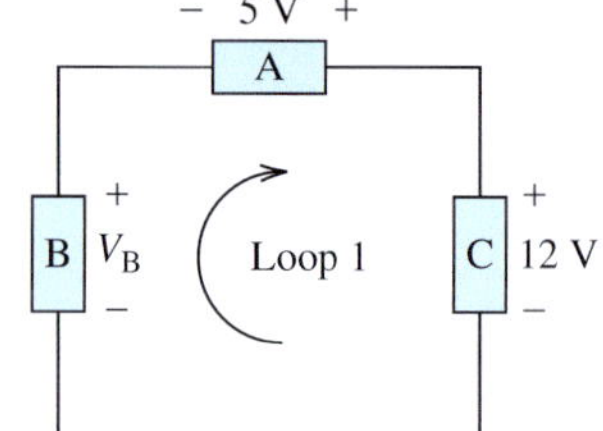

FIGURE 2.22 Circuit diagram for Example 2.6.

SOLUTION

For this circuit and the considered loop, the equation can be written:

$$-v_B - 5\text{ V} + 12\text{ V} = 0$$

Using this equation, we find:

$$v_B = 7\text{ V}$$

EXAMPLE 2.7 KVL

Find the unknown voltages v_C and v_D in Figure 2.23.

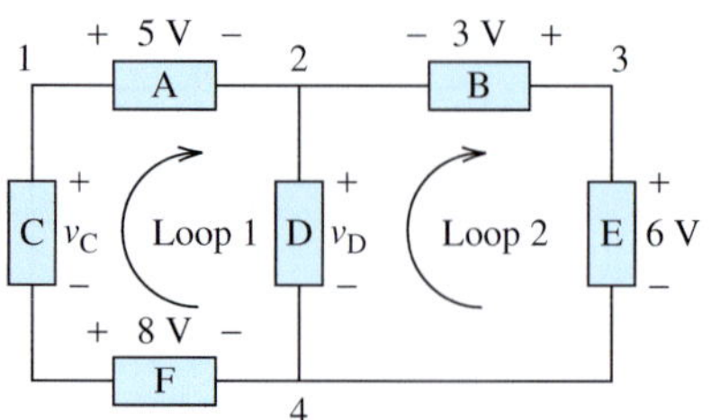

FIGURE 2.23 Circuit diagram for Example 2.7.

SOLUTION

In this circuit, there are two unknown voltages. Therefore, at least two KVL equations are needed to solve both unknowns. Two loops are shown in Figure 2.23. For Loop 1 the KVL equation can be written as:

$$-v_C + 5\text{ V} + v_D - 8\text{ V} = 0$$

For Loop 2, KVL leads to:

$$-v_D - 3\text{ V} + 6\text{ V} = 0$$

Solving these two equations, the following values are obtained for the unknown voltages:

$$v_D = 3\text{ V}$$
$$v_C = 0\text{ V}$$

APPLICATION EXAMPLE 2.8 Chemical Reaction

For this example, assume that the temperature of a chemical reaction needs to be monitored for the entire duration of an experiment. To monitor this temperature, an electric temperature sensor is connected to a computer that records the data (see Figure 2.24). The temperature sensor used is basically a resistor; its resistance changes with temperature (this type of sensor is known as a thermistor; see Chapter 11, Section 11.2 for a review on thermistors).

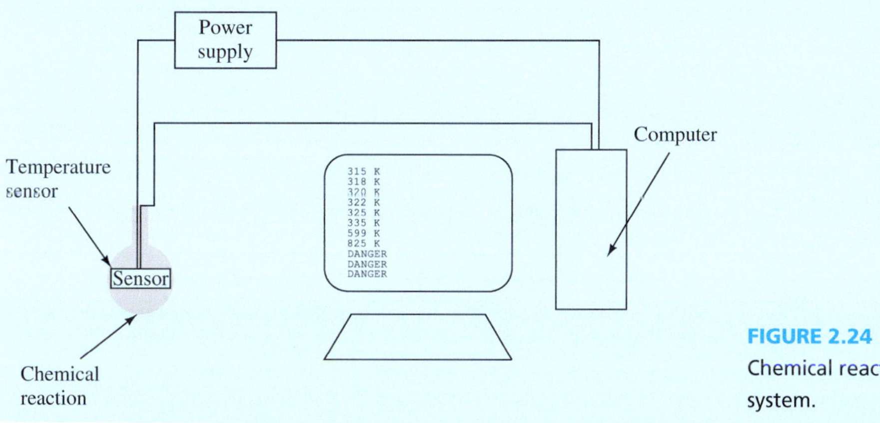

FIGURE 2.24 Chemical reaction system.

The computer can be modeled as an amplifier, which in turn can be modeled as a resistance. The entire system can be modeled as shown in Figure 2.25. The elements presented in this diagram are explained more fully later in this chapter. Find the voltage at the computer terminals.

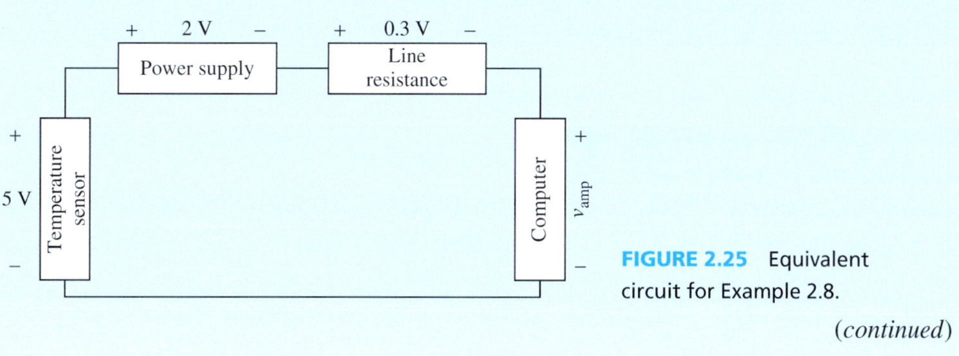

FIGURE 2.25 Equivalent circuit for Example 2.8.

(continued)

APPLICATION EXAMPLE 2.8 Continued

SOLUTION

Applying KVL, we see that:

$$0 = 2.0\text{ V} + 0.3\text{ V} + v_{\text{amp}} - 5\text{ V}$$

Next, we can calculate the voltage by solving the equation, resulting in:

$$v_{\text{amp}} = 2.7\text{ V}$$

APPLICATION EXAMPLE 2.9 Cooling Fan

An automobile cooling fan is connected to a 12 V car battery (Figure 2.26). The car battery supplies voltage to the car motor as well. The voltage drop across the fan motor consists of the voltage drop due to the internal resistance of the armature's windings. Back electromotive force (EMF) functions as a source. The equivalent circuit is depicted in Figure 2.27. Line resistance represents the resistance of the wire connecting the fan to the battery. What is the voltage drop over the line resistance, v_R?

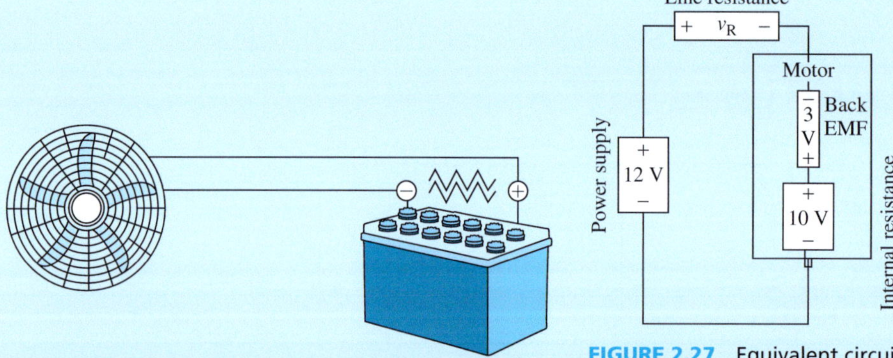

FIGURE 2.26 Cooling fan connected to car battery.

FIGURE 2.27 Equivalent circuit for Example 2.9.

SOLUTION

Applying KVL to the equivalent circuit shown in Figure 2.27, we see that:

$$-12\text{ V} + v_R - 3\text{ V} + 10\text{ V} = 0$$

Solving for the voltage using this equation, we see that:

$$v_R = 5\text{ V}$$

APPLICATION EXAMPLE 2.10 Loudspeaker

A basic loudspeaker is shown in Figure 2.28. The electromagnetic coil pushes and pulls on the magnet, causing the speaker cone to vibrate, producing sound. Figure 2.29 shows a simplified version of the electrical components of the speaker.

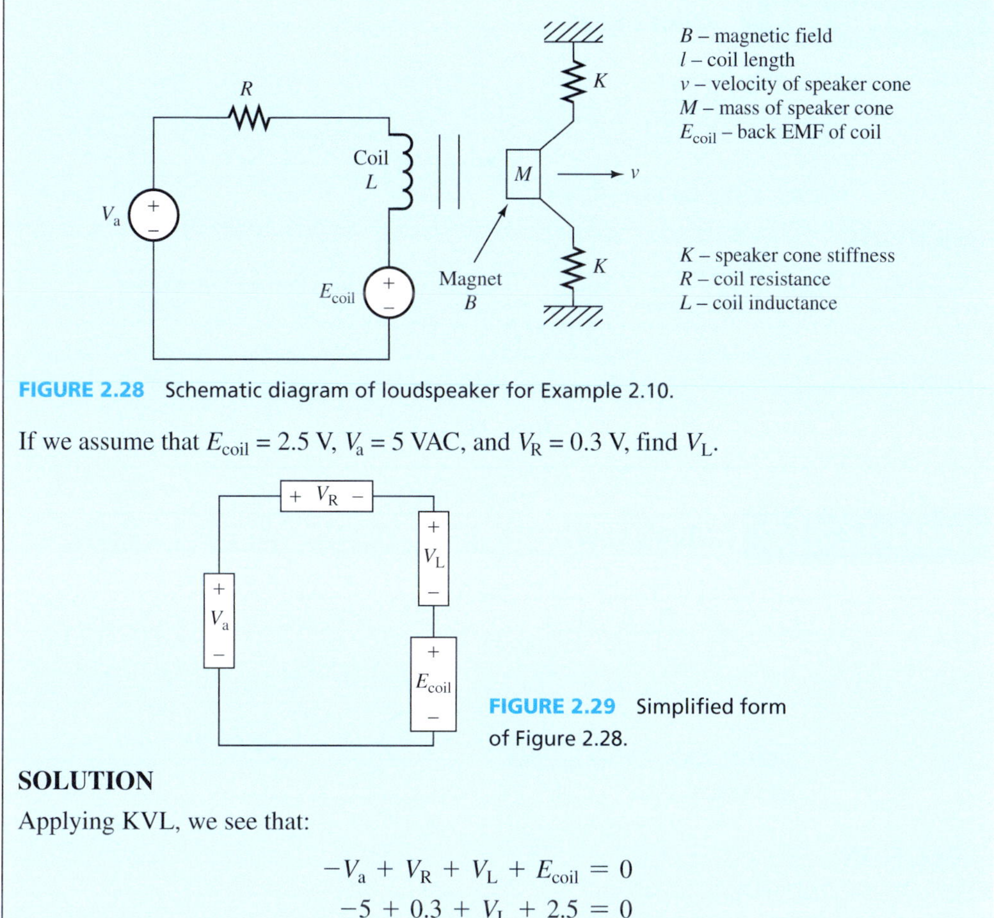

FIGURE 2.28 Schematic diagram of loudspeaker for Example 2.10.

If we assume that $E_{coil} = 2.5$ V, $V_a = 5$ VAC, and $V_R = 0.3$ V, find V_L.

FIGURE 2.29 Simplified form of Figure 2.28.

SOLUTION

Applying KVL, we see that:

$$-V_a + V_R + V_L + E_{coil} = 0$$
$$-5 + 0.3 + V_L + 2.5 = 0$$
$$V_L = 2.2 \text{ V}$$

2.7 OHM'S LAW AND RESISTORS

So far, we have introduced the circuit element variables, current, and voltage. The next important question to address is: what is the relationship between the current and the voltage of a passive element in a circuit?

The answer to this basic question depends on the properties of the element. The simplest passive element in an electric circuit is a resistor. In fact, every conducting material is a resistor. The voltage across the terminals of an ideal resistor is directly proportional to the current passing through that terminal. If we measure the ratio of the voltage and the current associated with an ideal resistor, we observe a constant value (at any given time). Accordingly, the relationship of the voltage and the current of a resistor can be stated as:

$$v = i \times R \tag{2.10}$$

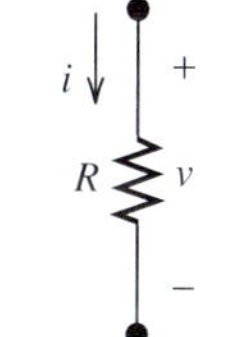

FIGURE 2.30 Ideal resistor symbol.

where the constant R is called the resistance and its unit is Ohm (Ω). Another important fact is that, 1 Ohm (Ω) = 1 volt/ampere (V/A). The Ohm is named after George Simon Ohm, who was probably the first physicist to measure the voltage and the current of a conductor and describe the constant voltage to current ratio, that is, the resistance.

Figure 2.30 shows the symbol commonly used to signify a resistor in an electric circuit. The voltage polarity and the direction of the current satisfy the sign conventions. The voltage–current or $v-i$ curve of an ideal (linear) resistor is a straight line as plotted in Figure 2.31. This curve is called the characteristic of a resistor.

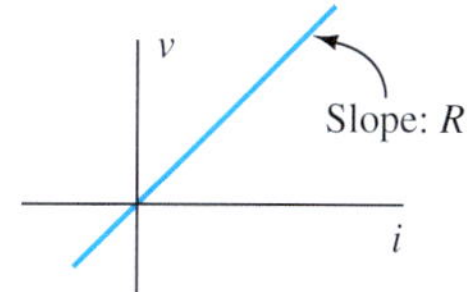

FIGURE 2.31 The linearity of an ideal resistor.

EXAMPLE 2.11 Ohm's Law

Find R in Figure 2.32.

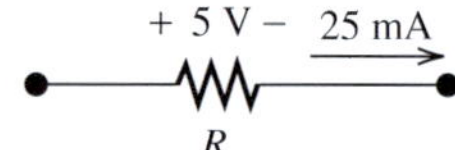

FIGURE 2.32 For Example 2.11.

SOLUTION

As shown in Figure 2.32, $v = 5$ V and $i = 0.025$ A. According to the Ohm's law:

$$5\text{ V} = R \times 0.025\text{ A}$$

Therefore:

$$R = 200\ \Omega$$

EXAMPLE 2.12 Ohm's Law

Find the current, i, in the circuit shown in Figure 2.33.

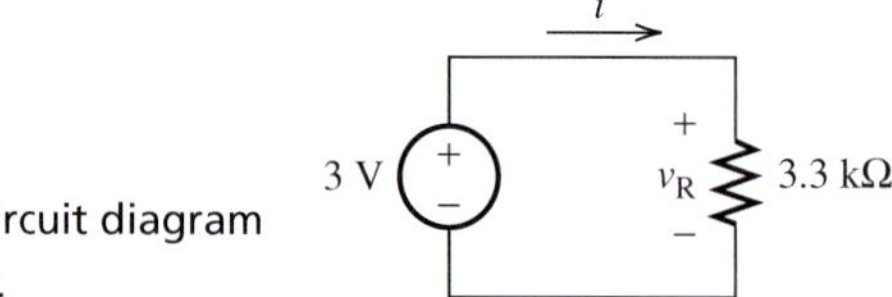

FIGURE 2.33 Circuit diagram for Example 2.12.

SOLUTION

The voltage across the terminals of the resistor equals the voltage across the voltage source, in this case, 3 V. Therefore, $v_R = 3$ V. Now, using Ohm's law, we can develop the equation:

$$3\text{ V} = 3300 \times i$$

Calculating i from this equation, we find:

$$i = 0.91\text{ mA}$$

EXERCISE 2.2

Assume that another 3.3 kΩ resistor is added to the circuit in series in Example 2.12. Calculate the current based on this revised circuit.

APPLICATION EXAMPLE 2.13 Electric Shock

An unfortunate bald utility worker climbs an aluminum ladder with bare feet and comes into contact with an overhead power line with the maximum voltage of 30 kV (Figure 2.34). The worker's body resistance is 315 kΩ. In this situation, his body can be modeled as a resistor connected between the terminals of a power supply (the wire) (Figure 2.35). How much current flows through his body? If the minimum current needed for a shock to be fatal is 100 mA, does the worker survive?

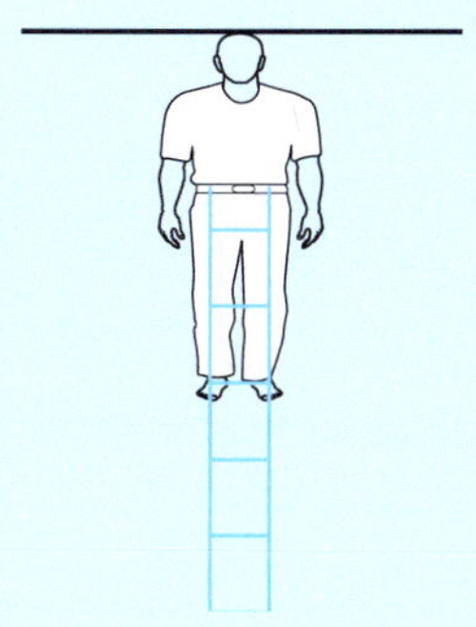

FIGURE 2.34 Application problem—electric shock.

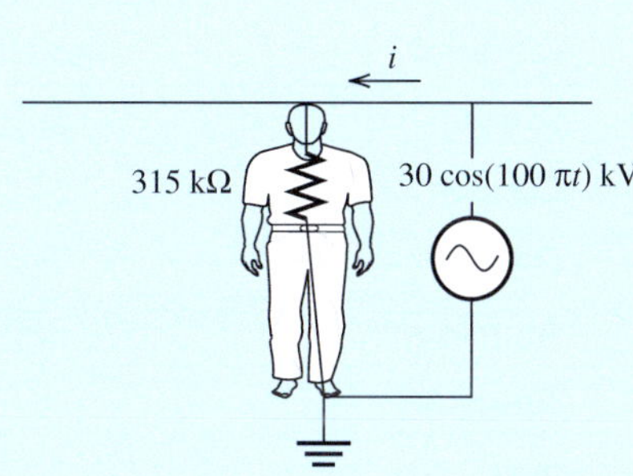

FIGURE 2.35 The equivalent circuit for Example 2.13.

SOLUTION

The equivalent circuit has been shown in Figure 2.35. Using Ohm's law, the maximum current through the worker's body is:

$$i_{\max} = \frac{30{,}000 \text{ V}}{315 \times 10^3 \ \Omega} = 95.2 \text{ mA}$$

Because the current passing through the worker is less than 100 mA, the worker survives, but barely! Note: As discussed in Chapter 15, to reduce the amount of current flow through the worker's heart, one hand should have been placed in a pocket. This creates a shortcut around the heart and reduces the current that goes through the heart.

2.7.1 Resistivity of a Resistor

The value of an ideal resistor depends on its physical characteristics. These physical characteristics include the conducting performance of the material and its physical and geometrical shape. A sample conductor, in the shape of a cylinder, is shown in Figure 2.36. The relationship, which expresses the resistance, R, of a conductor in terms of its physical parameters, corresponds to:

$$R = \rho\frac{L}{A} \tag{2.11}$$

where ρ is the resistivity of the material, expressed in units called ohm meters (Ωm). L and A, respectively, are the length and the area of the cylinder as shown in Figure 2.36.

Materials are usually categorized as conductors, semiconductors, and insulators. Conductors have the smallest resistivity and insulators have the largest resistivity. Semiconductors have values of resistivity that fall between conductors and insulators. Therefore:

$$\rho \text{ (conductors)} < \rho \text{ (semiconductors)} < \rho \text{ (insulators)}$$

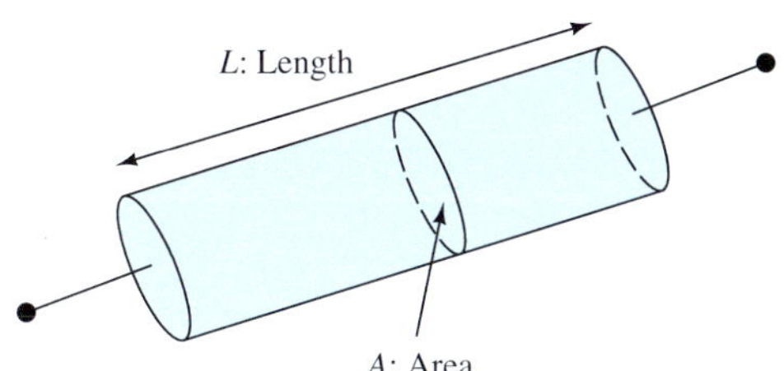

FIGURE 2.36 The geometry of a cylindrical conductor.

TABLE 2.1 The Resistivity Value of Some Materials

Material	Resistivity (Ωm)
Silver	1.59×10^{-8}
Copper	1.72×10^{-8}
Gold	2.44×10^{-8}
Aluminum	2.82×10^{-8}
Tungsten	5.6×10^{-8}
Iron	1×10^{-7}
Brass	0.8×10^{-7}
Silicon	6.4×10^{2}
Glass	10^{10} to 10^{12}

It should be noted that the value of the resistivity of a material may vary with time. For example, while aluminum is a conductor, aluminum contacts turn into aluminum oxide and aluminum oxide functions almost like an insulator. This is one reason why aluminum wire is discouraged for most home wiring settings. See Sections 9.2 and 9.3 for details.

The values of the resistivity at normal temperatures for selected materials are shown in Table 2.1. Materials such as silver and copper are good conductors. Silicon is a semiconductor and glass is an insulator.

APPLICATION EXAMPLE 2.14 Length of Tissue

Suppose that surgeons remove a cylindrical-shaped piece of tissue of unknown length from a patient. Their measurement for a piece of tissue shows that $\rho = 250\ \Omega\text{m}$ and the cross section is $6.875 \times 10^{-9}\ \text{m}^2$. When a voltage of 30 V is applied across the entire length of the tissue, the current flowing through it is 5.65 μA (see Figure 2.37). Find the length of the tissue.

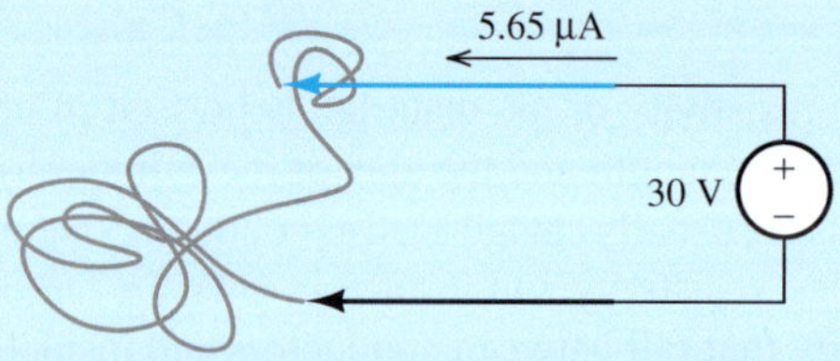

5.65 μA

30 V

FIGURE 2.37 Application problem (length of tissue).

SOLUTION

Assume R is the resistance of the tissue. Using Ohm's law, we can state that:

$$30 = R \times 0.00000565$$

Next, we can calculate that R is:

$$R = 5310000\ \Omega = 5.31\ \text{M}\Omega$$

According to Equation (2.11),

$$R = \frac{\rho L}{A} \Rightarrow 5.31\ \text{M}\Omega = \frac{250\ \Omega\text{m} \times L}{6.875 \times 10^{-9}\ \text{m}^2}$$

Using this equation, we can find the length $L = 146\ \mu\text{m}$. The equivalent circuit is shown in Figure 2.38.

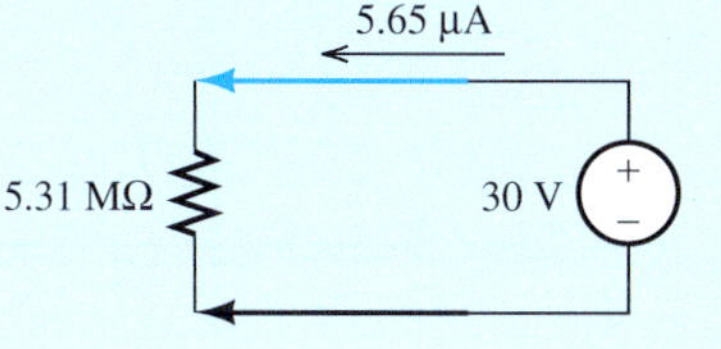

FIGURE 2.38 The equivalent circuit for Example 2.14.

EXAMPLE 2.15 Computing Resistance

Water conducts electricity. Electricity and water can be a fatal combination, as will be detailed in later chapters. As is also the case with other conductors, water resists the flow of electric current. Pure water has a resistivity of $\rho = 182$ kΩ-m. However, most water has some additive minerals or salt, and has a resistivity of about $\rho = 175$ kΩ-m.

In the water tank of Figure 2.39, find the resistance between (1) points A and B and (2) points X and Y, given $\rho = 175$ kΩ-m. Assume that the sides of the tank do not conduct current.

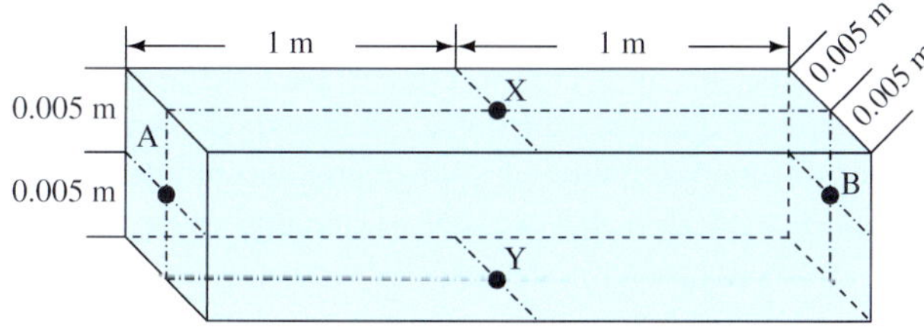

FIGURE 2.39 Water tank for Example 2.15.

SOLUTION

Using Equation (2.11), we know that:

$$R = \frac{\rho \times L}{A}$$

Next, we can apply the parameters outlined earlier to this equation.

a. When measuring the resistance between points A and B (i.e., the right and the left sides of the pool), we see that the length of the pool is 2 m and the cross section is 0.1 times 0.1. Thus:

$$R_{AB} = \frac{175{,}000 \times 2}{0.01 \times 0.01} = 3500\ \text{M}\Omega$$

b. When measuring the resistance between points X and Y, that is, the top and the bottom of the pool, the length is 0.01 and the cross section is 0.1 by 2 m. Thus:

$$R_{XY} = \frac{175{,}000 \times 0.01}{2 \times 0.01} = 87.5\ \text{k}\Omega$$

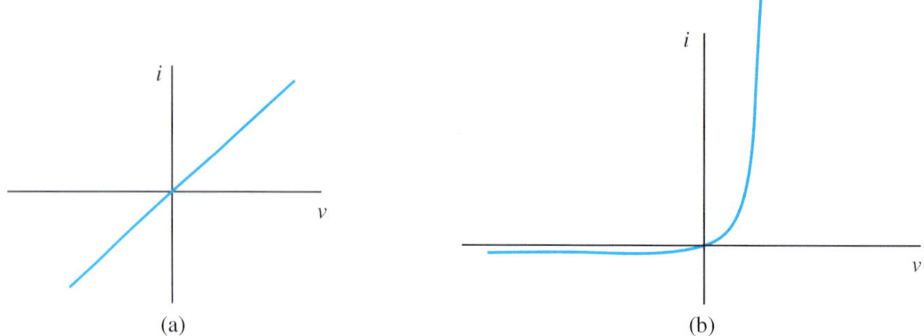

FIGURE 2.40 The v–i characteristics of a (a) linear and (b) nonlinear (diode) resistor.

2.7.2 Nonlinear Resistors

The linear relationship between voltage and current defined in Equation (2.10) is valid for linear resistors. In other words, as shown in Figure 2.40(a), for a linear resistor, if you sketch the voltage in terms of current or vice versa, the slope of this curve will not change with current and voltage. Therefore, for all currents and voltages, the resistance of a linear resistor is constant and remains unchanged.

However, note that in general, resistors are not linear. Therefore, the voltage–current relationship of the resistor is not linear. In other words, the resistance of a resistor may not be independent of the current that flows through it or the voltage across it. As a result, the slope of the characteristic curve of nonlinear resistors changes with current and/or voltage. Diodes are an example of nonlinear resistors (see Chapter 8).

The v–i characteristics of a diode are sketched in Figure 2.40(b). A diode is a device that allows the current to flow in only one direction in a circuit. Diodes are used in many applications, including voltage regulation. Chapter 8 discusses the details of how diodes are constructed. In Figure 2.40(b), we observe that for different amounts of voltage and current, the slope of the curve is different. Thus, the equivalent resistance is not constant. Typically, when the voltage is high enough, the slope tends toward infinity; therefore, the resistance tends toward zero. Inversely, when the voltage is negative, the slope is almost zero. In that case, the resistance is infinity.

2.7.3 Time-Varying Resistors

The magnitude of resistors may change with time due to environmental factors, such as temperature. As an example, resistivity of resistors changes with temperature. Because of this, in general, the resistance of a resistor varies with time due to the changing environmental factors. As a result, the characteristic curve of a time-varying resistor will not have the same slope at all times. In other words, the slope of the line in Figure 2.40(a) may change with time. In general, in this book, we consider only linear time-invariant resistors and we simply call them resistors.

2.8 POWER AND ENERGY

Electric circuits are used to generate energy for diverse applications. The electrical energy in power distribution systems is used to provide light, heat, and so on. The power and energy associated with a circuit element can be determined directly from the element's current, i, and voltage, V.

From the definition of voltage, we know that voltage is in fact the potential energy per charge and we know that current is the rate of charge passing through a circuit element. In other words, voltage and current correspond to:

$$\text{Voltage} = \frac{\text{Energy}}{\text{Charge}} \qquad \text{Current} = \frac{\text{Charge}}{\text{Time}} \tag{2.12}$$

TABLE 2.2 Notations for Voltage and Current

	DC	AC	Total: AC + DC	Amplitude of AC
Current	I_A	i_a	i_A	I_a
Voltage	V_A	v_a	v_A	V_a

Multiplying voltage and current results in the rate of energy transfer, also called power. Therefore, if power is denoted by p, the following equation can be developed:

$$p = v \cdot i \tag{2.13}$$

where the corresponding units for power are:

$$\text{Volt} \times \text{Ampere} = \frac{\text{Joule}}{\text{Coulomb}} \times \frac{\text{Coulomb}}{\text{Second}} = \frac{\text{Joule}}{\text{Second}} = \text{Watt} \tag{2.14}$$

Units of power are called watts (W), where 1 watt = 1 joule/second (J/s) = volts × amperes.

Note that lower case v is a general notation for a voltage that can be constant or time varying. If we know that the voltage is constant, usually, capital V is used. If we know that voltage is time varying then the notation $v(t)$ is used. The same notation is applied to current, i, i.e., i is a general notation for any current and $i(t)$ is for time-varying current. In addition, throughout this book, we consider two general types of currents and voltages: DC and AC (see Figure 2.7). The notations for AC and DC currents and voltages are listed in Table 2.2. It should be noted that in Table 2.2, "A" refers to a branch for current and a node for a voltage. That is, i_A is the total current through branch "A," and v_A is the total voltage at Node "A."

In Table 2.2, the total current (voltage) refers to the DC plus AC current (voltage) (see the third column of Table 2.2). These notations are used, because in general, a circuit includes both DC and AC sources. Similarly, upper case P is used for DC power (or for phasors in Chapter 6). In addition, lower case p is used for time-varying power. For example, in a radio (e.g., your cell phone), the DC source is the battery, and the AC source is the signal induced over the radio antenna. As detailed in Chapter 8, the DC source is used to bias the transistors of a radio system to allow the amplification of the weak AC signal induced on the antenna. Note that DC analysis of circuits is discussed in Chapter 3, while AC analysis of circuits is discussed in Chapter 6. As discussed in Chapter 6, an AC signal is represented using a sinusoid. Figure 2.7(a) represents a sinusoid. As shown in Figure 2.7, a sinusoid includes an amplitude. The last column in Table 2.2 represents the notation for the amplitude of the sinusoid. For details, refer to Chapter 6.

An electric circuit consists of one or more sources. A source supplies power and energy to other elements. A source is also called an **active element**. All power supplied by active elements in a circuit is consumed, dissipated, or converted by other elements in the circuit.

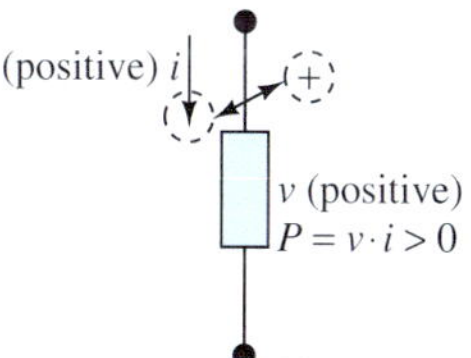

FIGURE 2.41 Passive element.

Elements that absorb energy are called **passive elements**. Passive and active elements in a circuit can be identified by considering the element's voltage polarity relative to the element's current direction. When the current direction is the same with the voltage drop direction (Figure 2.41), the element absorbs energy and it is passive. In this case, the power, $P = v \times i$, is a positive quantity. In other words, the element consumes positive power.

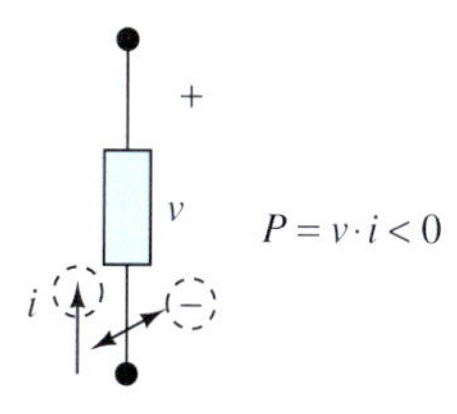

FIGURE 2.42 Active element.

On the other hand, if the current direction is not the same with the voltage drop direction (Figure 2.42), the element supplies energy and is considered an active element. In this case, the element consumes negative power (delivers positive power). It should be noted that in Figures 2.41 and 2.42, we consider the real (positive) directions/polarities of current/voltage. In contrast, the directions discussed in Figure 2.10 represent the relationship of current/voltage directions/polarities required for circuit analysis.

In general, if the current multiplied by the voltage for an element is negative, that element is an active element; otherwise it is a passive element (see Figures 2.41 and 2.42).

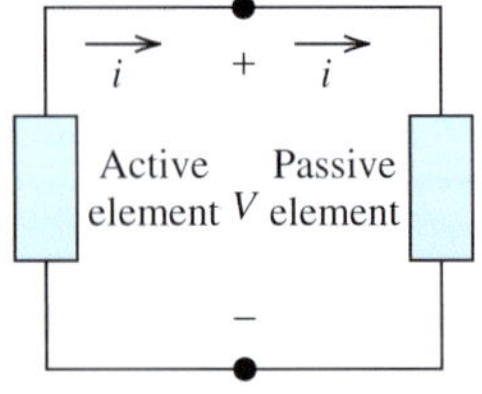

FIGURE 2.43 Passive and active elements connected to each other.

If active and passive elements are connected together as shown in Figure 2.43, the active element will supply energy to the passive element. As a result, the active element delivers energy and the passive element absorbs energy. The voltage polarity for both the elements is the same, but the current direction relative to the voltage polarity is different for each.

In general, if the element's current and voltage vary with time, the power of the element is a function of time as well. In this case, we can develop the following equation:

$$p(t) = v(t) \cdot i(t) \tag{2.15}$$

The total amount of energy supplied by a source element and consumed by a passive element can be determined by integrating the power time-dependent function over a specified time interval. As a result, the transferred energy, w, between time instants t_1 and t_2 corresponds to:

$$w = \int_{t_1}^{t_2} p(t)\mathrm{d}t \tag{2.16}$$

EXAMPLE 2.16 Voltage Direction Versus Current Direction

Assuming the directions of current and the polarities of voltages shown in Figure 2.44 refer to the true (positive) values, do the following voltage/current labels conform to how passive elements should be referenced?

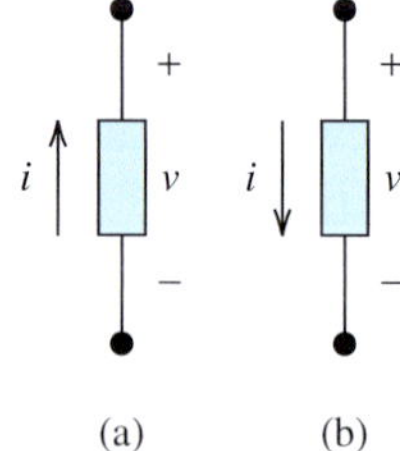

FIGURE 2.44 Configuration for Example 2.16.

SOLUTION

In passive elements, the current enters the positive polarity of the voltage; therefore, elements a and b in Figure 2.44 are active and passive, respectively.

APPLICATION EXAMPLE 2.17 Robotic Arm

A robotic arm—5 m long—is designed to load containers onto a cargo ship (Figure 2.45). The maximum mass that can be lifted by the robotic arm is 500 kg at a constant angular velocity of 0.2 rad/s. The power required to initiate the lifting is five times the power required for the lifting process. Assume that the voltage source for the robotic arm is 500 kV. Calculate the current flow through the circuit at the initial time, $t = 0$, and the current during the loading process. Assume the gravitational acceleration is $g = 10\ \text{ms}^{-2}$.

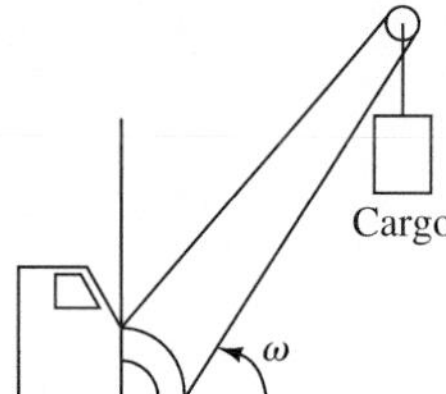

FIGURE 2.45 An example of a robotic arm used to load and unload a ship's cargo.

The equations needed to solve this problem include the following:

1. The relationship between the torque, distance (L), and force (F) is: $\tau = L \times F$.
2. The minimum force required to lift an object with mass m is: $F = m \times g$.
3. The relationship of power and torque is: $P = \tau \times \omega$.

SOLUTION

The torque required to lift up a 500 kg container is:

$$\begin{aligned}\tau = L \times F &= L \times \text{ m} \times g\\ &= 5\text{ m} \times 500\text{ kg} \times 10\text{ ms}^{-2}\\ &= 25{,}000\text{ Nm}\end{aligned}$$

The angular velocity is $\omega = 0.2$ rad/s; thus, the required power for the lifting process is:

$$P_{\text{process}} = \tau \times \omega = 25{,}000 \times 0.2 = 5000\text{ W}$$

The power required to initiate the lifting is five times the power required for the lifting process, thus:

$$P_{\text{initial}} = 5P_{\text{process}} = 25{,}000\text{ W}$$

Given voltage, V, is 500,000 V, the initial current is:

$$I_{\text{initial}} = \frac{P_{\text{initial}}}{V} = 0.05\text{ A}$$

The current during the lifting process is:

$$I_{\text{process}} = \frac{P_{\text{process}}}{V} = 0.01\text{ A}$$

APPLICATION EXAMPLE 2.18 Voltage Adapter

Many electronic devices are connected to their power source through an adapter. Because the "standard" voltage across different countries is not the same, universal device adapters have become popular. One example of this is adapters for laptop computers that enable the capability to connect to both 110 V and 220 V voltage sources without an additional step-down transformer.

Figure 2.46 represents a block diagram of a universal adapter. The power factor control (PFC) consists of a variable resistor that changes the resistance so that the power input to the transformer remains unchanged, regardless of the input voltage. Transformers will be discussed in more detail in Chapters 8 and 12.

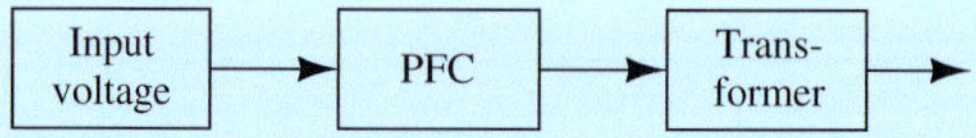

FIGURE 2.46 Block diagram of a universal adapter.

Assume that the desired adapter output is 15 V and 6 A, with 100% power efficiency. What is the value of the PFC resistor when:

a. input voltage is 110 V
b. input voltage is 220 V

a. First, calculate the output power:

$$P = IV$$
$$P = 6 \times 15$$
$$P = 90\,\text{W}$$

For a 100% power efficiency transformer, the power input would be the same as output power. In addition, we know that the input voltage is 110 V. Therefore, the required input current is:

$$I_{\text{input}} = \frac{P}{V_{\text{input}}} \rightarrow I_{\text{input}} = \frac{90}{110} \rightarrow I_{\text{input}} = 0.8182\,\text{A}$$

The required resistance for 110 V input voltage for the PFC is therefore:

$$R_{110} = \frac{V_{\text{input}}}{I_{\text{input}}} \rightarrow R_{110} = 134.44\,\Omega$$

b. We know that the input and output powers are the same in this problem. We have already calculated the power as 90 W. Thus, the resistance required for a 220 V input voltage can be determined directly using the definition below:

$$P = \frac{V^2}{R}$$

Where V is the input voltage and R is the required resistance. Thus, the required resistance is:

$$R_{220} = \frac{V^2}{P} \rightarrow R_{220} = \frac{220^2}{90} \rightarrow R_{220} = 537.78\,\Omega$$

2.8.1 Resistor-Consumed Power

Because resistors are passive elements, they always absorb energy. The absorbed electrical energy is converted to other types of energies such as heat or light. The power, P, absorbed by a resistor can be found by multiplying the current, I, passing through it and voltage, V, across its terminals (Section 2.8 discusses the power and energy in detail). Because in a resistor the ratio of voltage to current is expressed by the constant resistance value, R, we can state that:

$$P = V \times I = \frac{V^2}{R} = R \times I^2 \quad (2.17)$$

Because R is always a positive value, Equation (2.17) confirms that the power of a resistor (similar to the power of all other passive elements) is positive. As has been discussed, the value of the resistance of an ideal resistor or conductor is completely independent of the voltage across its terminals and the current passing through it.

EXAMPLE 2.19 Power Dissipation

Figure 2.47 represents a circuit that includes a dependent current source as detailed in Section 2.9. Here, the current of the current source is a function of the current in the 3 kΩ resistor. How much power is dissipated in the 33 kΩ resistor in Figure 2.47?

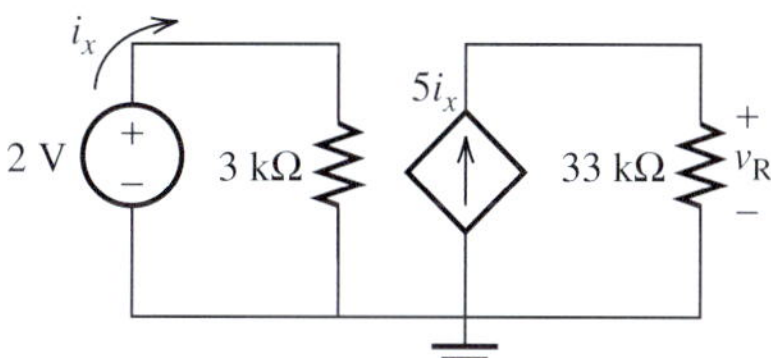

FIGURE 2.47 Circuit diagram for Example 2.19.

SOLUTION

The current source generates a current in the 33 kΩ resistor that depends on the current passing through the 3 kΩ resistor. To compute the voltage across the 33 kΩ resistor and then evaluate its power, we first find the value of i_x. Using Ohm's law, we can state:

$$i_x = \frac{2\ \text{V}}{3\ \text{k}\Omega} = 667\ \mu\text{A}$$

Therefore, $5i_x = 3.33$ mA. Then:

$$v_R = 3.33\ \text{mA} \times 33\ \text{k}\Omega = 110\ \text{V}$$

Next, we are able to find the value of the power dissipated in the 33 kΩ resistor. Using Equation (2.17), we see that:

$$P_R = 110 \times 3.33 \times 10^{-3} = 366.3\ \text{mW}$$

EXERCISE 2.3

Find the power dissipated in the 33 kΩ resistor if the dependent current source in Figure 2.47 is replaced by a dependent voltage source with its voltage a function of the voltage across the 3 kΩ resistor. (See Section 2.9 for a more detailed investigation of dependent and independent sources.)

APPLICATION EXAMPLE 2.20 Door Opener

A driveway gate shown in Figure 2.48 is driven by an electric motor that is connected to a voltage source of 200 V. In order to move the driveway gate, a force of 500 N or more is required. Assume that the motor drives the driveway gate at a constant speed of 0.2 ms^{-1} and that the gate path is 4 m long. Also, assume that the power required to start the motor is the same as the power required to move the gate. Calculate the total power that will be required for this process, and the possible resistance in the motor.

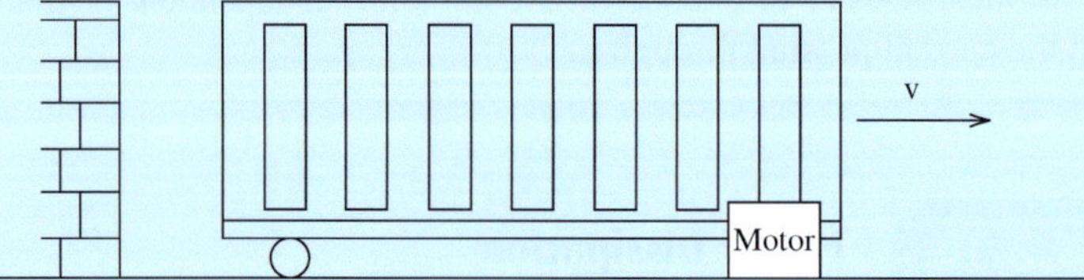

FIGURE 2.48 A simple illustration of a driveway gate driven by an electric motor.

SOLUTION

Here, force, $F = 500$ N, velocity, $V_e = 0.2$ ms^{-1}, and distance, $d = 4$ m. Thus, the total time of the process is $t = d/V_e = 20$ s. In addition, we can calculate the power needed as:

$$P = \frac{Fd}{t} = 100 \text{ W}$$

To calculate the resistance, we can use the following equation:

$$P = \frac{V^2}{R}$$

Given that $V = 200$ V, and the result of the power calculation ($P = 100$ W), we see that the resistance corresponds to $R = 400\ \Omega$.

2.9 INDEPENDENT AND DEPENDENT SOURCES

An *independent voltage source* is a voltage source (e.g., a battery or utility-connected electrical outlet) applied across a circuit that generates a current through the circuit. A voltage source maintains a constant or time-varying voltage across its terminal nodes.

An independent voltage source generates a predetermined voltage that is independent of the current flowing through and/or voltage across any given element of the circuit.

Figure 2.49(a) shows an ideal independent voltage source. The current passing through this element is determined by other circuit elements.

The voltage across the terminals of an independent DC voltage source (e.g., a battery) is constant and does not change with time. For example, the constant DC voltage in Figure 2.49(b) is 5 V.

EXERCISE 2.4

Sketch the voltage (y-axis) versus current (x-axis) diagram of an independent voltage source. Is it a line parallel to the (x- axis)? This diagram is called a v–i diagram. Resketch the voltage source v–i diagram when the voltage is zero. This represents the v–i diagram of a short circuit. In other words, a zero voltage source represents a short circuit.

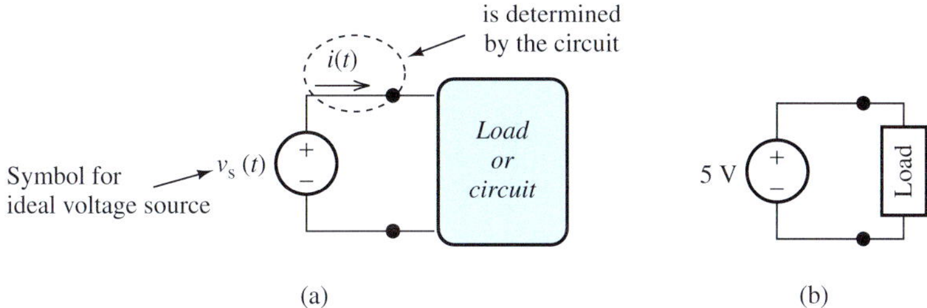

FIGURE 2.49 (a) Ideal voltage source symbol and (b) a 5 V DC source connected to a load.

In AC voltage sources, the voltage varies with time. An example of an AC voltage source is the signal that can be obtained from standard electrical outlets in a home. In Figure 2.50, the AC voltage is in the form of a sinusoid where the voltage changes periodically with the time. The symbol for an independent voltage source is a circle (see Figure 2.49). The sinusoidal waveform in the circle in Figure 2.50 represents the varying nature of the voltage. Note: standards for residential electric power vary from country to country. In this figure, the voltage source frequency is 50 Hz ($50 = 100\ \pi/2\pi$).

A *dependent or controlled voltage source* creates a voltage across its terminals. An automotive alternator is an example of a dependent voltage source—voltage is dependent on the revolutions per minute (RPM) of the engine. A dependent voltage is a function of the voltage across or current through a circuit element. In Figure 2.51, the voltage across the terminals of the voltage source is controlled by the voltage across the terminals of the element B. In this circuit, at a given time, the voltage delivered to the circuit by the source is four times the voltage across the element B at that time.

The voltage supplied by a dependent voltage source can also be controlled by a current passing through an element. Figure 2.52 shows a dependent voltage source. Here, the voltage

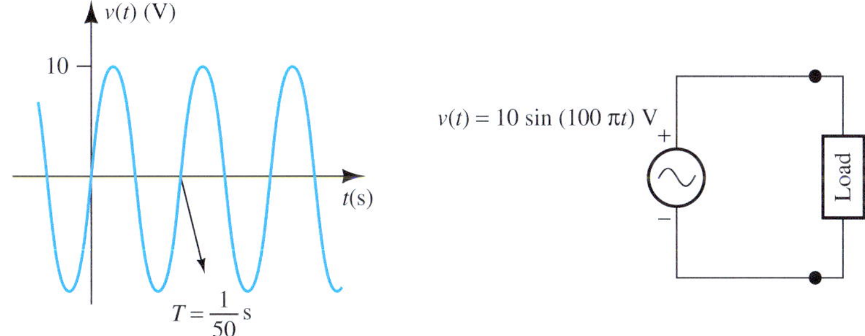

FIGURE 2.50 AC voltage source.

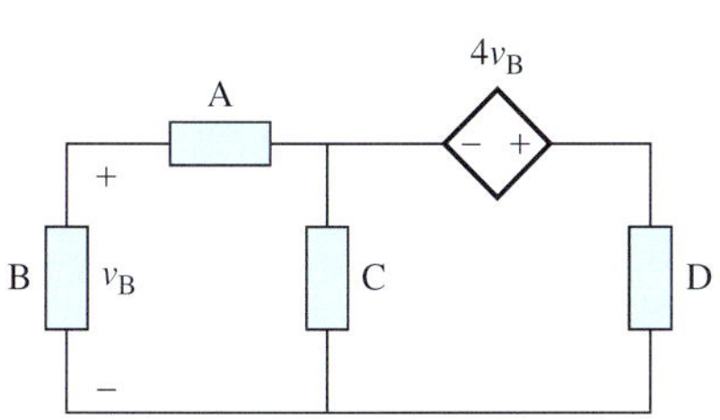

FIGURE 2.51 Voltage controlled voltage source.

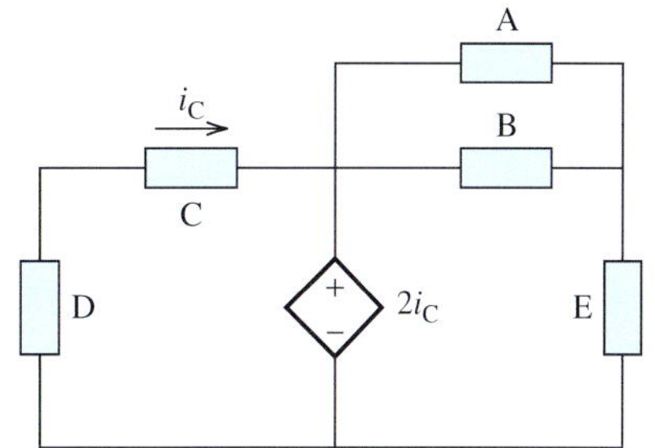

FIGURE 2.52 Current voltage source.

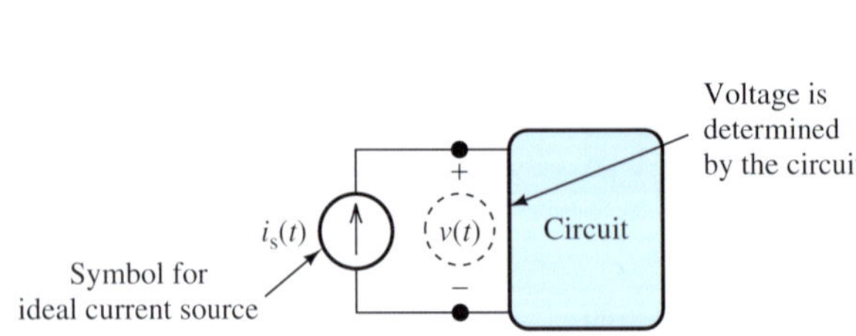

FIGURE 2.53 An ideal current source.

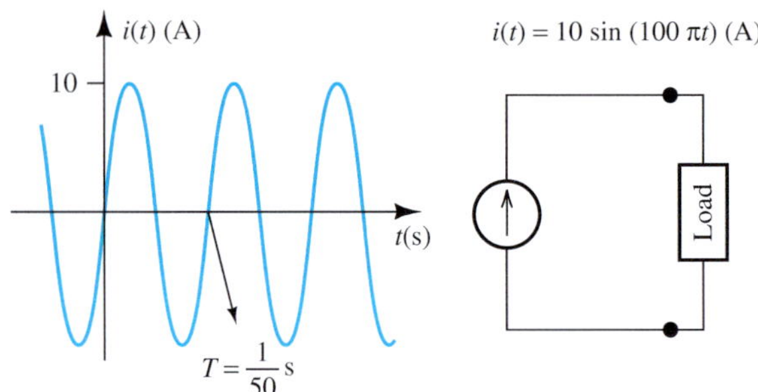

FIGURE 2.54 AC source.

provided by the source is two times the current passing through element C. A diamond is the common symbol for showing a current or voltage-controlled source in a circuit. The energy might be delivered to a circuit via a voltage source or a current source.

An *independent current source* is a current source that supplies energy to a circuit in the form of current. A current source maintains a specific current flow through its branch. Unlike voltage sources (e.g., batteries that can be observed physically) current sources are usually a model for electronic systems. Figure 2.53 shows an ideal independent current source that generates a current independent of the connected circuit.

EXERCISE 2.5

Sketch the voltage (y-axis) versus current (x-axis) diagram for an independent current source. Is it a line parallel to the (y-axis)? Resketch the current source v–i diagram when the current is zero. This represents the v–i diagram of an open circuit. As described, a zero-current source represents an open circuit.

The symbol commonly used for an independent current source is a circle with an arrow inside. The direction of the arrow represents the true direction of the flow of current. The DC and AC sources, respectively, force constant and time-varying currents in a circuit. Figure 2.54 represents an example of AC supply.

Dependent or controlled current sources generate currents that are determined by a current through or voltage across a circuit element. Voltage-dependent current sources establish a current in a circuit that is controlled by a voltage across the terminals of an element in the circuit. Current-dependent current sources create a current that is controlled by a current passing through an element. Figures 2.55 and 2.56 show some examples of current- and voltage-controlled current sources. A diamond with an arrow inside is the common symbol for dependent current sources.

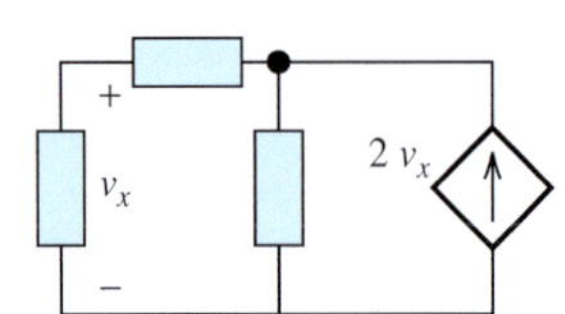

FIGURE 2.55 Voltage-controlled current source.

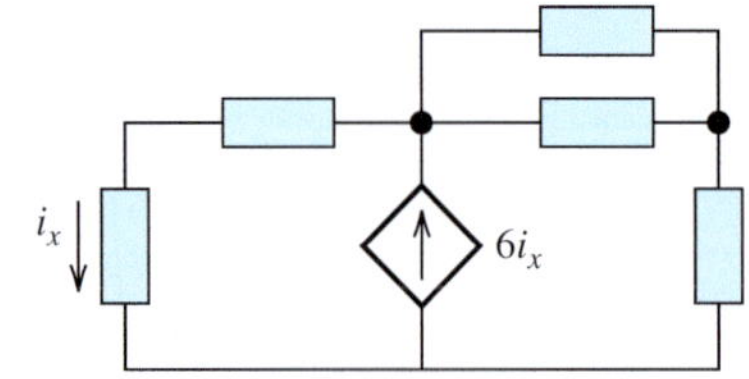

FIGURE 2.56 Current-controlled current source.

EXAMPLE 2.21 Dependent Source

Find the current, I_x, that is shown in Figure 2.57.

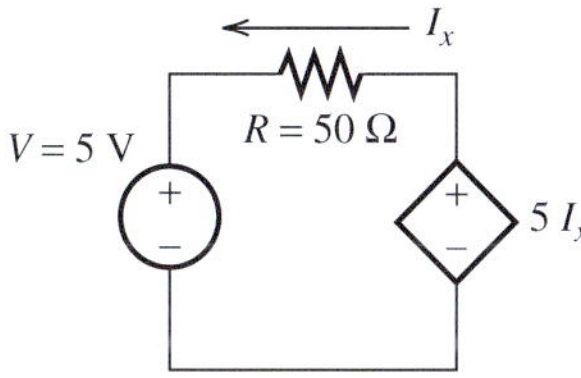

FIGURE 2.57 Circuit diagram for Example 2.21.

SOLUTION

Here, the current direction is arbitrarily selected to be counterclockwise, which is the opposite direction of the voltage source, 5 V. Using the values from Figure 2.57, the equation is:

$$5I_x = 50I_x + 5$$

Therefore, the current, I_x, is:

$$-45I_x = 5$$

$$I_x = -0.11 \text{ A}$$

The negative sign of the current shows that the correct current direction is clockwise instead of counterclockwise, as shown in Figure 2.57.

EXAMPLE 2.22 Dependent Source

Determine the voltage, V_A, across the 50-Ω resistor in Figure 2.58.

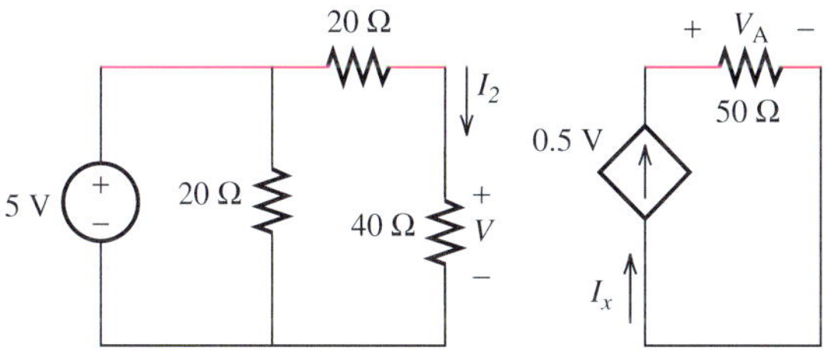

FIGURE 2.58 Circuit diagram for Example 2.22.

SOLUTION

V_2 is the voltage that corresponds to the current, I_2, which is also the voltage across both 20-Ω and 40-Ω resistors. Then, using the amounts from Figure 2.58, we can develop the equation:

$$I_2 = \frac{5}{40 + 20} = 0.0833 \text{ A}$$

The voltage across 40-Ω resistor, V, is:

$$V = 40I_2 = 3.33 \text{ V}$$

Then, the current, I_x, is:

$$I_x = 0.5 \text{ V} = 1.667 \text{ A}$$

The voltage, V_A, can then be calculated as:

$$V_A = 50I_x = 83.33 \text{ V}$$

2.10 ANALYSIS OF CIRCUITS USING PSPICE

This section provides an overview of the capabilities of PSpice software for circuit analysis. You will learn how to set up a PSpice circuit step by step. In order to analyze a circuit using PSpice, three stages are required:

a. Creating a new project file
b. Drawing circuit in the PSPICE schematic window
c. Running simulation analysis

This section explains the details of these stages.

a. Creating a new project file

1. Click on the PSpice icon and open the main page. Figure 2.59 shows a screenshot of the main PSpice page:
2. Go to "File" > "New" and choose Project.
3. A window (as shown in Figure 2.60) will pop up. Type a filename of your choice, for example, "Test01." Next, you will choose from among the different categories of projects allowed by PSpice, including the following:
   - **i.** ***Analog or Mixed A/D:*** Analog or digital circuits can be set up on the PSpice layout, and simulations can be run for circuit analysis.
   - **ii.** ***The PC Board:*** Used for setting up a PC Board layout.
   - **iii.** ***The Programmable Logic Wizard:*** Allows the users to prepare their design with the aid of a complex programmable logic device (CPLD) or a field programmable gate array (FPGA) tool. The libraries of the project are configured based on the manufacturer's type/part that is chosen.
   - **iv.** ***The Schematic:*** This option creates a project that contains only a design file.

   In this book, we mainly deal with analog or mixed A/D circuits. As such, choose the option "Analog or Mixed A/D project," as shown in Figure 2.60, and press OK. The file can be saved in a location of your choice.

4. As soon as "OK" is selected, the next window (shown in Figure 2.61) will pop up to ask whether to use a project created based on an existing project for the file. At this stage, you may use an existing project and modify the parameters or you may create a new one. In most cases, we will create a new project. Thus, we choose "Create a blank project" and press OK.
5. Next, the main PSpice schematic circuit board window will appear (as shown in Figure 2.62). Circuits can be easily created in this board.

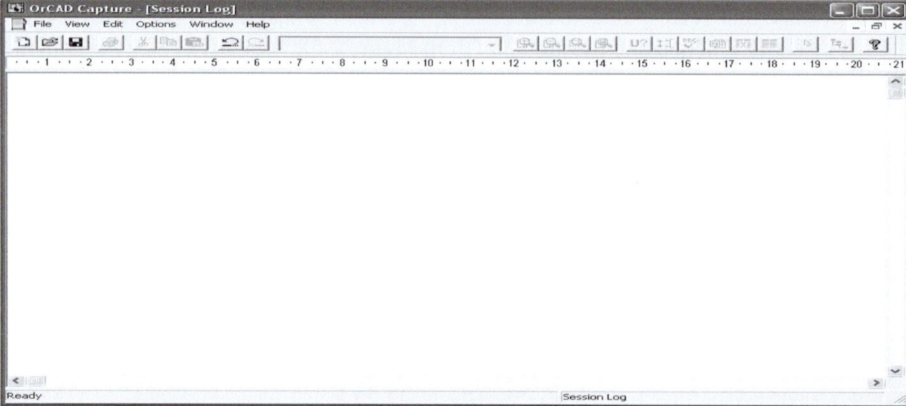

FIGURE 2.59 Main page of PSpice.

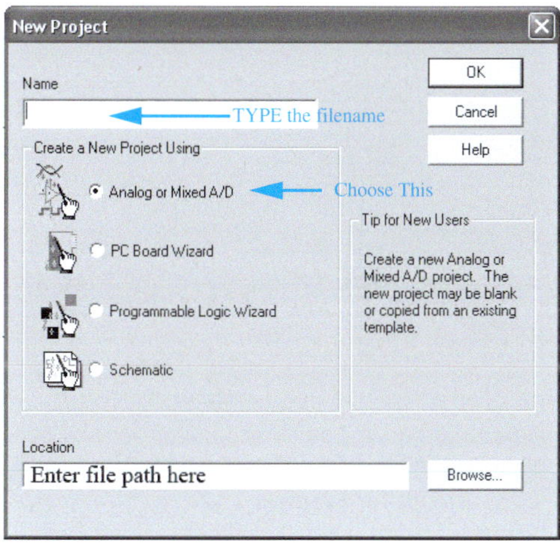

FIGURE 2.60 New project.

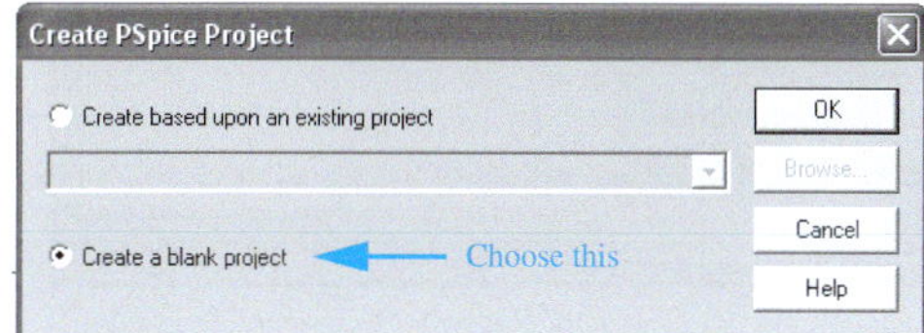

FIGURE 2.61 Create a PSPICE project.

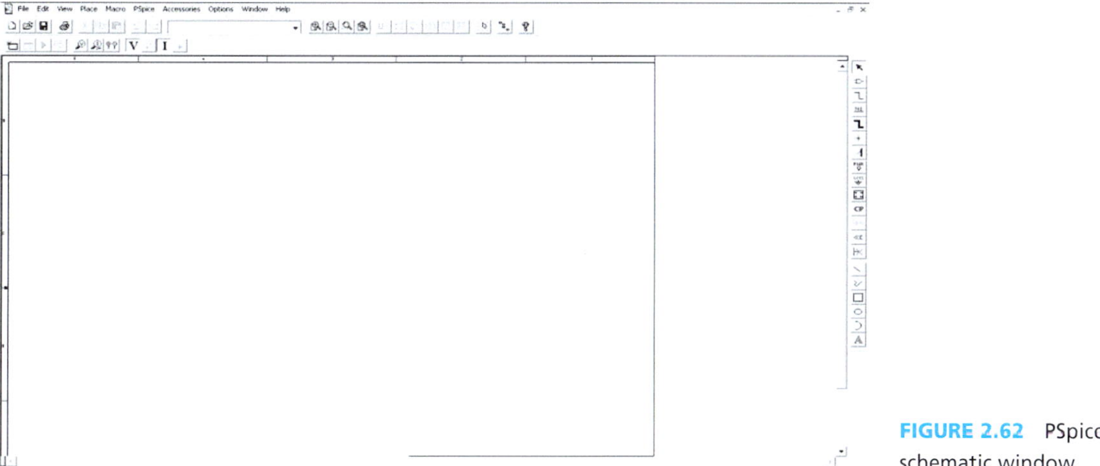

FIGURE 2.62 PSpice schematic window.

b. Drawing a circuit in the PSpice schematic window

PSpice consists of many tools to facilitate drawing circuits. Each category of these tools is allocated a specific icon located on the right-hand side of the schematic window. The following are some important examples of these icons:

- ***Place Part icon:*** This contains the various analog and digital parts that are needed to set up an electronic circuit.
- ***Wire icon:*** This connects the parts on the circuit board to create a complete circuit.
- ***Ground icon:*** All the analog circuits should be connected to a Ground to form a complete circuit.

Now, considering the circuit of Example 2.12 as an example; the basic PSpice schematic and simulation steps are shown below:

1. Select "Place" > "Part" or click on the icon . A window (as shown in Figure 2.63) will pop up.
2. The electronic parts are stored in a specific library file. Therefore, the library files are required to load into the Parts Database and search and place the particular electronic parts on the PSpice circuit board. To place the parts, choose Add Library (STEP A in Figure 2.63) and highlight all of the files or "Ctrl+A" to add all of the *.olb/*.lib files into the library.

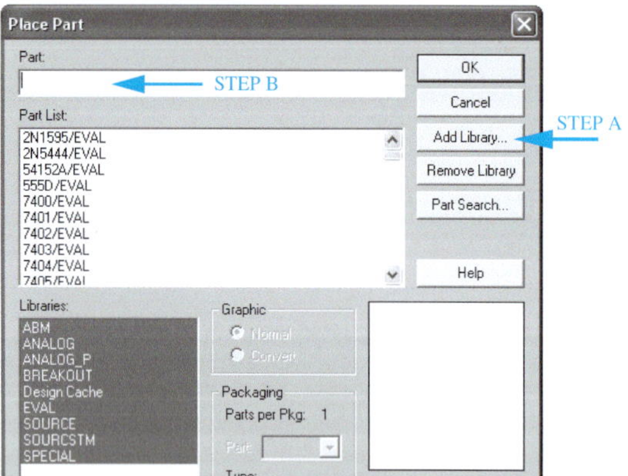

FIGURE 2.63 Place Part and Add Library.

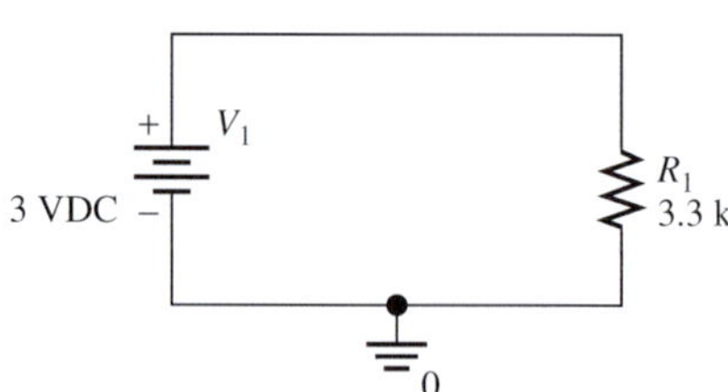

FIGURE 2.64 PSpice schematic circuit for Example 2.12.

3. Continuing with Example 2.12, to add a DC voltage source, type "Vdc" in the Part column (STEP B in Figure 2.63) and then click OK. Place the "Vdc" icon as shown in Figure 2.64, and then right click and choose "End Mode." *Press "r" to change the orientation or direction of the voltage source.*
4. To set the voltage source value, double-click on the "0 V DC" and type "3 V DC," where DC stands for direct current.
5. To place a resistor on the circuit board, go to "Place" > "Part" and type "R" in the Part column, then click OK. (Note: be sure to use R/analog and not R/discrete.) Place the resistor, "R1" as shown in Figure 2.64. *Press "r" to change the orientation of the resistor.*
6. To change the resistance of resistor, double click the "1 k" and change it into "3.3 k."
7. To connect the voltage and resistor, click "Place" > "Wire," or the icon on the vertical bar to the right. Left-click inside the square box of the voltage source (shown in Figure 2.65) to create the connection between the voltage source and the wire. For the other end of the wire, click on the empty square box of the resistor to create a connection between the voltage source and the resistor. A colored solid circle will appear during the connection if it is connected properly (shown in blue in Figure 2.66). Right-click and choose "End Wire" after connecting the wire to the resistor.
8. The direction of the wire can be changed by left-clicking once on the circuit board.
9. Most of the time, the parts and their square boxes appear small on the screen. To ease the process of wiring the electronic parts in the circuit, you can zoom in so that the electronic parts are shown larger on the screen. You may also highlight the electronic parts as shown in Figure 2.67. Note: hold the Ctrl" button down on the computer to select multiple parts simultaneously. We can change the orientation of the resistor by pressing "r."
10. Next, press the "zoom in" (solid arrow) icon as shown in Figure 2.68. If you wish to zoom out, press the "zoom out" (dashed arrow) icon in Figure 2.68.
11. Every circuit in PSpice is required to be connected to a ground to become a complete circuit. Click "Place" > "Ground" or the icon on the vertical bar to the right.

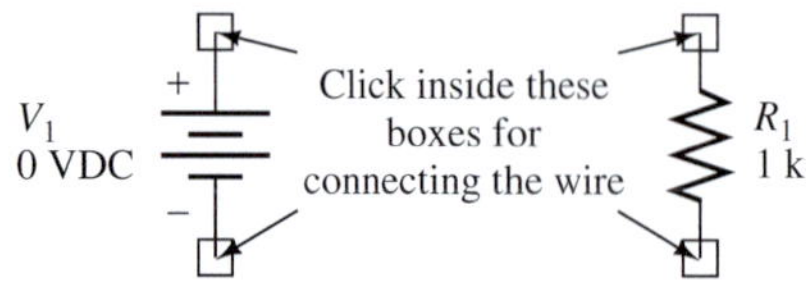

FIGURE 2.65 Empty square box on voltage source and resistor.

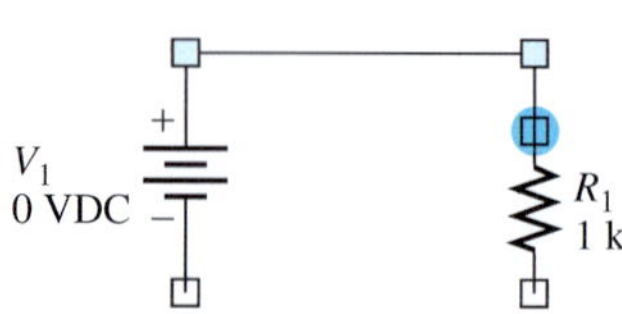

FIGURE 2.66 Colored circle for connection confirmation.

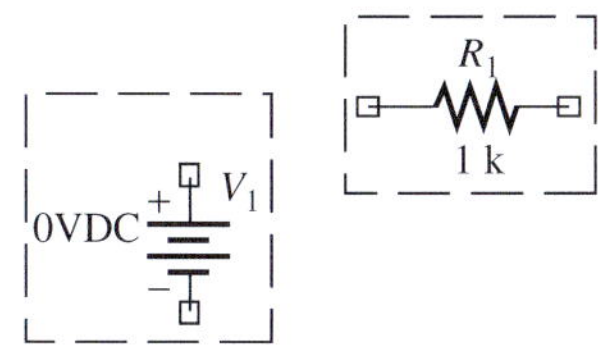

FIGURE 2.67 Highlight the electronic parts.

FIGURE 2.68 Zoom in and zoom out buttons.

12. A window that is similar to Place Part (Figure 2.63) will pop up. Choose add library, and choose source.olb/source.lib to add the ground into the library. Next, choose "0/Source" and place it as shown in Figure 2.64. Note: Always either press "Esc" or right click and choose "End" to complete the insertion.
13. You should be cautious not to choose and place any parts in the list of Figure 2.63 that have the suffix "Design Cache," that is, called "*partname*/Design Cache" in the PSpice circuit board. PSpice is unable to run any simulation analysis with those components.

c. **Running simulation analysis**

PSpice can run four types of simulation analyses, including the following:

i. ***Time Domain (Transient):*** The analysis of voltage and current over time.
ii. ***DC Sweep:*** The analysis of the circuit by changing voltage in linear or logarithmic steps.
iii. ***AC Sweep:*** Usually used when the voltage or current are time varying and also for frequency response analysis (e.g., Chapter 7).
iv. ***Bias Point:*** The analysis of voltage and current for a given (constant voltage or current), or when voltage or current varies around a bias voltage or current, for example, for the analysis of current and voltage in transistors and diodes (Chapter 8).

In this section, we will only look into Time Domain and Bias Point analysis. AC Sweep analysis will be discussed in Chapter 7.

Bias Point Analysis

1. First, go to "PSpice" > "New simulation profile" or click the bottom left icon that the **colored** arrow points to in Figure 2.69.
2. A window (shown in Figure 2.70) will pop up to prompt you to insert a simulation name. Type a preferred simulation name, for example,, "Bias Point" and then click "Create."
3. Another window, "Simulation Setting" will pop up, as shown in Figure 2.71. Go to the "Analysis" Tab, and choose the "Analysis Type" to be "Bias Point," and then click OK.
4. Go to "PSpice" > "Run" or click the "Play" icon (see the dotted arrow in Figure 2.69). The results are shown in Figure 2.72.
5. If either the voltage or current does not show up after running the simulation, click the "V" or "I" icon (for location, see the dashed arrow in Figure 2.69).

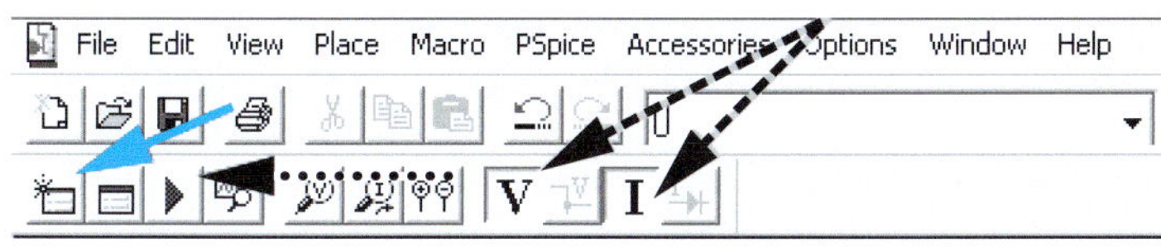

FIGURE 2.69 Create new simulation profile.

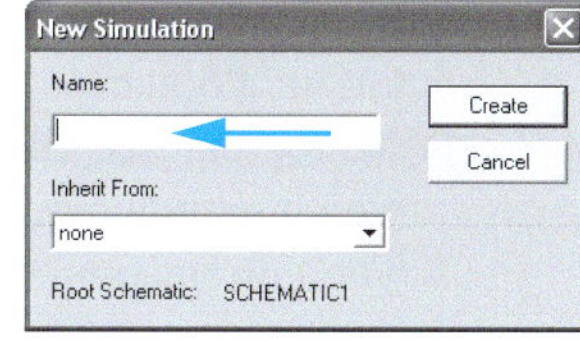

FIGURE 2.70 New simulation.

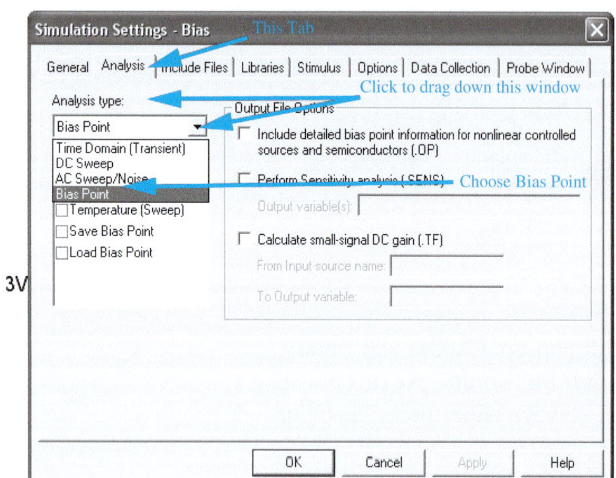

FIGURE 2.71 Simulation results for Example 2.12.

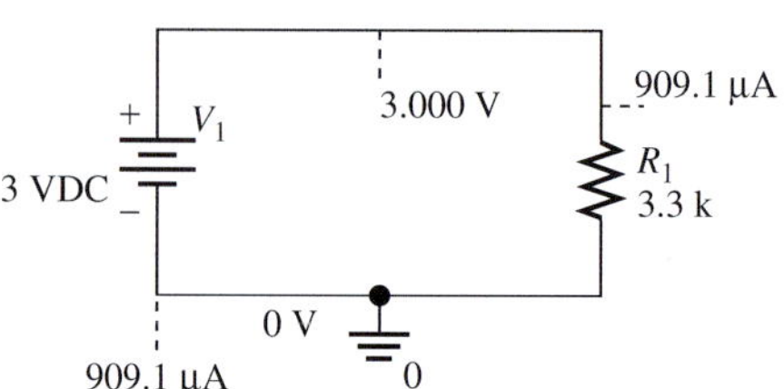

FIGURE 2.72 Simulation setting.

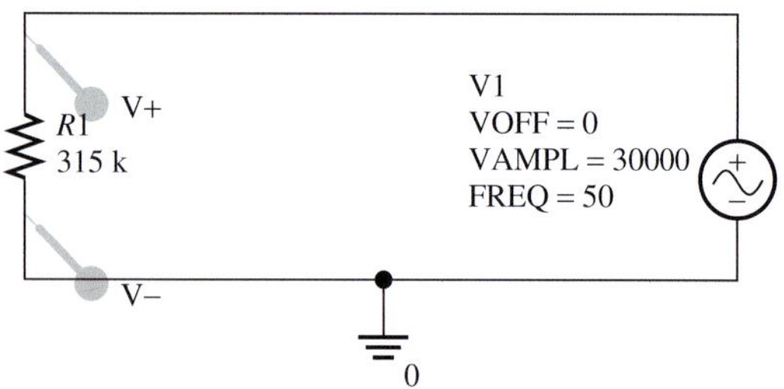

FIGURE 2.73 PSpice schematic circuit for Example 2.13.

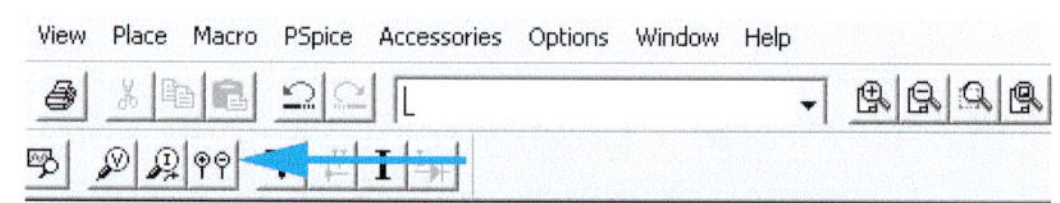

FIGURE 2.74 Voltage differential markers.

Time Domain (Transient) Analysis

Consider the circuit of Example 2.13 that is shown in Figure 2.73. Construct this circuit in PSpice using techniques similar to those shown earlier. It is clear that there is a set of "Voltage Differential Markers" at the resistance, R_1. The Voltage Differential Markers will allow the simulation to plot the desired observation automatically whenever the simulation is run. To place the Voltage Differential Markers on the circuit, click the "Voltage Differential Markers" (solid arrow) on the toolbars as shown in Figure 2.74. The first click (or odd number) of placing onto the circuit will be a positive Voltage Differential Marker, while the second click (or even number) of placing will be a negative Voltage Differential Marker.

Next, follow these steps to set up the time domain analysis simulation:

1. First, go to "PSpice" > "New simulation profile" or click the bottom left icon that is indicated by the solid arrow shown in Figure 2.69.
2. Insert a simulation name in the window when prompted. Type a preferred simulation name, for example, "Time Analysis" and then click "Create."
3. Another window, "Simulation Setting" will pop up in Figure 2.75. Go to the "Analysis" Tab, and choose the "Analysis Type" to be "Time Domain (Transient)."
4. Set the "Run to time" to 50 ms (milliseconds) and press OK.
5. Go to "PSpice" > "Run" or click the "Play" icon to run the simulation.
6. Notice that the Simulation Schematic Window will plot the voltage plot of the circuit, which is shown in Figure 2.76.
7. The current plot can be added using the "Add Trace" function. However, the amplitude of the current plot is much smaller than the voltage plot in most cases; therefore, we need to delete the voltage plot and replace it with the current plot. To do so, click the

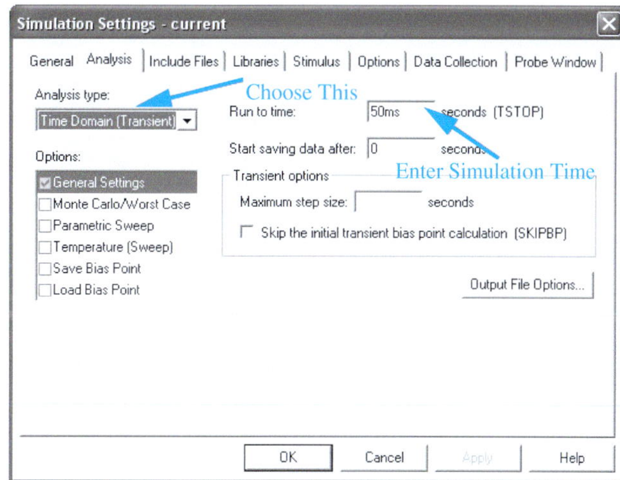

FIGURE 2.75 Set up a time domain (transient) simulation.

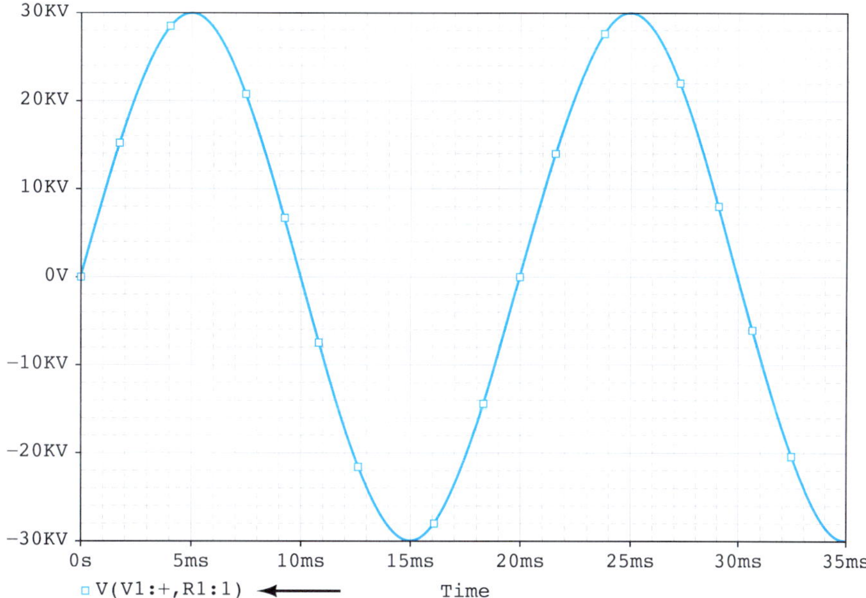

FIGURE 2.76 Voltage plot for time domain simulation.

"V(V1:+,R1:1)" that is located near the arrow in Figure 2.76, and press the "Delete" key on the keyboard.

8. Next, to add the current plot, go to "Trace" > "Add Trace."

9. A window will pop up as shown in Figure 2.77. The "add trace" window allows the user to perform mathematic summation, subtraction, multiplication, and so on on the "Simulation Output Variables" in the "Trace Expression" column for various plotting. However, in this section, we only intend to observe the current flow through the resistor R1; therefore, click "I(R1)" and choose OK, or double-click "I(R1)" in Figure 2.77.

10. The output of the plot for the current plot is shown in Figure 2.78.

Copy the Simulation Plot to the Clipboard to Submit Electronically

Sometimes, you may be requested to submit your work electronically. Unfortunately, the PSpice student/demo version does not allow the user to save the simulation plot in any type of file.

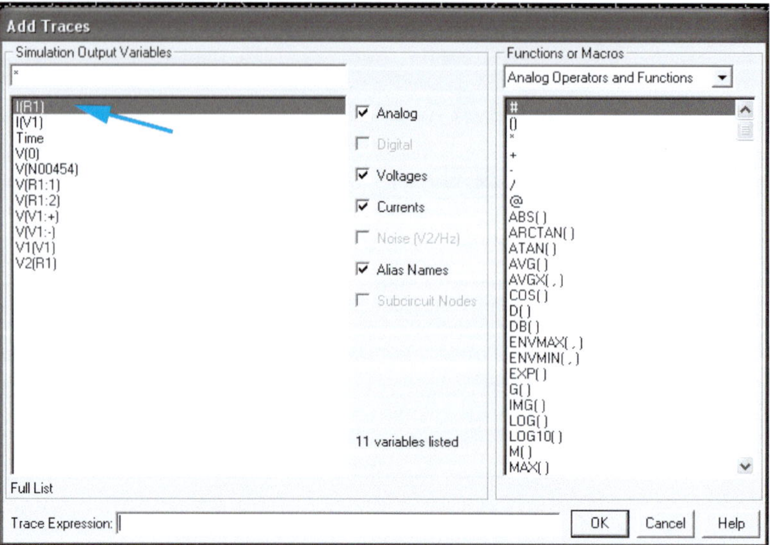

FIGURE 2.77 Add traces.

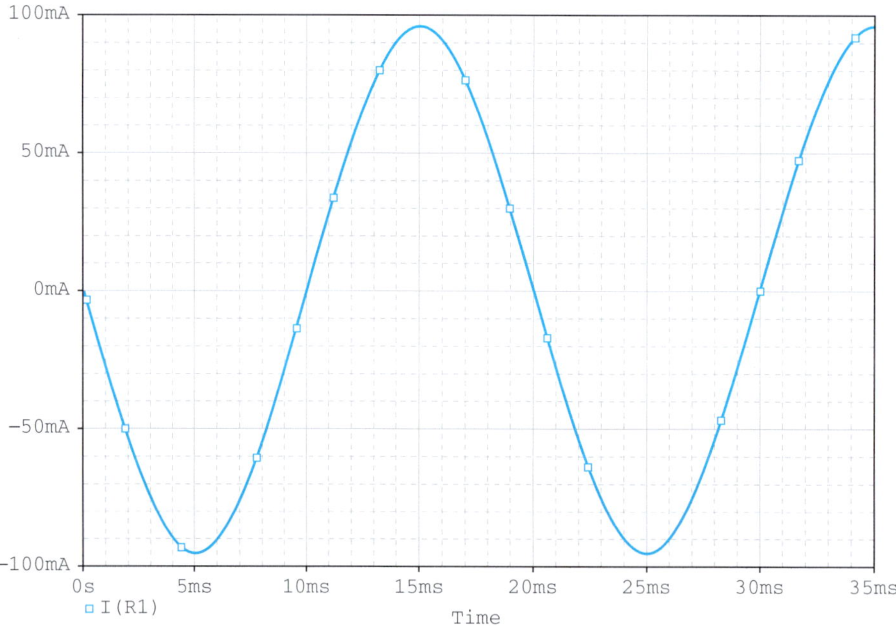

FIGURE 2.78 Current plot for time domain simulation.

Therefore, several steps are required to submit your work electronically. There are two ways to submit a PSpice plot electronically:

a. Use the Microsoft Windows *Print Screen* function, and paste it into image editor software, such as *Microsoft Paint*, *Photoshop*, and so on. However, this method requires an extra image editing process.

b. Use the "copy to clipboard" function and paste it into *Microsoft Paint*. This is a more straightforward method.

The following steps explain how to use the "copy to clipboard" function in PSpice. For these instructions, we will use the simulation plot for Example 2.13 that is shown in Figure 2.76.

a. In the schematic plot window, go to "Window" and choose "Copy to Clipboard" as shown in Figure 2.79.

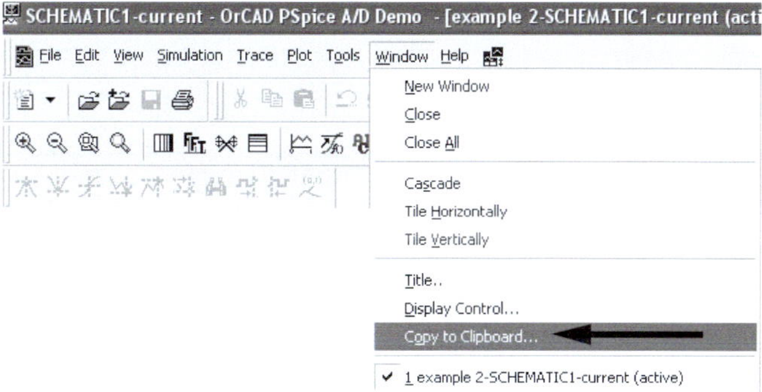

FIGURE 2.79 Copy to clipboard.

b. A copy to clipboard option window will pop up as shown in Figure 2.80. Choose either the option "change white to black" or "change all colors to black," then press OK.

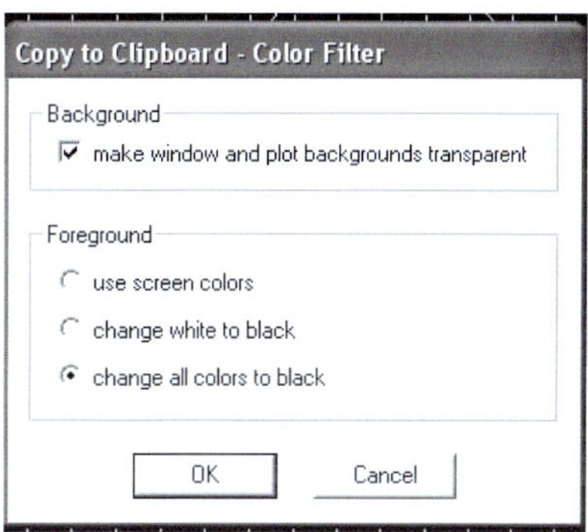

FIGURE 2.80 Copy to clipboard option.

c. Next, go to the Microsoft Windows Start Menu > All Programs > Accessories and click "Paint." Press "CTRL+V" or go to the "edit" menu and choose "Paste" to paste the plot.

d. Go to file > save to save the file with a specific file type, for example, JPG or JPEG.

e. The plot will be saved as shown in Figure 2.81.

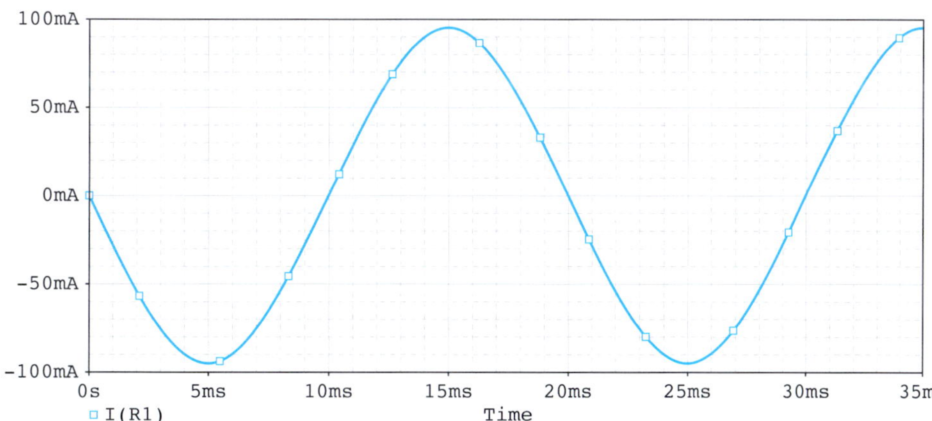

FIGURE 2.81 Simulation plot using copy to clipboard.

EXAMPLE 2.23 PSpice Example

Use PSpice to find the current, I, in the circuit of Example 2.12 (shown in Figure 2.82).

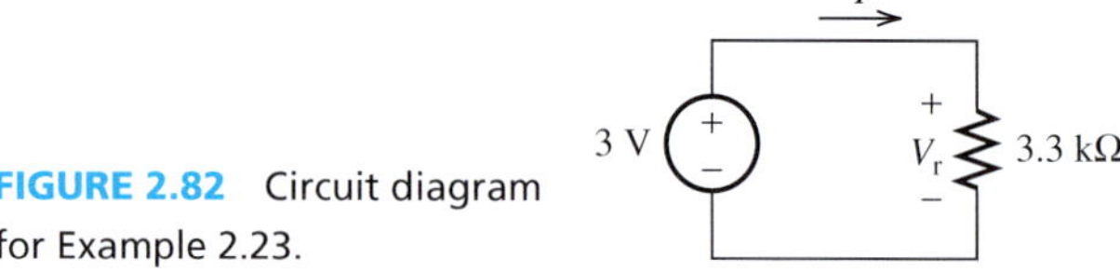

FIGURE 2.82 Circuit diagram for Example 2.23.

SOLUTION

Follow the steps in the PSpice tutorial to set up the PSpice circuit for Example 2.12, shown in Figure 2.82. The PSpice schematic solution is shown in Figure 2.83.

Next, set the simulation type to be "Bias Point" and run the simulation. The PSpice result is shown in Figure 2.84.

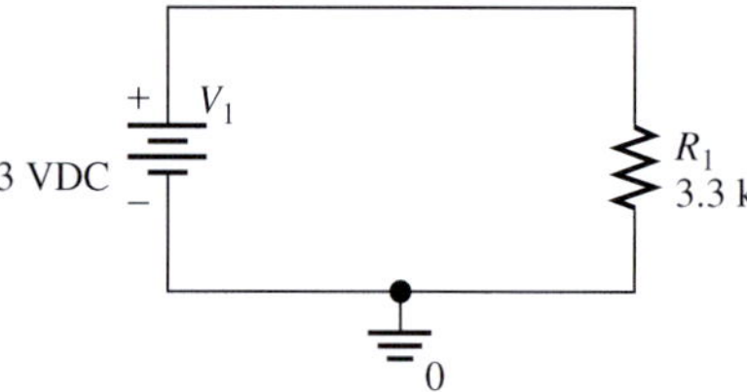

FIGURE 2.83 PSpice schematic for the circuit in Example 2.23.

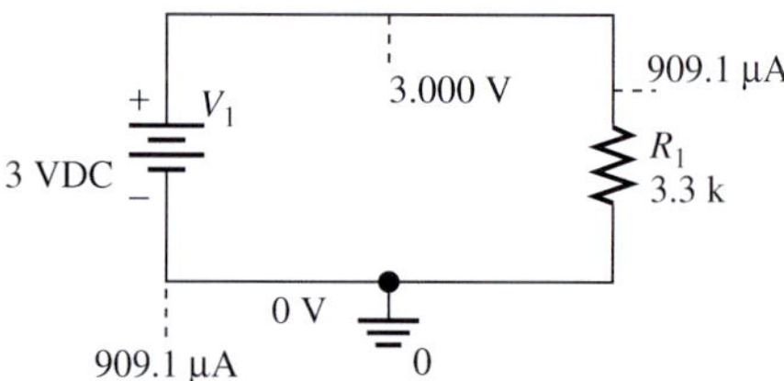

FIGURE 2.84 Simulation results for Example 2.23.

APPLICATION EXAMPLE 2.24 Electric Shock

Consider the unfortunate bald utility worker of Example 2.13 who climbs an aluminum ladder with bare feet and comes into contact with an overhead power line bearing 30 cos(100 πt) kV (Figure 2.86). The worker's body resistance is 315 kΩ. This situation can be modeled as a resistor (his body) connected between the terminals of a power supply (the wire) (Figure 2.85). How much current flows through his body? If the minimum current needed for a shock to be fatal is 100 mA, does the worker survive?

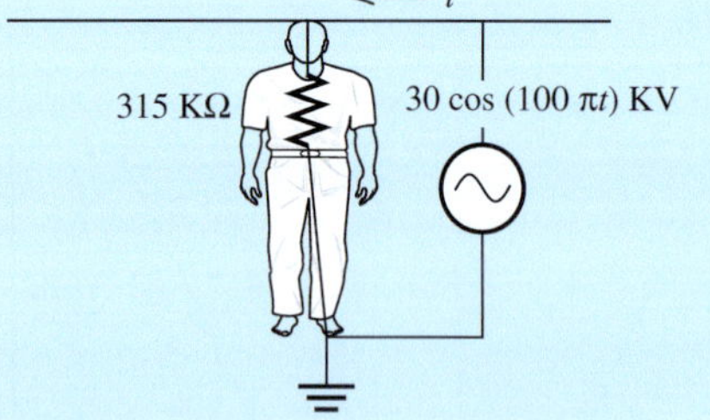

FIGURE 2.85 The equivalent circuit for Example 2.24.

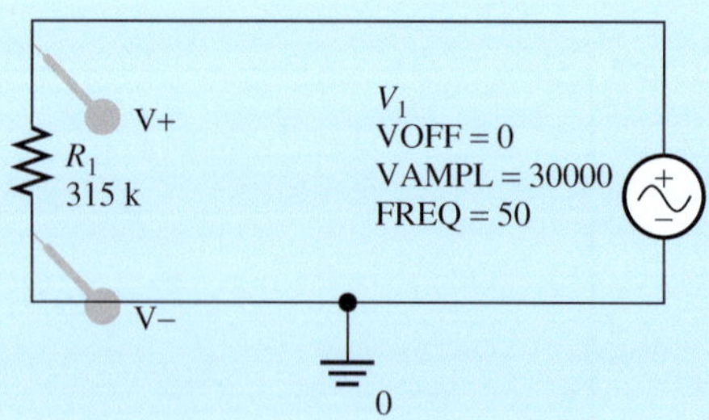

FIGURE 2.86 PSpice schematic circuit for Example 2.13.

SOLUTION

First, given the voltage source of 30 cos(100 πt) kV, we need to determine of the frequency of the voltage. The voltage source function is $V_S = V\cos(\omega t)$, and the frequency, f, is defined as

$f = \omega/2\pi$. This clearly shows that: $\omega = 100\pi$, and the frequency of the circuit is: $f = 100\pi/2\pi$ or $f = 50\,\text{Hz}$.

PSpice Schematic Setup:

1. Instead of using the Direct Current Voltage Source, a sinusoidal voltage source has been chosen. The sinusoidal voltage source for PSpice can be obtained by typing "VSIN" in the "Place Part" window; its amplitude and frequency are set as shown in Figure 2.86.
2. Follow the steps in the Tutorial section to set up a Time Domain Analysis Simulation. The current plot across the human body is shown in Figure 2.87.
3. Figure 2.87 shows the maximum (also known as amplitude) of the current is less than 100 mA. Therefore, the worker is able to survive the electric shock.

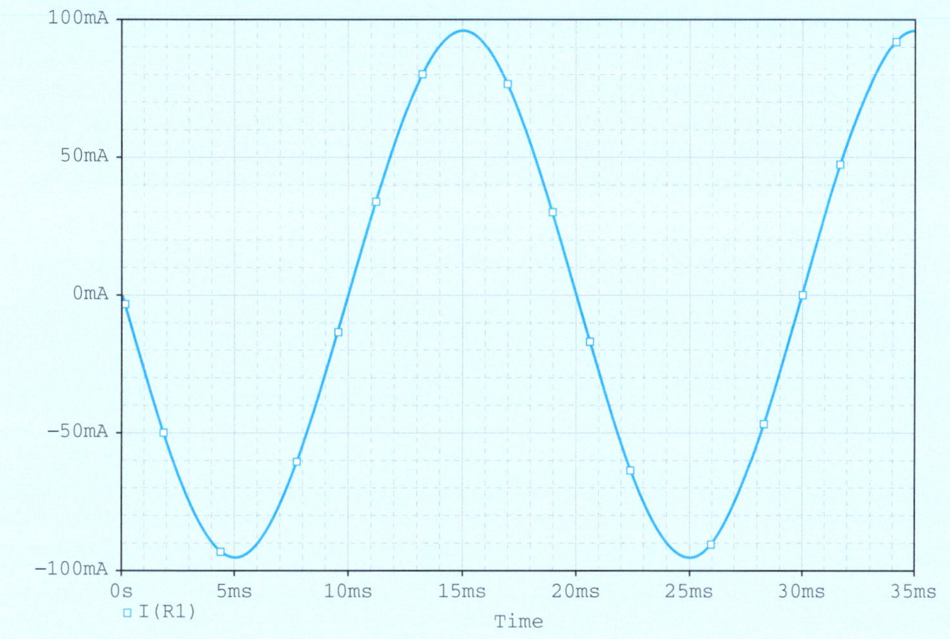

FIGURE 2.87 Simulation results for Example 2.24.

EXAMPLE 2.25 PSpice Example

Use PSpice to calculate the dissipated power in Example 2.16 (see Figure 2.88).

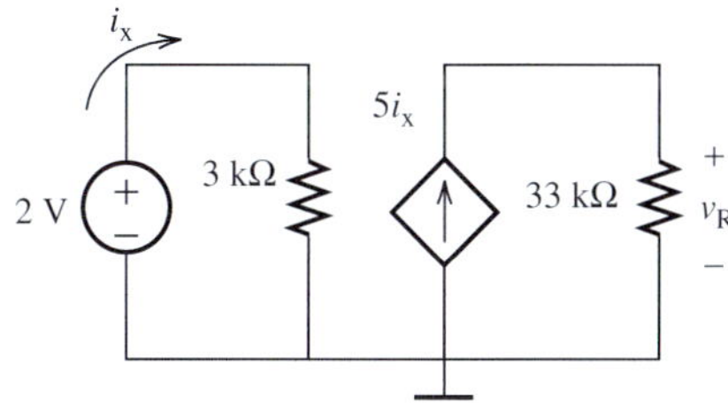

FIGURE 2.88 Circuit diagram for Example 2.25.

SOLUTION

You have already learned how to set up the PSpice Schematic. You will discover some other useful PSpice functions in other chapters. In this example, you will learn how to use the current-controlled current source (CCCS) gain in the PSpice software. After gaining familiarity with CCCS, you will

(continued)

EXAMPLE 2.25 Continued

be able to easily apply the same procedure to voltage-controlled voltage source (VCVS), voltage-controlled current source (VCCS), and current-controlled voltage source (CCVS).

1. First, set up the PSpice Schematic as shown in Figure 2.89.
2. Notice that there is an additional resistor with a very high resistance of 100 MΩ in the circuit. The reason for this is that an error will occur if the PSpice circuit is open. A high-resistance resistor is required and will act as a "bridge" to connect both sides of the circuit, while preventing any additional current flow through it.
3. To use the CCCS gain function in PSpice, go to "Place Part" and type "F." Choose "F/Analog."
4. After connecting the wires shown in Figure 2.89, double-click the F icon to open the schematic properties window shown in Figure 2.90.
5. Change the "GAIN" value from "1" to "5," and then click the "Display…" that is located near the arrow in Figure 2.90.
6. A "Display Properties" window will pop up as shown in Figure 2.91. In the "Display Format" section, choose to display "Name and Value," then click OK.
7. Then the CCCS gain block will show "GAIN = 5" as shown in Figure 2.92.

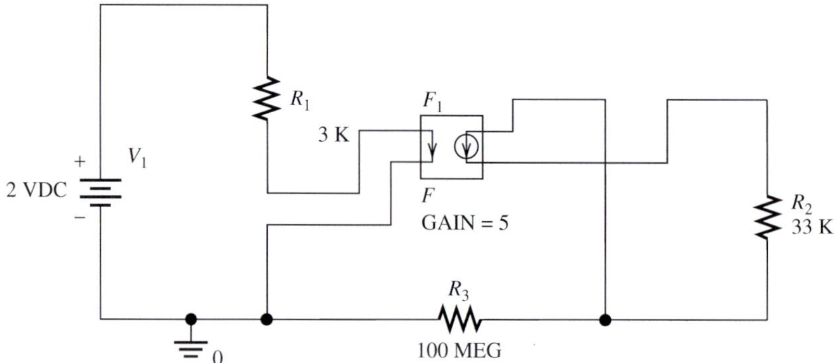

FIGURE 2.89 Spice schematic circuit for Figure 2.88.

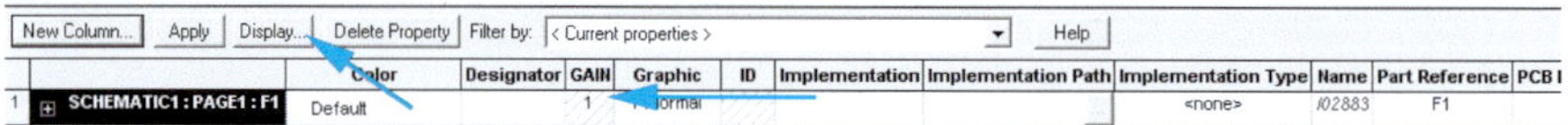

FIGURE 2.90 Schematic properties.

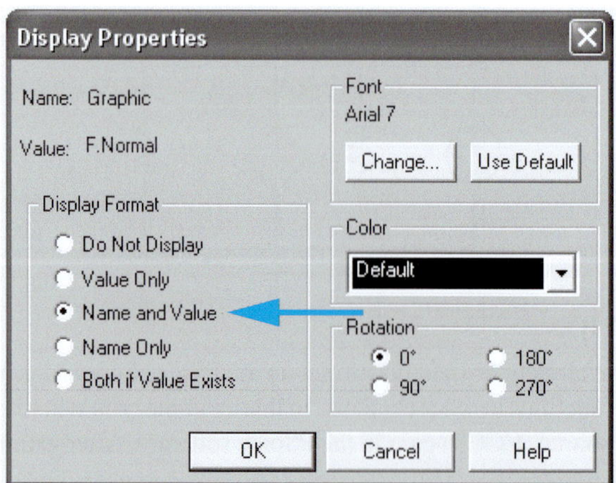

FIGURE 2.91 Display properties.

8. Set up the simulation profile with "Analysis Type" as "Bias Point" and run the simulation. The simulation result is shown in Figure 2.92.

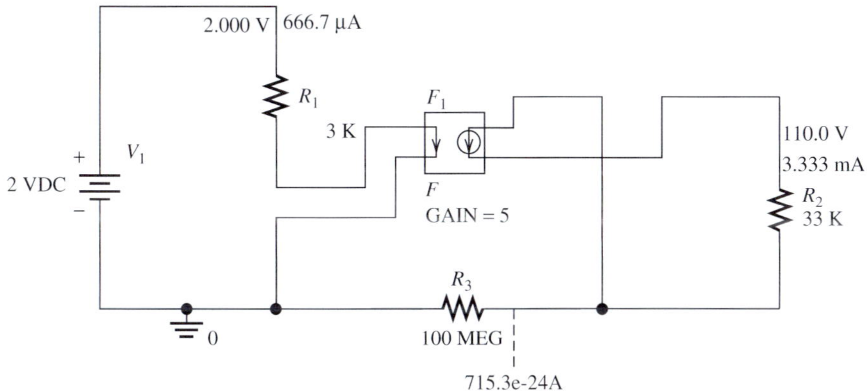

FIGURE 2.92 Simulation results.

2.11 WHAT DID YOU LEARN?

- The relationship between current and charges is expressed as [Equation (2.3b)]:

$$i(t) = \frac{dq(t)}{dt}$$

- Current direction is defined as the direction of the movement of *positive* charges.
- The voltage between two points in a circuit indicates the energy that is needed to move one coulomb (1 C) of charge and one volt (1 V); it is defined as:

$$1 \text{ volt (V)} = 1 \text{ joule/coulomb (J/C)}$$

- *Respective direction* of voltage and current (as shown in Figure 2.10) means that the direction of the current is from the positive sign of the voltage to the negative sign of the voltage.
- Kirchhoff's current law states that the currents entering a node are equal to the currents leaving the node [Equation (2.7)]:

$$\sum_{k=1}^{n} i_k = \sum_{k=n+1}^{N} i_k$$

- Kirchhoff's voltage law states that the algebraic sum of the voltages equals zero for any loop in an electric circuit.
- Ohm's law [Equation (2.10)] defines the relationship between voltage, v, current, i, and resistance, R, and can be written as:

$$v = i \times R$$

- The resistance, R, of a conductor is determined by its resistivity, ρ, length, L, and cross-sectional area, A, as shown in Equation (2.11):

$$R = \rho \frac{L}{A}$$

- The power consumed by an electric element is defined by [Equation (2.13)]:

$$p = v \times i$$

  If $p > 0$, the electric element absorbs energy and is called a *passive element*.
  If $p < 0$, the electric element supplies energy and is called an *active element*.
- Energy suppliers in a circuit are divided into dependent and independent sources.
- PSpice can be used to analyze simple circuits using the techniques outlined in the chapter.

Problems

*B refers to Basic, A refers to Average, H refers to Hard, and * refers to problems with answers.*

SECTION 2.1 INTRODUCTION

2.1 (B) Identify the device in Figure P2.1:

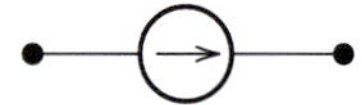

FIGURE P2.1 Image for Problem 2.1.

2.2 (B) Identify the device in Figure P2.2:

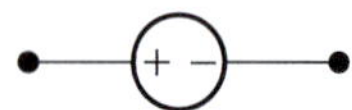

FIGURE P2.2 Image for Problem 2.2.

2.3 (B) Identify the device in Figure P2.3

FIGURE P2.3 Image for Problem 2.3.

2.4 (B) Identify the device in Figure P2.4:

FIGURE P2.4 Image for Problem 2.4.

2.5 (B) Identify the device in Figure P2.5:

FIGURE P2.5 Image for Problem 2.5.

SECTION 2.2 CHARGE AND CURRENT

2.6 (B) What is the definition of current direction?

2.7 (A) Find the charge as a function of time if:

$$i(t) = 2\sin(10\pi t)$$

2.8 (A) Find the number of electrons, n, as a function of time if:

$$i(t) = 12t$$

2.9 (A)* Direct current in steady state does not vary with time. If a DC currentof 5 A flows through a wire, find the charge through the wire with respect to time. The charge at $t = 0$ is 0.

2.10 (A)* In Problem 2.9, how many charges flow in 5 s?

2.11 (H)* Given that the number of charges measured at t_1 is $n_1 = 2 \times 10^{19}$, and the number of charges measured at t_2 is, $n_2 = 5.75 \times 10^{19}$. Find the current flow if the time interval between t_1 and t_2 is 2 s, with the assumption that the current, $i(t)$ is a linear function of charges, $q(t)$.

2.12 (H) Application: electroplating
A depth of 0.15 mm nickel is required to be coated on a metal surface by electroplating. The nickel density $\rho = 8.8 \times 10^3$ kg/m^3, the electrochemical equivalent $k = 0.304 \times 10^{-6}$ kg/C (which is the weight of the nickel produced by electrolysis during the flow of a quantity of electricity equal to 1 C), and the current per square meter $\rho = 35$ A/m^2. How long does it take to achieve the desired coating depth? (*Hint: Find the mass of the nickel.*)

SECTION 2.3 VOLTAGE

2.13 (B) The voltage across the device in Figure P2.13 is 1 V. What is the voltage polarity of Nodes a and b if the potential energy is reduced when a positive charge moves from a to b.

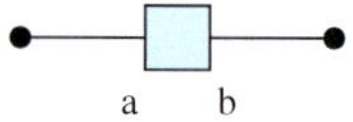

FIGURE P2.13 Image for Problem 2.13.

2.14 (A)* If a current of 3 A flows from a 5 V DC source for 1 s, how much energy, in Joules, is used?

2.15 (H) If 5×10^{16} electrons are pushed by the energy of 15 Joules, what is the voltage?

2.16 (H)* If an alternating current of $i(t) = 2\sin(3/2\pi t)$ flows from a voltage source of $v(t) = 5$, find the energy used in 1 s.

2.17 (H) 20 J is used from a 2 V source in 4 s. How much current, i, is flowing during this time?

SECTION 2.4 RESPECTIVE DIRECTION OF VOLTAGE AND CURRENT

2.18 (B) True/false: The notation in Figure P2.18 conforms to passive reference.

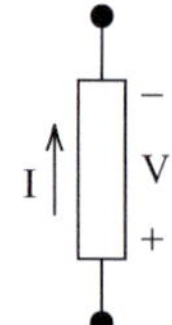

FIGURE P2.18 Image for Problem 2.18.

2.19 (B) True/false: The notation in Figure P2.19 conforms to passive reference.

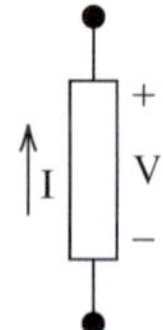

FIGURE P2.19 Image for Problem 2.19.

2.20 (B) True/false: The notation in Figure P2.20 conforms to passive reference.

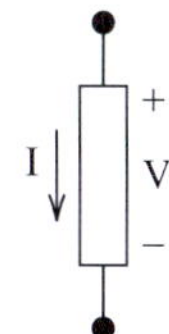

FIGURE P2.20 Image for Problem 2.20.

SECTION 2.5 KIRCHHOFF'S CURRENT LAW

2.21 (B) Use KCL to find i_2 in Figure P2.21.

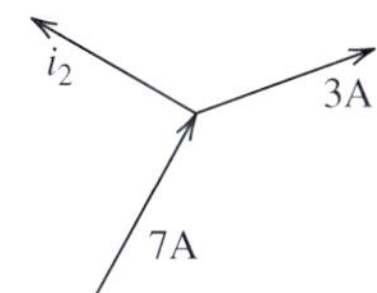

FIGURE P2.21 Image for Problem 2.21.

2.22 (B)* Use KCL to find i_4, i_5, and i_6 in Figure P2.22.

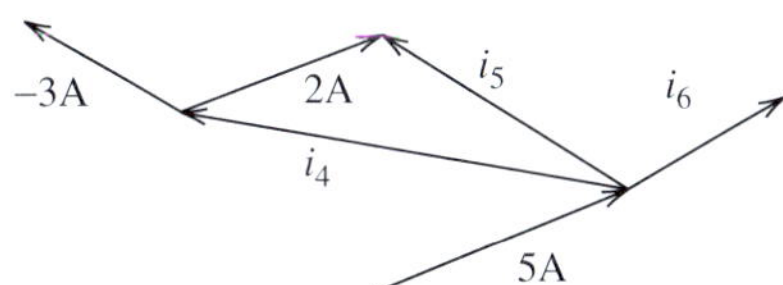

FIGURE P2.22 Image for Problem 2.22.

2.23 (A)* Find v in the circuit given in Figure P2.23.

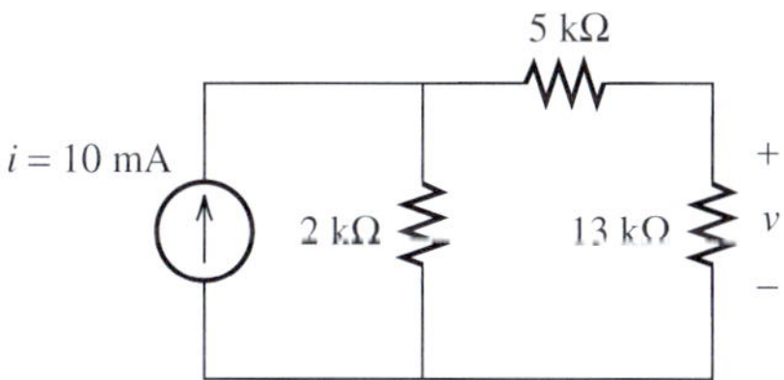

FIGURE P2.23 Circuit for Problem 2.23.

2.24 (A) Use KCL to show that i_1 is equal to i_4 in Figure P2.24.

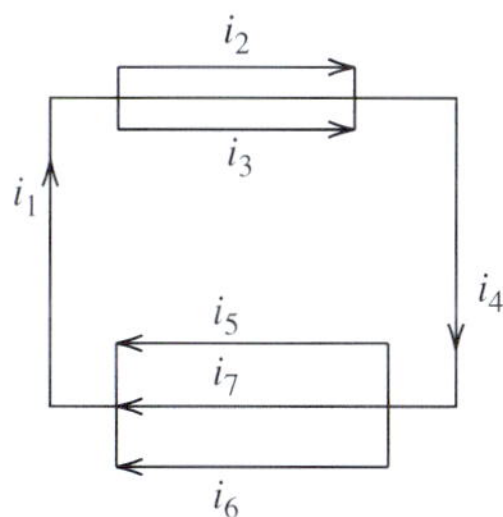

FIGURE P2.24 Circuit for Problem 2.24.

2.25 (H) In Figure 2.11, explain how KCL results from the law of conservation of charge.

2.26 (H)* Find the current, i, in the circuit shown in Figure P2.26.

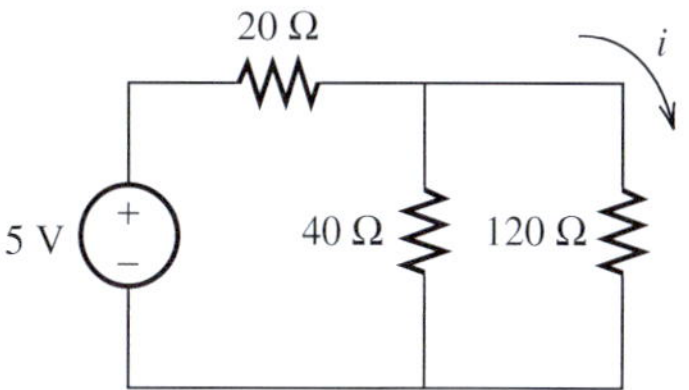

FIGURE P2.26 Circuit for Problem 2.26.

2.27 (H) Find I_x in this Figure P2.27.

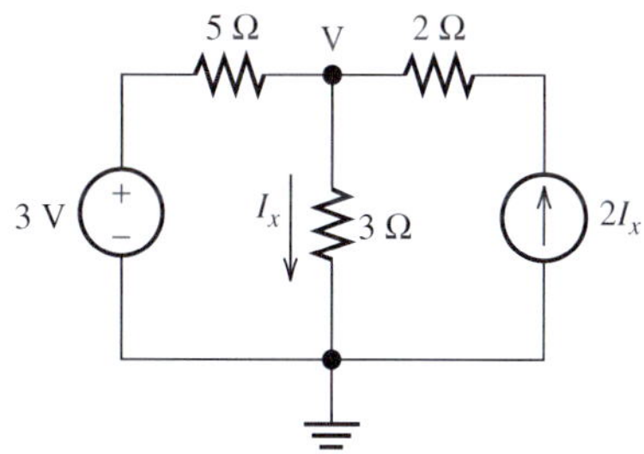

FIGURE P2.27 Circuit for Problem 2.27.

2.28 (H)* Use KCL to find the equivalent resistance between Node A and B, that is, $R_{AB} = V_{AB}/I_{AB}$.

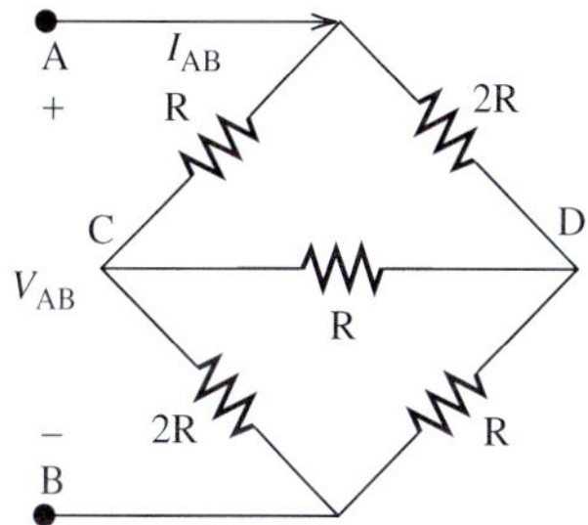

FIGURE P2.28 Circuit for Problem 2.28.

SECTION 2.6 KIRCHHOFF'S VOLTAGE LAW

2.29 (B) Find v_A in Figure P2.29:

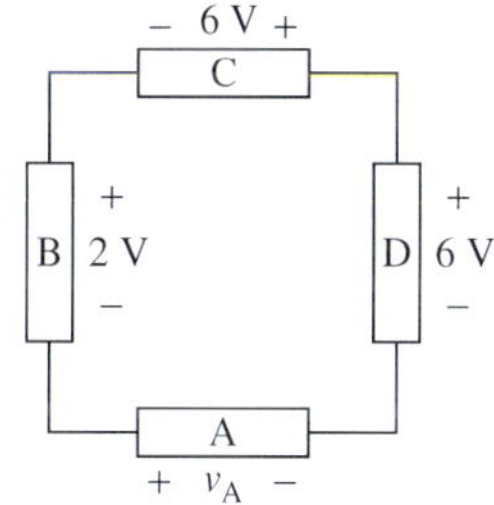

FIGURE P2.29 Circuit for Problem 2.29.

2.30 (B)* (Application) An automobile cooling fan is connected to the car's 12 V battery. The fan motor can be modeled as two voltage sources in series, a negative back EMF voltage, and a positive armature voltage, as shown in Figure P2.30.

Find the current through the line resistance, if the line resistance is 20 Ω.

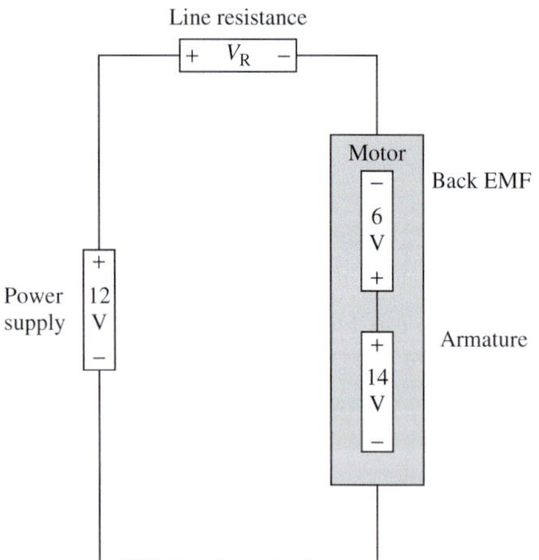

FIGURE P2.30 Cooling fan circuit for Problem 2.30.

2.31 (A) Find v_C and v_D in Figure P2.31:

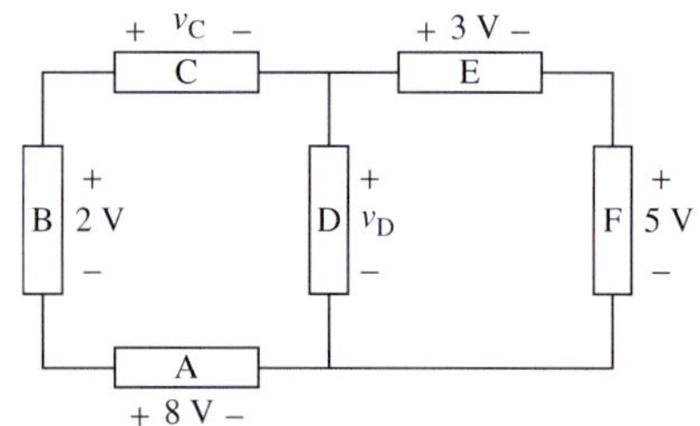

FIGURE P2.31 Circuit for Problem 2.31.

2.32 (A) Find v_C and v_E in Figure P2.32:

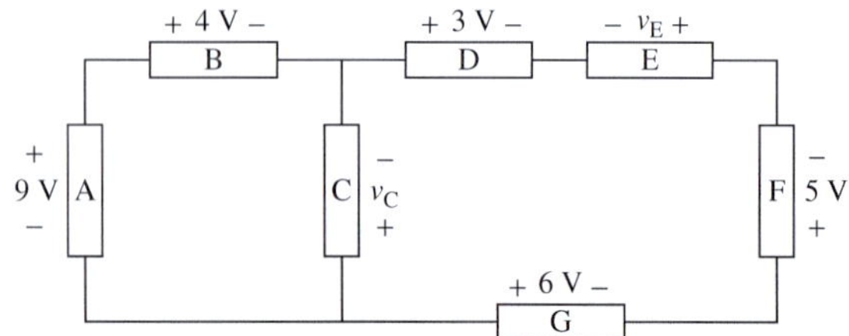

FIGURE P2.32 Circuit for Problem 2.32.

2.33 (A) Find v_C, v_D, and v_E in Figure P2.33:

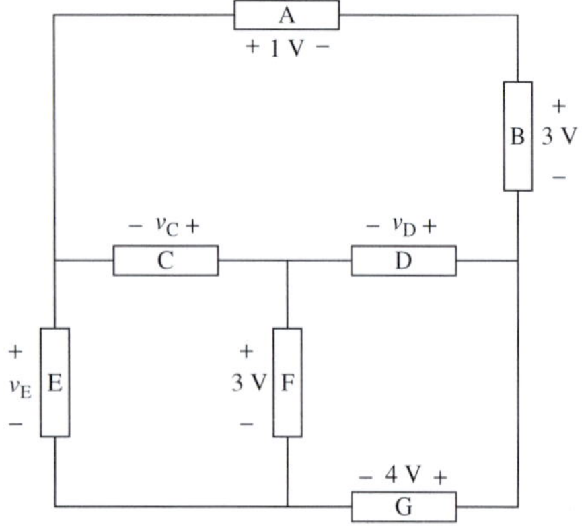

FIGURE P2.33 Circuit for Problem 2.33.

2.34 (H) Find v_A, v_B, v_D, and v_E in Figure P2.34:

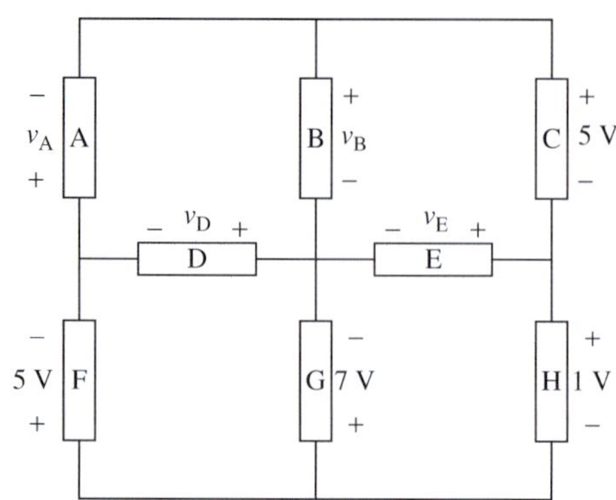

FIGURE P2.34 Circuit for Problem 2.34.

2.35 (H) Find V_C, V_D, V_H, and V_I in Figure P2.35:

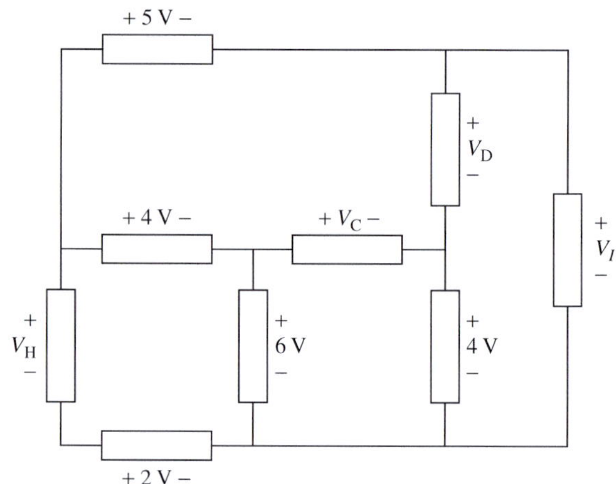

FIGURE P2.35 Circuit for Problem 2.35.

2.36 (H)* Shown in Figure P2.36 is a non-series-parallel connection known as a *Wheatstone bridge circuit.*

a. Given that $R_1 = 6\ \Omega$, $R_2 = 3\ \Omega$, find *i*.

b. When the current $I = 0$, we say the bridge is *balanced.* Under what condition (find an expression relating R_1 and R_2) will this bridge be balanced?

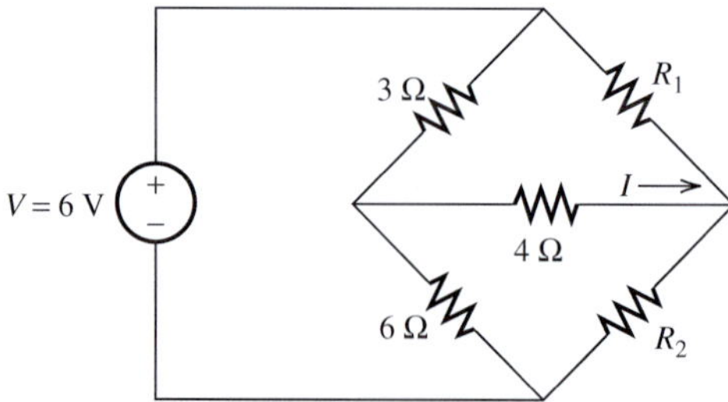

FIGURE P2.36 Bridge circuit.

2.37 (H)* Use KVL to find the equivalent resistance between Nodes A and B, $R_{AB} = V_{AB}/I_{AB}$.

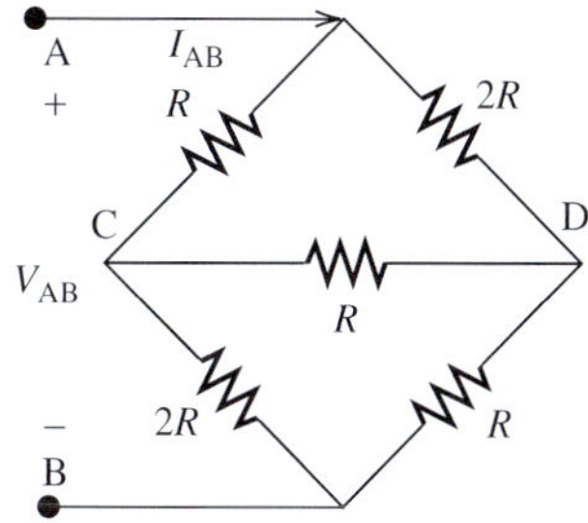

FIGURE P2.37 Circuit for Problem 2.37.

SECTION 2.7 OHM'S LAW AND RESISTORS

2.38 (B) Find V in the circuit of Figure P2.38.

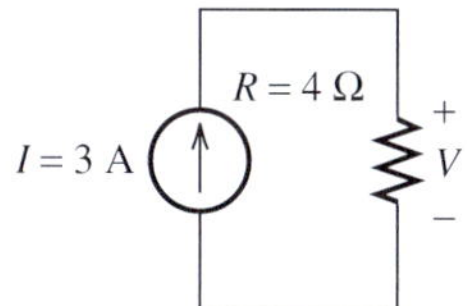

FIGURE P2.38 Circuit for Problem 2.38.

2.39 (B) Find R in the circuit shown in Figure P2.39.

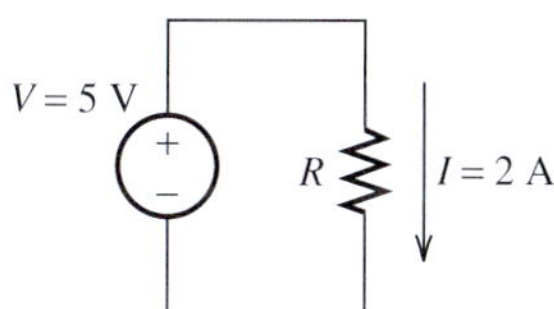

FIGURE P2.39 Circuit for Problem 2.39.

2.40 (B) Find I in the circuit shown in Figure P2.40.

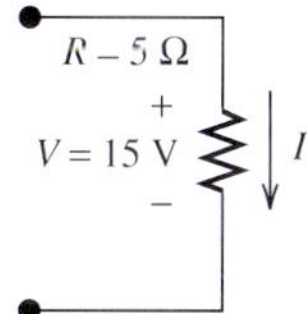

FIGURE P2.40 Circuit for Problem 2.40.

2.41 (A)* A 50-cm-long copper wire with a cross-sectional area $1.2286 \times 10^{-10}\,\text{m}^2$ is used to connect a voltage source and a device. What is the maximum voltage if the current is required to be not bigger than 1 A?

2.42 (A) (Application Problem)

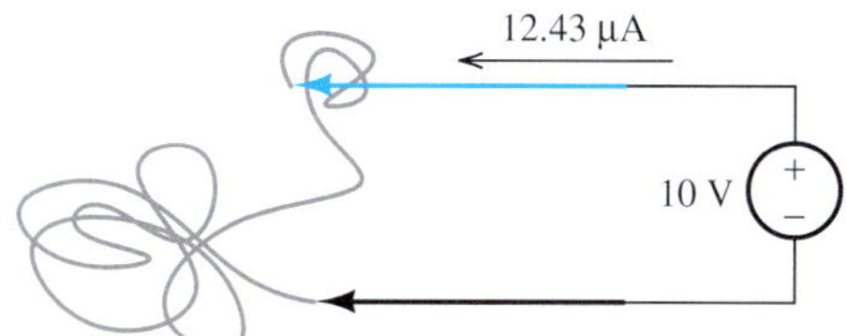

FIGURE P2.42 Circuit for Problem 2.42.

Surgeons remove a cylindrical-shaped piece of tissue of unknown length from a patient. They find that the piece of tissue has a length of 15.94 m and a cross-sectional area of $3.468 \times 10^{-3}\,\text{m}^2$. When a voltage of 10 V is applied across the entire length of the tissue as shown in Figure P2.42, a current of 12.43 μA flows through it. Find the resistivity of the tissue.

2.43 (H) Find an expression for i in Figure P2.43.

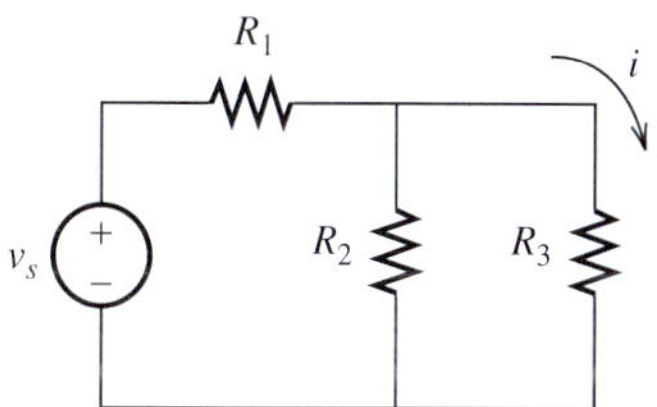

FIGURE P2.43 Circuit for Problem 2.43.

2.44 (H) Find V_A, V_B, V_C, and V_S in Figure P2.44.

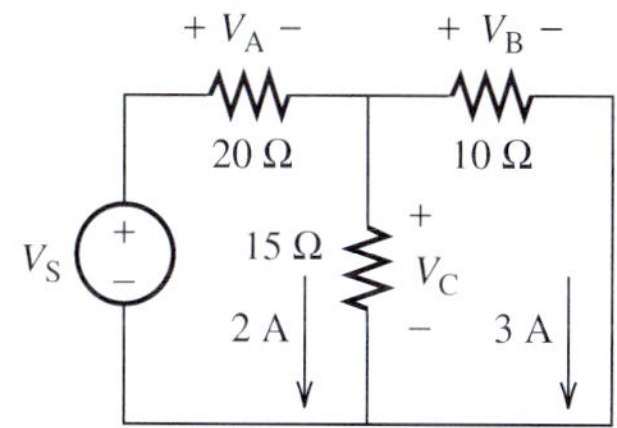

FIGURE P2.44 Circuit for Problem 2.44.

2.45 (A)* (Application Problem)

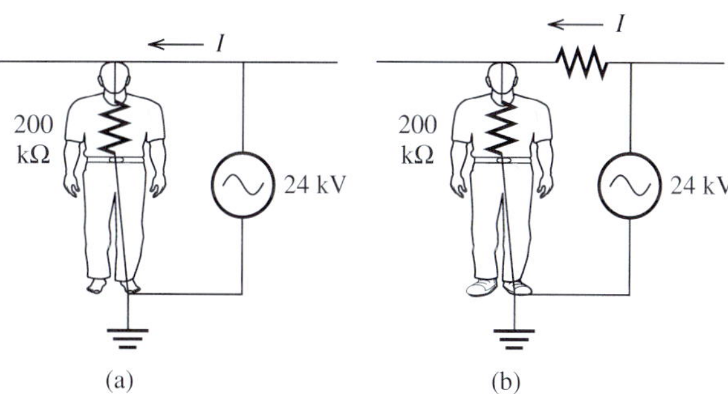

FIGURE P2.45 (a) Circuit for Problem 2.44. (b) Safety shoe included.

A careless utility worker climbs an aluminum ladder and comes into contact with an overhead power line bearing 24 kV, as shown in Figure P2.45a. His body resistance is 200 kΩ. How much current flows through his body? The worker wears a good safety shoe. The safety shoe can be modeled as a resistance that is located in series with the 240 V voltages (see Figure P2.45b). In this situation, the current drops to 40 mA. Calculate the resistance of the shoe.

2.46 (H) Find the equivalent resistance of the short circuit and the open circuit shown in Figure P2.46.

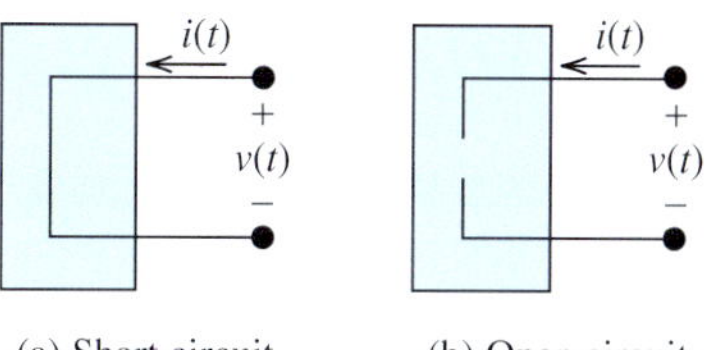

FIGURE P2.46 Circuit for Problem 2.46.

2.47 (H) Application: temperature meter
The circuit shown in Figure P2.47 is a bridge circuit. R_S is a resistive temperature sensor that varies with temperature. This circuit can be used as a thermometer by measuring the voltage V_{out}. To enable a high sensitivity in reading the temperature, the circuit should be designed to ensure that a small change in R_S would lead to a large change in V_{out}.

Determine the conditions for which the output voltage changes the most for a given change in the resistor R_S, that is, the derivative of V_{out} with respect to R_S, (dV_{out}/dR_S) would be maximum. (*Hint: Properly set* R_C.)

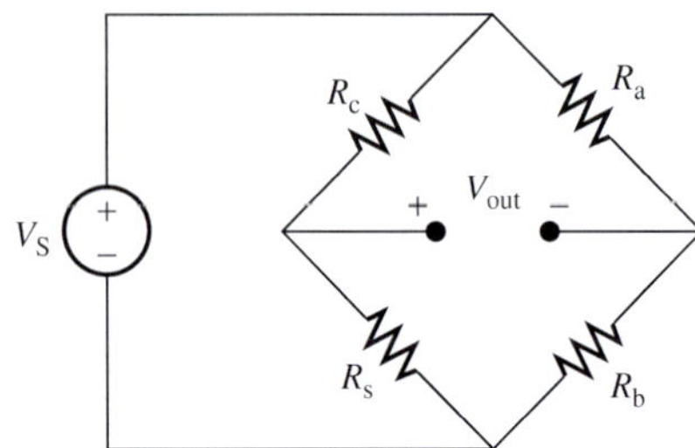

FIGURE P2.47 Circuit for Problem 2.47.

2.48 (H)* A thermistor is a temperature-dependent resistor. When the temperature changes, the resistance of the thermistor changes in a predictable way. The relationship of the absolute temperature and the thermistor's resistance is:

$$1/T = A + B\ln(R) + C[\ln(R)]^3$$

Here are some data points for a typical thermistor.

T (K)	*R* (Ω)
273	16,330
298	5000
323	1801

Find the coefficients *A*, *B*, and *C*.

SECTION 2.8 POWER AND ENERGY

2.49 (A) Identify the passive and active devices in Figure P2.49, and compute the power absorbed or supplied.

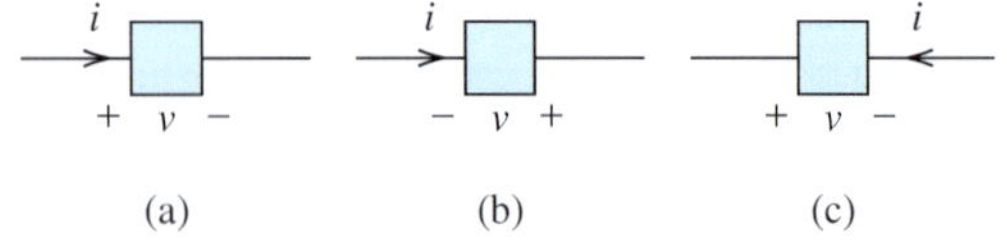

FIGURE P2.49 Circuit for Problem 2.49.

a. $v = -3$ V, $i = 1$ A
b. $v = -2$ V, $i = 2$ A
c. $v = 2$ V, $i = 3$ A

2.50 (H) Repeat Problem 2.49 with the following new assumptions.
a. $v = 20$ V, $i = 2.5e^{-2t}$ mA
b. $v = 20$ V, $i = 2\sin t$ mA

2.51 (A) Find the value of the resistance, *R*, in Figure P2.51.

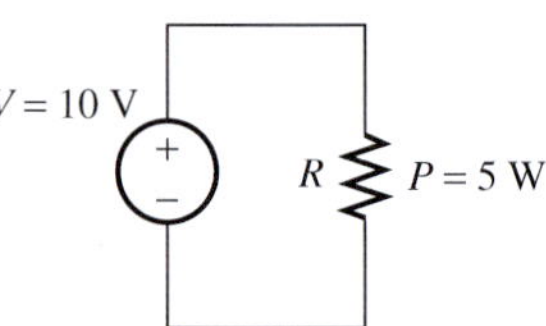

FIGURE P2.51 Circuit for Problem 2.51.

2.52 (A)* Find the power, *P*, dissipated by the 20-Ω resistor in Figure P2.52.

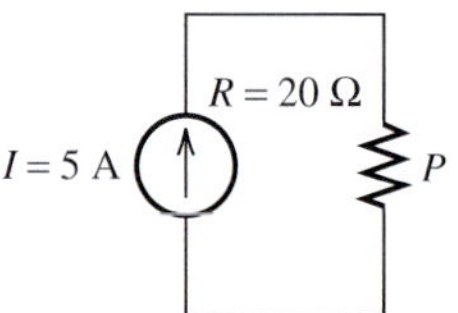

FIGURE P2.52 Circuit for Problem 2.52.

2.53 (H) Find the power P_1, P_2, and P_3 dissipated by all resistors in the circuit shown in Figure P2.53.

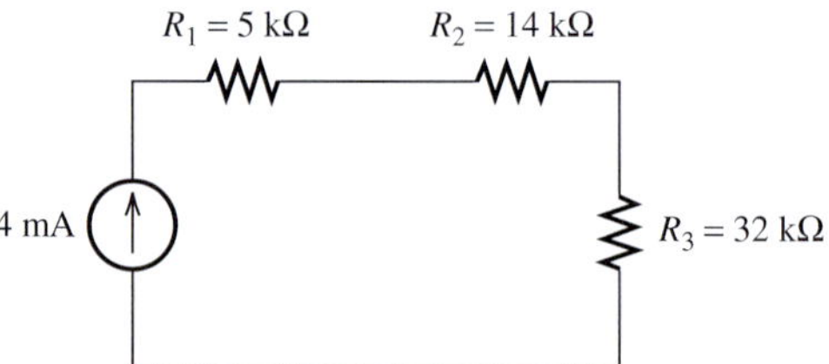

FIGURE P2.53 Circuit for Problem 2.53.

2.54 (H) Find *v* in the circuit shown in Figure P2.54.

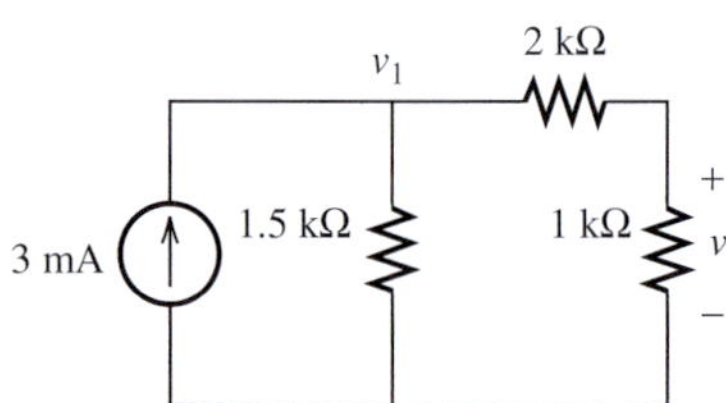

FIGURE P2.54 Circuit for Problem 2.54.

2.55 (H) Find the power dissipated in all the resistors of Figure P2.54.

2.56 (A) Show that the power absorbed in each resistor is a non-negative number.

2.57 (H) The voltage and the current across an electrical device are shown in Figure P2.57. The voltage and the current are associated in their direction. Draw the waveform of the power absorbed by the device and compute the energy consumed during the time interval [0, 2].

2.58 (A)* For the circuit shown in Figure P2.58:
a. Find the power provided by the −5 V voltage source.
b. In order to zero the power mentioned earlier, what will the current of the 4 A current source be changed to?

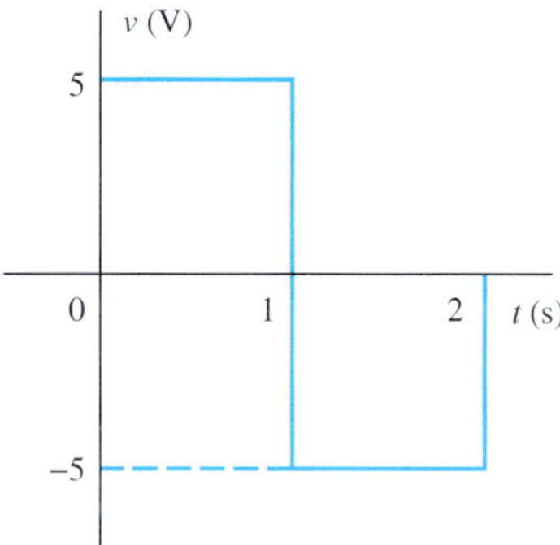

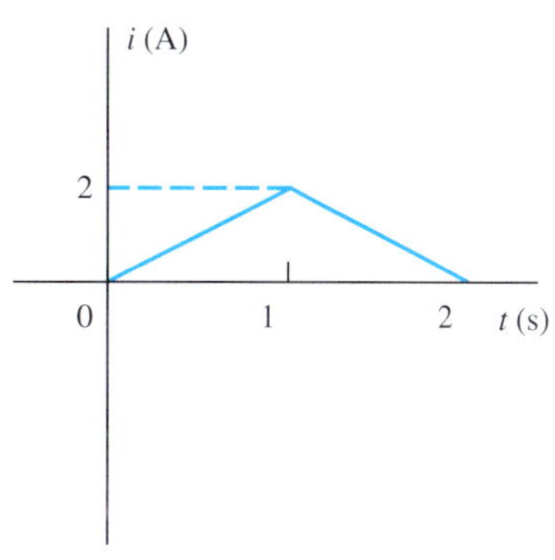

FIGURE P2.57 Voltage and current waveforms.

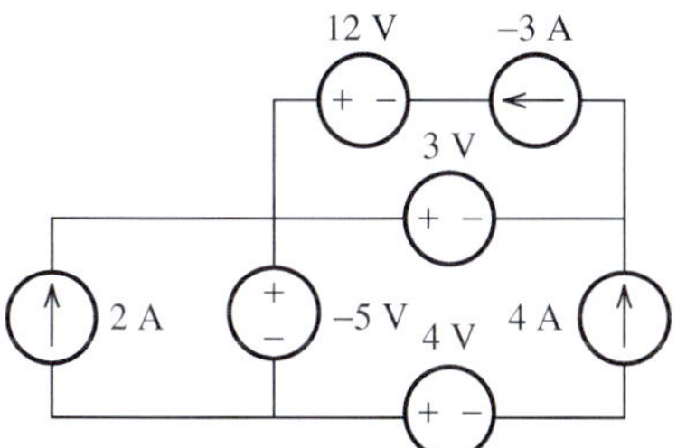

FIGURE P2.58 Circuit for Problem 2.58.

SECTION 2.9 INDEPENDENT AND DEPENDENT SOURCES

2.59 (B) Plot voltage versus current curves for ideal voltage source and current source.

2.60 (B) In Figure P2.60, what is the value that i can take?

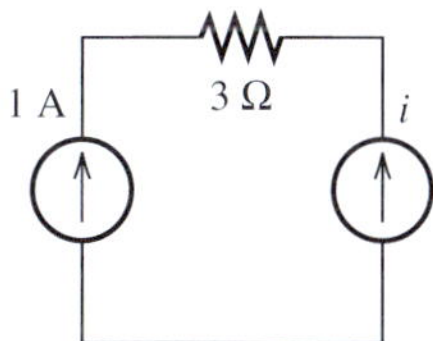

FIGURE P2.60 Circuit for Problem 2.60.

2.61 (A) In Figure P2.61, compute the voltage across the 3Ω resistor, the voltage across the current source, and the power of the current source.

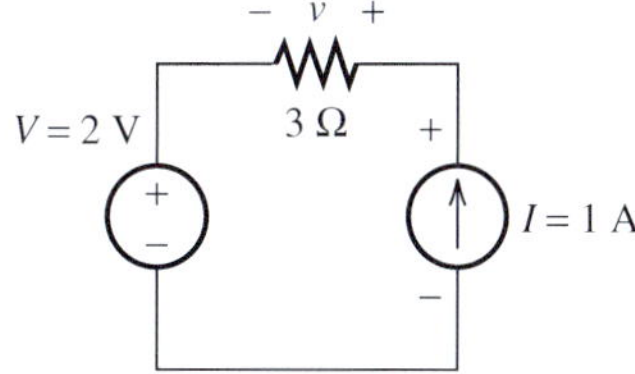

FIGURE P2.61 Circuit for Problem 2.61.

2.62 (A)* Find V_x in the circuit shown in Figure P2.62:

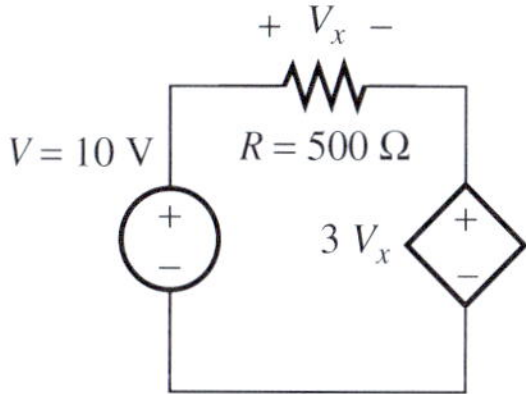

FIGURE P2.62 Circuit for Problem 2.62.

2.63 (A) Find the value of I_x in Figure P2.63.

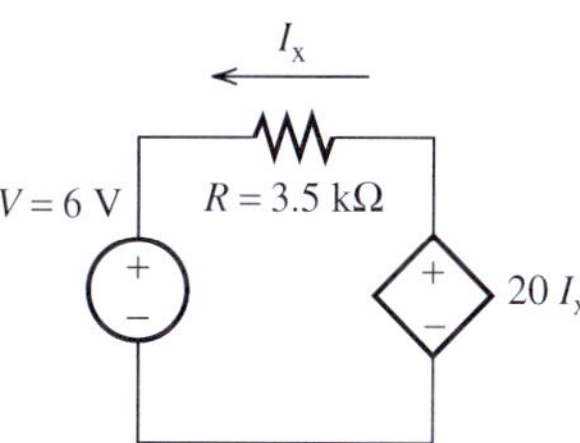

FIGURE P2.63 Circuit for Problem P2.63.

2.64 (A) In the circuit given in Figure P2.64, find the current of the dependent source.

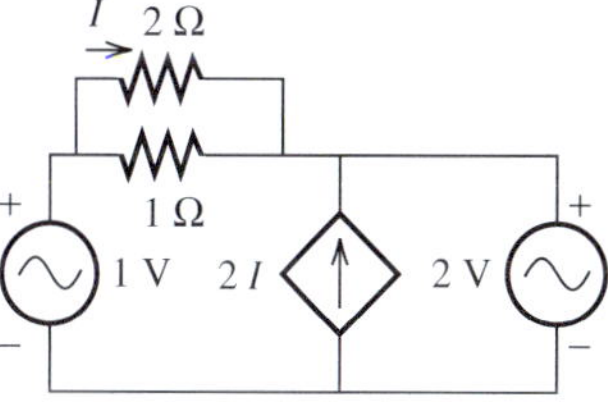

FIGURE P2.64 Circuit for Problem 2.64.

2.65 (H) Find the value of v_o with respect to v_S in Figure P2.65.

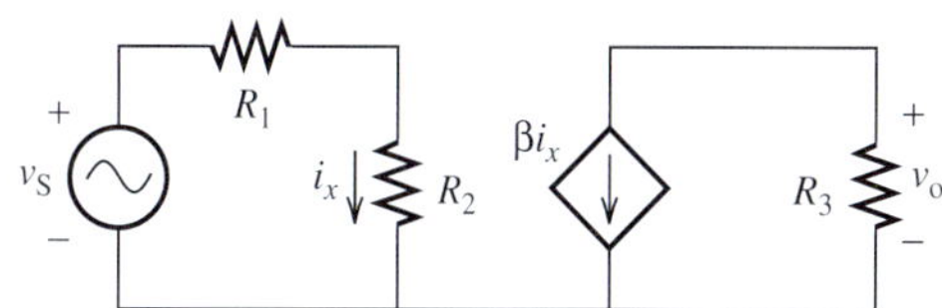

FIGURE P2.65 Circuit for Problem 2.65.

2.66 (H)* A current controlled current source (CCCS) circuit is given in Figure P2.66.
Find v_S and the power supplied by the dependent source.

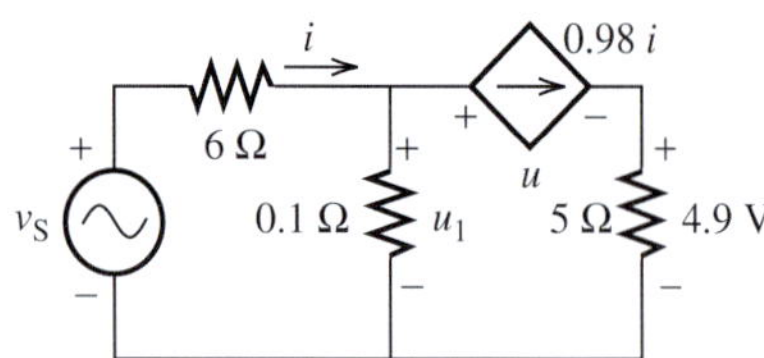

FIGURE P2.66 Circuit for Problem 2.66.

SECTION 2.10 ANALYSIS OF CIRCUITS USING PSPICE

2.67 (B) Use PSpice and find the current, i, that shown in Figure P2.66.

2.68 (A) In Figure P2.68, the current source A is 10 A, and resistors R_1, R_2, and R_3 are equal to 4 Ω, 3 Ω, and 2 Ω, respectively. Use PSpice and find the current i_2 and i_3. In addition, assume we replace the current source with a voltage source. Assuming i_1, i_2, and i_3 remain unchanged, find the magnitude of the voltage source.

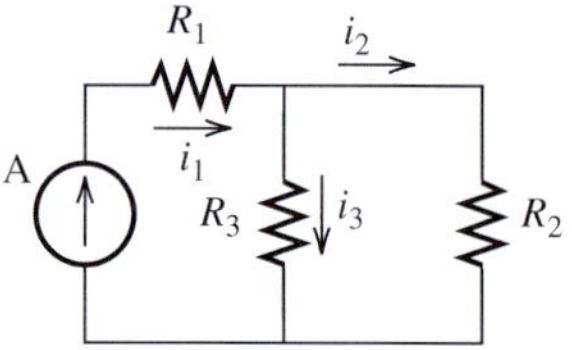

FIGURE P2.68 Circuit diagram for PSpice Problem 2.68.

2.69 (H) Given the circuit shown in Figure P2.69, assume a 10 V voltage source. Find the values of the resistor R_6 and the current i_5, assuming: $i_2 = 1$ A, $R_1 = R_5 = 2\ \Omega$, $R_2 = 4\ \Omega$, $R_3 = R_4 = 1\ \Omega$, $R_7 = 3\ \Omega$. Verify the answer through PSpice software using the value of R_6 obtained from calculation.

2.70 (H) Revise the PSpice Problem 2.70 by adding two new voltage sources v_2 and v_3 to the circuit as shown in Figure P2.70. Also, a new resistor R_8 has been added. Also, note that the values of the resistors R_1 through R_8 have been changed. Use PSpice to find the current at i_4 and i_6 with the resistance of resistors are given as below: $R_1 = 3\ \Omega$, $R_2 = R_3 = R_7 = 2\ \Omega$, $R_4 = 1\ \Omega$, $R_5 = 20\ \Omega$, $R_6 = 1.6\ \Omega$, and $R_8 = 8\ \Omega$.

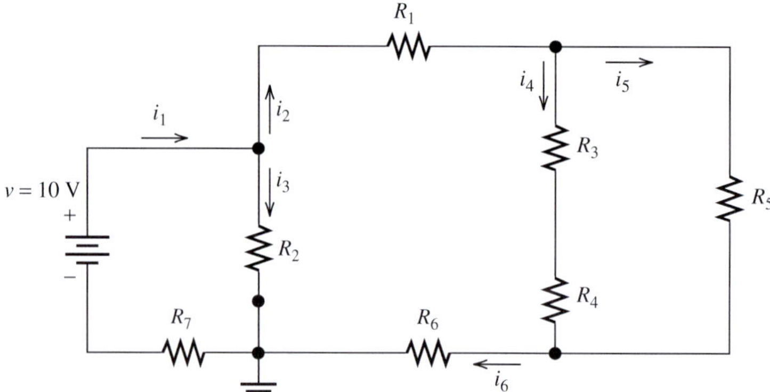

FIGURE P2.69 Circuit diagram for Problem 2.69.

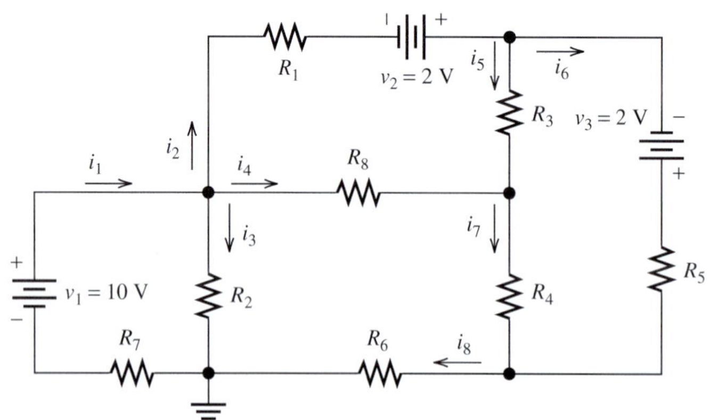

FIGURE P2.70 Circuit diagram for PSpice Problem 2.70.

TOPIC 3

Resistive Circuits

The content of this topic is compiled from:

Chapter 3 (pp. 61–134), *Electrical Engineering: Concepts and Applications*

By S.A. Reza Zekavat

CHAPTER 3

Resistive Circuits

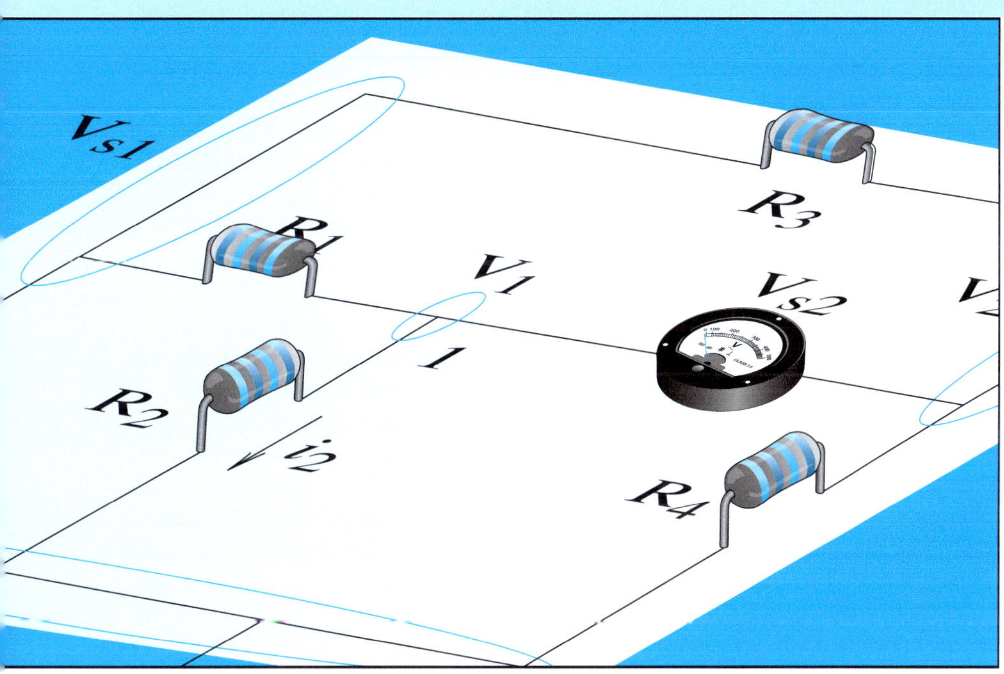

3.1 INTRODUCTION

Chapter 2 focused on the main variables of electric circuits: voltage, current, and power. Three elements of electric circuits were introduced: the voltage source, the current source, and resistors. This chapter outlines how to analyze complex electric circuits and calculate the voltage and current of circuit elements. This chapter focuses on circuits that consist of resistors and independent or dependent current and voltage sources. The strategies outlined here will enable you to find unknown current, voltage, power, and resistance values in terms of known circuit values. These techniques are applicable to any electric circuit, regardless of its complexity.

A resistive circuit consists of only resistors and sources. For example, the circuits shown in Figure 3.1 are resistive. The tools and techniques used for analyzing resistive circuits include Kirchhoff's voltage laws (KVL), Kirchhoff's current law (KCL), and Ohm's law, which were introduced in Chapter 2. To analyze a resistive circuit network with a large number of sources, resistors, nodes, and branches may seem extremely difficult; however, by applying the techniques outlined in this

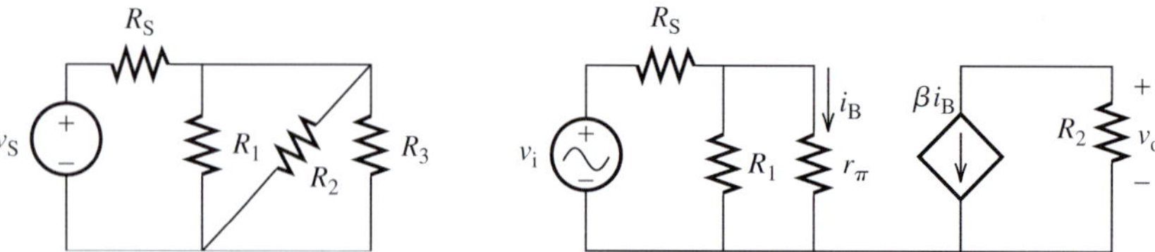

FIGURE 3.1 Resistive circuits.

chapter, you will learn to decrease the number of nodes and branches of a resistive circuit and therefore ease the process of resistive circuit analysis.

The first section of this chapter introduces *series* and *parallel* arrangements of resistors. In either case, an equivalent resistor can be developed. A network of resistors is usually a combination of parallel and series resistors. Therefore, by using an equivalent resistor for parallel and series configurations, several resistors in a network can be replaced by a single one. This technique reduces the complexity of a given circuit, simplifying the analysis.

This chapter also examines how voltages and currents are divided between the series and parallel resistors. Finding an equivalent circuit for a network may not always be easy or possible. This chapter outlines the techniques used to find unknown values in a circuit that are applicable to all circuits, regardless of their structural complexity. Nodal analysis will be discussed as a good tool for this goal. Nodal analysis applies Kirchhoff's rules to different nodes and loops of a circuit and finds unknown values by solving a series of linear equations.

Finally, this chapter summarizes Thévenin and Norton methods which are other useful techniques for finding an equivalent circuit for a resistive circuit in the presence of dependent or independent sources. Each of the techniques listed here are based on Kirchhoff's current and voltage laws. This chapter explains how these basic rules are applied to analyze circuits.

Many electric circuits can be modeled using a simple resistive circuit. For example, consider a 120-V power supply connected to a lamp, a fan, or a computer. The lamp and computer draw specific currents. Thus, a resistor can be used as a simple model to represent these circuits. Resistive circuit modeling allows engineers in different disciplines to use the techniques presented in this chapter to analyze systems. For instance, mechanical engineers may need to analyze resistive circuits to understand automotive system power requirements and ignition systems. Civil engineers may need to analyze resistive circuits formed by resistive strain gages. Resistive circuits are also vital to the study of building wiring, electric shock, and various other applications as detailed throughout the many applied examples in this chapter.

This chapter also offers examples to explain how circuit currents and voltages can be computed and analyzed using PSpice.

3.2 RESISTORS IN PARALLEL AND SERIES AND EQUIVALENT RESISTANCE

As discussed in Table 2.2, different notations for current and voltage might be used. In this chapter, lower case *i*/*v* (that is a general notation for current/voltage) and upper case *I*/*V* (that is used for DC current and voltage) are used interchangeably. The series elements in a circuit refer to those arranged in a chain. In other words, if any two branches share only one node, the elements are said to be arranged *in series*. Consider the two series elements in Figure 3.2. If KCL is applied to node a:

$$i_1 - i_2 = 0 \Rightarrow i_1 = i_2$$

As a result, currents that flow through two series elements are equal. Current flowing through two series elements has only one path to take. In practice, there might be more than two series elements in a circuit (see Figure 3.3 for example). For the current flowing through the series elements of Figure 3.3, by applying KCL to each node between the elements we arrive at the following equation:

$$i_1 = i_2 = i_3 = i_4 = \cdots = i \quad \textbf{(3.1)}$$

As shown, the same current flows through all of the series elements. Next, because this chapter studies resistive circuits, consider an example where the circuit element in Figure 3.3 is replaced

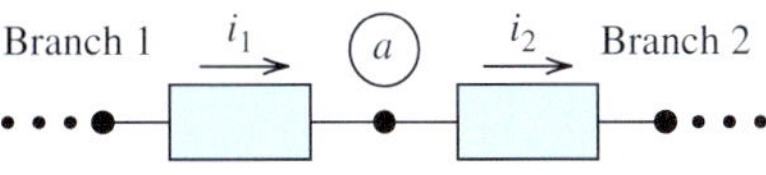

FIGURE 3.2 Two series elements.

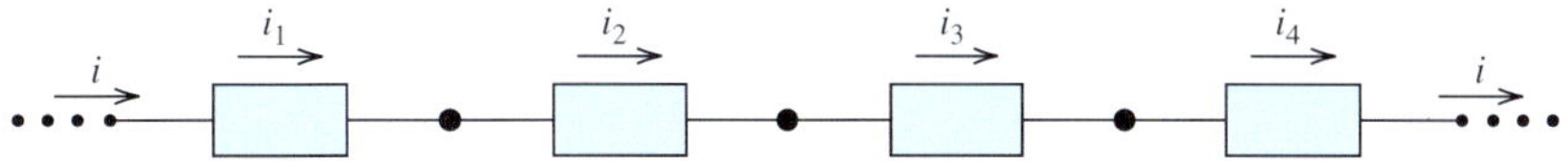

FIGURE 3.3 A circuit with several series elements.

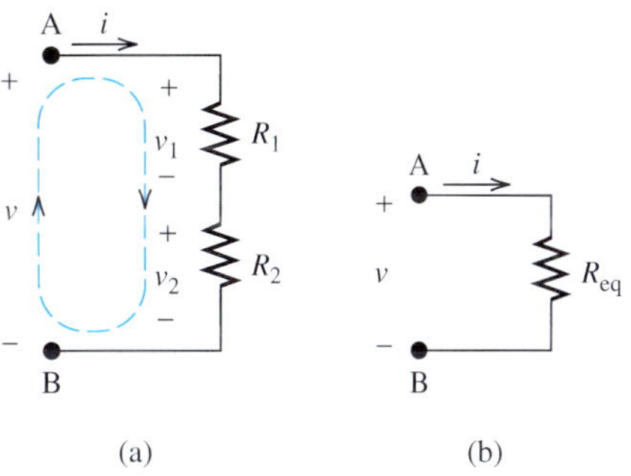

FIGURE 3.4 (a) Two series resistors. (b) The equivalent resistor.

by an ideal resistor. In Figure 3.4(a), a voltage, v, is applied to two series resistors. Applying KVL to the loop specified in Figure 3.4(a):

$$-v + v_1 + v_2 = 0 \tag{3.2}$$

Using Ohm's law for the two resistors and knowing that the currents passing through the two series elements are equal, that is, $i_1 = i_2 = i$, results in:

$$v_1 = iR_1 \tag{3.3}$$

In addition:

$$v_2 = iR_2 \tag{3.4}$$

Substituting Equations (3.3) and (3.4) into (3.2) yields:

$$v = v_1 + v_2 \Rightarrow v = iR_1 + iR_2 \Rightarrow v = i(R_1 + R_2)$$

Thus, the equivalent resistance through the terminals corresponds to:

$$R_{eq} = \frac{v}{i} = R_1 + R_2$$

Therefore, if the voltage, V, is applied to the equivalent resistor, R_{eq}, there will be no change in the current flowing through the circuit. The same approach can be applied to situations where multiple resistors are in series. In general, the equivalent resistance of N resistors in series is:

$$R_{eq} = \sum_{n=1}^{N} R_n \tag{3.5}$$

The current flowing through series resistors in a circuit is the same as the current flowing through their equivalent resistor. Therefore, the equivalent resistance (instead of series elements) can be used to analyze the circuit.

The other well-known configuration for circuit elements is the *parallel connection*. When two elements in a circuit are in parallel, their terminals are connected to each other. Parallel elements are arranged with their two sides (called *heads* and *tails*) connected together. In general, the voltages across two parallel elements are the same. Figure 3.5 shows two parallel elements.

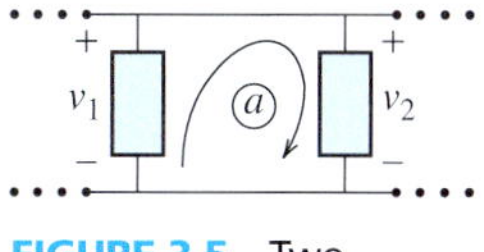

FIGURE 3.5 Two parallel elements.

Applying KVL to loop a in Figure 3.5 results in:

$$-v_1 + v_2 = 0 \Rightarrow v_1 = v_2$$

As shown above, the voltage across the two parallel elements is the same. In general, N circuit elements will be in parallel if they are connected at each end to the same point with no circuit elements in between (see Figure 3.6). KVL can be applied to each of the N loops shown in Figure 3.6. The voltage across all parallel elements is equal, that is, $v = v_1 = v_2 = v_3 = v_4 = \cdots$

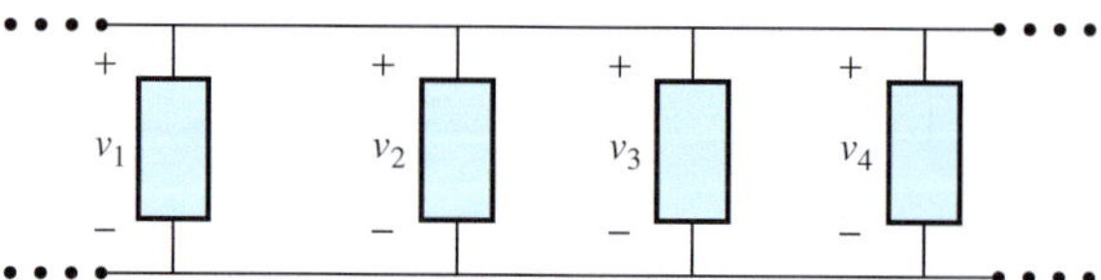

FIGURE 3.6 Parallel elements.

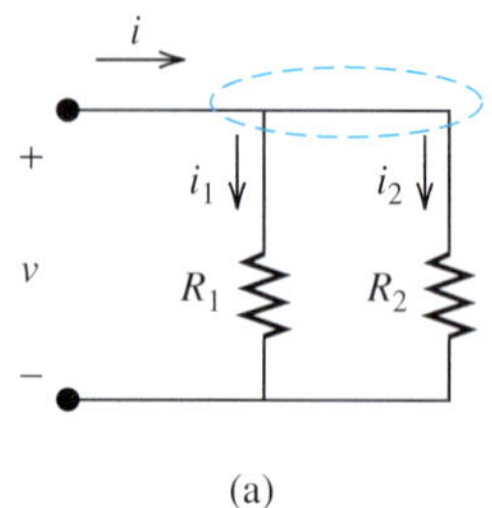

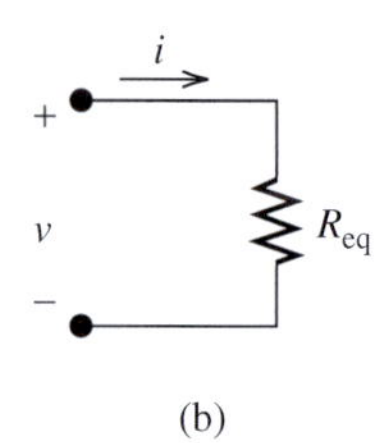

FIGURE 3.7 (a) Two parallel resistors. (b) The equivalent resistor.

Next, consider the situation where resistors are the circuit elements. Figure 3.7(a) shows two parallel resistors. A voltage, v, is applied to the two parallel resistors. The current of the circuit is broken up at the initial point of the two branches. Specifically, the current at the beginning of the source is divided between the two resistors. The point at which the two parallel resistors are connected possesses the same potential or voltage. This point, called a *node*, is represented by the dashed line in Figure 3.7(a). If KCL is applied to the node specified in the figure:

$$i - i_1 - i_2 = 0 \Rightarrow i = i_1 + i_2 \tag{3.6}$$

Next, Ohm's law can be used to define the currents of resistors in terms of their voltages and resistances. Because the voltages across parallel elements are the same:

$$i_1 = \frac{v}{R_1} \quad i_2 = \frac{v}{R_2} \tag{3.7}$$

If the equivalent resistor, R_{eq}, is attached to the voltage, V [Figure 3.7(b)], the same current is expected to flow through the circuit. This can be expressed as:

$$R_{eq} = \frac{v}{i} \Rightarrow i = \frac{v}{R_{eq}} \tag{3.8}$$

Now, substituting i_1 and i_2 in Equations (3.7) and (3.8) into (3.6):

$$i = i_1 + i_2 \Rightarrow \frac{v}{R_{eq}} = \frac{v}{R_1} + \frac{v}{R_2}$$

Therefore:

$$\frac{1}{R_{eq}} = \frac{1}{R_1} + \frac{1}{R_2} \tag{3.9}$$

Sometimes, the notation of two parallel lines || is used to show that two resistors are in parallel:

$$R_{eq} = R_1 || R_2 = \frac{R_1 R_2}{R_1 + R_2}$$

If R_{eq} is used instead of the two parallel resistors in a circuit, there will be no change in the voltage or the current of the circuit. Using the same approach, it is possible to find an equivalent resistance for N parallel resistors:

$$R_{eq} = \frac{1}{\frac{1}{R_1} + \frac{1}{R_2} + \cdots \frac{1}{R_N}} \tag{3.10}$$

Note that if $R_1 = R_2 = \cdots = R_N = R$, then:

$$R_{eq} = \frac{R}{N} \tag{3.11}$$

Sometimes, the process of circuit analysis can be simplified by reducing the number of circuit resistors. This can be done by using an *equivalent resistor*. Replacing several resistors in a resistive network with an equivalent resistor simplifies analysis but does not change the voltage or the current of the circuit.

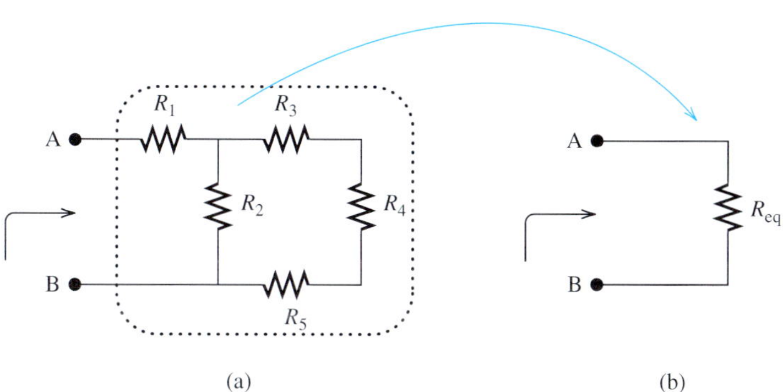

FIGURE 3.8 The equivalent resistor for five resistors in a circuit.

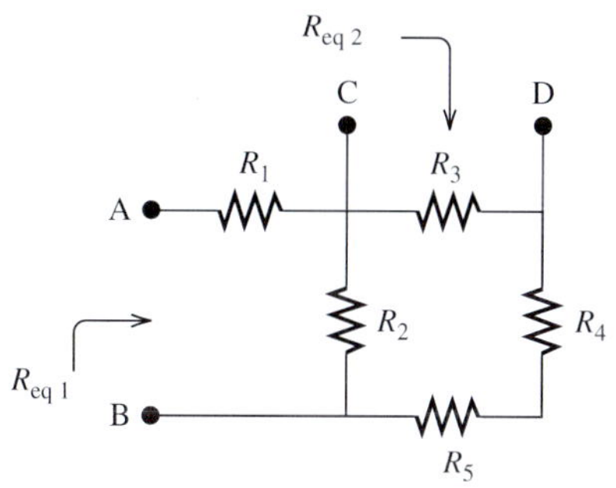

FIGURE 3.9 The equivalent resistors observed from two different locations in a circuit are not necessarily equal.

For example, in Figure 3.8, resistors R_1–R_5 can be replaced with the equivalent resistor R_{eq}. Now, by applying the same voltage across the two terminals A and B, the same current will flow in those terminals of the circuit.

> Therefore, the equivalent resistance observed across any two points in a circuit network can be found by applying a voltage across the two points, measuring the current flowing through the circuit, and computing the voltage to current ratio, which is the equivalent resistance.

Note that in general the equivalent resistances observed from two different locations in a resistive network are not the same. For instance, in Figure 3.9, R_{eq1} and R_{eq2} are not necessarily equal because they are seen through two different sets of terminals A–B and C–D.

Resistive circuits are usually a combination of parallel and series resistors. The equivalent resistance of a group of resistors can be found, step by step, by finding the equivalent resistances for series and parallel configurations. This approach of finding the equivalent resistance is illustrated in Figure 3.10.

Figure 3.10 shows the step-by-step process to find the equivalent resistance of the circuit in Figure 3.8(a). Each step shows a portion of the circuit being replaced by the equivalents of series and parallel resistor arrangements. First, observe that the resistors R_4 and R_5 are in series. Thus, they can be replaced with their equivalent resistance:

$$R_{54} = R_5 + R_4$$

Now, the resistance R_{54} is in parallel to the resistance R_3 (i.e., $R_{543} = R_3 || R_{54}$), and the resistance:

$$R_{543} = R_3 || R_{54} = \frac{R_{54}R_3}{R_{54} + R_3}$$

is equivalent. In the next step, R_{543} and i_2 are parallel and their equivalent resistance is:

$$R_{5432} = R_{543} || R_2 = \frac{R_{543}R_2}{R_{543} + R_2}$$

Finally, R_{5432} and R_1 are in series; therefore, their equivalent resistance is $R_{eq} = R_{54321} = R_{5432} + R_1$. This step-by-step process is reviewed in Figure 3.10.

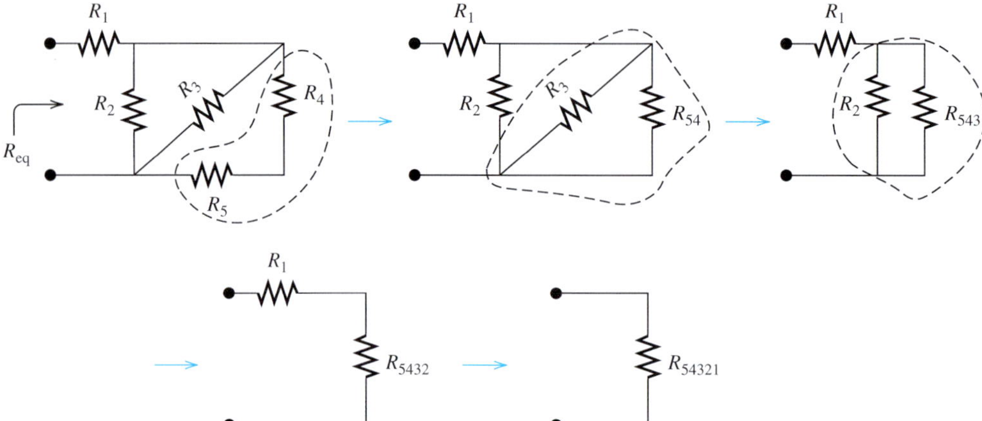

FIGURE 3.10 Step-by-step process to find an equivalent resistance.

EXAMPLE 3.1 Resistor Tolerance

In practice, the resistance of a given resistor varies within a given range. The accuracy of the stated value of a resistor is described as *resistor tolerance*. Resistor tolerance is depicted by colored strips on the resistor. Figure 3.11(a) represents different types of resistors. In the circuit shown in Figure 3.11(b), the resistances are given with a tolerance, $R_1 = 2.5\ \text{k}\Omega \pm 10\%$ and $R_2 = 7.5\ \text{k}\Omega \pm 5\%$. Find the nominal, minimum, and maximum equivalent resistances.

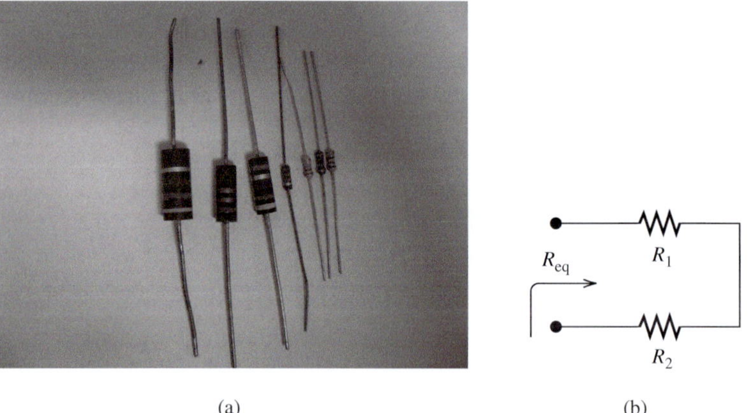

FIGURE 3.11 (a) Resistors; (b) the circuit for Example 3.1.

SOLUTION

The minimum and maximum values of R_1 and R_2 are:

$$\begin{aligned} R_1(\text{min}) &= 2.25\,\text{k}\Omega & R_1(\text{max}) &= 2.75\,\text{k}\Omega \\ R_2(\text{min}) &= 7.125\,\text{k}\Omega & R_2(\text{max}) &= 7.875\,\text{k}\Omega \end{aligned}$$

The two resistors are in series; therefore, their equivalent resistances are as follows:

$$R_{eq}(\text{nominal}) = R_1 + R_2 = 10.0\,\text{k}\Omega$$
$$R_{eq}(\text{min}) = R_1(\text{min}) + R_2(\text{min}) = 9.38\,\text{k}\Omega$$
$$R_{eq}(\text{max}) = R_1(\text{max}) + R_2(\text{max}) = 10.63\,\text{k}\Omega$$

EXAMPLE 3.2 Series and Parallel Resistors

Which resistors in Figure 3.12 are in series, and which are in parallel?

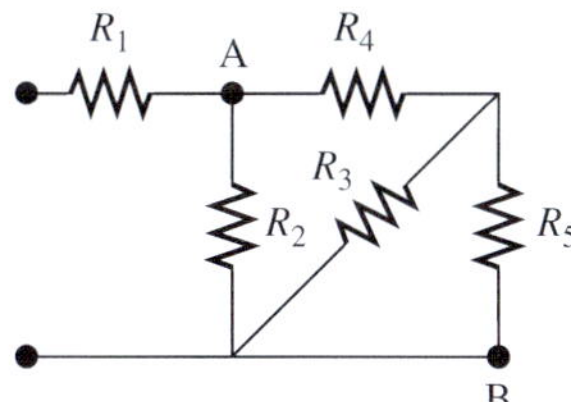

FIGURE 3.12 The circuit for Example 3.2.

SOLUTION

From Figure 3.12, it is clear that R_3 and R_5 are in parallel. The combination of these two is in series with R_4. R_2 is in parallel to the combination of R_3, R_4, and R_5. Finally, R_1 is in series with R_2–R_5.

EXCERCISE 3.1

In Example 3.2, if another resistor, R_6, is connected between nodes A and B, determine which resistors will be in parallel and which will be in series with others. Note: It may be that in some situations resistors are not in parallel and/or series with each other.

EXAMPLE 3.3 Equivalent Resistance

Find the equivalent resistance in Figure 3.13.

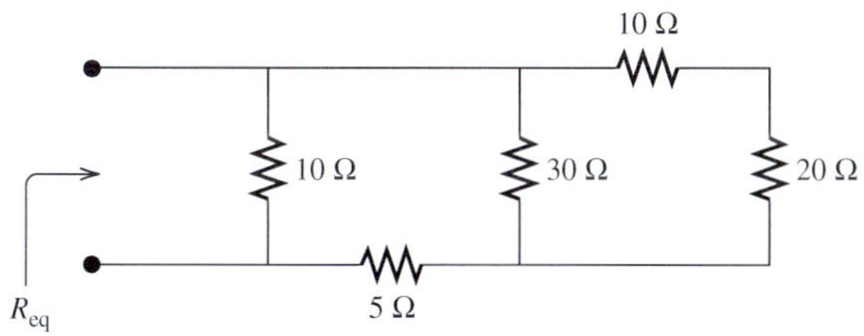

FIGURE 3.13 The circuit for Example 3.3.

SOLUTION

The equivalent resistance can be found using the step-by-step process of replacing the series and parallel resistor configurations with their equivalents. The steps are shown in Figure 3.14.

(continued)

EXAMPLE 3.3 Continued

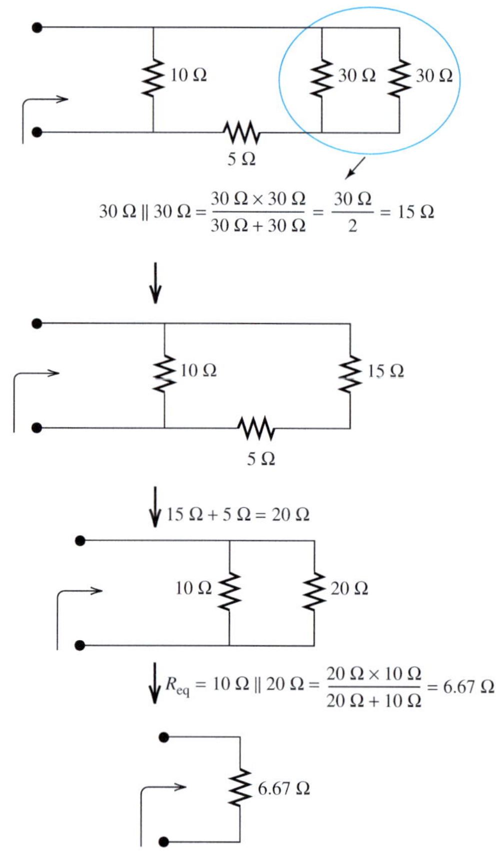

FIGURE 3.14 Steps for finding the equivalent resistance for Example 3.3.

Thus, the equivalent resistance is 6.67 Ω.

APPLICATION EXAMPLE 3.4 Is Building Wiring Parallel or Series?

Figure 3.15 shows a basic electric circuit with a power supply driving a single load. Two configurations are provided for study, series (Figure 3.16) and parallel (Figure 3.17). Assuming that $R_1 = 10\ \Omega$, $R_2 = 20\ \Omega$, and $R_3 = 30\ \Omega$:

a. Using the circuit in Figure 3.15, find V_1 and I_1.
b. For the circuit in Figure 3.16, find V_1, I_1, V_2, I_2, V_3, and I_3.
c. For the circuit in Figure 3.17, find V_1, I_1, V_2, I_2, V_3, and I_3.

Based on the results of this example, assume that in a building multiple devices, such as a computer, a television, and a lamp, are connected to the same circuit. Should these devices be connected in series or in parallel? Why?

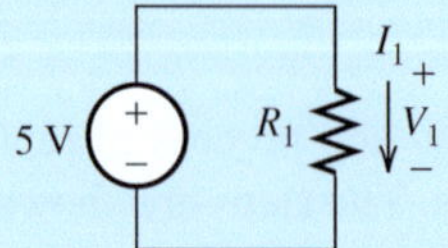

FIGURE 3.15 The basic electric circuit for Example 3.4.

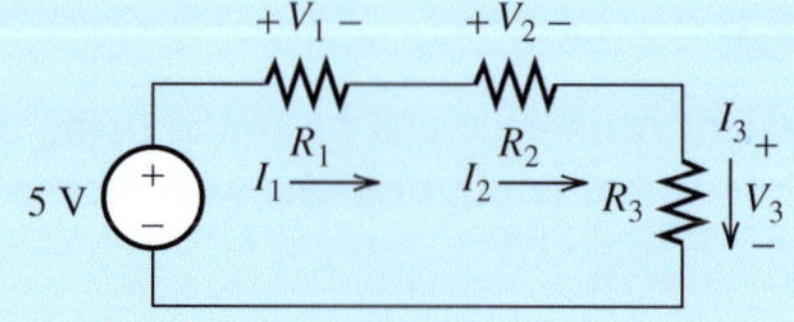

FIGURE 3.16 The circuit wired in series.

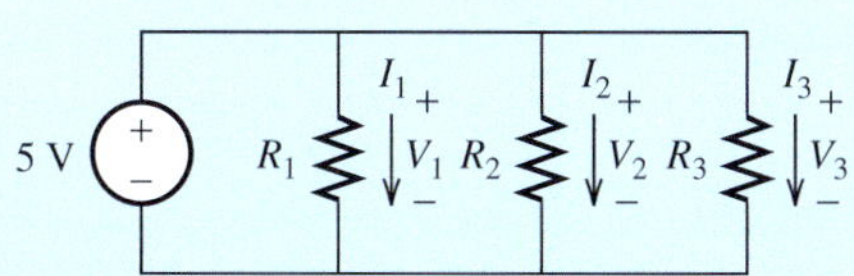

FIGURE 3.17 The circuit wired in parallel.

SOLUTION

(a)	$I_1 = 500$ mA	$V_1 = 5.000$ V
(b)	$I_1 = 83.3$ mA	$V_1 = 0.833$ V
	$I_2 = 83.3$ mA	$V_2 = 1.667$ V
	$I_3 = 83.3$ mA	$V_3 = 2.500$ V
(c)	$I_1 = 500$ mA	$V_1 = 5.000$ V
	$I_2 = 250$ mA	$V_2 = 5.000$ V
	$I_3 = 167$mA	$V_3 = 5.000$ V

When the circuit is wired in series, the current through and voltage over the elements changes as elements are added or removed. This poses several problems: (a) to use any single piece of equipment, all equipment must be turned on; (b) equipment must be designed to be powered by widely varying voltage levels; and (c) if an element burns out (i.e., is open circuited) all other elements will turn off. However, if the circuit is wired in parallel, the voltage across and the current through each element does not change as elements are added or removed. Thus, building wires are generally connected in parallel.

APPLICATION EXAMPLE 3.5 Shock and Water

Suppose that on a very rainy day a careless bald worker, wearing a hat and shoes, retrieves a dry wooden ladder from the garage. He climbs the ladder and begins working. Without warning, the old wooden ladder breaks and collapses below him as shown in Figure 3.18. Instinctively, he grabs the power line with his bare, wet hands. He grabs the soaked wooden power pole with his other hand, and feels a sharp tingle.

Due to the water on his hands and legs, his bodily resistance is lowered to 200 kΩ. The resistance of the wet telephone pole is 600 kΩ. The maximum power line voltage is 30 kV. What is the maximum current flow through his body?

FIGURE 3.18 Application problem: shock and water.

(*continued*)

APPLICATION EXAMPLE 3.5 Continued

SOLUTION

The situation shown in Figure 3.18 can be modeled as a series circuit as shown in Figure 3.19. Because it is a series circuit, current I_{max} is the same at all points in the circuit. Therefore, in order to find the current through the worker's body, compute the total current in the circuit. In order to do this, first compute the equivalent resistance of the circuit. For this simple series circuit, the equivalent resistance is 200 kΩ + 600 kΩ = 800 kΩ. Now, using Ohm's law, the maximum current is:

$$I_{max} = \frac{30\ \text{kV}}{800\ \text{k}\Omega} = 37.5\ \text{mA}$$

While this is not a fatal level of current, the worker may still fall hard to the ground.

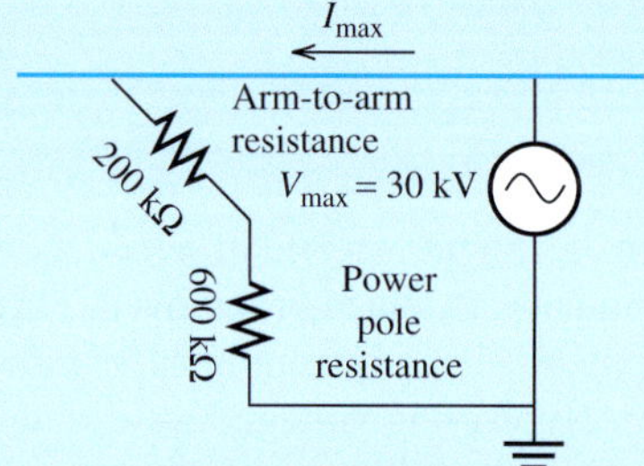

FIGURE 3.19 The circuit model for Example 3.5.

APPLICATION EXAMPLE 3.6 Solar Cells

The sunlight can be directly converted to electricity using solar cells. Solar cells are also called photovoltaic (PV) cells. Solar cells were first developed in the 1950s for the U.S. space satellites. Recently, they have become more widely used in daily life. As shown in Figure 3.20, solar cells can be used in homes for lights and appliances. The energy captured by these sources can be stored in batteries to power a lamp or an emergency roadside wireless telephone when no telephone wires are around (see Figure 3.20). They can also be used in vehicles to directly power electric motors.

Solar cells are usually arranged in a PV module and modules are wired together in a PV array to produce the required output power, as shown in Figure 3.21(a). The current produced by a solar cell is proportional to the solar illumination level and the produced voltage is almost invariable to the solar illumination level.

Assume that there are two panels of 10 cells each. One panel has 10 cells connected in series and the other has parallel connection. Each cell creates 0.50 V and 1 A; thus, the output power of each cell is 0.5 W. Because each panel consists of 10 cells, the total power output is 5.0 W ($10 \times 0.50 \times 1$).

a. If a single cell is completely shaded in both panels, what is the power output of each panel?
b. Which type of cell connection [series-connected and parallel-connected; see Figure 3.21(b)] is more robust ?

SOLUTION

a. In a series-connected panel, if one cell is completely shaded, the current produced by this cell drops to zero. Therefore, the output current will be limited by the current of the shaded cell and it also drops to zero. In this case, the output power is:

$$P_s = V_{out} \times I_{out} = 0$$

FIGURE 3.20 Sample applications of solar cells. (Used with permission from © Martin Shields/Alamy and (b) Sebastien Burel/Shutterstock.com.)

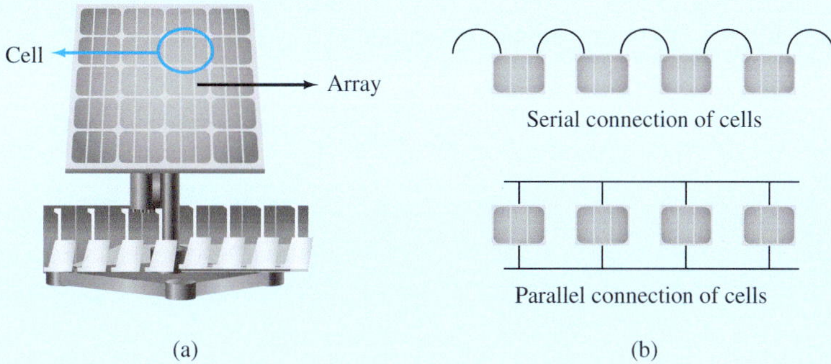

FIGURE 3.21 Solar cell module and array.

In the parallel-connected panels, if the current of one cell drops to zero, then the output current becomes:

$$I_{out} = 9 \times 1\,A = 9\,A$$

and the output voltage remains. Therefore the power output is:

$$P_{out} = V_{out} I_{out} = 0.5\,V \times 9\,A = 4.5\,W$$

b. From the computation shown in part (a), we can see that the parallel-connected panel is more robust. In a series network of solar cells, no output power is created when any single cell fails; however, in a parallel network, the failure of one cell only results in a partial decrease in power.

In practice, cells are usually put in series to meet minimum system voltage requirements, then multiple strings of series-connected cells are paralleled to attain the required system current.

3.3 VOLTAGE AND CURRENT DIVISION/DIVIDER RULES

3.3.1 Voltage Division

In order to determine the current and voltage of the parallel- and series-configured resistors in a circuit, it must first be determined how the voltage or current is divided across series and parallel resistors. The current passing through the two series resistors in Figure 3.22 is the same but the voltage across each of them is a fraction of the source voltage, V_s.

Applying KVL to the loop shown in the Figure 3.22:

$$-V_s + V_1 + V_2 = 0 \quad \textbf{(3.12)}$$

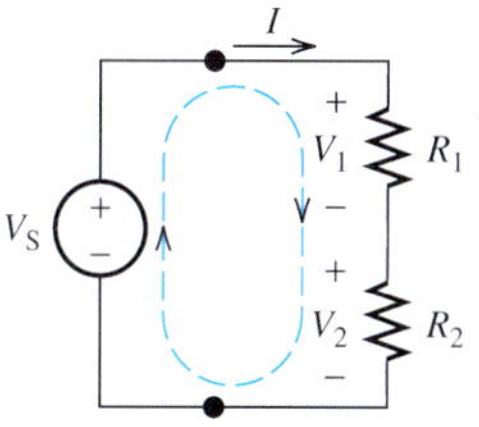

FIGURE 3.22 Voltage divider.

Next, Ohm's law can be used to maintain the relationship of current in the loop and voltage across the two resistors. Because $V_1 = R_1 \cdot I$ and $V_2 = R_2 \cdot I$:

$$I = \frac{V_1}{R_1} = \frac{V_2}{R_2}$$

This equation can be used to maintain the relationship of the two voltages V_1, and V_2, which is:

$$V_2 = \frac{R_2}{R_1} V_1 \quad \textbf{(3.13)}$$

Substituting V_2 from Equation (3.13) into Equation (3.12):

$$V_1 + \frac{R_2}{R_1} V_1 = V_s$$

Factorizing V_1:

$$V_1 = \frac{R_1}{R_1 + R_2} V_s \quad \textbf{(3.14)}$$

Equivalently, if $V_1 = \frac{R_1}{R_2} V_2$ is substituted into Equation (3.12):

$$V_2 = \frac{R_2}{R_1 + R_2} V_s \quad \textbf{(3.15)}$$

Similarly, the voltage across each of the N series-configured resistors is obtained using:

$$V_k = \frac{R_k}{R_1 + R_2 + \cdots + R_N} V_s; \quad k = 1, 2, \ldots N \quad \textbf{(3.16)}$$

Therefore, the voltage of each resistor in a series combination is a fraction of the total voltage. This fraction is the ratio of the resistor's individual resistance to the total series resistance. Larger resistance leads to larger voltage drop. Equation (3.16) is called *voltage division rule.*

EXCERCISE 3.2

Can we use voltage division rule depicted in Equation (3.16) to find the value of an unknown resistance? How?

Do you feel that this method of measuring an unknown resistance involves measurement errors?

Note that a voltage source such as a battery usually has an internal resistance. This resistance increases as the battery ages. Thus, at a given time, the resistance of a battery source is not constant. A better method for measuring the resistance of a resistor is Wheatstone bridge as discussed in Example 3.22.

APPLICATION EXAMPLE 3.7 Rocket Launch Setup

A simple rocket launch setup is shown in Figure 3.23. A controller with a voltage source of 24 V is used to control the switch and the current flow through the igniter. Switch 1 is opened to serve as a safety backup while the igniter is set up. The indicator serves as a high-resistance controller so that it lowers the current flow through the igniter to avoid launching rocket when switch 2 is opened and switch 1 is closed. When both switch 1 and switch 2 are closed, the

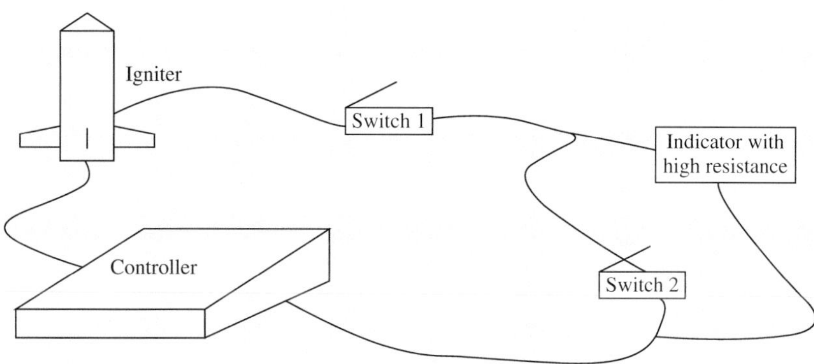

FIGURE 3.23 A simple rocket launch setup.

current flow through the igniter will become high enough to initiate the launch. Assume that a voltage source is in series with the controller, and it supplies 24 V. The resistances of the controller, igniter, and indicator are 500 Ω, 10 Ω, and 1 MΩ, respectively. Calculate the current flow through the igniter and the voltage across it when:

a. Switch 1 is closed and switch 2 is opened.
b. Both switch 1 and switch 2 are closed.

SOLUTION

The equivalent circuit for the rocket launch setup is shown in Figure 3.24.

a. If switch 1 is closed and switch 2 is opened, the total resistance is:

$$R_{\text{total}} = R_{\text{controller}} + R_{\text{igniter}} + R_{\text{indicator}}$$
$$R_{\text{total}} = 500 + 10 + 10^6 = 1000510\ \Omega$$

The current flow through the igniter is:

$$I = \frac{V_{\text{source}}}{R_{\text{total}}}$$

$$I = \frac{24}{1{,}000{,}510} = 2.3988 \times 10^{-5}\ \text{A}$$

The voltage across the igniter is:

$$V_{\text{igniter}} = I \times R_{\text{igniter}} = 0.23988\ \text{mV}$$

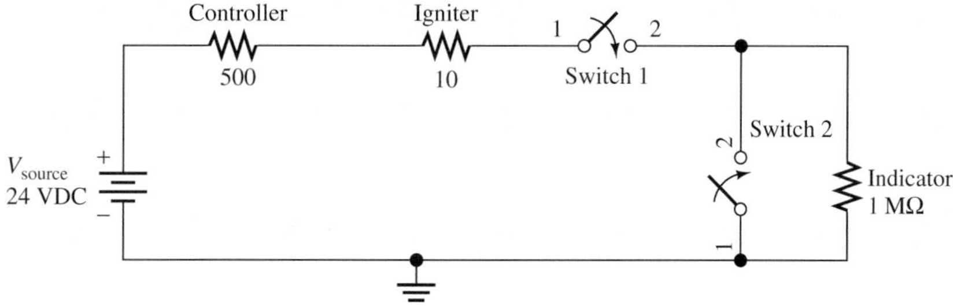

FIGURE 3.24 The equivalent circuit for the rocket launch setup.

(continued)

APPLICATION EXAMPLE 3.7 Continued

This problem can also be solved using voltage division rule of Equation (3.16):

$$V_{\text{igniter}} = \frac{R_{\text{igniter}}}{R_{\text{total}}} \times V_{\text{source}}$$

$$= \frac{10}{1{,}000{,}510} \times 24 = 0.23988 \text{ mV}$$

b. If both switch 1 and switch 2 are closed, a short circuit is created by switch 2. Therefore, the current does not flow through the indicator. The total resistance is:

$$R_{\text{total}} = R_{\text{controller}} + R_{\text{igniter}}$$

$$R_{\text{total}} = 500 + 10 = 510\ \Omega$$

In this case, the current flow through the igniter is:

$$I = \frac{V_{\text{source}}}{R_{\text{total}}}$$

$$I = \frac{24}{510} = 0.0471 \text{ A}$$

The voltage across the igniter is:

$$V_{\text{igniter}} = I \times R_{\text{igniter}} = 0.471 \text{ V}$$

Again, using voltage division rule of Equation (3.16):

$$V_{\text{igniter}} = \frac{R_{\text{igniter}}}{R_{\text{total}}} \times V_{\text{source}}$$

$$= \frac{10}{510} \times 24 = 0.471 \text{ V}$$

3.3.2 Current Division

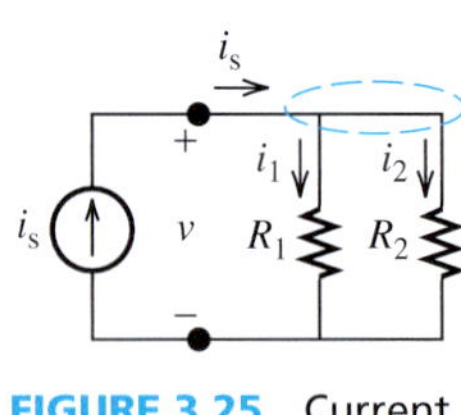

FIGURE 3.25 Current divider.

The current entering a configuration with parallel resistors is divided between the resistors. Therefore, a parallel resistor combination is also called a current divider. Consider Figure 3.25 as a current divider. Applying KCL:

$$i_s - i_1 - i_2 = 0 \qquad \textbf{(3.17)}$$

Using Ohm's law:

$$v_1 = R_1 i_1 = R_2 i_2 \qquad \textbf{(3.18)}$$

Equation (3.18) can be used to find the relationship of the currents in the two parallel branches:

$$i_2 = \frac{R_1}{R_2} i_1 \qquad \textbf{(3.19)}$$

Substituting Equations (3.19) into (3.17):

$$i_1 + \frac{R_1}{R_2} i_1 = i_s$$

Factorizing i_1:

$$i_1 = \frac{R_2}{R_1 + R_2} i_s \tag{3.20}$$

Similarly, using Equation (3.18) to write i_1 in terms of i_2 results in:

$$i_2 = \frac{R_1}{R_1 + R_2} i_s \tag{3.21}$$

Equations (3.20) and (3.21) are *current division rules.*

EXERCISE 3.3

Prove Equation (3.21).

Each parallel resistor receives a fraction of the total current. For two resistors, this fraction equals the ratio between the other resistance in the circuit and the sum of the two resistances. In this case, a smaller amount of current flows through larger resistor. Recall that resistors resist the flow of current.

EXAMPLE 3.8 Voltage Division in Resistive Circuits

In Figure 3.26, find the resistor voltage drops for V_1, V_2, and V_3.

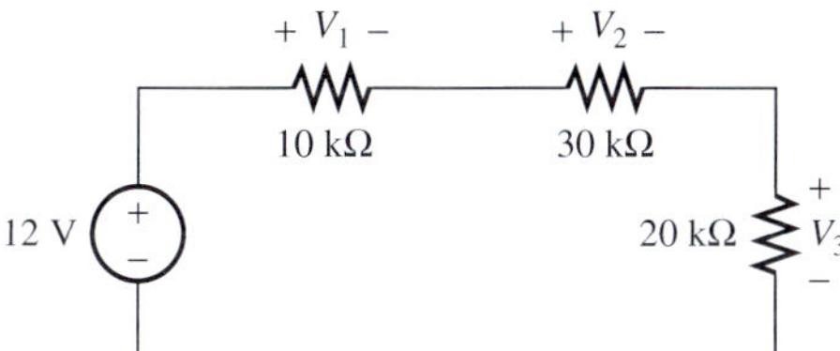

FIGURE 3.26 Figure for Example 3.8.

SOLUTION

According to the voltage division principle of Equation (3.16):

$$V_1 = 12\text{ V} \times \frac{10\text{K}}{10\text{K} + 30\text{K} + 20\text{K}} = 2\text{V}$$

$$V_2 = 12\text{ V} \times \frac{30\text{K}}{10\text{K} + 30\text{K} + 20\text{K}} = 6\text{V}$$

$$V_3 = 12\text{ V} \times \frac{20\text{K}}{10\text{K} + 30\text{K} + 20\text{K}} = 4\text{V}$$

As expected from Kirchhoff's voltage law, the total resistance drop is equal to the applied voltage, that is, 2 + 4 + 6 = 12 V.

EXAMPLE 3.9 Current Division in Resistive Circuits

Figure 3.27 is a current divider with a dependent source. Find I in this circuit.

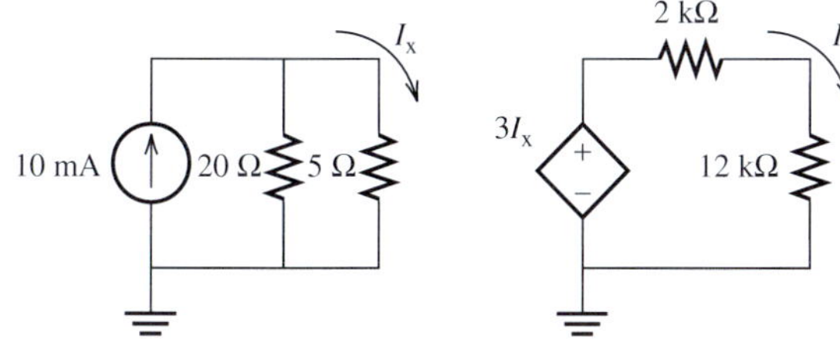

FIGURE 3.27 A current divider with a dependent source.

(continued)

EXAMPLE 3.9 Continued

SOLUTION

This circuit has a 10-mA independent current source and a dependent voltage source equal to $3I_x$. In the section of the circuit located in the right-hand side:

$$I = \frac{3I_x}{2\,\text{k}\Omega + 12\,\text{k}\Omega} = \frac{3I_x}{14\,\text{k}\Omega}$$

To find I, first find I_x. According to the current division principle (see Equation (3.20)):

$$I_x = 10\text{ mA} \times \frac{20\,\Omega}{20\,\Omega + 5\,\Omega} = 8\,\text{mA}$$

Now, replacing I_x with 8 mA:

$$I = \frac{3 \times 8\text{ mA}}{14\text{k}\Omega} = 1.71\,\mu\text{A}$$

APPLICATION EXAMPLE 3.10 Strain Gage

A strain gage is a variable resistor whose resistance varies by the strain on the device to which it is attached. Strain gages can be used as sensors in "smart" buildings. Smart buildings are equipped with sensors that enable load specifications to be monitored and can inform authorities if there is any probability of building damage.

In this example, consider a strain gage attached to a beam (see Figure 3.28). This strain gage is connected to a computer that records the data over a given period of time. The computer has an equivalent resistance of 10 kΩ. The parameters of the strain gage are:

$R_{min} = 1.285$ kΩ
$R_{max} = 20.1$ kΩ
$I_{max} = 325$ μA (maximum current the gage can handle)
$P_{max} = 2.9$ mW (maximum power the gage can dissipate).

This system can be modeled as shown in Figure 3.29. Given $I_{max} = 325$ μA for the strain gage, determine if the current-limiting resistance meets the I_{max} and P_{max} requirements for the strain gage. If not, how can this problem be addressed?

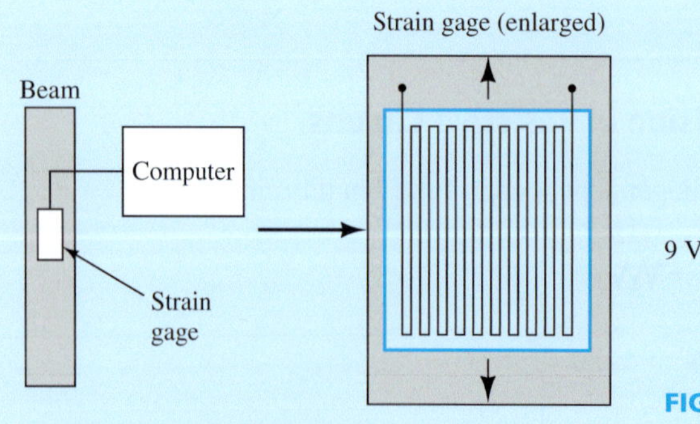

FIGURE 3.28 Strain gage.

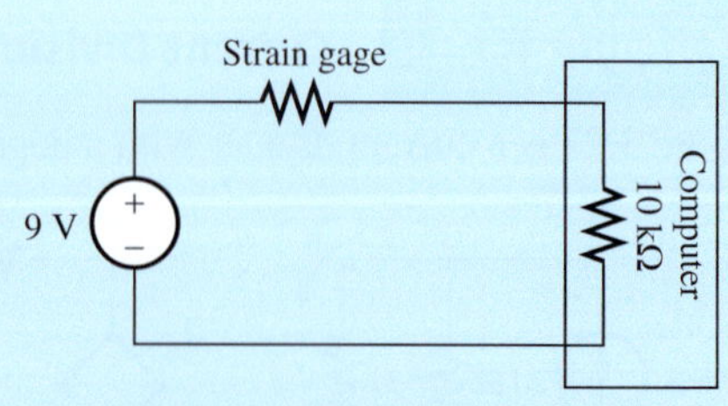

FIGURE 3.29 Equivalent circuit for Figure 3.28.

SOLUTION

No, the resistance does not meet the I_{max} and P_{max} requirements because at the minimum resistance, the current through the circuit is $9/(1.285\text{k}\Omega + 10\text{k}\Omega) = 797\ \mu\text{A}$, which is greater than I_{max} (325 μA).

In order to fix this problem, a resistor must be added in series with the gage to limit the current. Now, calculate the value of the required current-limiting resistor to ensure all of the strain gage specifications are met.

Because $I_{max} = 325\ \mu\text{A}$, the resistor of the circuit should be [see Figure 3.30(a)]:

$$\frac{9}{325\ \mu\text{A}} = 1.285\,\text{k}\Omega + R + 10\,\text{k}\Omega$$

$$\Rightarrow R = 16.4\,\text{k}\Omega$$

To allow for resistor tolerances, adjust its value to 17 kΩ as shown in Figure 3.30.

Including this resistor, the maximum voltage drop across the gage for the two situations of R_{max} and R_{min}, respectively, equals:

$$V_{R_{max}} = \frac{20.1\ \text{k}\Omega}{20.1\ \text{k}\Omega + 17\ \text{k}\Omega + 10\ \text{k}\Omega} \times 9\,\text{V} = 3.84\ \text{V}$$

$$V_{R_{min}} = \frac{1.285\ \text{k}\Omega}{1.285\ \text{k}\Omega + 17\ \text{k}\Omega + 10\ \text{k}\Omega} \times 9\,\text{V} = 0.409\ \text{V}$$

In addition, their corresponding currents are:

$$I_{R_{max}} = \frac{V_{R_{max}}}{20.1\,\text{k}\Omega} = 191\ \mu\text{A}$$

$$I_{R_{min}} = \frac{V_{R_{min}}}{1.285\,\text{k}\Omega} = 318\ \mu\text{A}$$

$$P_{max} = 9 \times 318 = 2.868\,\text{mW} < 2.9\,\text{mW}$$

Note that 2.9 mW is the maximum power allowable. Thus, this circuit with its current-limiting resistance meets the design requirements for the strain gage.

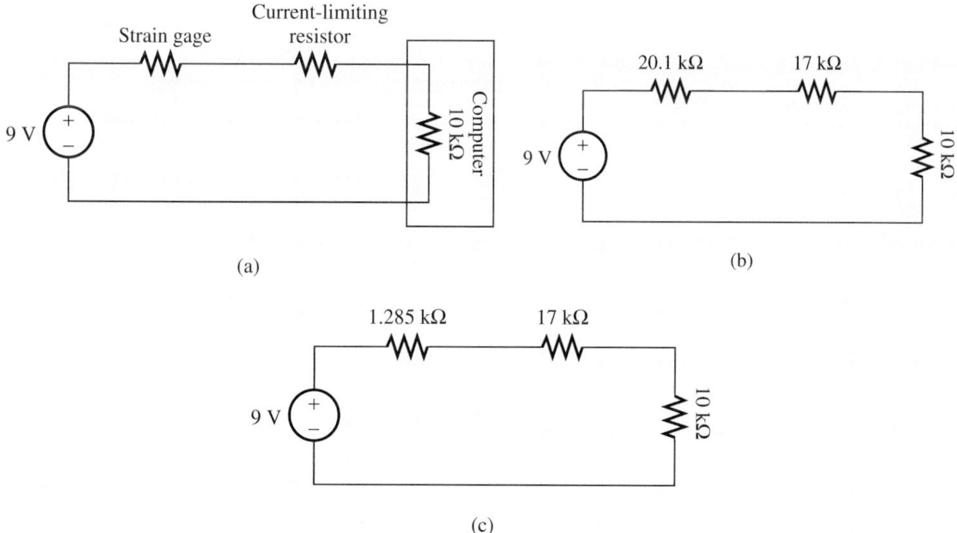

FIGURE 3.30 (a) Calculating the resistor that is needed to limit the current. (b) Equivalent circuit for R_{max}. (c) Equivalent circuit for R_{min}.

APPLICATION EXAMPLE 3.11 Automotive Power System

To design an automotive power supply appropriately, it is important to properly compute the amount of current that is required by its electrical system. The components of an automotive electrical system are connected to the power supply (battery–alternator combination) in parallel, as shown in Figure 3.31. Each additional component causes more current to be drawn from the supply. Find the total current i in Figure 3.31 given that $i_1 = 3$ A, $R_2 = 2\ \Omega$, $R_3 = 6\ \Omega$, $R_4 = 10\ \Omega$, and $i_5 = 1$ A.

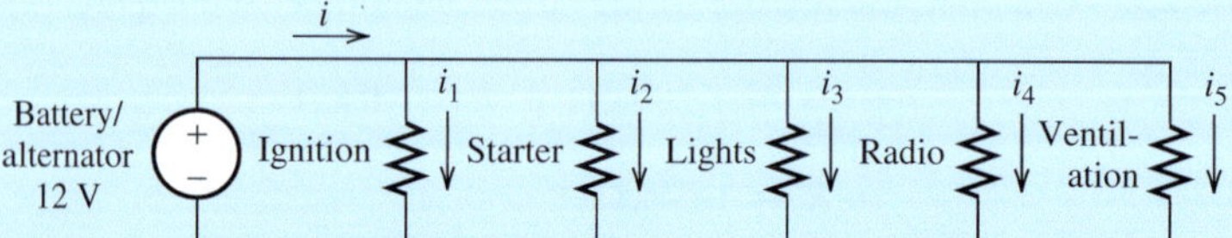

FIGURE 3.31 Circuit for Example 3.11.

SOLUTION

The voltage across all resistors is constant and is equal to 12 V; thus:

$$i_2 = \frac{12}{2} = 6\,\text{A} \quad i_3 = \frac{12}{6} = 2\,\text{A} \quad i_4 = \frac{12}{10} = 1.2\,\text{A}$$

Now, using KCL, and given the two known currents, i_1 and i_5:

$$-i + i_1 + i_2 + i_3 + i_4 + i_5 = 0$$

Replacing for the known currents results in:

$$-i + 3 + 6 + 2 + 1.2 + 1 = 0$$

$$i = 13.2\ \text{A}$$

EXERCISE 3.4

Why are the components of an automotive electric system connected in parallel?

APPLICATION EXAMPLE 3.12 Impedance Plethysmography

Biomedical engineers must measure biological features without invasive techniques. For example, measuring the volume of a heart chamber is not easily accomplished by conventional means, as it could require the removal of the heart from the body. Other less invasive techniques are therefore preferred.

One technique to measure the left ventricular (LV) volume is impedance plethysmography. This involves measuring the electrical resistance (impedance) of the blood in the heart chamber and using formulas to convert this measurement into a volume.

Electrodes are implanted along the length of the ventricle at an evenly spaced distances. A known voltage is applied across the two outer electrodes as shown in Figure 3.32. The current and the voltages between each pair of electrodes are measured. From this, the cross-sectional area of the cylinder between each electrode pair (as shown in the figure) can be calculated using the formula: $R = \rho L/A$, where ρ is the resistivity, L is the length of the material, and A is its cross-sectional area. Then, the volume of each cylinder can be easily found and added to get an approximate total volume.

Suppose five electrodes are inserted into the left ventricle with a spacing of 1 cm. $\rho = 150\ \text{m}\Omega$. The resistance between the electrodes A and E is found to be 7.5 kΩ. With

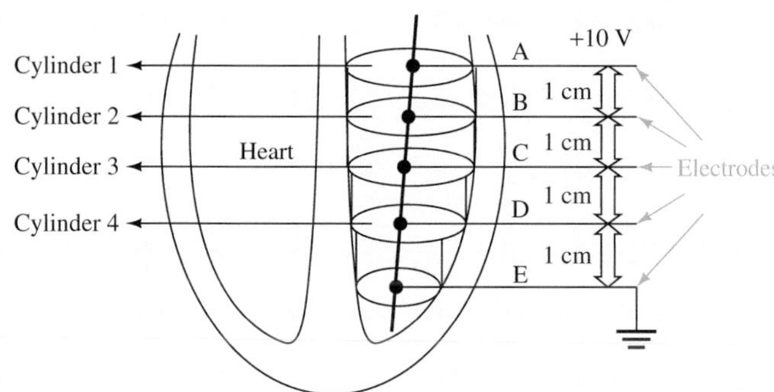

FIGURE 3.32 Impedance plethysmography.

a voltage of 10 V applied between A and E, the electrode voltages (with respect to the ground) are:

Electrode	Voltage
Electrode A	$V_A = 10$ V
Electrode B	$V_B = 8$ V
Electrode C	$V_C = 6.2$ V
Electrode D	$V_D = 3.8$ V
Electrode E	$V_E = 0$ V

Find the resistance of each cylinder, and from this, find the cross-sectional area and volume for each cylinder.

SOLUTION

This system can be modeled as the circuit shown in Figure 3.33. The resistance voltages can be calculated to be $V_{AB} = V_A - V_B = 2$ V, $V_{BC} = V_B - V_C = 1.8$ V, $V_{CD} = V_C - V_D = 2.4$ V and $V_{DE} = V_D - V_E = 3.8$ V. Using these values, the resistances can be calculated:

$$V_{AB} = 2\,\text{V} = \frac{10\,\text{V} \times R_1}{7.5\,\text{k}\Omega} \rightarrow R_1 = 1.5\text{k}\Omega$$

$$V_{BC} = 1.8\,\text{V} = \frac{10\,\text{V} \times R_2}{7.5\,\text{k}\Omega} \rightarrow R_2 = 1.35\,\text{k}\Omega$$

$$V_{CD} = 2.4\,\text{V} = \frac{10\,\text{V} \times R_3}{7.5\,\text{k}\Omega} \rightarrow R_3 = 1.8\,\text{k}\Omega$$

$$V_{DE} = 3.8\,\text{V} = \frac{10\,\text{V} \times R_4}{7.5\,\text{k}\Omega} \rightarrow R_4 = 2.85\,\text{k}\Omega$$

These resistances can then be used to calculate the cross-sectional area of the cylinders. The electrode spacing is 1 cm. Using $A = \dfrac{\rho L}{R}$:

$$A_1 = \frac{150\,\Omega\text{m} \times 0.01\,\text{m}}{1.5\,\text{k}\Omega} = 10\,\text{cm}^2$$

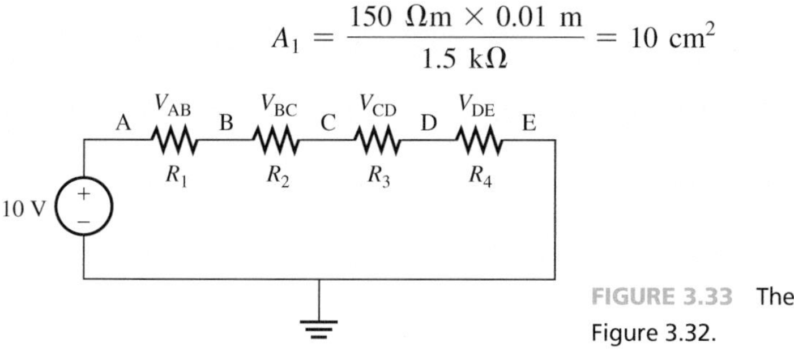

FIGURE 3.33 The circuit model for Figure 3.32.

(continued)

APPLICATION EXAMPLE 3.12 Continued

$$A_2 = \frac{150\ \Omega\text{m} \times 0.01\ \text{m}}{1.35\ \text{k}\Omega} = 11.1\ \text{cm}^2$$

$$A_3 = \frac{150\ \Omega\text{m} \times 0.01\,\text{m}}{1.8\ \text{k}\Omega} = 8.33\ \text{cm}^2$$

$$A_4 = \frac{150\ \Omega\text{m} \times 0.01\,\text{m}}{2.85\,\text{k}\Omega} = 5.26\ \text{cm}^2$$

Now, find the volume by adding the volume of each cylinder:

$$\text{Total volume} = 10\,\text{cm}^2 \times 1\,\text{cm} + 11.1\,\text{cm}^2 \times 1\,\text{cm} + 8.33\,\text{cm}^2 \times 1\,\text{cm} + 5.26\,\text{cm}^2 \times 1\,\text{cm}$$
$$= 34.69\,\text{cm}^3$$

APPLICATION EXAMPLE 3.13 Automobile Ignition System

The ignition system of an automobile generates a high voltage (>20,000 V) from the 12-V battery. Figure 3.34 represents a very simplified ignition system. When the switch, S (known as the "points") is opened, a high voltage is generated in the secondary coil, which is directed to the appropriate spark plug. The function of the ignition system will be detailed in future chapters.

a. If the switch is closed, and $V_R = 5.2$ V, what is V_{SW}?
b. What is V_L?
c. If $R = 10\ \Omega$, what is the coil resistance R_L?
d. What is the voltage V_{SW} when the switch opens?

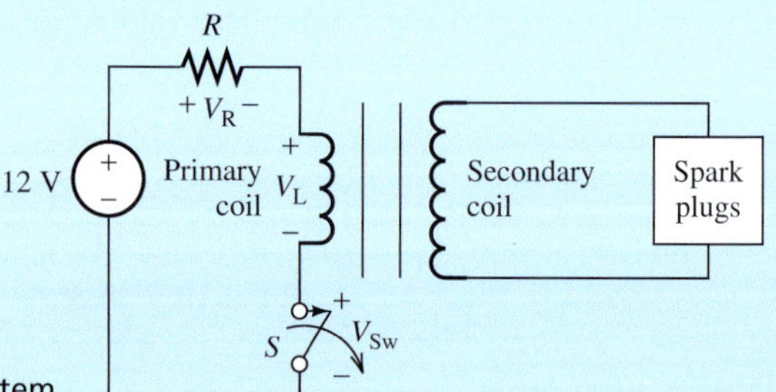

FIGURE 3.34 A simplified ignition system.

SOLUTION

a. Because the switch is closed, the resistance across its contacts is close to zero. According to Ohm's law, $V = I \times R$. Using the equation, if the resistance is zero, the voltage will also be zero. Therefore:

$$V_{SW} = 0\,\text{V}$$

b. According to KVL:

$$-12 + 5.2 + V_L + 0 = 0$$

This equation can be used to calculate the voltage, $V_L = 6.8\,\text{V}$

c. According to the voltage divider formula:

$$6.8 = 12 \times \frac{R_L}{10 + R_L + 0}$$

Now, the resistance can be calculated as $R_L = 13.08\ \Omega$.

d. When the switch is open, the resistance across its contacts is nearly infinite. According to the voltage divider formula:

$$V_{SW} = 12 \times \frac{R_{SW}}{R + R_L + R_{SW}}$$

As R_{SW} approaches infinity, V_{SW} approaches 12 V. Therefore:

$$V_{SW} = 12\text{V}$$

3.4 NODAL AND MESH ANALYSIS

In some circuits, equivalent resistance and voltage/current divider rules are not enough for resistive circuit analysis. For example, consider the circuit in Figure 3.35. None of the resistors in the circuit is in series or parallel to each other.

Therefore, a different analysis method is needed that works regardless of circuit configuration and size. Nodal voltage analysis and mesh current analysis are examples of such systems. In the following sections, nodal and mesh analysis are introduced.

3.4.1 Nodal Analysis

In the nodal analysis method, the nodes in the circuit are determined first. In a resistive circuit, a node is the connecting point of two or more circuit resistors. The nodes in Figure 3.35 are specified in Figure 3.36. Four nodes are observed in the figure.

In the next step, a node is picked from the specified nodes as the *reference node*. The voltages of the other nodes are then measured with respect to this node.

As shown in Figure 3.37, a voltage is assigned to each node. Any node in the circuit can be used as the reference node. However, the circuit analysis can be simplified by selecting one of the end points of a voltage source as the reference node. As can be observed in Figure 3.37, ground is the symbol used for the reference node. The reference node is assigned zero voltage. The voltage of some nodes in the circuit can be determined easily. For example, in Figure 3.37, there is only one voltage source between Node 1 and the reference node. As a result, the difference of the voltage between Node 1 and the zero voltage of the reference node is V_s and V_s is assigned to I_1.

The main step in node voltage analysis is to write KCL equations in terms of the node voltages. After writing the KCL equations corresponding to each node, solve the system of equations to find unknown voltages. For example, assume that Figure 3.38 is a part of the circuit in

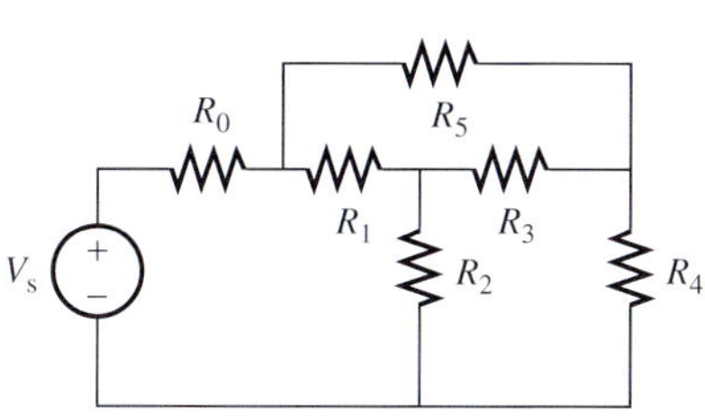

FIGURE 3.35 A resistive circuit.

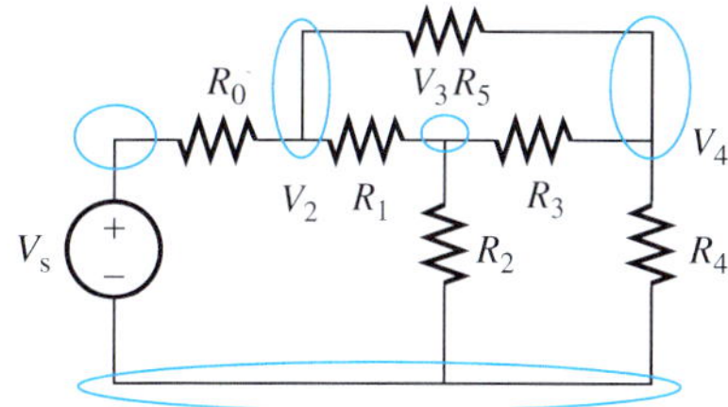

FIGURE 3.36 The nodes of the circuit of Figure 3.35.

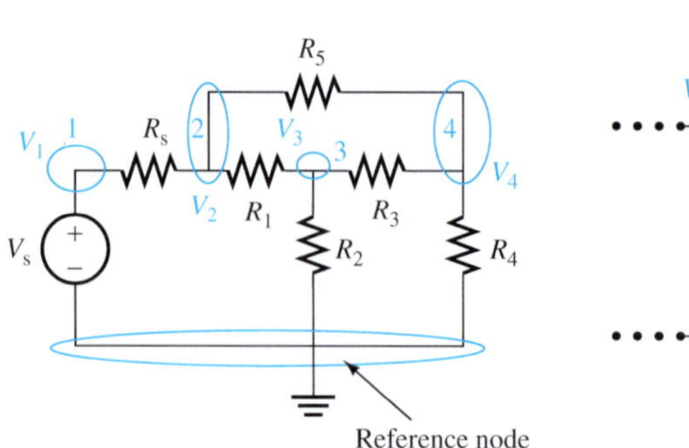

FIGURE 3.37 A section of Figure 3.35.

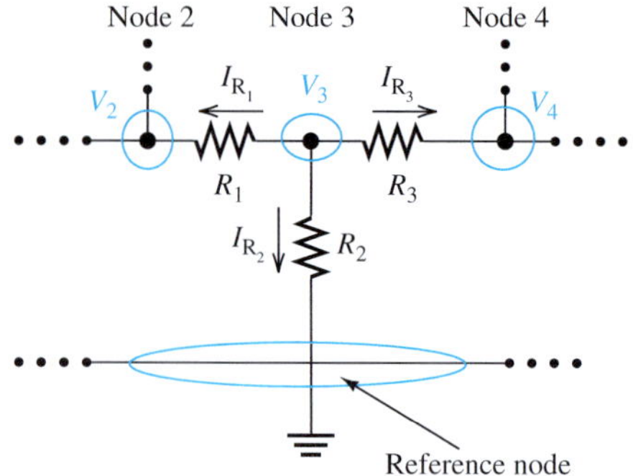

FIGURE 3.38 Different nodes in a resistive circuit.

Figure 3.37. Ohm's law states that the current of each resistor is the difference of the potentials of their terminals divided by their resistances, that is:

$$I_{R_1} = \frac{V_3 - V_2}{R_1}$$

$$I_{R_2} = \frac{V_3 - 0}{R_2}$$

$$I_{R_3} = \frac{V_3 - V_4}{R_3}$$

Actually, for each node, it is assumed that the current is leaving the node. Therefore, according to KCL, the sum of all the leaving currents is zero. Therefore, in Figure 3.38 (Node 3):

$$I_{R_1} + I_{R_2} + I_{R_3} = 0, \quad I_{R_1} = \frac{V_3 - V_2}{R_1}, \quad I_{R_2} = \frac{V_3}{R_2}, \quad I_{R_3} = \frac{V_3 - V_4}{R_3}$$

Similar equations can be generated for Nodes 2 and 4.

EXCERCISE 3.5

Write the equations for Nodes 1 and 2 of Figure 3.37.

Writing the leaving current from each node in terms of the node voltages and the resistance enables generation of KCL equations. Using KCL for the three nodes of Figure 3.37 and noting that $V_1 = V_s$:

KCL for Node 2:

$$\frac{V_2 - V_s}{R_s} + \frac{V_2 - V_3}{R_1} + \frac{V_2 - V_4}{R_5} = 0$$

KCL for Node 3:

$$\frac{V_3 - V_2}{R_1} + \frac{V_3}{R_2} + \frac{V_3 - V_4}{R_3} = 0$$

KCL for Node 4:

$$\frac{V_4 - V_3}{R_3} + \frac{V_4}{R_4} + \frac{V_4 - V_2}{R_5} = 0$$

Algebraic manipulation results in the following system of equations for the node voltages:

$$\begin{cases} \left(\dfrac{1}{R_s}+\dfrac{1}{R_1}+\dfrac{1}{R_5}\right)V_2-\dfrac{V_3}{R_1}-\dfrac{V_4}{R_5}=\dfrac{V_s}{R_S} \\ -\dfrac{V_2}{R_1}+\left(\dfrac{1}{R_1}+\dfrac{1}{R_2}+\dfrac{1}{R_3}\right)V_3-\dfrac{V_4}{R_3}=0 \\ -\dfrac{V_2}{R_5}-\dfrac{V_3}{R_3}+\left(\dfrac{1}{R_3}+\dfrac{1}{R_4}+\dfrac{1}{R_5}\right)V_4=0 \end{cases} \tag{3.22}$$

EXCERCISE 3.6

Show the details of generating the above set of equations.

The system of linear equations can also be shown in matrix form. The matrix form of the above node voltage equations is:

$$\begin{pmatrix} \dfrac{1}{R_5}+\dfrac{1}{R_1}+\dfrac{1}{R_5} & -\dfrac{1}{R_1} & -\dfrac{1}{R_5} \\ -\dfrac{1}{R_1} & \dfrac{1}{R_1}+\dfrac{1}{R_2}+\dfrac{1}{R_3} & -\dfrac{1}{R_3} \\ -\dfrac{1}{R_5} & -\dfrac{1}{R_3} & \dfrac{1}{R_3}+\dfrac{1}{R_4}+\dfrac{1}{R_5} \end{pmatrix}\begin{pmatrix} V_2 \\ V_3 \\ V_4 \end{pmatrix}=\begin{pmatrix} \dfrac{V_s}{R_S} \\ 0 \\ 0 \end{pmatrix}$$

Matrix form facilitates solving the problem and obtaining the voltages using a calculator or software such as MATLAB®. The technique used to solve these matrix equations and evaluate the unknown parameters is discussed in Appendix A.

EXAMPLE 3.14 Reference Node

Which node in Figure 3.39 is the best choice for a reference node?

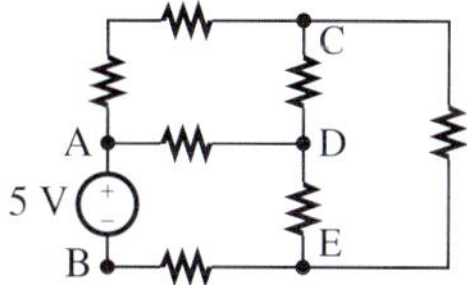

FIGURE 3.39 The circuit for Example 3.14.

SOLUTION

The negative terminal of the voltage source (in this case, node B) in the circuit is the best choice for the reference node.

EXAMPLE 3.15 Reference Node

Which node is the best choice for a reference node in Figure 3.40?

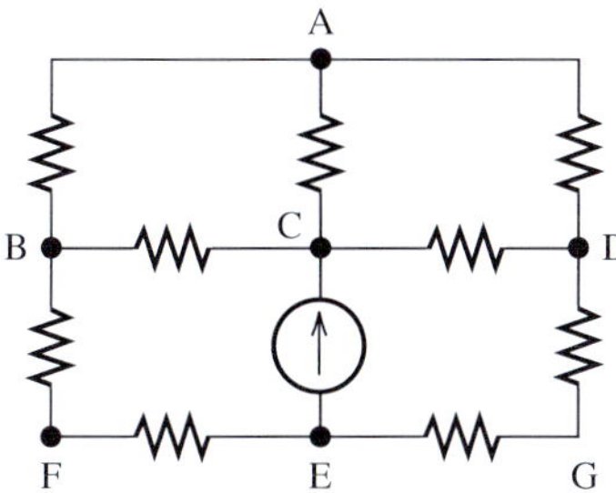

FIGURE 3.40 The circuit for Example 3.15.

(*continued*)

EXAMPLE 3.15 Continued

SOLUTION

There is no voltage source in the circuit. Therefore, any of the nodes in Figure 3.40 can be used as a reference node.

EXCERCISE 3.7

In Figure 3.40, if all resistors are equivalent to R, find the voltage of nodes B and D. Are these voltages equivalent? If two points (nodes) have the same voltage, they can be overlapped. This simplifies the topology of the circuit. Use this method to find the equivalent resistance across nodes C and E.

EXAMPLE 3.16 Nodal Analysis

Find V_1, V_2, and V_3 in Figure 3.41 using nodal equations.

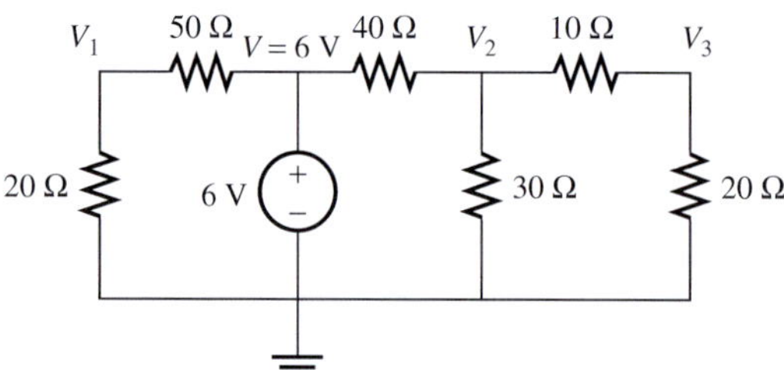

FIGURE 3.41 The circuit for Example 3.16.

SOLUTION

If the sum of the current leaving from each node is set to zero, the following set of linear equations can be obtained in terms of node voltages.

$$\begin{cases} \dfrac{V_1}{20} + \dfrac{V_1 - 6}{50} = 0 \\ \dfrac{V_2 - 6}{40} + \dfrac{V_2}{30} + \dfrac{V_2 - V_3}{10} = 0 \\ \dfrac{V_3 - V_2}{10} + \dfrac{V_3}{20} = 0 \end{cases} \longrightarrow \begin{cases} \left(\dfrac{1}{20} + \dfrac{1}{50}\right)V_1 = \dfrac{6}{50} \\ \left(\dfrac{1}{10} + \dfrac{1}{30} + \dfrac{1}{40}\right)V_2 - \dfrac{V_3}{10} = \dfrac{6}{40} \\ \left(\dfrac{-1}{10}\right)V_2 + \left(\dfrac{1}{10} + \dfrac{1}{20}\right)V_3 = 0 \end{cases}$$

Solving this system of equations:

$$V_1 = 1.71 \text{ V}$$
$$V_2 = 1.64 \text{ V}$$
$$V_3 = 1.09 \text{ V}$$

EXCERCISE 3.8

Use the voltage divider method to find the voltages V_1, V_2, and V_3 in Example 3.16.

APPLICATION EXAMPLE 3.17 Automobile Acceleration

The inventor of a new sports car intends to verify if its initial acceleration from 0 to 100 mph is indeed better than any competitors. She has gathered the acceleration data from the leading competitors to compare to her own. To record her own data, she connects an accelerometer to her initial prototype. The accelerometer is connected to a PDA located in the car, which logs the data. The PDA has an equivalent resistance of 50 Ω which is considerably lower than the resistances of the resistors in the network. Therefore, the current through it would be large.

To avoid having the PDA overload the accelerometer's resistive network, it is isolated from the accelerometer using a buffer, modeled by the dependent voltage source. The accelerometer gives the acceleration as a varying output voltage, V_A. The voltage, V_A, ranges from 0 to V_{max} = 5 V. The circuit model is shown in Figure 3.42. To calibrate the accelerometer and the PDA, the inventor needs to find the circuit parameters as follows:

a. Find V_o if V_A = 0.3 V
b. Find I_A when $V_A = V_{A,\,max}$
c. How much power is dissipated by the PDA?

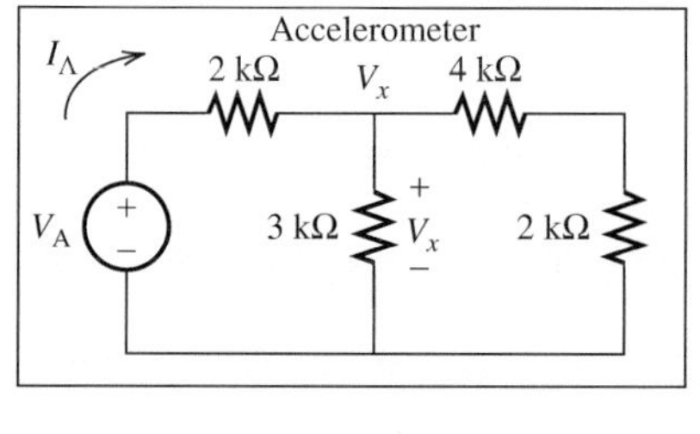

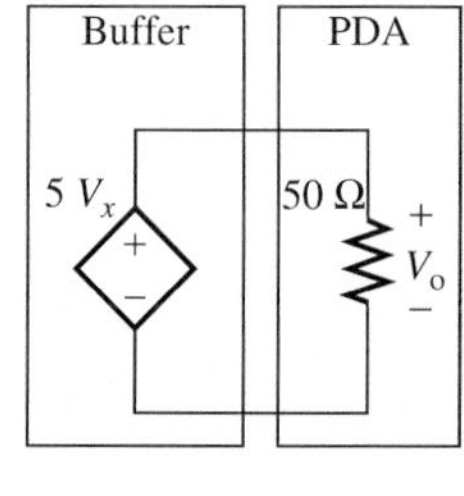

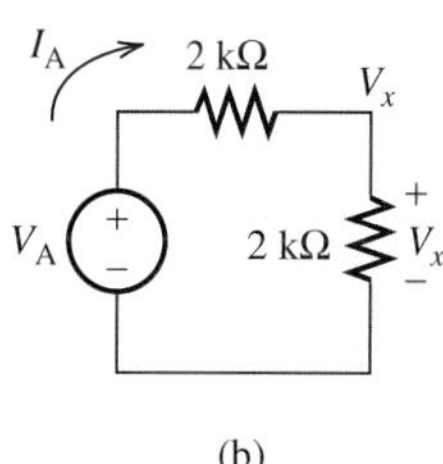

(a) (b)

FIGURE 3.42 (a) The circuit model for the accelerometer and the PDA; (b) equivalent circuit for the accelerometer.

SOLUTION

a. Using KCL for the node with the voltage V_x:

$$\frac{V_x - V_A}{2\text{ k}} + \frac{V_x}{3\text{ k}} + \frac{V_x}{4\text{ k} + 2\text{ k}} = 0$$

If V_A = 0.3 V:

$$V_x = 0.15\text{ V}$$

Therefore (see Buffer–PDA circuit):

$$V_o = 5V_x = 0.75\text{ V}$$

b. The equivalent resistor of the accelerometer is 4 kΩ and V_{max} = 5 V. As a result:

$$I_A = \frac{5\text{V}}{4\text{K}} = 1.25\,\text{mA}$$

c. Figure 3.42(b) shows the equivalent circuit for the accelerator obtained by finding the equivalent resistance of 3||(2 + 4) = 2 kΩ. Using simple voltage division, and taking $V_A = V_{max}$:

$$V_x = \frac{1}{2}V_A = \frac{1}{2} \times 5 = 2.5\text{ V}$$

(*continued*)

APPLICATION EXAMPLE 3.17 Continued

In addition, the voltage of the dependent voltage source of the buffer is $V_o = 2.5 \times 5 = 12.5$ V and the power dissipated at the load resistance will be:

$$P = \frac{V_o^2}{R} = \frac{(12.5)^2}{50} = 3.125\,\text{W}$$

APPLICATION EXAMPLE 3.18 Chemical Process

The temperature of a chemical process needs to be monitored in order to control the process (Figure 3.43). This requires a device that can electrically measure temperature to allow it to be monitored by an electric controller. There are two types of electric temperature sensors: thermocouples and thermistors. A thermistor is simply a resistor whose value changes with temperature. A thermistor is simpler to use than a thermocouple. The approximate temperature-to-resistance relationship of a thermistor is given by the following equation, known as the *Steinhart–Hart Equation*:

$$T = \frac{1}{a + b\ln(R_T) + c(\ln(R_T))^3}$$

where

T = temperature in Kelvin
a, b, c = parameters specific to the thermistor device
R_T = normalized resistance of the thermistor; the normalization is with respect to the resistance in ambient temperature.

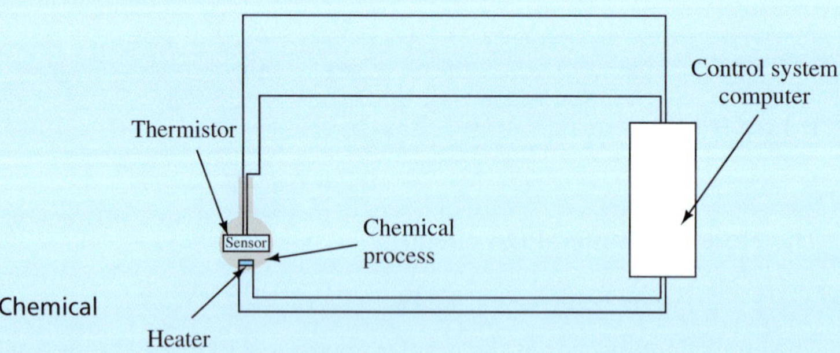

FIGURE 3.43 Chemical thermistor.

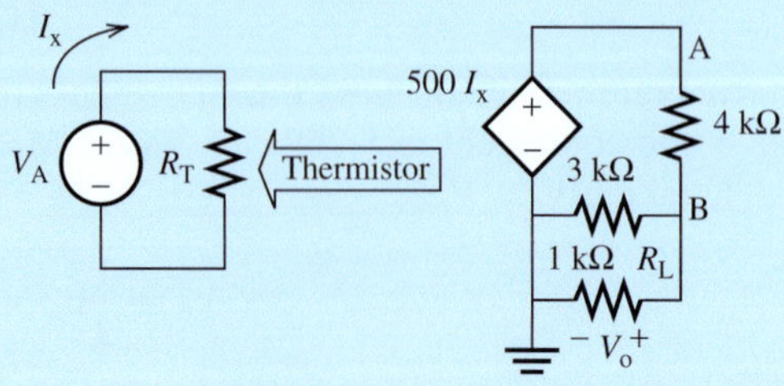

FIGURE 3.44 Circuit model for the controlled thermistor.

For this example, assume the associated ambient temperature is 1 Ω and that the thermistor has the following device parameters:

$$a = 0.0160717; \quad b = -0.00286422; \quad c = 0.0000193965$$

The thermistor is connected as shown in Figure 3.44. The output voltage, V_o, is measured to be 0.73 V. This is the voltage seen by the monitoring computer. The voltage, V_A, is used to convert the varying resistance of the thermistor to the varying current, I_x. This is isolated from the computer by the dependent source.

a. Find the voltage at node A.
b. What is the temperature at the thermistor, in kelvin?

SOLUTION

a. Based on Figure 3.44, the output voltage is given, therefore, $V_o = V_B = 0.73$ V. Now, using node voltage analysis applied to node B:

$$\frac{0.73 - V_A}{4\,\text{k}} + \frac{0.73}{3\,\text{k}} + \frac{0.73}{1\,\text{k}} = 0$$

Through calculation:

$$V_A = 4.623 \text{ V}$$

b. To solve part (b), start with the result from above:

$$V_A = 500\, I_x = 4.623 \text{ V}$$

Calculating the current:

$$I_x = 9.247 \text{ mA}$$

Now, see that:

$$R_T = \frac{V_A}{I_x} = \frac{5\text{ V}}{9.247\,\text{mA}} = 540.7\,\Omega$$

Using the *Steinhart–Hart* Equation:

$$T = \frac{1}{a + b\ln(R_T) + c(\ln(R_T))^3}$$

$$T = 347 \text{ K}$$

The node voltage analysis procedure can be summarized in the following steps:

1. Determine all nodes in the circuit.
2. Pick a node as a reference node.
3. Assign voltage variables to each node.
4. For each node, write KCL equations in terms of node voltages.
5. Solve the system of equations to find unknown voltages.

3.4.2 Mesh Analysis

In addition to nodal analysis, mesh analysis can be used to find the circuit parameters in Figure 3.35. The procedures are outlined as follows.

A mesh is defined as a closed loop of a circuit. The first step in mesh analysis is to identify the meshes in the given circuit. In the circuit of Figure 3.45, three meshes are identified and then the mesh current variables i_1, i_2, i_3 are assigned to them, respectively. To avoid confusion in writing circuit equations, unknown mesh currents are defined exclusively clockwise.

The major step is to write KVL equations for each mesh. Beginning with mesh 1, note that the voltages around the mesh have been assigned in Figure 3.45 *consistent with* the direction of the mesh current, i_1. Now it is important to observe that while mesh current i_1 is equal to the current flowing through resistor R_0, it is not equal to the currents through R_1 and R_2. The branch current through R_1 is the difference between the two mesh currents, that is, $i_1 - i_2$. Similarly, the current through R_2 is $i_1 - i_3$. Note that the current direction is determined consistent with the polarity of the voltage across R_2 which is from the positive pole toward the negative one. Here, the direction of i_1 is consistent with the polarity of v_2 while the direction of i_3 is not consistent with this polarity. Therefore, the current through R_2 would be $i_1 - i_3$.

According to the polarity of voltage v_1, the voltage, v_1, is given by:

$$v_1 = (i_1 - i_2)R_1$$

Similarly, the voltage, v_2, corresponds to:

$$v_2 = (i_1 - i_3)R_2$$

Therefore, the KVL equation for mesh 1 is:

$$v_S - i_1 R_0 - (i_1 - i_2) R_1 - (i_1 - i_3) R_2 = 0$$

The same approach can be applied to mesh 2. Figure 3.46 depicts the voltage assignment around mesh 2, which follows the clockwise direction mesh current i_2.

The mesh current i_2 is the branch current that passes through resistor R_5. However, the current that passes through resistor R_1 that is shared by mesh 1 and mesh 2 is equal to $i_2 - i_1$, and the current through resistor R_3 that is shared by mesh 2 and mesh 3, is equal to $i_2 - i_3$. Note that the current direction is determined consistent with the polarity of the voltage across R_3 which is from the positive pole toward the negative one. Here, the direction of i_2 is consistent with the polarity of v_3 while the direction of i_2 is not consistent with this polarity. Therefore, the current through R_3 would be $i_2 - i_3$.

Therefore, the voltages across R_1 and R_3 are:

$$v_1 = (i_2 - i_1)R_1$$
$$v_3 = (i_2 - i_3)R_3$$

Accordingly, the KVL equation for mesh 2 is:

$$i_2R_5 + (i_2 - i_3)R_3 + (i_2 - i_1)R_1 = 0$$

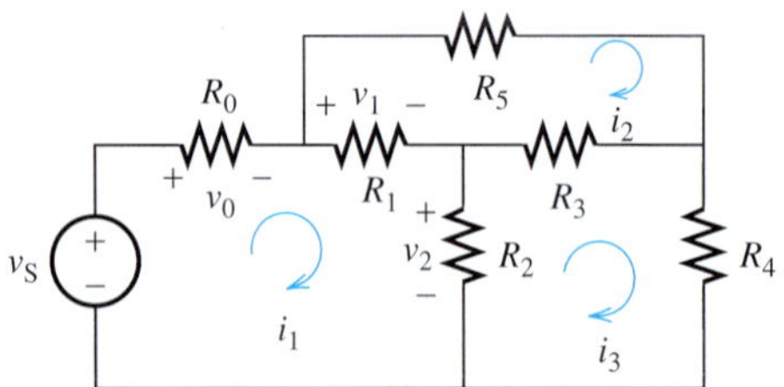

FIGURE 3.45 Assignment of currents and voltages around mesh 1 for the circuit in Figure 3.35.

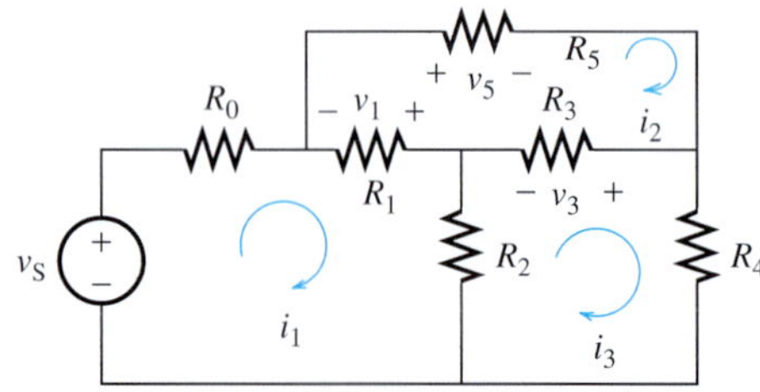

FIGURE 3.46 Assignment of currents and voltages around mesh 2.

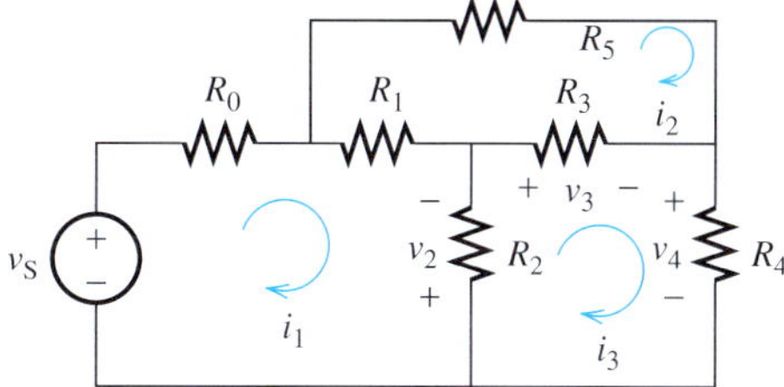

FIGURE 3.47 Assignment of currents and voltages around mesh 3.

Figure 3.47 depicts that the voltage assignment around mesh 3, which follows the clockwise direction for mesh current i_3.

Following the voltage polarity shown in Figure 3.47, voltages v_2 and v_3 can be expressed by:

$$v_2 = (i_3 - i_1)R_2$$
$$v_3 = (i_3 - i_2)R_3$$

Then, the KVL equation for mesh 3 is:

$$(i_3 - i_1)R_2 + (i_3 - i_2)R_3 + i_3R_4 = 0$$

Combining the KVL equations for each mesh 1, 2, and 3, we obtain the following system of equations:

$$(R_0 + R_1 + R_2)i_1 - R_1i_2 - R_2i_3 = v_s$$
$$-R_1i_1 + (R_1 + R_3 + R_5)i_2 - R_3i_3 = 0$$
$$-R_2i_1 - R_3i_2 + (R_2 + R_3 + R_4)i_3 = 0$$

This system of equation can be solved to find the mesh currents i_1, i_2, and i_3.

The mesh current analysis procedure can be summarized in the following steps:

1. Define each mesh current (usually exclusively clockwise).
2. Assign current variables to each mesh.
3. To write the KVL for each mesh, determine the voltage across the components of that mesh based on the mesh current direction.
4. For each mesh, write the KVL equation.

EXAMPLE 3.19 Mesh Determination

Find the meshes in the circuit of Figure 3.48.

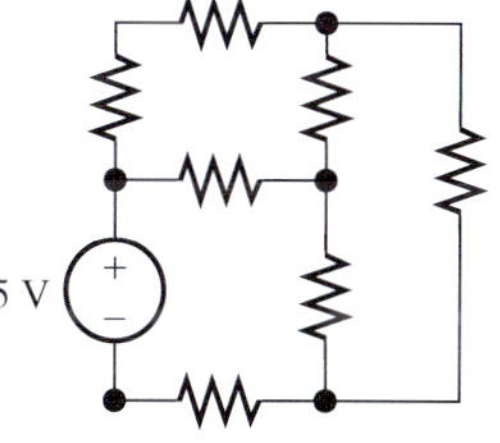

FIGURE 3.48 Find the meshes.

SOLUTION

By inspection there are three meshes in the circuit of Figure 3.48. The mesh currents are defined clockwise and denoted in Figure 3.49.

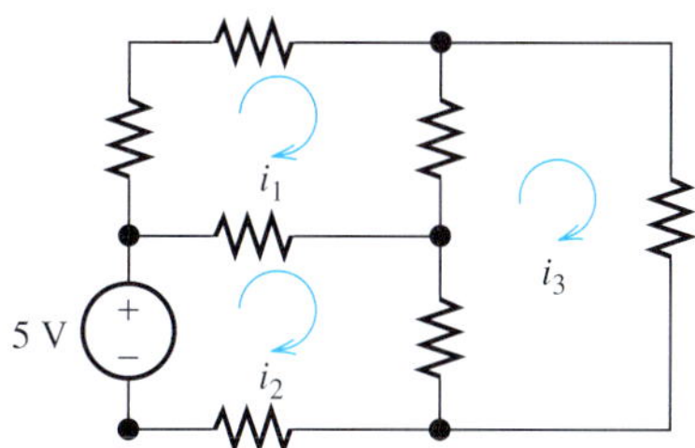

FIGURE 3.49 Find the meshes.

EXAMPLE 3.20 Mesh Analysis

Use mesh analysis to find the current, i, flowing through the 6-V voltage source in Figure 3.50.

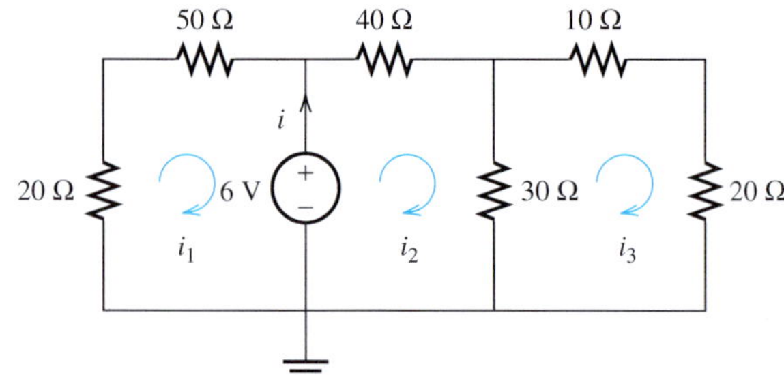

FIGURE 3.50 The circuit for Example 3.20.

SOLUTION

First, three meshes are identified in the circuit of Figure 3.50. The current flowing through the 6-V voltage source is:

$$i = i_2 - i_1$$

We need to find the mesh currents i_1 and i_2. Thus, we write KVL equations for each mesh to find i_1 and i_2.

Mesh 1:

$$20i_1 + 50i_1 + 6 = 0$$

Mesh 2:

$$40i_2 + 30(i_2 - i_3) - 6 = 0$$

Mesh 3:

$$30(i_3 - i_2) + 10i_3 + 20i_3 = 0$$

Rearranging the above three equations, we obtain the system of equations:

$$i_1 = -0.086$$
$$70i_2 - 30i_3 = 6$$
$$-30i_2 + 60i_3 = 0$$

Solving the system of equations:

$$i_1 = -0.086\text{A}, \quad i_2 = 0.109\text{A}, \quad i_3 = 0.055\text{A}$$

Therefore:

$$i = i_2 - i_1 = 0.195\text{A}$$

EXAMPLE 3.21 Mesh Analysis with Current Sources

Find the mesh currents in the circuit shown in Figure 3.51, when $i_s = 0.5$ A, $v_s = 6$ V, $R_1 = 3\ \Omega$, $R_2 = 8\ \Omega$, $R_3 = 2\ \Omega$.

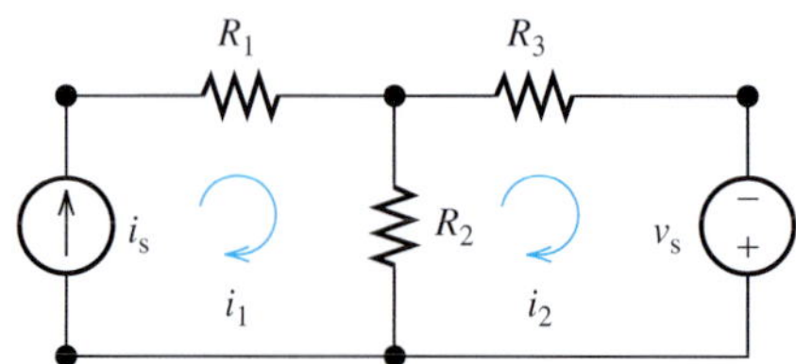

FIGURE 3.51 Mesh analysis with current source.

SOLUTION

For mesh 1, the mesh current is equal to the current source i, that is, $i_1 = i_s = 0.5$ A. Therefore, it is not necessary to write the KVL equation for mesh 1.

For mesh 2, the KVL equation is:

$$(i_2 - i_1)R_2 + i_2R_3 - 6 = 0$$

And i_2 can be expressed by:

$$i_2 = (6 + i_1R_2)/(R_2 + R_3)$$

Substituting i_1 and the resistor values, we can solve the mesh currents:

$$i_1 = 0.5\,\text{A}, \quad i_2 = 1\,\text{A}$$

EXAMPLE 3.22 Wheatstone Bridge

The circuit in Figure 3.52 is known as a Wheatstone bridge. This circuit is commonly used to identify the value of unknown resistor. The application of this bridge has already been introduced in Chapter 1. Chapter 14 also reveals the application of this bridge in measurement devices. Example 3.38 represents another application of Wheatstone bridge.

As shown in Figure 3.52, the circuit of this bridge consists of four elements (in this case, resistors). In the circuit, R_1 and R_2 are known resistors; resistor R_4 is known and adjustable; resistor R_3 is unknown and three meshes are identified.

a. List the KVL equations for all meshes.

b. In this bridge, the value of resistor R_4 can be adjusted to maintain zero current flow between the two nodes A and B. In this case, find the unknown resistor R_3 in terms of R_1, R_2, and R_4.

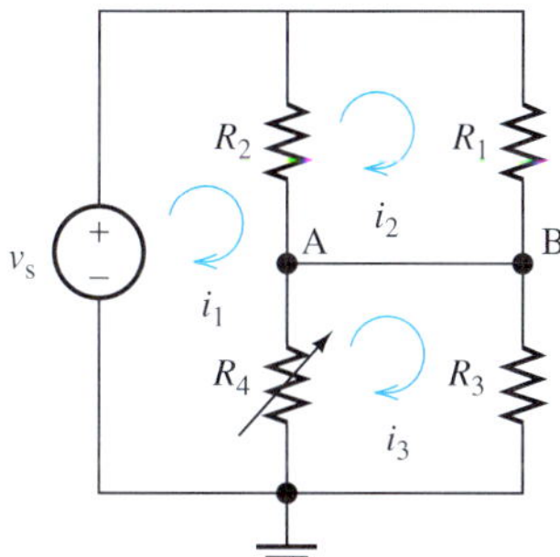

FIGURE 3.52 Wheatstone bridge.

SOLUTION

a. For mesh 1, the KVL equation is:

$$R_2(i_1 - i_2) + R_4(i_1 - i_3) - v_s = 0$$

For mesh 2, the KVL equation is:

$$R_1i_2 + R_2(i_2 - i_1) = 0$$

For mesh 3, the KVL equation is:

$$R_3i_3 + R_4(i_3 - i_1) = 0$$

b. When the current that flows from node A to node B is zero, $i_2 - i_3 = 0$. In this case, the KVL equation for mesh 2 becomes:

$$R_1/R_2 = (i_1 - i_2)/i_2 = (i_1 - i_3)/i_3$$

(continued)

EXAMPLE 3.22 Continued

In addition, the KVL equation for mesh 3 is:

$$R_3/R_4 = (i_1 - i_3)/i_3$$

Equating the left-hand sides of above two equations:

$$R_3 = (R_1R_4)/R_2$$

Using this equation, it is clear that in the Wheatstone bridge the voltage source does not impact the measurement of the resistance R_3. Therefore, the weakness of the battery does not have any effect on the accuracy of the resistance measurement.

3.5 SPECIAL CONDITIONS: SUPER NODE

Recall again the circuit in Figure 3.37. Now, consider this question: what happens if a KCL equation is written for the reference node in addition to the three specified nodes?

The KCL equation for the reference node in this circuit is:

$$\frac{V_s - V_2}{R_s} + \frac{0 - V_3}{R_2} + \frac{0 - V_4}{R_4} = 0$$

Combining this with the system of Equations of (3.22) results in a system of four equations that corresponds to:

$$\begin{cases} \left(\frac{1}{R_s} + \frac{1}{R_1} + \frac{1}{R_5}\right)V_1 - \frac{V_2}{R_1} - \frac{V_3}{R_5} = \frac{V_s}{R_s} \\ -\frac{V_1}{R_1} + \left(\frac{1}{R_1} + \frac{1}{R_2} + \frac{1}{R_3}\right)V_2 - \frac{V_3}{R_3} = 0 \\ -\frac{V_1}{R_5} - \frac{V_2}{R_3} + \left(\frac{1}{R_3} + \frac{1}{R_4} + \frac{1}{R_5}\right)V_3 = 0 \\ \frac{V_2}{R_s} + \frac{V_3}{R_2} + \frac{V_4}{R_4} = \frac{V_s}{R_s} \end{cases}$$

Now, add up the first three equations. The result is the last equation. What conclusion can be drawn from this? We can conclude that these four equations are *linearly dependent*, that is, by combining three of them, the next one is found. As a result, only three equations are sufficient in order to find the unknown voltages. In contrast, any three equations in this set are *linearly independent* from each other; considering any three equations in the set, you cannot combine two of them in order to find the third one. Therefore, any linearly independent set of equations can be used to find the unknown variables. In general, for n nodes in a circuit, $n - 1$ independent equations are needed to find the unknown voltages.

Sometimes, the current in a circuit branch cannot be expressed in terms of its terminal node voltages. In such cases, the aforementioned nodal voltage analysis methods cannot be applied. For example, consider Figure 3.53. The two circuits in this figure are the same. There are two voltage sources. In (a) and (b), two different reference nodes are shown at the negative end of voltage sources. Applying KCL to Node 1 in the circuit of Figure 3.53(a) results in:

$$\frac{V_1 - V_{s1}}{R_1} + \frac{V_1}{R_2} + I_a = 0$$

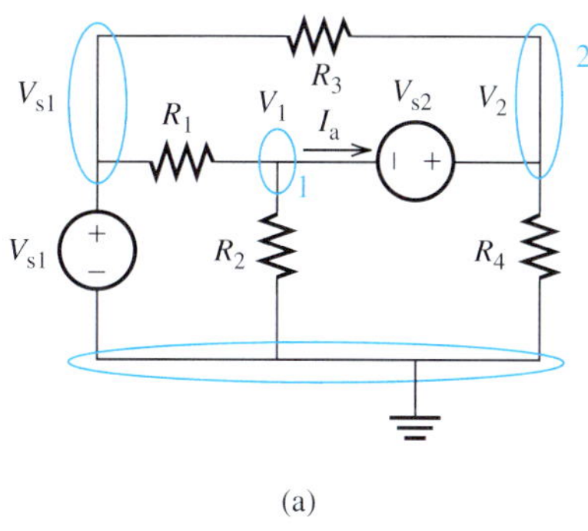

(a)

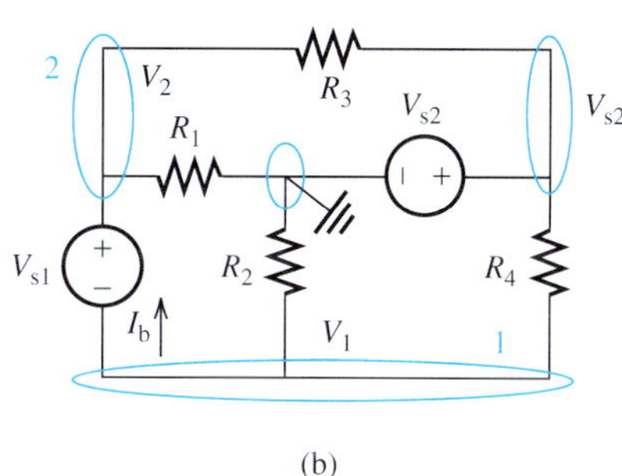

(b)

FIGURE 3.53 Two examples in which the current cannot be written in terms of single node voltages.

Accordingly, the current I_a cannot be expressed in the circuit shown in Figure 3.53(a) in terms of its node voltages, V_1 and V_2. Similarly, the current I_b in the circuit shown in Figure 3.53(b) can be found using:

$$\frac{V_1 - V_{s2}}{R_4} + \frac{V_1}{R_2} + I_b = 0$$

The same problem exists for another reference node as well: the current I_b cannot be written in terms of its node voltages.

In order to solve this problem, in these situations a *super node* is defined instead of a single node. A super node is a closed surface in a circuit that includes several nodes. As described in Chapter 2, KCL can be applied to any closed surface in a circuit. Recall that a node is actually a closed surface with a zero-limit area. Considering the circuit shown in Figure 3.54, the super node shown contains two single nodes.

If KCL is applied to the super node's closed surface:

$$I_1 + I_2 + I_3 + I_4 = 0$$

This can be written in terms of voltages as:

$$\frac{V_1 - V_{s1}}{R_1} + \frac{V_1}{R_2} + \frac{V_2 - V_{s1}}{R_3} + \frac{V_2}{R_4} = 0$$

In cases where super nodes are used instead of single nodes, a KVL equation must also be written for the super node's closed surface:

$$-V_1 - V_{s2} + V_2 = 0$$

This equation represents the KVL for the loop formed by the super node. Therefore, the following system of equations is derived for the circuit in Figure 3.54:

$$\begin{cases} \left(\frac{1}{R_1} + \frac{1}{R_2}\right)V_1 + \left(\frac{1}{R_3} + \frac{1}{R_4}\right)V_2 = \left(\frac{1}{R_1} + \frac{1}{R_3}\right)V_{s1} \\ -V_1 + V_2 = V_{s2} \end{cases}$$

Note that there is another approach for finding the voltage or current in circuits that include more than one independent source. This approach is called superposition and will be discussed in Section 3.7.

Nodal voltage analysis is also useful for cases where a dependent voltage or current sources exist in the circuit. In these situations, similar to the previous cases, a reference node is assigned to one of the end points (usually the negative point) of the independent sources. Figure 3.55 shows an example of a circuit with a dependent current source. In this example, the KCL equations at nodes 1 and 2, respectively, are:

$$\frac{V_1 - V_{s1}}{R_1} + \frac{V_1}{R_2} + \frac{V_1 - V_2}{R_3} = 0 \tag{3.23}$$

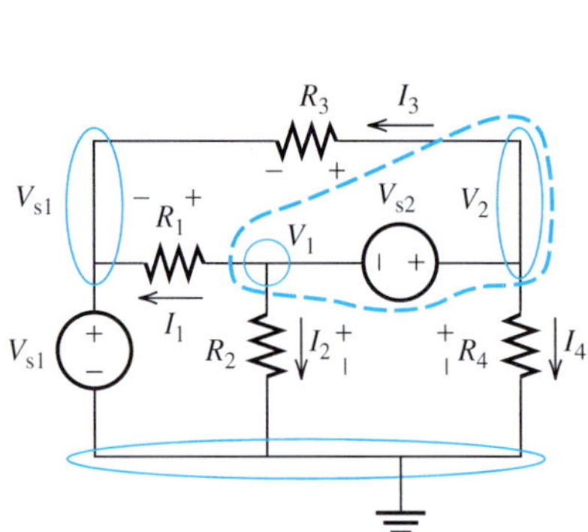

FIGURE 3.54 A super node in a circuit.

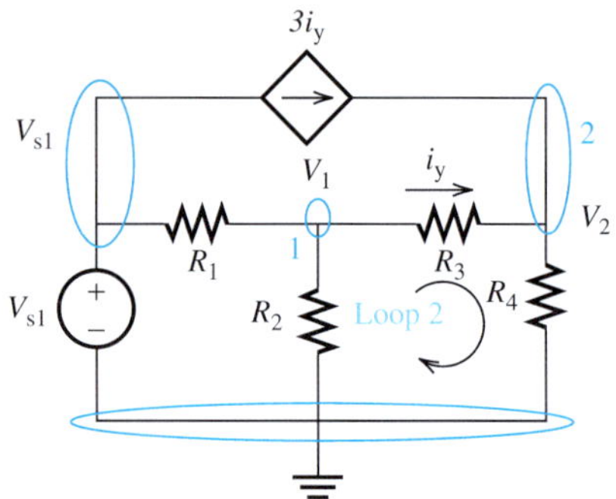

FIGURE 3.55 Node voltage analysis in a circuit with a dependent current source.

and

$$\frac{V_2 - V_1}{R_3} + \frac{V_2}{R_4} - 3i_y = 0 \tag{3.24}$$

The KCL equation at Node 2 is in terms of the current i_y as well. This term has appeared due to the dependent current source. Because i_y flows through the resistor R_3, the KVL in loop 2 shown in Figure 3.55 can be written as:

$$i_y = \frac{V_1 - V_2}{R_3} \tag{3.25}$$

Replacing i_y in Equation (3.24) with that of Equation (3.25), results in the following system of equations for this circuit:

$$\begin{array}{r} \text{Equation (3.23)}: \\ \text{Modified Equation (3.24)}: \end{array} \left\{ \begin{array}{l} \dfrac{V_1 - V_{s1}}{R_1} + \dfrac{V_1}{R_2} + \dfrac{V_1 - V_2}{R_3} = 0 \\ \dfrac{V_2 - V_1}{R_3} + \dfrac{V_2}{R_4} - 3\dfrac{V_1 - V_2}{R_3} = 0 \end{array} \right.$$

Now, consider Figure 3.56 as a circuit with a dependent voltage source. In this case, if node voltage analysis is applied with single nodes, it will not be possible to express the current passing through the dependent voltage source in terms of voltage. As a result, a super node is used which includes the nodes of the dependent voltage source. The KCL equations are:

$$\frac{V_1}{R_1} + \frac{V_1 - V_3}{R_2} - I_s + \frac{V_2 - V_3}{R_3} = 0$$

$$\frac{V_3 - V_1}{R_2} + \frac{V_3 - V_2}{R_3} + \frac{V_3}{R_4} = 0$$

KVL must also be applied to the super node's closed surface node:

$$-V_1 - 2V_x + V_2 = 0$$

where:

$$V_x = V_3$$

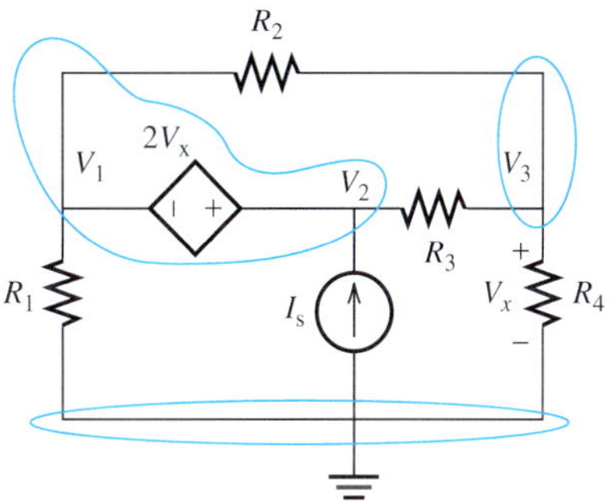

FIGURE 3.56 Voltage analysis in a circuit with a dependent voltage source.

Accordingly, the following set of node voltage equations is derived:

$$\begin{cases} \left(\frac{1}{R_1} + \frac{1}{R_2}\right)V_1 + \frac{V_2}{R_3} - \left(\frac{1}{R_2} + \frac{1}{R_3}\right)V_3 = I_s \\ -\frac{V_1}{R_2} - \frac{V_2}{R_3} + \left(\frac{1}{R_2} + \frac{1}{R_3} + \frac{1}{R_4}\right)V_3 = 0 \\ -V_1 + V_2 - 2V_3 = 0 \end{cases}$$

EXAMPLE 3.23 Node Analysis

Find V_o in the circuit shown in Figure 3.57.

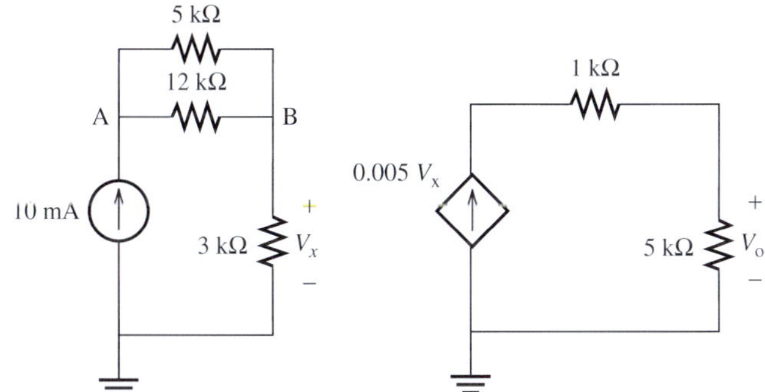

FIGURE 3.57 The circuit for Example 3.23.

SOLUTION

The following are the KCL equations for nodes A and B, respectively:

$$\begin{aligned} -10\text{ mA} + \frac{V_A - V_B}{12\text{ k}} + \frac{V_A - V_B}{5\text{ k}} &= 0 \\ \frac{V_B - V_A}{12\text{ k}} + \frac{V_B - V_A}{5\text{ k}} + \frac{V_B}{3\text{ k}} &= 0 \end{aligned} \longrightarrow \begin{cases} \left(\frac{1}{12} + \frac{1}{5}\right)V_A - \left(\frac{1}{12} + \frac{1}{5}\right)V_B = 10 \\ -\left(\frac{1}{12} + \frac{1}{5}\right)V_A + \left(\frac{1}{12} + \frac{1}{5} + \frac{1}{3}\right)V_B = 0 \end{cases}$$

Solving this set of equations:

$$V_B = 30\text{ V}$$

In addition, $V_x = V_B$. Using these voltages, the output voltage will be:

$$V_o = 0.005 \times V_B \times 5\text{ k}\Omega = 750\text{ V}$$

EXAMPLE 3.24 Super Node

Find the nodal voltages V_1 and V_2 for the circuit shown in Figure 3.58.

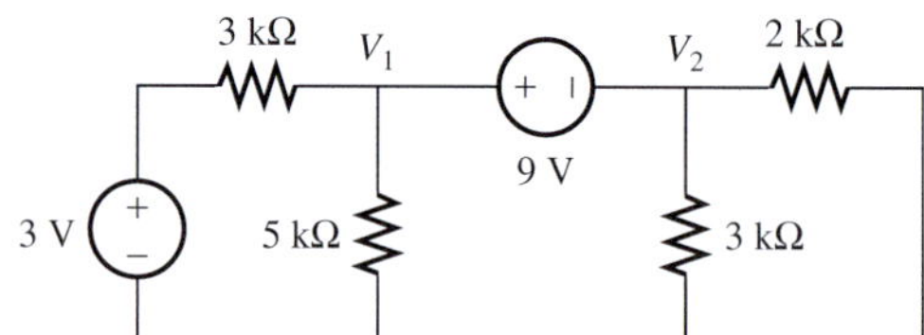

FIGURE 3.58 The circuit for Example 3.24.

SOLUTION

The current that flows through the 9-V source cannot be expressed in terms of its end voltages. Therefore, as shown in Figure 3.59, the 9-V source nodes are selected as a super node. Applying KCL at the super node results in:

$$\frac{V_1 - 3}{3\,\text{k}} + \frac{V_1}{5\,\text{k}} + \frac{V_2}{3\,\text{k}} + \frac{V_2}{2\,\text{k}} = 0$$

$$\left(\frac{1}{3} + \frac{1}{5}\right)V_1 + \left(\frac{1}{3} + \frac{1}{2}\right)V_2 = 1$$

Applying KVL to the loop formed by the super node loop, that is, the loop formed by the two voltages across the node, as shown in Figure 3.59 results in:

$$V_2 = V_1 - 9\text{ V}$$

These two equations are used to find V_1 and V_2:

$$V_1 = 6.22\text{ V}$$
$$V_2 = -2.78\text{ V}$$

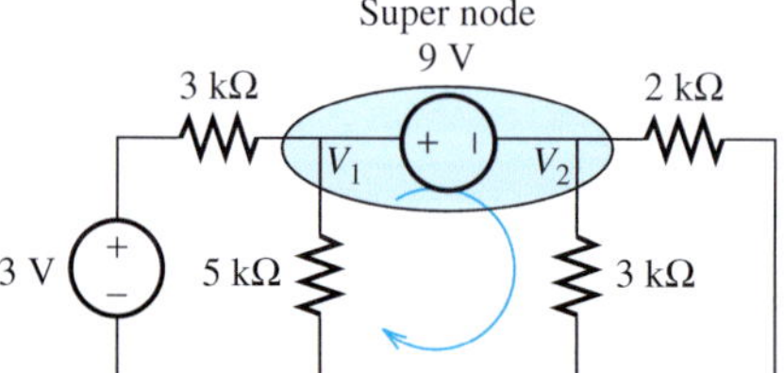

FIGURE 3.59 Circuit for Example 3.24.

EXAMPLE 3.25 Super Node

Find the nodal voltages in Figure 3.60.

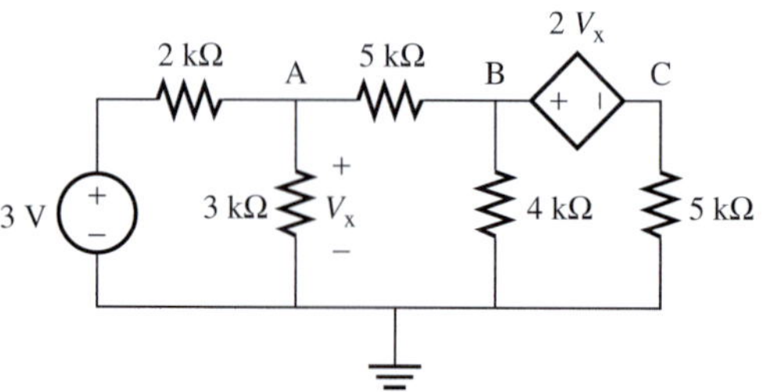

FIGURE 3.60 The circuit for Example 3.25.

SOLUTION

A current that flows through a dependent source cannot be expressed in terms of its end voltages. Therefore, as shown in Figure 3.61, the two nodes across the dependent source are selected as a super node.

Next, write two KCL equations for node A and super node "B, C" and one KVL equation for the super node loop. The system of equations corresponds to:

$$\frac{V_A - 3}{2\,k} + \frac{V_A}{3\,k} + \frac{V_A - V_B}{5\,k} = 0$$

$$\frac{V_B - V_A}{5\,k} + \frac{V_B}{4\,k} + \frac{V_C}{5\,k} = 0$$

$$V_B - V_C = 2V_x = 2V_A$$

The system of equations can be represented in matrix form as follows:

$$\begin{bmatrix} 31 & -6 & 0 \\ -4 & 9 & 4 \\ 2 & -1 & 1 \end{bmatrix} \begin{bmatrix} V_A \\ V_B \\ V_C \end{bmatrix} = \begin{bmatrix} 45 \\ 0 \\ 0 \end{bmatrix}$$

This form helps us to find the voltages using an inverse matrix, as discussed in the Appendix A:

$$\begin{bmatrix} V_A \\ V_B \\ V_C \end{bmatrix} = \begin{bmatrix} 1.767\text{ V} \\ 1.631\text{ V} \\ -1.903\text{ V} \end{bmatrix}$$

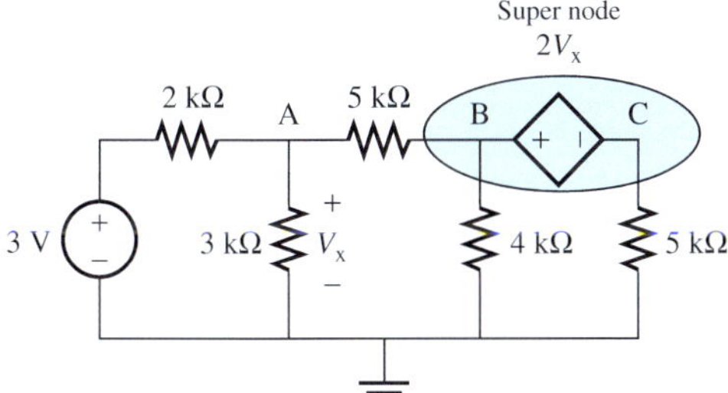

FIGURE 3.61 The circuit for Example 3.25.

APPLICATION EXAMPLE 3.26 Sports Car Inventor/Accelerometer

The inventor of the sports car discussed earlier has been informed by an anonymous source that connecting an accelerometer to a PDA is a good way to measure initial acceleration data. The 6-V source gives the signal a small amount of amplification. A circuit model for this example is shown in Figure 3.62. Remember, for this accelerometer, $0 \leq V_A \leq 5$ V.

a. If the accelerometer produces its maximum output voltage, 5 V, what voltage is delivered to the PDA?

b. In addition, if the maximum input voltage the PDA can handle is 5 V, will this circuit damage the PDA?

(continued)

APPLICATION EXAMPLE 3.26 Continued

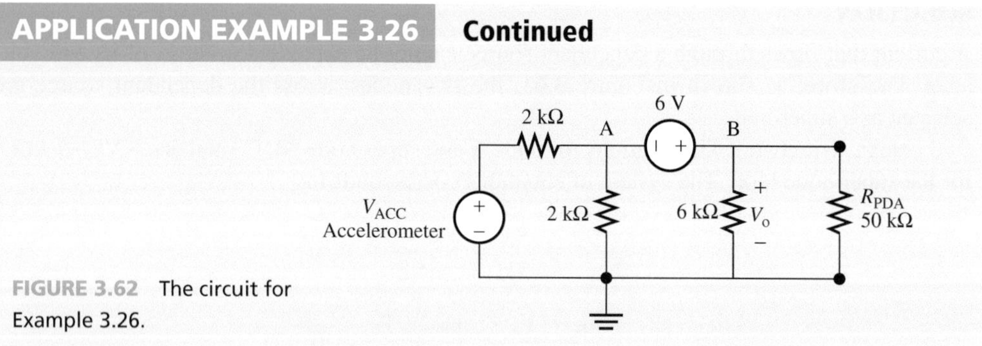

FIGURE 3.62 The circuit for Example 3.26.

SOLUTION

In Figure 3.63 the nodal voltage equations are:

$$\text{KCL for the super node } A,B\text{: } \frac{V_A - 5}{2\text{k}} + \frac{V_A}{2\text{k}} + \frac{V_B}{6\text{k}} + \frac{V_B}{50} = 0$$

$$\text{KVL for the super node loop: } V_B = V_A + 6$$

Solving these two equations results in $V_o = V_B = 0.402V$. Thus, this circuit is feasible, because the maximum voltage it can produce is not higher than the maximum voltage the PDA can handle.

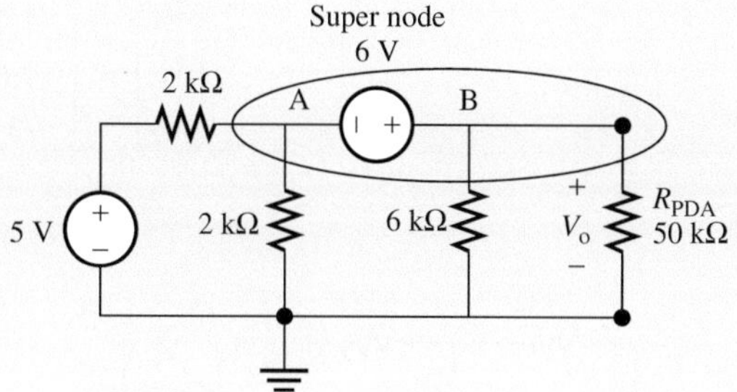

FIGURE 3.63 The circuit for Example 3.26.

APPLICATION EXAMPLE 3.27 A Chemical Sensor

A titration sensor is connected to measure the PH level of a solution. The solution is kept at a constant temperature by a heater connected to a thermostat. The sensor produces a voltage output and requires the circuit shown in Figure 3.64(b) to connect it to an LCD display. If the sensor voltage $V_s = 0.201$ V, find the voltage V_o seen by the LCD display.

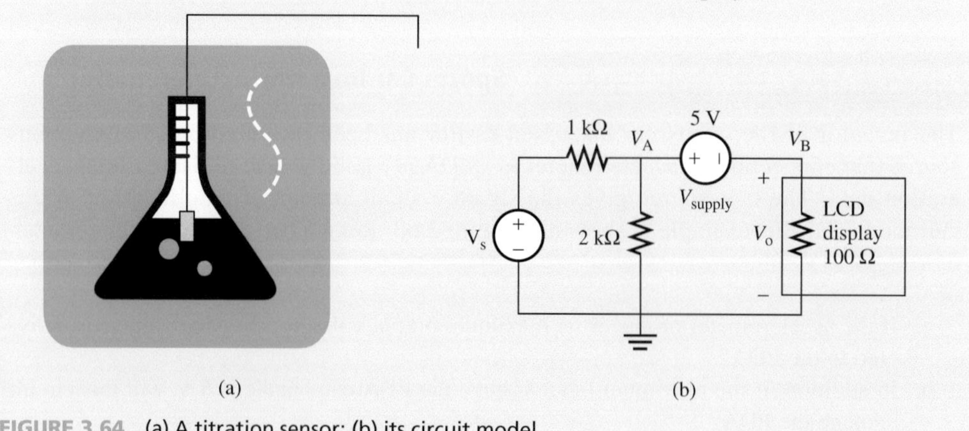

FIGURE 3.64 (a) A titration sensor; (b) its circuit model.

SOLUTION

First, write the KCL and KVL equations for the super node shown in Figure 3.65.

$$\frac{V_A}{2\text{ k}} + \frac{V_A - V_s}{1\text{ k}} + \frac{V_B}{0.1\text{k}} = 0$$

$$V_A - V_B = 5\text{ V}$$

This system of equations can be rearranged and written again in the matrix format and solved for the voltages to find $V_A = 4.3653$ V and $V_o = V_B = -0.6347$ V. The mathematical details of calculating these voltages are left to the student.

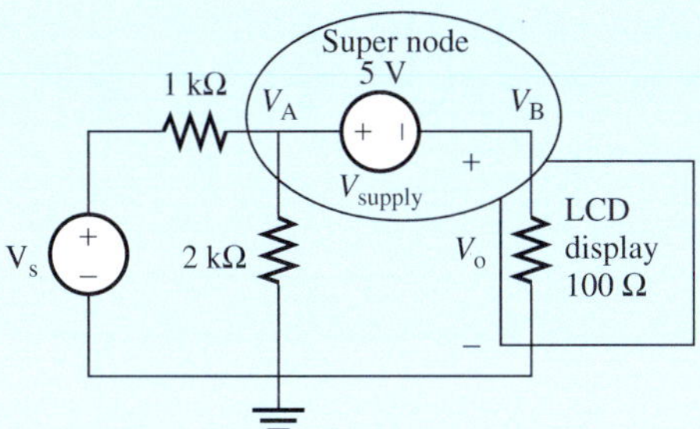

FIGURE 3.65 The circuit for Example 3.27.

3.6 THÉVENIN/NORTON EQUIVALENT CIRCUITS

This section introduces two theorems which facilitate the possibility of replacing a complex system or circuit containing resistors and sources with a simple equivalent circuit. This replacement eases the process of evaluating the current and voltage of a load when it is connected to the system. For example, consider a computer that is connected to a robot to control the movement of the robot. How much current and, accordingly, how much power is delivered to the robot. Is that power sufficient enough to control the robot's movement? If not, then what kind of amplifier (see Chapter 8 for more information about amplifiers) should be connected to the output of the computer to enable the desired task?

The first theorem, known as the Thévenin theorem, states that any resistive circuit regardless of its complexity can be represented by a voltage source in series with a resistance. The equivalent circuit is called the *Thévenin equivalent circuit*. In this equivalent circuit, the voltage source is referred to as the Thévenin voltage (V_{th}) while the resistance is called the Thévenin resistance (R_{th}). The second theorem, known as the Norton theorem, states that any resistive circuit regardless of its complexity can be represented by a current source in parallel with a resistance. The equivalent circuit is called a *Norton equivalent circuit*. In this equivalent circuit, the current source is referred to as the *Norton voltage* (V_n) while the resistance is called the *Norton resistance* (R_n).

Figure 3.66 shows an example of Thévenin and Norton equivalent circuits. If the resistance, R_L, is connected to the equivalent Thévenin and Norton circuits instead of the resistive circuit (shown at left), there will be no change in its current and voltage. Therefore, compared to the original circuit, the voltage (current) of R_L remains unchanged.

Consider the circuit in Figure 3.67 and its Thévenin equivalent where terminals A and B are open-circuited. The Thévenin equivalent circuit doesn't change the current through or the voltage across the terminals A and B when they are connected to a load. As a result, V_{oc1} and V_{oc2} will be equal. Thus, Thévenin and Norton equivalent circuits are used to represent the internal structure of complex circuits simply by using a source and a resistor independent of their complexity.

Now, if KVL is applied to the open-circuited Thévenin equivalent circuit sketched in Figure 3.67 (right):

$$-V_{th} + R_{th}i + V_{oc2} = 0$$

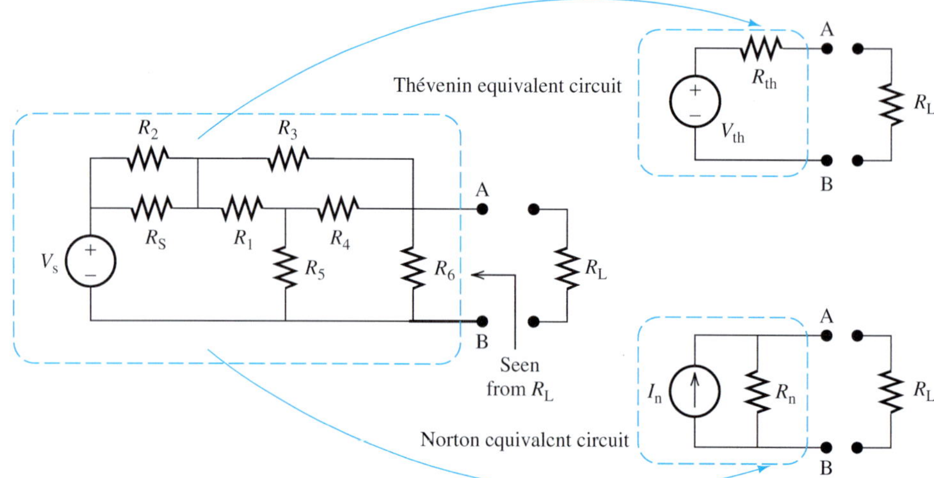

FIGURE 3.66 Thévenin and Norton equivalent circuits.

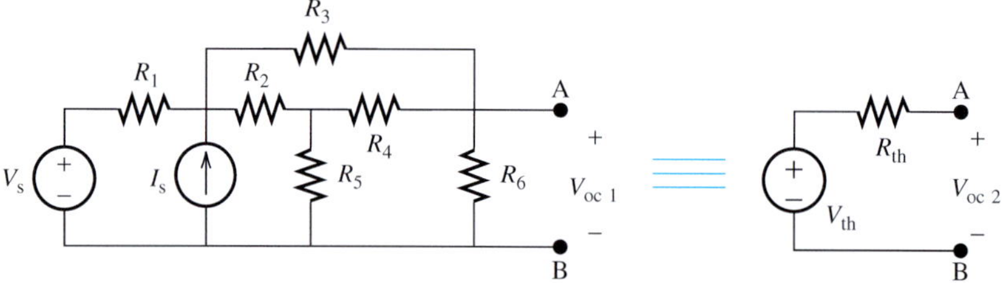

FIGURE 3.67 A Thévenin equivalent circuit with open-circuited terminals.

The circuit is open and therefore $i = 0$, so:

$$V_{th} = V_{oc2} = V_{oc1}$$

Accordingly, the Thévenin voltage will be equal to the open-circuit voltage of the original circuit, and in order to find it the terminals are opened, through which the Thévenin circuit can be found. Note that V_{th} is always proportional to V_s. Thus, if $V_s = 0$, V_{th} will be equal to zero. In that case, the resistor seen through the terminals A and B will be equal to R_{th}. This is an easy approach for finding R_{th}.

Now, consider the circuit shown in Figure 3.68 with its Norton equivalent. In this case, the terminals are short-circuited. According to the Norton theorem, $I_{sc1} = I_{sc2}$. In addition, by writing KCL for node A in the Norton equivalent short-circuit of Figure 3.68 (right) we find that:

$$-I_n + I + I_{sc2} = 0$$

Now, because $V = 0$, then $I = V/R_n = 0$. Thus:

$$I_n = I_{sc2} = I_{sc1}$$

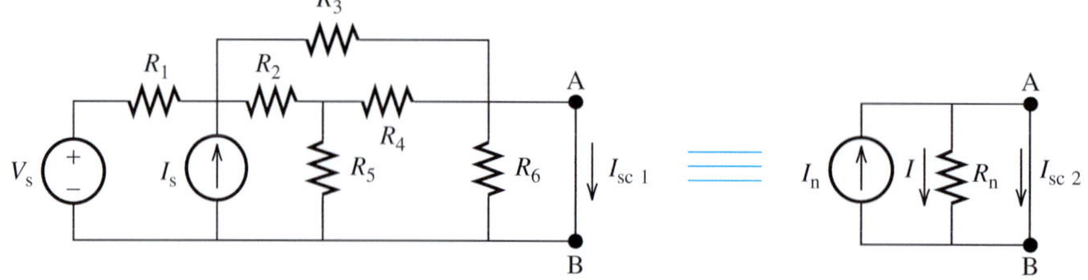

FIGURE 3.68 Norton equivalent circuit with short-circuited terminals.

Therefore, the Norton current source is the same as the short-circuit current of the original circuit. In order to find the Thévenin and Norton resistances, consider the open-circuited Norton and short-circuited Thévenin equivalent circuits in Figures 3.69(a) and (b).

From Figure 3.69(a) and (b):

$$R_n = \frac{V_{oc}}{I_n}$$

and

$$R_{th} = \frac{V_{th}}{I_{sc}}$$

But, as shown in Figure 3.67, because no current flows through the circuit when it is open-circuited, $V_{th} = V_{oc}$. In addition, as seen in Figure 3.69(a), if terminals A and B are short-circuited, all I_n current will flow through these terminals, and $I_n = I_{sc}$; therefore:

$$R_{th} = R_n = \frac{V_{oc}}{I_{sc}} \quad \textbf{(3.26)}$$

Consequently, the Thévenin and Norton resistances are the same. In addition, in practice they can be found by first opening the circuit to find V_{oc} and then shortening the circuit to find I_{sc}. Then, by dividing V_{oc} and I_{sc} [Equation (3.26)] R_{th} can be found. Note that R_{th} or R_n is indeed the equivalent resistance seen through the same terminals.

(a)

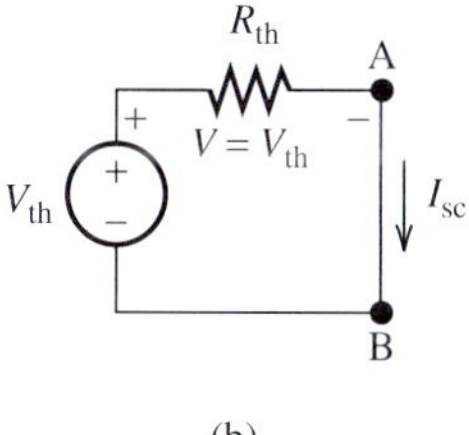

(b)

FIGURE 3.69 (a) Open-circuited Norton equivalent circuit. (b) Short-circuited Thévenin equivalent circuit.

Thus, Equation (3.26) represents another approach for finding the equivalent resistance of a circuit seen through any terminal: simply open-circuit the terminal and find the voltage seen through that open-circuited terminal; then, short-circuit that terminal and find the current through the short-circuited terminal; the division of the open-circuited voltage and short-circuited current is the equivalent resistance.

As stated earlier, $R_{th} = R_n$ can also be found by setting all independent sources in the main circuit to zero. This point is depicted in Figure 3.70.

Therefore, there are two ways to calculate R_{th}:

1. Using Equation (3.26), that is, finding V_{oc} and I_s.
2. Setting all sources to zero and finding the equivalent resistance through the terminals as discussed in Section 3.2.

Note that:

- An independent voltage source with zero voltage is equivalent to a short circuit.
- An independent current source with zero current is equivalent to an open circuit.

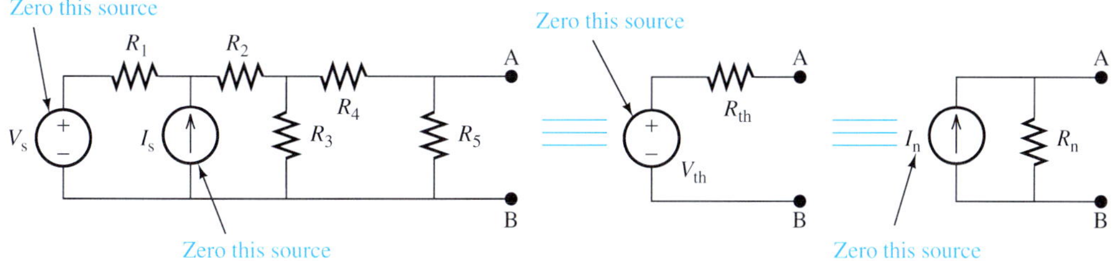

FIGURE 3.70 By zeroing the sources, the Thévenin and Norton theorems will still be true.

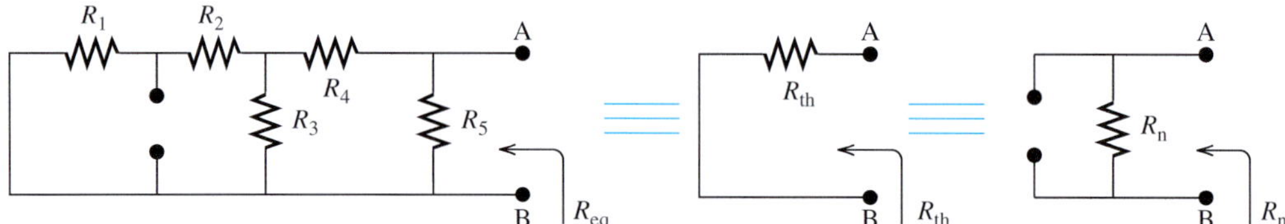

FIGURE 3.71 Zeroing the sources in order to find the Thévenin and Norton resistances.

Therefore, in order to zero the independent sources in a circuit, replace the voltage sources with a short circuit and the current sources with an open circuit. For example, consider Figure 3.70. If all the sources are zeroed, the circuit and its equivalents will correspond to Figure 3.71. After zeroing the sources, the equivalent resistance between the terminals will be the same as the Thévenin or Norton equivalent resistance. In this case, this equivalent resistance is:

$$R_{th} = R_n = R_{eq} = \{[(R_1 + R_2)||R_3] + R_4\} || R_5$$

Here, the notation || represents parallel resistors.

It is *important to note* that the source zeroing technique can only be used for independent, *not dependent*, sources. For circuits that include only dependent sources, the first technique [i.e., Equation (3.26)] is the only method for finding the equivalent resistance of the circuit.

EXAMPLE 3.28 Thévenin Equivalent Circuit

Find the Thévenin equivalent of the circuit in Figure 3.72(a) as seen by R_L.

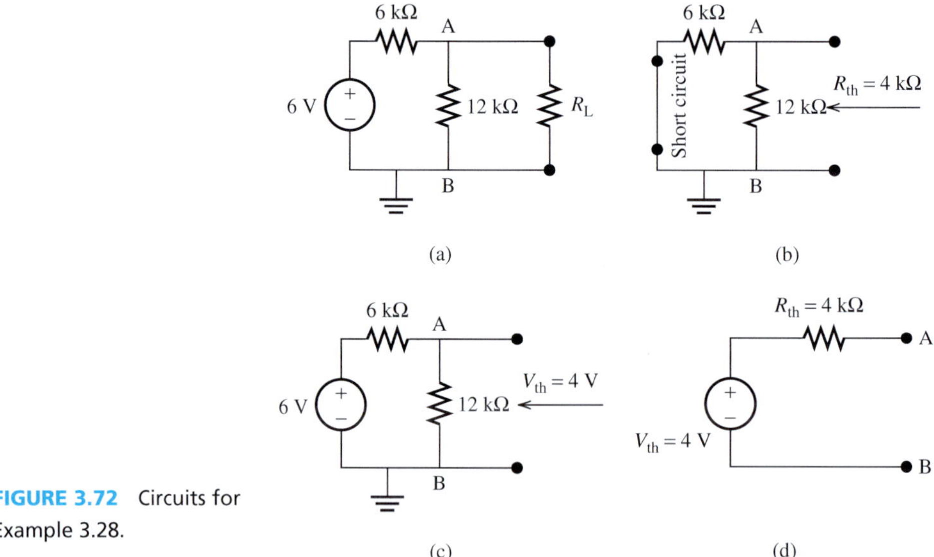

FIGURE 3.72 Circuits for Example 3.28.

SOLUTION

Equating the independent sources to zero, the 6-V voltage source is replaced by a short circuit. Recall that by equating a voltage source to zero, it will be replaced by a short circuit. In addition, equating a current source to zero, it will be replaced by an open circuit. The Thévenin resistance is the equivalent resistance between the A and B terminals when these terminals are open-circuited, which is shown in Figure 3.72(b):

$$R_{th} = R_{eq} = 6\text{ k}\Omega||12\text{ k}\Omega = 4\text{ k}\Omega$$

The Thévenin voltage is the open-circuit voltage of node B as shown in Figure 3.72(c). Using the voltage dividing technique:

$$V_{th} = \frac{12}{6 + 12} \times 6 = 4\text{ V}$$

The Thévenin equivalent circuit of Figure 3.72(c) is sketched in Figure 3.72(d).

EXAMPLE 3.29 Norton Equivalent Circuit

Find the Norton equivalent of the circuit in Figure 3.73(a) as seen by R_L.

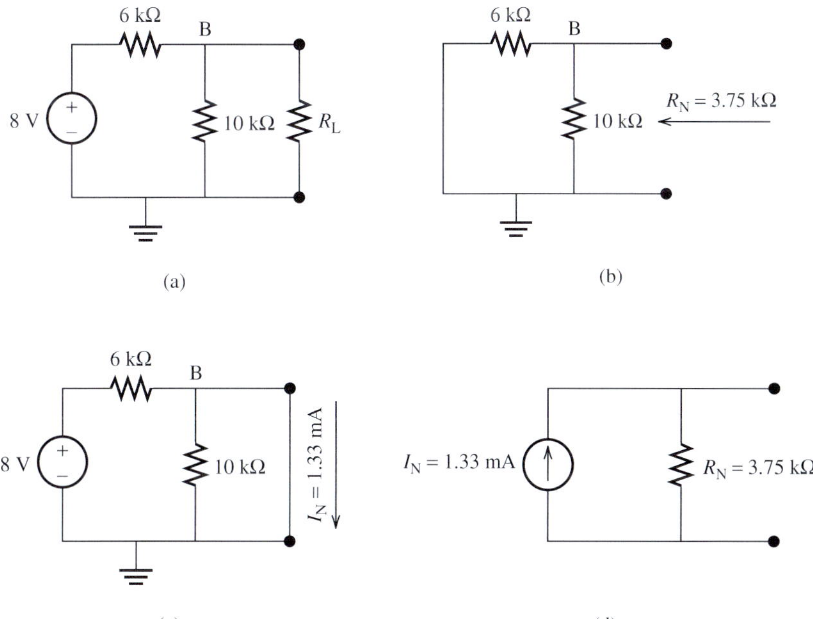

FIGURE 3.73 Circuits for Example 3.29.

SOLUTION

Equating the independent voltage sources to zero, the 8-V voltage source will be replaced by a short circuit. The Norton resistance is the equivalent resistance between the terminals in Figure 3.73(b), when they are open-circuited, which is:

$$R_n = R_{eq} = 10\,\text{k}\Omega \,\|\, 6\,\text{k}\Omega = 3.75\,\text{k}\Omega$$

The Norton current is the current passing from the short-circuited terminals in Figure 3.73(c). In this case, the current through the 10-kΩ resistance is zero. This can be verified simply by using the current division technique. As a result:

$$I_n = \frac{8\text{V}}{6\,\text{k}\Omega} = 1.33\,\text{mA}$$

The Norton equivalent circuit is sketched in Figure 3.73(d).

EXCERCISE 3.9

Consider current I is applied to two parallel resistances: one has a nonzero resistance of R and the other one has a zero resistance (i.e., short circuit). Calculate the current through each resistor.

EXAMPLE 3.30 Thévenin Equivalent Circuit through a Resistor

In Figure 3.74, find the Thévenin equivalent as seen by R_L.

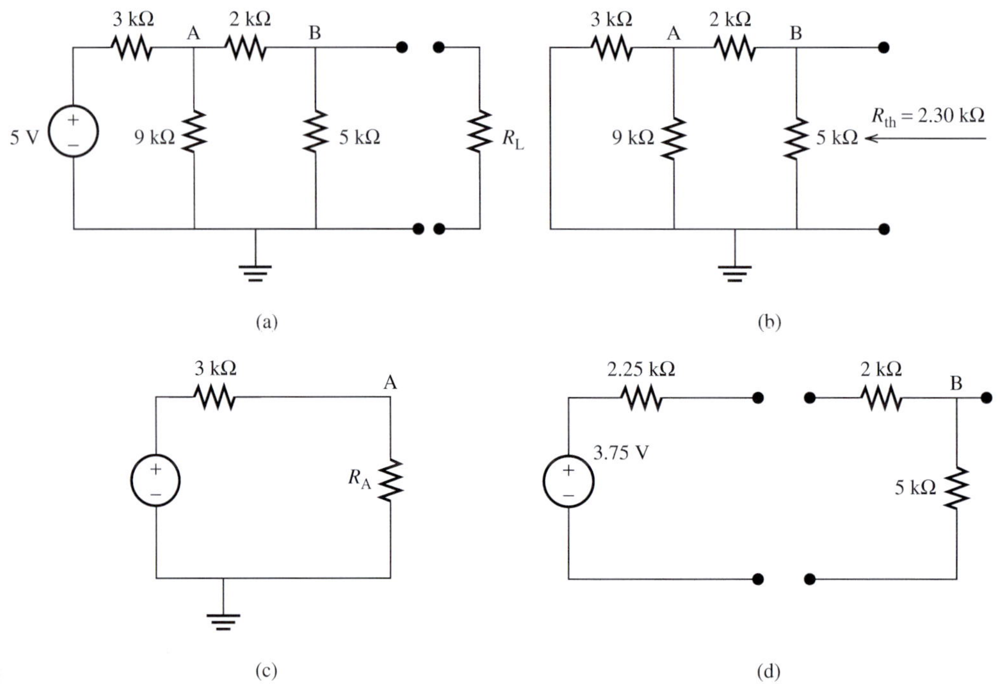

FIGURE 3.74 Circuits for Example 3.30.

SOLUTION

Replace the 3-V voltage source with a short circuit. The Thévenin resistance is the equivalent resistance between the terminals in Figure 3.74(b), which is:

$$R_{th} = R_{eq} = 5\text{ k}\Omega \parallel (2\text{ k}\Omega + (9\text{ k}\Omega \parallel 3\text{ k}\Omega)) = 2.30\text{ k}\Omega$$

The Thévenin voltage is the open-circuit voltage of node B as shown in Figure 3.74(a). To find this voltage, any of the following approaches can be selected:

Approach 1: Writing node voltage equations for nodes A and B results in:

$$\begin{cases} \dfrac{V_A - 5}{3\text{ K}} + \dfrac{V_A}{9\text{ K}} + \dfrac{V_A - V_B}{2\text{ K}} = 0 \\[2ex] \dfrac{V_B - V_A}{2\text{ K}} + \dfrac{V_B}{5\text{ K}} = 0 \end{cases} \rightarrow \begin{cases} \left(\dfrac{1}{3} + \dfrac{1}{2} + \dfrac{1}{9}\right)V_A - \dfrac{1}{2}V_B = \dfrac{5}{3} \\[2ex] -\dfrac{1}{2}V_A + \left(\dfrac{1}{2} + \dfrac{1}{5}\right)V_B = 0 \end{cases}$$

Solving for V_A and V_B, the results are $V_A = 2.838$ V and $V_B = 2.027$ V. Therefore, $V_{th} = V_B = 2.027$ V.

Approach 2: The voltage dividing concept can be used. First, find the voltage of node A and then the voltage of node B. The resistance at node A, R_A corresponds to:

$$R_A = 9\text{ k}\Omega \parallel (2\text{ k}\Omega + 5\text{ k}\Omega) = 3.9375\text{ k}\Omega$$

Therefore, the equivalent circuit from the point of view of node A is the one sketched in Figure 3.74(c). Accordingly, the voltage at V_A is:

$$V_A = \frac{R_A}{R_A + 3} \times 5 = 2.838$$

Now, once more using voltage dividing concept:

$$V_{th} = V_B = \frac{5}{5 + 2} \times V_A = 2.027 \text{ V}$$

Approach 3: Find the Thévenin equivalent circuit seen through node A first by disconnecting the 2-kΩ and 5-kΩ resistors. Now the Thévenin equivalent circuit across terminal A can be easily calculated. Its voltage can be found using voltage division:

$$3.75 \text{ V} = \frac{9}{3 + 9} \times 5 \text{ V}$$

In addition, the equivalent resistance will be $2.25\,\text{k}\Omega = 3\,\text{k}\Omega \parallel 9\,\text{k}\Omega$. Thus, Figure 3.74(d) shows the equivalent Thévenin circuit of the circuit seen through terminal A. Now, by connecting it to the rest of the circuit [i.e., 2-kΩ and 5-kΩ resistors, as shown in Figure 3.74(d)], the equivalent circuit seen through terminal B is found. Now, using voltage division rule:

$$V_{th} = V_B = \frac{5}{5 + 2 + 2.25} \times 3.75 = 2.027 \text{ V}$$

EXCERCISE 3.10

In Example 3.30, use approach 3 to find the equivalent Thévenin resistance and compare it with the computed value.

EXAMPLE 3.31 Norton Resistance

Find the Norton resistance as shown in Figure 3.75 by equating the sources to zero.

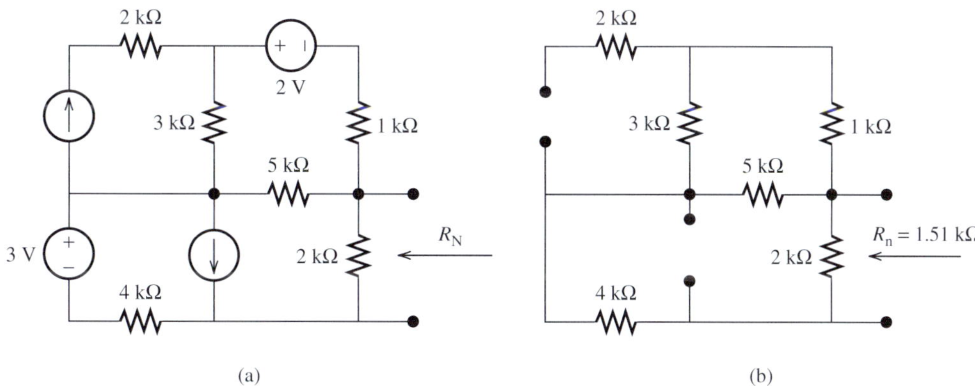

FIGURE 3.75 Circuits for Example 3.31.

SOLUTION

After equating the current and voltage sources to zero, the equivalent circuit between the terminals will be as shown in Figure 3.75(b). Accordingly, it can be verified that:

$$R_n = R_{eq} = [\{(1\text{ k}\Omega + 3\text{ k}\Omega \parallel (5\text{ k}\Omega)\} + 4\text{ k}\Omega] \parallel (2\text{ k}\Omega) = 1.5135\ \Omega$$

EXAMPLE 3.32 Norton Resistance

Find the Norton equivalent resistance as seen by R_L in Figure 3.76(a).

SOLUTION

The Norton equivalent resistance is $R_n = V_{oc}/I_{sc}$. It is important to determine both the open-circuit voltage and the short-circuit current of the terminals. Therefore, two separate circuits need to be analyzed. The open-circuit scenario is shown in Figure 3.76(b). Here, no current flows through the 2-kΩ resistor.

a. KCL in node A: Here, the current in the 3-kΩ resistor leaves the node. The voltage across this resistor is $V_A - 4$. Thus, the current in it will be $(V_A - 4)/3$ k. In addition, the current that enters this node is 0.004 V_x. Thus:

$$\frac{V_A - 4}{3\text{k}} - 0.004\ V_x = 0 \rightarrow 12\ V_x = V_A - 4$$

b. KVL in loop A:

$$V_x = V_A - 4$$

Now, observe that $12\ V_x = V_A - 4$ and $V_x = V_A - 4$, thus, $V_x = 0$ and:

$$V_A = V_{oc} = 4\text{ V}$$

To find the short-circuit current, short-circuit the output terminal [see Figure 3.76(c)].

a. KCL in node A:

$$\frac{V_A - 4}{3\text{k}} - 0.004\ \text{V}_x + \frac{V_A}{2\text{k}} = 0 \rightarrow \left(\frac{1}{2} + \frac{1}{3}\right)V_A - 4\ \text{V}_x = \frac{4}{3}$$

b. KVL in loop A:

$$V_x = V_A - 4$$

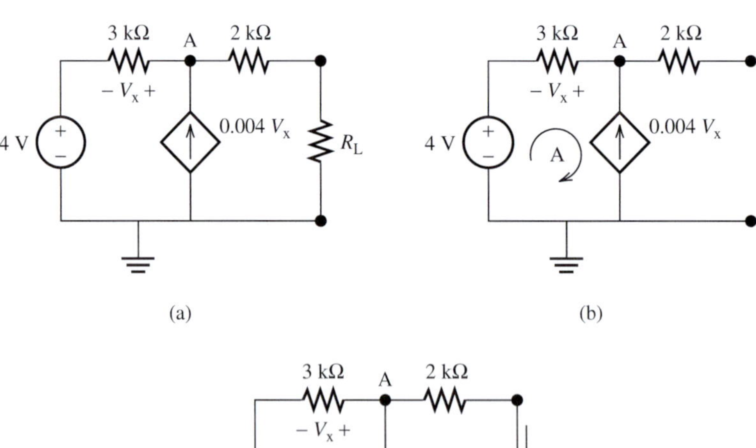

(c)

FIGURE 3.76 Circuits for Example 3.32.

Combining these two equations results in: $V_A = 4.63157$ V and $I_{sc} = \frac{V_A}{2\text{ K}} = 2.3157\text{mA} = I_n$. Therefore, the Thévenin/Norton resistance corresponds to:

$$R_n = \frac{V_{oc}}{I_{sc}} = \frac{4}{2.3157\text{ mA}} = 1.727\text{ k}\Omega$$

APPLICATION EXAMPLE 3.33 Aerial Fireworks

Aerial fireworks are used in various celebration events. In order to synchronize the firework ignition process, a computer is required. Figure 3.77(a) shows a simple setup of the computer and the igniter for an aerial fireworks show.

Assume a voltage source of 20 V is connected to the computer in series; the computer has a resistance of 500 Ω, and the igniter has a resistance of 50 Ω when it receives a signal from the computer to order it to launch. Assume the same ground for all igniters and the computer voltage source. The equivalent circuit for the computer and igniters is shown in Figure 3.77(b).

a. Sketch the Thévenin and Norton equivalent circuits of the computer.

b. Calculate the current flow through each igniter when it receives the order to ignite the firework. Use two methods: (i) calculate the total resistance and then the current through the computer first, and (ii) use the Norton equivalent circuit of computer and current division.

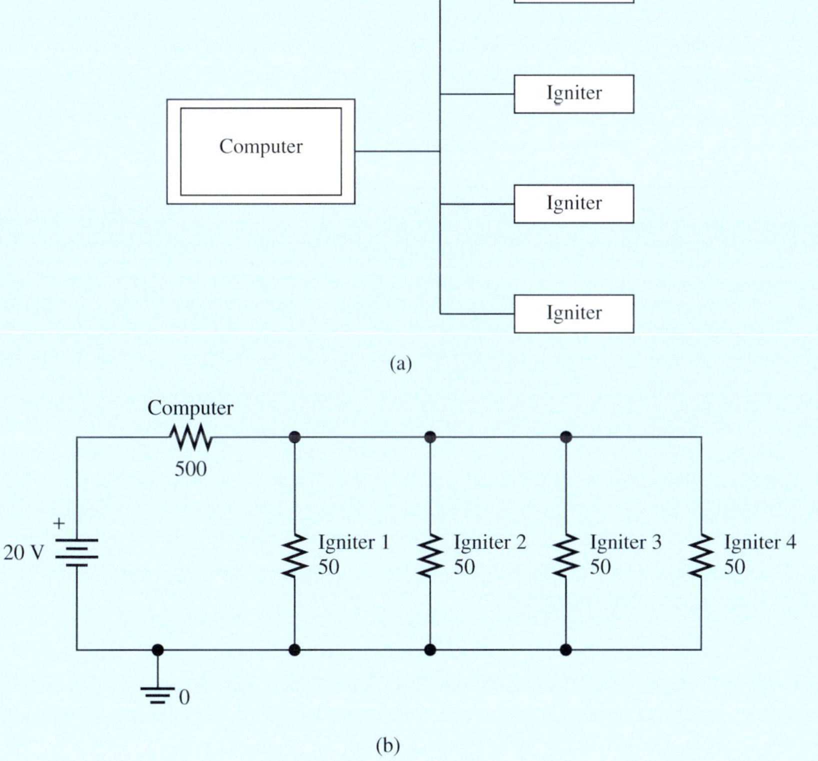

FIGURE 3.77 (a) A simple computer setup and igniters; (b) its equivalent circuit.

(continued)

APPLICATION EXAMPLE 3.33 Continued

SOLUTION

a. From Figure 3.77(b), the computer's equivalent circuit is a voltage source in series with a 500-Ω resistor. Indeed, this is the Thévenin equivalent circuit of the computer. The Norton equivalent circuit of the computer will be a current source in parallel to the same resistor. The Thévenin and Norton equivalent circuits are shown in Figure 3.78(a) and (b), respectively.

b. The total resistance from the igniters can be calculated as follows:

$$\frac{1}{R_{\text{parallel}}} = \frac{1}{R_1} + \frac{1}{R_2} + \frac{1}{R_3} + \frac{1}{R_4}$$

$$\frac{1}{R_{\text{parallel}}} = \frac{1}{50} + \frac{1}{50} + \frac{1}{50} + \frac{1}{50}$$

$$R_{\text{parallel}} = 12.5\ \Omega$$

The total resistance is:

$$R_{\text{total}} = R_{\text{computer}} + R_{\text{parallel}}$$

$$R_{\text{total}} = 500 + 12.5 = 512.5\ \Omega$$

Therefore, the current of the computer is:

$$I = \frac{V}{R_{\text{total}}}$$

$$I = \frac{20}{512.5} = 39\text{ mA}$$

Because the resistance of each igniter is equivalent, the current flow in each igniter is:

$$I_{\text{igniter}} = \frac{1}{4}I$$

$$I_{\text{igniter}} = 9.75\text{ mA}$$

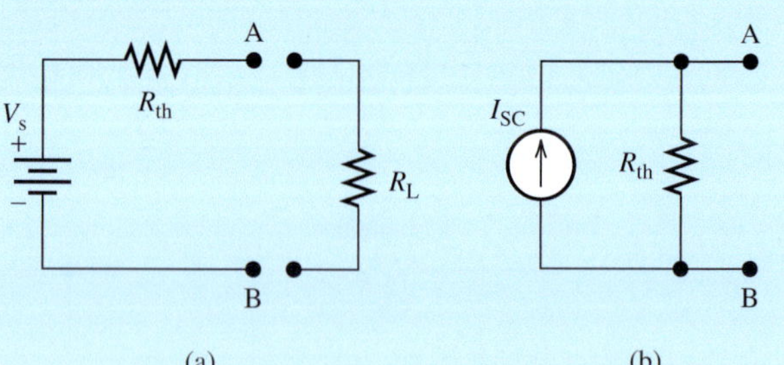

FIGURE 3.78 (a) Thévenin and (b) Norton equivalent circuit for the circuit of Figure 3.77.

3.6.1 Source Transformation

We have been exclusively working with ideal voltage and current sources which provide constant voltage or current regardless of the load resistor. A more accurate source model contains an internal resistance which accounts for observed reduction in voltage or current. The practical voltage and current sources are given in Figure 3.79(a) and (b), respectively.

We will see that the practical voltage and current sources may be interchangeable without affecting the remainder of the circuit. Such practical sources are defined as being equivalent if they produce identical values of I_L and V_L when they are connected to the identical load R_L, which is shown in Figure 3.79. Here, we intend to find the relationship of R_{th} and R_n, and the relationship of V_{th} and I_n.

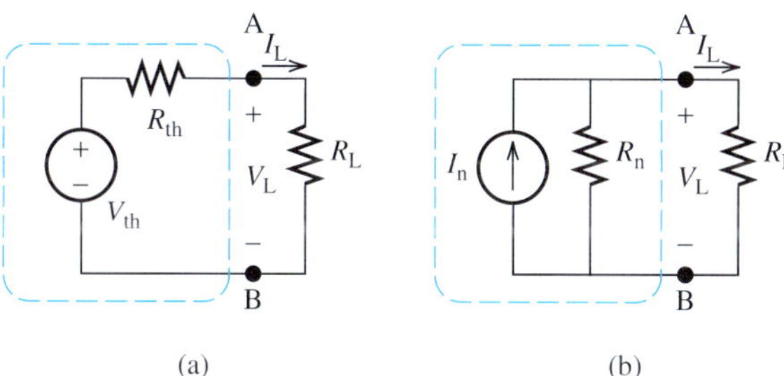

FIGURE 3.79 Equivalent sources.

When $R_L = 0$, the circuit between nodes A and B is short, that is, the total voltage of the source is across R_{th} in Figure 3.79(a), and no current flows through R_n in Figure 3.79(b). Therefore, I_L for Figure 3.79 (a) and (b) is:

$$I_L^a = V_{th}/R_{th}$$
$$I_L^b = I_n$$

When $R_L = \infty$, the circuit between nodes A and B is open, that is, there is no current through R_{th} in Figure 3.79(a) and the total current of the source flows through R_n in Figure 3.79(b).

Therefore, V_L for Figure 3.79(a) and (b) is:

$$V_L^a = V_{th}$$
$$V_L^b = I_n R_n$$

Because of equivalency, $V_L^a = V_L^b$ and $I_L^a = I_L^b$, then:

$$I_n = V_{th}/R_{th}$$
$$I_n = V_{th}/R_n$$

Therefore:

$$R_{th} = R_n$$
$$I_n = V_{th}/R_n$$

As a conclusion we can convert a Norton circuit to Thévenin and vice-versa. This approach can be used to calculate the equivalent Thévenin and Norton circuit for more complicated circuits (see Examples 3.34 and 3.35).

EXAMPLE 3.34 Equivalent Thévenin and Norton Using Source Transformation

The circuit of Example 3.28 has been re-sketched in Figure 3.80(a). Find the equivalent Thévenin and Norton circuit seen across the load R_L using source transformation.

(continued)

EXAMPLE 3.34 **Continued**

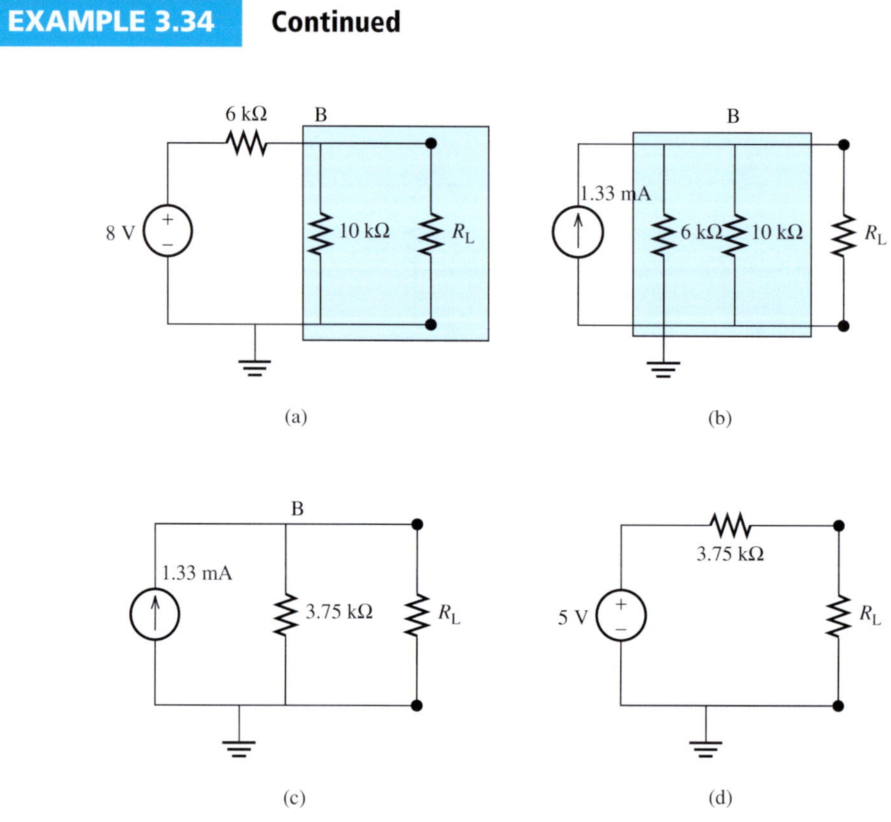

FIGURE 3.80 The figure of Example 3.34.

SOLUTION

Here, first, for simplicity, we assume everything beyond node B as the circuit load. Next, we find the equivalent Thévenin circuit of the 8-V source that is in series with the 6-kΩ resistor. The equivalent current source is:

$$I = \frac{8 \text{ V}}{6 \text{ k}\Omega} = 1.33 \text{ mA}$$

This equivalent circuit is sketched in Figure 3.80(b). Now, there are two parallel resistors of 6 kΩ, and 10 kΩ. The equivalent resistance of these two is 3.75 kΩ. This equivalent circuit has also been depicted in Figure 3.80(c). Now, the 1.33 mA in parallel to 3.75 kΩ form the equivalent Norton circuit that is seen through the terminals of the load resistor. We can equivalently find the corresponding Thévenin resistor seen through the terminals of the load resistor which is sketched in Figure 3.80(d).

EXAMPLE 3.35 **Equivalent Thévenin and Norton Using Source Transformation**

In the circuit shown in Figure 3.81, find the equivalent Norton and Thévenin circuit observed across the load resistor using source transformation.

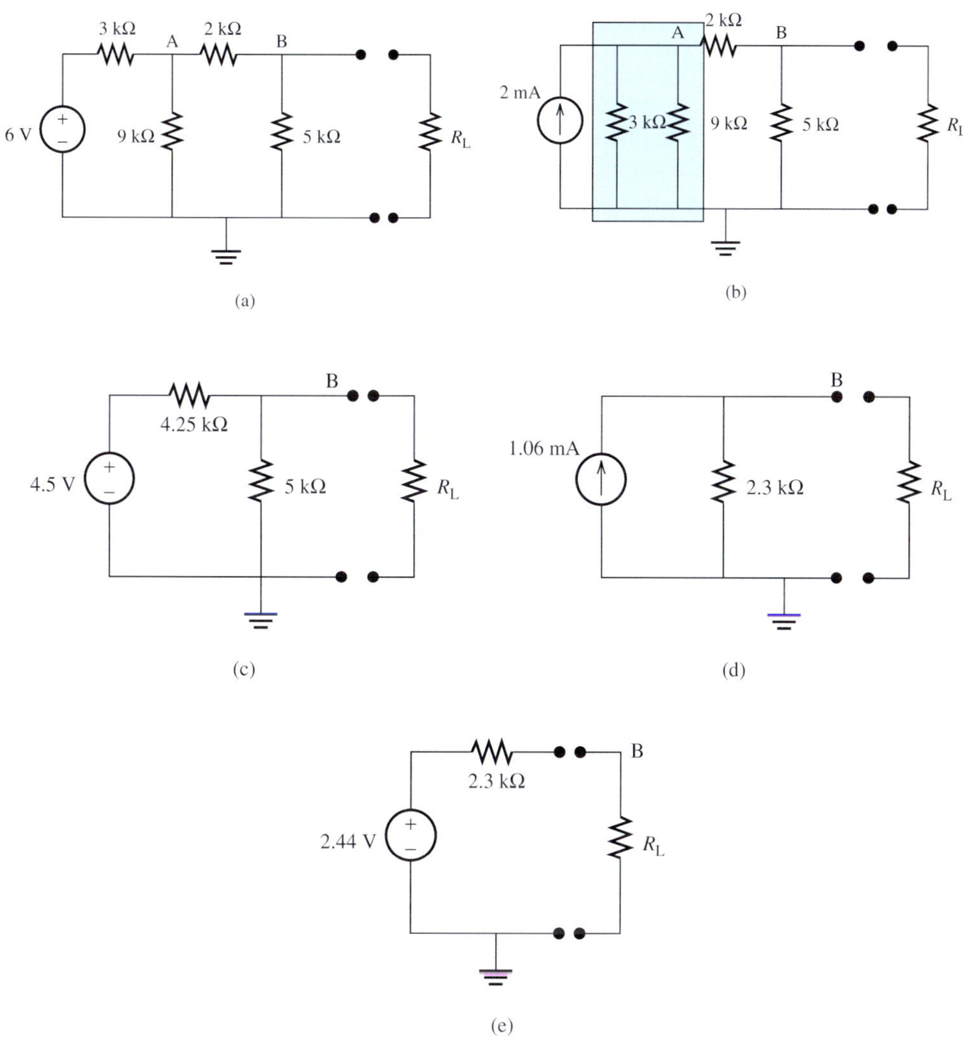

FIGURE 3.81 The figure of Example 3.35.

SOLUTION

The step-by-step process has been detailed in Figure 3.81(b–d). In Figure 3.81(b), we have converted the Thévenin circuit of 6 V voltage and the series resistor of 3 kΩ to its Norton equivalent circuit. Now there is another resistor of 9 kΩ. The equivalent resistance of 3 kΩ || 9 kΩ is 2.5 kΩ. Thus, the 2-mA source would be indeed in parallel to the 2.5-kΩ resistor.

Now, the 2-mA current source and the parallel 2.5-kΩ resistance create a new Norton circuit. Converting this Norton circuit into Thévenin, we will have the 5-V source in series with the 2.5-kΩ resistor. In this condition, the 2.5-kΩ resistor will be in series with the 2-kΩ between the terminals A and B. Thus, the equivalent resistance will be 4.5 kΩ, as shown in Figure 3.81(c).

Now, again, the 5-V source is in series with the 4.5-kΩ resistor which forms another Thévenin circuit. Converting it to its Norton equivalent, the current source will be 4.5 V/4/25 kΩ = 1.06 mA. In addition, we will have two parallel resistances of 4.5 kΩ || 5 kΩ, which is equivalent to 2.368 kΩ. Thus, the total Norton circuit across the load would be as shown in Figure 3.81(d). The Thévenin converted of this load is shown in Figure 3.81(d).

3.7 SUPERPOSITION PRINCIPLE

A circuit may consist of multiple sources. An example is a vehicle where the electrical systems might be supplied by both the car's battery and its generator. When multiple sources are available, it might seem difficult to find the voltage or current of the resistor. In a resistive circuit, the voltage or the current of a resistor can be considered a response to independent sources. The principle of superposition states that the total voltage (current) in any part of a linear circuit equals the algebraic sum of the voltages (currents) produced by each source, when other sources are set to zero.

It is *important to note* that superposition only applies to *linear* systems. The resistors, capacitors, and inductors described in this book are considered linear elements and the circuits they make up are considered linear systems. However, not all resistors, capacitors, or inductors are not linear systems. A linear system is the one, whose input, x, and output, $f(x)$, relationship can be maintained using a linear function. A function, $f(x)$, is called linear if it possesses the two conditions of additivity and homogeneity that are defined as:

$$\text{Additivity: } f(x_1 + x_2) = f(x_1) + f(x_2)$$
$$\text{Homogeneity: } f(ax) = af(x)$$

For example for a linear resistor, input–output relationship is defined by Ohm's law, that is, $v = R \cdot i$. It is clear that if we replace the current, i, with $i_1 + i_2$, we will have:

$$v = R \cdot (i_1 + i_2) = R \cdot i_1 + R \cdot i_2 = v_1 + v_2$$

In addition, if we replace i, with $\alpha\, i$, we will have:

$$v = R \cdot \alpha i = \alpha(R \cdot i) = \alpha v$$

Thus, a linear resistor fulfills both additivity and homogeneity conditions and can be considered as a linear system. Note that, in general the current–voltage relationship might not be linear.

EXCERCISE 3.11

Show that if voltage–current relationship is represented by $v = Ri + \beta i^2$, the system would be nonlinear.

In order to apply the superposition principle in a circuit, first find the response (voltage or current) of a resistor to a source by setting other sources in the circuit to zero. Recall that the voltage and current sources can be equated to zero by replacing them with short and open circuits, respectively. Note that, in applying the superposition principle, the dependent sources are not set to zero.

For example, consider the circuit in Figure 3.82(a). The goal is to determine the current, I, flowing through the resistor R_2. By applying nodal voltage analysis, and writing KCL for node A in Figure 3.82(b):

$$\frac{V - V_s}{R_1} + \frac{V}{R_2} - I_s = 0$$

Rearranging to find the voltage, V:

$$V = \frac{R_1 R_2}{R_1 + R_2}\left(\frac{V_s}{R_1} + I_s\right) = \frac{R_1 R_2}{R_1 + R_2}\frac{V_s}{R_1} + \frac{R_1 R_2}{R_1 + R_2} I_s \quad \textbf{(3.27)}$$

Now, the current through R_2, is equal to its voltage divided by the resistance, that is:

$$I = \frac{V}{R_2} = \frac{R_1}{R_1 + R_2}\left(\frac{V_s}{R_1} + I_s\right) = \frac{V_s}{R_1 + R_2} + \frac{R_1 I_s}{R_1 + R_2} \quad \textbf{(3.28)}$$

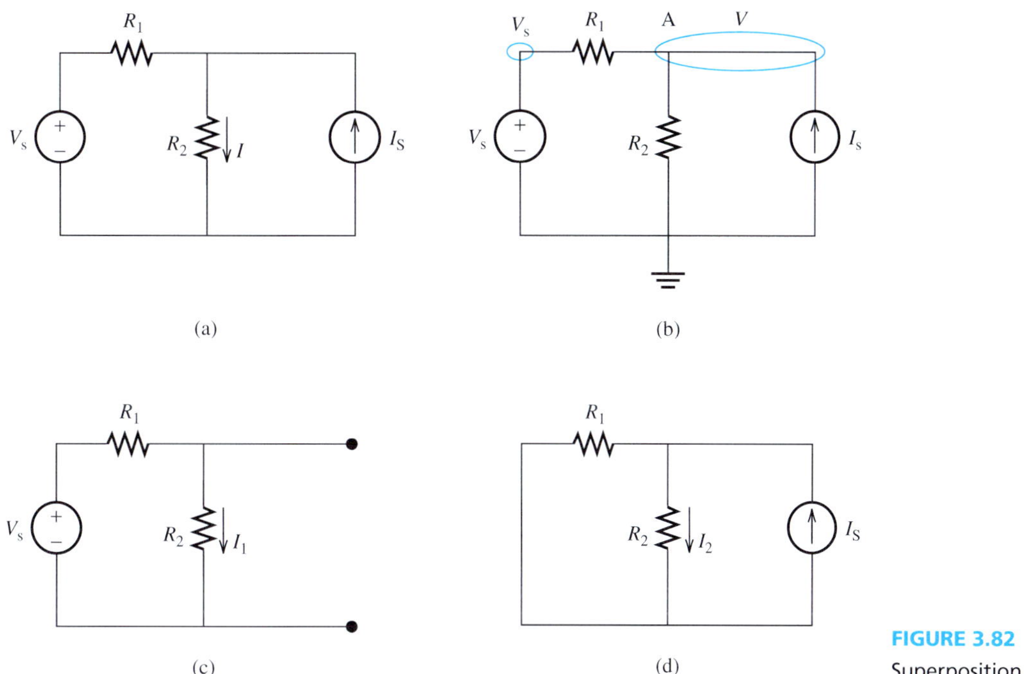

FIGURE 3.82
Superposition principles.

In Equations (3.27) and (3.28), the first term is the response to the voltage source (setting $I_s = 0$) and the second term is the response to the current source (setting $V_s = 0$). Thus, the superposition principle can also be used to find the current, I, by first setting $V_s = 0$ and finding V that is due to I_s, and then setting $I_s = 0$ and finding a second V that is due to V_s, and finally, adding these two voltages. Note that (Chapter 2, Exercises 2.4 and 2.5):

> A zero-current source is equivalent to an open circuit.
> A zero-voltage source is equivalent to a short circuit.

The current due to the voltage source and in the absence of the current source [Figure 3.82(c)] is:

$$I_1 = \frac{V_s}{R_1 + R_2}$$

The current due to the current source while the voltage source is zero [Figure 3.82(d)] will be:

$$I_2 = \frac{R_1 I_s}{R_1 + R_2}$$

Applying the superposition principle, the current i, in the original circuit is:

$$I = I_1 + I_2 = \frac{R_1}{R_1 + R_2}\left(\frac{V_s}{R_1} + I_s\right)$$

which is equivalent to Equation (3.28).

EXAMPLE 3.36 Superposition

Find the voltage V_1 in Figure 3.83 using the superposition technique.

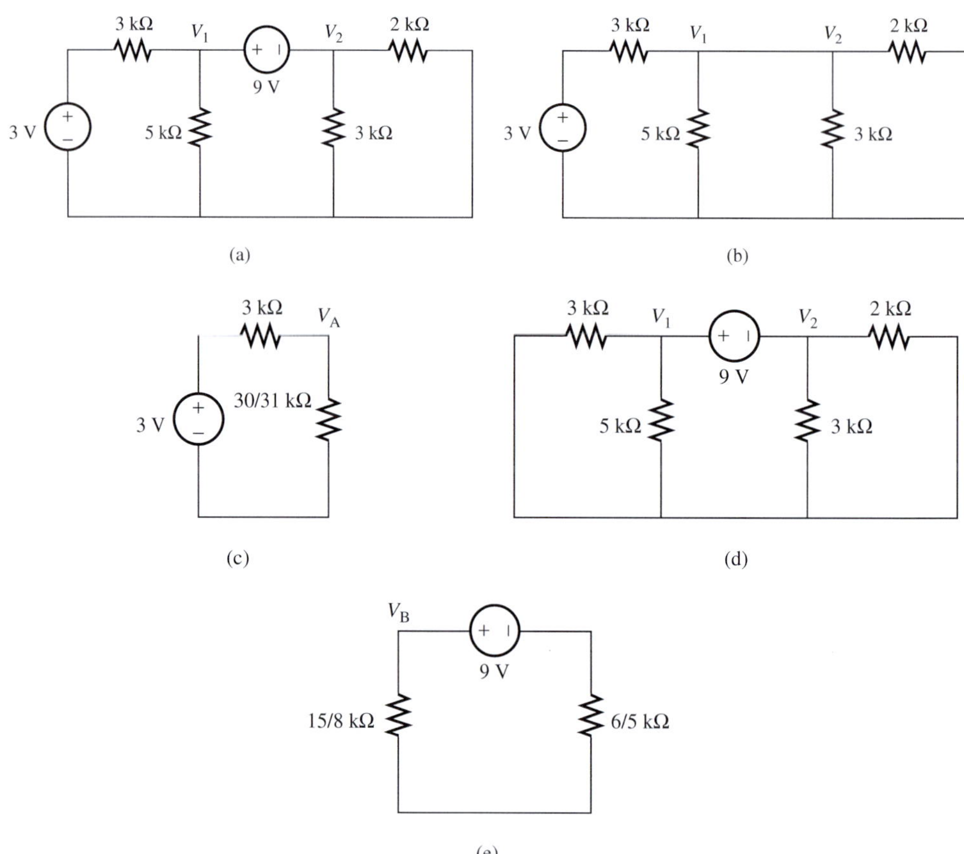

FIGURE 3.83 Circuits for Example 3.37. (a) The original circuit. (b,c) Setting 9 V to zero. (d, e) Setting 3 V to zero.

SOLUTION

First, equate the 9-V and 3-V voltage sources to zero, respectively. The resulting circuits and their equivalents are shown in Figure 3.83. According to the equivalent circuit in Figure 3.83(c), which is for the case when the 9-V voltage source is equated to zero [see Figure 3.83(b)], the node voltage due to the 3-V voltage source (V_A) is:

$$V_A = \frac{\frac{30}{31}}{3 + \frac{30}{31}} \times 3 = 0.7317$$

Similarly, according to the equivalent circuit in Figure 3.83(e), which is for the case when the 3-V voltage source is equated to zero [see Figure 3.83(d)] the node voltage due to the 9-V voltage source (V_B) is:

$$V_B = \frac{\frac{15}{8}}{\frac{15}{8} + \frac{6}{5}} \times 9 = 5.4878 \text{ V}$$

Applying the superposition principle to the node voltage V_1 in Figure 3.83(a) is the summation of the voltages V_A and V_B, that is:

$$V_1 = V_A + V_B = 6.22 \text{ V}$$

APPLICATION EXAMPLE 3.37 Solar-Powered Heater

Figure 3.84 shows a solar-powered shower heating system. In order to reduce the cost of energy, solar energy is produced during the daytime to heat the water that is used for shower purposes.

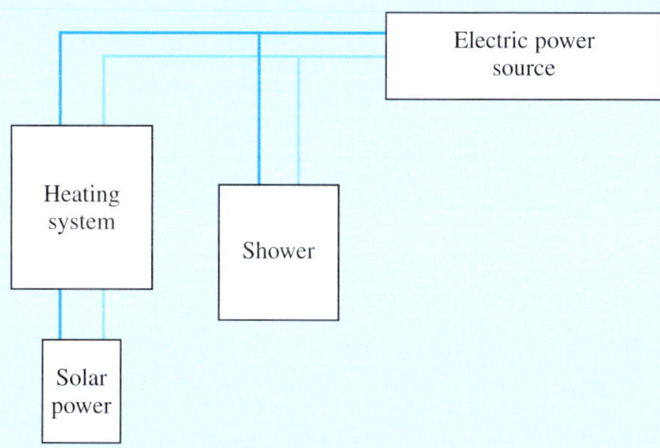

FIGURE 3.84 A solar-powered shower system.

Assume the solar power produces a voltage source of 100 V, the electric power supply is 110 V, the heater resistance is 100 Ω and the shower resistance is 80 Ω. Calculate the current flow through the shower.

SOLUTION

The equivalent circuit for the system in Figure 3.84 is shown in Figure 3.85.

Using the superposition concept, first, equate the electric voltage source to zero as shown in Figure 3.85(b). In this case, the shower is short-circuited. Thus, the current through the shower, I_1, is zero.

Next, consider the case when there is no solar power source [Figure 3.85(c)]. In this case, the heater and the shower become parallel resistors. The current flow through the shower, I_2, can be calculated as:

$$I_2 = \frac{V_{\text{electric}}}{R_{\text{shower}}}$$

$$I_2 = \frac{110}{80} = 1.375 \text{ A}$$

Therefore, the current flow through the shower is:

$$I_{\text{shower}} = I_1 + I_2$$

$$I_{\text{shower}} = 0 + 1.375 = 1.375 \text{ A}$$

So far superposition has been discussed for use across independent sources. However, superposition can also be applied to circuits that include both independent and dependent sources. The following examples clarify this point.

(continued)

APPLICATION EXAMPLE 3.37 Continued

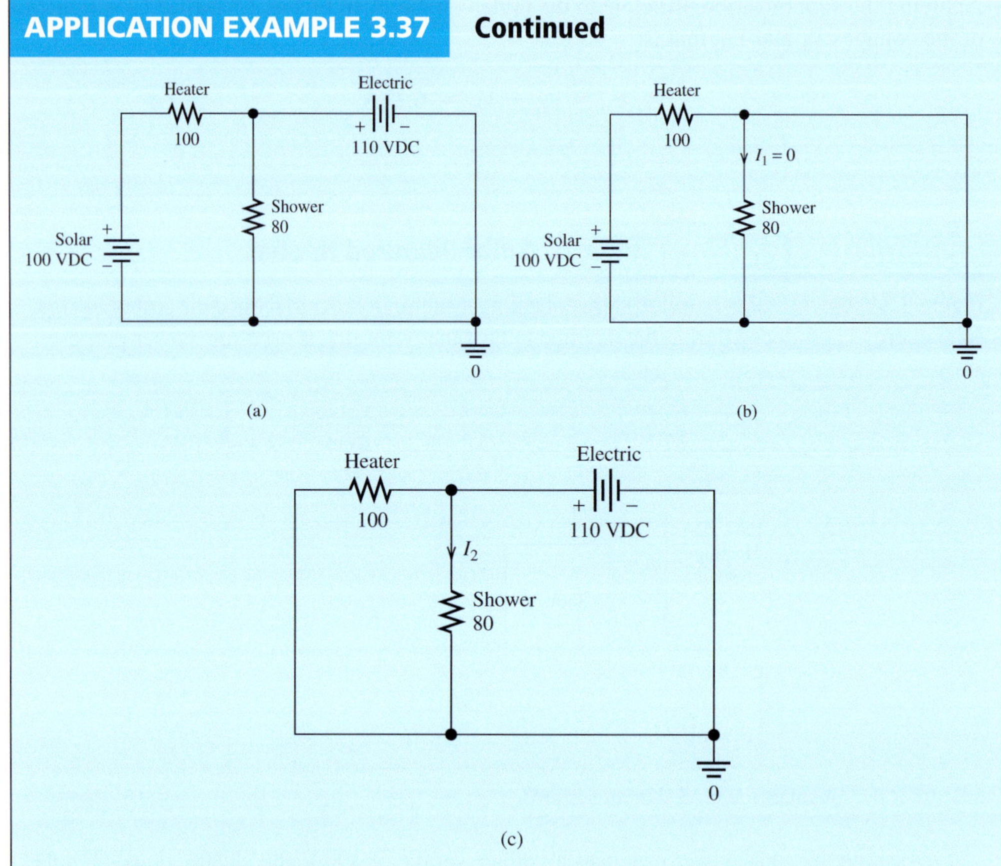

FIGURE 3.85 (a) The equivalent circuit of Figure 3.84; (b) equating the electric voltage source to zero; (c) equating the heater to zero.

EXAMPLE 3.38 Superposition

Find the current, *I*, in Figure 3.86 using the superposition technique.

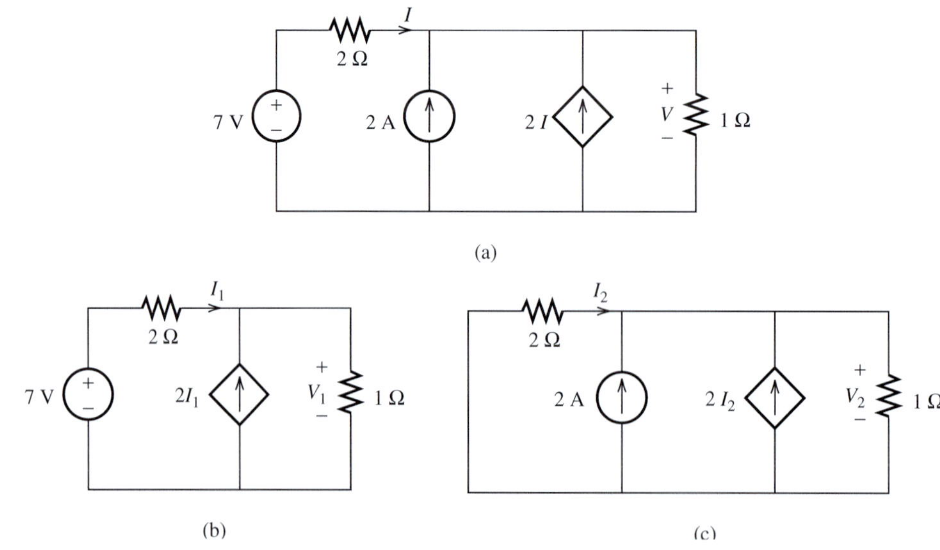

FIGURE 3.86 Circuits for Example 3.38. (a) The original circuit. (b) Setting 2 A to zero. (c) Setting 7 V to zero.

SOLUTION

Equate the 7-V voltage sources and the 2-A current sources to zero, respectively. The resulting circuits and their equivalents are shown in Figure 3.86.

According to the equivalent circuit in Figure 3.86(b), the node voltage due to the 7-V voltage source (V) is:

$$V_1 = 7 - 2I_1 = (I_1 + 2I_1) \times 1$$

$$I_1 = \frac{7}{5}\text{A} = 1.4 \text{ A}$$

Similarly, according to the equivalent circuit in Figure 3.86(c), the node voltage due to the 2-A current source (V) is:

$$V_2 = -2I_2 = (3I_2 + 2) \times 1$$

$$I_2 = -\frac{2}{5}\text{A} = -0.4 \text{ A}$$

According to the superposition principle, the current, I, in Figure 3.86(a) is the summation of the currents I_1 and I_2. Thus:

$$I = I_1 + I_2 = 1 \text{ A}$$

EXAMPLE 3.39 Superposition

Find the voltage, V, in Figure 3.87 using the superposition technique.

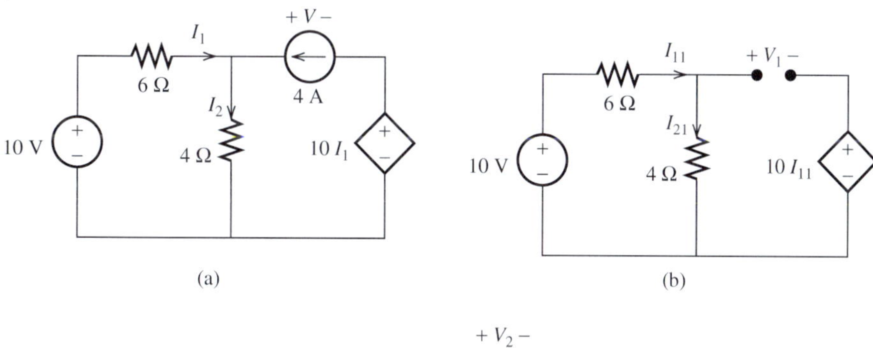

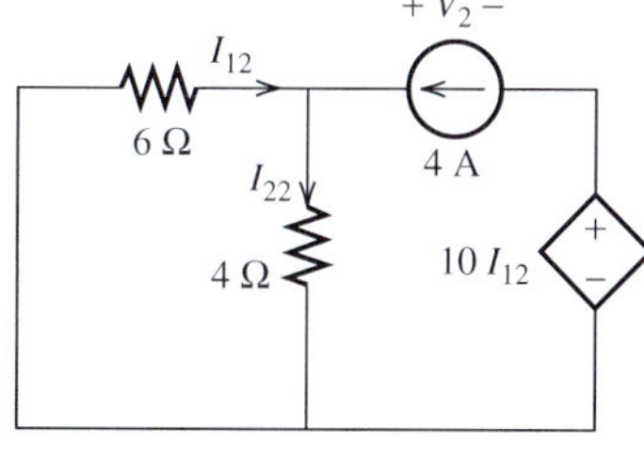

FIGURE 3.87 Circuits for Example 3.39. (a) The original circuit. (b) Setting 4 A to zero. (c) Setting 10 V to zero.

(continued)

EXAMPLE 3.39 Continued

SOLUTION

First, zero the 10-V voltage sources and 4-A current sources, respectively. The resulting circuits and their equivalents are shown in Figure 3.87 (b) and (c). According to the equivalent circuit in Figure 3.87(b), the node voltage due to the 10-A voltage source, I_{11}, is:

$$I_{11} = \frac{10}{6 + 4} = 1\text{ A}$$
$$V_1 = -10I_{11} + 4I_{11} = -6\text{ V}$$

Similarly, according to the equivalent circuit in Figure 3.87(c), the node voltage due to the 2-A voltage source, I_{12}, is:

$$I_{12} = -\frac{4}{6 + 4} \times 4 = -1.6\text{ A}$$
$$I_{22} = 4 + I_{12} = 2.4\text{ A}$$
$$V_2 = -10I_{12} + 4I_{22} = 25.6\text{ V}$$

According to the superposition principle, the current I in Figure 3.87(a) is the summation of the voltages V_1 and V_2. Thus:

$$V = V_1 + V_2 = 19.6\text{ V}$$

3.8 MAXIMUM POWER TRANSFER

As demonstrated by the Thévenin theorem, every two-terminal resistive circuit can be replaced by a Thévenin equivalent circuit that consists of a voltage source and a resistance. Suppose a load resistance, R_L, is connected to the terminals of the Thévenin equivalent circuit as shown in Figure 3.88.

In this section, the goal is to find the load resistance that absorbs the maximum power from the two-terminal resistive circuit. The current flowing through the load resistance is given by:

$$I_L = \frac{V_{th}}{R_L + R_{th}}$$

The power delivered to the load is:

$$P_L = R_L I_L^2$$

Therefore:

$$P_L = \frac{V_{th}^2 R_L}{(R_{th} + R_L)^2}$$

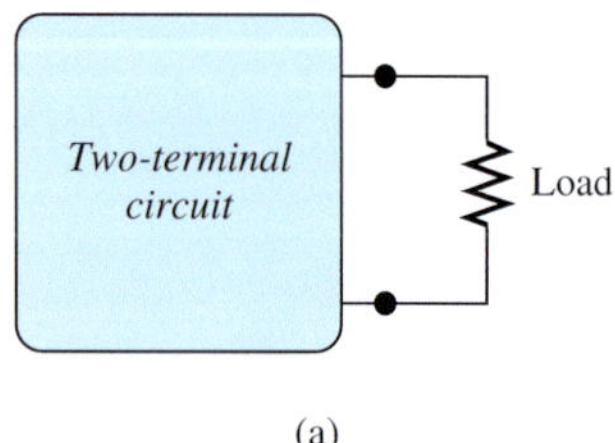

(a)

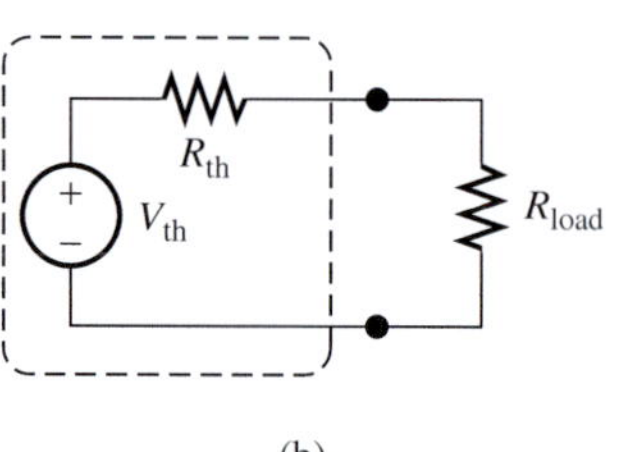

(b)

FIGURE 3.88 (a) A two-terminal circuit and (b) its Thévenin equivalent.

In order to determine the load resistance such that the maximum possible power is delivered to the load, the derivative of the power, P_L, with respect to R_L, is set to zero. This is a general concept in mathematics.

To find the maximum or minimum of a function, $f(x)$, with respect to the variable x, find the derivative of $f(x)$ with respect to x, that is, $f'(x)$, and set it to zero. Solving $f'(x) = 0$, for x, will result in the value of x, namely x_m, that maximizes or minimizes $f(x)$. In addition, from the concepts learned in mathematics, replacing x in the second derivative, that is, $f''(x_m)$, with x_m, $f''(x_m) < 0$, then x_m represents a maxima; otherwise, it represents a minima.

Solving $dP_L/dR_L = 0$ with respect to R_L, it can be determined that:

$$\textit{for } P_{max}: R_L = R_{th} \tag{3.29}$$

As a result, the load resistance that absorbs the maximum power is equal to the Thévenin equivalent resistance. In this case, the power transferred to the load will be:

$$P_L = \frac{V_{th}^2}{4R_{th}} \tag{3.30}$$

EXCERCISE 3.12

Work out the details of differentiation $dP_L/dR_L = 0$ and show that indeed $R_L = R_{th}$ leads to the maximum (and not the minimum) power transfer.

EXAMPLE 3.40 Maximum Power Transfer

In Figure 3.89, find the maximum power transferred to a speaker, when the resistor R is equal to the speaker resistance. Here, $V = 10$ V and $R_{th} = 100\ \Omega$.

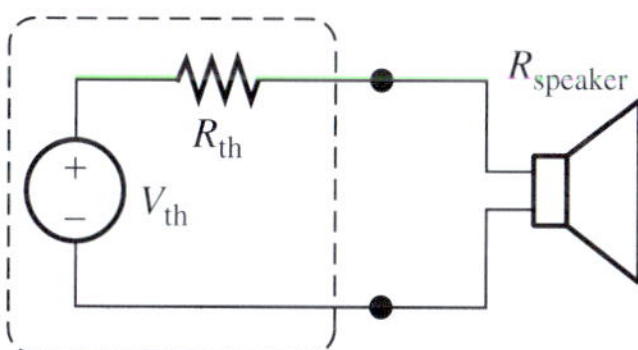

FIGURE 3.89 Figure for Example 3.40.

SOLUTION

Using Equation (3.30), and assuming $R_{speaker} = R_{th}$ we have:

$$P_L = \frac{V_{th}^2}{4R_{th}} = \frac{10^2}{4 \times 100} = 0.25 \text{ W}$$

APPLICATION EXAMPLE 3.41 Strain Gage

Strain gage has already been introduced and discussed in Example 3.10. Resistive strain gages are often connected in groups of four, as shown in Figure 3.90(a), known as quad cells. These groupings are to eliminate the effect of the surrounding temperature on the strain reading. Only two of the gages are connected to the beam.

a. Find the Thévenin equivalent voltage and resistance between V_A and V_B;
b. Find the power dissipated in a 20.8-kΩ resistance installed between nodes A and B.

(continued)

APPLICATION EXAMPLE 3.41 Continued

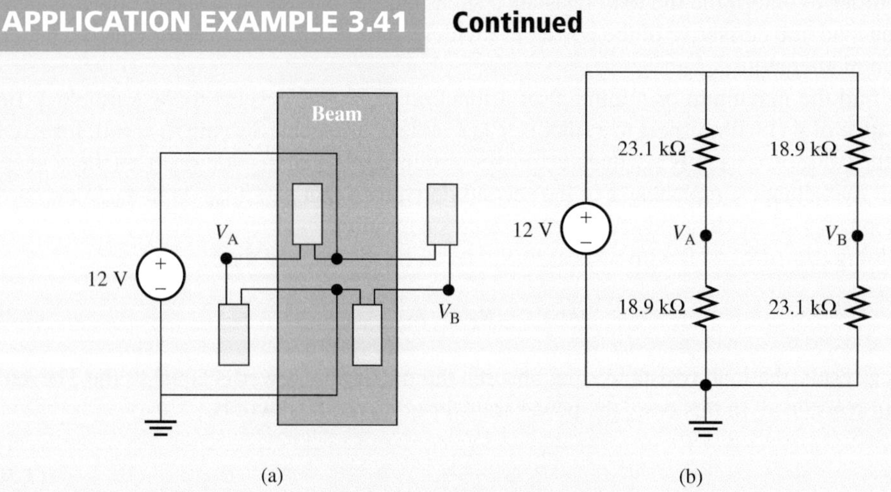

FIGURE 3.90 (a) The strain gage and (b) its equivalent circuit.

SOLUTION

a. The equivalent circuit of Figure 3.90(a) is sketched in Figure 3.90(b). This structure is called a *Wheatstone bridge*. Here, the approach of $R_{th} = V_{oc}/I_{sc}$ is used. The open-circuit voltage V_{oc} is across the terminals A and B, that is, when A and B are detached (open-circuited), and I_{sc} is for the scenario that these two terminals are connected (short-circuited).

To find the short-circuit current between nodes A and B, short-circuit these two nodes as shown in Figure 3.91 and write the nodal voltage equations as:

$$\frac{V_A - 12}{23.1\,\text{K}} + \frac{V_A}{18.9\,\text{K}} + I_{sc} = 0$$
$$\frac{V_B - 12}{18.9\,\text{K}} + \frac{V_B}{23.1\,\text{K}} - I_{sc} = 0$$
$$V_A = V_B$$

Solving for I_{sc}, $I_{sc} = -57.72\ \mu\text{A}$.

Now, find the open-circuit voltage of the nodes $V_{oc} = V_A - V_B$. Based on the voltage division rule, the node voltages can be simply calculated:

$$V_A = 12\text{ V} \times \frac{18.9\text{ k}\Omega}{23.1\text{ k}\Omega + 18.9\text{ k}\Omega} = 5.4\text{ V}$$

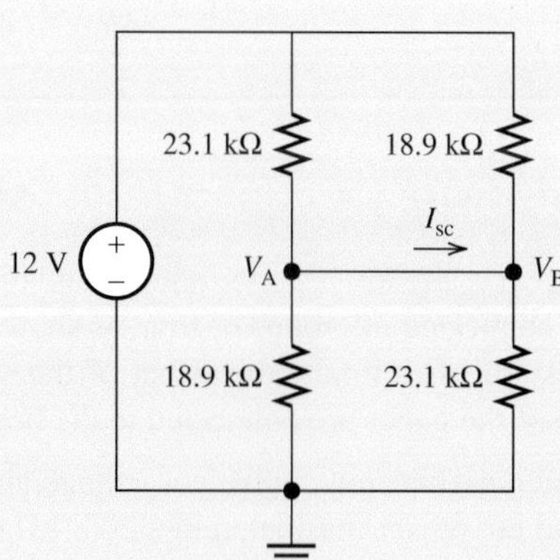

FIGURE 3.91 Circuit for Example 3.41.

$$V_B = 12\text{ V} \times \frac{23.1\text{ k}\Omega}{23.1\text{k}\Omega + 18.9\text{ k}\Omega} = 6.6\text{V}$$

Next, see that:

$$V_{th} = V_{oc} = V_A - V_B = -1.2\text{ V}$$

Therefore, the Thévenin equivalent resistance will be:

$$R_{th} = \frac{V_{oc}}{I_{sc}} = 20.8\text{ k}\Omega$$

b. The Thévenin equivalent circuit was calculated seen through terminals A and B. It can be used to calculate the maximum power transfer for a load connected between terminals A and B. Incorporating Equation (3.30), it corresponds to:

$$P_L = \frac{V_{th}^2}{4R_{th}} = \frac{1.44}{4 \times 20.8\text{ k}\Omega} = 17.3\mu\text{W}$$

APPLICATION EXAMPLE 3.42 Sports Car Inventor

Back to our inventor and the hot rod acceleration test (see Example 3.17). Determine which of the circuits shown in Figure 3.92(a) or (b) match output resistance with the 50-Ω input resistance of the PDA. For this accelerometer, $0 \leq V_{ACC} \leq 5$ V.

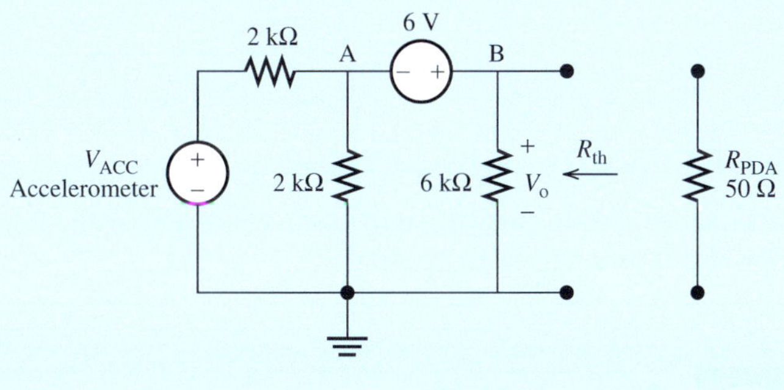

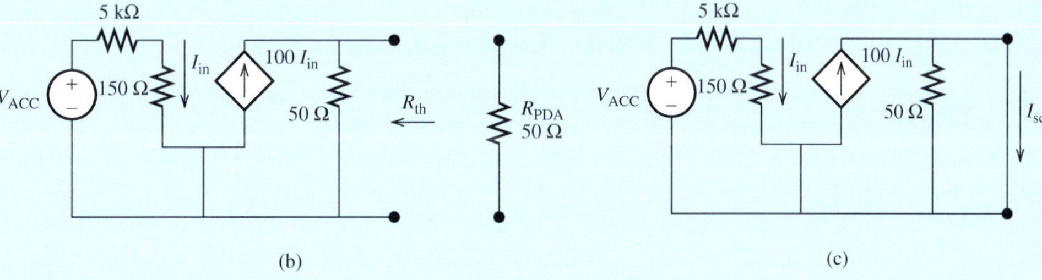

FIGURE 3.92 The circuits for Example 3.42.

SOLUTION

To find if the 50-Ω resistance has a good match with the circuit, first find the Thévenin equivalent circuit seen through terminal B (with respect to the ground), and check if $R_{th} = R_{PDA}$. For the circuit of Figure 3.92(a), in order to determine the Thévenin equivalent resistance,

(continued)

APPLICATION EXAMPLE 3.42 Continued

equate all sources to zero and find the equivalent resistance between the terminals. In this case, the equivalent resistance will be $R_{th} = 2\text{ k}\Omega \| 2\text{ k}\Omega \| 6\text{ k}\Omega = 857\ \Omega$, but $R_{PDA} = 50\ \Omega$. Thus, $R_{th} \neq R_{PDA}$ and this circuit does not deliver maximum energy to the output (PDA).

For the second circuit in Figure 3.92(b), $V_{ACC,max} = 5$ V and $I_{in} = 5/(5000 + 150)$. The open-circuit voltage between the terminals is given by:

$$V_{oc,max} = 100 I_{in} \times 50 = 100 \times \frac{5}{5150} \times 50 = 4.85\text{ V}$$

As a result, the Thévenin voltage will be:

$$V_{th,max} = V_{oc,max} = 4.85\text{ V}$$

Moreover, the short-circuit current between the terminals [Figure 3.91(c)] will be obtained simply as:

$$I_{sc,max} = 100 I_{in} = 100 \times \frac{5}{5150} = 97.1\text{ mA}$$

Using the open-circuit voltage and the short-circuit current between the terminals:

$$R_{th} = \frac{V_{oc,max}}{I_{sc,max}} = \frac{4.85\text{ V}}{97.1\text{ mA}} = 49.94\ \Omega$$

Therefore, in this case, $R_{th} = R_{PDA}$, and the maximum power is absorbed by the PDA resistance.

3.9 ANALYSIS OF CIRCUITS USING PSPICE

Here, examples are provided to help students learn to solve problems using PSpice. Students are encouraged to review the PSpice tutorial in Section 2.9.

EXAMPLE 3.43 PSpice Analysis

In Example 3.4, assuming $R_1 = 10\ \Omega$, $R_2 = 20\ \Omega$, $R_3 = 30\ \Omega$, compute and compare the current flow through the circuits in Figure 3.16 and 3.17 using PSpice software.

SOLUTION

The PSpice solution for a series circuit is shown in Figure 3.93. Figure 3.94 shows the solution for a parallel circuit.

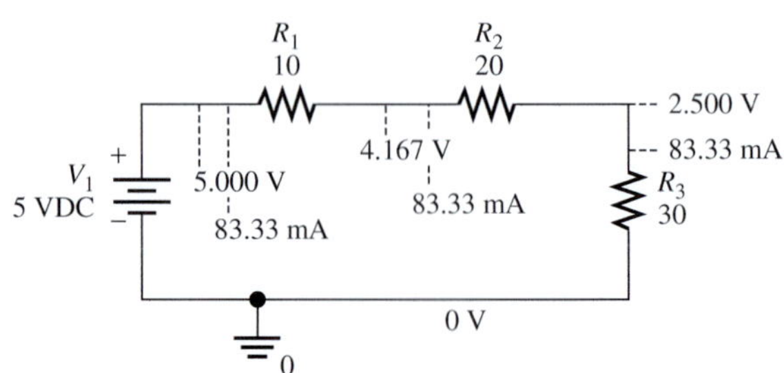

FIGURE 3.93 PSpice solution for Example 3.4, series circuit.

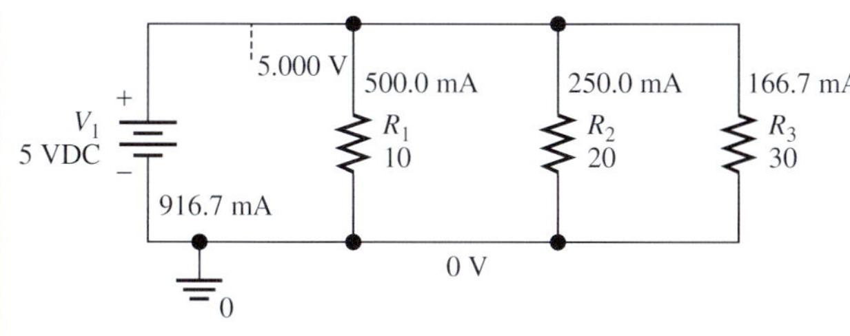

FIGURE 3.94 PSpice solution for Example 3.4, parallel circuit.

EXAMPLE 3.44 PSpice Analysis

Generate the results of Example 3.17 using PSpice.

SOLUTION

a. Find V_o if $V_A = 0.3$ V.

1. To add a voltage-controlled voltage source into the PSpice schematic board go to "Parts" (see Figure 2.63 in Chapter 2) and type "E."
2. Double-click the "E" icon to get into properties menu. Change the Gain to "5," and choose "display setting" while the "Gain" column is highlighted. Choose "Name and Value" for the display format and click OK.
3. Run the simulation for Bias Point Analysis.

The result is shown in Figure 3.95(a).

b. Find I_A when $V_A = V_{Amax}$. The result is shown in Figure 3.95(b).

c. How much power is dissipated by the PDA?

$$P = \frac{V_o^2}{R} = \frac{(12.5)^2}{50} = 3.125W$$

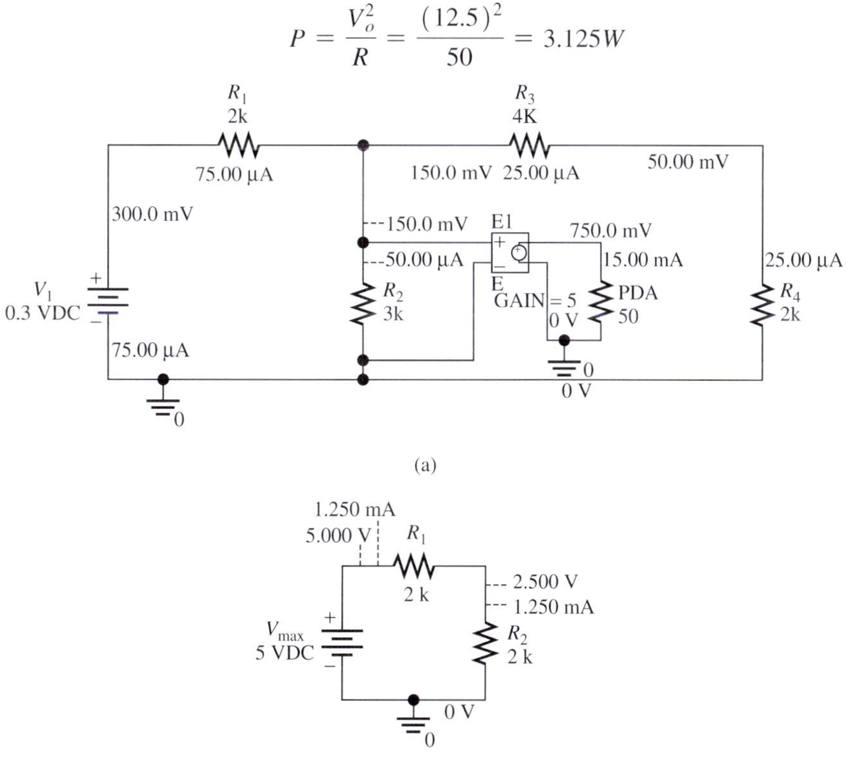

FIGURE 3.95 PSpice solution for Example 3.17 (a) Part (a); (b) Part (b).

EXAMPLE 3.45 PSpice Analysis

Use PSpice to find the nodal voltages V_1 and V_2 in Example 3.20.

SOLUTION

To study the voltage at a particular node, click and drag the voltage bar to the particular node as shown in Figure 3.96.

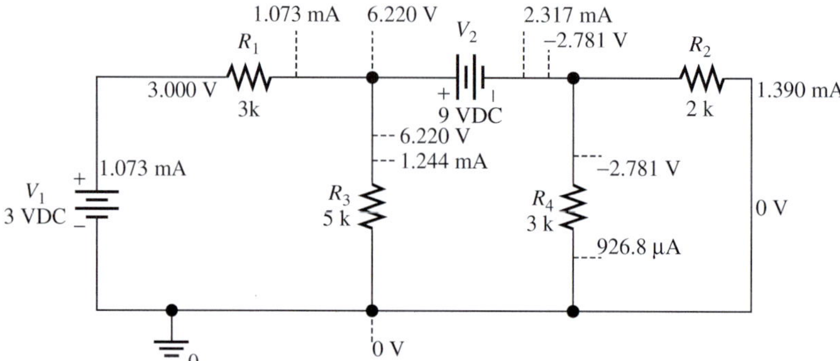

FIGURE 3.96 PSpice solution for Example 3.20.

EXAMPLE 3.46 PSpice Analysis

Use PSpice to find the Thévenin equivalent of the circuit in Example 3.28 [Figure 3.72(a)] seen by R_L.

SOLUTION

Because PSpice cannot run the simulation with an open circuit, the process instead re-defines the value of resistor R_L. In order to study the Thévenin equivalent of voltage, we set the value of R_L to be a very large resistance, which is shown in Figure 3.97.

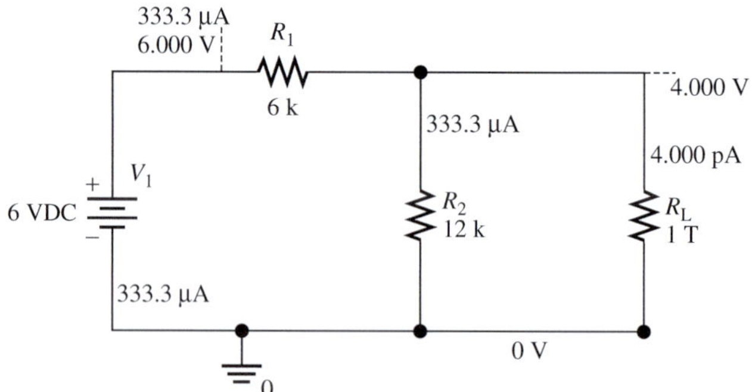

FIGURE 3.97 PSpice solution for Example 3.28.

EXAMPLE 3.47 PSpice Analysis

Find the Norton equivalent of the circuit in Example 3.29 as seen by R_L.

SOLUTION

To study the Norton equivalent current, the value of resistor, R_L needs to be selected as very small to represent a short circuit. Thus, the corresponding circuit corresponds to Figure 3.98.

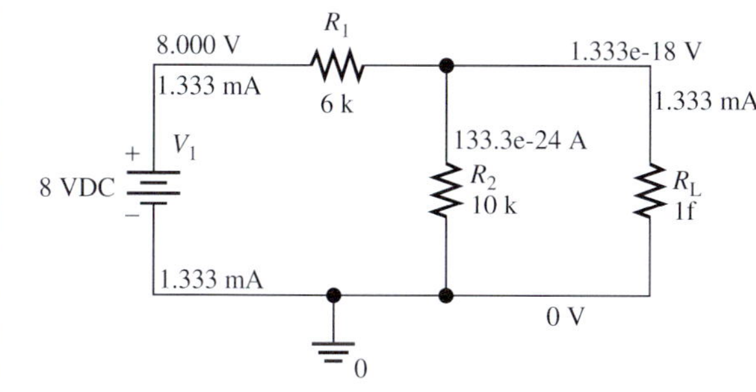

FIGURE 3.98 PSpice solution for Example 3.29.

3.10 WHAT DID YOU LEARN?

- The equivalent resistance of N resistors in series is shown in Equation (3.5) and is equal to:

$$R_{eq} = \sum_{n=1}^{N} R_n$$

- The equivalent resistance of N resistors in parallel is shown in Equation (3.10) and is equal to:

$$R_{eq} = \frac{1}{\dfrac{1}{R_1} + \dfrac{1}{R_2} + \cdots + \dfrac{1}{R_N}}$$

- The *voltage division rule* states that voltage is divided between N series resistors in direct proportion to their resistance [Equation (3.16)]:

$$v_k = \frac{R_k}{R_1 + R_2 + \cdots + R_N} v_s \; ; \; k = 1, 2, \ldots N$$

- The *current division rule* states that current is divided between two parallel resistors in inverse proportion to their resistance [see Equations (3.20) and (3.21)]:

$$i_1 = \frac{R_2}{R_1 + R_2} i_s$$

$$i_2 = \frac{R_1}{R_1 + R_2} i_s$$

- The nodal analysis procedure involves the following steps:
  1. Determine all nodes in the circuit.
  2. Pick a node as a reference node.
  3. Assign voltage variables to each node.
  4. For each node, write KCL equations in terms of node voltages.
  5. Solve the system of equations to find unknown voltages.
- The mesh analysis procedure involves the following steps:
  1. Define each mesh current (usually exclusively clockwise).
  2. Assign current variables to each mesh.
  3. To write the KVL for each mesh, determine the voltage across the components of that mesh based on the mesh current direction.
  4. For each mesh, write the KVL equation.
  5. Solve the system of equations to find unknown mesh currents.
- A *super node* is a closed surface in a circuit that includes several nodes. KCL can be applied to super nodes.
- Thévenin and Norton equivalent techniques can be used to simplify complex resistive circuits (see Figure 3.66).

  The *Thévenin theorem* states that any resistive circuit can be represented by a voltage source in series with a resistance.

  The *Norton theorem* states that any resistive circuit can be represented by a current source in parallel with a resistance.

Source transformation can be used to find the Thévenin or Norton equivalent of a complex circuit.

- The *principle of superposition* simplifies the solution of circuits with multiple sources. It states that the total voltage (or current) in any part of a linear circuit equals the algebraic sum of the voltages (or currents) produced by each source, when other sources are set to zero.
- *Maximum power transfer* means that the load resistance that absorbs the maximum power from a resistive network is equal to the Thévenin equivalent resistance of the resistive network [Equation (3.29)].

$$R_L = R_{th}$$

- PSpice can be used to analyze complex resistive circuits to find the voltage across and the current through all elements.

Problems

*B refers to Basic, A refers to Average, H refers to Hard, and * refers to problems with answers.*

SECTION 3.1 INTRODUCTION

3.1 (B) Which of the following is not resistive circuit?

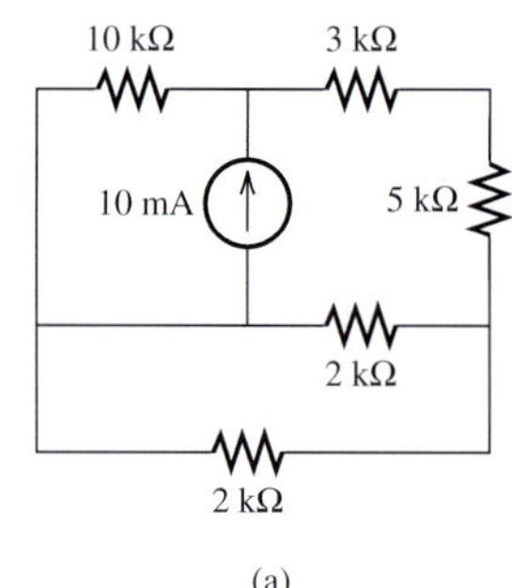

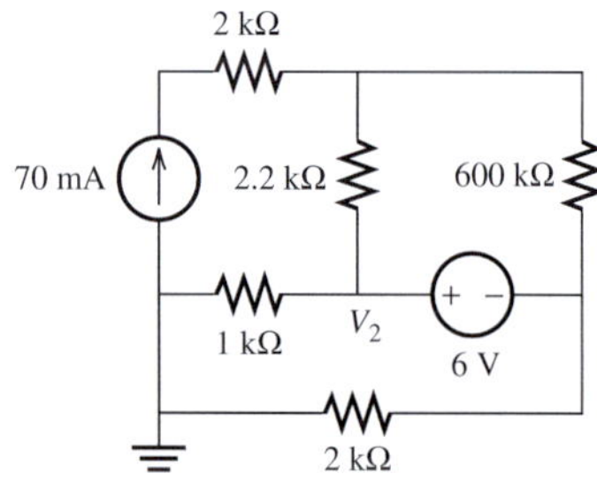

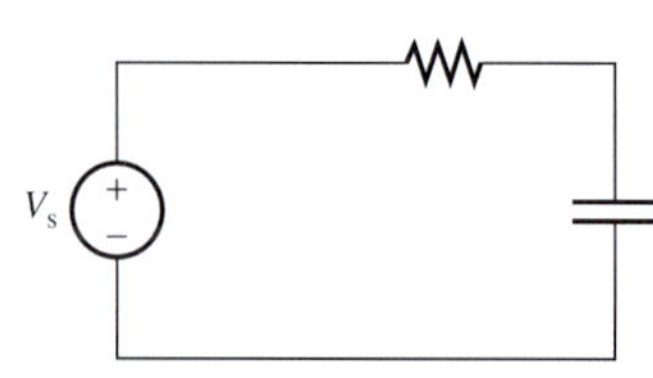

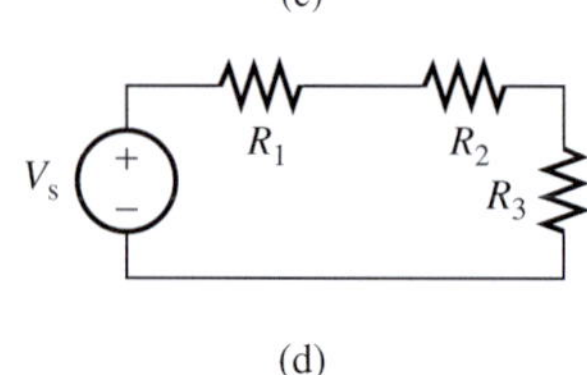

FIGURE P3.1 Circuit for Problem 3.1.

SECTION 3.2 RESISTORS IN PARALLEL AND SERIES AND EQUIVALENT RESISTANCE

3.2 (B) Identify which resistors are in parallel and which are in series in Figure P3.2.

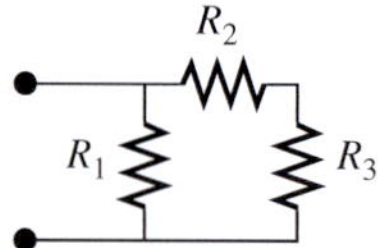

FIGURE P3.2 Circuit for Problem 3.2.

3.3 (B) Identify which resistors are in parallel and which are in series in Figure P3.3.

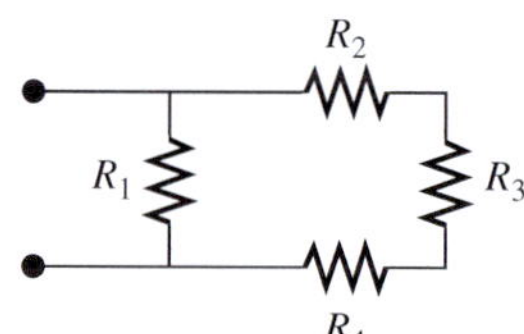

FIGURE P3.3 Circuit for Problem 3.3.

3.4 (B) Find R_{eq} in terms of R_1, R_2, R_3 in Figure P3.4.

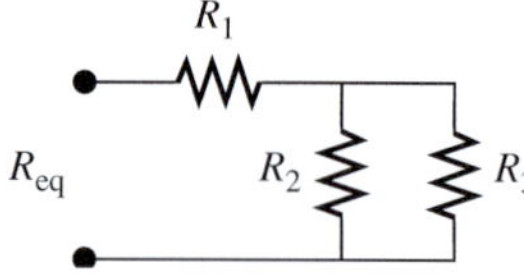

FIGURE P3.4 Circuit for Problem 3.4.

3.5 (A) Identify which resistors are in parallel and which are in series in Figure P3.5.

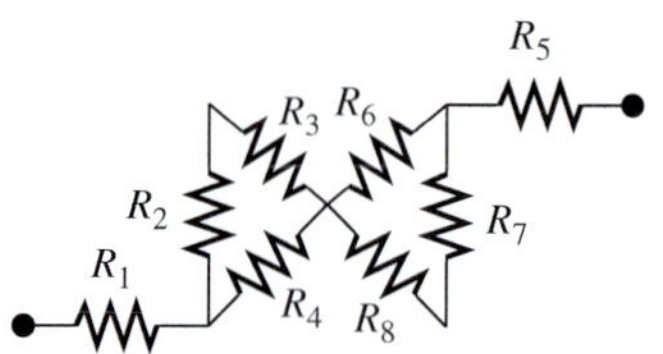

FIGURE P3.5 Circuit for Problem 3.5.

3.6 (A)* Find R_2 in Figure P3.6, given $R_{eq} = 17\ \Omega$, $R_1 = 5\ \Omega$, $R_3 = 15\ \Omega$.

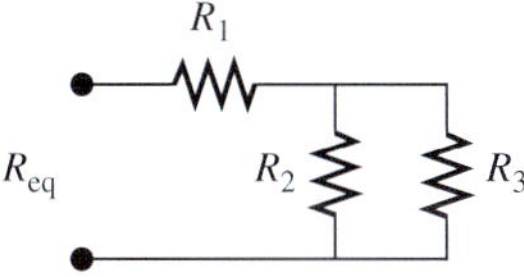

FIGURE P3.6 Circuit for Problem 3.6.

3.7 (A) Find R_{eq} in Figure P3.7, given

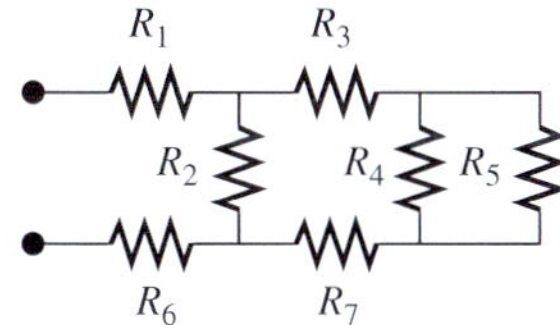

FIGURE P3.7 Circuit for Problem 3.7.

3.8 (A)* Find the R_{eq} in Figure P3.8, given that every resistor has an equal resistance, R (the center crossing is connected).

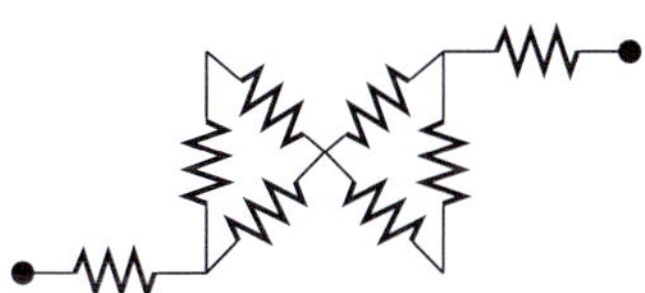

FIGURE P3.8 Circuit for Problem 3.8.

3.9 (H) Find R_{AB} in Figure P3.9:

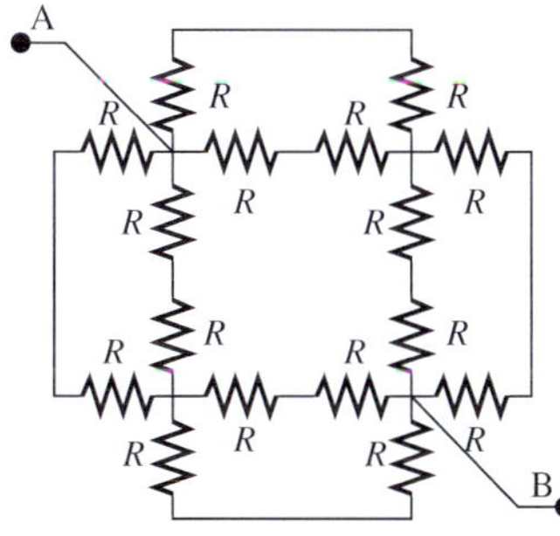

FIGURE P3.9 Circuit for Problem 3.9.

3.10 (H) Find R_{AB} and R_{CD} in Figure P3.10. Let node C and D open when calculating R_{AB} and let node A and B open when calculating R_{CD}.

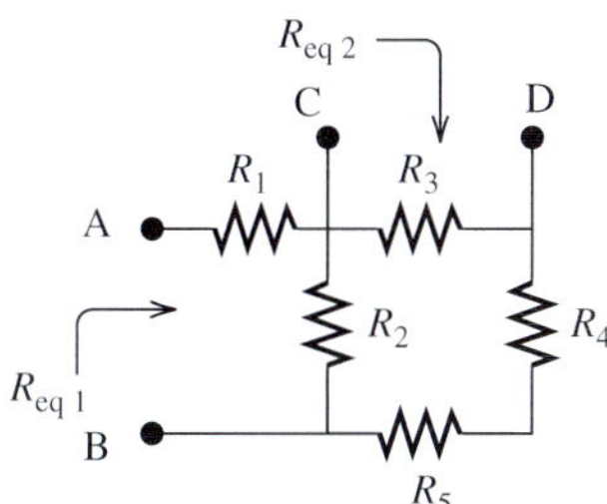

FIGURE P3.10 Circuit for Problem 3.10.

3.11 (H)* For Figure P3.11, find (a) the equivalent resistance, R_{AB} and (b) the equivalent resistance R_{CD}.

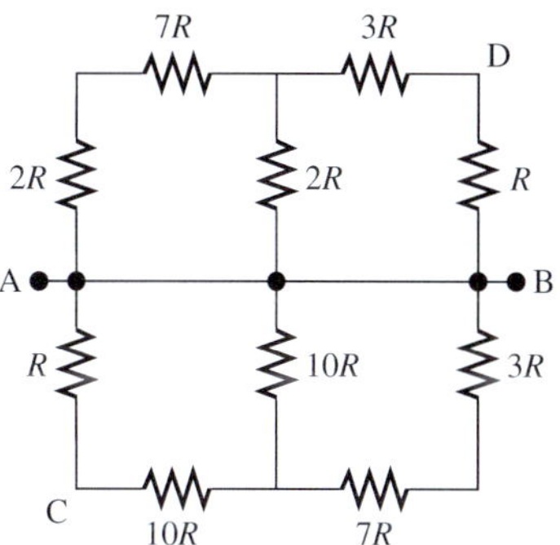

FIGURE P3.11 Resistor network for Problem 3.11.

3.12 (H) **Δ-Y transformation**

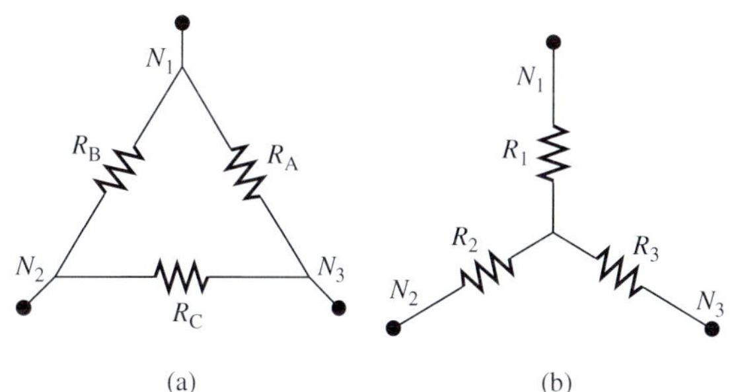

FIGURE P3.12 Δ-Y transformation.

The resistant network consisting of three resistors R_A, R_B, and R_C as shown in Figure P3.12 (a) forms a delta (Δ) connection and the resistors R_1, R_2, and R_3 form a Y connection shown in Figure P3.12 (b). Verify that the two circuits are equivalent provided that:

$$R_1 = R_A R_B/(R_A + R_B + R_C)$$
$$R_2 = R_B R_C/(R_A + R_B + R_C)$$
$$R_3 = R_C R_A/(R_A + R_B + R_C)$$

3.13 (H)* Find R_{AB} in Figure P3.13 using Δ-Y transformation shown in Problem 3.12.

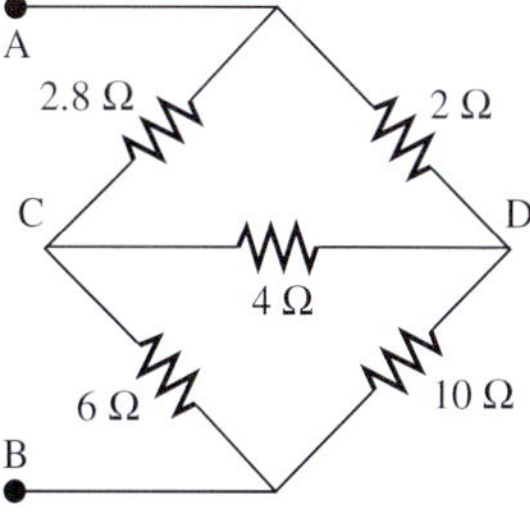

FIGURE P3.13 Bridge circuit.

3.14 (H) Find the resistance seen through the terminals A and B (R_{AB}) in Figure P3.14. The circuit in the box repeats infinitely.

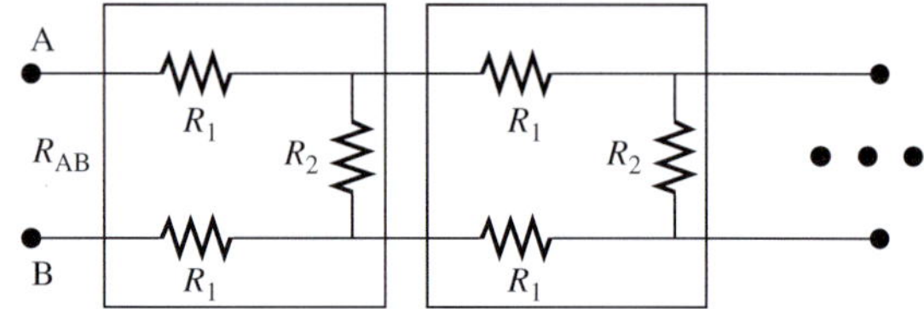

FIGURE P3.14 Circuit for Problem 3.14.

3.15 (H)* Each segment in Figure P3.15 represents a resistor R. Find R_{AB}, that is, the resistance between terminals A and B. (*Hint*: Let a current source I connect terminals A and B, and find the current through each resistor by symmetry of the circuit.)

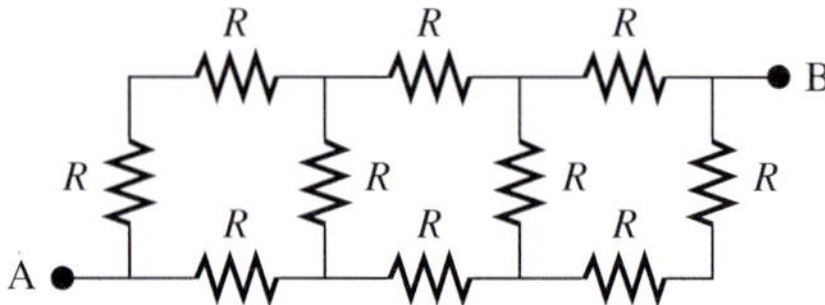

FIGURE P3.15 Circuit for Problem 3.15.

SECTION 3.3 VOLTAGE AND CURRENT DIVISION/DIVIDER RULES

3.16 (B) Find V_1, V_2, V_3 in Figure P3.16.

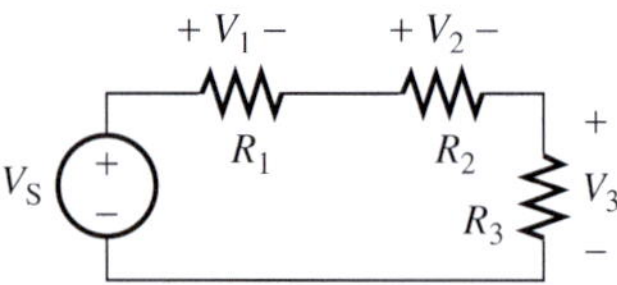

FIGURE P3.16 Circuit for Problem 3.16.

3.17 (A) Find V_s given V in Figure P3.17.

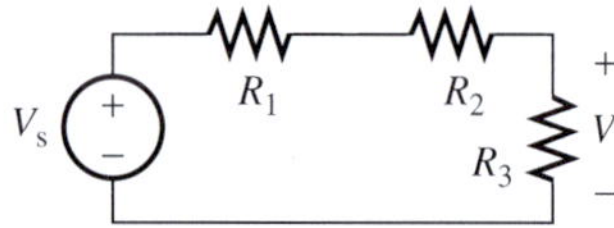

FIGURE P3.17 Circuit for Problem 3.17.

3.18 (A) Find I_1 and I_2 in Figure P3.18.

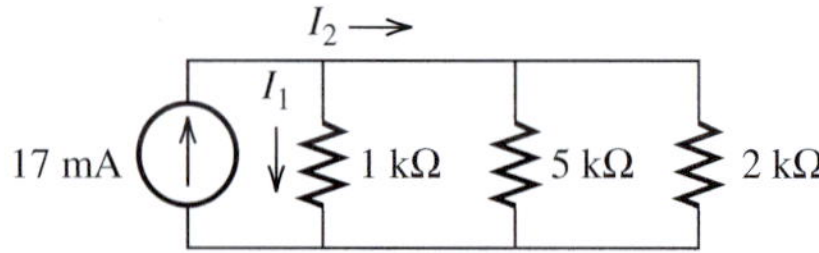

FIGURE P3.18 Circuit for Problem 3.18.

3.19 (A)* Find i_1, i_2, i_3, and i_4 in Figure P3.19 supposing $i_5 = 2$ A.

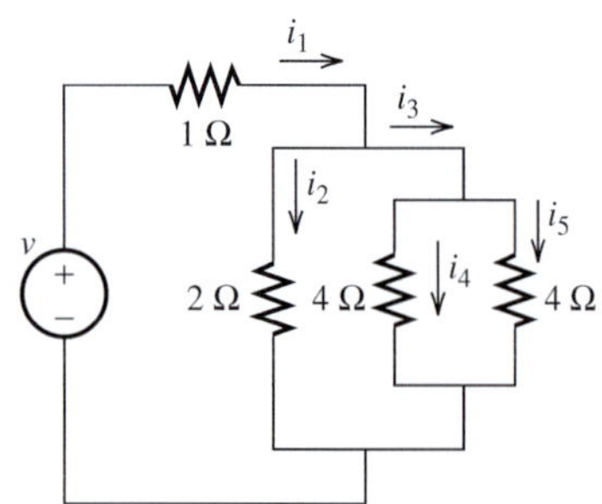

FIGURE P3.19 Circuit for Problem 3.19.

3.20 (H) Find i_2, i_3, i_4, and i_5 in Figure P3.19 assuming $i_1(t) = 4_{\text{cost}}$ A.

3.21 (H) **Current Division in Conductance**
Given a resistor R, we define its conductance G to be $G = 1/R$. Therefore in Figure P3.21, the conductances of R_1 and R_2 are $G_1 = 1/R_1$ and $G_2 = 1/R_2$, respectively. Prove that the current division formulas in terms of conductance are:

$$i_1 = \frac{G_1}{G_1 + G_2} i_s,\ i_2 = \frac{G_2}{G_1 + G_2} i_s$$

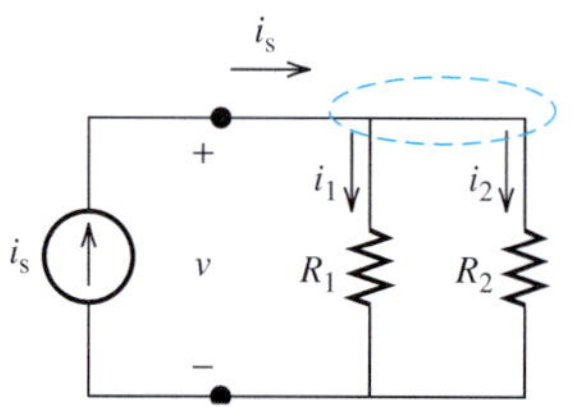

FIGURE P3.21 Circuit for Problem 3.21.

3.22 (A)* Using the current division formulas in Problem 3.21 to recompute the currents through 1 kΩ, 5 kΩ, and 2 kΩ in Figure P3.18.

3.23 (A) Find the voltage drop across all resistors in Figure P3.23.

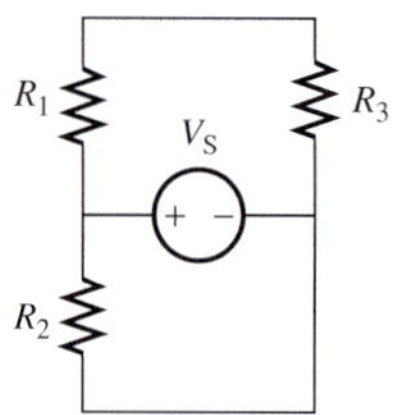

FIGURE P3.23 Circuit for Problem 3.23.

3.24 (A) Given V_o, find R_3 in Figure P3.24.

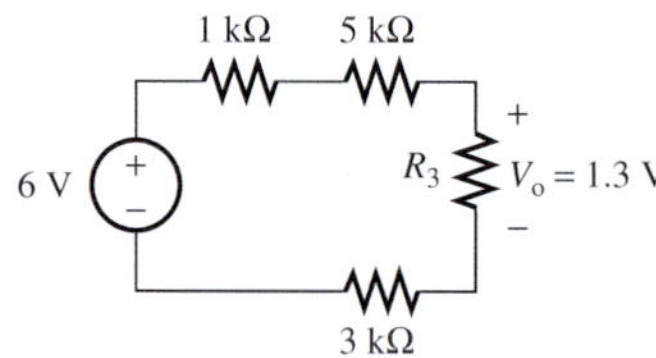

FIGURE P3.24 Circuit for Problem 3.24.

3.25 (H) Find I_1, I_2, and I_3 in Figure P3.25.

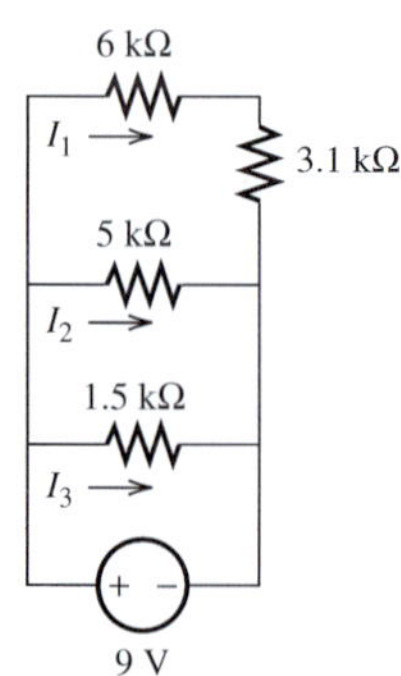

FIGURE P3.25 Circuit for Problem 3.25.

3.26 (H) Assume that you need to power a system from a V_s power supply, but the system requires a voltage of V_{sys}. Use voltage division as shown in Figure P3.26, express R_1 as a function of R_2, V_s, V_{sys}, and R_L. ($V_{sys} < V_s$)

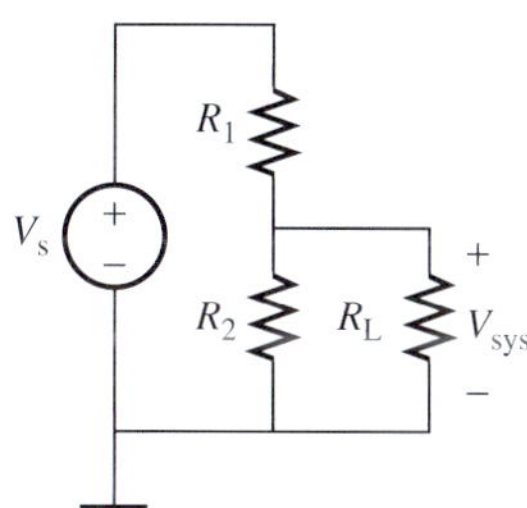

FIGURE P3.26 Circuit for Problem 3.26.

3.27 (A) Find I_1, I_2, I_3, and I_4 in Figure P3.27.

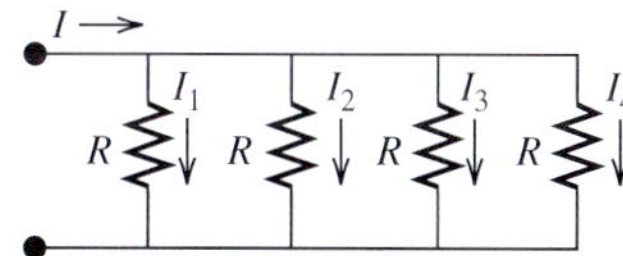

FIGURE P3.27 Circuit for Problem 3.27.

3.28 (H)* Find V_o in Figure P3.28.

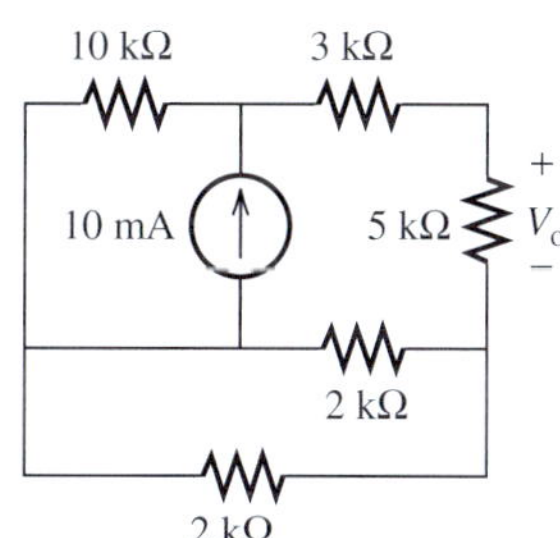

FIGURE P3.28 Circuit for Problem 3.28.

SECTION 3.4 NODAL AND MESH ANALYSIS

3.29 (B) Which node is the best choice for a reference node in the circuit in Figure P3.29?

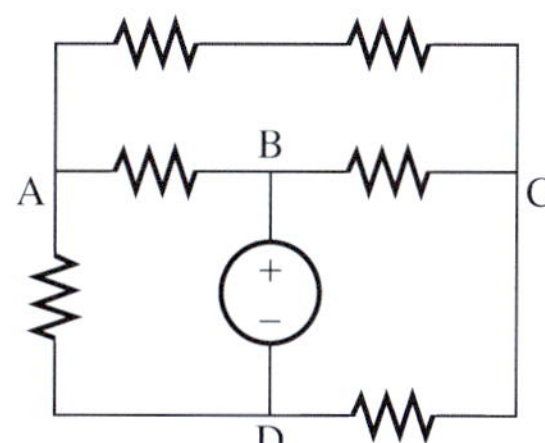

FIGURE P3.29 Circuit for Problem 3.29.

3.30 (A)* Which node is the best choice for a reference node in the circuit in Figure P3.30?

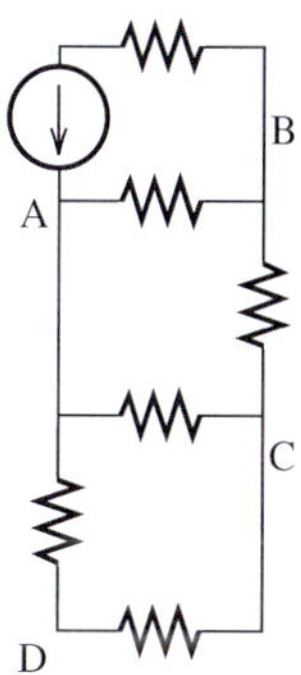

FIGURE P3.30 Circuit for Problem 3.30.

3.31 (A) Use MATLAB® to find V_1, V_2, and V_3 if $2\,V_1 + 2\,V_2 - 4\,V_3 = -2$, $3\,V_1 + 2\,V_2 = 1$, $-2\,V_1 - 3\,V_2 + 5\,V_3 = -3$.

3.32 (A) Use Cramer's rule to find V_1, V_2, and V_3 in Problem 3.31 (MATLAB® command "det" may be used to compute determinant).

3.33 (H) Determine the best choice of node as a reference node for the circuit in Figure P3.33. Then determine the current between B and D.

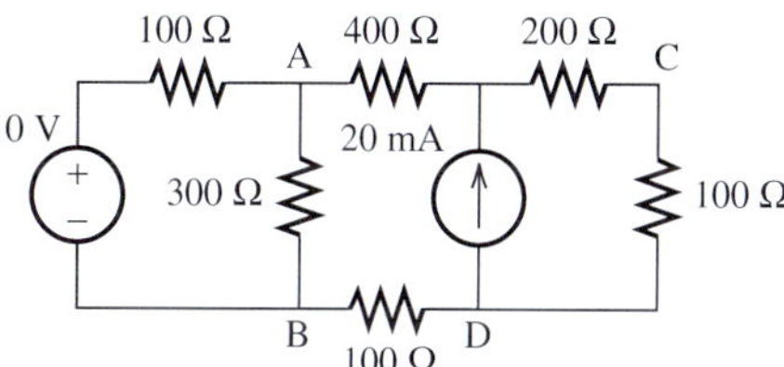

FIGURE P3.33 Circuit for Problem 3.33.

3.34 (B) Find V_1 and V_2 in Figure P3.34.

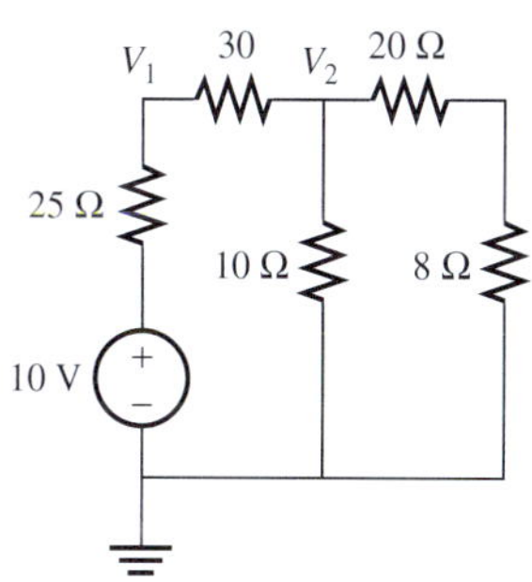

FIGURE P3.34 Circuit for Problem 3.34.

3.35 (B)* Find V_1 and V_2 in Figure P3.35.

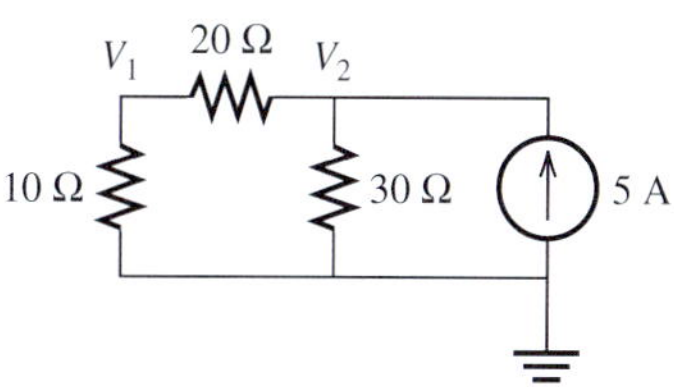

FIGURE P3.35 Circuit for Problem 3.35.

3.36 (A) Find V_1, V_2, and V_3 in Figure P3.36:

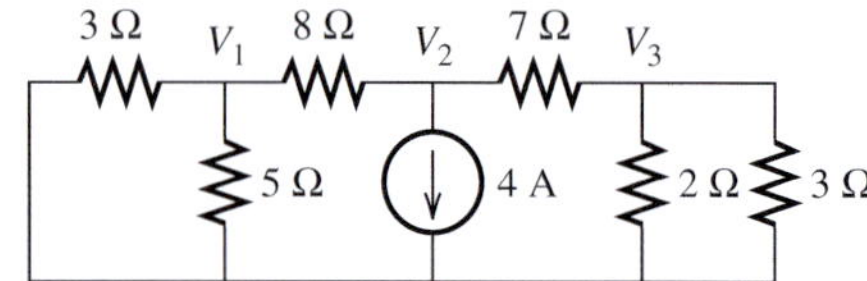

FIGURE P3.36 Circuit for Problem 3.36.

3.37 (A) Find V_1, V_2, and V_3 in Figure P3.37:

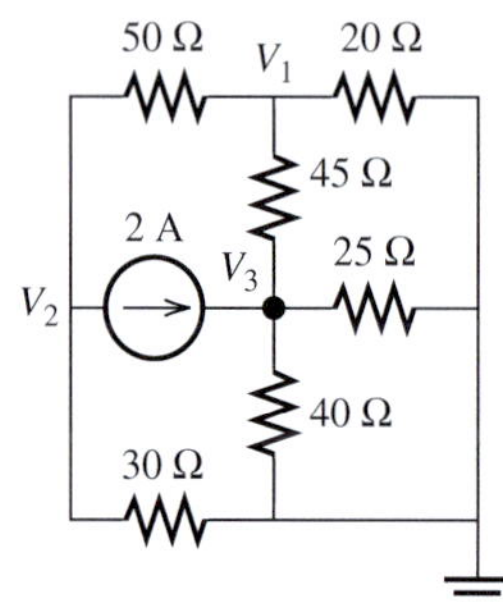

FIGURE P3.37 Circuit for Problem 3.37.

3.38 (A)* Find V_1, V_2, and V_3 in Figure P3.38:

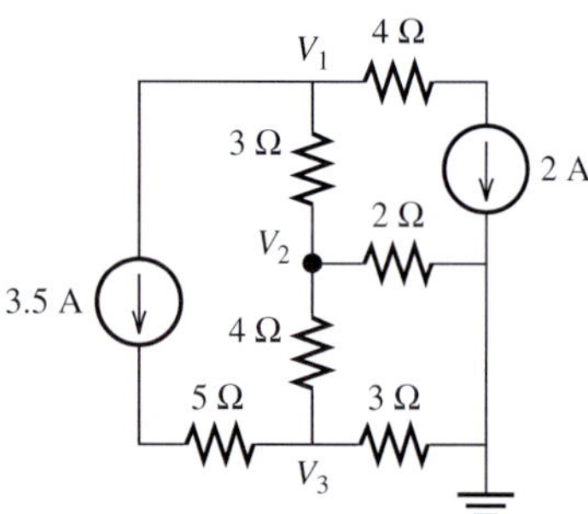

FIGURE P3.38 Circuit for Problem 3.38.

3.39 (H) Find I_x in Figure P3.39.

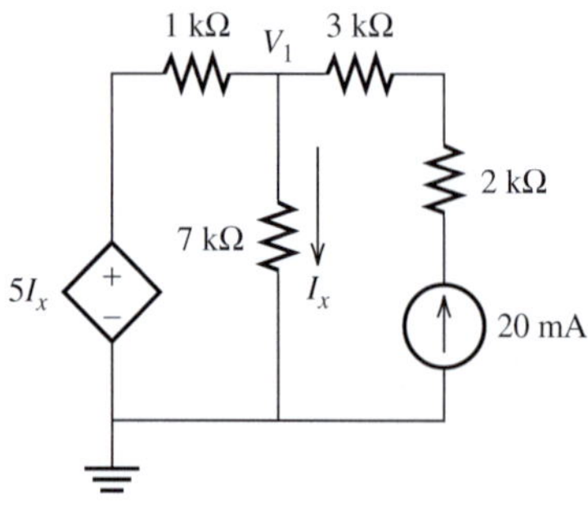

FIGURE P3.39 Circuit for Problem 3.39.

3.40 (H) Find I_x in Figure P3.40.

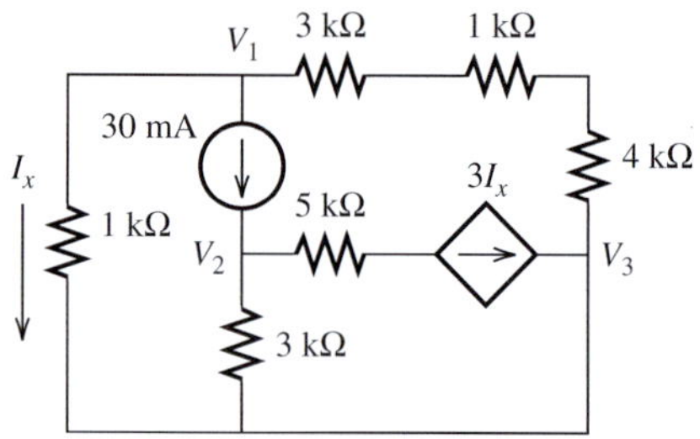

FIGURE P3.40 Circuit for Problem 3.40.

3.41 (H) The circuit shown in Figure P3.41 is a simple bipolar junction transistor amplifier. The portion of the circuit in the shaded box is an approximate T-model of a transistor in the common-emitter configuration.

Use nodal analysis to find the voltage gain V_2/V_1 when $R_B = 80\ \text{k}\Omega$, $R_C = R_E = 2\ \text{k}\Omega$.

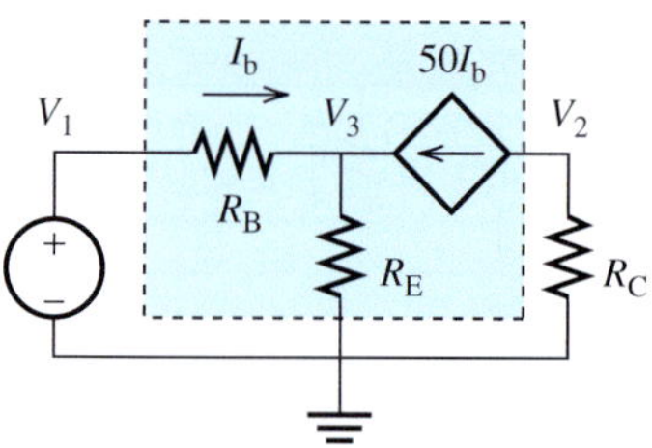

FIGURE P3.41 Circuit for Problem 3.41.

3.42 (B) Identify the meshes and mark the mesh currents in Figure P3.29.

3.43 (B) Identify the meshes and mark the mesh currents in Figure P3.30.

3.44 (A) Find the mesh currents in the circuit shown in Figure 3.44, when $I = 1$ A, $V = 8$ V, $R_1 = 3\ \Omega$, $R_2 = 4\ \Omega$, $R_3 = 20\ \Omega$.

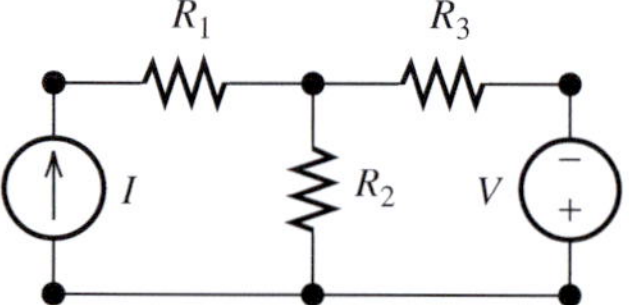

FIGURE P3.44 Circuit for Problem 3.44.

3.45 (A) Use mesh analysis to find the current through the 8-Ω resistor in Figure P3.34.

3.46 (H) Find the mesh currents in Figure P3.46.

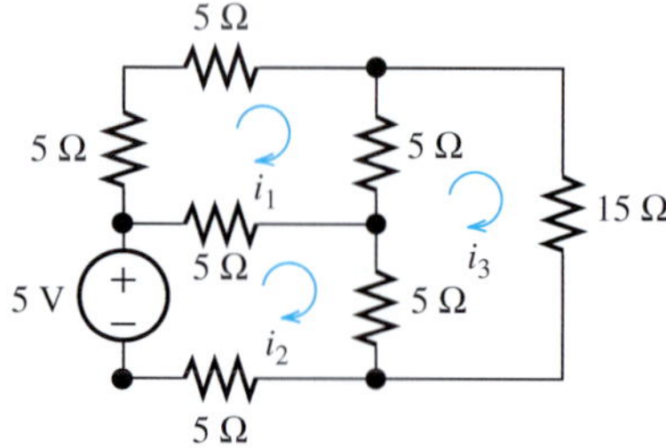

FIGURE P3.46 Circuit for Problem 3.46.

SECTION 3.5 SPECIAL CONDITIONS: SUPER NODE

3.47 (B) Which of the following circuits requires super node analysis?

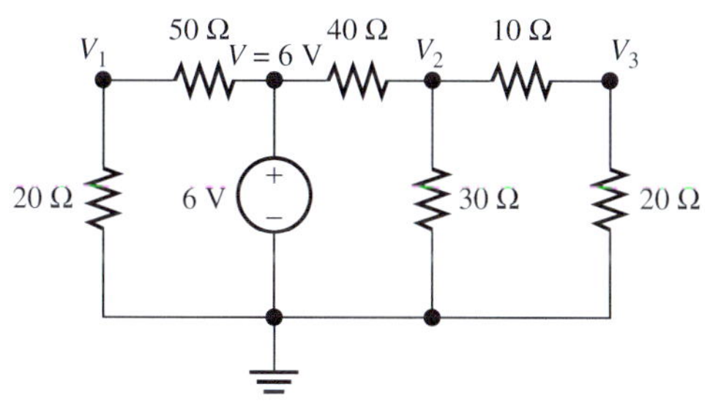

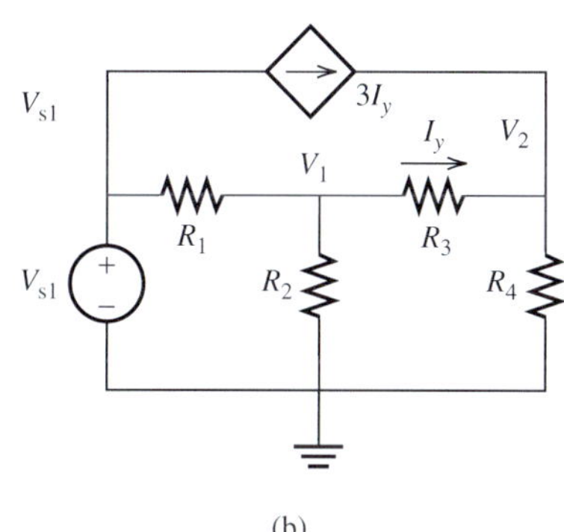

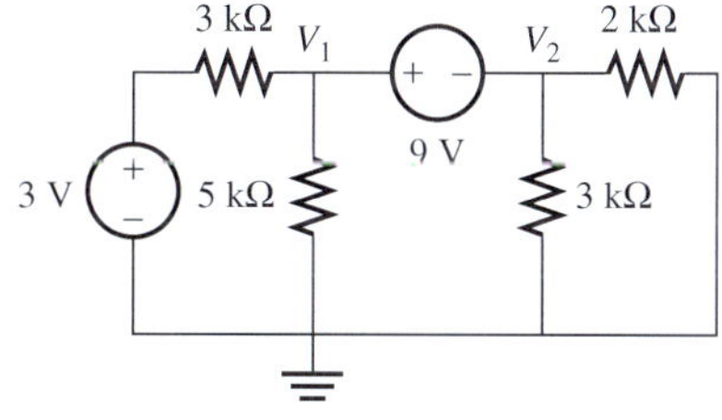

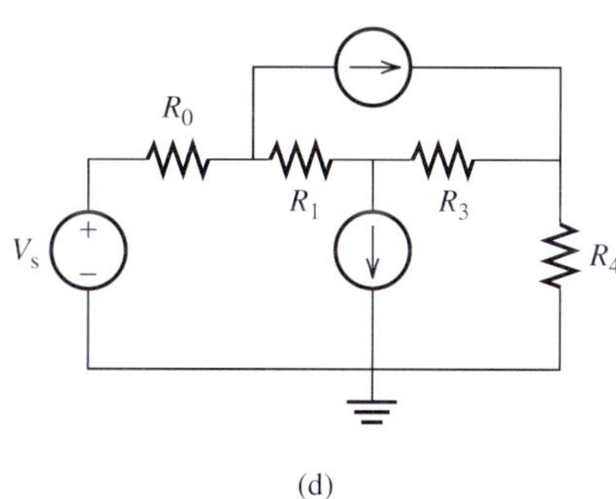

FIGURE P3.47 Circuit for Problem 3.47.

3.48 (A) Under what circumstances is a super node required in nodal analysis?

3.49 (H)* Find V_1, V_2, and V_3 in Figure P3.49:

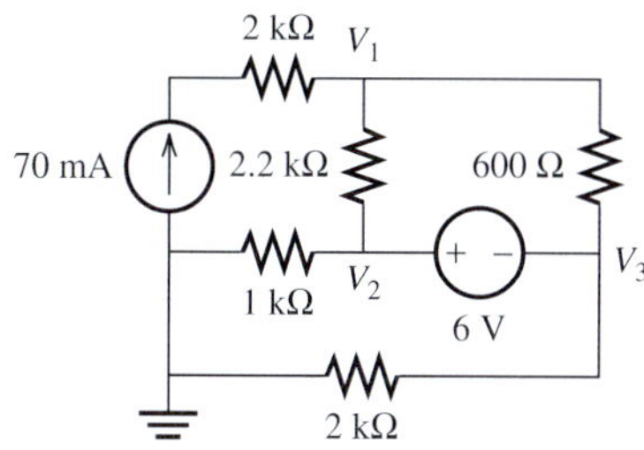

FIGURE P3.49 Circuit for Problem 3.49.

3.50 (H) Find V_1 and V_2 in Figure P3.50:

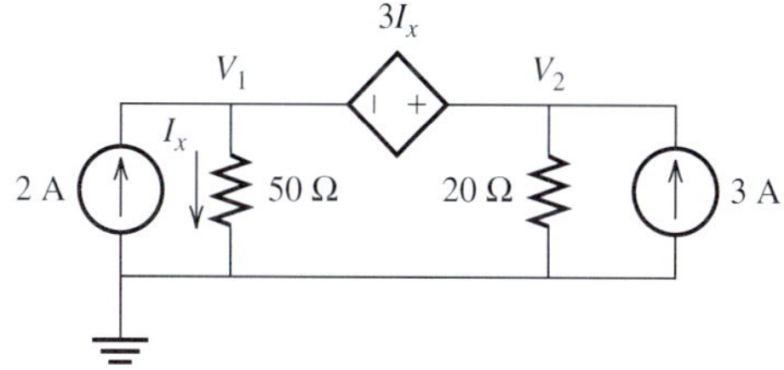

FIGURE P3.50 Circuit for Problem 3.50.

3.51 (H) Find I in Figure P3.51:

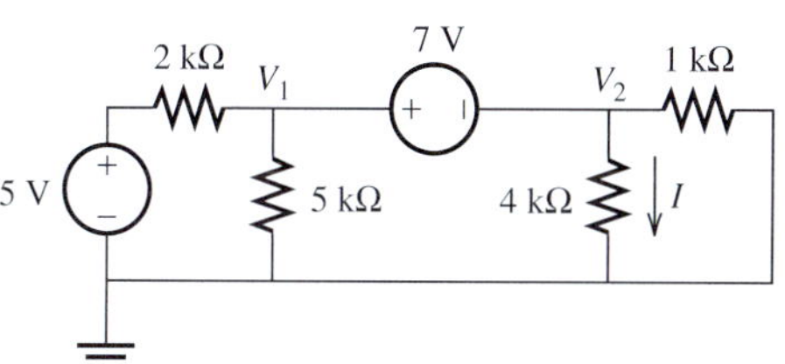

FIGURE P3.51 Circuit for Problem 3.51.

SECTION 3.6 THÉVENIN/NORTON EQUIVALENT CIRCUITS

3.52 (B) Find R_{th} in Figure P3.52:

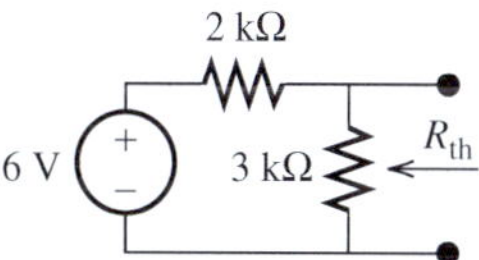

FIGURE P3.52 Circuit for Problem 3.52.

3.53 (B)* Find the Thévenin resistance for the circuit in Figure P3.53:

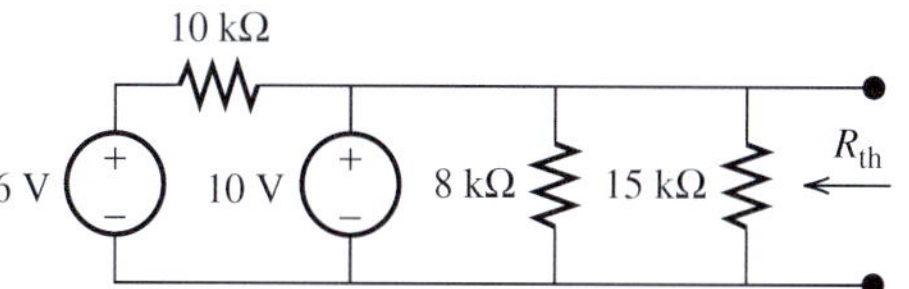

FIGURE P3.53 Circuit for Problem 3.53.

3.54 (A) Find the Thévenin equivalent resistance and voltage as seen by R_L for the circuit in Figure P3.54.

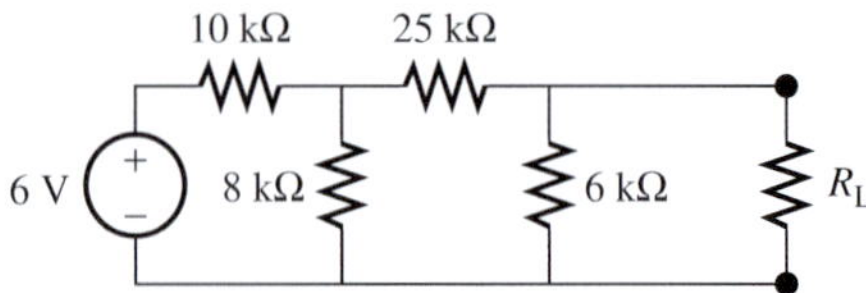

FIGURE P3.54 Circuit for Problem 3.54.

3.55 (A) Find the Norton equivalent as seen by R_L in the circuit in Figure P3.55.

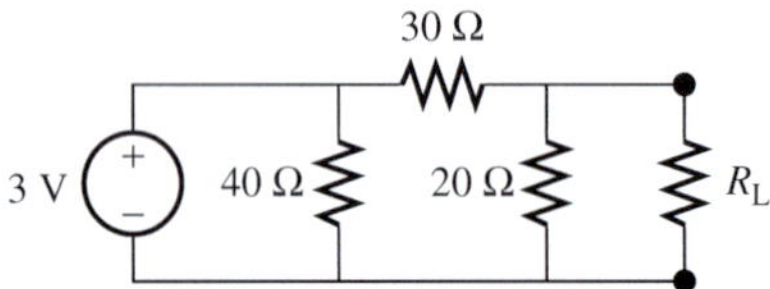

FIGURE P3.55 Circuit for Problem 3.55.

3.56 (A) Find the Thévenin equivalent circuit as seen by R_L in Figure P3.56.

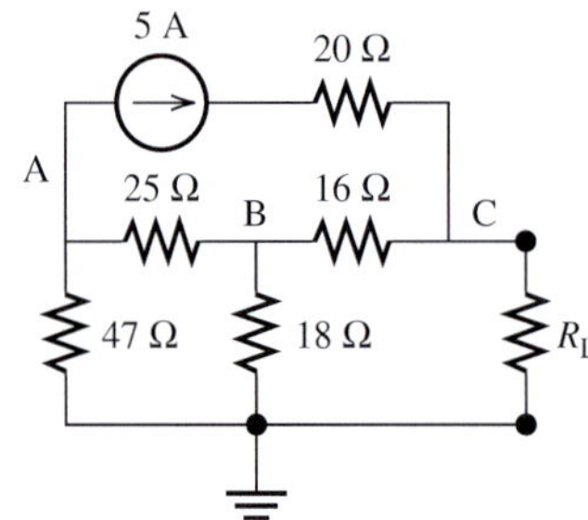

FIGURE P3.56 Circuit for Problem 3.56.

3.57 (A) Find the Thévenin/Norton equivalents for Figure P3.57.

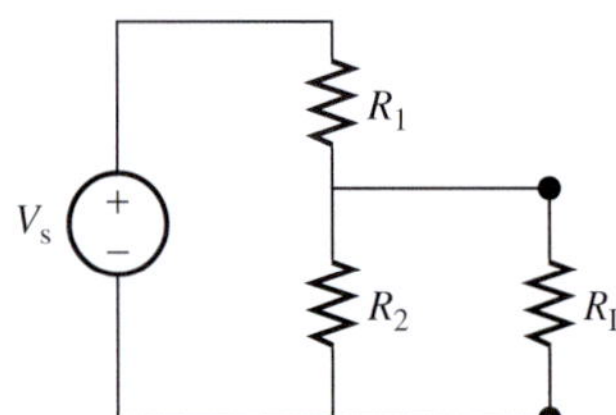

FIGURE P3.57 Circuit for Problem 3.57.

3.58 (H)* Find the Thévenin/Norton equivalents as seen by R_L in Figure P3.58.

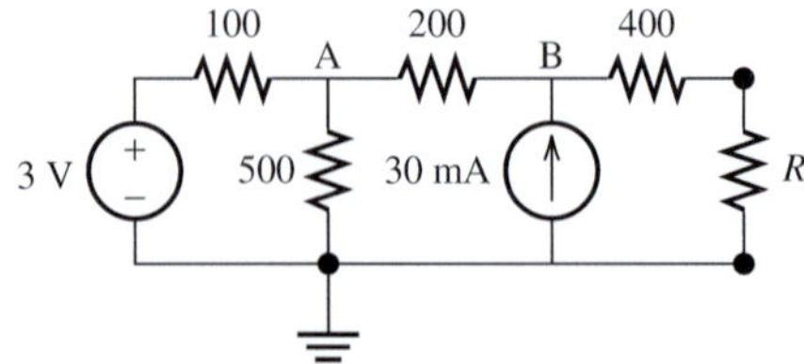

FIGURE P3.58 Circuit for Problem 3.58.

3.59 (H) **Source transformation**
In Figure P3.59(a) and (b) use the methods introduced in Examples 3.34 and 3.35 to calculate the total Thévenin and Norton Circuit observed across the load terminals A and B.

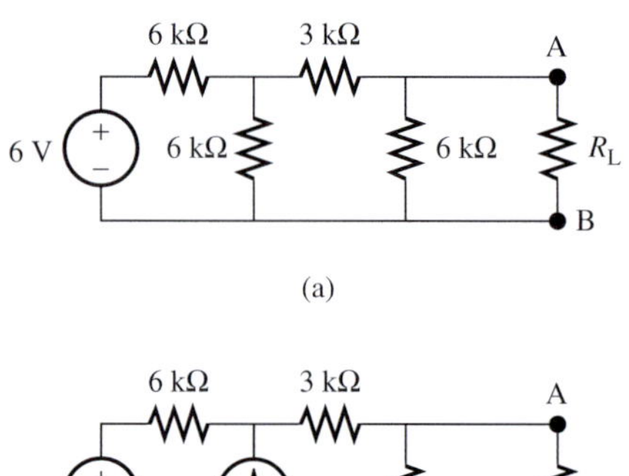

FIGURE P3.59 Circuits of Problem 3.59.

3.60 (A) Repeat Problem 3.52 using source transformation.

3.61 (H) Assume that there are 10 nonideal 1.5-V voltage sources. Each has an internal resistance of 1.5 Ω. Those voltage sources are divided into $10/n$ groups so that n voltage sources are connected in series and the $10/n$ groups are connected in parallel. A 5-Ω external resistor is driven by the voltage source combination. Find the maximum current through the 5-Ω resistors and determine the value of n.

3.62 (H)* In Figure 3.62, $\varepsilon = 100$ V, $\varepsilon' = 40$ V, $R = 10$ Ω, $R' = 30$ Ω. Find the current through R'.

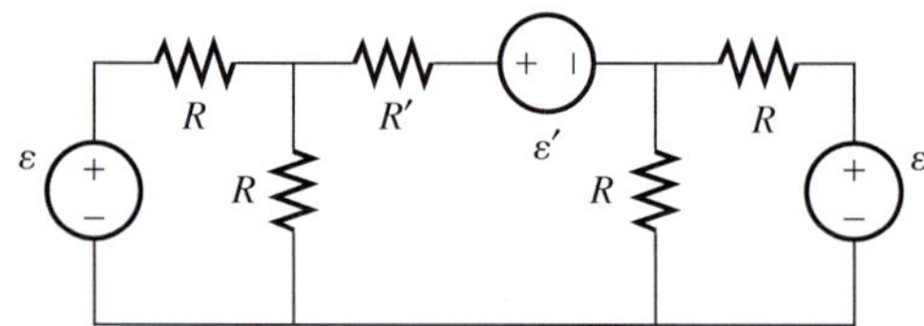

FIGURE P3.62 Circuit for Problem 3.62.

SECTION 3.7 SUPERPOSITION PRINCIPLE

3.63 (B) Find the current I_2 in Figure P3.63 using the superposition technique.

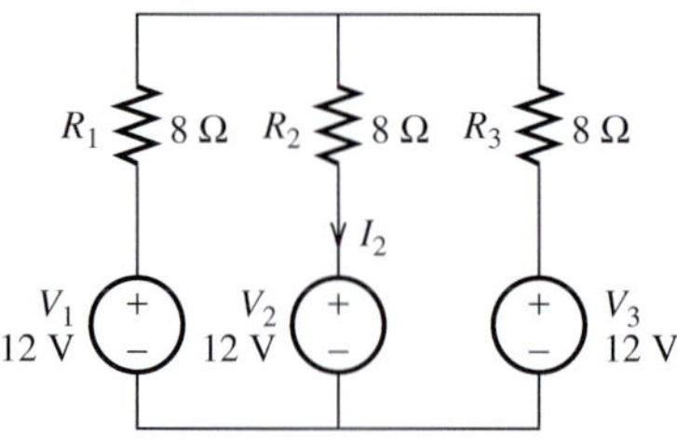

FIGURE P3.63 Circuit for Problem 3.63.

3.64 (A) Find I_x in Figure P3.64 using the principle of superposition.

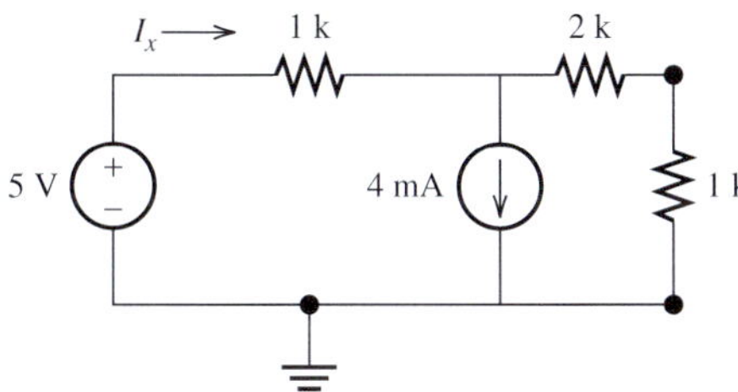

FIGURE P3.64 Circuit for Problem 3.64.

3.65 (A) Use the principle of superposition to find the power dissipated in the 2-kΩ resistor in Figure P3.64.

3.66 (H) If $R_L = 0$, which source in Figure P3.66 contributes the most to the power dissipated in the 500-Ω resistor? Which contributes the least? What is the power dissipated in the 500-Ω resistor? Does it equal the sum of power dissipation contributed by each source?

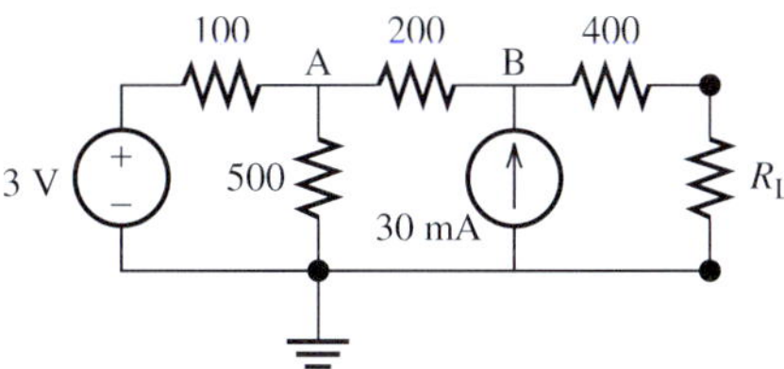

FIGURE P3.66 Circuit for Problem 3.66.

3.67 (A) Repeat Problem 3.33 using the superposition principle.

3.68 (A) Repeat Problem 3.51 using the superposition principle.

3.69 (A)* Find the current I in Figure P3.69 using the superposition principle.

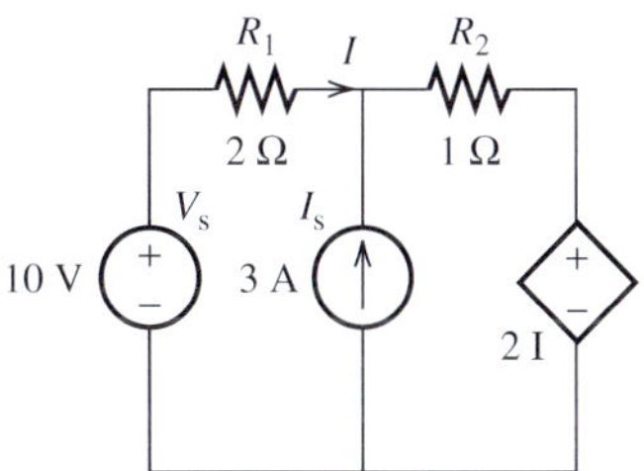

FIGURE P3.69 Circuit for Problem 3.69.

3.70 (A) Repeat Problem 3.50 using the superposition principle.

3.71 (H) When $V_s = 4$ V and $I_s = 2$ A, then $V_x = 8$ V; when $V_s = 7$ V and $I_s = 1$ A, then $V_x = 6$ V. Find the voltage V_x in Figure P3.71 using the superposition principle when $V_s = 2$ V and $I_s = 6$ A.

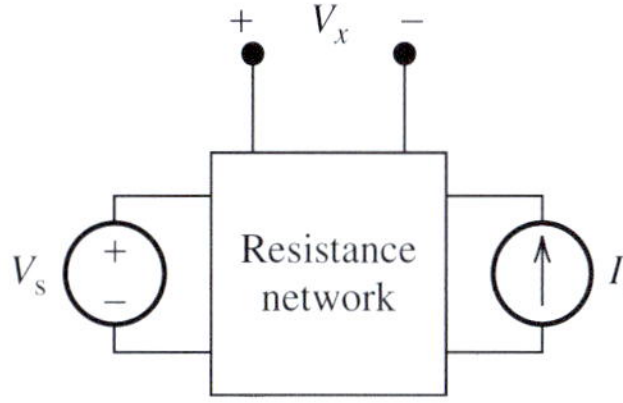

FIGURE P3.71 Circuit for Problem 3.71.

3.72 (H) Using the superposition principle, find the Thévenin equivalence seen by R_L in Figure P3.72.

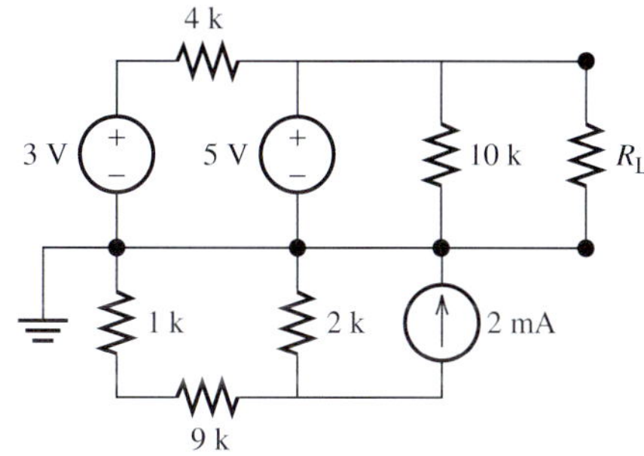

FIGURE P3.72 Circuit for Problem 3.72.

SECTION 3.8 MAXIMUM POWER TRANSFER

3.73 (B) Find the resistance R_1 in Figure P3.73 to transfer maximum power to R_2.

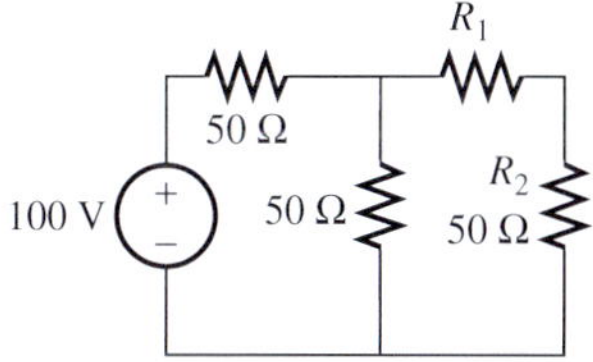

FIGURE P3.73 Circuit for Problem 3.73.

3.74 (A)* Find the value of R where R_L equals 50 Ω and draws the maximum power in Figure P3.74.

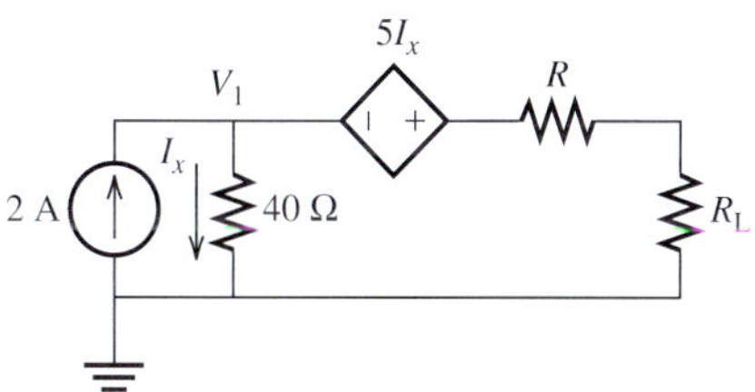

FIGURE P3.74 Circuit for Problem 3.74.

3.75 (A) Determine the maximum power transfer for the circuit shown in Figure P3.75.

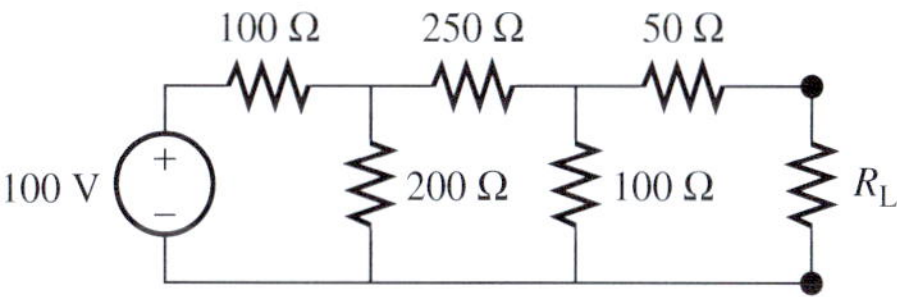

FIGURE P3.75 Circuit for Problem 3.75.

3.76 (A)* In Figure 3.76, find $R_L$which draws the maximum power from the source.

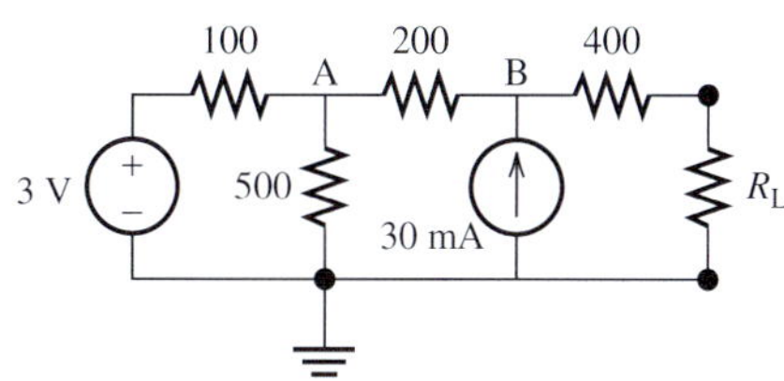

FIGURE P3.76 Circuit for Problem 3.76.

3.77 (H) For the circuit shown in Figure P3.77, draw a graph of the maximum output power of R_L when R_s changes from 1 Ω to 100 Ω. R_L is adjustable.

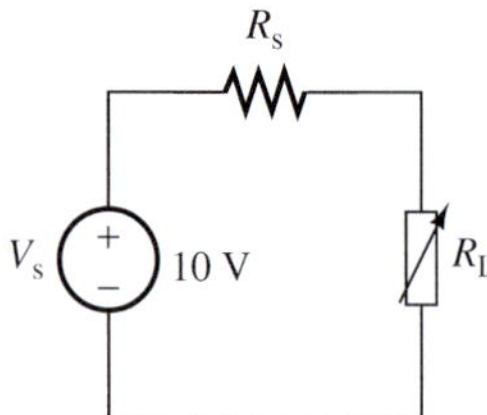

FIGURE P3.77 Circuit for Problem 3.77.

3.78 (H) The light bulb shown in Figure P3.78 has a resistance of $R_0 = 2\ \Omega$. The working voltage $V_0 = 4.5$ V. $V = 6$ V. Find the conditions under which the power efficiency η (defined as the ratio of the power consumed by the bulb and the power provided by the source) is maximum.

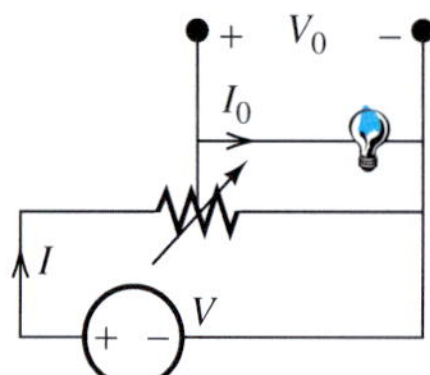

FIGURE P3.78 Circuit for Problem 3.78.

SECTION 3.9 ANALYSIS OF CIRCUITS USING PSPICE

3.79 (B)* Use PSpice to find I_1, I_2, and I_3 in Figure P3.25.

3.80 (B)* Find I_x in Figure P3.39 using PSpice.

3.81 (A) For the circuit shown in Figure 3.81, run the Bias Point simulation analysis for three cases in which the negative terminal of each voltage source is selected as a reference node. Assume $R_1 = 3\ \Omega$, $R_2 = R_3 = R_7 = 2\ \Omega$, $R_4 = 1\ \Omega$, $R_5 = 20\ \Omega$, $R_6 = 1.6\ \Omega$, $R_8 = 8\ \Omega$. (*Hint*: Apply a ground (GND) at the negative terminal of the voltage source that is selected as reference node.)

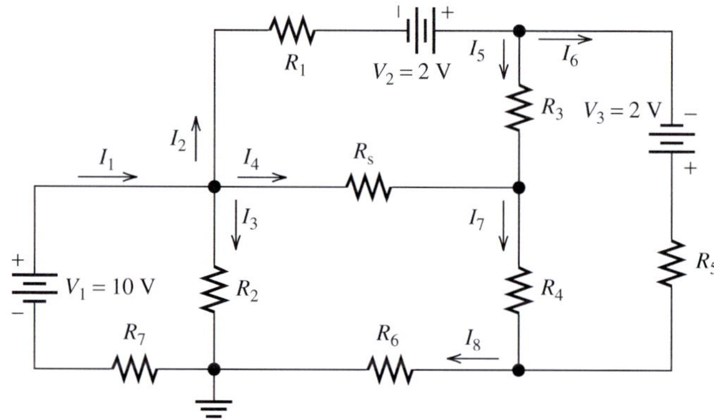

FIGURE P3.81 Circuit of PSpice Problem 3.81.

3.82 (A) Use PSpice to find the Thévenin equivalent resistance and voltage as seen by R_L for the circuit shown in Figure P3.54.

TOPIC 4

Electronic Circuits

The content of this topic is compiled from:

Chapter 8 (pp. 316–394), *Electrical Engineering: Concepts and Applications*

By S.A. Reza Zekavat

CHAPTER 8

Electronic Circuits

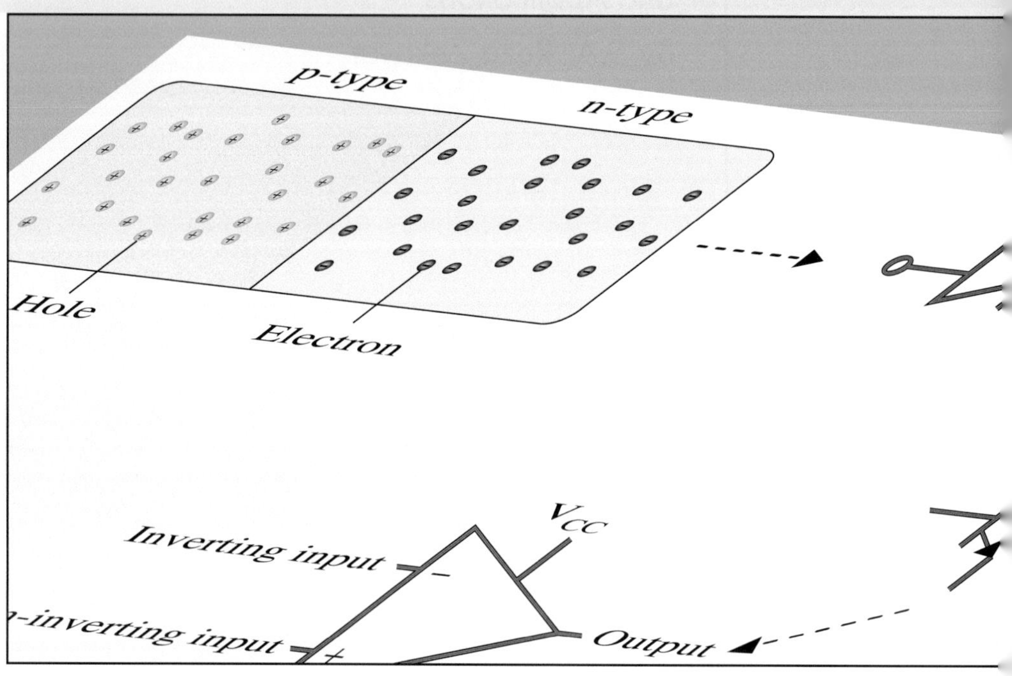

8.1 INTRODUCTION

All electronic systems and circuits, including radio, television, cell phone, and others consist of many electronic components. In general, electronic components are divided into two main categories: passive and active. The nature of the element—active or passive—is determined by the element's energy, $w(t)$, which is defined as:

$$w(t) = \int_{-\infty}^{t} p(\alpha)\, d\alpha = \int_{-\infty}^{t} v(\alpha) \cdot i(\alpha)\, d\alpha$$

Passive components are those that only absorb energy. In this case, the energy $w(t)$ $[v(t) \cdot i(t)]$ is always positive. In contrast, active elements are those that may deliver energy. In this case, their energy is negative. Resistors, capacitors, and inductors are examples of passive elements. Examples of active elements include diodes and transistors.

Active elements are an important part of electronic systems. For example, they are used to make amplifiers. Amplifiers may be used to increase the power of signals; in this case, they are called power amplifiers. Amplifiers may also increase the voltage of signals; in this case, they are called voltage amplifiers. Finally, amplifiers may also be used to increase current; in this case, they are called current amplifiers. Amplifiers are integral parts of many electronic systems, such as two-way radios, cell phones, radios, and television sets. Power-assisted brakes and power steering in automobiles are mechanical examples of amplification.

Chapters 2 to 7 introduced circuits that included only passive elements. This chapter examines circuits that include active elements such as diodes and transistors. This chapter also discusses operational amplifiers (op amps), an important element of circuits. Op amps (made up of transistors) are part of many electronic circuits. For example, op amps are part of electronic amplifiers, which are used to invert signals (i.e., apply 180° phase), or to isolate one part of a circuit from another (this type of system is called a *buffer*). Chapter 7 investigated passive filters. In this chapter, op amps are discussed as an important part of active filters. Active filters are discussed in Chapter 11. Each of the elements described earlier is an essential part of electronic systems; thus, Chapter 8 is titled Electronic Circuits.

Currently, all active elements in a circuit are made of semiconductors. Semiconductors are a group of materials, which possess an electric conductivity that is intermediate between conductors (e.g., metals) and insulators (e.g., dielectrics and plastic). The variability of the electric properties of semiconductors makes these materials a natural choice for investigation of electronic devices.

The conductivity of these materials varies with temperature, illumination, and the semiconductor's impurity content. Semiconductor materials with no impurities are called *pure* or *intrinsic* semiconductors. Applying impurity atoms to semiconductors leads to the formation of two different categories of semiconductors: *p-type* and *n-type*. The category of semiconductor is specified by the nature of the impurity atoms. The process of adding impurity atoms to a pure semiconductor is called doping, which is introduced in Section 8.2.

Most semiconductor devices contain at least one junction between p-type and n-type materials. These p–n junctions are fundamental to the performance of functions, such as rectification, amplification, switching, and other operations in electronic circuits. This chapter discusses examples of these electronic devices and analyzes their operation in electric circuits.

In this chapter, direct current (DC) parameters are denoted by uppercase letters, such as "V" and "I," and AC parameters are denoted by lower case letters, such as "v" and "i."

8.2 P-TYPE AND N-TYPE SEMICONDUCTORS

This section briefly introduces the two categories of semiconductors, p-type and n-type. In addition, this section introduces two popular materials used in the construction of diodes—germanium (Ge) and silicon (Si).

Figure 8.1 shows a simplified two-dimensional representation of an intrinsic or a pure silicon crystal (i.e., does not have impurities). Specifically, Figure 8.1(a) shows the covalent bonds that form the silicon structure. Each silicon atom in the crystal is surrounded by four

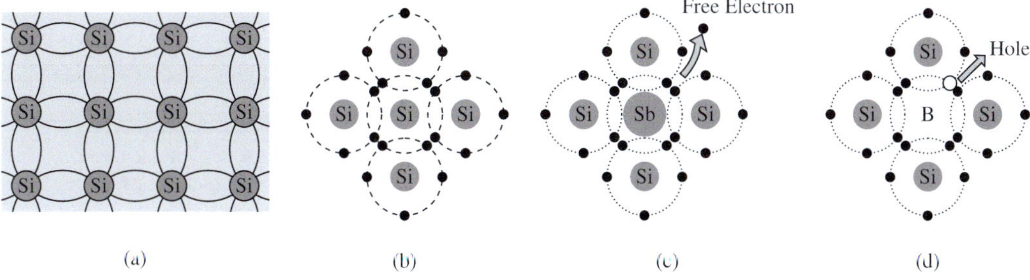

FIGURE 8.1 Crystal structure of silicon. (Courtesy of Wayne Storr, Electronics-Tutorials, www.electronics-tutorials.ws.)

other silicon atoms. Each atom includes four valence electrons (i.e., the electrons in the outermost shell of the atom), which are shared with its neighbors to form a covalent bond. At a temperature of zero kelvin (0 kelvin), Si works as a dielectric. As temperature increases, more electrons are freed and the conductivity increases. While germanium and other materials are also suitable for semiconductors, silicon is by far the most widely used, due to its cost and versatility.

Another technique used to increase the conductivity of semiconductors is *doping,* which is the addition of foreign atoms to the regular crystal lattice of silicon or germanium. The addition of pentavalent impurities (i.e., atoms with five valence electrons) such as antimony, arsenic, or phosphorous, produces n-type semiconductors by contributing extra electrons. Note that silicon can maintain a covalent bond using only four valance electrons of impurity, while the impurity has five electrons. Therefore, free electrons are produced in the n-type semiconductors, as shown in Figure 8.1(c).

Conversely, the addition of trivalent impurities (i.e., atoms with three valence electrons) such as boron, aluminum, or gallium produces p-type semiconductors by contributing extra holes. In other words, only three electrons contribute in forming the covalent bonds. Thus, one valence connection remains electron-less [see Figure 8.1(d)]. In this case, it is said that a *hole* is generated.

It should be noted that in spite of the existence of the free electrons that appear in n-type and the free holes that appear in p-type impure semiconductors, these materials are electrically neutral (i.e., the total number of electrons and protons in these materials are equivalent. However, the addition of impurity increases the conductivity).

Now, suppose that a p-type block of silicon is placed in perfect contact with an n-type block forming a p–n junction shown in Figure 8.2. In this case, free electrons from the n-type region diffuse across the junction to the p-type side where they recombine with some of the many holes in the p-type material. Similarly, holes diffuse across the junction in the opposite direction and recombine, as shown in Figure 8.2. Remember that before connecting p-type and n-type semiconductors, each was neutral. After connecting the two materials and replacing holes and electrons, the material is no longer neutral at the junction.

In addition, the recombination of free electrons and holes in the vicinity of the junction creates a narrow region on either side of the junction that contains no mobile charges. This narrow region, which has been depleted of mobile charge, is called the depletion layer. It extends into both the p-type and n-type regions as shown in Figure 8.3.

The diffusion of holes from the p-type side of the depletion layer generates fixed negative charges (the acceptor ions) in that region. Similarly, fixed positive charges (donor ions) are generated on the n-type side of the depletion layer. There is then a separation of charges: negative fixed charges on the p-type side of the depletion layer and positive fixed charges on the n-type side. Similar to a capacitor, this separation of charges creates an electric field across the depletion layer. Existence of an electric field represents the existence of a voltage difference across the depletion layer. This voltage difference is usually called *built-in* voltage and it varies according to many parameters, including temperature. It is approximately 0.7 for silicon and 0.3 for germanium.

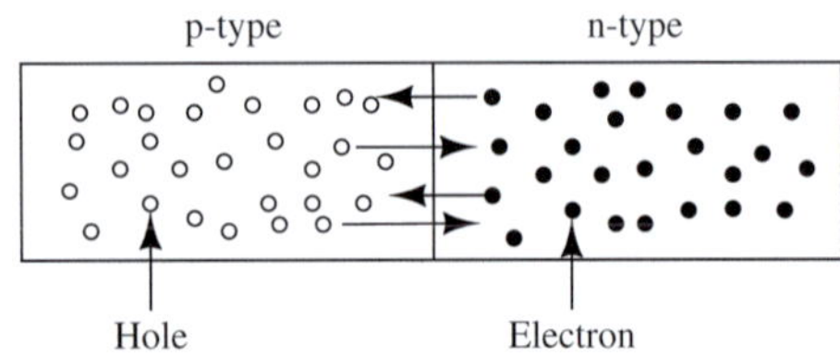

FIGURE 8.2 Electron and hole diffusion in a p–n junction.

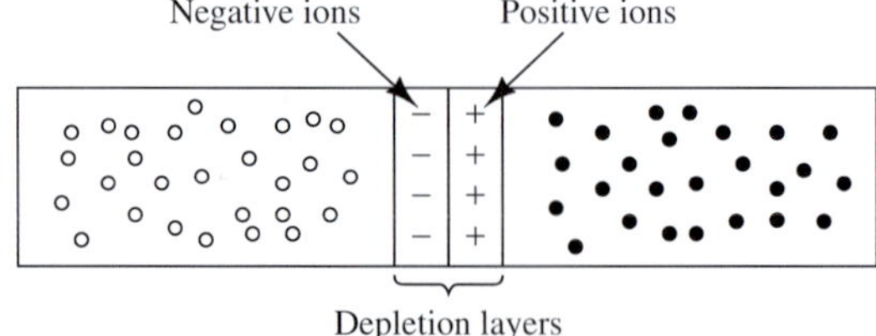

FIGURE 8.3 Depletion layer formation in a p–n junction.

8.3 DIODES

A diode is a p–n junction that allows current to pass in only one direction—from the p-type (called an *anode*) toward the n-type (called a *cathode*) semiconductor. A diode is similar to a "check valve" in a water system that allows water to pass in only one direction. The circuit diagram of a diode is shown in Figure 8.4. Equation (8.1) represents the relationship between voltage across and current through the diode.

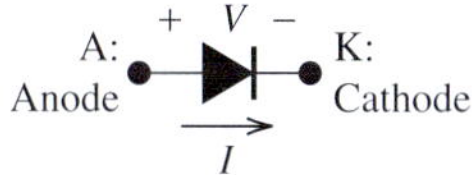

FIGURE 8.4 Circuit diagram of a diode.

$$I = I_o(e^{V/V_T} - 1) \quad \textbf{(8.1)}$$

The current, I_o, in Equation (8.1) is called a reverse saturation current, and the voltage, V_T, is called thermal voltage and corresponds to:

$$V_T = \frac{KT}{q} \quad \textbf{(8.2)}$$

where K is the Boltzmann's constant, T is the absolute temperature in kelvin, and q is the magnitude of the electron charge. The value of V_T at room temperature (i.e., $T = 300$ K) equals 0.0259 V. The reverse saturation current, I_o, is very low (typically in the microampere range). The plot of a diode's characteristics is shown in Figure 8.5.

If the voltage across the diode is negative (i.e., $V < 0$), the diode is said to be in *reverse bias*. The current that passes through the diode in reverse bias is called I_o. The reverse bias occurs when the voltage across the diode is $V < -4V_T$ or, typically, $V < -0.1$ V at room temperature. The direction of the reverse saturation current is from the cathode to the anode. The diode is said to be OFF when the voltage across it is $V < -4V_T$.

The diode is said to be in *forward bias* if its voltage is positive (i.e., $V > 0$) and it starts to pass current as the voltage increases beyond the built-in voltage or the *threshold voltage*. This threshold voltage (V_γ) is shown in Figure 8.5. The threshold voltage varies with many parameters, and it is approximately 0.7 for silicon diodes and 0.3 for germanium diodes.

Instead of using the exact relationship between voltage and current to analyze circuits that contain diodes, approximate models are used. These models simplify the process of analysis of diode circuits. One of these models is the *practical model,* which mimics practical situations. The *I–V* characteristics of the practical model are shown in Figure 8.6. In this figure, the effect of reverse saturation current is ignored. As discussed earlier, this current is very small (on the order of microamperes) and can be easily ignored in many practical problems. Figure 8.7 represents the diode and its voltage and current. In this figure, the voltage on the anode of the diode and cathode are V_1 and V_2, respectively. Therefore, the voltage across the diode corresponds to: $V = V_1 - V_2$.

Accordingly, the current–voltage (*I–V*) model of Figure 8.6 is described as follows:

1. $V < 0 \rightarrow$ Reverse bias and the diode is OFF $\rightarrow I = 0$
2. $0 < V < V_\gamma \rightarrow$ Forward bias and the diode is OFF $\rightarrow I = 0$
3. $V > V_\gamma \rightarrow$ Forward bias and the diode is ON $\rightarrow V = V_\gamma$

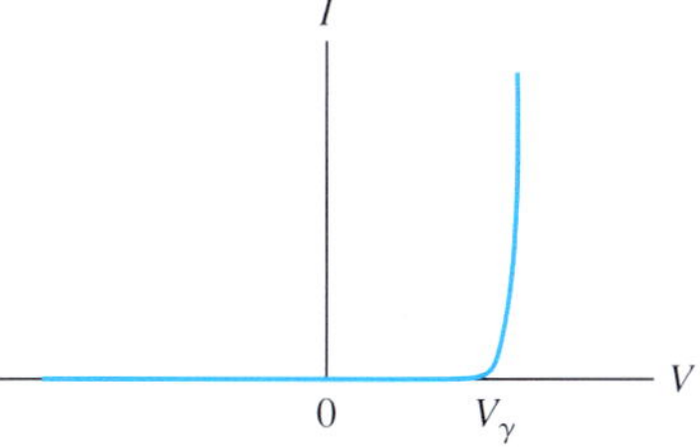

FIGURE 8.5 Diode characteristics: the relationship between voltage and current of a diode.

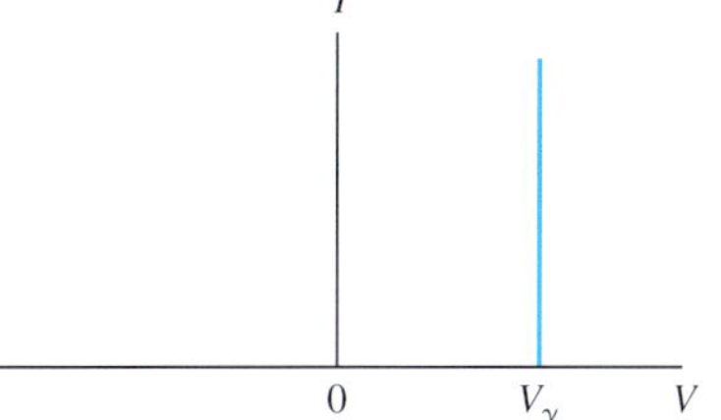

FIGURE 8.6 Diode characteristics: practical model.

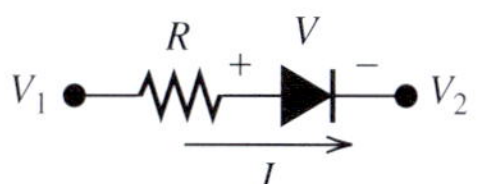

FIGURE 8.7 A simple diode circuit.

Note:

All diodes used in this chapter are assumed to be silicon diodes.

EXAMPLE 8.1 Current and Voltage of a Diode

For the circuit shown in Figure 8.8, find I and V when:

a. $V_S = -10$ V
b. $V_S = -1$ V
c. $V_S = 0.5$ V
d. $V_S = 8$ V

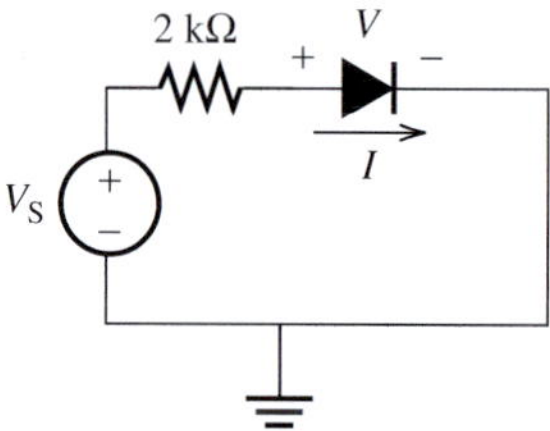

FIGURE 8.8 Circuit for Example 8.1.

SOLUTION

a. Referring to Figure 8.7, $V_S = -10$ V means $V_1 - V_2 = -10$ V. Therefore, the diode is reverse biased (OFF) and the current flowing through it $I = 0$ and the voltage across the diode $V = -10$ V.

b. Similar to part a, $I = 0$ and $V = -1$ V.

c. For the silicon diode, $V_\gamma = 0.7$

$$0 < V_1 - V_2 = V_S = 0.5 < V_\gamma = 0.7$$

Therefore, the diode is forward biased but it is OFF because there is no sufficient voltage to turn it ON. Therefore, $I = 0$ and $V = 0.5$ V.

d. Next, $V_1 - V_2 = V_S = 8\text{ V} > V_\gamma = 0.7$.
Therefore, the diode is forward biased and it is ON. Accordingly:

$$V = 0.7 \text{ V}$$

$$I = \frac{V_S - V_\gamma}{R} = \frac{8 - 0.7}{2} = 3.65 \text{ mA}$$

EXAMPLE 8.2 The Current of Diode

Find the current, I, flowing through the diode of the circuit shown in Figure 8.9 when:

a. $R = 1$ kΩ
b. $R = 10$ kΩ

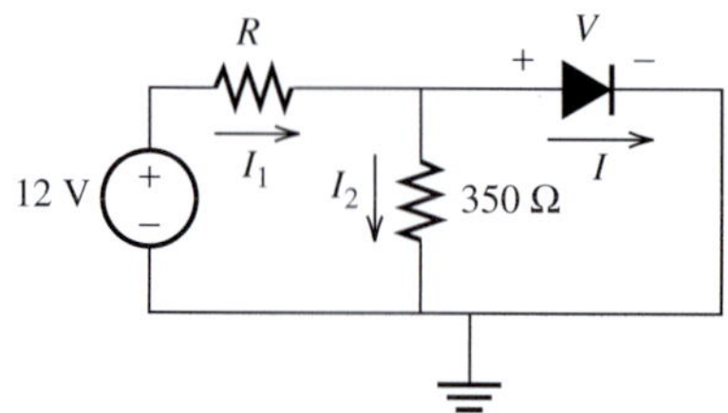

FIGURE 8.9 Circuit for Example 8.2.

SOLUTION

This problem can be solved using either of two methods.

Method 1: Because the voltage available to directly drive the diode is unknown (i.e., $V_1 - V_2$ of Figure 8.7), an assumption must be made of the status of the diode (ON or OFF). This assumption is used to solve the problem, and then the validity of the assumption can be checked by observing the direction of the diode current. For example, when assuming the diode is ON, the assumption will be valid if the current through the diode flows from the anode to the cathode, because that situation is consistent with an ON diode.

a. $R = 1\ \text{k}\Omega$

Assume the diode is ON. Therefore, the voltage across it is:

$$V = V_\gamma = 0.7\ \text{V}$$

Accordingly, the voltage across the 350-Ω resistor is 0.7 V and the current through it corresponds to:

$$I_2 = \frac{0.7}{350} = 2\ \text{mA}$$

The current through R corresponds to:

$$I_1 = \frac{12 - 0.7}{1} = 11.3\ \text{mA}$$

Applying KCL, the current through the diode corresponds to:

$$I = I_1 - I_2 = 11.3 - 2 = 9.3\ \text{mA}$$

This current flows through the diode from anode to the cathode, which confirms the validity of the assumption.

b. $R = 10\ \text{k}\Omega$

Again, assume the diode is ON. Therefore, the voltage across it is:

$$V = V_\gamma = 0.7\ \text{V}$$

Accordingly, the voltage across the 350-Ω resistor is 0.7 V and the current through it corresponds to:

$$I_2 = \frac{0.7}{350} = 2\ \text{mA}$$

The current through R corresponds to:

$$I_1 = \frac{12 - 0.7}{10} = 1.13\ \text{mA}$$

Applying KCL, the current through the diode corresponds to:

$$I = I_1 - I_2 = 1.13 - 2 = -0.87\ \text{mA}$$

The negative sign means that the current flows in an opposite direction to that shown in Figure 8.9 (i.e., from cathode to anode). The diode passes current only in one direction (i.e., from anode to cathode). Thus, the assumption of the diode being ON is not valid.

Now, consider another assumption: the diode is OFF. In this case: $I = 0$. Therefore, the currents I_1 and I_2 are equal and are given by:

$$I_1 = I_2 = \frac{12}{10 + 0.35} = 1.1594\ \text{mA}$$

(*continued*)

EXAMPLE 8.2 Continued

The voltage across the diode corresponds to:

$$V = 1.1594 \times 0.35 = 0.41 \text{ V}$$

It is notable that the voltage across the diode is less than V_γ. This confirms that the diode is OFF.

Method 2: For simplicity, find the Thévenin equivalent circuit between the diode terminals.

a. $R = 1 \text{ k}\Omega$

The Thévenin voltage corresponds to:

$$V_{\text{th}} = 12 \times \frac{0.35}{1 + 0.35} = 3.11 \text{ V}$$

In addition, the Thévenin resistance corresponds to:

$$R_{\text{th}} = \frac{1 \times 0.35}{1 + 0.35} = 0.2593 \text{ k}\Omega$$

The Thévenin equivalent circuit is shown in Figure 8.10. Because 3.11 V is greater than V_γ, the diode is ON and the current through it corresponds to:

$$I = \frac{3.11 - 0.7}{259.3} = 9.3 \text{ mA}$$

b. $R = 10 \text{ k}\Omega$

The Thévenin voltage corresponds to:

$$V_{\text{th}} = 12\frac{0.35}{10 + 0.35} = 0.41 \text{ V}$$

In addition, the Thévenin resistance is:

$$R_{\text{th}} = \frac{1 \times 0.35}{10 + 0.35} = 0.3382 \text{ k}\Omega$$

The Thévenin equivalent circuit is shown in Figure 8.11.
Because 0.41 V is less than $V_\gamma = 0.7$ V, the diode is OFF and no current flows through it. Thus, $I = 0$ and $V = 0.41$ V.

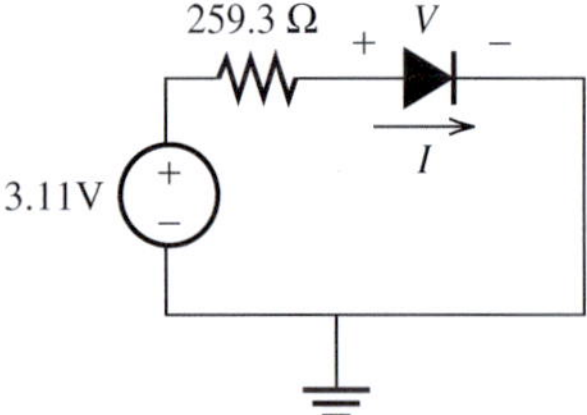

FIGURE 8.10 Thévenin equivalent of the circuit of Figure 8.9 with $R = 1 \text{ k}\Omega$.

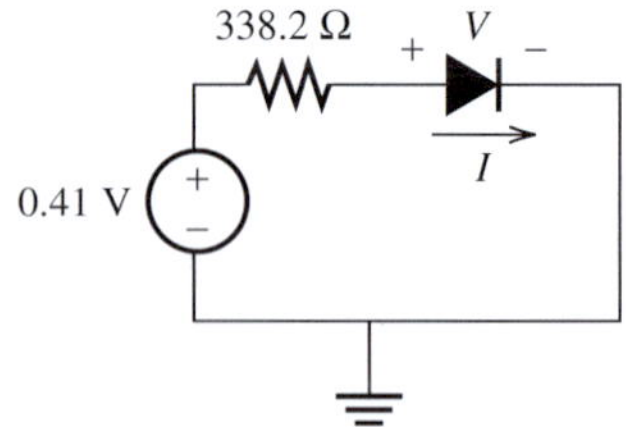

FIGURE 8.11 Thévenin equivalent of the circuit of Figure 8.9 with $R = 1 \text{ k}\Omega$.

EXERCISE 8.1

In Figure 8.9, find R to maintain 0.7 V across the diode. Find the diode current in this case.

8.3.1 Diode Applications

In some cases, diodes (or their circuits) are designed for applications such as rectification, capacitance [e.g., voltage controlled oscillators (VCOs) used in communication systems], tunneling, and light emission and detection. This section focuses only on rectifier diodes and some of their applications. As discussed in the previous section, a diode allows current flow in only one direction. Accordingly, the diode is usually called a rectifier diode or simply, a rectifier.

8.3.1.1 HALF-WAVE RECTIFIER

Consider the circuit shown in Figure 8.12. Here, the input voltage is a sinusoidal waveform with the peak voltage $V_m > 0.7$. The goal is to evaluate the voltage across the load resistance, R_L, or the output voltage, $v_o(t)$. The input voltage changes continuously between $-V_m$ and V_m. Therefore, the diode status varies between OFF and ON.

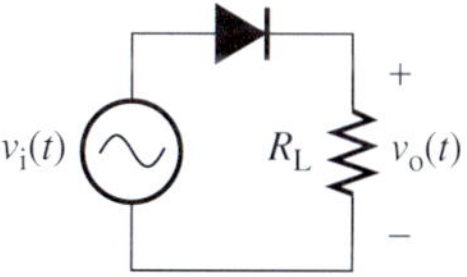

FIGURE 8.12 A half-wave rectifier.

When the value of the input voltage is less than 0.7, the diode is OFF and the current flowing through it is $i = 0$. Therefore, the output voltage corresponds to:

$$v_o(t) = i \times R = 0 \tag{8.3}$$

As the value of the input voltage increases beyond 0.7, the current starts to flow and the voltage drop across the diode is 0.7 V. Therefore, applying KVL to the loop, the output voltage corresponds to:

$$v_o(t) = v_i(t) - 0.7 \tag{8.4}$$

The transfer characteristic (i.e., the output voltage versus the input voltage) is shown in Figure 8.13(a). Plots of the input and output voltage waveforms are shown in Figures 8.13(b) and 8.13(c), respectively. Note that the maximum value of the output voltage is 0.7 V less than that of the input voltage. While the input voltage has positive and negative values, the output voltage is always positive. Therefore, the current flowing through the load resistance, $i = v_o/R$, is always positive. In other words, the current passes through the load in only one direction. Therefore, the term *rectifier* is used. The term *half wave* is used because the current through the load exists only during the positive half cycle.

8.3.1.2 BRIDGE RECTIFIER

The *bridge* rectifier circuit is shown in Figure 8.14. A bridge rectifier allows the current to pass through the load resistance in one direction during the two half cycles. The positive sign above the AC source indicates that the voltage of the upper terminal is positive with respect to the bottom terminal during the positive half cycle. Now, suppose the input voltage of the circuit shown in Figure 8.14 is a sinusoidal waveform with a peak value $V_m > 1.4$ V. During the positive half cycle, diodes D_1 and D_3 are forward biased and diodes D_2 and D_4 are reverse biased. Because diodes D_2 and D_4 are reverse biased, the current flowing through them is zero and each can be replaced with an open circuit, as shown in Figure 8.15.

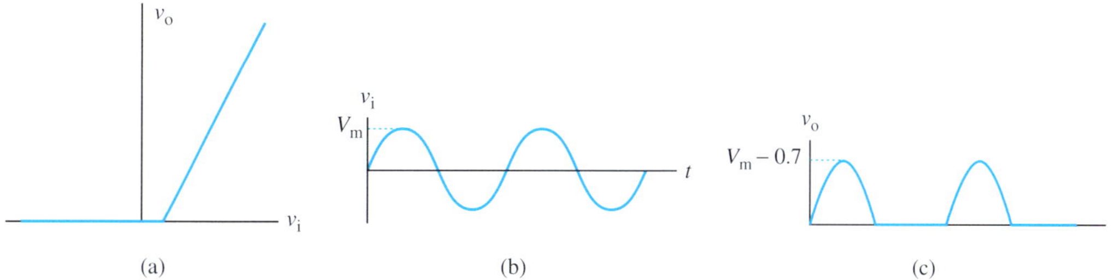

FIGURE 8.13 (a) Transfer characteristics for the circuit shown in Figure 8.12; (b) the input voltage; (c) the output voltage.

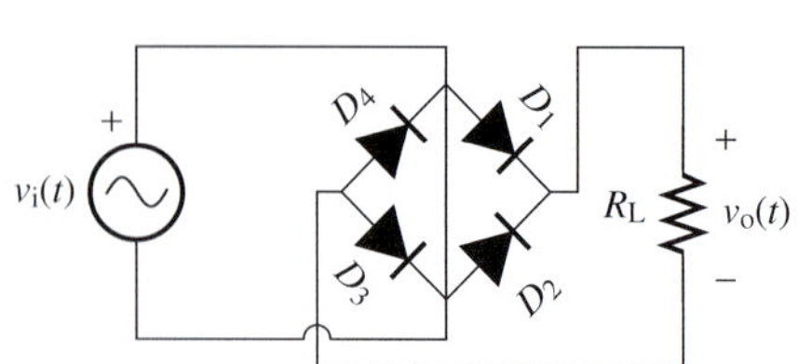

FIGURE 8.14 A bridge rectifier.

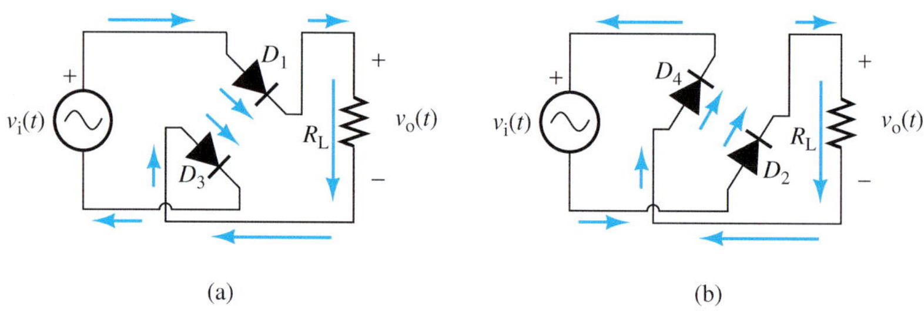

FIGURE 8.15 The equivalent circuit during (a) the positive half cycle; (b) the negative half cycle.

To turn both diodes of the circuit shown in Figure 8.15 ON, the value of the input voltage must be greater than 1.4 V (0.7 V is required for each diode). Thus, when the value of the input voltage increases beyond 1.4 V, D_1 and D_3 turn ON and the current will pass in the direction shown in Figure 8.15(a). Because the current passes through the load resistance from the top to the bottom, the output voltage has a positive value and corresponds to (consider a KVL through the loop):

$$v_o(t) = v_i(t) - 1.4 \tag{8.5}$$

When the value of the input voltage is less than 1.4 V, both diodes will turn OFF and the current flowing through the circuit will be zero. Therefore, the output voltage $v_o(t) = 0$.

During the negative half cycle, diodes D_2 and D_4 are forward biased and diodes D_1 and D_3 are reverse biased. Similarly, because diodes D_1 and D_3 are reverse biased, each can be replaced with an open circuit as shown in Figure 8.15(b). The magnitude of the input voltage value must be greater than 1.4 V to turn both diodes ON in the circuit. In other words, the input voltage must be less than −1.4 V to enable the current to pass in the direction shown in Figure 8.15(b). In this case, the current passes through the load resistance from the top to the bottom in the same direction as in the case of positive half cycle. Referring to the circuit shown in Figure 8.15(b), and writing a simple KVL, the output voltage corresponds to:

$$v_o(t) = -v_i(t) + 1.4 \tag{8.6}$$

When the value of the input voltage is greater than −1.4 V, both diodes turn OFF and the current flowing through the circuit will be zero. Therefore, the output voltage $v_o(t) = 0$. Plots of transfer characteristics, input voltage, and output voltage are shown in Figure 8.16.

8.3.1.3 DIODE LIMITERS

A *limiter* is a device that limits a waveform from exceeding a specified value. Limiters have applications in wave shaping and circuit protection. Many circuits cannot tolerate voltages beyond a limit; limiters protect circuits from voltages that could harm the circuit. The circuit of a limiter is similar to that of a rectifier; however, a limiter's circuit usually includes a constant (independent) voltage source as well. This section discusses examples of diode limiters, and explains how to plot their transfer characteristics as well as their input and output voltage waveforms.

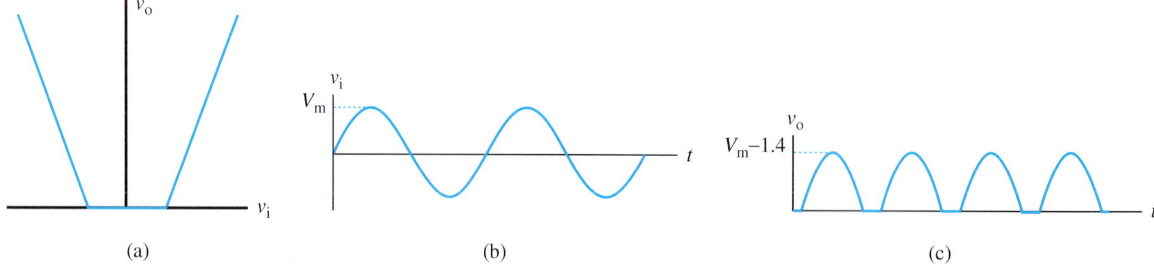

FIGURE 8.16 (a) Transfer characteristics for the circuit of Figure 8.14; (b) its input voltage; (c) its output voltage.

EXAMPLE 8.3 One-Sided Limiter

Find the output voltage of the circuit shown in Figure 8.17.

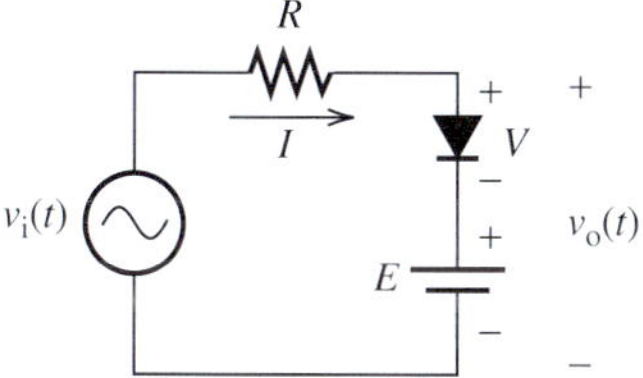

FIGURE 8.17 The limiter circuit for Example 8.3.

SOLUTION

The output voltage can be calculated by applying *KVL*. If the voltage across the diode is known, the output voltage will correspond to:

$$v_o(t) = V + E \tag{8.7}$$

On the other hand, if the current flowing through the diode is known, the output voltage will correspond to:

$$v_o(t) = v_i(t) - I \times R \tag{8.8}$$

The knowledge of v or I depends on whether the diode is ON or OFF, respectively. If the diode is ON, the voltage across the diode $v = 0.7$, and based on Equation (8.7):

$$v_o(t) = 0.7 + E \tag{8.9}$$

If the diode is OFF, the current flowing through the diode $I = 0$, and based on Equation (8.8):

$$v_o(t) = v_i(t) \tag{8.10}$$

If the voltage difference between $v_i(t)$ and E exceeds 0.7 V, the diode will turn on. In other words, to turn the diode on, the following must be true:

$$v_i(t) > E + 0.7 \tag{8.11}$$

Accordingly, the output voltage corresponds to Equation (8.9). On the contrary, the diode will be turned off if:

$$v_i(t) < E + 0.7 \tag{8.12}$$

In this case, the output voltage will correspond to Equation (8.10).

As a summary, the output voltage is given by:

$$v_o(t) = \begin{cases} E + 0.7 & v_i(t) > E + 0.7 \\ v_i & v_i(t) < E + 0.7 \end{cases} \tag{8.13}$$

Plots of the transfer characteristic, input voltage, and output voltage for the limiter circuit of Figure 8.17 are shown in Figure 8.18. As shown in the figure, it is evident that the output voltage doesn't exceed $E + 0.7$. The upper limit of the output voltage can be controlled by adjusting the value of E.

(continued)

EXAMPLE 8.3 Continued

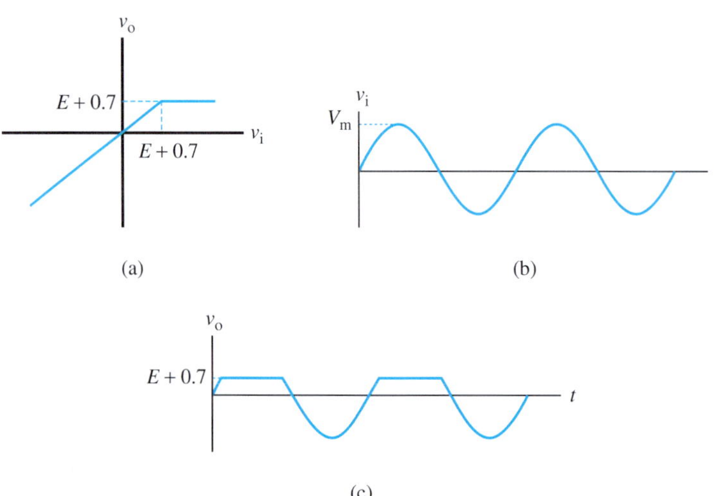

FIGURE 8.18 (a) Transfer characteristics for the circuit shown in Figure 8.17; (b) its input voltage; (c) its output voltage.

EXAMPLE 8.4 One-Sided Limiter

Another limiter circuit is shown in Figure 8.19. Find the output voltage.

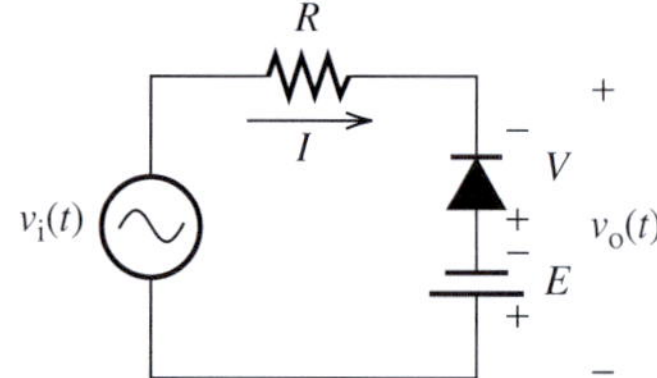

FIGURE 8.19 The limiter circuit for Example 8.4.

SOLUTION

When the voltage across the diode terminals is known (i.e., the diode is ON):

$$v_o(t) = -V - E \quad \textbf{(8.14)}$$

If the diode is OFF or the current flowing through it is zero:

$$v_o(t) = v_i(t) - I \times R \quad \textbf{(8.15)}$$

The diode is turned on as the voltage difference between $-E$ and $v_i(t)$ exceeds 0.7 V. In other words, to turn the diode on, the following must be true:

$$-E - v_i(t) > 0.7 \quad \textbf{(8.16)}$$

Or equivalently:

$$v_i(t) < -E - 0.7 \quad \textbf{(8.17)}$$

Accordingly, the output voltage corresponds to:

$$v_o(t) = -E - 0.7 \quad \textbf{(8.18)}$$

On the contrary, the diode is turned off if:

$$v_i(t) > -E - 0.7 \tag{8.19}$$

In this case, the output voltage will correspond to:

$$v_o(t) = v_i(t) \tag{8.20}$$

In summary, the output voltage is given by:

$$v_o(t) = \begin{cases} -E - 0.7 & v_i(t) < -E - 0.7 \\ v_i & v_i(t) > -E - 0.7 \end{cases} \tag{8.21}$$

Plots of the transfer characteristic, input voltage, and output voltage for the limiter circuit of Figure 8.19 are shown in Figure 8.20. As shown in the figure, it is notable that the output voltage doesn't drop below $-E - 0.7$. The lower limit of the output voltage can be controlled by adjusting the value of E.

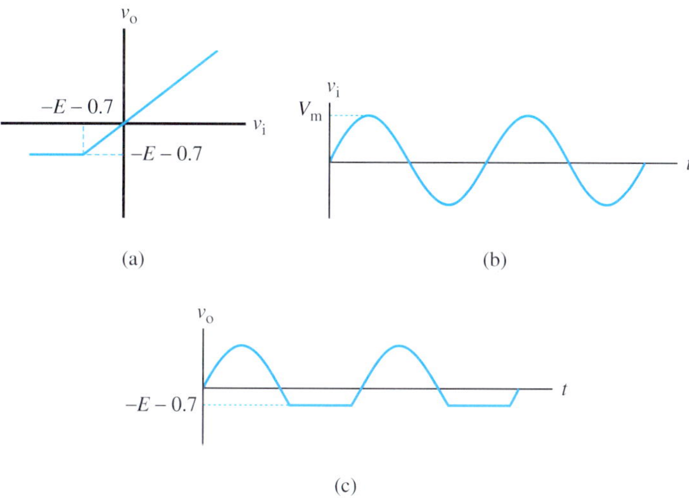

FIGURE 8.20 (a) Transfer characteristics for the circuit shown in Figure 8.19; (b) its input voltage; (c) its output voltage.

EXAMPLE 8.5 Two-Sided Limiter

For the circuit shown in Figure 8.21, sketch the transfer characteristics, input voltage, and output voltage. Assume the input voltage is a sinusoidal waveform with a peak value of 10 V and $E_1 = E_2 = 4$ V.

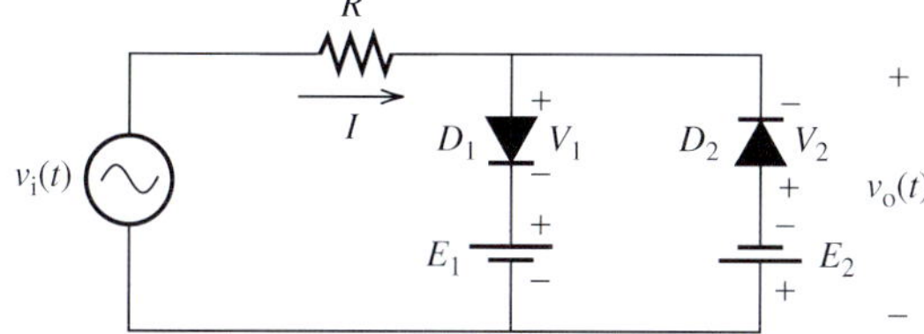

FIGURE 8.21 Circuit for Example 8.5.

(continued)

EXAMPLE 8.5 Continued

SOLUTION

The value of the output voltage depends on the status of diodes D_1 and D_2. Based on the distribution shown, both diodes cannot be turned ON simultaneously because:

- To turn on D_1, the value of the input voltage must be greater than E_1 by at least 0.7 V.
- To turn on D_2, the value of the input voltage must be less than $-E_2$ by at least 0.7 V.

Table 8.1 summarizes the status of diodes, the required value of the input voltage, and the corresponding output voltage.

For the given value of E_1 and E_2, the output voltage corresponds to:

$$v_o(t) = \begin{cases} 4.7\text{ V} & v_i(t) > 4.7\text{ V} \\ -4.7\text{ V} & v_i(t) < -4.7\text{ V} \\ v_i(t) & -4.7\text{ V} < v_i(t) < 4.7\text{ V} \end{cases}$$

Plots of the transfer characteristic, input voltage, and output voltage for the limiter circuit of Figure 8.21 are shown in Figure 8.22. As shown in the figure, it is notable that the output voltage is bounded between 4.7 V and −4.7 V.

TABLE 8.1 Analysis of the Circuit in Example 8.5

D_1	D_1	Criterion	$v_o(t)$
ON	OFF	$v_i(t) > E_1 + 0.7$	$E_1 + 0.7$
OFF	ON	$v_i(t) < -E_2 - 0.7$	$-E_2 - 0.7$
OFF	OFF	$-E_2 - 0.7 < v_i(t) < E_1 + 0.7$	$v_i(t)$

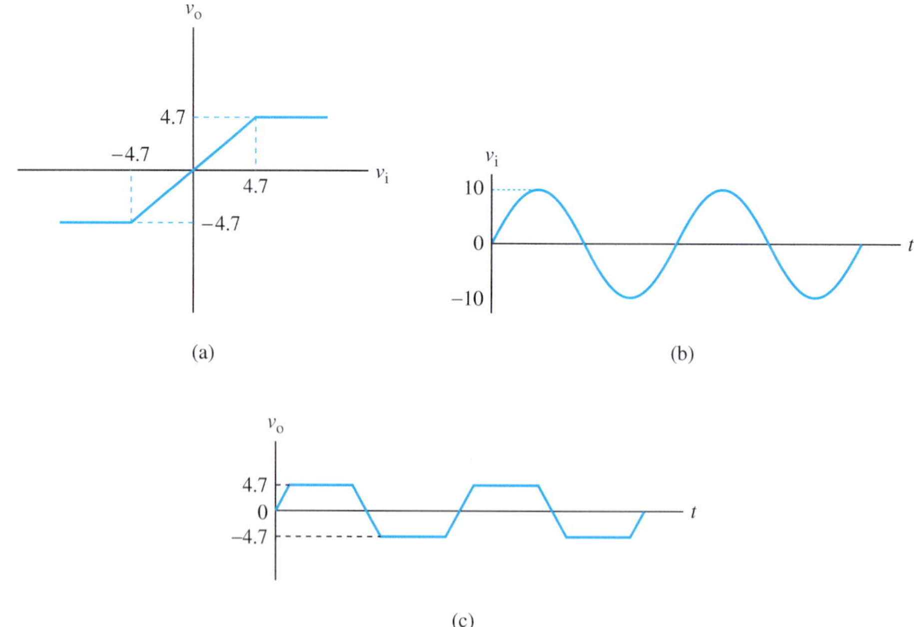

FIGURE 8.22 (a) Transfer characteristics for the circuit shown in Figure 8.21; (b) its input voltage; (c) its output voltage.

APPLICATION EXAMPLE 8.6 Protection of Electrical Devices

Electronic and electrical devices are designed to operate at a certain maximum supply voltage: They can be seriously harmed if their voltage exceeds that threshold. An excessive increase in voltage sources may be from a natural source (e.g., lightening) or a man-made source (e.g., electromagnetic induction when switching on or off inductive loads). In addition, the devices can be harmed if the polarity of the input voltage is reversed (i.e., some electronic devices cannot tolerate negative voltages).

A limiter circuit can protect devices from excess voltage. Assume a limiter circuit for the input of an electronic device needs to be designed with the following ratings: $V_{max} = 9$ V and $V_{min} = 0$.

SOLUTION

Because the input voltage to the device needs to be bounded between two values, a double limiter must be used to protect it. Considering the limiter circuit discussed in Example 8.5, and by referring to Table 8.1, the upper limit of the voltage is controlled by diode D_1 and the voltage source, E_1. The value of E_1 is calculated as follows:

$$E_1 + 0.7 = V_{max} = 9\text{ V} \rightarrow E_1 = 8.3\text{ V}$$

The lower limit is controlled by diode D_2 and the voltage source, E_2. The value of E_2 is calculated as follows:

$$-E_2 - 0.7 = V_{min} = 0 \rightarrow E_2 = -0.7\text{ V}$$

The negative sign indicates that the battery should be reversed. The limiter circuit is connected between the input terminals of the device as shown in Figure 8.23.

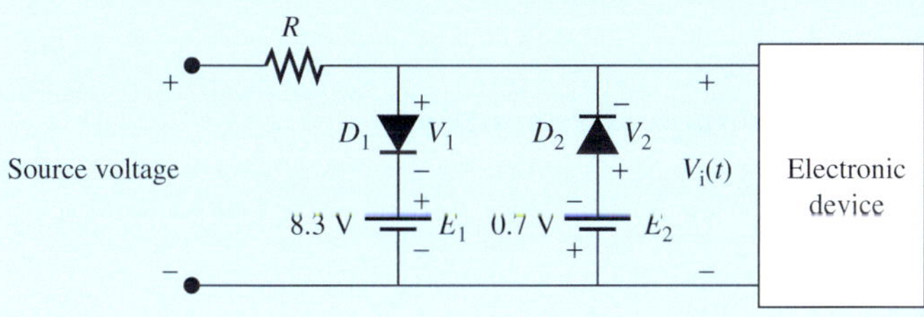

FIGURE 8.23 The limiter circuit for Example 8.6.

EXERCISE 8.2

Sketch the transfer characteristics and the output voltage for the circuit shown in Figure 8.23, assuming the input source is a sinusoid with an amplitude of 5 V.

8.3.2 Different Types of Diodes

The function of diodes is not limited to rectification. There are many types of diodes; each has a different function. Diodes are designed to exploit specific junction properties, such as capacitance or charge storage. This section provides an overview of common types of diodes and their applications.

8.3.2.1 ZENER DIODES

A *Zener diode* or a *breakdown diode* is similar to a forward-biased rectifier diode, like those discussed in Section 8.1. However, the rectifier diode does not conduct current when it is reverse biased. The Zener diode allows current to pass in the reverse direction if the reverse voltage

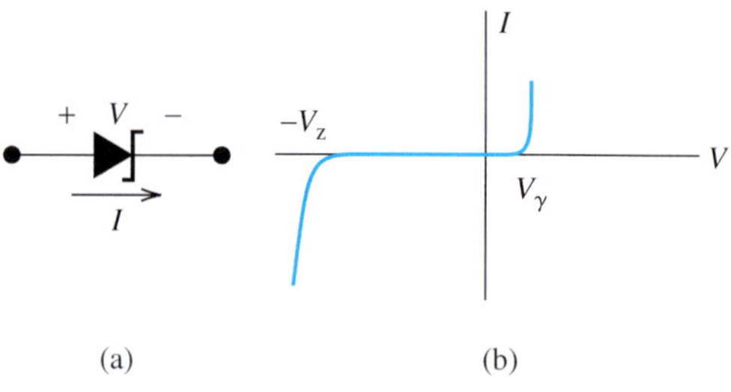

FIGURE 8.24 (a) Symbol of a Zener diode. (b) Characteristics of a Zener diode.

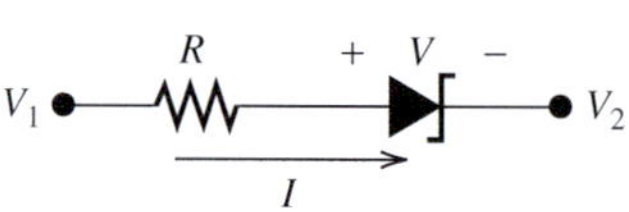

FIGURE 8.25 Circuit with a Zener diode.

TABLE 8.2 Operation of a Zener Diode

	Diode Status	*V*	*I*
$V_1 - V_2 > V_\gamma$	ON	V_γ	Positive
$-V_Z < V_1 - V_2 < V_\gamma$	OFF	$V_1 - V_2$	$=0$
$V_1 - V_2 < -V_Z$	Zener region	$-V_Z$	Negative

exceeds a precisely defined voltage called the Zener voltage, V_Z. Zener voltage ranges from 3.3 to 75 V. The circuit diagram of a Zener diode and its characteristics are shown in Figure 8.24.

To explain the operation of a Zener diode, consider the circuit shown in Figure 8.25. Assuming a practical model, Table 8.2 summarizes the operation of the Zener diode. Observe that if the reverse voltage across the diode is more than V_Z, the Zener diode will be equivalent to a constant voltage source with the value of $V = -V_Z$.

EXAMPLE 8.7 Current and Voltage of Zener Diode

For the circuit shown in Figure 8.26, the Zener diode has $V_Z = 3.6$ V. Find *I* and *V* when:

a. $V_S = -1$ V
b. $V_S = 8$ V
c. $V_S = -10$ V

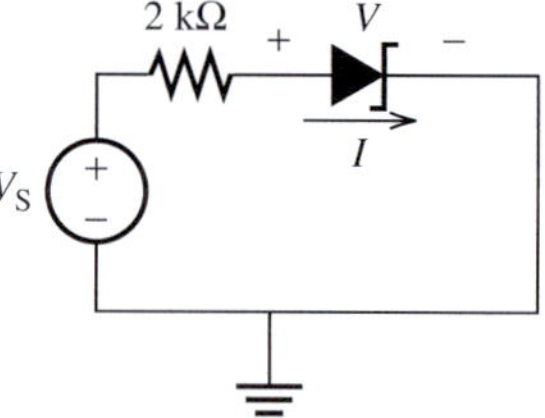

FIGURE 8.26 Circuit for Example 8.7.

SOLUTION

a. Refer to Figure 8.25, $V_S = -1$ V means $-3.6\text{ V} < V_1 - V_2 < 0$ V. Therefore, the diode is reverse biased, it is OFF, and the current flowing through it is $I = 0$. Applying KVL to Figure 8.26, the voltage across the diode corresponds to:

$$V = V_S = -1\text{ V}$$

b. Because $V_1 - V_2 = V_S > V_\gamma = 0.7$, the diode is forward biased and it is ON. Therefore, $V = V_\gamma = 0.7$ and the current corresponds to:

$$I = \frac{8 - 0.7}{2} = 3.65\text{ mA}$$

c. In this example, $V_1 - V_2 = V_S = -10\text{ V} < -V_z = -3.6\text{ V}$. Therefore, the diode is reverse biased and it is in the Zener region. Therefore, the voltage across the diode corresponds to:

$$V = -V_z = -3.6\text{ V}$$

In addition, the current corresponds to:

$$I = \frac{V_S + V_z}{R} = \frac{-10 + 3.6}{2} = -3.2\text{ mA}$$

EXAMPLE 8.8 Using Zener Diodes in Limiter Circuits

A Zener diode can be used to replace with DC sources in limiter circuits, as shown in the circuit in Figure 8.27. For this circuit, sketch the transfer characteristics, input voltage, and output voltage. Assume $V_z = 5.6$ V and that the input voltage is a sinusoidal waveform with a peak voltage of 12 V.

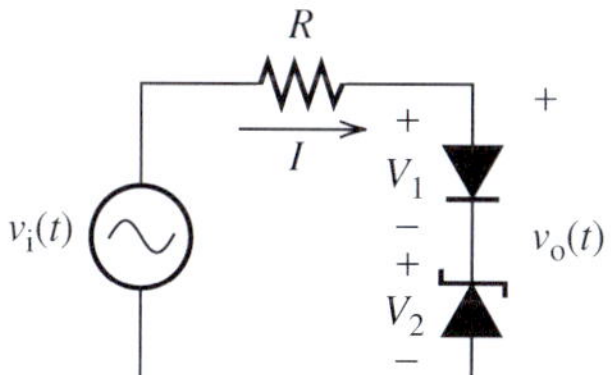

FIGURE 8.27 Limiter using a Zener diode.

SOLUTION

During the positive half cycle, the normal diode is forward biased and the Zener diode is reverse biased. To turn the rectifier diode on, and push the Zener diode to its Zener region, the input voltage must satisfy:

$$v_i(t) > V_\gamma + V_z = 0.7 + 5.6 = 6.3\text{ V}$$

Now, because both diodes allow the flow of current: $V_1 = 0.7$ V, and, $V_2 = 5.6$ V. Therefore, the output voltage corresponds to:

$$v_o(t) = V_1 + V_2 = 6.3\text{ V}$$

If the input voltage $v_i(t) < 6.3$ V, at least one of the diodes avoids the flow of current, which leads to:

$$v_o(t) = v_i(t)$$

During the negative half cycle, the normal diode is reverse biased and the Zener diode is forward biased. However, the Zener diode can pass current and the normal diode cannot. Therefore, $v_i(t) = v_o(t)$. In general, the output voltage corresponds to:

$$v_o(t) = \begin{cases} 6.3\text{ V} & v_i(t) > 6.3\text{ V} \\ v_i(t) & v_i(t) < 6.3\text{ V} \end{cases}$$

Plots of transfer characteristics, input voltage, and output voltage are shown in Figure 8.28.

The Zener diode has an important application in the design of power supplies (i.e., AC-to-DC converters) where it is used as a voltage regulator. AC-to-DC conversion is presented in Section 8.3.3.

(continued)

EXAMPLE 8.8 Continued

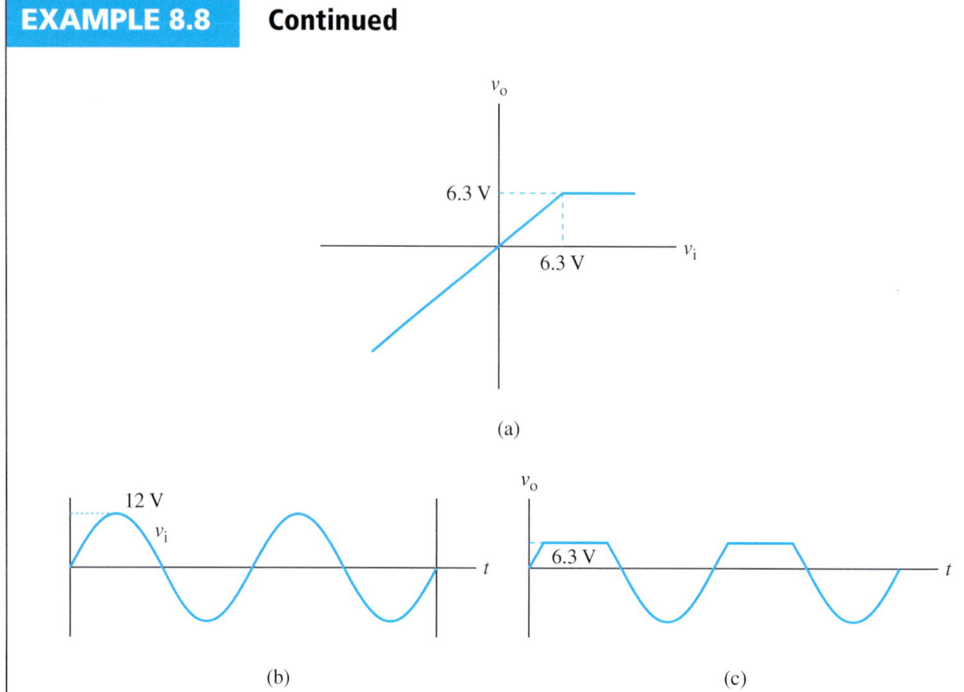

FIGURE 8.28 (a) Transfer characteristics for the circuit shown in Figure 8.27; (b) its input voltage; (c) its output voltage.

8.3.2.2 VARACTOR DIODE

The term *varactor* is a shortened form of *variable reactor*. Varactor diodes are typically used in a reverse bias configuration. The varactor, when biased in this fashion, acts as a variable capacitor. The capacitance of the varactor decreases with applied reverse voltage and it ranges from a few pico farads (pF) to over 100 pF. The symbol and characteristics (i.e., the capacitance versus the voltage across the diode) of a varactor are shown in Figure 8.29.

Varactors are commonly used in voltage-to-frequency converters or voltage control oscillators (VCOs). The VCO is an oscillator that changes the frequency of its output signal according to the value of the input voltage. The VCO is a widely used device in Frequency Modulation (FM) circuits. Another popular application of the varactor is in electronic tuning circuits (e.g., radio and television tuners). In these tuners, the DC control voltage varies the capacitance of the varactor, retuning the resonant circuit. A simple tuning circuit is shown in Figure 8.30.

In Figure 8.30, $v_S(t)$ is a small signal, (i.e., its amplitude is in the range of millivolts or less) that is much smaller than the DC control voltage. The small signal is a model for the received

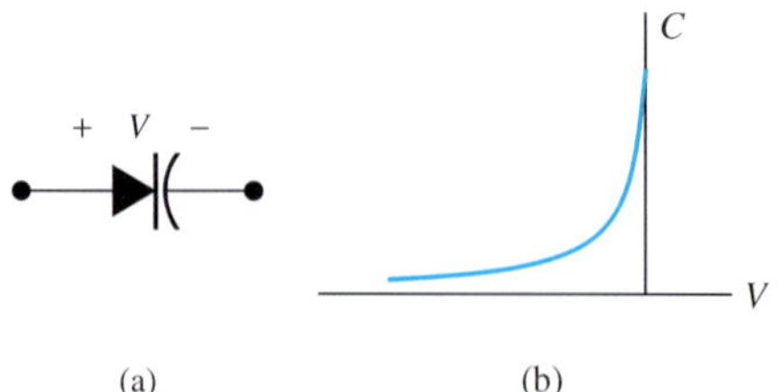

FIGURE 8.29 (a) The symbol for a varactor diode; (b) C-V characteristics of a varactor diode.

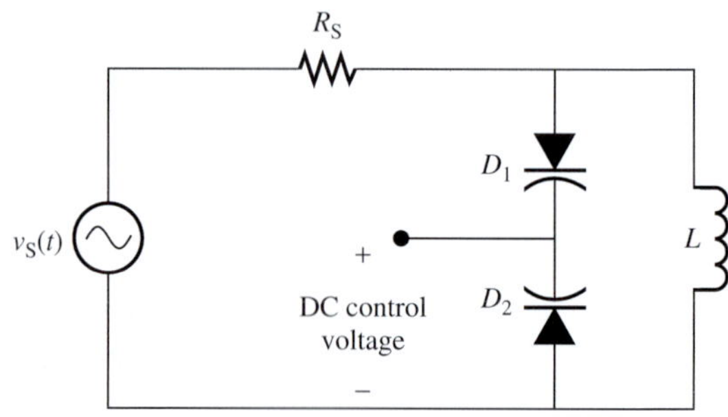

FIGURE 8.30 A tuning circuit using a varactor diode.

signal by the antenna. If the applied DC control voltage is positive, both varactors will be reverse biased. If the equivalent capacitance of both diodes is labeled C, the resonant frequency will correspond to:

$$f_r = \frac{1}{\sqrt{2\pi LC}} \tag{8.22}$$

8.3.2.3 LIGHT-EMITTING DIODE

A *light-emitting diode* (LED) is another type of diode that emits light when current passes through it. LEDs operate in the forward bias and the emitted light intensity increases as the current flowing through it increases. The color of the light mainly depends on the material used for diode fabrication. LEDs with visible light (e.g., red, blue, yellow, and green) are commonly used as indicators and are used in many devices, such as TV power indicators, digital clocks, and seven-segment displays.

The threshold voltage, V_γ, for an LED is about 1.6 V, as it is fabricated by materials different from Si and Ge (usually gallium arsenide, or GaAs). The symbol of the LED is shown in Figure 8.31.

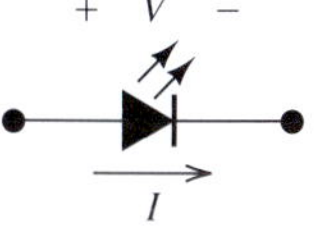

FIGURE 8.31 The symbol of an LED.

In many applications, the light from an LED is invisible to the human eye. Infrared emitters are well-suited to optical communication systems. For example, an LED might be used in conjunction with a photodiode (discussed in the next section) or other photosensitive devices to transfer information optically between locations.

8.3.2.4 PHOTODIODES

In contrast to LEDs, photodiodes are semiconductor devices that function as photodetectors. There are two major modes of operation for photodiodes: photovoltaic mode and photoconductive mode. In the photovoltaic mode, the diode operates under zero bias, but when light is emitted on the diode, it creates a current that leads to forward biasing. Series and parallel connection of photodiodes operating in photovoltaic mode are used in *solar cells*. A solar cell is a device that converts light or solar energy into electrical energy. Solar cells have many applications in power systems for remote areas, Earth-orbiting satellites and space probes, and consumer systems, for example, handheld calculators, wrist watches, and remote radiotelephones.

Considering the scarcity and the cost of oil resources on the earth, a significant amount of research is ongoing to develop and improve solar cells capable of efficiently converting solar energy to power. One approach that is under study by many investigators worldwide is to launch low-orbit satellites equipped with solar cells capable of efficiently collecting solar energy in outer space and transmitting it to the earth via wireless communications.

To use a photodiode in its photoconductive mode, the photodiode is reverse biased; the photodiode then allows current flow when it is illuminated. The conductivity of the photo diode varies directly with the intensity of the light. Photodiodes that operate in this mode have many applications, including smoke detectors and motor speed control.

APPLICATION EXAMPLE 8.9 Motor Speed Control

In many applications, the speed of motors needs to be kept constant. One example is the rotation of radar antennas. Radar stands for Radio Detection And Ranging. Accordingly, these devices use radio waves to detect targets and find their location. In order to locate targets in full 360° around the radar position, the radar directional antenna (usually a parabolic antenna) is rotated via a motor. The speed of this motor must be precisely monitored to avoid errors in localizing targets. In order to control the speed of a motor, the speed must be accurately measured. The speed must be corrected if it is not consistent with the desired value.

An *optical encoder* is a commonly used circuit to measure motor speed. An optical encoder produces a signal with a frequency proportional to the speed of the motor. This frequency is

(*continued*)

APPLICATION EXAMPLE 8.9 Continued

converted to a voltage using another circuit called a frequency-to-voltage converter. This voltage is used to control the speed of the motor. This example focuses on the optical encoder circuit.

An optical encoder is a circuit that uses light to generate a periodic signal whose frequency is proportional to the motor speed. Here, the motor speed is "encoded" in the frequency of the signal. Figure 8.32 illustrates the operation of an optical encoder. An opaque disk is mounted on a motor shaft and turns with the shaft. Slots (in this case, 500) are cut into the disk. An LED is positioned on one side of the disk. A light sensor (photodiode) on the opposite side of the disk senses light from the LED and produces a voltage when one of the slots is positioned exactly between the LED and the sensor, allowing light to pass. When the light is blocked by the solid part of the disk between slots, the sensor's output voltage drops to zero. As the motor rotates, the sensor produces a series of pulses that take the form of a sinusoidal signal, as shown in Figure 8.32. The frequency of this signal is determined by the speed of rotation. Specifically, 500 periods of this waveform are produced during each 360° rotation of the motor. The signal is sinusoidal instead of square because the light is not interrupted instantaneously by the disk.

Assume the output signal has a frequency $f = 200$ Hz, the radius of the circle passing through the centers of slots $r = 10$ cm, and the disc has 500 slots. Find the speed of the disc at the holes.

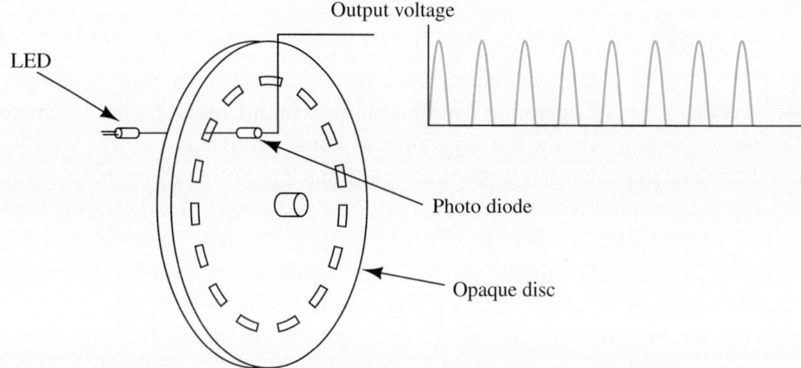

FIGURE 8.32 Optical encoder configuration.

SOLUTION

First, calculate the distance between two successive slots. The perimeter of the circle passing through the slot's centers corresponds to:

$$p = 2\pi r = 2\pi \times 0.1 = 0.6287 \text{ m}$$

Because the disc has 500 slots, the distance between two successive slots corresponds to:

$$d = \frac{p}{500} = 0.1257 \text{ cm}$$

The time between two successive peaks in the output signal is equivalent to the time to move from one slot to the next one, and it corresponds to:

$$T = \frac{1}{f} = \frac{1}{200} = 0.005 \text{ s}$$

Therefore, the speed of the encoder disc at the holes corresponds to:

$$V = \frac{d}{T} = 0.1257 \times 10^{-2} \times 200 = 0.2514 \text{ m/s}$$

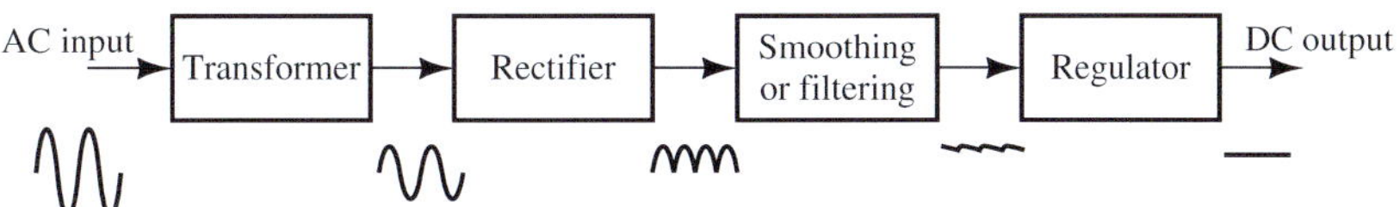

FIGURE 8.33 Block diagram of an AC-to-DC converter.

8.3.3 AC-to-DC Converter

An AC-to-DC converter is a device that converts AC signals into DC signals. Sometimes, it is called an adapter. An adapter is a DC power supply that receives 120 V/60 Hz AC signal as an input and generates a DC voltage output. Many electronic systems such as Labtop, Electronic tooth brush, radio and printer, need a DC supply (independent voltage source) for their operation. A DC supply can be offered by a battery or by converting AC source to DC. The block diagram of an AC-to-DC converter is shown in Figure 8.33. The second block, that is, the rectifier, has been discussed in the previous sections. This section examines the other three blocks.

8.3.3.1 TRANSFORMER

A transformer is a device that steps-up or steps-down an input voltage (see Section 12.4 for additional details). In DC power supplies, step-down transformers are typically used because the input voltage is 120 V AC and most power supplies have a maximum DC voltage output of 30 V. Figure 8.34 shows the circuit block diagram of the transformer. A transformer is a two-port device, with the two ports called the *primary* port and the *secondary* port. The input voltage is applied to the primary port and the output voltage is taken from the secondary port. The relationship between the input and the output voltages is given by:

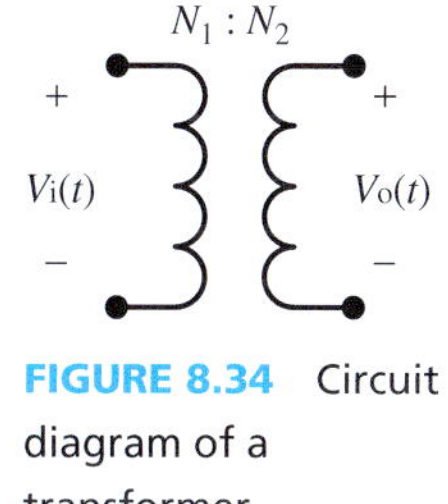

FIGURE 8.34 Circuit diagram of a transformer.

$$\frac{v_o(t)}{v_i(t)} = \frac{N_2}{N_1} \tag{8.23}$$

where N_2/N_1 is called the transformer's *turns ratio*. If the turns ratio is greater than unity, the transformer is called a step-up transformer and the output voltage will be greater than the input. If the turns ratio is less than unity, the transformer is called a step-down transformer and the output voltage will be less than the input.

8.3.3.2 SMOOTHING OR FILTERING

In general, the voltage output from a rectifier has an invariable polarity. It also has a nonzero average but it does not have constant magnitude (DC means size and direction are constant over time). A capacitor can be used to smooth the rectifier's output, as shown in Figure 8.35. In this case, the capacitance of the capacitor must be large enough to increase the discharge time. For an input voltage of a sinusoidal waveform, the operation of the filter shown in the circuit shown in Figure 8.35 is described as follows.

During the first half cycle, the capacitor charges through the rectifier output across zero resistance—neglecting the diode resistance—toward the peak voltage. When the rectifier output starts to decrease; the capacitor discharges through the load resistor, R_L. If the time constant, R_LC, is large enough compared to the input signal *period*, the capacitor will take a long time to discharge. Before losing its charge, the capacitor starts to charge again through the next half period maintaining its voltage near the peak value. The time constant of charging is relatively low because the resistance of the diodes and the transformer is very small. Plots of input voltage, secondary voltage, and output voltage are shown in Figure 8.36.

The voltage across the transformer's secondary port is related to the input by Equation (8.23). Therefore:

$$V_{sm} = \frac{N_2}{N_1} \times V_{im} \tag{8.24}$$

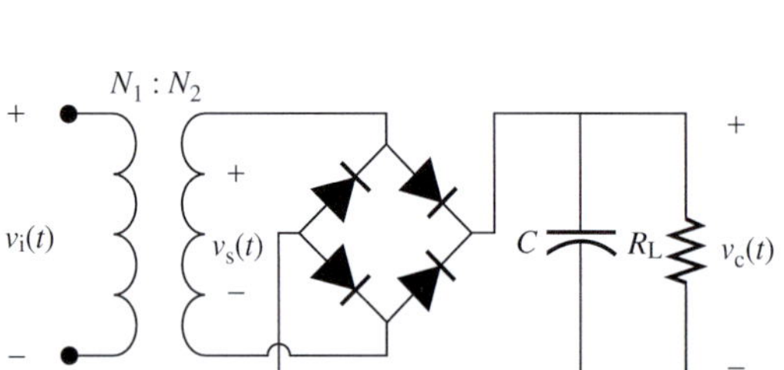

FIGURE 8.35 Using an RC filter to smooth the rectifier output.

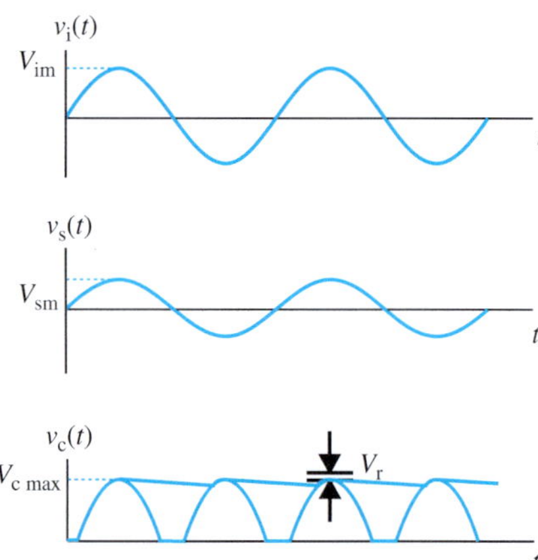

FIGURE 8.36 Voltages at different points for the circuit shown in Figure 8.35.

As discussed in the previous section, the voltage drop across the two forward biased diodes in the bridge rectifier is 1.4 V. Therefore, the peak value of the output voltage corresponds to:

$$V_{c\,max} = V_{sm} - 1.4 \tag{8.25}$$

The difference between the maximum and the minimum values of the output voltage is called the *ripple voltage*, V_r. As the capacitance increases, the discharge time increases, and therefore the ripple voltage decreases leading to more smooth output voltage. The minimum value of the output voltage corresponds to:

$$V_{c\,min} = V_{c\,max} - V_r \tag{8.26}$$

As explained in Chapter 4, the voltage drop corresponds to:

$$v(t) = V_{c\,max} \cdot \left(1 - e^{-\frac{t}{R_L C}}\right)$$

Applying Taylor series to the exponential term and ignoring higher order terms:

$$v(t) \cong V_{c\,max} \cdot \left(\frac{t}{R_L C}\right)$$

Now, in a bridge rectifier, the voltage drops until about $t = T/2$, ($T = 1/f$ is the period of the sine wave). Therefore, the ripple voltage corresponds to:

$$v_r = \frac{V_{c\,max}}{2f R_L C} \tag{8.27}$$

Note that in a half-wave rectifier, the ripple voltage is twice the value of Equation (8.27). Here, f is the frequency of the input signal (in North America, f = 60 Hz, elsewhere it is commonly 50 Hz).

EXERCISE 8.3

Start with Equation (8.26) and using the guidelines provided, prove Equation (8.27).

The ripple voltage is a function of many variables. Some are controllable, including the capacitance: as the capacitance increases, the ripple voltage decreases. In contrast, other variables are uncontrollable, including signal frequency and load resistance. Because the output voltage has a ripple voltage that cannot be completely eliminated, another circuit is required to stabilize the value of the output voltage. This circuit is called a voltage regulator.

8.3.3.3 VOLTAGE REGULATORS

A *voltage regulator* is a circuit used for voltage stabilization. A simple regulator is shown in Figure 8.37. If the Zener diode is reverse biased, it will turn on. In this case, the output voltage corresponds to:

$$v_o(t) = V_z \tag{8.28}$$

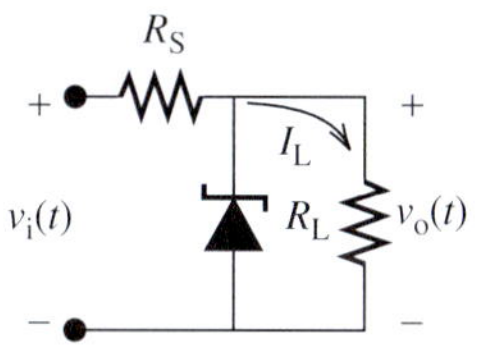

FIGURE 8.37 A simple regulator made by Zener diode.

To turn the Zener diode on in the reverse mode, the output voltage must be positive and the input voltage must satisfy:

$$v_i(t)\frac{R_L}{R_L + R_s} \geq V_z \tag{8.29}$$

or:

$$v_i(t) \geq V_z\frac{R_L + R_s}{R_L} = V_z\left(1 + \frac{R_s}{R_L}\right) \tag{8.30}$$

In other words, the minimum value voltage introduced in Equation (8.26):

$$V_{c\,min} \geq V_z\left(1 + \frac{R_s}{R_L}\right)$$

Therefore, the minimum value of the input voltage required to turn the regulator on corresponds to:

$$V_{on} = V_z\left(1 + \frac{R_s}{R_L}\right) \tag{8.31}$$

The complete AC-to-DC converter circuit is shown in Figure 8.38.

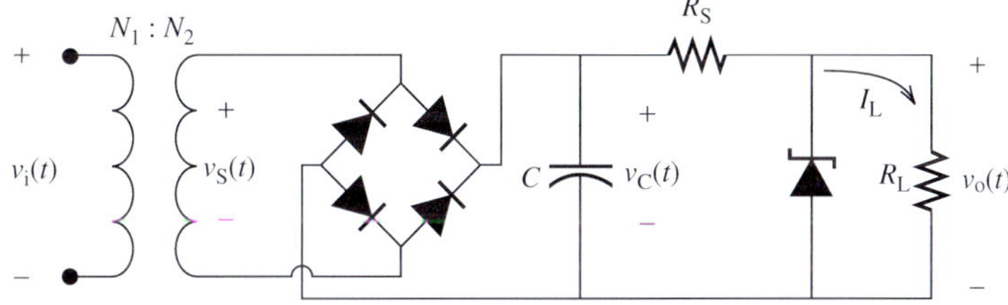

FIGURE 8.38 An AC-to-DC converter with a voltage stabilizer.

EXAMPLE 8.10 DC Power Supply: Adapter

Design a DC power supply that supplies the DC voltage of 10 V to a load resistance of 1 kΩ. The AC input is 120 V/60 Hz. Choose suitable values for the capacitor and the transformer turns ratio. The ripple voltage should be less than 10% of the output voltage and $R_S = 200\ \Omega$.

SOLUTION

The circuit shown in Figure 8.38 is used for power supply design. Because the power supply output voltage is 10 V, the Zener diode must have $V_z = 10$ V. First, propose a value for the minimum voltage across the capacitor $V_{c\,min} > V_{on}$. Here, V_{on} is the voltage required to turn on the Zener diode and is given by Equation (8.31).

$$V_{on} = 10\left(1 + \frac{200}{1000}\right) = 12\ \text{V}$$

Let's select:

$$V_{c\,min} = 15\ \text{V}$$

(*continued*)

EXAMPLE 8.10 Continued

The ripple voltage $V_r < 0.1\ V_o = 1$ V. Select:

$$V_r = 0.5 \text{ V}$$

The maximum capacitor voltage is calculated using Equation (8.26):

$$\begin{aligned} V_{c\,max} &= V_{c\,min} + V_r \\ &= 15 + 0.5 \\ &= 15.5 \text{ V} \end{aligned}$$

The ripple voltage is calculated using Equation (8.27):

$$\frac{15.5}{2 \times 60 \times 1000 \times C} = 0.5$$

Therefore:

$$C = 2600\ \mu\text{F}$$

To calculate the transformer turns ratio, first calculate the maximum voltage across the transformer's secondary port using Equation (8.25):

$$\begin{aligned} V_{s\,max} &= V_{c\,max} + 1.4 \\ &= 15.5 + 1.4 \\ &= 16.9 \text{ V} \end{aligned}$$

Using Equation (8.23), the transformer turns ratio corresponds to:

$$\frac{N_1}{N_2} = \frac{V_{i\,max}}{V_{s\,max}} = \frac{120\sqrt{2}}{16.9} = 10.42$$

8.4 TRANSISTORS

The previous section outlined diodes that consist of two terminals and are formed by connecting two types of semiconductors. A transistor is a semiconductor device that has three terminals and can be used in electrical circuits as an amplifier or as a switch. There are two major types of three-terminal semiconductor devices: the bipolar junction transistor (BJT) and the field-effect transistor (FET). This section briefly introduces the BJT and FET.

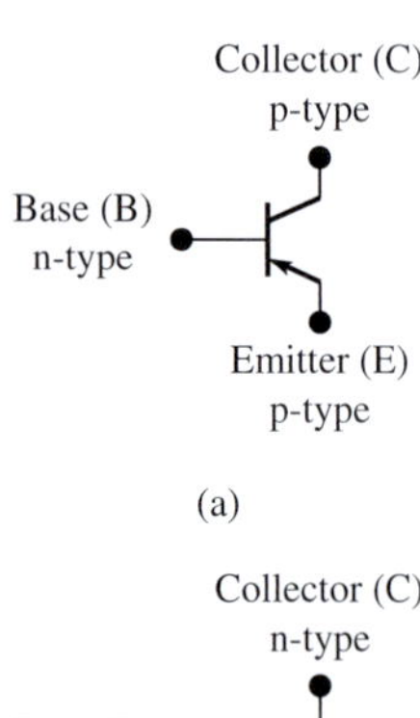

FIGURE 8.39 Symbols of a BJT transistor (a) PNP type; (b) NPN type.

8.4.1 Bipolar Junction Transistor

A BJT has three layers of p-type and n-type semiconductors. Based on the order of these layers, there are two types of BJT: PNP and NPN transistors, named after the order of their layers. The circuit diagrams of both types are depicted in Figure 8.39.

The three terminals of a BJT are called the base, the emitter, and the collector and are abbreviated as B, E, and C, respectively. As discussed in Section 8.1, the two types of semiconductors (i.e., p-type and n-type) are connected in a junction. Accordingly, the BJT has two junctions: the first junction is between the base and the emitter (BE junction), and the second junction is between the base and the collector (BC junction).

Biasing of BE and BC junctions (i.e., applying proper voltages to the junctions) determines the region of operation of the transistor (Table 8.3). As seen in Table 8.3, three regions are defined for the operation of the transistor: cutoff, active, and saturation.

TABLE 8.3 Regions of Operation of an NPN Transistor

Region of Operation	BE	BC	V_{BE}	I_B	I_C	V_{CE}	Application
Cutoff	Reverse	Reverse	<0.7 V	0	0	Computed by the circuit	Switch
Active	Forward	Reverse	$=0.7$ V	Computed by the circuit	$=\beta I_B$	Computed by the circuit >0.2 V	Amplifier
Saturation	Forward	Forward	$=0.7$ V	Computed by the circuit	Computed by the circuit	$=0.2$ V	Switch

The arrow on the emitter refers to the direction of the current. Therefore, the current flows into (toward) the emitter if the transistor is PNP (i.e., from the p-type and toward an n-type as is the case in a diode as well). However, the current flows out of the emitter in NPN transistors. The current and voltage representation for the PNP and NPN transistor is shown in Figure 8.40. Because the characteristics and operation of PNP and NPN transistors are similar, in this chapter we mainly discuss the analysis of NPN transistors.

To evaluate circuits consisting of transistors, the relationship between currents and voltages must be known. First, applying KCL to the transistor shown in Figure 8.40:

$$I_E = I_C + I_B \tag{8.32}$$

The value of I_C is determined according to the region of operation of the transistor. BJT transistors operate in three regions. Table 8.3 presents these three regions of operation of the transistor and their properties. In the active region, the value of I_C depends on the value of I_B and they are related by β, which is defined as the DC gain. Many transistor circuits are used for amplification of the input signal. In this case, the transistor is operated in the active region. In the saturation region, the collector emitter voltage has its minimum value of $V_{CESat} = 0.2$ V.

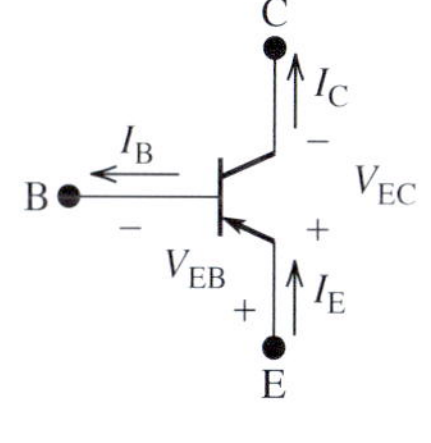

(a)

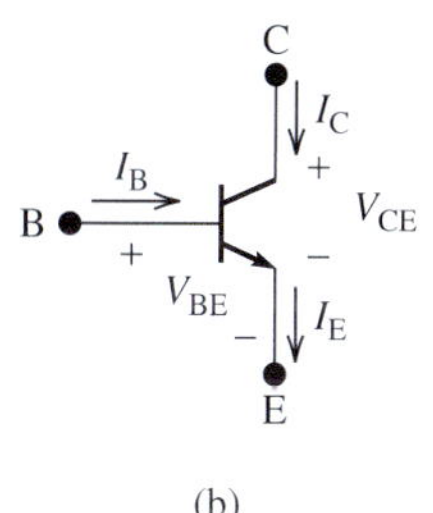

(b)

FIGURE 8.40 Current directions and voltage polarities of a BJT (a) PNP type; (b) NPN type.

8.4.2 Transistor as an Amplifier

One of the most important functions of a transistor is *amplification* of AC signals. An amplifier is a device with an output signal equal to a factor multiplied by the input signal. This factor is called *voltage gain* or *current gain*. These gains are defined based on whether the signal to be amplified is voltage or current. Because the relationship between the output and the input is linear (e.g., $v_o(t) = A_v\, v_{in}(t)$; where A_v is the voltage gain), the transistor must operate in the active region when used as an amplifier. As it is observed in Table 8.3, in the saturation region the collector–emitter voltage is constant, that is, it does not vary with the input; in the cutoff region, $I_B = I_C = 0$. Thus, transistors cannot be operated as amplifiers in those regions.

Transistors must be properly biased in order to operate in the active region and function as an amplifier. Transistor bias is adjusted by applying proper DC voltage to transistor nodes. One important term, called the *operating point*, is defined as a DC voltage and current that determines the operation of transistors in the desired bias. Another well-known term for the operating point is *quiescent point* or *Q-point*. Analysis of transistor circuits is simplified by using the superposition theorem studied in Chapter 3. First, only consider DC sources to determine the operating point or the Q-point. Next, consider AC sources to determine the AC voltage or current gains. To minimize confusion, the following symbols are defined for voltage and current notations:

DC only: Capital letters for both variable and subscript → V_{BE}, I_B, I_C, and V_{CE}
AC only: Small letters for both variable and subscript → v_{be}, i_b, i_c, and v_{ce}
DC plus AC: Small letter for variable and capital letter for subscript → v_{BE}, i_B, i_C, and v_{CE}

where:

$$v_{BE} = V_{BE} + v_{be} \quad \textbf{(8.33)}$$

$$i_B = I_B + i_b \quad \textbf{(8.34)}$$

$$v_{CE} = V_{CE} + v_{ce} \quad \textbf{(8.35)}$$

$$i_C = I_C + i_c \quad \textbf{(8.36)}$$

Based on Equations (8.33) to (8.36), the analysis of an amplifier made from a transistor circuit can be completed by:

1. *DC analysis*: Specifying its operation point and ensuring that the transistor is in its active region
2. *AC analysis*: Analyzing its voltage gain assuming it is properly adjusted in the active region

One of the most famous transistor amplifier circuits is shown in Figure 8.41. The DC voltage source, V_{CC}, together with resistors R_1, R_2, R_C, and R_E are used to determine the Q-point or the region of operation. The AC voltage signal to be amplified, v_{in}, is applied to the base and the AC amplified voltage signal, v_o, is considered across the load resistor, R_L. The capacitors C_1 and C_2 are called *coupling capacitors* while C_e is called the *bypass capacitor*.

As discussed in Chapter 4, if a DC signal is applied to a capacitive circuit, all capacitors will be equivalent to an open circuit after a short time period. Therefore, capacitors isolate the source from the transistor for DC analysis. In addition, when the signal variation is high, that is, the signal frequency is high, the capacitor will be equivalent to a short circuit. Note that the capacitor reactance corresponds to $X_c = 1/(2\pi fC)$. Therefore, for AC analysis (when f is high), the capacitors will be equivalent to a short circuit. Thus, the purpose of coupling capacitors (C_1) is to allow the AC signal to pass into and from the transistor circuit while keeping the DC circuit isolated. In order for a capacitor to function as a coupling capacitor, its capacitance must be large enough. That is, $X_c = 1/\omega C$ must be very small to maintain the capacitor as a short circuit.

A nonzero resistance of R_E reduces the amplification gain of the transistor. Thus, the purpose of the bypass capacitor, C_e, is to create an AC ground at the emitter, while the Q-point can be readjusted using R_E without affecting the AC circuit.

The DC equivalent circuit for the amplifier circuit shown in Figure 8.41 is depicted in Figure 8.42(a). Note that all capacitors are removed from the DC equivalent circuit as they act as an open circuit. To obtain the AC equivalent circuit, first replace all capacitors with short circuits. Moreover, according to the superposition theorem, the independent DC voltage source, V_{CC}, is short-circuited for AC analysis. The AC equivalent circuit is shown in Figure 8.42(b).

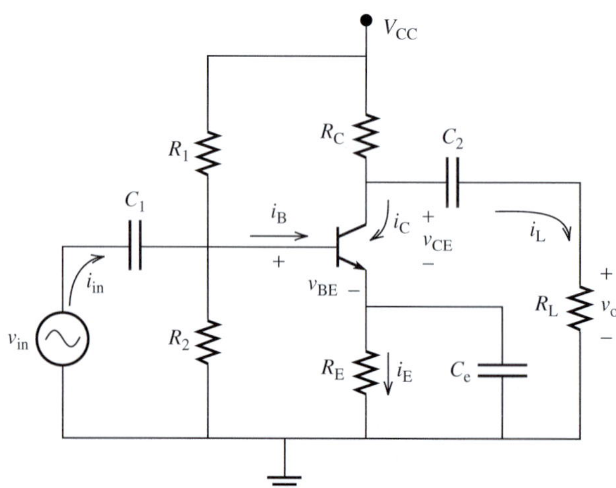

FIGURE 8.41 Circuit for a common emitter amplifier.

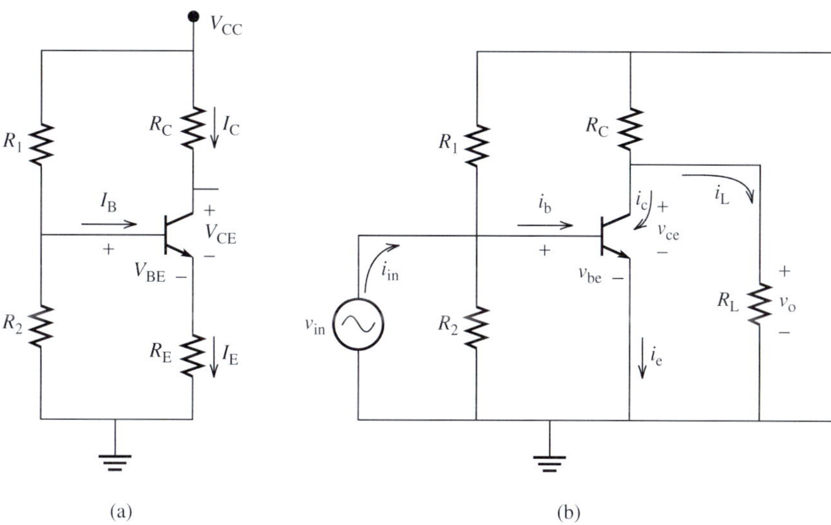

FIGURE 8.42 (a) DC; (b) AC equivalent circuits of the amplifier circuit shown in Figure 8.41.

8.4.2.1 DC ANALYSIS

In order to initiate a DC analysis, first generate a simplified version of Figure 8.42(a) and let $R_E = 0$, as shown in Figure 8.43. Create the simplified version by calculating the Thévenin equivalent circuit of Figure 8.42(a) seen through the base of the transistor. Based on Figure 8.42(a), the parameters in Figure 8.43 correspond to (see Example 8.14 for details):

$$R_B = \frac{R_1 R_2}{R_1 + R_2}$$

$$V_{BB} = \frac{R_1}{R_1 + R_2} \cdot V_{CC}$$

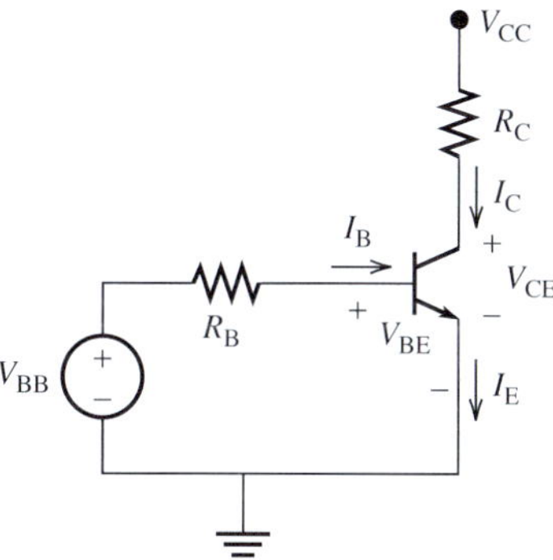

FIGURE 8.43 A basic transistor biasing circuit.

To determine the region of operation, or the Q-point (i.e., the value of voltages and currents), consider the circuit shown in Figure 8.43 and use the following procedure:

1. Replace all capacitors with open circuits.
2. Remove all independent AC sources by replacing voltage sources with short circuits and current sources with open circuits.
3. Check the biasing of the BE junction. If $V_{BB} < 0.7$ V the transistor operates at the cutoff region (OFF) and: $I_B = I_C = 0$, $V_{BE} = V_{BB}$, and $V_{CE} = V_{CC}$.
4. If $V_{BB} > 0.7$ V the transistor is ON and operates in one of two regions: the active region or the saturation region. In either region, the base emitter voltage $V_{BE} = 0.7$ V. To determine the

region of operation, make an assumption. Assume the active region and calculate I_B, I_C, and V_{CE} as follows:

Writing KVL for the input loop yields:

$$V_{BB} - I_B R_B - 0.7 = 0 \tag{8.37}$$

Thus, the current, I_B, is given by:

$$I_B = \frac{V_{BB} - 0.7}{R_B} \tag{8.38}$$

Referring to Table 8.3, with the assumption that the transistor operates in the active region:

$$I_C = \beta I_B \tag{8.39}$$

Writing KVL for the input loop yields:

$$V_{CC} - I_C R_C - V_{CE} = 0 \tag{8.40}$$

Thus, the voltage, V_{CE}, is given by:

$$V_{CE} = V_{CC} - I_C R_C \tag{8.41}$$

To check the validity of the assumption, compare V_{CE} and 0.2 V:

- If $V_{CE} > 0.2$ V, the transistor operates in the active region and the assumption is valid.
- If $V_{CE} < 0.2$ V, the transistor operates in the saturation region and the assumption is not valid.

In the saturation region, the collector–emitter voltage is constant (i.e., $V_{CE} < 0.2$ V). Thus, the value of I_C needs to be recalculated based on the properties of the saturation region. Using Equation (8.41), the collector current can be calculated as follows:

$$I_C = \frac{V_{CC} - 0.2}{R_C} \tag{8.42}$$

As shown in Table 8.3, when a transistor is used as a switch (ON–OFF), there is a transition between high voltage and low voltage. For example, using Figure 8.43, it is observed that when the transistor is in the cutoff region, $I_C = 0$; thus, $V_{CE} = V_{CC}$. When, it is in the saturation region, $V_{CE} = 0.2$, that is, the output voltage is almost zero. See Section 8.4.2 for additional details.

EXERCISE 8.4

For the circuit shown in Figure 8.43, use KVL in the output loop to validate that in the cutoff region: $V_{CE} = V_{CC}$.

EXAMPLE 8.11 Bias Point of a Transistor

For the circuit shown in Figure 8.44, find V_{BE}, I_B, I_C, and V_{CE}, if:

a. $V_{BB} = 0.3$ V
b. $V_{BB} = 2.7$ V
c. $V_{BB} = 6.7$ V

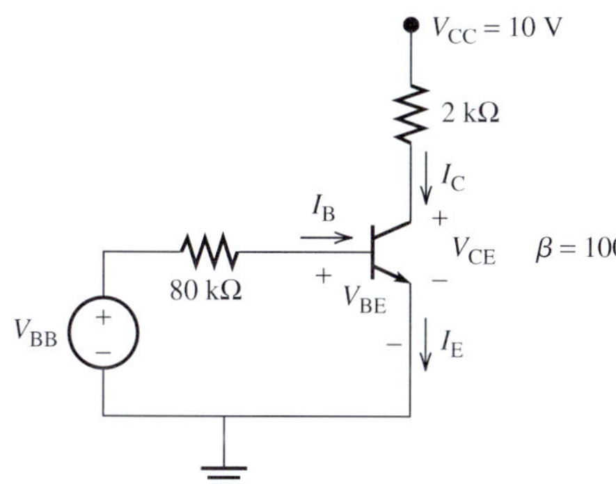

FIGURE 8.44 The circuit for Example 8.11.

SOLUTION

a. Because the value of the base–emitter voltage $V_{BB} = 0.3 \text{ V} < 0.7 \text{ V}$, the transistor operates in the cutoff region. Therefore:

$$I_B = I_C = 0$$

Applying KVL to the input loop:

$$V_{BB} - I_B \times R_B - V_{BE} = 0$$

Therefore:

$$V_{BE} = V_{BB} = 0.3 \text{ V}$$

Applying KVL to the output loop:

$$V_{CC} - I_C \times R_C - V_{CE} = 0$$

Thus:

$$V_{CE} = V_{CC} = 10 \text{ V}$$

b. Because $V_{BB} > 0.7$ V, the transistor is ON and $V_{BE} = 0.7$. Applying KVL to the input loop:

$$2.7 - I_B \times 80 - 0.7 = 0$$

The current, I_B, is given by:

$$I_B = \frac{2.7 - 0.7}{80} = 0.025 \text{ mA}$$

Assuming the active region:

$$I_C = \beta I_B = 100 \times 0.025 = 2.5 \text{ mA}$$

Applying KVL to the output loop yields:

$$V_{CE} = V_{CC} - I_C R_C = 10 - 2.5 \times 2 = 5 \text{ V}$$

Because, the value of the collector–emitter voltage $V_{CE} > 0.2$ V, the transistor does operate in the active region and the assumption is valid.

(continued)

EXAMPLE 8.11 Continued

c. Again, because $V_{BB} > 0.7$ V the transistor is on and, therefore, $V_{BE} = 0.7$. Applying KVL to the input loop:

$$6.7 - I_B \times 10 - 0.7 = 0$$

The current, I_B, is given by:

$$I_B = \frac{6.7 - 0.7}{80}$$
$$= 0.075 \text{ mA}$$

Assuming the active region:

$$I_C = \beta I_B$$
$$= 100 \times 0.075$$
$$= 7.5 \text{ mA}$$

Applying KVL to the output loop yields:

$$V_{CE} = V_{CC} - I_C R_C$$
$$= 10 - 7.5 \times 2$$
$$= -5 \text{ V}$$

Because the value of the collector–emitter voltage $V_{CE} < 0.2$ V, the transistor operates in the saturation region and the assumption was not valid. In the saturation region:

$$V_{CE} = 0.2 \text{ V}$$

Therefore, the collector current corresponds to:

$$I_C = \frac{10 - 0.2}{2}$$
$$= 4.9 \text{ mA}$$

EXAMPLE 8.12 Bias Point of a Transistor

For the circuit shown in Figure 8.45, find V_{BE}, I_B, I_C, and V_{CE}.

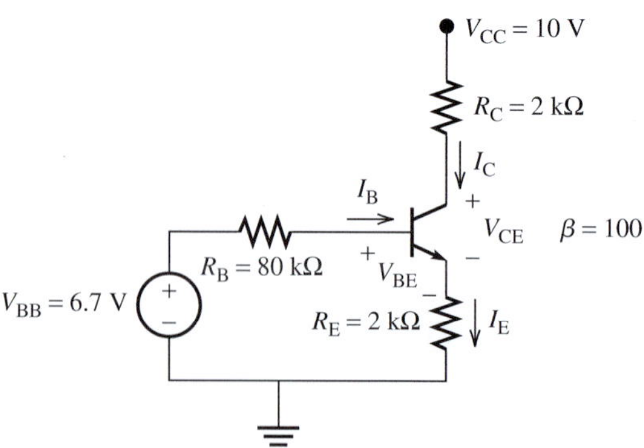

FIGURE 8.45 Circuit for Example 8.12.

SOLUTION

Because $V_{BB} = 6.7 \text{ V} > 0.7$ V, the transistor is ON. Therefore, $V_{BE} = 0.7$. Applying KVL to the input loop:

$$V_{BB} - I_B R_B - V_{BE} - I_E R_E = 0$$

Assuming the active region where $I_C = \beta I_B$, and substituting Equation (8.39) into Equation (8.32):

$$I_E = I_B + \beta I_B = (1 + \beta)I_B$$

Thus, the input loop equation is:

$$V_{BB} - I_B R_B - V_{BE} - (1 + \beta)I_B R_E = 0$$

The base current corresponds to:

$$I_B = \frac{V_{BB} - V_{BE}}{R_B + (1 + \beta)R_E} = \frac{6.7 - 0.7}{80 + 101 \times 2} = 0.0213 \text{ mA}$$

In addition, the collector current corresponds to:

$$I_C = 100 \times 0.0213 = 2.13 \text{ mA}$$

Finally, the collector–emitter voltage corresponds to:

$$V_{CE} = V_{CC} - I_C R_C - I_E R_E = 10 - 2.13 \times 2 - 101 \times 0.0213 \times 2 = 1.45 \text{ V}$$

Because $V_{CE} > 0.2$ V, the transistor operates in the active region. Comparing Examples 8.11 and 8.12, it is notable that adding the resistor, R_E, prevents the transistor from saturating with higher DC voltages.

EXAMPLE 8.13 Transistor Computations

For the circuit shown in Figure 8.46, find the value of the resistor, R_B, which adjusts the collector–emitter voltage to be midway between V_{CC} and ground (i.e., $V_{CE} = V_{CC}/2$).

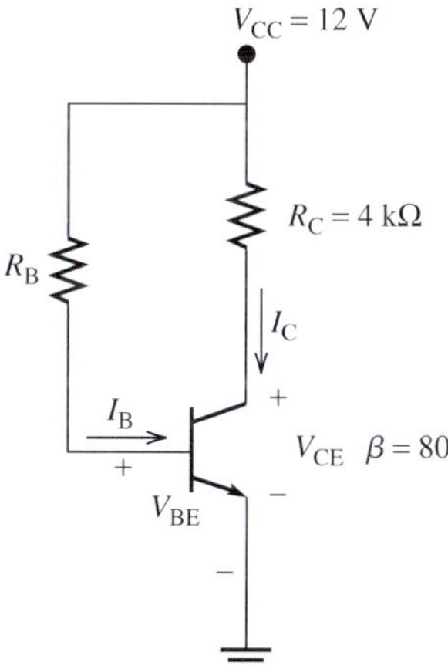

FIGURE 8.46 Circuit for Example 8.13.

SOLUTION

Applying KVL to the output loop yields:

$$12 - I_C \times 4 - V_{CE} = 0$$

(continued)

EXAMPLE 8.13 Continued

But, in this example, $V_{CE} = V_{CC}/2 = 6$ V. Therefore, the collector current corresponds to:

$$I_C = \frac{12 - 6}{4} = 1.5 \text{ mA}$$

Because $V_{CE} > 0.2$ V, the transistor operates in the active region. Therefore, the base current corresponds to:

$$I_B = \frac{I_C}{\beta} = \frac{1.5}{80} = 0.01875 \text{ mA}$$

Applying KVL to the input loop:

$$V_{CC} - I_B R_B - V_{BE} = 0$$

In addition, because the transistor operates in the active region, $V_{BE} = 0.7$ V and the resistor, R_B, corresponds to:

$$R_B = \frac{12 - 0.7}{0.01875} = 602.67 \text{ k}\Omega$$

EXAMPLE 8.14 Bias Point of a Transistor

For the circuit shown in Figure 8.47, find V_{BE}, I_B, I_C, and V_{CE}.

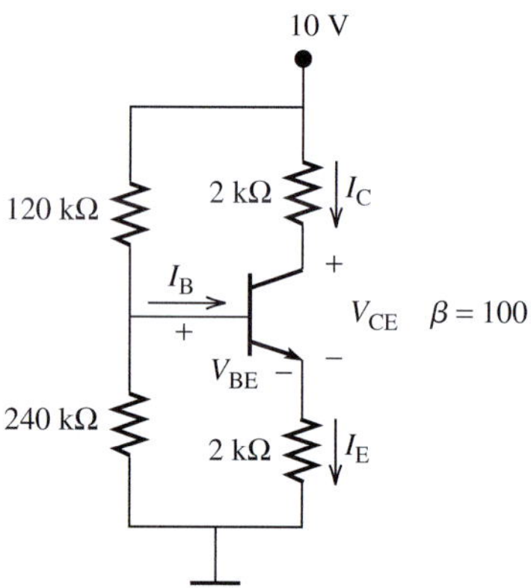

FIGURE 8.47 Circuit for Example 8.14.

SOLUTION

The circuit can be represented as shown in Figure 8.48(a). Applying the Thévenin theorem to find the equivalent circuit between A and B in the circuit at left results in:

$$V_{BB} = V_{th} = 10 \times \frac{240}{240 + 120} = 6.7 \text{ V}$$

and:

$$R_B = R_{th} = 240 \| 120 = \frac{240 \times 120}{240 + 120} = 80 \text{ k}\Omega$$

The simplified circuit is shown in Figure 8.48(b). The solution can be completed easily as discussed in Example 8.12.

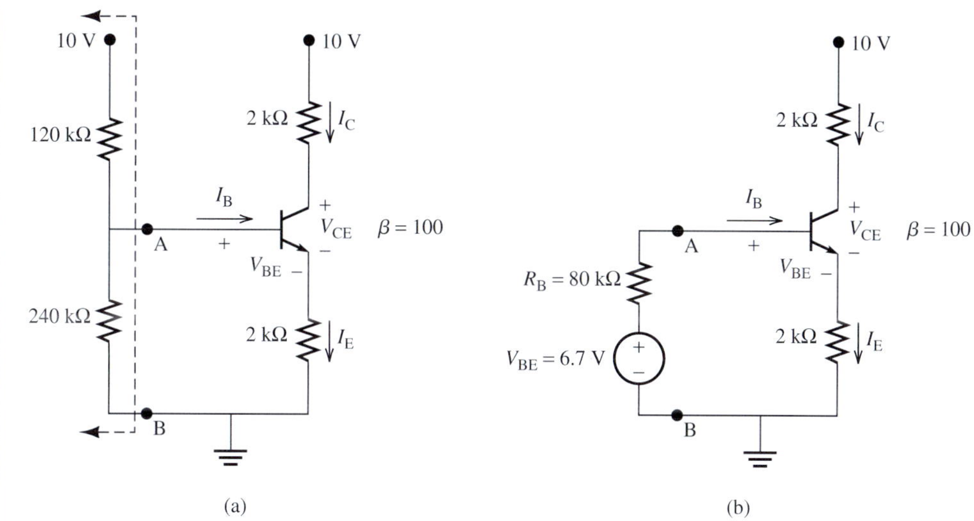

FIGURE 8.48 A simplified representation of the circuit shown in Figure 8.47.

EXERCISE 8.5

Prepare a detailed solution for Example 8.14.

APPLICATION EXAMPLE 8.15 Gyro Sensor

Gyno sensor and its applications was discussed in Example 5.8. Figure 8.49 shows the circuit for a single-axis gyro sensor. The gyro sensor senses a single-axis orientation of a particular moment, and sends the signal to another device that processes the information. In this problem, a gyro sensor is studied without its processing device. The circuit consists of a PNP transistor, with a current gain, β, of 200. The circuit also has two capacitors, one gyro sensor, and a resistor with a fair resistance to prevent short circuits from occurring.

Assume the voltage source is 5 V, the resistance of the resistor is 5 kΩ, the resistance of the gyro sensor is 500 Ω, the capacitance of C_1 is 33 μF, and C_2 is 0.33 μF. Determine the current and voltage across the gyro sensor.

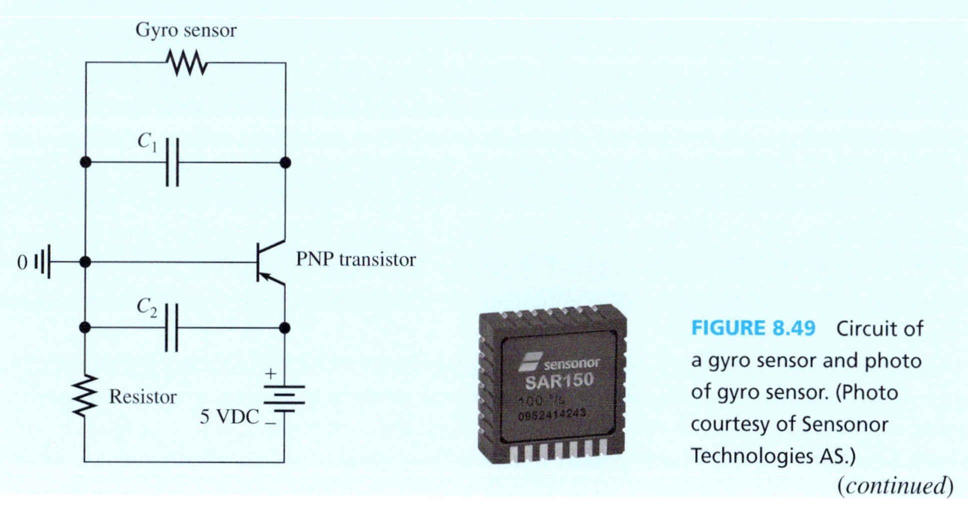

FIGURE 8.49 Circuit of a gyro sensor and photo of gyro sensor. (Photo courtesy of Sensonor Technologies AS.)

(continued)

APPLICATION EXAMPLE 8.15 Continued

SOLUTION

The circuit analysis of PNP transistors is similar to that of NPN transistors, but the directions of node currents and voltage polarities are different [see Figure 8.40(a)]. When a PNP transistor is in the active region, $V_{EB} = 0.7$ V.

Because the voltage source is 5 V > 0.7 V, the transistor is ON. In addition, the two capacitors function as open circuits. Applying KVL, the emitter current across the resistor is:

$$I_E R = 5\text{ V} - 0.7\text{ V}$$

$$I_E = \frac{5 - 0.7}{5000}$$

$$I_E = 0.86\text{ mA}$$

Next, applying KCL to the transistor node, with $I_C = \beta I_B$, the base current is:

$$I_E = I_B + I_C$$

$$I_E = I_B + \beta I_B$$

$$(1 + \beta)I_B = I_E$$

$$I_B = \frac{I_E}{1 + \beta}$$

$$I_B = \frac{0.86}{1 + 200}$$

$$I_B = 0.0043\text{ mA}$$

In addition, the collector current across the gyro sensor is:

$$I_C = 0.8557\text{ mA}$$

Thus, the voltage across the gyro sensor is:

$$V_C = I_C \times R_{gyro}$$

$$V_C = 0.8557 \times 500$$

$$V_C = 0.4279$$

8.4.2.2 AC ANALYSIS

The previous section outlined how to determine the region of operation and how to design a circuit to adjust the Q-point (see Example 8.13). This section explains how to perform AC analysis to determine voltage and current amplification, input impedance, and output impedance of a transistor circuit. Because a transistor is used to amplify AC signals with very low amplitudes, AC signals are usually called *small signals*. One example is the amplification of radio signals in a radio system. The signal received on the radio antenna is in the order of a millivolt, while the batteries used to maintain the operating point of the transistor may have a voltage of 3 V. Comparing a 1-mV signal with 3 V, the first signal is considered to be small. Therefore, these signals are called small signals.

In AC analysis, the transistor is replaced by an equivalent circuit called the *small-signal model*. This small-signal model is appropriate for a transistor if it operates in its active region. It should be noted that the transistor is used as a voltage and/or current amplifier only in its active region. Therefore, it is important to make sure that the transistor operates in this region prior to initiating this analysis. A small-signal model for an NPN transistor is shown in Figure 8.50.

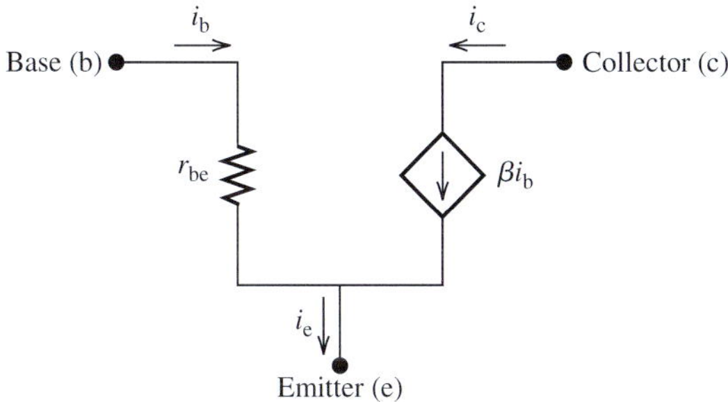

FIGURE 8.50 AC equivalent circuit (small-signal model) of an NPN transistor.

The resistance, r_{be}, is called the base–emitter resistance and it corresponds to:

$$r_{be} = \frac{0.0259}{I_{BQ}} \tag{8.43}$$

Here, I_{BQ} is the value of the base current at the Q-point. The AC collector current is represented by a dependent current source (i.e., a current-controlled current source as discussed in Chapter 3) and it corresponds to:

$$i_c = \beta i_b \tag{8.44}$$

Here, β is the AC gain. For simplicity, this text assumes that both DC and AC gains are equal and both have the same symbol, β.

The following procedure is used to perform AC analysis:

1. First, short-circuit all capacitors
2. Remove all independent DC sources by replacing voltage sources with short circuits and current sources with open circuits
3. Calculate the base-emitter resistance, r_{bc}, using the DC bias current, I_{BQ}, calculated using DC analysis, as shown in Equation (8.43)
4. Replace the transistor with its AC equivalent circuit depicted in Figure 8.50
5. Evaluate the following components:

Voltage gain: Amplification of the voltage:

$$A_v = \frac{v_o}{v_{in}}$$

Current gain: Amplification of the current:

$$A_i = \frac{i_L}{i_{in}}$$

Input impedance, R_i: The equivalent impedance seen by the input source.
Output impedance, R_o: The equivalent impedance seen by the load.

Note: Impedance seen across two terminals is calculated using one of the two methods below:

1. Equate all independent sources to zero, and find the ratio of the voltage to current at the desired terminal
2. Leave all sources, and use the following equation (discussed in Chapter 4):

$$R = \frac{v_{oc}}{i_{sc}}$$

Where, v_{oc} is the voltage seen across the two terminals when it is open circuited, and i_{sc} is the current through that terminal when it is short circuited.

EXAMPLE 8.16 Voltage and Current Gain-Input and Output Resistance

Find A_v, A_i, R_i, and R_o for the circuit shown in Figure 8.51.

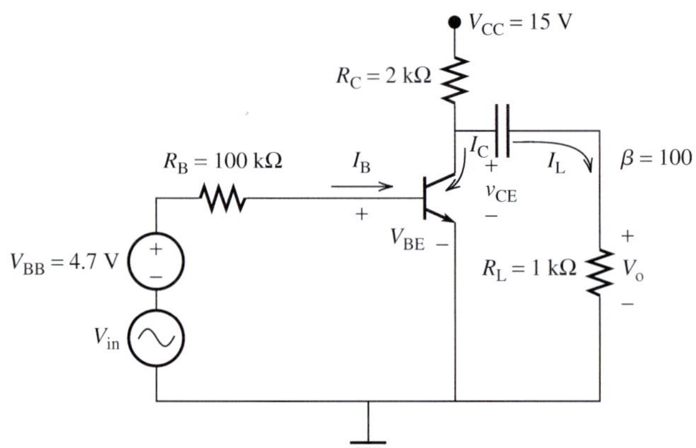

FIGURE 8.51 Circuit for Example 8.16.

SOLUTION

First, calculate the Q-point from the DC analysis to check the region of operation. The DC equivalent circuit is shown in Figure 8.52(a). Applying the same procedure as in the previous section:

$$V_{BE} = 0.7 \text{ V}, I_B = \frac{4.7 - 0.7}{100} = 0.04 \text{ mA}, I_C = 100 \times 0.04 = 4 \text{ mA, and}$$
$$V_{CE} = 15 - 4 \times 2 = 7 \text{ V}$$

The transistor operates in the active region because $V_{CE} > 0.2$ V.

Now, replace C, V_{BB}, and V_{CC} with short circuits, which leads to the AC equivalent circuit shown in Figure 8.52(b). To complete the AC analysis, replace the transistor with the small-signal model given in Figure 8.50. The resistance, r_{be}, is calculated using Equation (8.43):

$$r_{be} = \frac{0.0259}{0.04 \times 10^{-3}} = 647.5\ \Omega$$

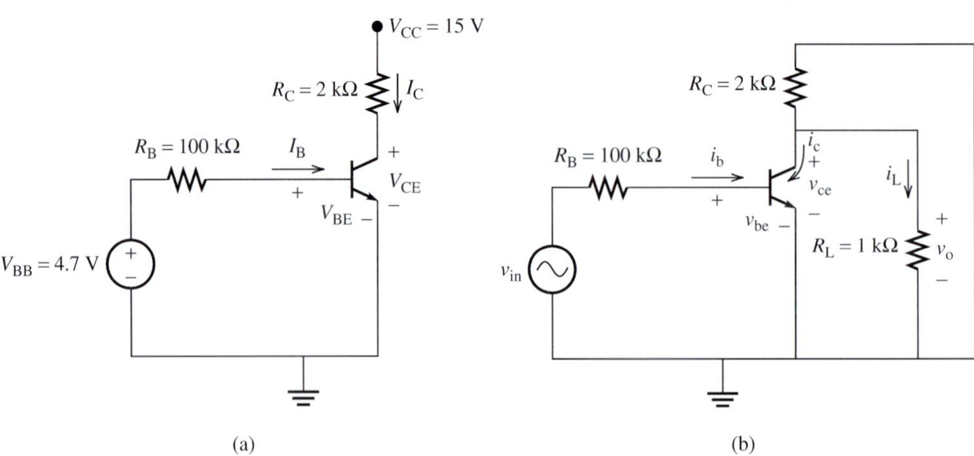

FIGURE 8.52 (a) DC and (b) AC equivalent circuits for the circuit shown in Figure 8.51.

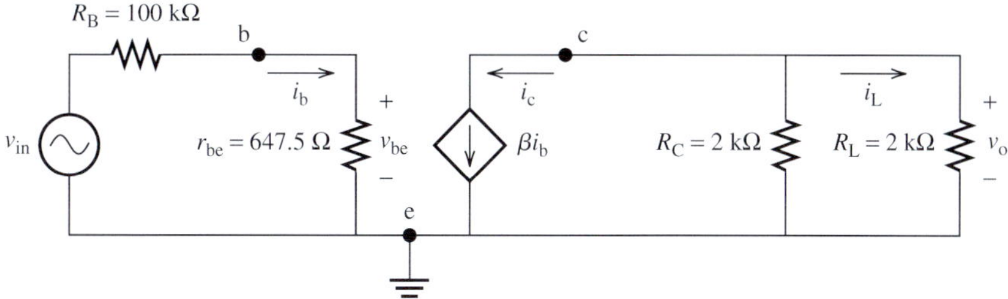

FIGURE 8.53 AC equivalent circuit for the circuit shown in Figure 8.51.

The simplified AC equivalent circuit is shown in Figure 8.53. The output voltage is given by:

$$v_o = i_L R_L$$

Applying the current division law, the load current corresponds to:

$$i_L = -i_c \times \frac{R_C}{R_C + R_L} = -\beta \frac{R_C}{R_C + R_L} \times i_b$$

Applying KVL to the input loop, the base current can be calculated:

$$i_b = \frac{v_{in}}{R_B + r_{be}}$$

Therefore, the output voltage corresponds to:

$$v_o = -\frac{\beta \cdot R_C R_L}{R_C + R_L} \times \frac{v_{in}}{R_B + r_{be}}$$

The voltage gain corresponds to:

$$\begin{aligned} A_v &= \frac{v_o}{v_{in}} \\ &= -\frac{\beta R_C R_L}{(R_C + R_L)(R_B + r_{be})} \\ &= -\frac{100 \times 2000 \times 1000}{(2000 + 1000) \times (1 \times 10^5 + 647.5)} = -0.66 \end{aligned}$$

The negative sign indicates that the input and output voltage are out of phase. In other words, there is a 180° phase angle between the input voltage (at the base) and the output voltage (at the collector). The current gain is:

$$\begin{aligned} A_i &= \frac{i_L}{i_{in}} = -\frac{\frac{\beta R_C}{Rc + R_L} i_b}{i_b} \\ &= -\frac{\beta R_C}{Rc + R_L} \\ &= -\frac{100 \times 2000}{2000 + 1000} \\ &= -66.67 \end{aligned}$$

The negative sign indicates that the direction of the load current must be reversed.
The input impedance seen by the input source is:

$$R_i = R_B + r_{be} = 100.647 \text{ k}\Omega$$

(continued)

EXAMPLE 8.16 Continued

The output impedance seen by the load corresponds to:

$$R_o = R_C = 2\ \text{k}\Omega$$

Note that the voltage gain is less than unity, which means that the transistor attenuates the input voltage signal instead of amplifying it. The reason for that is the existence of the large resistance, R_B, in the denominator. The resistance, R_B, cannot be eliminated or reduced because it is used to adjust the Q-point and limit the base current. However, its effect can be eliminated by connecting it in parallel with the input source (see the next example).

EXAMPLE 8.17 Voltage and Current Gain-Input and Output Resistance

Considering the circuit shown in Figure 8.54, calculate voltage gain, current gain, input impedance, and output impedance.

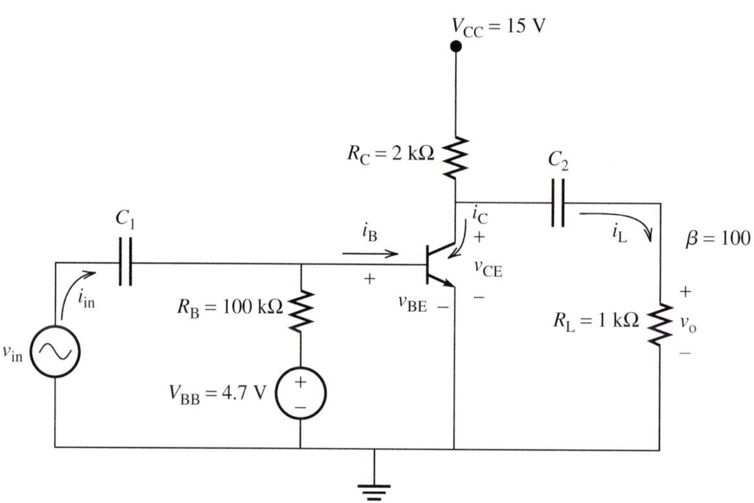

FIGURE 8.54 Circuit for Example 8.16.

SOLUTION

The DC circuit and the operating point are the same as the previous example. The AC equivalent circuits are shown in Figure 8.55. The base–emitter resistance is calculated in the previous example and is $r_{be} = 647.5\ \Omega$. Referring to Figure 8.55(b), the output voltage corresponds to:

$$v_o = -\frac{\beta R_L R_C}{R_L + R_C} i_b$$

The base current corresponds to:

$$i_b = \frac{v_{in}}{r_{be}}$$

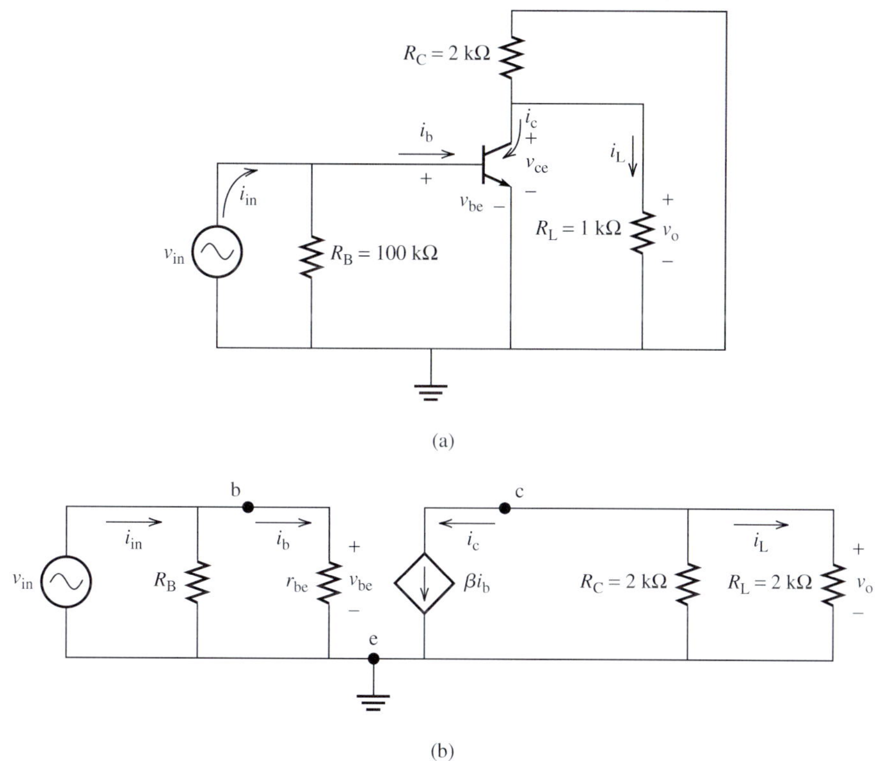

FIGURE 8.55 AC equivalent circuits for the circuit shown in Figure 8.54.

Therefore, the output voltage can be written as:

$$v_o = -\frac{\beta R_L R_C}{R_L + R_C} \frac{v_{in}}{r_{be}}$$

Moreover, the voltage gain corresponds to:

$$\begin{aligned} A_v &= -\frac{\beta R_L R_C}{(R_L + R_C) r_{be}} \\ &= -\frac{100 \times 1000 \times 2000}{(1000 + 2000) \times 647.5} \\ &= -102.96 \end{aligned}$$

Next, the current gain is:

$$\begin{aligned} A_i &= \frac{i_L}{i_{in}} \\ &= -\frac{\dfrac{\beta R_C}{Rc + R_L} i_b}{i_{in}} \\ &= -\frac{\beta R_C}{Rc + R_L} \frac{R_B}{R_B + r_{be}} \text{(using current division)} \\ &= -\frac{100 \times 2000}{2000 + 1000} \times \frac{10^5}{10^5 + 647.5} \\ &= -66.24 \end{aligned}$$

$$R_o = R_C = 2\ k\Omega$$

EXAMPLE 8.18 Voltage and Current Gain-Input and Output Resistance

For the circuit shown in Figure 8.56 find A_v, A_i, R_i, and R_o.

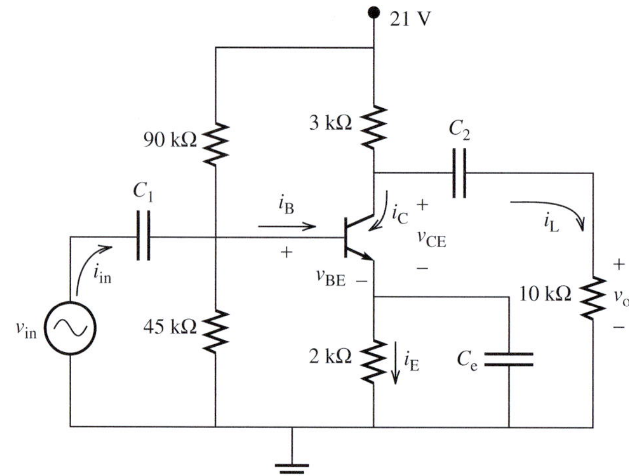

FIGURE 8.56 Circuit for Example 8.17.

SOLUTION

DC analysis

The DC equivalent circuits are shown in Figure 8.57.

The Thévenin voltage, V_{BB}, corresponds to:

$$V_{BB} = 21 \times \frac{45}{45 + 90} = 7 \text{ V}$$

Also, the Thévenin resistance corresponds to:

$$R_B = \frac{45 \times 90}{45 + 90} = 30 \text{ k}\Omega$$

Referring to Figure 8.57(b), and assuming the active region, the base current corresponds to:

$$I_B = \frac{V_{BB} - 0.7}{R_B + (1 + \beta)R_E}$$
$$= \frac{7 - 0.7}{30 + 101 \times 2}$$
$$= 0.0272 \text{ mA}$$

The collector current is given by:

$$I_C = \beta I_B = 100 \times 0.0272 = 2.72 \text{ mA}$$

The collector–emitter voltage corresponds to:

$$V_{CE} = V_{CC} - I_C R_C - I_E R_E$$
$$= 21 - 2.72 \times 3 - (2.72 + 0.0272) \times 2$$
$$= 7.37 \text{ V}$$

Because $V_{CE} > 0.2$ V, the transistor does operate in the active region; therefore, the assumption was correct.

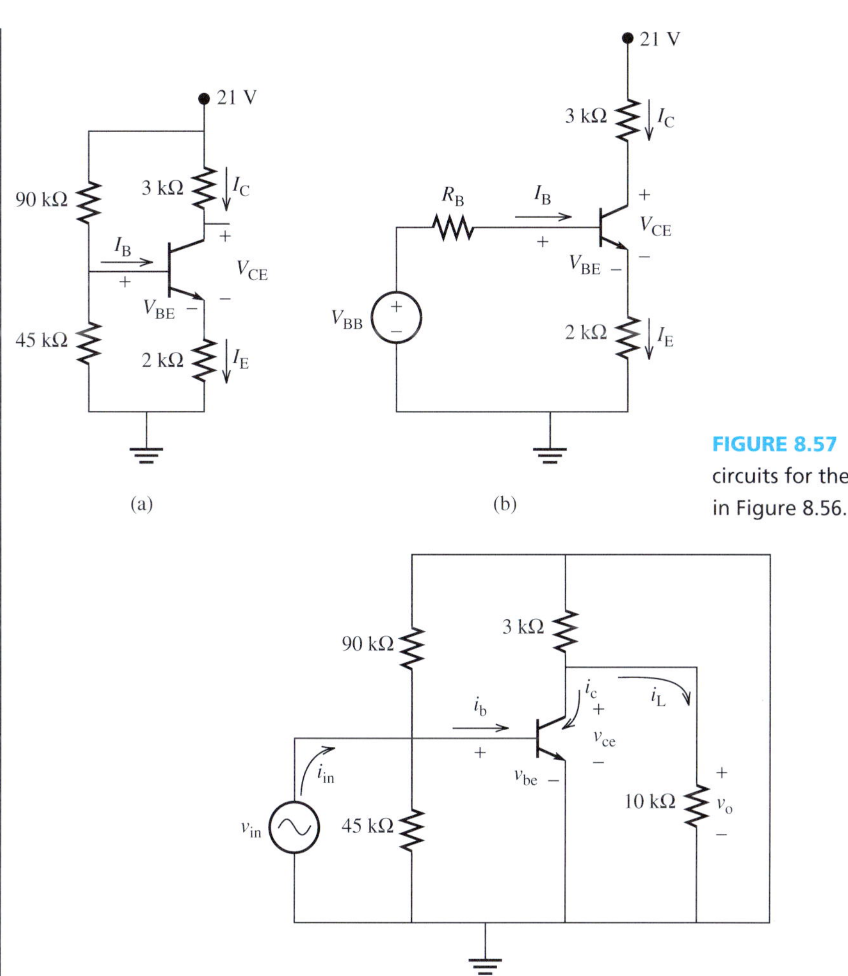

FIGURE 8.57 DC equivalent circuits for the circuit shown in Figure 8.56.

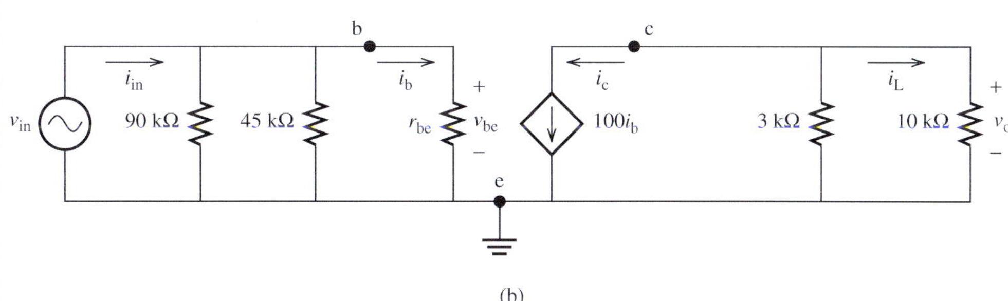

FIGURE 8.58 AC equivalent circuits for the circuit shown in Figure 8.56.

AC analysis

The AC equivalent circuits are shown in Figure 8.58. The base–emitter resistance corresponds to:

$$r_{be} = \frac{0.259}{0.0272 \times 10^{-3}} = 953.78\ \Omega$$

Following the same procedure as in the previous examples:

$$A_v = \frac{-100 \times 3000 \times 10{,}000}{(3000 + 10{,}000) \times 953.78} = -241.95$$

(continued)

EXAMPLE 8.18 Continued

$$A_i = -241.95 \times \frac{90{,}000 \| 45{,}000 \| 953.78}{10{,}000} = -22.37$$

$$R_i = 90{,}000 \| 45{,}000 \| 953.78 = 924.39\ \Omega$$

$$R_o = R_C = 3\ \text{k}\Omega$$

8.4.3 Transistors as Switches

A transistor can function as a switch between the collector and the emitter terminals where the base terminal is used to control the status of the switch. As discussed in the previous sections, when the biasing voltage applied to the transistor base is less than 0.7 V, the transistor operates in the cutoff region, no current passes through the collector (i.e., $I_C = 0$), and the switch will be open. When the value of the biasing voltage is large enough, the transistor saturates, the voltage across the transistor, V_{CE}, is at the minimum (i.e., $V_{CE} = 0.2$ V), and the switch will be closed. The transistor that operates between cutoff and saturation is called an electronic switch. The switch is controlled by an electric voltage on the transistor base.

EXAMPLE 8.19 Transistor Characteristic

Consider the circuit shown in Figure 8.59. Sketch the transfer characteristics (i.e., the relationship between the input voltage, V_{BB}, and the collector–emitter voltage, V_{CE}, and calculate the minimum value of V_{BB} required such that the transistor operates in the saturation region (the switch is ON).

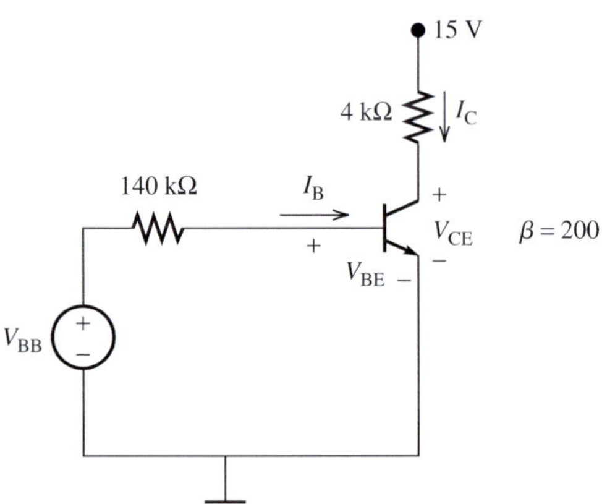

FIGURE 8.59 The circuit for Example 8.18.

SOLUTION

Three cases are possible:

1. When the input voltage $V_{BB} < 0.7$ V, the transistor operates in the cutoff region. Therefore, $I_B = I_C = 0$ and the voltage. V_{CE}, corresponds to:

$$V_{CE} = V_{CC} = 15\ \text{V}$$

2. As the value of V_{BB} increases beyond 0.7 V, both the base current and the collector current begin to flow. Therefore, the voltage, V_{CE}, will decrease. Because V_{CE} is still high (i.e., $V_{CE} > 0.2$ V), the transistor operates in the active region and the relationship between V_{BB} and V_{CE} is computed as follows:

Applying KVL to the output loop, the collector–emitter voltage corresponds to:

$$V_{CE} = V_{CC} - I_C R_C$$

Replacing I_C with βI_B yields:

$$V_{CE} = V_{CC} - \beta I_B R_C$$

Find I_B by applying KVL to the input loop. Substituting I_B by $(V_{BB} - 0.7)/R_B$:

$$\begin{aligned} V_{CE} &= V_{CC} - \frac{\beta R_C(V_{BB} - 0.7)}{R_B} \\ &= 15 - \frac{200(V_{BB} - 0.7)}{140} \\ &= 15 - \frac{10V_{BB}}{7} + 1 \\ &= 16 - 1.43V_{BB} \end{aligned}$$

This equation shows that the value of V_{CE} decreases linearly with V_{BB}. But, the transistor will operate in the saturation region as the value of V_{CE} reaches 0.2 V (see Figure 8.60). In other words, the transistor saturates when:

$$16 - 1.43V_{BBsat} = 0.2$$

where V_{BBSat} is the value of the input voltage required to operate the transistor in the saturation region and it corresponds to:

$$V_{BBsat} = \frac{16 - 0.2}{1.43} = 11.06 \text{ V}$$

3. When $V_{BB} > V_{BBSat} = 11.06$ V, the transistor operates in the saturation region. Therefore, $V_{CE} = 0.2$ V. As a summary, the collector–emitter voltage corresponds to:

$$V_{CE} = \begin{cases} 15 \text{ V} & V_{BB} < 0.7 \text{ V} \\ 16 - 1.43\,V_{BB} & 0.7 \text{ V} < V_{BB} < 11.06 \text{ V} \\ 0.2 \text{ V} & V_{BB} > 11.06 \text{ V} \end{cases}$$

A plot of the transfer characteristics is shown in Figure 8.60. Note that the transistor or the switch is considered OFF if $V_{BB} < 0.7$ V. Moreover, it will be ON if $V_{BB} > 0.7$ V. Therefore, the minimum voltage required to turn the switch ON is the same as the minimum voltage required to saturate the transistor and corresponds to $V_{BBSat} = 11.06$ V.

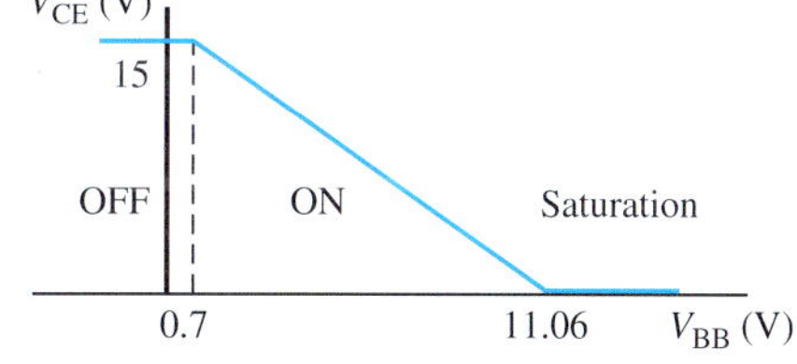

FIGURE 8.60 Transfer characteristics for the circuit shown in Figure 8.59.

8.4.4 Field-Effect Transistors

This section introduces field-effect transistors (FETs). The operational principle of the FET is based on the formation of a channel between two terminals. This channel is called the conduction channel and is formed by an electric field created between two terminals of the transistor. Therefore, it is called a *field-effect transistor*. There are many different kinds of FET (e.g., MOSFET, JFET, MESFET, MODFET, etc.), but the most commonly used FET is

(a) (b) (c)

FIGURE 8.61 (a) Top view of an integrated circuit (Used with permission from Stockex/Alamy.). (b) Packaged IC microchip with large pins, ready to be mounted on a printed circuit board (PCB) (Used with permission from Timothy Hodgkinson/Shutterstock.com.). (c) Scanning electron microscope image of the cross-sectional view of a single MOSFET device, the most elemental unit of an IC (Image courtesy of MuAnalysis.).

the metal–oxide–semiconductor field-effect transistor (MOSFET). Complex digital integrated circuits (ICs) such as microprocessors and memory chips are primarily made up of MOSFETs [see Figure 8.61(a) and 8.61(b)]. The operational principles of all FETs are approximately the same; therefore, this section will study MOSFETs as a case study to understand FETs.

MOSFETs are the basic building block of very large-scale integration (VLSI) circuits [see Figure 8.61(c)]. MOSFETs can operate at a much lower current than the BJTs discussed earlier in the chapter. Therefore, they consume less power ($P = I \times V$) and generate less heat per area. A MOSFET also occupies a smaller chip area and requires a lower number of fabrication steps compared to a BJT. Portable handheld devices such as MP3 players, computer game players, and laptops have limited battery lifetime. Therefore, low-power circuit design using FETs is absolutely essential to enable minimum power consumption.

An analogy to understand the hierachy in an IC is to think of the MOSFET as the brick that builds a house. If a few "bricks" are stacked together, a new digital module called a *logic gate* (NOT, OR, AND, etc.) is built. Logic gates are discussed in detail in Chapter 10. When a few logic gates are connected together, other more complex digital modules can be built (e.g., latches, flip-flops). These modules enables larger digital modules like adders and multiplexers. The functionality of a circuit becomes more complex as its size increases. The scale of an IC is analogous to a city. ICs are connected to each other using a PCB, which is similar to the highway system that connects two cities. This section focuses on the "brick" (FETs) and does not discuss more advanced digital modules. More advanced courses in electrical engineering like digital design or very large-scale integration (VLSI) design are recommended for further understanding of these systems. Chapter 10 offers a summary of digital circuits.

8.4.4.1 BASIC OPERATION OF MOSFETS

Similar to BJTs, FETs are used to make amplifiers and logic switches. The amplifying function of a MOSFET is similar to that of a BJT, which was explained at length in the previous section. This section focuses on using MOSFETs as logic switches, which are used in the world of digital design. MOSFET analysis can be very complex. This chapter first explains MOSFET structure in simple (but accurate) terms and then outlines the details of FET analysis.

Figure 8.62(a) shows a cross-sectional view of the physical layout of a single MOSFET device. A MOSFET has three terminals: (a) drain, (b) source, and (c) gate. The drain and source junctions define the direction of the electron flow, which is opposite to the direction of the current flow. There are two kinds of MOSFET, PMOS (p-type MOSFET) and NMOS (n-type MOSFET). These types are defined based on the nature of the device: if the source drain junctions are n-type, the MOSFET is called NMOS; if the source and drain junctions are p-type, the MOSFET is called PMOS. Figure 8.62(b) shows the symbols that are normally used to represent these MOSFETs.

As shown in Figure 8.62(c), an NMOS is fabricated on a p-type silicon substrate. The drain and source junctions of the NMOS are formed by an n-type semiconductor. When a positive

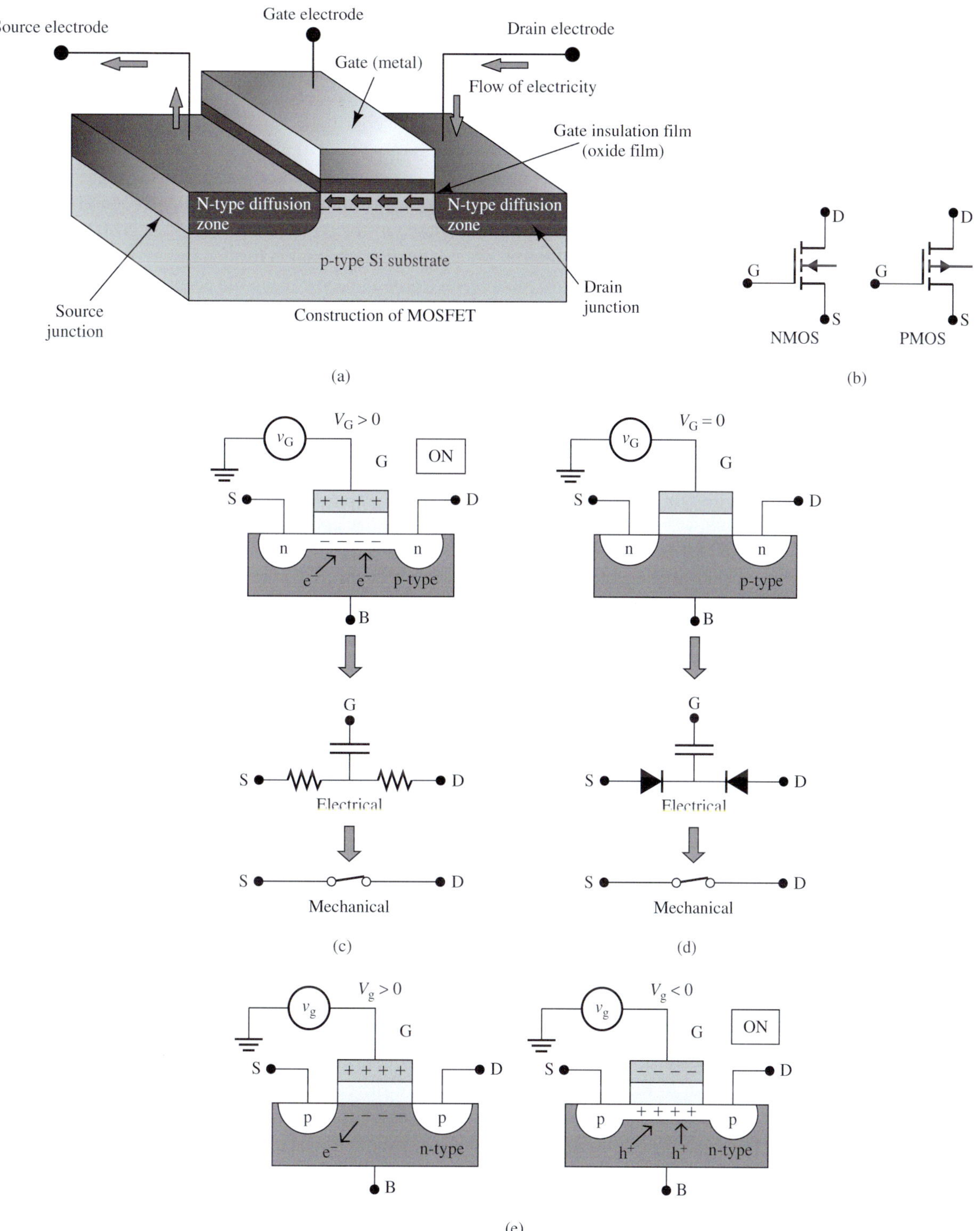

FIGURE 8.62 (a) Physical structure of an n-type MOSFET (NMOS). For PMOS, the silicon substrate needs to be n-type. (b) Symbolic representation of NMOS and PMOS. (c) When V_G is positive, electrons in the p-type substrate are attracted to the oxide–silicon interface, and form an n-type conduction channel. The electrical model is represented by resistors in series. The transistor is in its ON state. (d) When $V_G = 0$, the area underneath the oxide layer is still p-type, which forms a "back-to-back" diode with the n region, as shown in the electrical representation. The transistor is in its OFF state; (e) For PMOS, the substrate is n-type. The p-type inversion channel can be formed by applying a negative bias to the gate electrode, turning the transistor from OFF (left) to ON (right).

voltage bias is applied to the gate of an NMOS, an electric field is generated across the insulating silicon dioxide layer. The electric field generated by the positive charges attracts negative charges (electrons) from the silicon substrate to the oxide–silicon interface. The accumulation of free electrons forms an n-type conduction channel. This conduction channel is called an n-type because it has excess free electrons, similar to the concept discussed earlier in the diode section (Section 8.2).

Note that although the p-type substrate is rich with free-hole carriers, there are still some minority electron carriers available in the p-type substrate. With the conduction channel formed, current ($I = dQ/dt$) flows freely from the source to the drain junction and the NMOS is said to be in its ON state. In electrical terms, the conduction channel is formed by inverting the original p-type substrate between the two junctions to form an n-type using the accumulation of free electrons. Thus, it is also called the *inversion channel.* Conceptually, using very crude electrical terms, under positive gate bias, the drain–source channel can now be modeled by a resistor. A closed switch is a mechanical analogy of the NMOS in its ON state, as illustrated in Figure 8.62(c). A water pipe controlled by the water tap (gate) is another analogy. When the water tap is turned ON, the water (electrons) will start flowing.

When the voltage applied to the gate of the NMOS is zero, there are no electrons accumulating at the oxide–silicon interface, that is, in between the source and the drain junctions. The current flow from source to drain is now impeded and thus, the NMOS is in its OFF state. Under these voltages, hole carriers (discussed in Section 8.2) are accumulated at the oxide–silicon interface. A rough but easy-to-understand electrical model is shown in Figure 8.62(d), where the drain–source channel can now be seen as two p–n diodes connected serially in reverse direction (i.e., back-to-back diodes). This simply means that the channel is now under reverse-biased condition and there is minimal (if not zero) current flow. Again, a good mechanical analogy of this device at its OFF state is an open switch, or a water pipe with a closed tap. The electrical behavior of a PMOS is just the opposite of an NMOS. Negative gate voltage is needed to turn the PMOS to an ON state, and a positive gate voltage is needed to turn it to its OFF state, as illustrated in Figure 8.62(e).

8.4.4.2 DESIGN OF DYNAMIC RANDOM ACCESS MEMORY (DRAM) USING MOSFET

The simplest example of a practical MOSFET application is its use in the design of dynamic random access memory (DRAM). DRAM chips are used in almost every computer. An example is shown in Figure 8.63(a). A DRAM is an IC chip used to store data temporarily while the microprocessor computes the data. The beauty of DRAM is that it only requires one transistor and one capacitor to represent one bit of data (either logic 0 or logic 1) as shown in Figure 8.63(b) and 8.63(c). Because the transistor and capacitor can be made very small in size, the DRAM structure can achieve very high density compared to other kinds of memory, such as static-RAM (SRAM), which consists of six transistors.

The capacitor stores electron charges in the DRAM cell structure and discharges electrons over time; therefore, the DRAM needs to be recharged repeatedly. DRAM loses its stored data once the computer is turned off. Therefore, it is considered to be one kind of volatile memory chip. As shown in Figure 8.63(c), the voltage of the capacitor of one DRAM cell is originally zero. The data represented by zero capacitor voltage is logic 0. Once V_G is turned to high voltage ($V_G > 0$), the NMOS transistor is turned ON. Then, turning V_S to high voltage ($V_S > 0$) enables current to flow through the conduction channel of the NMOS, and then, through the capacitor. This increases the voltage of the capacitor. The high voltage of the capacitor represents one bit of data, called logic 1. The single DRAM cell in Figure 8.63(c) can be used to build huge arrays of DRAM. Figure 8.63(d) shows an example of a 4×4 array, which can store 16 bits of data.

8.4.4.3 DESIGN OF A MOSFET AS A LOGIC GATE IN A MICROPROCESSOR

The microprocessor is the heart of all electronic devices. It is capable of compiling a large number of calculations in a very short time period depending on the frequency of the processor. The frequency of the processor is related to how fast a transistor can switch from ON state to OFF state or vice versa. A 2.0-GHz processor is composed of transistors with the switching

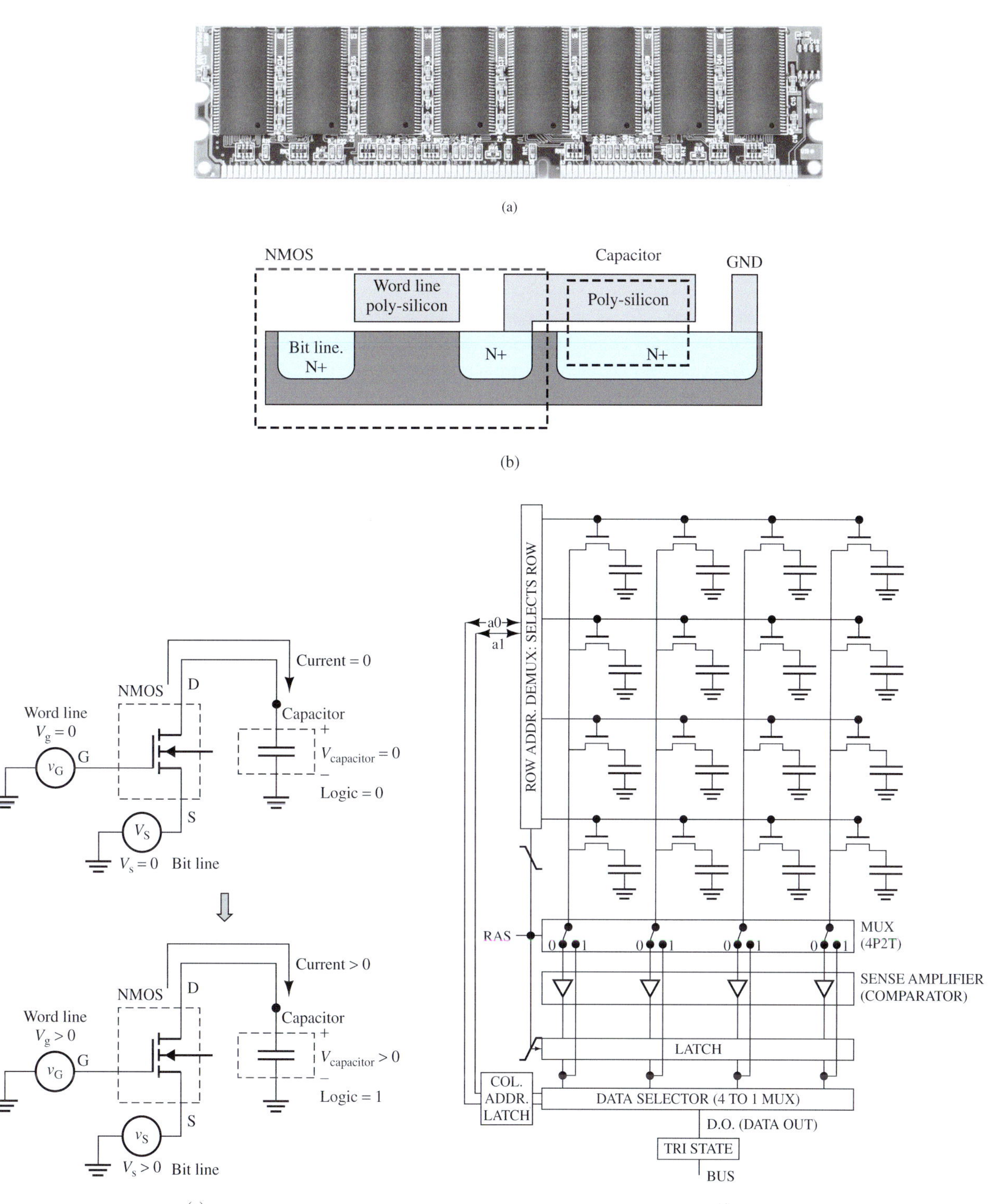

FIGURE 8.63 (a) An EDO DRAM memory module found in a personal computer. A 4-megabyte DRAM chip consist of 32 million DRAM cells. (Used with permission from Timothy Hodgkinson/Shutterstock.com.) (b) Cross-sectional view of the physical layout of a single DRAM cell. Note that the empty gap between both highly conductive poly-silicon and the n+ region of the capacitor is an insulator. (c) Electrical representation of a single DRAM cell, made of one MOSFET and one capacitor. Turning V_G and V_S to high voltage turns the data stored in the DRAM cell from logic 0 to logic 1. (d) The principles of operation of a DRAM read, for a simple 4 × 4 array. Accessing an individual DRAM cell is enabled by selecting one specific bit line and one word line. (Drawings adapted with permission from Steven Mann, University of Toronto.)

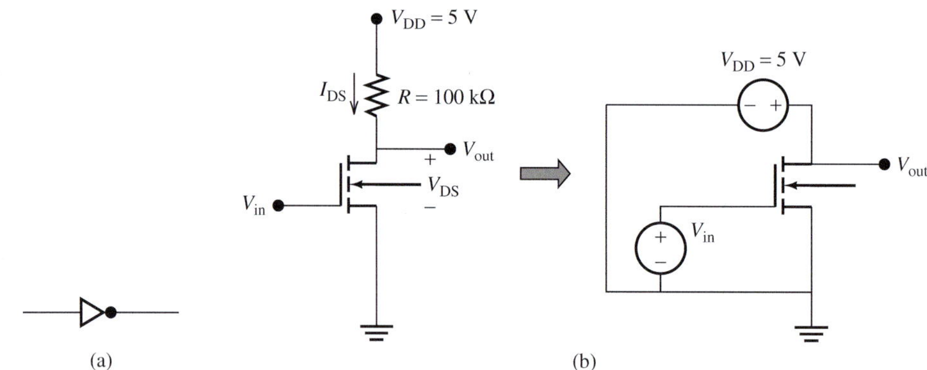

FIGURE 8.64 (a) Symbol of a NOT gate, also called an inverter. (b) A NOT gate made with one NMOS and one resistor. The equivalent electrical representation is illustrated on the right.

speed of 2×10^9 times in 1 s. As mentioned earlier, microprocessors are made of several millions of transistors. These transistors enable a very basic digital module logic gate. Examples of basic logic gates are the NOT gate, OR gate, and AND gate. These individual gates are generally interconnected in a high-density large network of IC to perform complex computations.

It is helpful to start with the simplest example of the NOT gate, also called an *inverter* because this gate produces an inverted signal. The symbolic representation of the NOT gate is shown in Figure 8.64(a) (see Chapter 10 for details). There is more than one way to design a NOT gate; the simple way is to use one NMOS and one resistor, as shown in Figure 8.64(b). The function of the NOT gate is very straightforward; it produces a voltage $V_{out} = V_{high}$ when an input voltage $V_{in} = V_{low}$ is applied to its input, and vice versa. The truth table of the NOT gate is listed in Table 8.4.

The operation of this inverter can be analyzed using the KVL rule. From Figure 8.64(b), $V_{out} = V_{DS} = V_{DD} - I_{DS}R$. When the input voltage, V_{in}, is low or zero, the NMOS is in its OFF state, and it behaves as an open circuit. The circuit shown in Figure 8.64(b) can be redrawn and represented with the crude model shown in Figure 8.65(a). In this case, $I_{DS} = 0$A because an NMOS is an open switch. Therefore, the voltage drop across the resistor, R, is zero, and $V_{out} = V_{DD} - (0)R = V_{DD} = 5$ V. However, when V_{in} is high (assume, for example, $V_{in} = 5$ V), as shown in Figure 8.65(b); the NMOS is in its ON state, and it behaves similar to a closed circuit. In this case:

$$I_{DS} = \frac{V_{DD}}{R} = \frac{5}{100 \times 10^3} = 0.05 \text{ mA}$$

In addition, V_{out} is directly connected to the ground through the closed circuit of the NMOS, so $V_{out} = 0$ V. Note that the result of this analysis is consistent with the characteristics in the truth table of the NOT gate in Table 8.4.

TABLE 8.4 Truth Table of the NOT Gate

V_{in}	V_{out}
High voltage (logic 1)	Low voltage (logic 0)
Low voltage (logic 0)	High voltage (logic 1)

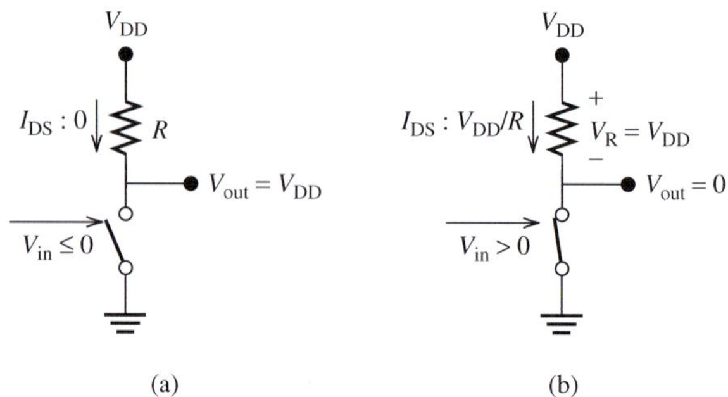

FIGURE 8.65 (a) Equivalent circuit of Figure 8.64(b) when V_{in} = Low is applied to the gate of the NMOS and it acts as an open circuit. (b) Equivalent circuit of Figure 8.64(b) when V_{in} = High is applied to the gate of the NMOS and it acts as a short circuit.

EXAMPLE 8.20 NOT Gate Analysis

The NOT gate analysis in Figure 8.65(a) and 8.65(b) assumes an ideal NMOS, that is, there is no resistance in the switch. In reality, there is a nonzero voltage drop across the NMOS channel because the resistance of the NMOS conduction channel (R_{DS}) is nonzero. Assume $V_{DD} = 5$ V, $V_{in} = 5$ V and $R = 100$ kΩ, $R_{DS} = 1$ Ω. Draw the simplified equivalent electrical circuit and calculate V_{out} and the current flowing through the transistor in its ON state.

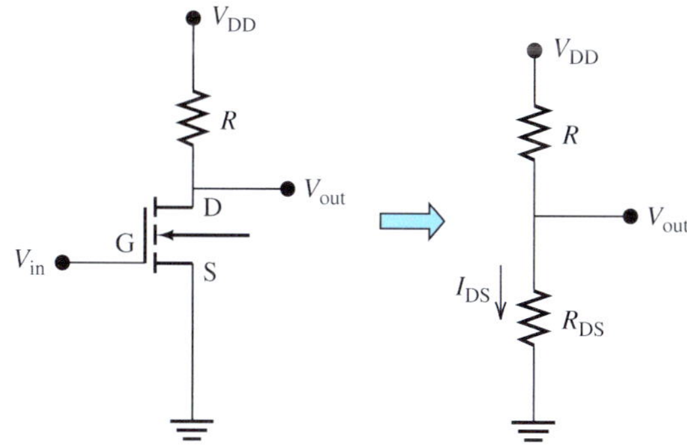

FIGURE 8.66 A simplified model of an NMOS NOT gate when $V_{in} > 0$.

SOLUTION

For an NMOS to be in its ON state, V_{in} has to be at a high voltage (5 V). In reality, as shown in Figure 8.66, the closed switch can be replaced with a resistor, representing the resistance of the conduction channel. This is a simple voltage divider problem, where:

$$V_{out} = V_{DS} = V_{DD} \times \frac{R_{DS}}{R + R_{DS}}$$

Therefore:

$$V_{out} = 5 \times \frac{1}{1 + 100 \times 10^3} = 0.05 \text{ mV}$$

which is very close to 0 V, representing logic = 0.

In addition, writing KVL for Figure 8.66:

$$I_{DS} = \frac{V_{DD}}{R + R_{DS}} = \frac{5}{100 \times 10^3 + 1} = 0.049995 \text{ mA}$$

EXAMPLE 8.21 Power Consumption of NOT Gate

Calculate the power consumption of one NOT logic gate when V_{in} = high.

SOLUTION

From the solution of Example 8.20, when V_{in} is high, NMOS is in its ON state. In this case, current flows through the resistor and also the NMOS conduction channel.
Therefore, the power consumption is:

$$P = V_{DD} \times I_{DS} = 5 \times 0.0499995 \text{ mA} = 0.249975 \text{ mW}$$

which is approximately 0.25 mW.

EXAMPLE 8.22 Logic Gate Power Consumption

A microprocessor consists of 1 million logic gates. Assume that this microprocessor is placed in a package that dissipates a maximum power of 4 W and all logic gates have the same power consumption. For this scenario:

a. Compute the maximum power that is required for each logic gate. Assume that all logic gates are in the ON state.

b. When the microprocessor functions normally, all logic gates switch between ON and OFF states. If the supply voltage is reduced to $V_{DD} = 1$ V, calculate the current that should be used by each gate. Assume that half of the logic gates are in conducting state (ON) at any given time.

SOLUTION

a. Note that the maximum power is the largest power consumption a single logic gate can afford. If the power dissipation of each logic gates exceeds the maximum power, the total power dissipation of the IC will exceed the limit that the package can withstand and may cause heat damage to the chip. Therefore, the maximum power required for each logic gate is 4 W/1 million = 4 μW.

b. Some logic gates are in their ON and some are in their OFF state. Thus, approximately 1 million/2 = 500,000 logic gates are active (ON state) at any given time. Power consumption for each logic gate is:

$$4\text{W}/0.5 \text{ million} = 0.008 \text{ mW} = 8 \ \mu\text{W}$$

$$P = V_{DD} \times I_{DS}$$

Thus:

$$8 \ \mu\text{W} = 1 \ \text{V} \times I_{DS}$$

Thus, the current that should be used by each logic gate is $I_{DS} = 8$ μA.

8.4.4.4 DESIGN OF MOSFETS AS AMPLIFIERS

MOSFETs can also be used as amplifiers. Understanding a more accurate model of the MOSFET and the current–voltage characteristics of MOSFET will help to explain how they can be used as amplifiers. The electrical model in Figure 8.66 is oversimplified. An accurate model for an NMOS can be created by considering a back-to-back pair of diodes due to the n–p–n structure from the drain junction to the channel and to the source junction, as illustrated in Figure 8.67.

When the model of Figure 8.67 is considered, the current–voltage characteristic of a MOSFET corresponds to:

$$I_{DS} = 0, \text{ when } V_{GS} < V_T \text{ (cuttoff region)}$$

$$I_{DS} = K\left[2(V_{GS} - V_T)V_{DS} - V_{DS}^2\right], \text{ when } V_{DS} < V_{GS} - V_T \text{ and } V_{GS} > V_T \text{ (triode region)}$$

$$I_{DS} = K(V_{GS} - V_T)^2, \text{ when } V_{DS} \geq V_{GS} - V_T \text{ and } V_{GS} > V_T \text{ (saturation region)} \quad \textbf{(8.45)}$$

Here, V_{DS} is the drain–source voltage ($V_{DS} = V_D - V_S$) and V_{GS} is the gate–source voltage ($V_{GS} = V_G - V_S$). Here, V_S is the source voltage and normally it is connected to ground, therefore:

$$V_{DS} = V_D - 0 = V_D$$

$$V_{GS} = V_G - 0 = V_G$$

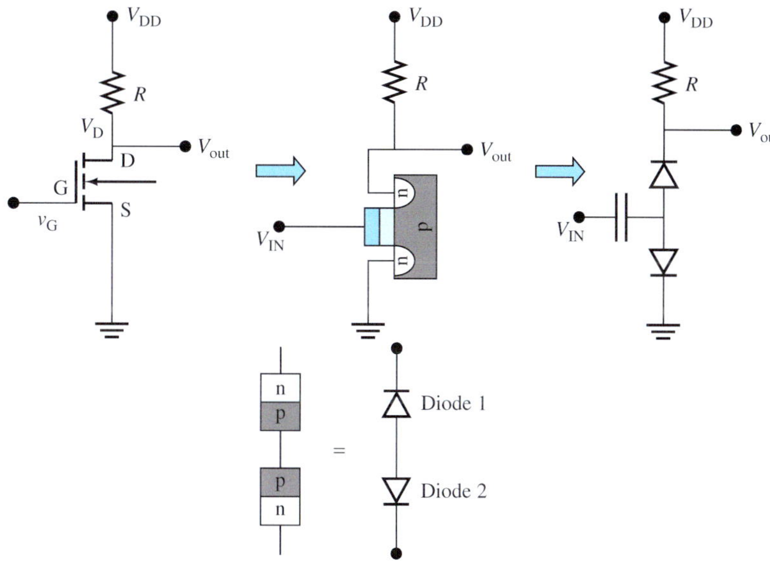

FIGURE 8.67 From left, the symbolic representation of an NMOS NOT gate, its physical layout connection, and its electrical model.

In addition, in Equation (8.45) V_T is the threshold voltage and K is a parametric constant value with a unit of A/V^2. One can see that now, the current is not only a function of V_G, but also V_D. There will be a current flow only when V_G is greater than V_T, or else $I_{DS} = 0$.

An NMOS transistor behaves somewhat similarly to, if not entirely like, a water pipe. Think of electrons as water, as water flow can be controlled by the water tap, so the electron flow (electric current) is controlled by the gate voltage. When the water tap is tightly closed, no water is flowing; this is similar to the cutoff region of a BJT transistor, where current is zero. A force that is slightly greater than the initial tightness of the water tap is needed to open up the water pipe, this scenario mirrors the fact that a gate voltage (V_G) that is greater than the threshold voltage (V_T) is required for any current to start flowing ($V_G > V_T$, I_{DS} is not zero). As the water starts dripping slowly when the water tap is slightly opened, turning the water tap a few more times will increase the water flow; this reflects the *I–V* characteristic shown in Figure 8.68, where drain current (I_{DS}) is always higher with higher V_G.

Drain voltage (V_D) is similar to the water pressure in the water pipe; when the water pressure increases slowly, so will the water flow for a given tap opening. The triode region of a transistor fits this scenario, as long as $V_D < V_G - V_T$. However, there is always a point where the transistor current comes to saturation, where current flow does not increase any more no matter how high V_D is. Naturally, this region is called the saturation region ($V_D > V_G - V_T$).

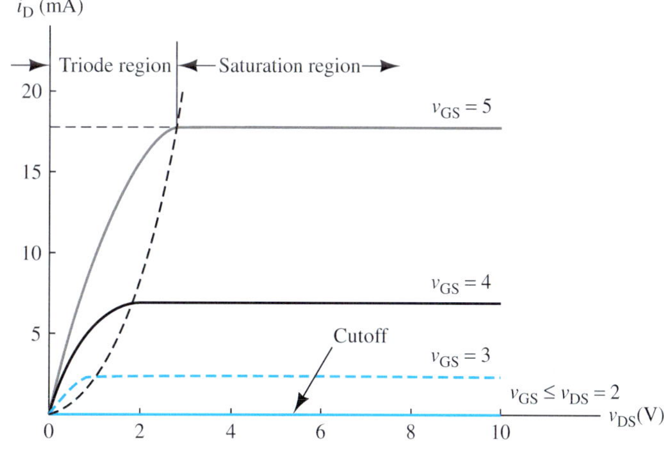

FIGURE 8.68 Current–voltage characteristics of an NMOS transistor.

Figure 8.68 represents the most important current–voltage characteristic of a MOSFET. The cutoff region actually represents the OFF state of a transistor, where I_{DS} is zero. The saturation region represents the ON state of the transistor, where I_{DS} is nonzero for a given V_G. The triode region is just a transition region between the ON and OFF state. Normally, the cutoff and saturation regions of a transistor are used to represent logic 0 and 1, respectively, as illustrated in Figure 8.63(c). Also, in the saturation region, I_{DS} is directly proportional to V_G, where different magnitude of I_{DS} can be produced by applying different V_G. Therefore, it is useful as an amplifier, a device used to increase the amplitude of the input signal.

EXAMPLE 8.23 Current-Voltage Characteristics of NMOS

Assume an enhancement-mode NMOS is used as an amplifier. It has $K = 2\ \text{mA/V}^2$ and $V_T = 2$ V. Plot the current–voltage characteristic curve for $V_{GS} = 0, 1, 2, 3$, and 4 V. For this example, use the NMOS represented by the model shown in Figure 8.67.

SOLUTION

Using the model in Figure 8.67, first study what happens when $V_{GS} = 0, 1$, and 2 V.

When $V_{GS} < V_T$ that is, V_{GS} is smaller than V_T (2 V), the current $i_D = 0$.

When $V_{GS} = 3$ V, there are two choices:

$$I_{DS} = K[2(V_{GS} - V_T)V_{DS} - V_{DS}^2], \text{ when } V_{DS} < V_{GS} - V_T$$
$$I_{DS} = K(V_{GS} - V_T)^2, \text{ when } V_{DS} \geq V_{GS} - V_T$$

V_D normally starts from zero and slowly increases. Let us plot $0\text{ V} < V_D < 10$ V.

Because $V_{GS} - V_T = 3 - 2 = 1$ V, for $0 < V_D < 1$, V_D is smaller than $V_{GS} - V_T$, and the following equation can be used to plot the colored dotted curve that forms the triode region of the $I-V$ curve:

$$I_{DS} = K[2(V_{GS} - V_T)V_{DS} - V_{DS}^2]$$

For $1 < V_D < 10$, V_D is greater than $V_{GS} - V_T$, therefore, use the equation:

$$I_{DS} = K(V_{GS} - V_T)^2 = 2\text{ mA}$$

to plot the colored dashed line, which is the saturation region. As shown in Figure 8.69, both colored dotted and dashed lines are for $V_{GS} = 3$ V. The solid colored line is for $V_{GS} = 0, 1$, and 2 V, representing the cutoff region.

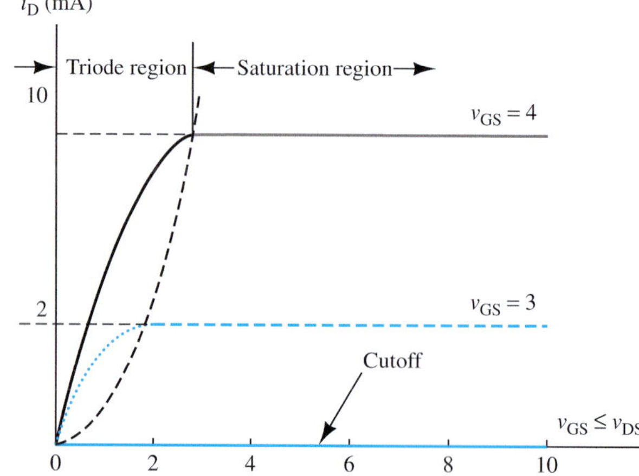

FIGURE 8.69 Current–voltage characteristic plot for Example 8.23.

When $V_{GS} = 4$ V:

$$V_{GS} - V_T = 4 - 2 = 2 \text{ V}$$

Thus, for $0 < V_D < 2$, V_D is smaller than V_{GS}. Therefore, use the equation:

$$I_{DS} = K\left[2(V_{GS} - V_T)V_{DS} - V_{DS}^{2}\right]$$

To plot the black curve that forms the triode region of the *I–V* curve, assuming $1 < V_D < 10$, V_D is greater than $V_G - V_T$, so use the equation:

$$I_{DS} = K(V_{GS} - V_T)^2 = 8 \text{ mA}$$

to plot the grey line, which is the saturation region. Both black and grey lines in Figure 8.69 are for $V_{GS} = 4$ V.

Using the solution in Example 8.23, and based on the setup shown in Figure 8.70(a), a simple NMOS amplifier circuit can be made.

Here, an alternating input voltage source (v_{in}) is in series with the DC source V_G. v_{in} is a sinusoid that is:

$$v_{in}(t) = 1 \times \sin(\omega t)$$

v_{in} has the amplitude of 1 V and the frequency of f. The amplifier operates similar to a BJT, except that for MOSFETs, $I_G = 0$. Therefore, a MOSFET is said to be modulated by the input voltage instead of the current.

In Figure 8.70(b), an alternating source combines with a constant DC source. This DC source produces a DC offset in the alternating sine function. When the input voltage is a combination of AC and DC components, the KVL loop produces the output voltage, V_{out}, that corresponds to:

$$v_{out}(t) = V_{DD} - i_D(t)R_D$$

Figure 8.70(c) illustrates the relationship between output and input voltages. Based on this figure, observe how the equivalent input voltage, that is:

$$v_{GS}(t) = V_{GS} + v_{in}(t)$$

creates an alternative output voltage. The straight line is called the *load line*. The load line is created by connecting two points. The first point assumes $V_{DS} = 0$, the second point assumes $I_D = 0$.

When $V_{DS} = 0$:

$$I_D = V_{DD}/R_D = 20/1 \text{ k}\Omega = 20 \text{ mA}$$

When $I_D = 0$:

$$V_{DS} = V_{DD} - 0 = V_{DD} = 20 \text{ V}$$

In general, as the current $i_D(t)$ varies, the output voltage will change. On the other hand, as explained in the previous subsections, $i_D(t)$, that is, the current flowing through the channel is controlled by the gate voltage. Thus, the load line maintains the relationship of $V_{GS}(t)$ and $V_{out}(t)$, which is V_{DS} in this case.

8.4.5 Design of NOT Gates Using NMOS Only for High-Density Integration

The design of NOT logic gates with one resistor and one NMOS transistor has been shown in Figures 8.64 to 8.67. Although, the resulting structure is very simple, there is a big disadvantage

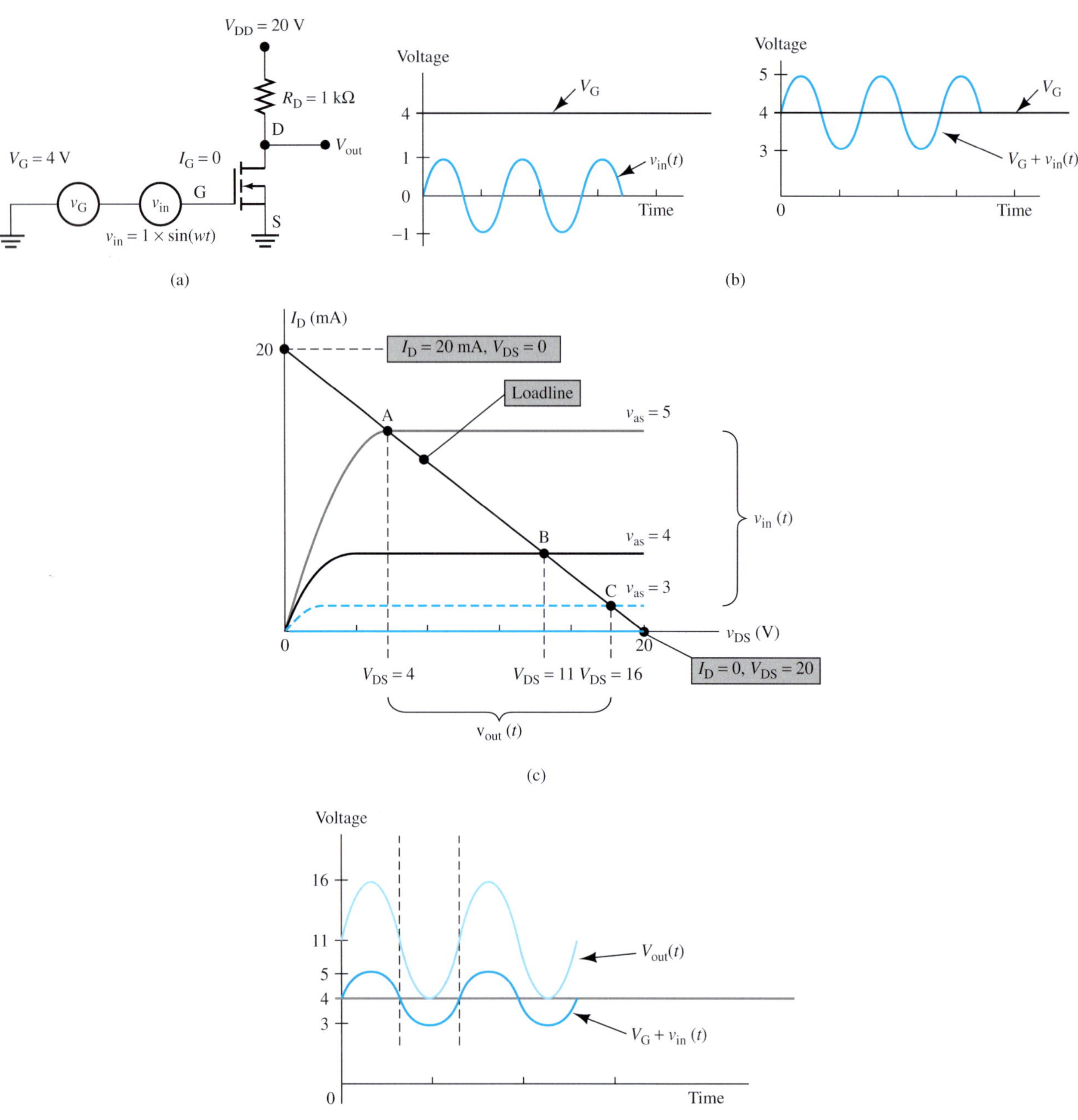

FIGURE 8.70 (a) Circuit of an NMOS amplifier. (b) Relationship between V_G and V_{in}. (c) Load-line analysis and relationship between V_{out} and V_{in}. (d) Comparison of the amplified $V_{out}(t)$ and the original signal $V_{in}(t)$.

of using resistors in logic gate design. Resistors need considerable space to be implemented on a silicon substrate using MOSFET technology. The resistance can be computed by:

$$R = \rho \times L/A$$

where ρ is the resistivity (Ω cm), L is the length of the conductor, and A is the cross-sectional area of the conductor. As an example, in Figure 8.71, a 100-kΩ resistor requires more than 10,000 μm^2 area, compared to a mere 2 μm^2 space for the NMOS. Assuming ρ is 0.001 Ω cm, for a 100-kΩ resistor, L is 10,000 μm, and A = width × depth = 1 μm × 1 μm = 1 μm^2.

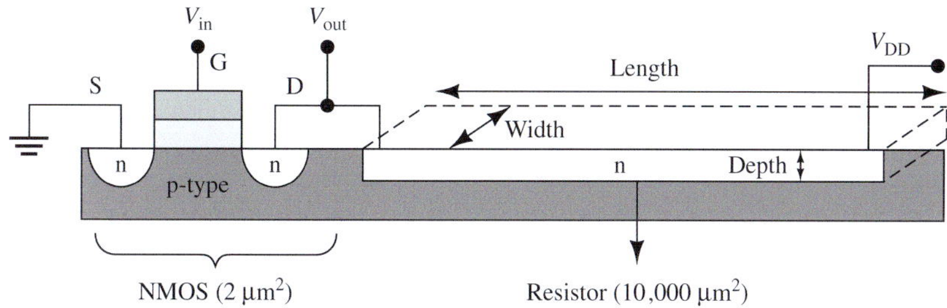

FIGURE 8.71 Physical layout of an NMOS NOT gate using one NMOS and one resistor, resulting in a low-density IC because the resistor needs a large amount of space.

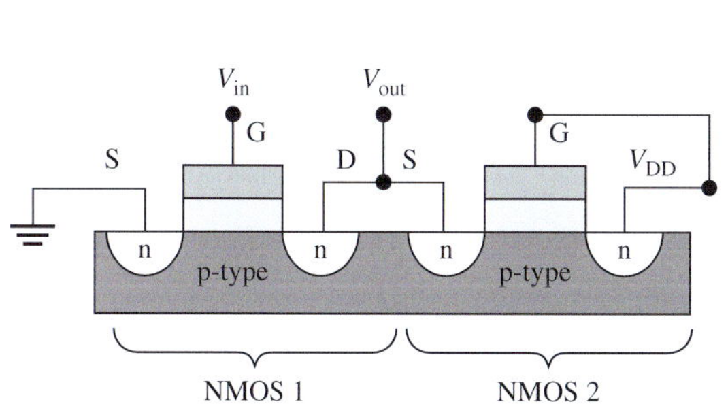

FIGURE 8.72 Physical layout of an NMOS NOT gate using NMOS only. The result is much a smaller layout area per logic gate. Therefore, a higher density IC is possible for a given chip area, for example 1 cm^2.

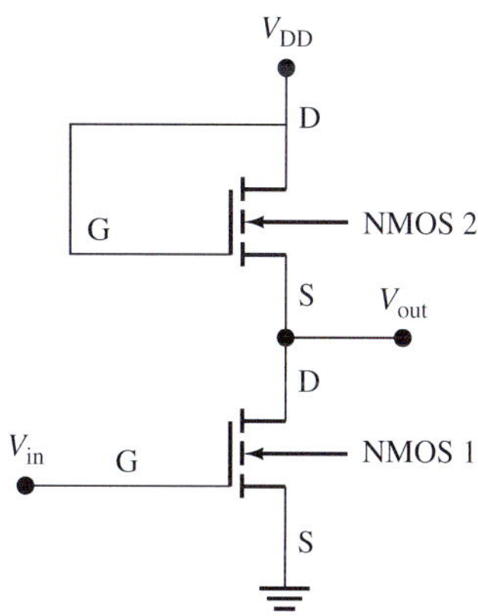

FIGURE 8.73 Symbol representation of a NOT gate using only NMOS transistors.

An alternative that allows implementation of a NOT gate without a resistor is the circuit shown in Figure 8.72. The design in this figure allows the NMOS logic gate to be implemented on a high-density IC.

Figure 8.73 represents the circuit model for the circuit shown in Figure 8.72. In this design, NMOS 2 is always ON. If V_{in} is low, then NMOS 1 will be OFF, and $V_{out} = V_{DD}$. If V_{in} is high, then V_{out} is the voltage drop across NMOS 1, which is very close to zero. NMOS 2 has to be designed such that its conduction channel resistance will be much greater than NMOS 1. In this case, the main voltage drop is across the V_{DS} of NMOS 2 instead of NMOS 1.

8.4.6 Design of a Logic Gate Using CMOS

Complementary MOSFET (CMOS) is the standard of today's ICs. It is called complementary because it has both PMOS and NMOS on the same silicon substrate instead of only NMOS or only PMOS. Its existance is due to the relatively high power consumption of NMOS-only (see Figure 8.74) and PMOS-only ICs. As the technology has progressed through the years, higher density ICs have become available. This has led to increases in the power dissipated per area. Since the surface area of a single microchip is fixed, there is a limit to how much heat can be generated due to the power consumption of each transistor before the entire circuit breaks down due to overheating.

Why does CMOS use less power than NMOS-only or PMOS-only ICs? The answer is simple. First, consider a NOT logic gate made using CMOS. In this case, there is only one transistor that is turned ON on both occasions where the input voltage, V_{in}, is low and high. There is no current path between V_{DD} and ground. In this case, the current, I_{DS}, that flows through the V_{DD} will not be too high. However, considering an NMOS-only NOT gate (see Figure 8.73), the top NMOS 1 is always turned ON. When V_{in} is high, NMOS 2 turns ON as well. In this case, a direct current path is generated between V_{DD} and the ground through these two NMOSs. Because the resistance of the two NMOSs in their ON status is low, I_{DS} flowing through them will be high.

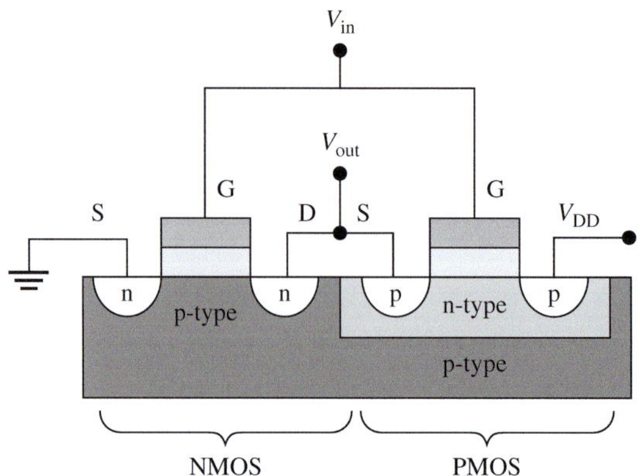

FIGURE 8.74 A cross-sectional view of the physical layout of a single CMOS. The NMOS transistor is on the left and the PMOS transistor is on the right. Note that PMOS is fabricated in a n-type region, where the p-type substrate is lightly doped with n-type dopant, thus turning it into a "minimized n-type substrate" (a so-called n-well) that is required for the PMOS to exist on a p-type substrate.

Thus, the total power consumption ($V_{DD} \times I_{DS}$) will be high and the NMOS-only logic gate leads to higher static power consumption as compared to CMOS.

As a result of overheating, a MOSFET transistor may lose its discreteness, that is, logic 1 and 0 might become indistinguishable. Digital calculation is not possible when the transistor is always displaying logic 1 when it actually should alternate between logic 1 and 0.

EXAMPLE 8.24 CMOS Inverter

Draw the equivalent circuit of the CMOS inverter when V_{in} is low and when it is high. Assume $V_{DD} = 5$ V. What is V_{out} for both occasions?

SOLUTION

The equivalent circuit is shown in Figure 8.75. When V_{in} is low, NMOS turns OFF, and PMOS turns ON. Applying a KVL loop leads to the conclusion that V_{out} is high. When V_{in} is high, NMOS is ON and PMOS is OFF. In this case, V_{out} is low.

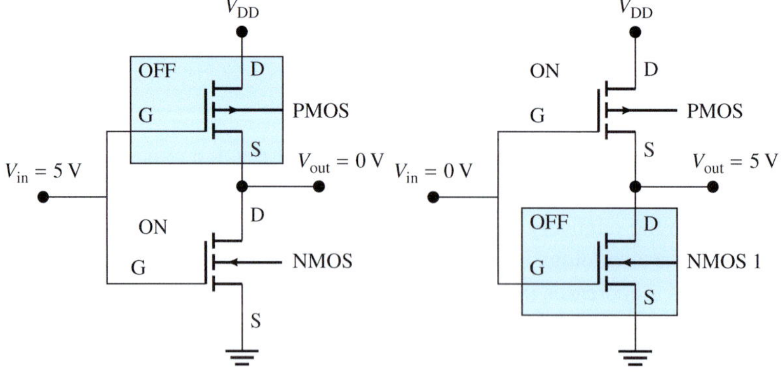

FIGURE 8.75 The equivalent circuit of the CMOS shown in Figure 8.73.

EXAMPLE 8.25 Truth Table

Develop the truth table of the two-input gate for the circuit shown in Figure 8.76(a). That is, determine the output voltage, V_{out}, in terms of inputs A and B when they are high (1) or low (0), for example, when both are 0, both are 1, and one is 0 and the other is 1. What is this operation?

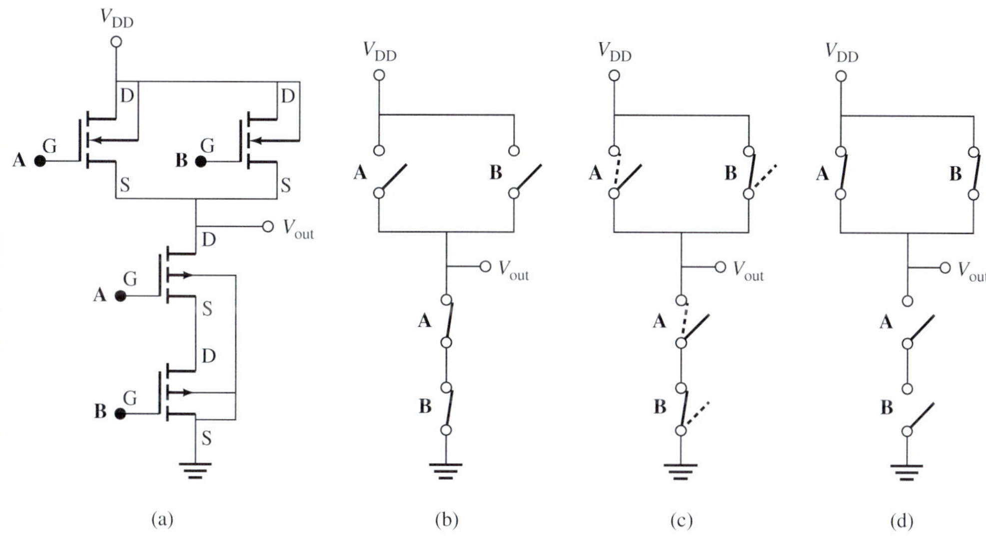

FIGURE 8.76 The logic gate for Example 8.25.

SOLUTION

It is clear that when both A and B are low (0), the NMOS will be OFF and the PMOS will be ON, thus, the output will be connected to the ground, i.e., the output will be low (0). This scenario is shown in Figure 8.76(b). When one of these inputs is high (1) and the other is low (0), the two series PMOSs block the output to be zero while one of the two parallel NMOSs allows connection to the V_{DD}. This scenario is shown in Figure 8.76(c). When both A and B are high (1), the output will be connected to V_{DD}, while the two PMOSs avoid connection to the ground. Thus, the output will be high (1) [see Figure 8.76(d)]. The solution is shown in Table 8.5. Based on this table, the output will be an OR gate (see Chapter 10).

TABLE 8.5 The Truth Table for Figure 8.76(a)

A	*B*	*A* + *B*
0	0	0
0	1	1
1	0	1
1	1	1

8.5 OPERATIONAL AMPLIFIERS

Operational amplifiers (or *op amps*) are high-gain voltage amplifiers with two differential inputs and a single output. Op amps have applications in almost all electronic circuits and are used to build high-power amplifiers, inverters, and buffers. Op amps are also an important part of active filters.

Compared to active filters (discussed in Chapter 11), passive filters (discussed in Chapter 7) have some disadvantages. First, the inductors needed are large and expensive for the audio frequency range in which most mechanical measurements are made. Second, passive filters restrict the sharpness of the cutoff because they do not have power amplification, which leads to loading in the measurements.

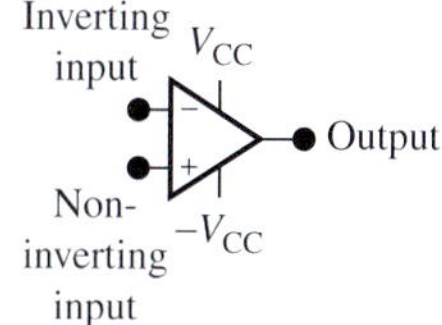

FIGURE 8.77 Circuit diagram of an operational amplifier (op amp).

The circuit diagram of the op amp and its important terminals are shown in Figure 8.77. It should be noted that the op amp itself is made up of a combination of many transistors. Here, for simplicity in presentation, the internal structure of an op amp is not presented. The two DC voltage sources, V_{CC} and $-V_{CC}$, are used to bias the transistors. The output voltage is limited by the values of the voltage sources where:

$$-V_{CC} \leq V_o \leq V_{CC} \tag{8.46}$$

An op amp can be used to amplify AC signals as well as DC signals.

An ideal op amp has the following characteristics:

1. Infinite open-loop voltage gain, $A_o = \infty$
2. Infinite input impedance (i.e., the impedance between the two inputs), $R_{in} = \infty$
3. Zero output impedance (i.e., the impedance seen between the output and the ground), $R_o = 0$
4. Infinite bandwidth (i.e., its gain doesn't change with the input signal frequency)
5. Zero-input offset voltage (i.e., the output voltage is exactly zero if the input is zero)

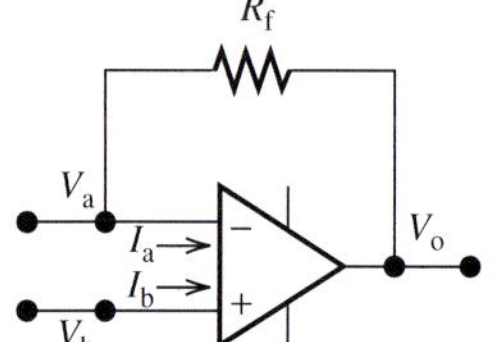

FIGURE 8.78 An op amp circuit with feedback.

Because an op amp has a very high open-loop voltage gain, negative feedback is usually considered to control the output voltage and to limit the voltage gain. The voltage gain of the op amp with negative feedback is called closed-loop gain (or simply voltage gain) and is referred to by A_v. An op amp circuit with negative feedback is shown in Figure 8.78. To achieve negative feedback, a feedback resistance, R_f, is connected between the output and the inverting input terminals.

Five properties stated above lead to two important rules called "golden rules." Referring to Figure 8.78, the golden rules are:

1. There is no input current to the op amp, that is in Figure 8.78, $I_a = I_b = 0$.
2. The voltage difference between the two inputs is zero, that is, $V_a - V_b = 0$ or $V_a = V_b$.

These two rules are very important, and are always used to analyze op amp circuits.

A simple procedure is followed to analyze any op amp circuit:

1. Calculate the voltage V_b.
2. Based on the second golden rule, $V_a = V_b$.
3. Write the *KCL* for the op amp input nodes to find the output voltage.

Note that KCL should not be written for the output node because the op amp's output current isn't known. Op amps have many applications, which are discussed in the following examples.

EXAMPLE 8.26 An Inverting Amplifier

Consider the circuit shown in Figure 8.79 and calculate the voltage gain, $A_v = V_o/V_i$.

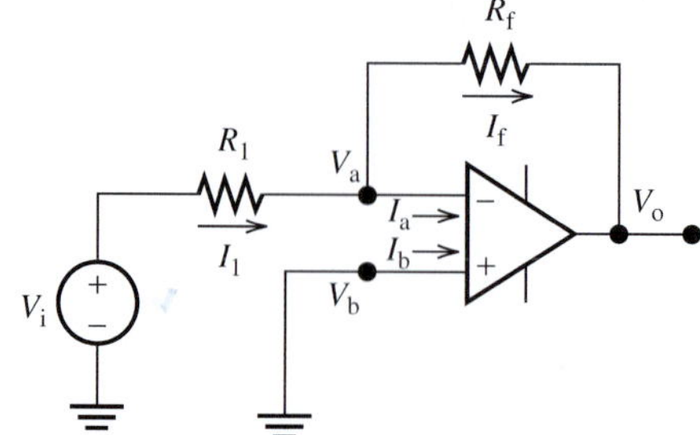

FIGURE 8.79 An inverting amplifier.

SOLUTION

Following the given procedure, first calculate V_b. Because the noninverting input is grounded:

$$V_b = 0$$

According to the second golden rule:

$$V_a = V_b = 0 \tag{8.47}$$

Applying KCL to node "a" yields:

$$I_1 = I_a + I_f \tag{8.48}$$

According to the first golden rule:

$$I_a = 0 \tag{8.49}$$

Therefore:

$$I_1 = I_f \tag{8.50}$$

Based on Ohm's law:

$$\begin{aligned} I_1 &= \frac{V_i - V_a}{R_1} \\ &= \frac{V_i - 0}{R_1} \\ &= \frac{V_i}{R_1} \end{aligned} \tag{8.51}$$

and:

$$\begin{aligned} I_f &= \frac{V_a - V_o}{R_f} \\ &= \frac{0 - V_o}{R_f} \\ &= -\frac{V_o}{R_f} \end{aligned} \tag{8.52}$$

Substituting Equations (8.51) and (8.52) into Equation (8.50) yields:

$$\frac{V_i}{R_1} = -\frac{V_o}{R_f} \tag{8.53}$$

Reordering Equation (8.53), the output voltage corresponds to:

$$V_o = -\frac{R_f}{R_1} V_i \tag{8.54}$$

And the voltage gain corresponds to:

$$A_v = \frac{V_o}{V_i} = -\frac{R_f}{R_1} \tag{8.55}$$

The negative sign in the voltage gain indicates that there is 180° of phase difference between the input and the output. Accordingly, this amplifier circuit is called an *inverting amplifier*.

Note that the gain and accordingly the output voltage are controlled by resistors R_1 and R_f. Although the output voltage $V_o = A_v V_i$, it is also limited by V_{CC} and $-V_{CC}$ (see Equation 8.46).

(*continued*)

EXAMPLE 8.26 Continued

The amplifier is said to be saturated if the absolute value of the output voltage exceeds V_{CC}. As a summary, the output voltage corresponds to:

$$V_o = \begin{cases} -\dfrac{R_f}{R_1}V_i & -V_{CC} < V_o < V_{CC} \\ V_{CC} & V_o > V_{CC} \\ -V_{CC} & V_o < -V_{CC} \end{cases}$$

This same concept can be applied for all applications.

EXAMPLE 8.27 A Noninverting Amplifier

For the circuit shown in Figure 8.80, follow the same procedure as the previous example to calculate the output voltage and determine the voltage gain.

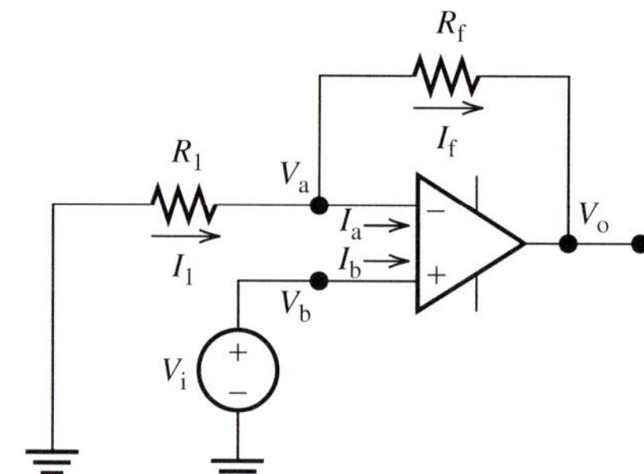

FIGURE 8.80 A noninverting amplifier.

SOLUTION

At node "b":

$$V_b = V_i \tag{8.56}$$

The second golden rule yields:

$$V_a = V_b = V_i \tag{8.57}$$

Applying KCL at node "a," and taking into account that $I_a = 0$:

$$I_1 = I_f \tag{8.58}$$

or:

$$\frac{0 - V_a}{R_1} = \frac{V_a - V_o}{R_f} \tag{8.59}$$

Substituting for V_a from Equation (8.57) into Equation (8.59):

$$-\frac{V_i}{R_1} = \frac{V_i - V_o}{R_f} \tag{8.60}$$

$$-\frac{R_f}{R_1}V_i = V_i - V_o \tag{8.61}$$

Therefore, the output voltage corresponds to:

$$V_o = V_i + \frac{R_f}{R_1} V_i \tag{8.62}$$

or:

$$V_o = \left(1 + \frac{R_f}{R_1}\right) V_i \tag{8.63}$$

and the voltage gain corresponds to:

$$A_v = \left(1 + \frac{R_f}{R_1}\right) \tag{8.64}$$

Note that there is no negative sign in the voltage gain. Thus, this circuit is called a *noninverting amplifier*. The minimum gain of the noninverting amplifier is unity, and this will happen if $R_f = 0$ and/or $R_1 = \infty$. A noninverting amplifier with unity gain is called a *buffer* or *voltage follower* and is discussed in the following example.

EXAMPLE 8.28 Buffers or Voltage Followers

An op amp voltage follower is shown in Figure 8.81. This circuit is similar to the circuit of Figure 8.80 after replacing R_1 with an open circuit (infinity resistance) and R_f with a short circuit (zero resistance). Using Equation (8.63), the output voltage corresponds to:

$$V_o = \left(1 + \frac{0}{\infty}\right) V_i = V_i \tag{8.65}$$

Therefore, the voltage gain corresponds to:

$$A_v = 1 \tag{8.66}$$

The result can be obtained following the same procedure as in the previous examples. Referring to Figure 8.81:

$$V_b = V_i \tag{8.67}$$

The second golden rule yields:

$$V_a = V_b = V_i \tag{8.68}$$

Because the output is directly connected to node "a" via a short circuit:

$$V_o = V_a = V_i \tag{8.69}$$

Buffers have applications in connecting cascaded amplifiers. Because their output resistance is zero, they reduce the impact of the previous amplifier on the next one. In addition, because their input resistance is infinity, they reduce the impact of the next amplifier on the previous one.

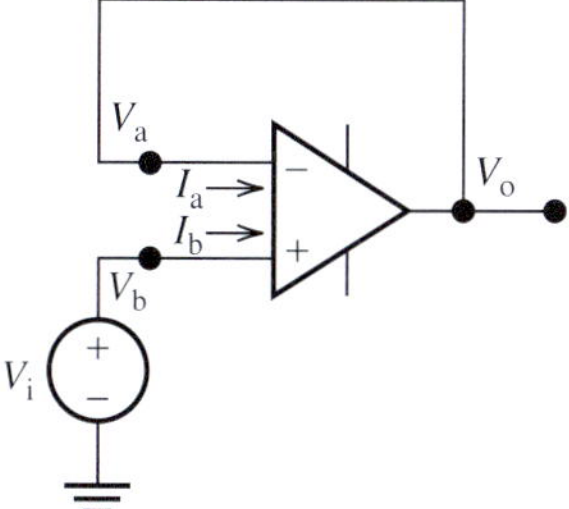

FIGURE 8.81 Buffer or voltage follower.

APPLICATION EXAMPLE 8.29 An Audio System

An audiophile has connected his new speakers as shown in Figure 8.82. This is only one channel; other channels have the same connections (with the exception of the subwoofer). After verifying that the speakers are up to his standards, he wants to know how much the speakers will vibrate when he blasts them.

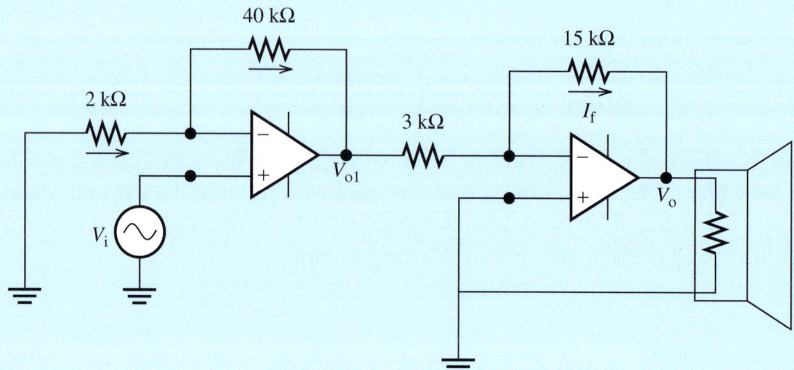

FIGURE 8.82 The circuit for Example 8.28.

SOLUTION

The input voltage is amplified using two amplifier stages. The first stage is a noninverting amplifier followed by an inverting amplifier. The output voltage from the first stage corresponds to:

$$V_{o1} = A_{v1}V_i$$

where, A_{v1} is given by Equation (8.64). In addition, the output voltage from the second stage corresponds to:

$$V_o = A_{v2}V_{o1}$$

where, A_{v2} is calculated using Equation (8.55). Therefore, the relationship between the output and the input voltage corresponds to:

$$V_o = A_{v1}A_{v2}V_i$$

Finally, the overall gain corresponds to:

$$A_v = \frac{V_o}{V_i} = A_{v1}A_{v2}$$

Using Equations (8.55) and (8.64):

$$A_v = \left(1 + \frac{40}{2}\right) \times -\frac{15}{3} = -105$$

EXERCISE 8.6

For the circuit shown in Figure 8.83, write the output phasor V_o in terms of input phasor V_i and show that the V_o/V_i represents a filter (see Chapter 7). What type of filter does it represent?

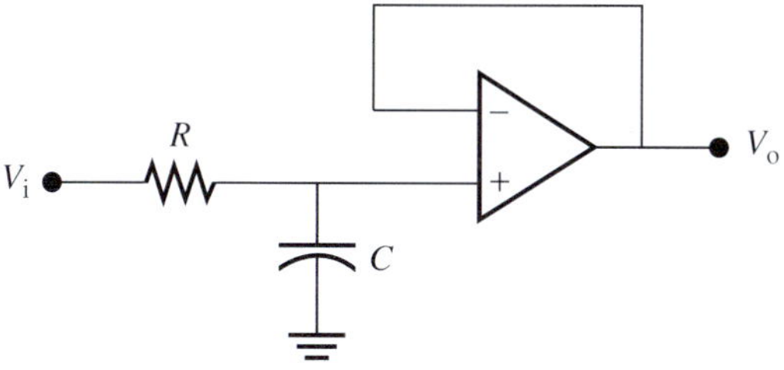

FIGURE 8.83 The circuit for Exercise 8.6.

8.6 USING PSPICE TO STUDY DIODES AND TRANSISTORS

This tutorial explains the process of editing the PSpice model to study diodes. This knowledge will also aid in any future PSpice circuit design that uses a variety of electronic devices with specific parameters. PSpice can be used to create electronic devices with the specifications of interest. For example, PSpice can be used to create a transistor with a specific DC gain, or a diode with different forward bias voltages.

EXAMPLE 8.30 PSpice Analysis

Use PSpice to study the circuit shown in Figure 8.9.

a. $R = 1\ \text{k}\Omega$
b. $R = 10\ \text{k}\Omega$

SOLUTION

The voltage source is given: $V(t) = 12V$.

a. $R = 1\ \text{k}\Omega$

A simple PSpice circuit that consists of a diode can be set up using the following steps:

1. Go to "Place Part" (see Figure 2.63), under the "BREAKOUT" library, search for the part name "Dbreak."
2. Set up the PSpice schematic shown in Figure 8.84 and run the simulation analysis as "Bias Point" (see Figure 2.71).

Notice that there is a slight difference between the current flowing through the diode compared to its value calculated theoretically in Example 8.2. This is due to the forward bias resistor of the diode, R_f, which is nonzero. In Example 8.2, it was assumed to be zero.

b. $R = 10\ \text{k}\Omega$

Next, change the resistance, R1, to 10 kΩ and run the bias point analysis again. The PSpice schematic solution is shown in Figure 8.85. The small amount of current across the diode shows that the diode has been turned off.

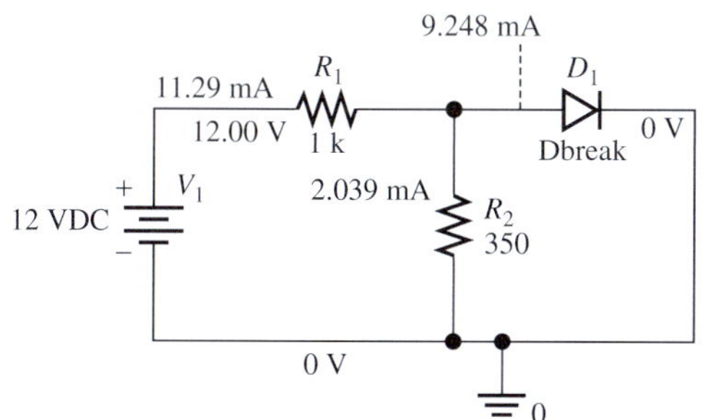

FIGURE 8.84 PSpice schematic for Example 8.30(a).

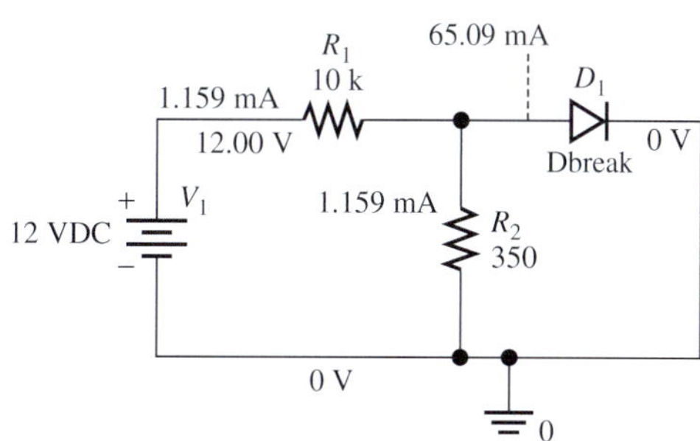

FIGURE 8.85 PSpice schematic solution for Example 8.30(b).

EXAMPLE 8.31 PSpice Analysis

Use PSpice to find the output of the voltage, $v_o(t)$, for the circuit shown in Figure 8.17 (see Example 8.3) using the following values: $v_i(t) = 5 \sin(120\pi t)\text{V}$, $R = 1\ \text{k}\Omega$, $E = 2\ \text{V}$.

SOLUTION

The voltage source is $v_i(t) = 5 \sin(120\pi t)\text{V}$. First, determine the frequency of the voltage source.

$$f = \frac{\omega}{2\pi}$$

$$f = \frac{120\pi}{2\pi}$$

$$f = 60\ \text{Hz}$$

Follow the solution of Example 2.24 to setup the VSIN voltage source. Then, the limiter, E, can be setup by placing a direct source voltage in the direction opposite from the voltage source. The PSpice schematic solution is shown in Figure 8.86. A time analysis simulation (see Figure 2.75) has been run to observe the voltage output across the diode and limiter. This simulation is shown in Figure 8.87.

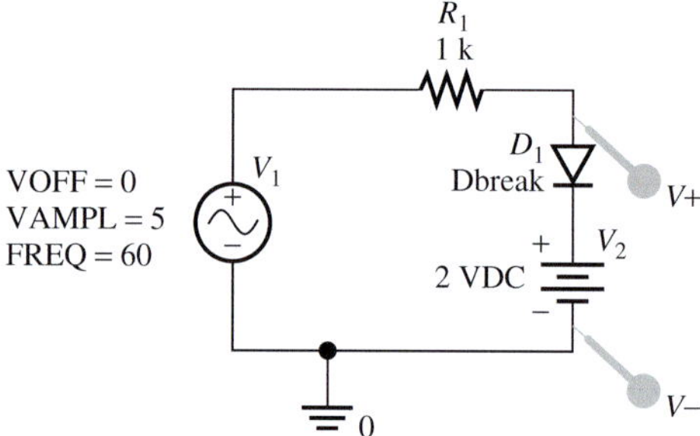

FIGURE 8.86 PSpice schematic solution for Example 8.31.

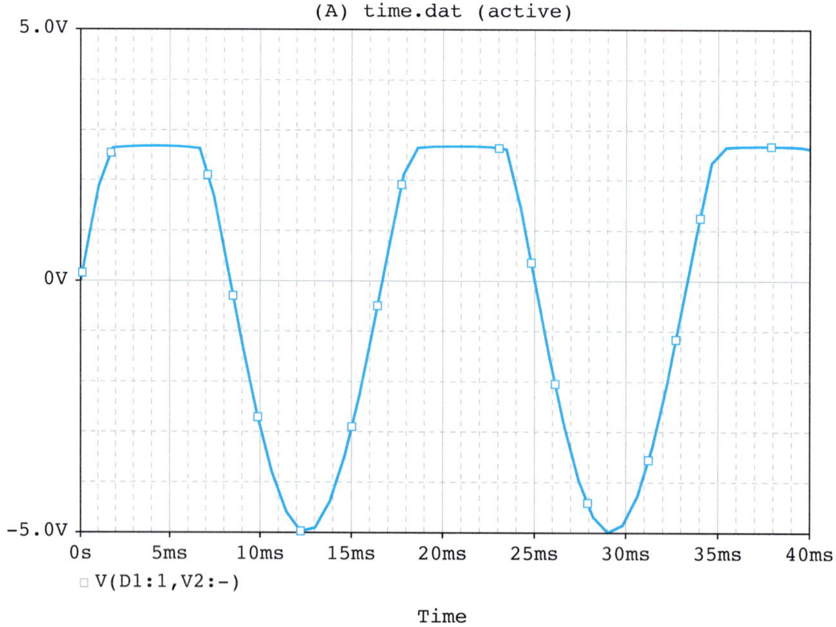

FIGURE 8.87 Voltage output across the diode and limiter for Example 8.31.

EXAMPLE 8.32 PSpice Analysis

Use PSpice to study the circuit shown in Figure 8.26 in Example 8.7. The Zener diode has $V_Z = 3.6$ V. Find I and V when:

a. $V_S = -1$ V
b. $V_S = 8$ V
c. $V_S = -10$ V

SOLUTION

The PSpice model editor is required to set up a Zener diode that meets the requirements for this circuit.

1. To place a Zener diode in the circuit, go to "Place Part" (see Figure 2.65) and type "Dbreakz," which is in the "BREAKOUT" library.
2. Place the Zener diode as shown in Figure 8.88.
3. Next, set the required reverse voltage for Zener diode.
4. Click on the Zener diode or Dbreakz, and go to the menu "Edit" → "PSpice Model" (see the arrow) as shown in Figure 8.89.
5. A model window with specific parameters will appear, as shown in Figure 8.90.
6. Change the parameters in the window to the value shown in Figure 8.91 and save it. "BV" is the breakaway voltage or the reversed voltage, and "IBV" is the breakaway current.

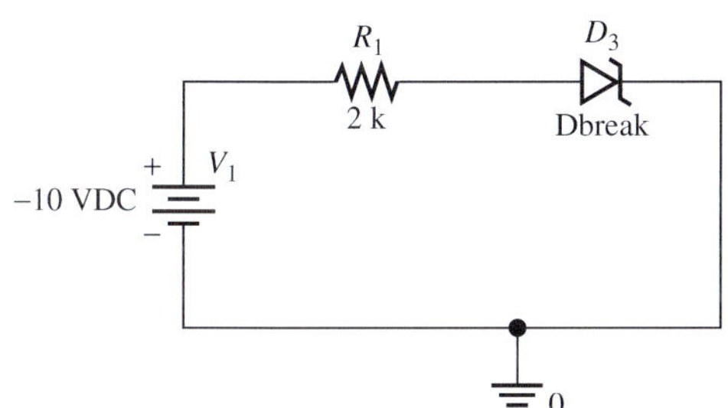

FIGURE 8.88 Place the Zener diode into the PSpice circuit.

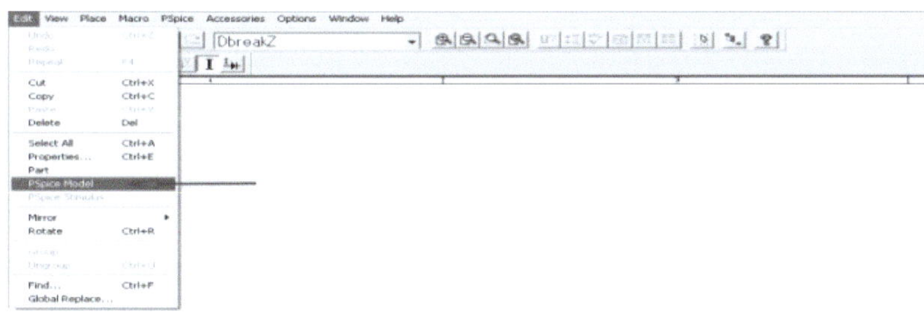

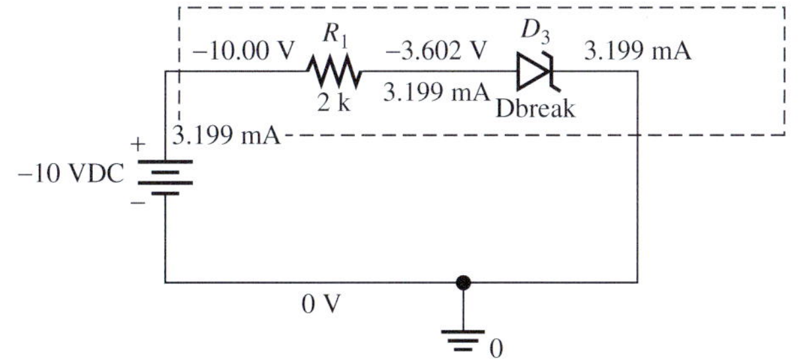

FIGURE 8.89 Enter the PSpice model editor for the Zener diode.

(continued)

EXAMPLE 8.32 Continued

7. Run the bias point analysis (see Figure 2.73). The PSpice schematic solutions for each voltage source are shown in Figures 8.92 to 8.94.
   a. $V_S = -1$ V
   b. $V_S = 8$ V
   c. $V_S = -10$ V

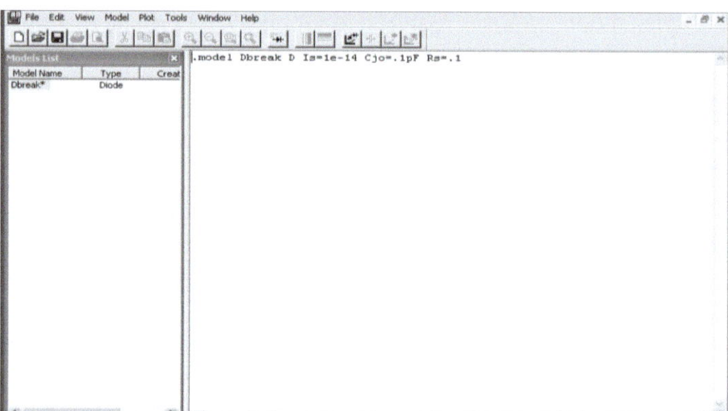

FIGURE 8.90 Default model parameters for the Zener diode.

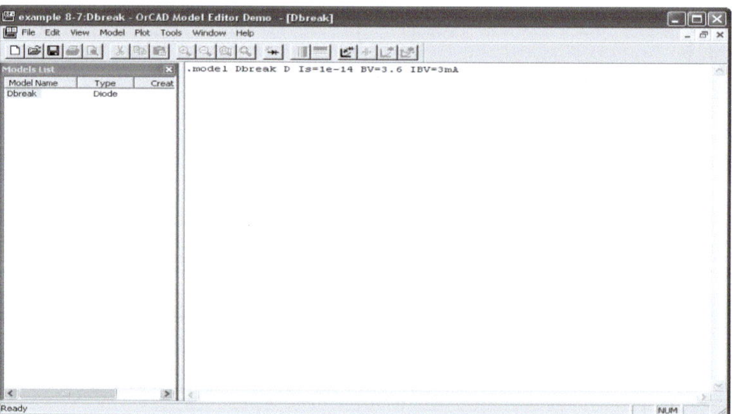

FIGURE 8.91 Desired parameters for the Zener diode in Example 8.32.

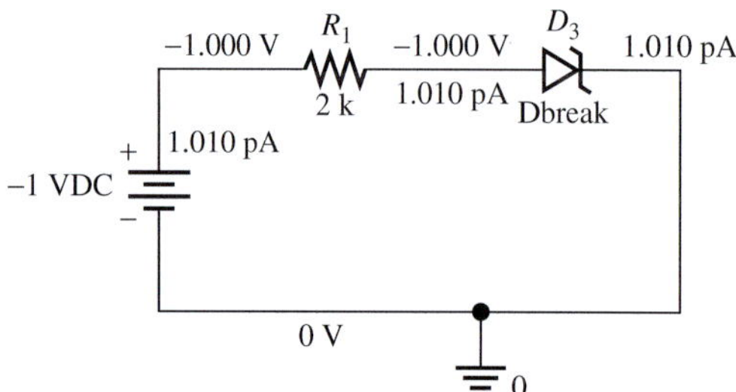

FIGURE 8.92 PSpice schematic solution for Example 8.32(a).

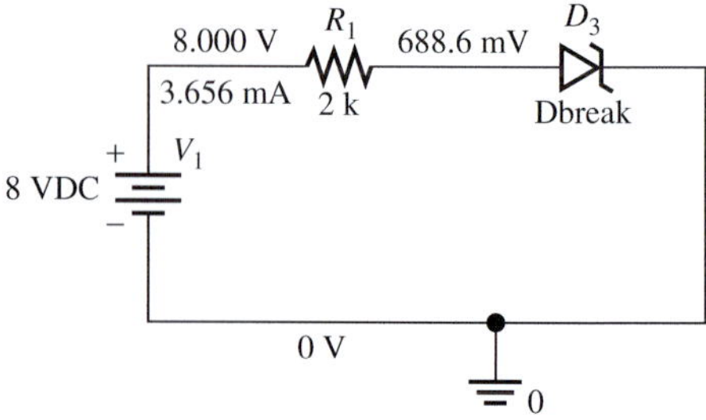

FIGURE 8.93 PSpice schematic solution for Example 8.32(b).

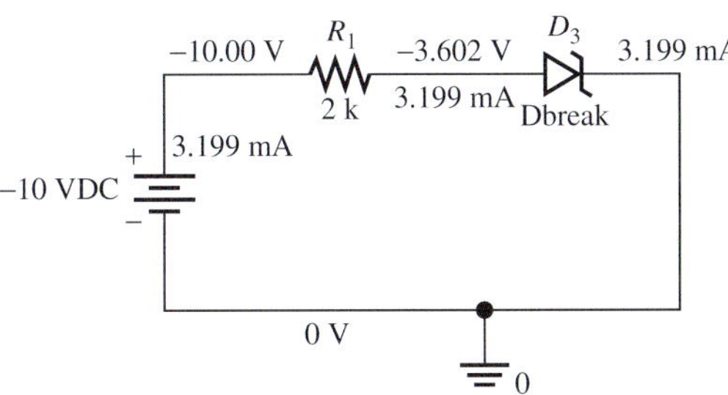

FIGURE 8.94 PSpice schematic solution for Example 8.32(c).

EXAMPLE 8.33 PSpice Analysis

Use PSpice to find V_{BE}, I_B, I_C, and V_{CE} in the circuit shown in Figure 8.44, if:

a. $V_{BB} = 0.3$ V
b. $V_{BB} = 2.7$ V
c. $V_{BB} = 6.7$ V

Use the PSpice model editor to set up the transistor to meet the requirements for the circuit.

SOLUTION

1. Go to "Place Part" (see Figure 2.63) and type "Qbreakn," which is in the "BREAKOUT" library to place a transistor in the circuit.
2. Place the transistor and set up the PSpice circuit as shown in Figure 8.95.
3. Click on the transistor or Qbreakn, and go to the menu "Edit" → "PSpice Model" (see the arrow) as shown in Figure 8.96.
4. Change the model parameters of the transistor to those shown in Figure 8.97. Here, Bf represents the amplifier, β.
5. Run the bias point analysis by varying the voltage sources (see Figure 2.71). The results are shown in Figures 8.98 to 8.100.
   a. $V_{BB} = 0.3$ V
   b. $V_{BB} = 2.7$ V
   c. $V_{BB} = 6.7$ V

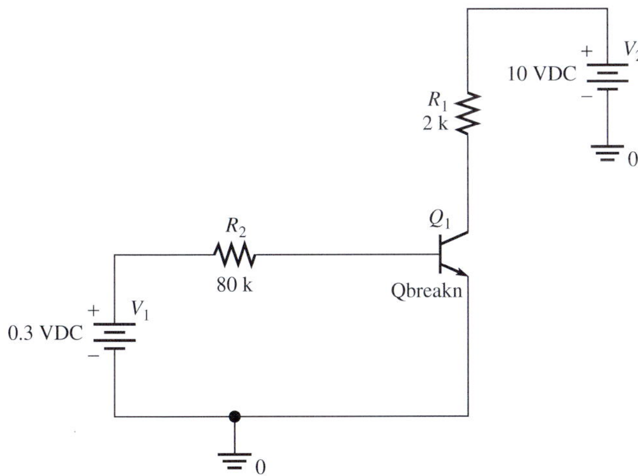

FIGURE 8.95 Placing the transistor.

(*continued*)

EXAMPLE 8.33 Continued

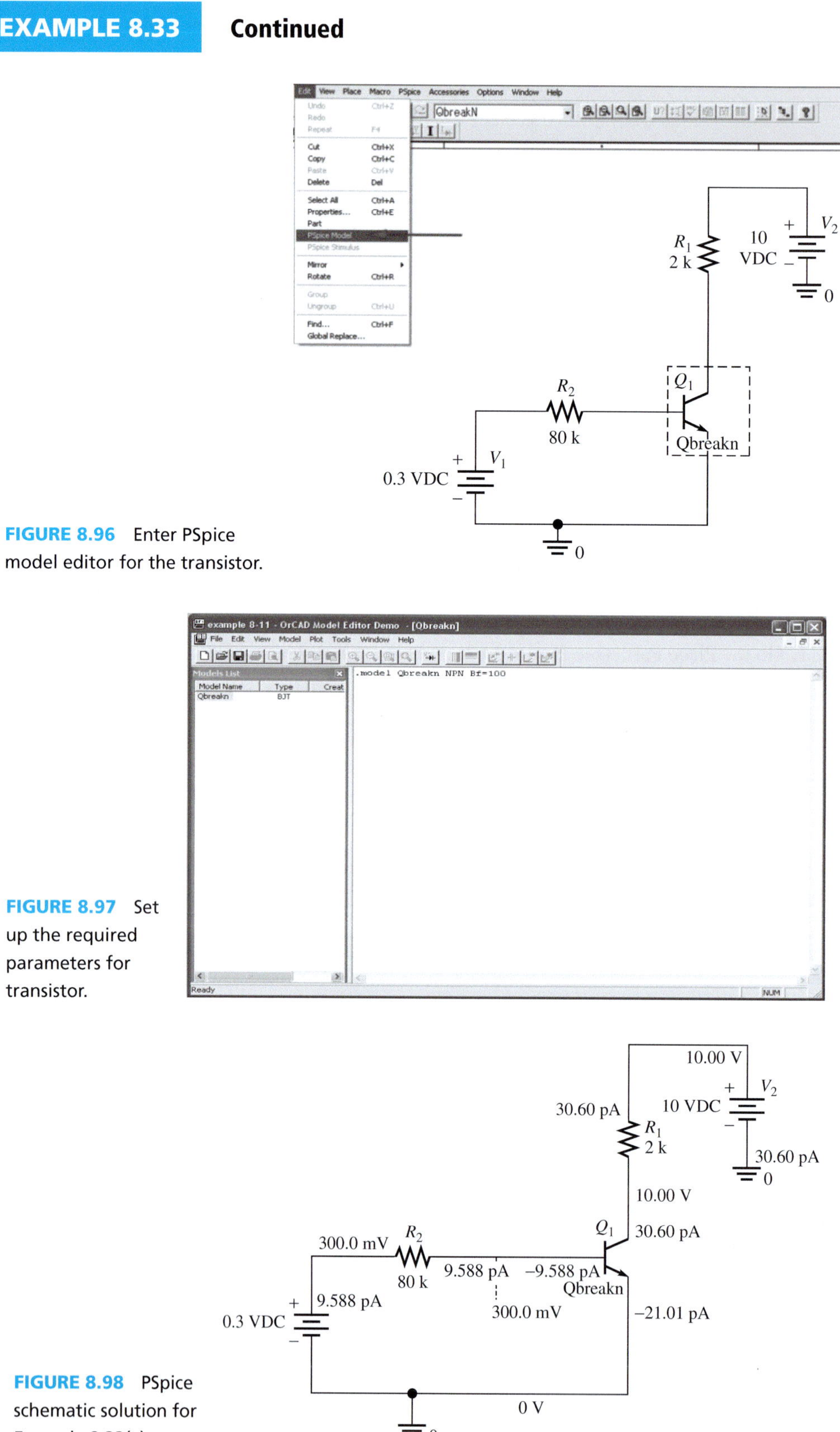

FIGURE 8.96 Enter PSpice model editor for the transistor.

FIGURE 8.97 Set up the required parameters for transistor.

FIGURE 8.98 PSpice schematic solution for Example 8.33(a).

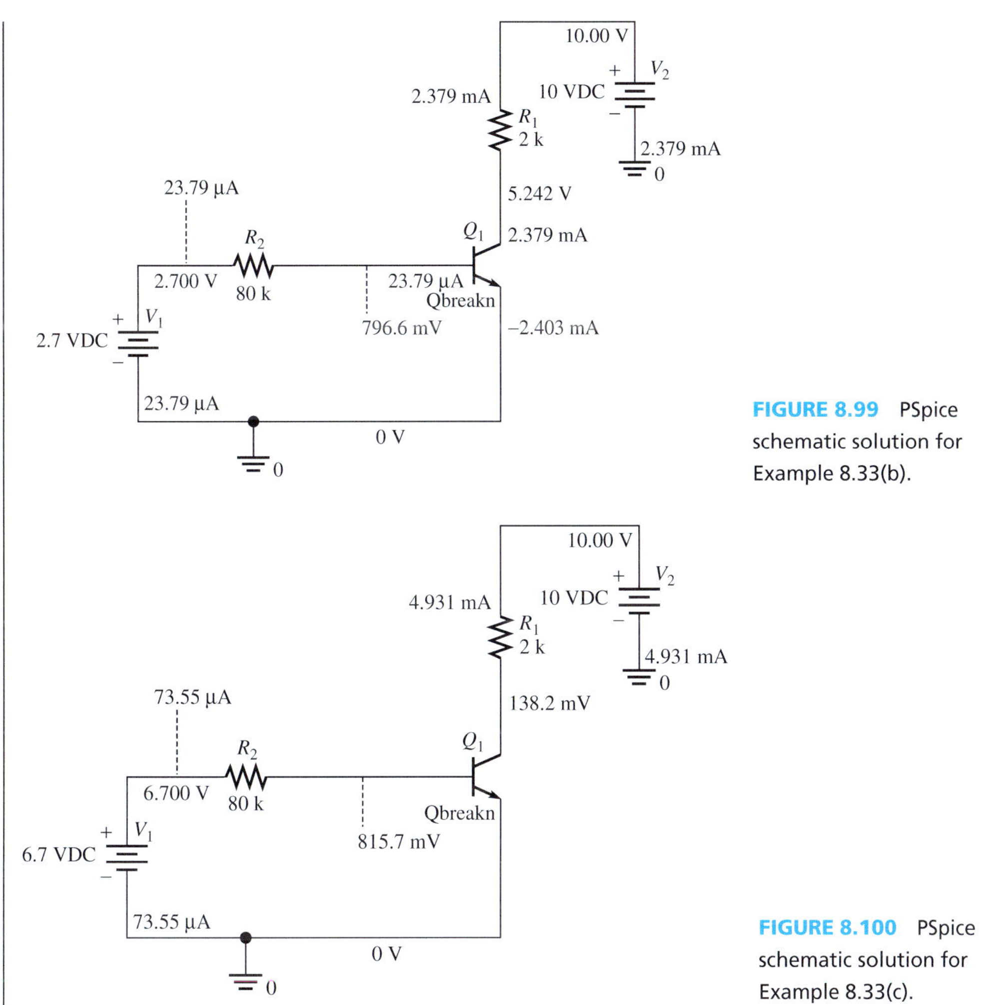

FIGURE 8.99 PSpice schematic solution for Example 8.33(b).

FIGURE 8.100 PSpice schematic solution for Example 8.33(c).

APPLICATION EXAMPLE 8.34 Microwave Oven

Figure 8.101 shows the simplified circuit of a microwave oven. A transformer at the voltage source increases the voltage input to 2800 V and charges the capacitor in the circuit. When the capacitor is fully charged, it supplies the voltage to the magnetron (power amplifier)

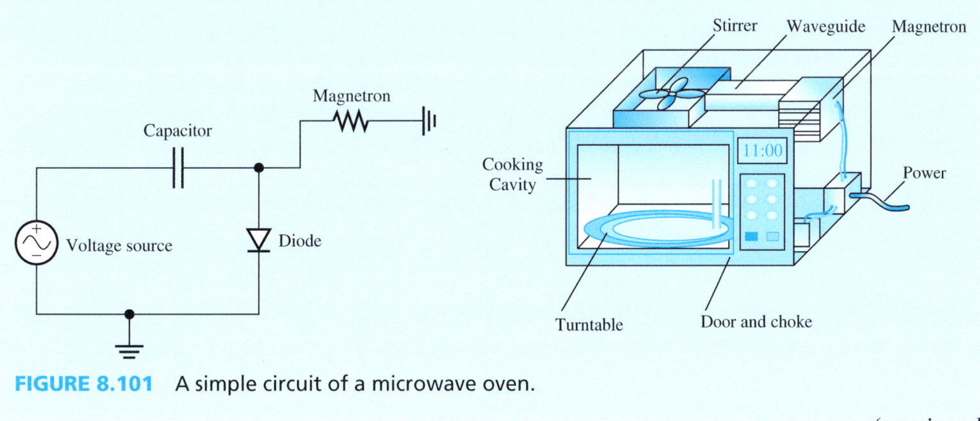

FIGURE 8.101 A simple circuit of a microwave oven.

(continued)

APPLICATION EXAMPLE 8.34 Continued

together with the voltage source from the transformer. A very high input voltage is supplied to the magnetron in order to generate the microwave energy to heat the food.

The voltage input after increasing by the transformer is 2800 sin $(120\pi t)$, the capacitor has the capacitance of 100 mF, and the resistance of the magnetron is 20 kΩ. Use PSpice to find the voltage across the magnetron during the charging and heating process.

SOLUTION

First, given the voltage source, $v_S(t) = 2800 \sin(120\pi t)$, the frequency is 60 Hz. Follow Example 2.24 to set up the VSIN voltage source, and follow Example 8.25 to place a diode in the circuit. The PSpice schematic solution is given in Figure 8.102.

Next, run the simulation as time transient analysis (see Figure 2.75) for 1 s, with the voltage differential markers located before and after the magnetron (see Figure 8.102). The voltage across the magnetron is shown in Figure 8.103.

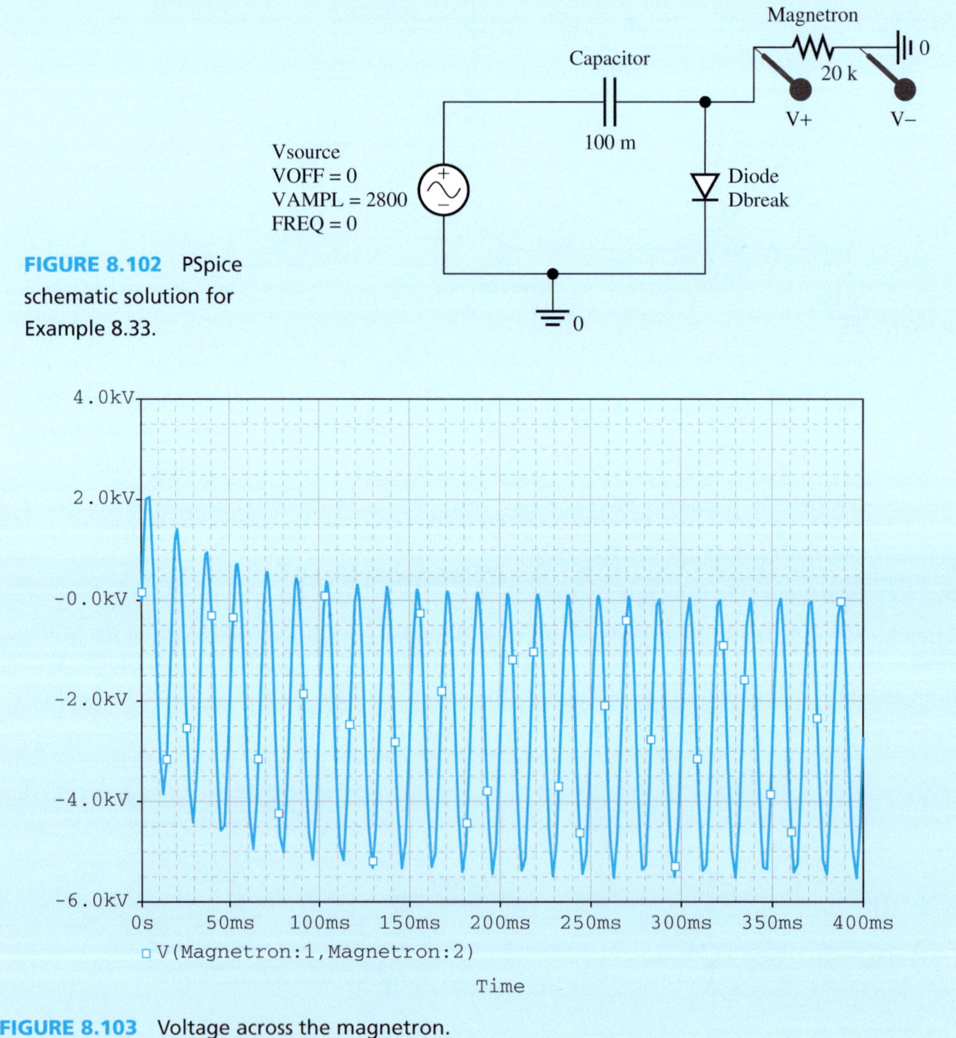

FIGURE 8.102 PSpice schematic solution for Example 8.33.

FIGURE 8.103 Voltage across the magnetron.

8.7 WHAT DID YOU LEARN?

- There are two different categories of semiconductors: p-type and n-type. The addition of pentavalent impurities (i.e., atoms with five valence electrons) produces n-type semiconductors by contributing extra electrons. The addition of trivalent impurities (i.e., atoms with three valence electrons) produces p-type semiconductors by contributing extra holes.
- A diode is a p–n junction that allows current to pass in only one direction from the p-type semiconductor toward the n-type semiconductor.
- The relationship between voltage across the diode and current through the diode is shown in Equation (8.1):

$$I = I_o(e^{V/V_T} - 1)$$

- A diode is said to be in *forward bias* if its voltage is positive and it starts to pass current as the voltage increases beyond the threshold voltage.
- Applications of diodes include rectifiers, diode limiters, and voltage regulators.
- Different types of diodes include Zener diodes, varactor diodes, light-emitting diodes, and photodiodes.
- A bipolar junction transistor (BJT) has three terminals: base, collector, and emitter.
- BJTs can operate in three regions: the active region, the saturation region, and the cutoff region. The characteristics of the three regions are summarized in Table 8.3.
- To be used as an amplifier, a BJT must be biased in the active region.
- DC analysis is used to ensure the BJT is in the active region. The DC equivalent circuit is shown in Figure 8.42a.
- AC analysis is used to find the voltage gain. The AC equivalent circuit is shown in Figure 8.50.
- The transistor that operates between cutoff and saturation is called an electronic switch.
- A MOSFET is the primary device that is used in complex digital ICs.
- The physical structure of an n-type MOSFET (NMOS) is shown in Figure 8.62(a).
- MOSFETs operate in cutoff, in triode, or in saturation (Figure 8.68).
- MOSFET applications include dynamic random access memory (DRAM), logic gates, and amplifiers.
- Complementary MOSFET (CMOS) contains both PMOS and NMOS on the same silicon substrate. It is the standard of today's ICs because it consumes less power than NMOS-only circuits.
- Op amps are characterized by having: high input resistance, low output resistance, and very high gain.
- Golden rules of op amps: (1) There is no input current to an op amp; (2) The voltage difference between the two inputs is zero.
- Voltage followers represent one application of op amps.

Further Reading

Ben Streetman and Sanjay Banerjee. 1999. *Solid state electronic devices*, 5th ed. Upper Saddle River, NJ: Prentice Hall.
Richard C. Jaeger. 1997. *Microelectronics circuit design*. New York: Mcgraw Hill.
Allan R. Hambley. 2000. *Electronics*, 2nd ed. Upper Saddle River, NJ: Prentice Hall.

Problems

*B refers to basic, A refers to average, H refers to hard, and * refers to problems with answers.*

SECTION 8.2 P-TYPE AND N-TYPE SEMICONDUCTORS

8.1 (B) Explain the difference between p-type and n-type semiconductors.

8.2 (B)* Which of the following is neutral?

a. p-type semiconductor
b. n-type semiconductor
c. p–n junction.

8.3 (B) What is the approximate built-in voltage for silicon and germanium?

SECTION 8.3 DIODES

8.4 (B) Given that the Boltzmann's constant is 8.617343×10^{-5} eV/K, with the thermal voltage at room temperature is 0.0259V determine the thermal voltage if a diode is at temperature:

a. 200 K
b. 40°C
c. 400 K

8.5 (B) Provide an application of a limiter diode and explain how it works.

8.6 (B)* What is the minimum number of standard AA batteries required to light a standard LED?

8.7 (B) What will happen to the variable reactor's resonant frequency when:

a. The inductance in circuit increased by nine times
b. The capacitance in the diode reduced by half

8.8 (B) Provide an application of a photovoltaic diode and explain how it works.

8.9 (B) Explain how a photoconductive diode works.

8.10 (B) For the circuit shown in Figure P8.10, find I and V when:

a. $V_S = -5$ V
b. $V_S = 0.3$ V
c. $V_S = 12$ V

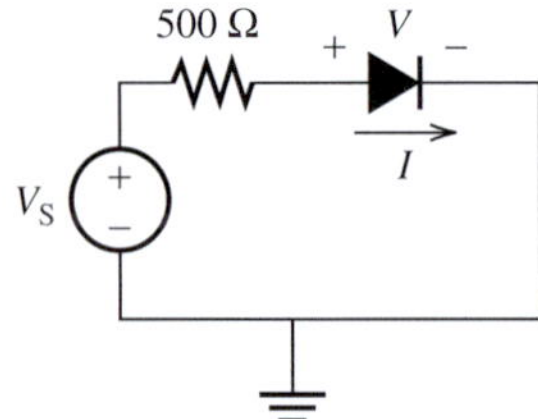

FIGURE P8.10 Circuit for Problem 8.10.

8.11 (A)* Find the current, I, flowing through the diode for the circuit shown in Figure P8.11 when:

(a) $R = 2$ kΩ
(b) $R = 8$ kΩ

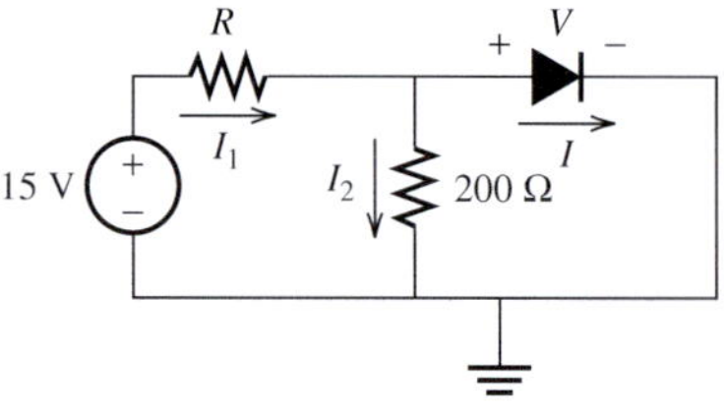

FIGURE P8.11 Circuit for Problem 8.11.

8.12 (A) Write the expression of I_B in the terms of voltage across the 10-Ω resistor, V_A, that is shown in Figure P8.12. Assume that both diodes are on and that V_S is 10 V.

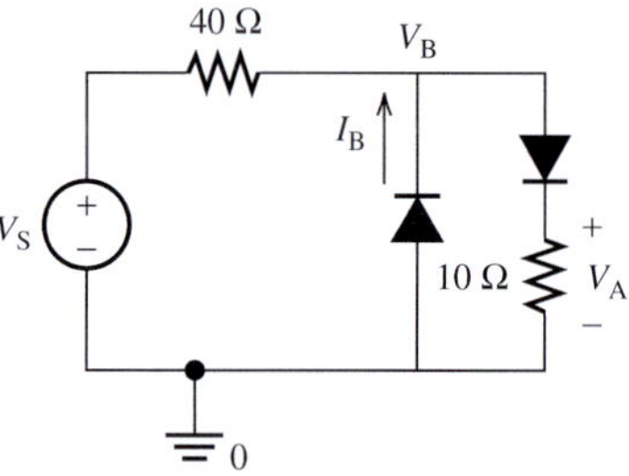

FIGURE P8.12 Circuit for Problem 8.12.

8.13 (H) Find I_D in Figure P8.13 when

a. $V_S = 6$ V
b. $V_S > 0$ V all the time, find the range of V_S when the diode is off.

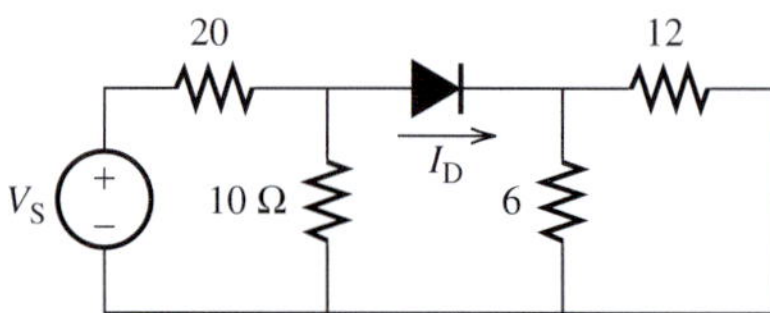

FIGURE P8.13 Diode circuit.

8.14 (H)* Determine the current through the diode for the circuit shown in Figure P8.14.

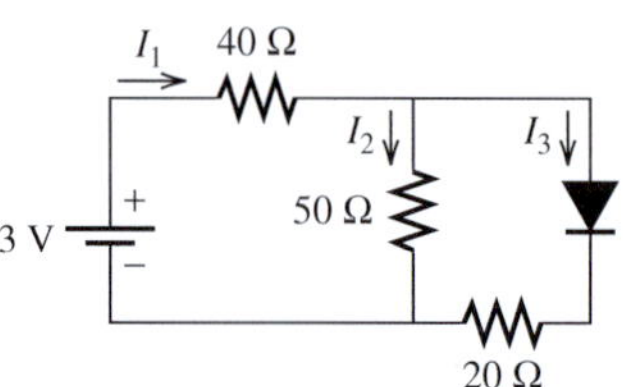

FIGURE P8.14 Circuit for Problem 8.14.

8.15 (B) For the circuit shown in Figure P8.15:

a. Sketch the transfer characteristics when $E = 4$ V.
b. Sketch the output voltage if the input voltage is $v_i(t) = 15\ \sin(10\pi t)$V.

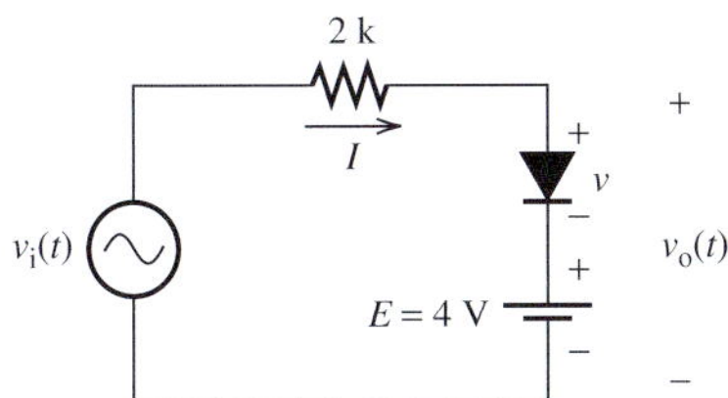

FIGURE P8.15 Circuit for Problem 8.15.

8.16 (B) For the circuit shown in Figure P8.16:

a. Sketch the transfer characteristics when $E = 4$ V.

b. Sketch the output voltage if the input voltage is $v_i(t) = 15\ \sin(10\pi t)$V.

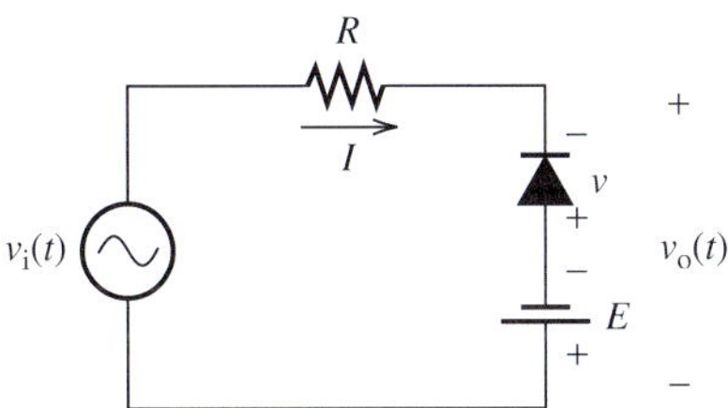

FIGURE P8.16 Circuit for Problem 8.16.

8.17 (A) For the circuit shown in Figure P8.17, sketch the transfer characteristics, input voltage, and output voltage. Assume the input voltage is a sinusoidal waveform with a peak value of 20 V and $E_1 = E_2 = 8$ V.

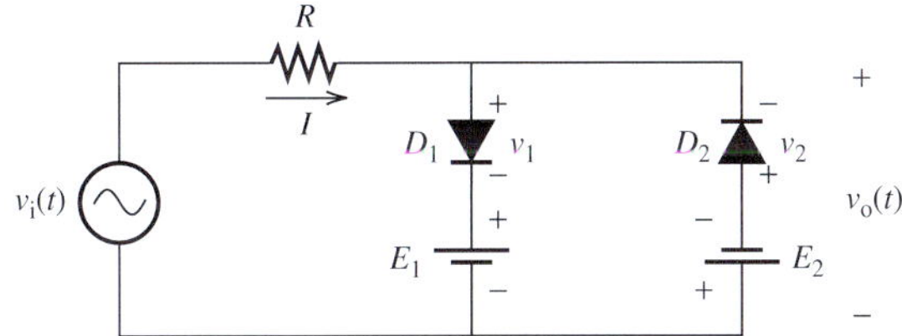

FIGURE P8.17 Circuit for Problem 8.17.

8.18 (A)* The output voltage for the circuit shown in Figure P8.18 (a) is shown in Figure P8.18(b). Given that the voltage source, v_S, is 10 V, and the resistance of the resistor is 20 Ω, determine:

a. E_1 and E_2.

b. The current flow through resistor, R, when the output voltage is at the upper and lower limit bounds.

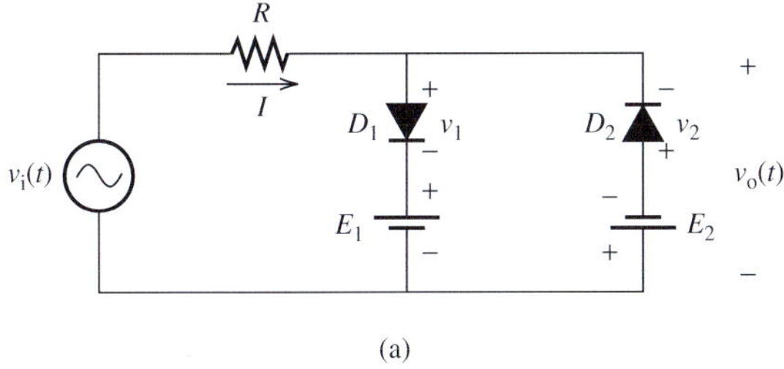

(a)

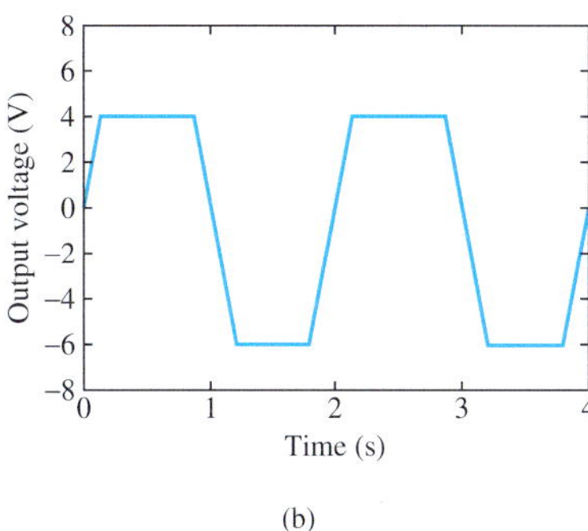

(b)

FIGURE P8.18 (a) Circuit for Problem 8.18. (b) Voltage output.

8.19 (H) Find the transfer characteristic for the circuit shown in Figure P8.19 when $E = 4.3$ V.

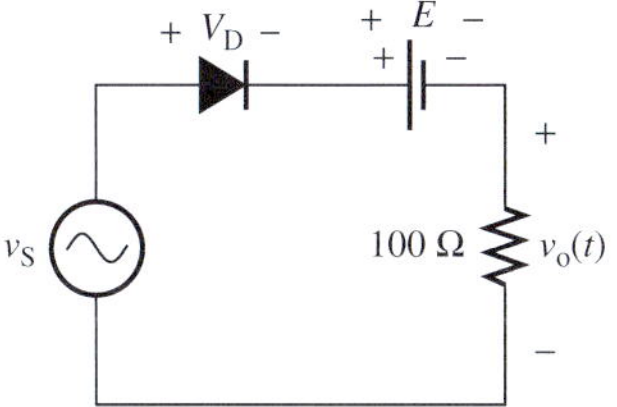

FIGURE P8.19 Voltage limiter.

8.20(H)* Find $v_o(t)$ with respect to $v_S(t)$ for Figure P8.20 if $E = 2.3$ V.

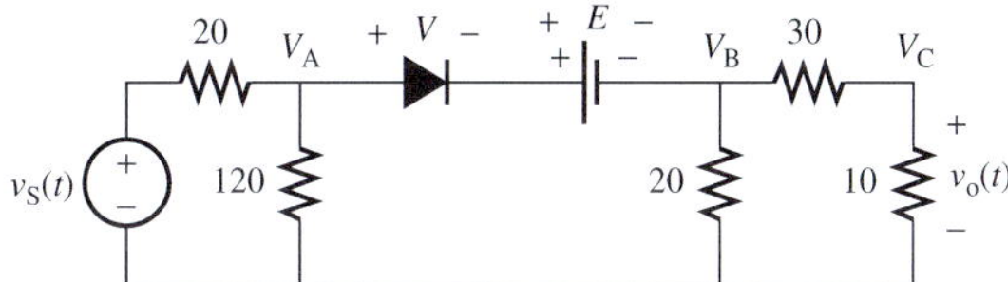

FIGURE P8.20 Voltage limiter.

8.21 (H)* Find I_1, I_2, and I_3 for the circuit shown in Figure P8.21.

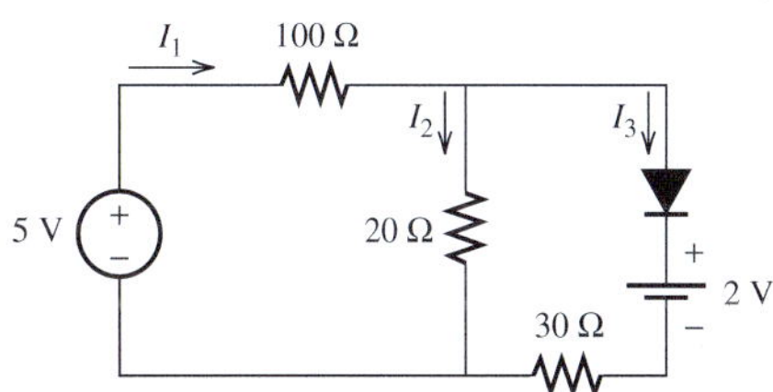

FIGURE P8.21 Circuit for Problem 8.21.

8.22 (B) Figure P8.22 shows the output of a Zener diode. From the plot, what information about the Zener diode can be obtained?

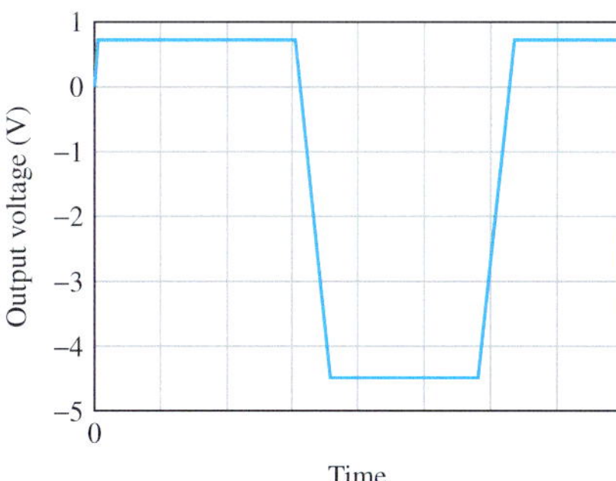

FIGURE P8.22 Zener diode output for Problem 8.22.

8.23 (A) For the circuit shown in Figure P8.23 the Zener diode has $V_Z = 6.8$ V. Find I and V when:

a. $V_S = -3$ V
b. $V_S = 6$ V
c. $V_S = -20$ V

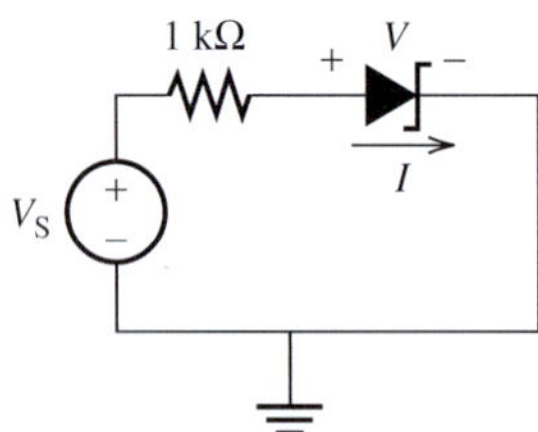

FIGURE P8.23 Circuit for Problem 8.23.

8.24 (A) Given that $V_z = 5.5$ V, find I_1 and I_2 in the circuit shown in Figure P8.24 when:

a. $V_S = -5$ V
b. $V_S = 3$ V
c. $V_S = -12$ V

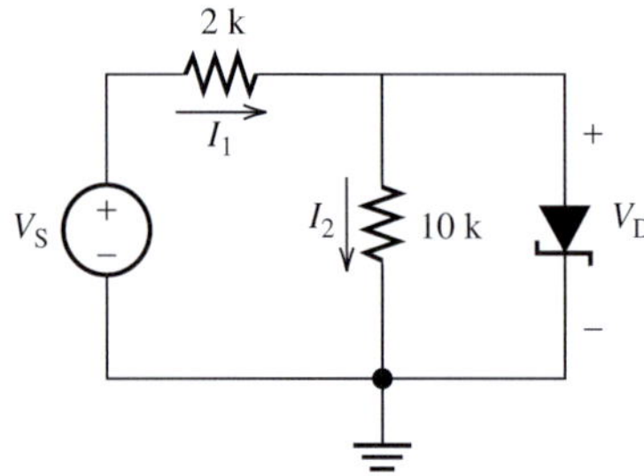

FIGURE P8.24 Circuit of the Zener diode.

8.25 (H) Given that the Zener voltage is 5.3 V for the circuit shown in Figure P8.25, find I_1 and I_2 if the input voltage is:

a. $V_S = -5$ V
b. $V_S = 8$ V
c. $V_S = 16$ V

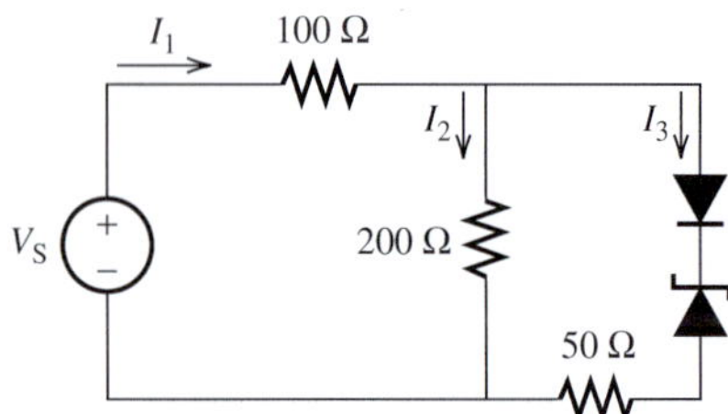

FIGURE P8.25 Circuit for Problem 8.25.

8.26 (H) Find the characteristic curve of the voltage across the diode (V_D) in Problem 8.24.

8.27 (B)* What is the minimum input voltage required for a full-wave rectifier if the required output voltage is 5 V?

8.28 (A) Determine the current through the LEDs when R is:

a. 400 Ω
b. 0 Ω

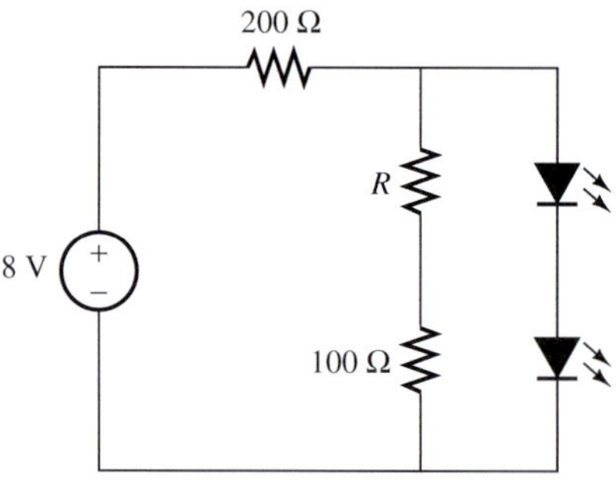

FIGURE P8.28 Circuit for Problem 8.28.

8.29 (H)* What is the maximum resistance, R, allowable to let a light-emitting diode work in the circuit shown in Figure P8.29?

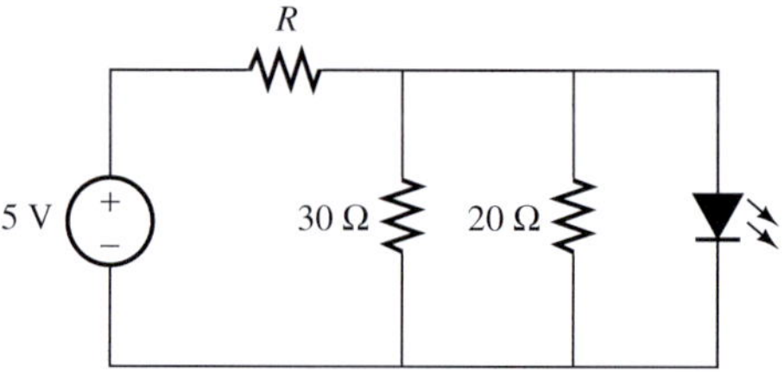

FIGURE P8.29 Circuit for Problem 8.29.

8.30 (B) Assume that the voltage input to the voltage regulator in Figure P8.30 is 10 V, with the Zener voltage of 4.5 V. Determine if the voltage regulator will turn on with the following combinations of R_L and R_S.

a. $R_L = 500\ \Omega$ and $R_S = 20\ \Omega$
b. $R_L = 40\ \Omega$ and $R_S = 90\ \Omega$
c. $R_L = 30\ \Omega$ and $R_S = 33\ \Omega$

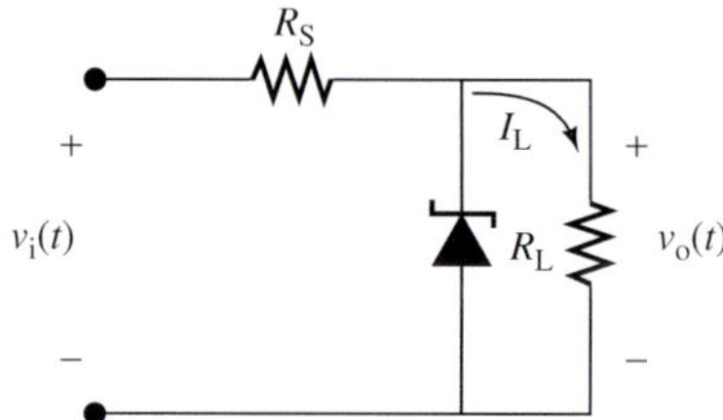

FIGURE P8.30 Circuit for Problem 8.30.

8.31 (A)* Design a DC power supply that supplies a DC voltage of 20 V to a load resistance of 500 Ω. Consider an AC input of 120 V/60 Hz. Choose suitable values for the capacitor and the transformer turns ratio. The ripple voltage should be less than 10% of the output voltage. Consider $R_S = 100\ \Omega$.

8.32 (H) A transformer has 200 turns of coil at its input stage and 50 turns of coil at its output stage. A voltage regulator is connected to the transformer (see Figure P8.32) to drive the load, $R_L = 300\ \Omega$. Given that the transformer is connected to a source, $V_S = 120 \sin(100t+60°)$ V, and the Zener voltage is 5 V, determine the capacitance required in the voltage regulator to meet the design requirements.

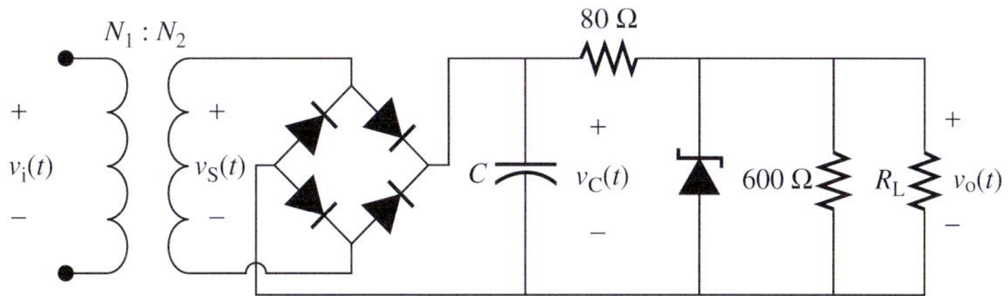

FIGURE P8.32 Circuit for Problem 8.32.

SECTION 8.4.1 BIPOLAR JUNCTION TRANSISTORS

8.33 (B) What is the direction of the current in both PNP and NPN transistors?

8.34 (H) What are the disadvantages of silicon transistors compared to electron tubes?

8.35 (A)* When a transistor, $\beta = 80$, is working in its saturation region, which one of the following statements is correct?

a. $I_C = 80 I_B$
b. $I_C > 80 I_B$
c. $I_C < 80 I_B$
d. $V_{CE} > 0.2$ V

8.36 (B) Figure P8.36 shows the relationship between the collector current and the base–emitter voltage of a transistor. Provide the name of each region, OA, AB, BC, and CD.

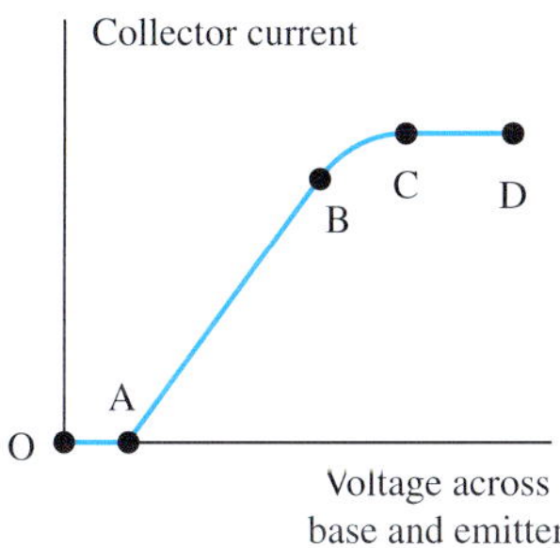

FIGURE P8.36 Collector current and the base–emitter voltage curve.

8.37 (H)* Assume that the voltage across a transistor's collector and emitter has been measured at 3 V, as shown in Figure P8.37. Other parameters are also given in Figure P8.37. Calculate the current gain, β.

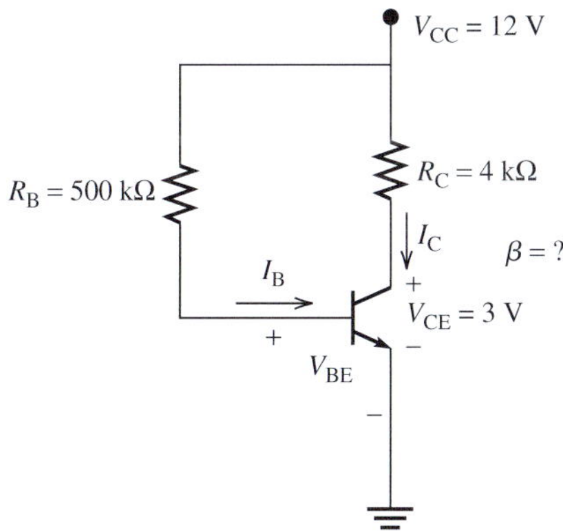

FIGURE P8.37 A simple transistor circuit.

8.38 (A) For the circuit shown in Figure P8.38, find V_{BE}, I_B, I_C, and V_{CE} when:

a. $V_{BB} = 0.5$ V
b. $V_{BB} = 1.7$ V
c. $V_{BB} = 3.7$ V

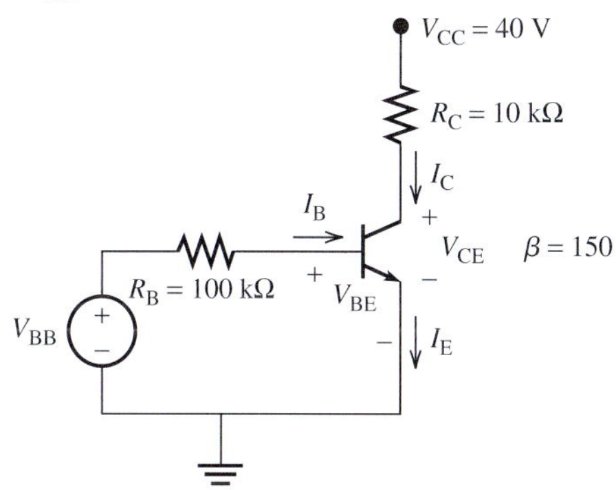

FIGURE P8.38 The circuit for Problem 8.38.

8.39 (A) For the circuit shown in Figure P8.39, find V_{BE}, I_B, I_C, and V_{CE}.

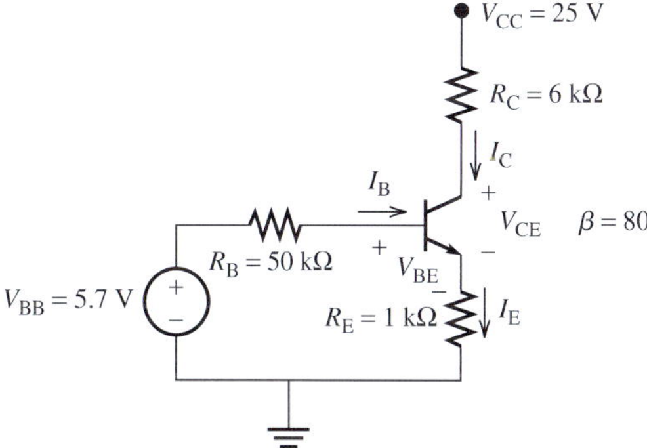

FIGURE P8.39 Circuit for Problem 8.39.

8.40 (A)* Determine the value of R_B in Figure P8.40 such that the output voltage is 3.93 V.

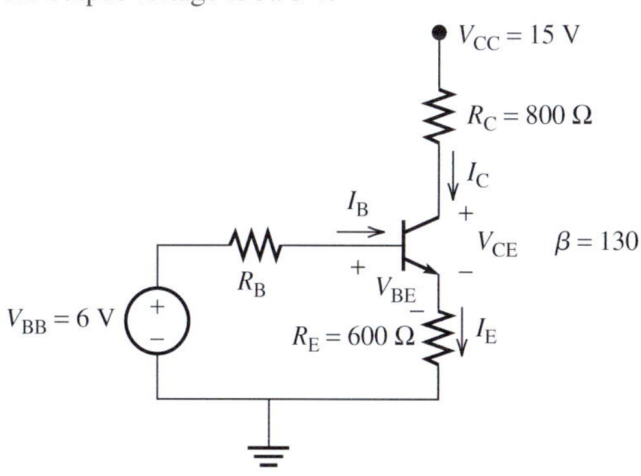

FIGURE P8.40 Transistor circuit for Problem 8.40.

8.41 (A) For the circuit shown in Figure P8.41, find the value of the resistor, R_B, which adjusts the collector–emitter voltage to be 6 V.

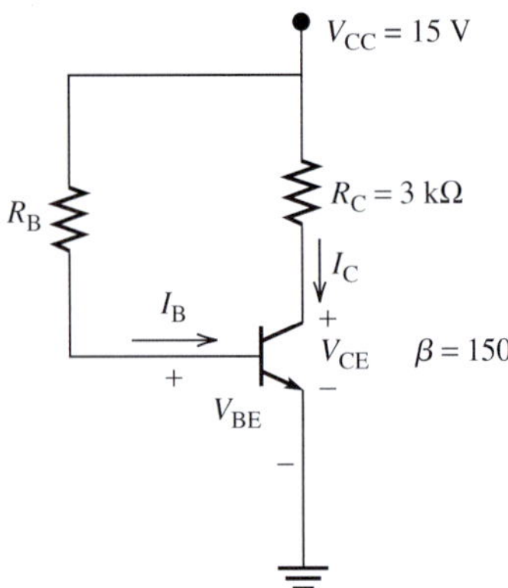

FIGURE P8.41 Circuit for Problem 8.41.

8.42 (A) For the circuit shown in Figure P8.42, find V_{BE}, I_B, I_C, and V_{CE}.

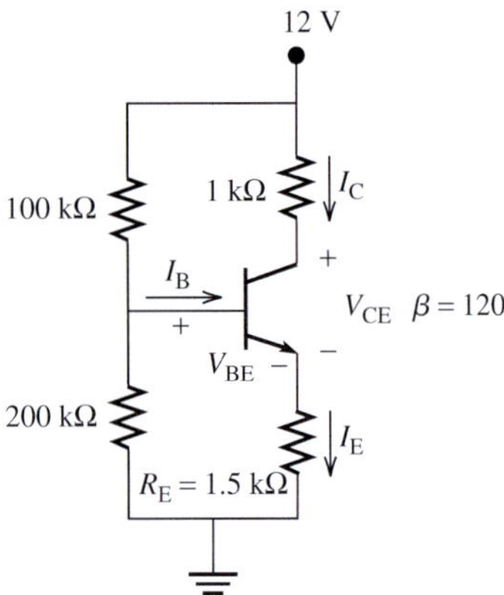

FIGURE P8.42 Circuit for Problem 8.42.

8.43 (B) Find the equivalent circuit of the one shown in Figure P8.42. (Hint: the equivalent circuit looks like the one shown in Figure P8.44.)

8.44 (A)* For the circuit shown in Figure P8.44, find V_B such that $I_C = 6$ mA. $\beta = 200$.

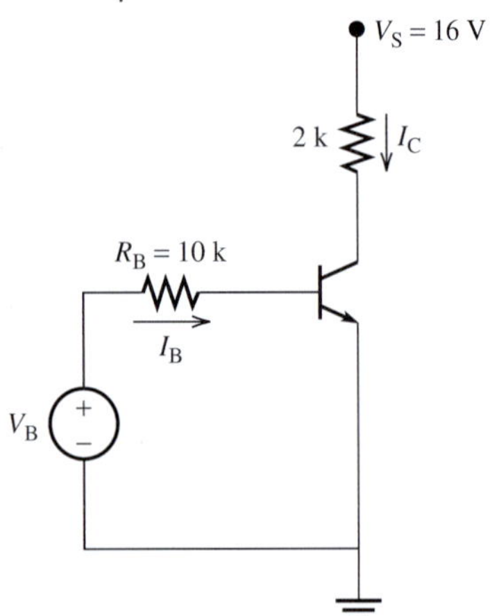

FIGURE P8.44 Transistor circuit for Problems 8.44 and 8.45.

8.45 (H) If $V_B = 4$ V for the circuit shown in Figure P8.44, find an equivalent circuit replacing V_B and R_B, having the source from V_S. (Hint: The circuit looks like Figure P8.45.)

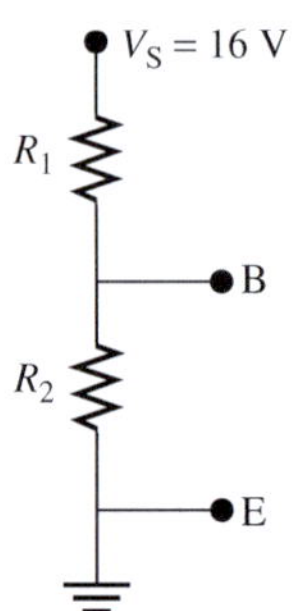

FIGURE P8.45 Hint circuit for Problem 8.45.

8.46 (B) Find the equivalent AC circuit for the one shown in Figure P8.46.

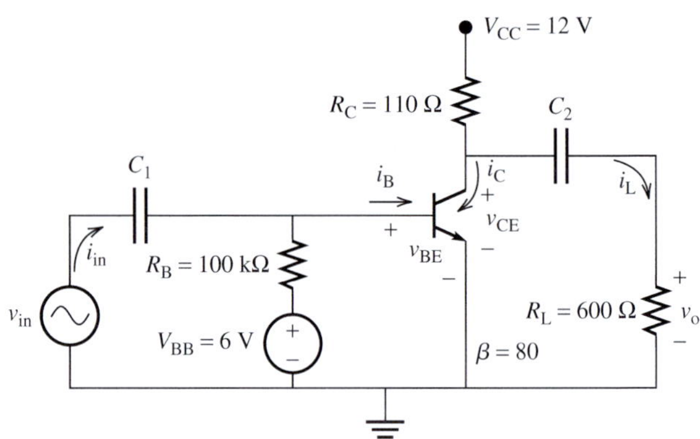

FIGURE P8.46 Biased transistor circuit.

8.47 (H) Find the relationship between R_{B1} and R_{B2}, and the range of R_C to make the circuit given in Figure P8.47 operate in the:

a. cutoff region
b. active region
c. saturate region

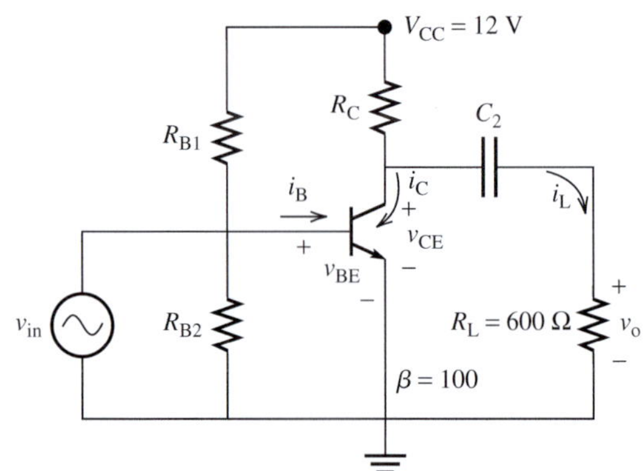

FIGURE P8.47 Circuit for Problem 8.47.

8.48 (H)* Find A_v, A_i, R_i, and R_o for the circuit shown in Figure P8.48.

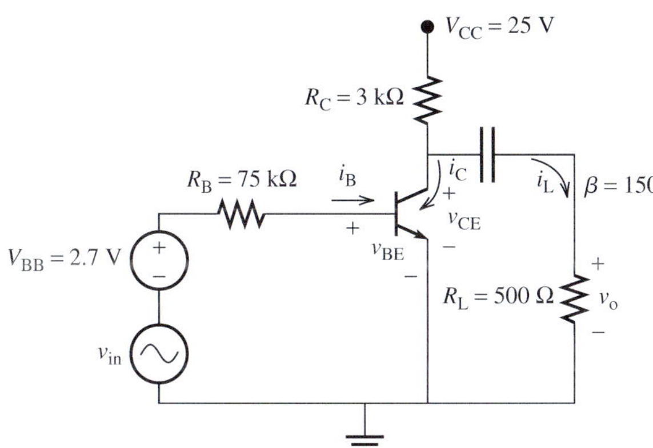

FIGURE P8.48 Circuit of Problem 8.48.

8.49 (H) For the circuit shown in P8.49, calculate voltage gain, current gain, input impedance, and output impedance.

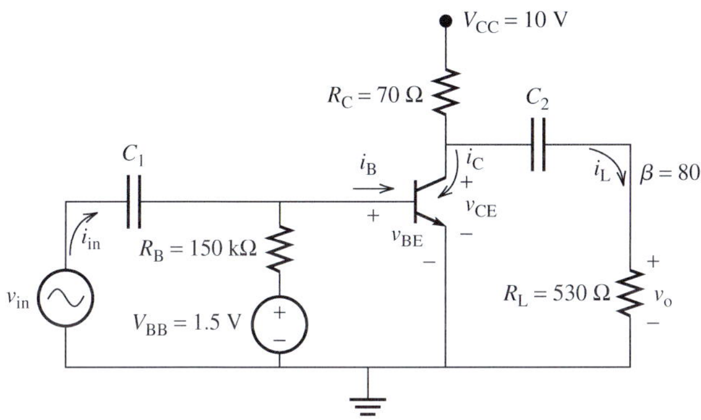

FIGURE P8.49 Circuit for Problem 8.49.

8.50 (H) For the circuit shown in Figure P8.50, find A_v, A_i, R_i, and R_o.

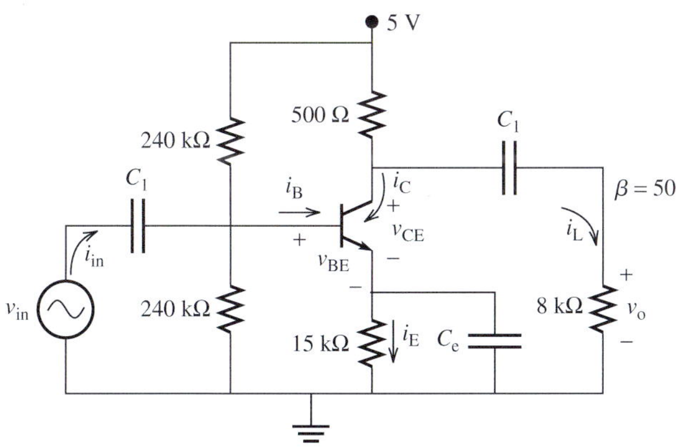

FIGURE P8.50 Circuit for Problem 8.50.

8.51 (B) Provide at least three advantages of using a transistor switch rather than a mechanical switch.

8.52 (A) A transistor switch (operating in the saturation region) is controlled by the output of a digital logic gate as shown in Figure P8.52 . The output of the logic gate is 5 V. A resistor (R_B) is needed to limit the current out from the digital logic gate to 5 mA to protect it, and a resistor (R_L) is also needed to limit the current through the photodiode to 10 mA. Calculate the value of the two protecting resistors.

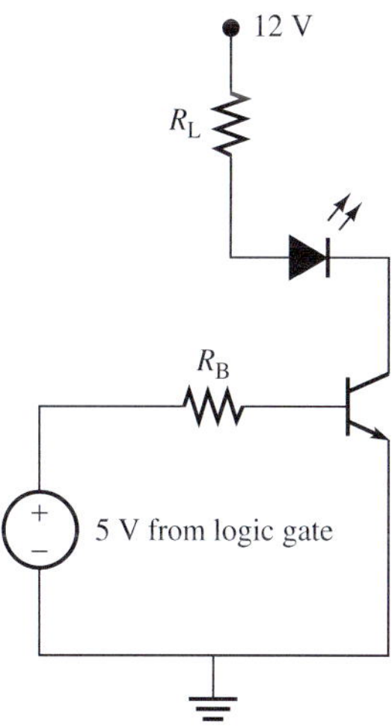

FIGURE P8.52 Logic gate-controlled photodiode.

8.53 (A)* For the circuit shown in Figure P8.51, find the minimum required value of V_{BB} to turn on the switch.

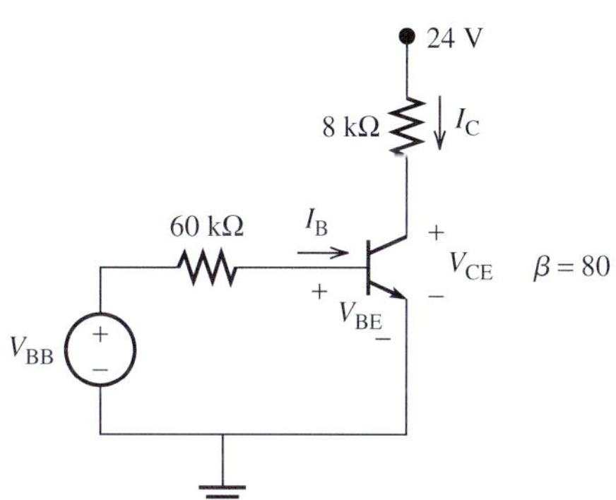

FIGURE P8.53 Circuit for Problem 8.53.

SECTION 8.4.4 FIELD-EFFECT TRANSISTORS

8.54 (B) What is the major difference between a BJT and a MOSFET?

8.55 (B) What are the three regions in which MOSFETS operate?

8.56 (A) If FETs are used as amplifiers, which region do FETs operate in? Explain the reason.

8.57 (B) Sketch the physical structure of an n-channel enhancement MOSFET. Label the channel length (L), the width (W), the terminals, and the channel region. Draw the corresponding circuit symbols.

8.58 (A) Draw a crude model of an NMOS at both its ON and OFF state, using a mechanical switch.

8.59 (A) Describe the conduction channel in an NMOS transistor and explain what causes it to form.

8.60 (A) Describe the conduction channel in a PMOS transistor and explain what causes it to form.

8.61 (H) When an NMOS transistor is in its OFF state, the drain–source channel is not conductive and it can be considered a back-to-back diode pair. Explain why this diode pair will impede the current flow.

8.62 (H) When the PMOS transistor is in its OFF state, the drain–source channel is not conductive. The channel is still a diode pair, but no longer back-to-back like NMOS. Describe the diode pair model for PMOS and explain why this diode pair will impede the current flow.

8.63 (H) When the NMOS transistor is in its ON state, the drain source channel is conductive and it can be thought of as a resistor with very low resistance. The conduction channel formation is a phenomenon that will change upon external bias. Explain (1) why the accumulation of electrons will change the p-type channel into an n-type, and (2) why an n-type channel will make the channel conductive.

8.64 (A)* Sketch the truth table of a two-input OR gate for the circuit shown in Figure P8.64. That is, determine the output voltage, V_{out}, in terms of inputs A and B when they are high (1) or low (0), for example, when both are 0, when both are 1, and when one is 0 and the other is 1. What is this operation?

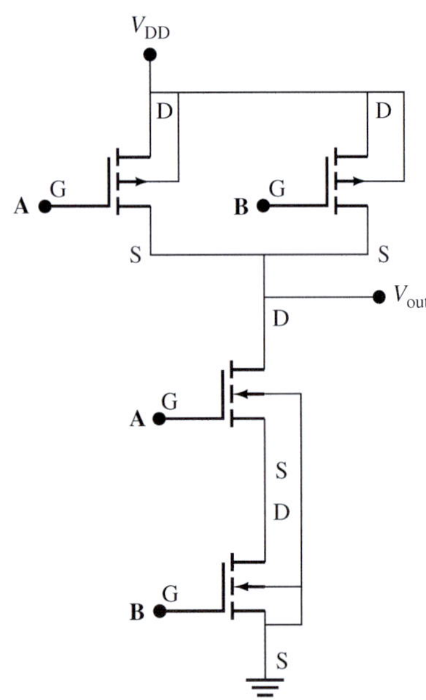

FIGURE P8.64 A CMOS circuit that functions as a gate.

8.65 (A) * Sketch the truth table of a two-input OR gate for the circuit shown in Figure P8.65. That is, determine the output voltage, V_{out}, in terms of inputs A and B when they are high (1) or low (0), for example, when both are 0, when both are 1, and when one is 0 and the other is 1. What is this operation?

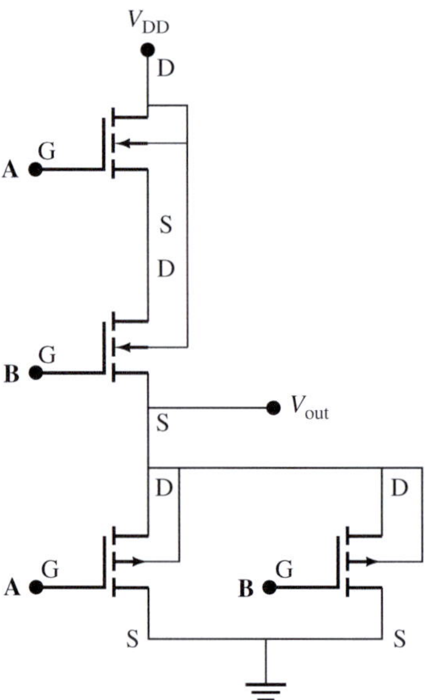

FIGURE P8.65 A CMOS circuit that functions as a gate.

8.66 (A) Repeat Problem 8.65 if PMOS is replaced by NMOS and vice versa.

8.67 (A) Design an AND gate for three inputs A, B, and C using CMOS technology. The truth table of an AND gate is shown in Table P8.67.

TABLE P8.67 The Truth Table of an AND Gate for Three Inputs

A	B	C	ABC
0	0	0	0
0	0	1	0
0	1	0	0
0	1	1	0
1	0	0	0
1	0	1	0
1	1	0	0
1	1	1	1

8.68 (A) Design an OR gate for three inputs A, B, and C using CMOS technology. The truth table of an OR gate is shown in Table P8.68.

TABLE P8.68 The Truth Table of an AND Gate for Three Inputs

A	B	C	A + B + C
0	0	0	0
0	0	1	1
0	1	0	1
0	1	1	1
1	0	0	1
1	0	1	1
1	1	0	1
1	1	1	1

SECTION 8.5 OPERATIONAL AMPLIFIERS

8.69 (B) Using the circuit shown in Figure P8.69, find the following:
- **(a)** The voltage gain in the circuit
- **(b)** The output voltage for the following input voltages, where V_{CC} is 15 V:
  - **i.** $V_i = 4$ V
  - **ii.** $V_i = -3\sin(\omega t)$V
  - **iii.** $V_i = -7$ V

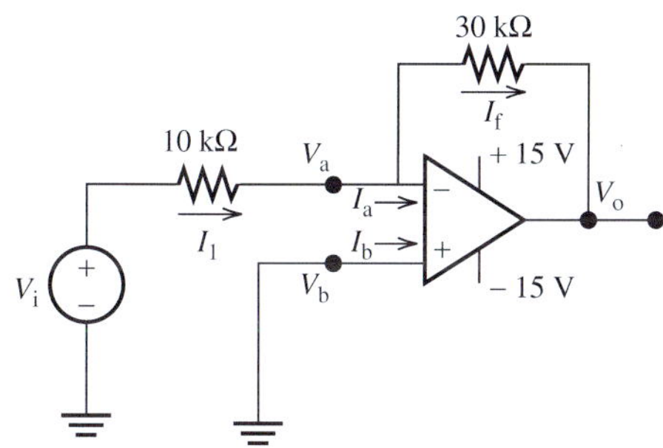

FIGURE P8.69 Circuit for Problem 8.69.

8.70 (B)* Consider the circuit shown in Figure P8.70. Find the following:
- **a.** The voltage gain
- **b.** The output voltage if the input voltage is 4 V

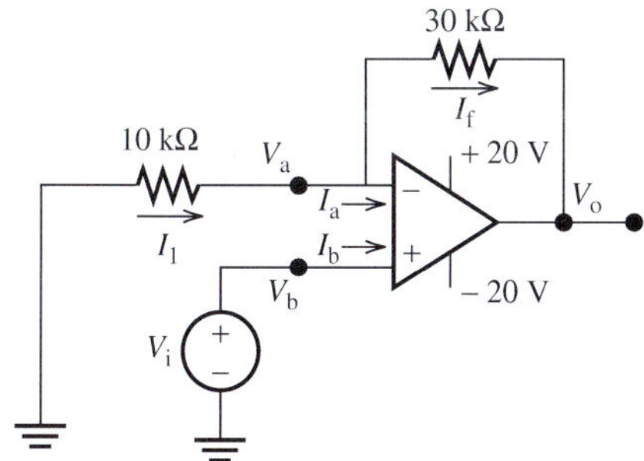

FIGURE P8.70 Circuit for Problem 8.70.

8.71 (A)* Find the output voltage for the circuit shown in Figure P8.71.

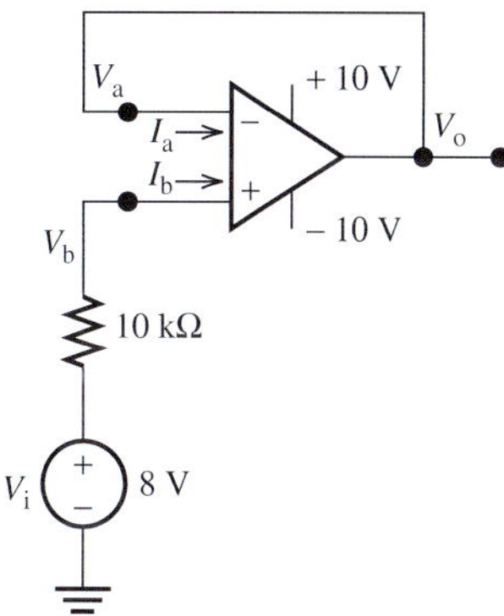

FIGURE P8.71 Circuit for Problem 8.71.

8.72 (A) Design a noninverting amplifier having a gain of 50, using a single op amp.

8.73 (H)* Find the formula for the voltage gain for the amplifier shown in Figure P8.73.

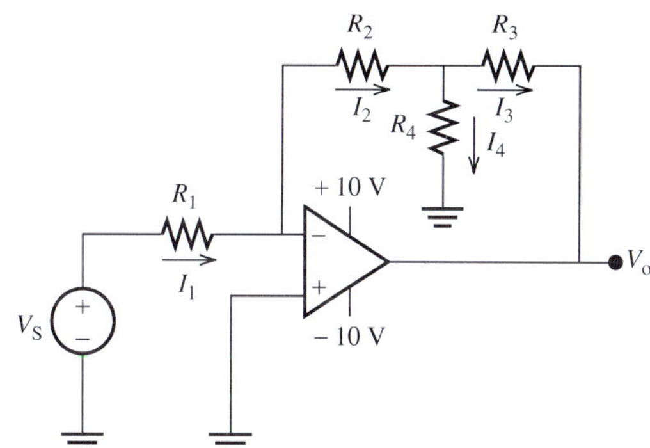

FIGURE P8.73 Advanced op amp circuit.

8.74 (H) The input voltage of the circuit shown in Figure P8.74 is 20 V. Determine the output voltage of both op amps, V_{o1} and V_{o2}.

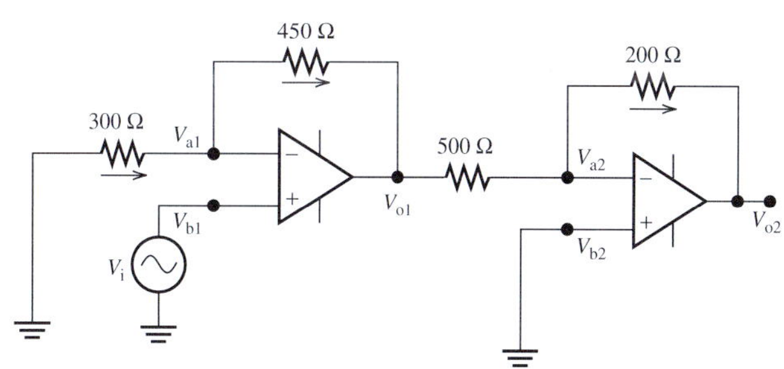

FIGURE P8.74 Circuit for Problem 8.74.

8.75 (B)* For the circuit shown in Figure P8.75, the input voltages are $V_1 = 15$ V and $V_2 = 10$ V. Determine the current, I_f, in the figure.

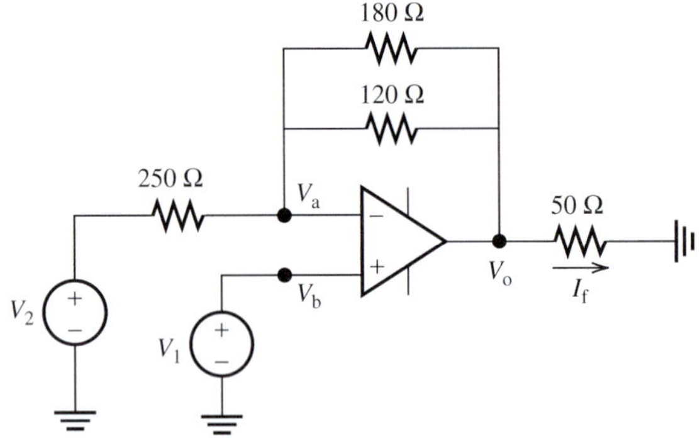

FIGURE P8.75 Circuit for Problem 8.75.

8.76 (A) Summing amplifier
For the circuit shown in Figure P8.76, verify that V_o can be represented in terms of the addition of the two input voltages, V_1 and V_2, if $R_1 = R_2$.

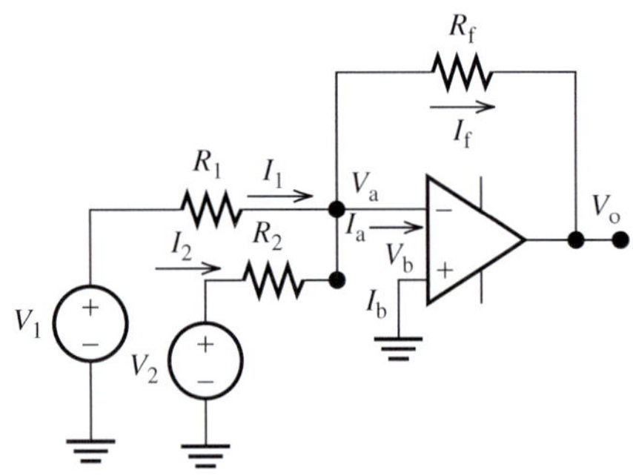

FIGURE P8.76 A summing amplifier.

8.77 (A) Difference amplifier
For the circuit shown in Figure P8.77, verify that V_{out} can be expressed in terms of the difference of V_1 and V_2 if $R_1 = R_2$, $R_f = R_g$.

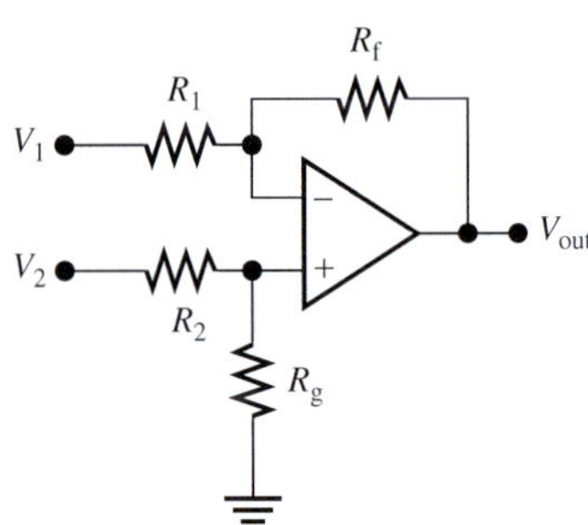

FIGURE P8.77 A difference amplifier.

8.78 (H)* Integrator
For the circuit shown in Figure P8.78, verify that the output voltage, V_{out}, is the integral of the input voltage, V_{in} (multiplied by some constants).

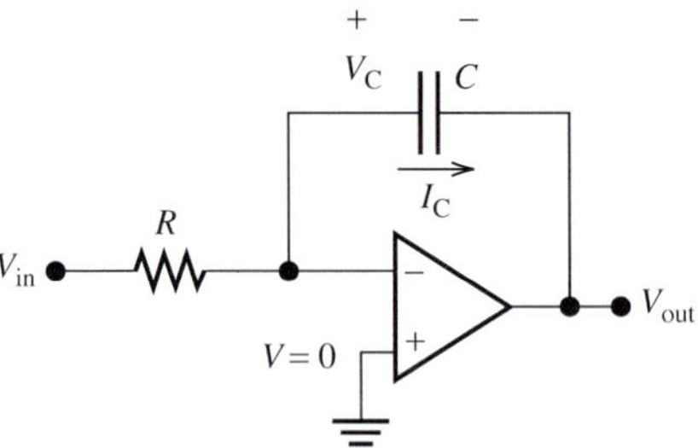

FIGURE P8.78 An integrator.

8.79 (H) Differentiator
For the circuit shown in Figure P8.79, verify that the output voltage, V_{out}, is the derivative of the input voltage V_{in} (multiplied by some constants).

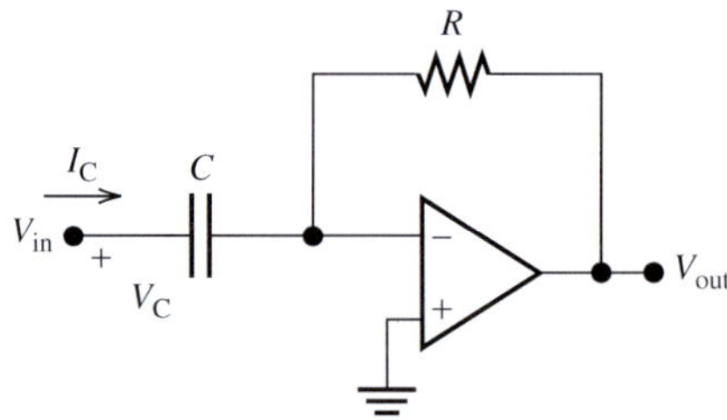

FIGURE P8.79 A differentiator.

SECTION 8.6 USING PSPICE TO STUDY DIODES AND TRANSISTORS

8.80 (B) Use PSpice to find the current, I, flowing through the diode of the circuit shown in the Figure P8.1 when:

a. $R = 2$ kΩ
b. $R = 8$ kΩ

8.81 (B) Consider the circuit shown in Figure P8.15. Use PSpice to plot the voltage output, $V_o(t)$, if the input voltage is $V_i(t) = 15\ \sin(10\pi t)$ V.

8.82 (H) Use PSpice to find V_{BE}, I_B, I_C, and V_{CE} for the circuit shown in Figure P8.39.

TOPIC 5

Fundamentals of Logic Circuits

The content of this topic is compiled from:

Chapter 10 (pp. 440–487), *Electrical Engineering: Concepts and Applications*

By S.A. Reza Zekavat

CHAPTER 10

Fundamentals of Logic Circuits

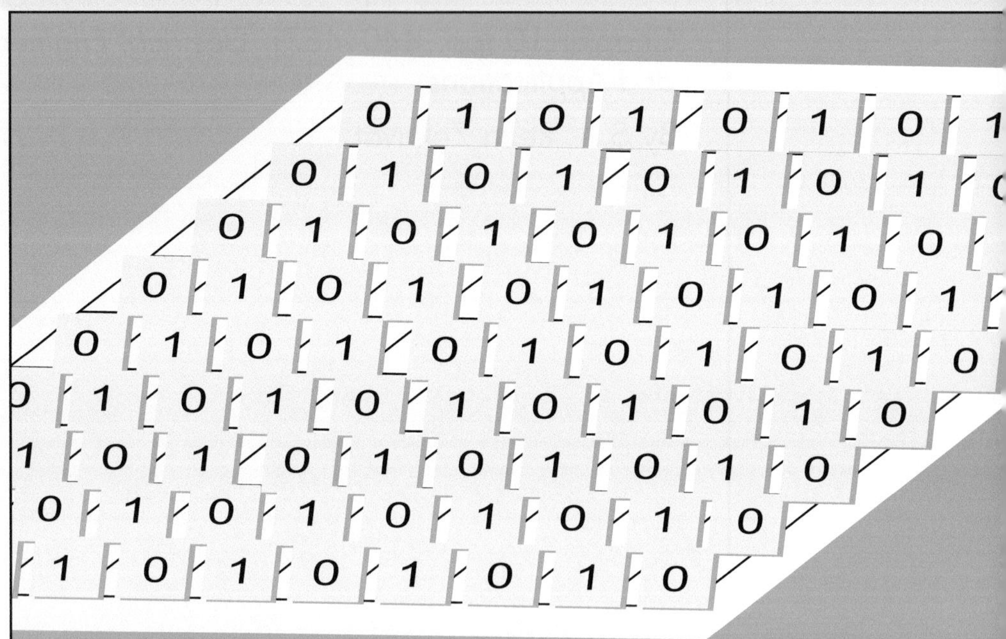

10.1 INTRODUCTION

So far, this book has examined circuits that process analog signals. As you have learned, a signal represents the variation of amplitude with time. For example, the amplitude of a sinusoid varies periodically with time. An analog signal is continuous in both the time domain and the amplitude domain. An example of such a signal is shown in Figure 10.1. In this example, no discontinuity is observed in amplitude and time. A discrete signal or a discrete-time signal (Figure 10.2) is a time series that is sampled from an analog signal. Note that a discrete-time signal may achieve any value for its amplitude. Thus, while it is discrete in time, it is not discrete in amplitude. On the other hand, a digital signal (Figure 10.3) is a discrete-time signal that attains only a discrete set of values. Thus, it is discrete in both the amplitude domain and the time domain.

The process of converting an analog signal to a digital signal is called *digitization*. Digitization includes two steps: sampling and quantization. In the *sampling* stage, the continuous signal is sampled at regular intervals. This creates a time series (discrete signals) as shown in Figure 10.2. *Quantization* refers to the adjustment of sampled levels to some predefined signal levels (Figure 10.3). As an example, consider three predefined signal levels of +5 V,

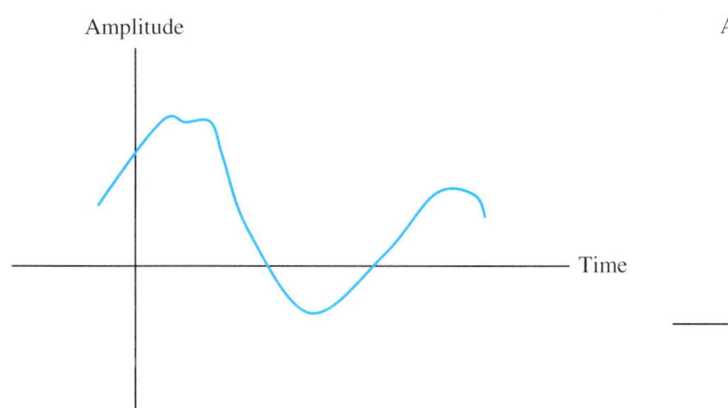

FIGURE 10.1 Analog signal.

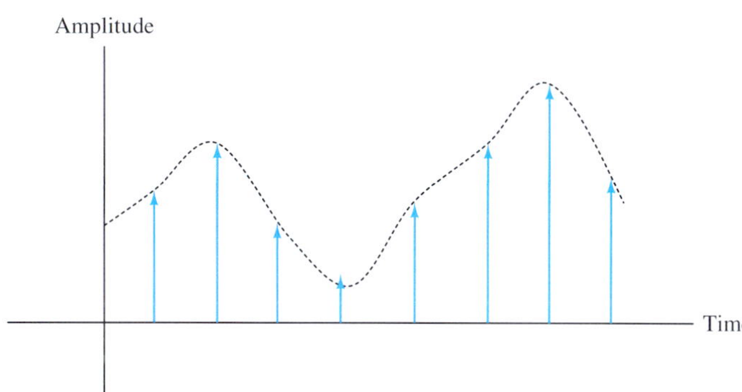

FIGURE 10.2 Discrete time signal.

0 V, and −5 V. If the sampled voltage level is 3.2 volts, it will be quantized to 5 V because that is the closest predefined level. Likewise, if the sampled signal is closest to the 0 V level (e.g., 1.2 V), it will be quantized as 0 V. Although, in general, a digital signal may have multiple voltage levels, this chapter is interested in two-level digital signals, namely '0' and '1,' as shown in Figure 10.4. These two levels refer to different voltage levels. For example, 1 may refer to +5 V and 0 may refer to 0 V.

Note that a multi-level digital signal may also be formed by a two-level one. For instance, each level of an eight-level quantized signal can be represented by three (0 or 1) digits. As an example, if the eight levels start at 1 V and go to 8 V, 1 V can be represented by 000, 2 V by 001, 3 V by 010, and 8 V can be represented by 111. The process of converting these eight predefined voltages to 0-1 (digital) is called analog-to-digital conversion. In general, the devices that convert analog signals to digital are called ADC (analog-to-digital convertor). For the remainder of this chapter, digital signals will be referred to as two-level (digital) signals.

In order to process analog signals using digital signal processors (DSP), analog signals must be converted to digital. In order to apply the signal to a DSP, the original analog (continuously varying) signal is first sampled, then quantized, and then converted to digital (0 or 1).

Today, almost all systems are digital. Examples include digital cell phones, digital TV, digital cameras, and so on. Digital representation of signals offers many advantages over analog, including: fewer components, stability, and a wide range of applications. In addition, signal processing of analog signals is not as straightforward as processing of digital signals. In digital communication systems, analog signals are first converted to digital. For example, in today's cellular systems, after reception at the radio's RF front-end, all signals are converted to digital to allow further processing. This eases the reconstruction process of transmitted signals that are altered by interference and noise effects. Thus, it improves performance and supports transmission of data (e.g., an image).

In digital logic circuits, the binary numbers 1 and 0 are used to represent the voltage or current. The logic 1 is also called high, true, or ON; logic 0 is also called low, false, or OFF. However, in daily life, decimal numbers are most commonly used to represent quantities.

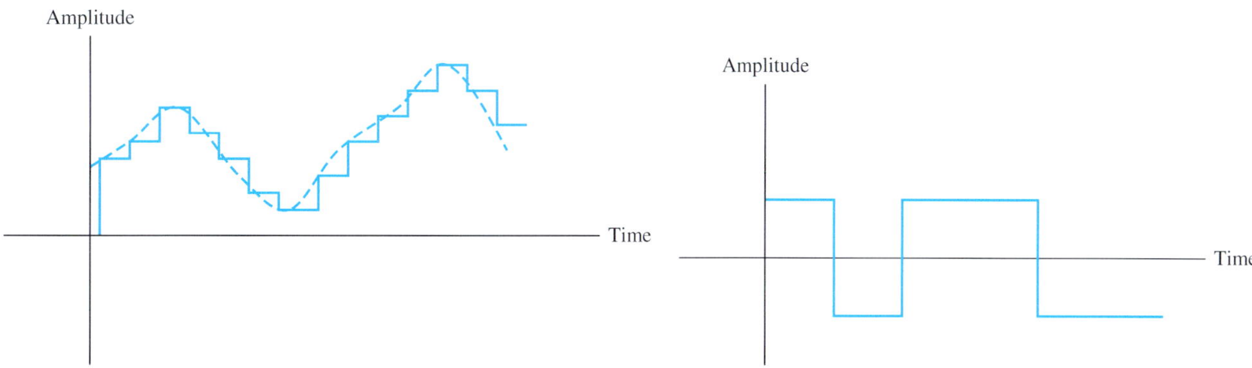

FIGURE 10.3 Multi-level digital signal.

FIGURE 10.4 Two-level digital signal.

Therefore, this chapter first discusses how to convert decimal numbers into binary. The chapter introduces number systems, Boolean algebra, basic logic gates, and sequential logic circuits that are used in digital circuits. PSpice examples are provided to help construct digital systems.

10.2 NUMBER SYSTEMS

10.2.1 Binary Numbers

Currently, all computer data is stored in binary format (using only the numbers 0 or 1). Examples of this type of data include the following:

- Data on a floppy disk, hard disk, and so on (magnetic storage)
- Data in flash memory (semiconductor storage)
- Data written to a CD, DVD, and so on (optical disc storage).

Therefore, most numerical data is represented in the binary format (base 2). However, in daily life, decimal numbers (base 10) are more commonly used to represent numerical data. Therefore, it is important to define the relationship between decimal and binary numbers.

First, let's see how decimal numbers can be converted to binary. The decimal number 1450 can be represented as:

$$1450 = 1 \times 10^3 + 4 \times 10^2 + 5 \times 10^1 + 0 \times 10^0$$

In this representation, the base is 10 (ten). The number 1450 can also be represented in a similar way in base 2:

$$1450 = 1 \times 2^{10} + 0 \times 2^9 + 1 \times 2^8 + 1 \times 2^7 + 0 \times 2^6 + 1 \times 2^5 + 0 \times 2^4 + 1 \times 2^3 + 0 \times 2^2 + 1 \times 2^1 + 0 \times 2^0$$

The coefficients of each term form the binary format of the number 1450 (see Figure 10.5).

Therefore, the binary form of the number 1450 is 10110101010. Two methods are available for converting a decimal number to binary form that are detailed in the next subsections.

10.2.1.1 CONVERTING A DECIMAL INTEGER TO BINARY FORM

Method 1: Figure 10.6 summarizes Method 1 using 1450 as the number to be converted. Therefore, 1450 can be represented as:

$$1450 = 1 \times 2^{10} + 0 \times 2^9 + 1 \times 2^8 + 1 \times 2^7 + 0 \times 2^6 + 1 \times 2^5 + 0 \times 2^4 + 1 \times 2^3 + 0 \times 2^2 + 1 \times 2^1 + 0 \times 2^0$$

Thus, the binary form of 1450 is 10110101010.

In summary, to use Method 1 to convert a decimal number to binary, first select the largest power of 2 (e.g., 1024, 512, 256, ...) for which the result is smaller than the original decimal number. This prevents obtaining negative values during conversion. Subtract the largest power of 2 from the original decimal number and then repeat this process with the remaining number. This process is repeated until the subtraction results in 0. In each round, one of the digits of the binary number is generated. The decimal number is then represented as the summation of numbers each with a power of 2. The corresponding coefficients of the numbers to the power of 2 form the binary form of this decimal number.

10th	9th	8th	7th	6th	5th	4th	3th	2th	1th	0th
1	0	1	1	0	1	0	1	0	1	0

FIGURE 10.5 Coefficients of binary form of 1450 (decimal).

$1450 - 1024 = 426$ $(1024 = 2^{10})$
$426 - 256 = 170$ $(256 = 2^8)$
$170 - 128 = 42$ $(128 = 2^7)$
$42 - 32 = 10$ $(32 = 2^5)$
$10 - 8 = 2$ $(8 = 2^3)$
$2 - 2 = 0$ $(2 = 2^1)$

FIGURE 10.6 Converting 1450 to binary form (Method 1).

	Quotient	Remainder
1450/2 =	725	0
725/2 =	362	1
362/2 =	181	0
181/2 =	90	1
90/2 =	45	0
45/2 =	22	1
22/2 =	11	0
11/2 =	5	1
5/2 =	2	1
2/2 =	1	0
1/2 =	0	1

FIGURE 10.7 Converting 1450 to binary form (Method 2).

Method 2: Method 2 is summarized in Figure 10.7, again for the decimal number of 1450.

In Method 2, the quotient (the first quotient is the original number) is repeatedly divided by 2 until quotient is 0. The reverse sequence of the remainders forms the binary number (10110101010) of the decimal number.

10.2.1.2 CONVERTING A DECIMAL FRACTION TO BINARY FORM

Method 1: To explain this method we will use 0.2111 as a sample number (see Figure 10.8).

As shown in Figure 10.8, the process outlined above can continue to the required precision to convert decimal numbers. The process is stopped when it reaches the precision requirement.

$$0.2111 \approx 0 \times 2^{-1} + 0 \times 2^{-2} + 1 \times 2^{-3} + 1 \times 2^{-4} + 0 \times 2^{-5} + 1 \times 2^{-6} + 1 \times 2^{-7}$$

Therefore, the binary form for 0.2111 is 0.0011011.

In Method 1, when converting a decimal fraction number to binary, choose the largest negative number of power of 2 (e.g., 0.5, 0.25, 0.125 . . .), that is smaller than the value you are subtracting from. This prevents obtaining negative values during conversion. The process is repeated until 0 is obtained, or the value of the binary form of the fraction number approximates the original number with the required precision. Finally, the decimal fraction number is represented as the summation of numbers of negative power of 2. The corresponding coefficients of power of 2 form the binary number consistent with the desired decimal number.

Method 2: Method 2 is summarized in Figure 10.9. As shown in this figure, the binary form of 0.2111 is 0.0011011.

In Method 2, the fraction part of the number is repeatedly multiplied by 2 (the first fraction part to be multiplied is the original number) until the fraction part becomes 0 or until the desired precision is reached. As shown, the equivalent binary number in our example corresponds to 0.0011011.

In order to convert a decimal number consisting of both integer and fraction parts, each part is separately converted to binary.

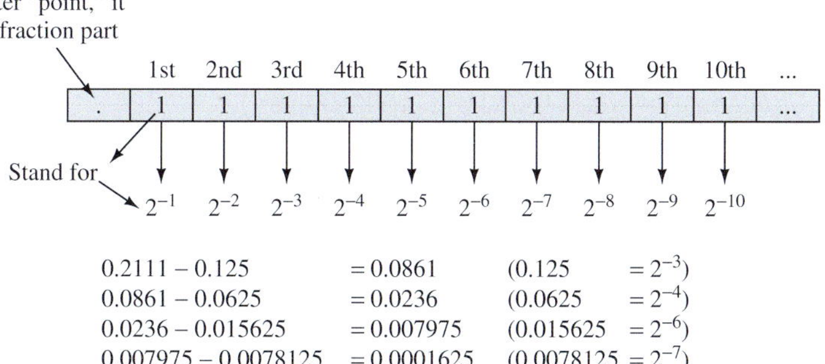

FIGURE 10.8 Converting 0.2111 to binary (Method 1).

	Integer part	Fraction part
0.2111 × 2 =	0	0.4222
0.4222 × 2 =	0	0.8444
0.8444 × 2 =	1	0.6888
0.6888 × 2 =	1	0.3776
0.3776 × 2 =	0	0.7552
0.7552 × 2 =	1	0.5104
0.5104 × 2 =	1	0.0208

FIGURE 10.9 Converting 0.2111 to binary form (Method 2).

EXAMPLE 10.1 Decimal to Binary Conversion

Convert the decimal integer number 626 into the binary form.

SOLUTION

The approach has been summarized in Figure 10.10.

According to the results shown in Figure 10.10, 626 (decimal) = 1001110010 (Binary).

		Quotient	Remainder
626/2	=	313	0 ↑
313/2	=	156	1
156/2	=	78	0
78/2	=	39	0
39/2	=	19	1
19/2	=	9	1
9/2	=	4	1
4/2	=	2	0
2/2	=	1	0
1/2	=	0	1

FIGURE 10.10 Converting 626 to binary form.

EXAMPLE 10.2 Decimal to Binary Conversion

Convert the decimal fraction number 0.47 into binary (stop when the fraction part has 8 bits).

SOLUTION

The process has been summarized in Figure 10.11.

Accordingly: 0.47(decimal) = 0.01111000 (binary).

	Integer part	Fraction part
$0.47 \times 2 =$	0	0.94
$0.94 \times 2 =$	1	0.88
$0.88 \times 2 =$	1	0.76
$0.76 \times 2 =$	1	0.52
$0.52 \times 2 =$	1	0.04
$0.04 \times 2 =$	0	0.08
$0.08 \times 2 =$	0	0.16
$0.16 \times 2 =$	0 ↓	0.32

FIGURE 10.11 Converting 0.47 to binary form.

EXAMPLE 10.3 Decimal to Binary Conversion

Convert the decimal number 626.47 into binary form (stop when the fraction part has 8 bits).

SOLUTION

Based on the results of Example 10.1 and 10.2:

626.47(decimal) = 1001110010.01111000 (binary).

EXERCISE 10.1

Determine the binary form of the decimal number 627.47. Can you use the results from Example 10.3 to solve this problem?

10.2.1.3 CONVERTING A BINARY NUMBER TO A DECIMAL NUMBER

Here, we use the number 110110.011011(binary) as an example. The solution is shown in Figure 10.12.

As shown in Figure 10.12:

$$110110.011011 = 1\times2^5+1\times2^4+0\times2^3+1\times2^2+1\times2^1+0\times2^0+0\times2^{-1}+1\times2^{-2}+1\times2^{-3}+0\times2^{-4}+1\times2^{-5}+1\times2^{-6}$$
$$= 54.421875 \text{ (decimal)}$$

To convert a binary number to a decimal, each bit (0 or 1) is multiplied by the corresponding power of 2. Then, the results are added together. The summation will be the desired decimal number.

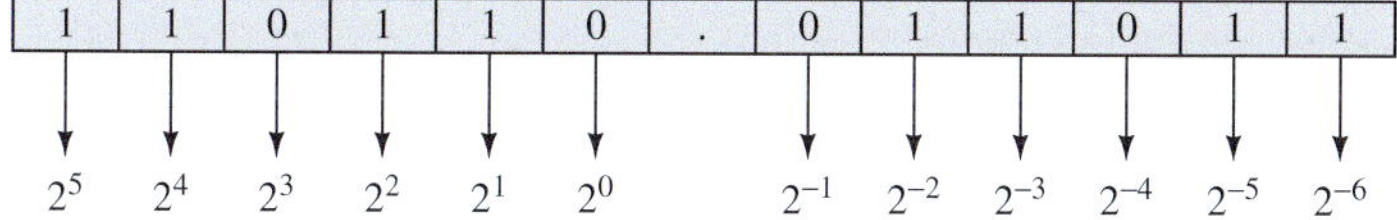

FIGURE 10.12 Converting a binary number to decimal form.

APPLICATION EXAMPLE 10.4 Computing Device

An engineer intends to design his own computing device for calculating the daily income from his store. He wants the device to have a memory to store the input data. The maximal incoming data is 1000.00, and he wants the device to have 2-degrees of precision. If one memory cell only can store one binary bit, how many memory cells are required for his computing device? (No memory cell is required to store the point '.' in the fraction number. In addition, to minimize the error and use less memory cells, make the absolute value of the error between the decimal value and the corresponding binary form less than 0.005.)

SOLUTION

For the integer part (using Method 1), converting 1000 to binary as shown in Figure 10.13 results in: 1000 (decimal) = 1111101000 (binary). Thus, 10 memory cells are required for the integer part.

For the fraction part (using Method 2): Assuming two-degrees of precision, the maximal value for the fraction part will be 0.99, and the minimal valid fraction part is 0.01. The conversion process of 0.99 to binary is shown in Figure 10.14.

For the fraction part, based on Figure 10.14, if 6 cells are used, the absolute value of error will be $0.99-(2^{-1}+2^{-2}+2^{-3}+2^{-4}+2^{-5}+2^{-6})=0.005625$, which is greater 0.005. If 7 cells are used, the absolute value of error is $|-0.0021871| < 0.005$. Considering the minimal valid fraction of 0.01, the last bit (0000001) corresponds to 2^{-7}; thus, the absolute value of error is $|0.01-2^{-7}|=0.0021875 < 0.005$.

Therefore, 7 memory cells are required to store the fraction part. In total, the engineer needs 17 memory cells for his computing device.

	Quotient	Remainder
1000/2 =	500	0
500/2 =	250	0
250/2 =	125	0
125/2 =	62	1
62/2 =	31	0
31/2 =	15	1
15/2 =	7	1
7/2 =	3	1
3/2 =	1	1
1/2 =	0	1

FIGURE 10.13 Converting 1000 to binary.

0.99–0.5	= 0.49	($0.5 = 2^{-1}$)
0.49–0.25	= 0.24	($0.25 = 2^{-2}$)
0.24–0.125	= 0.115	($0.125 = 2^{-3}$)
0.115–0.0625	= 0.0525	($0.0625 = 2^{-4}$)
0.0525–0.03125	= 0.02125	($0.03125 = 2^{-5}$)
0.02125–0.015625	= 0.005625	($0.015625 = 2^{-6}$)
0.005625–0.0078125	= – 0.0021875	($0.0078125 = 2^{-7}$)

FIGURE 10.14 Converting 0.99 to binary form.

10.2.1.4 BINARY ARITHMETIC

Addition. The addition of binary numbers is similar to that of decimal numbers. The rules for the addition of binary bits are summarized in Table 10.1.

TABLE 10.1 Addition Rules for Binary Bits

	Carry	Sum
0 + 0 =	0	0
0 + 1 =	0	1
1 + 1 =	1	0
1+1+1=	1	1

EXAMPLE 10.5 Binary Addition

Use binary addition to calculate 12 (decimal) + 14 (decimal).

SOLUTION

The solution has been summarized in Figure 10.15.

To conduct simple addition using binary numbers, first convert the decimal numbers to binary and then use the binary addition rules. In this example, the decimal answer should be 26, which will serve as a check by converting the binary answer back to decimal form.

Figure 10.16 shows how the answer is converted to decimal form to check the accuracy of the binary addition.

Thus, the final answer is correct.

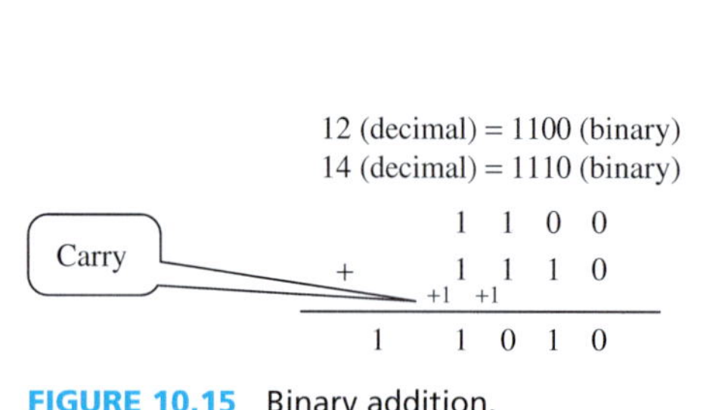

FIGURE 10.15 Binary addition.

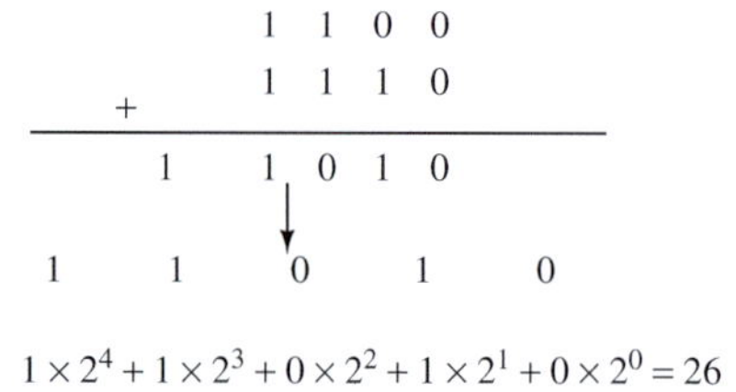

$1 \times 2^4 + 1 \times 2^3 + 0 \times 2^2 + 1 \times 2^1 + 0 \times 2^0 = 26$

FIGURE 10.16 Converting the answer to decimal form.

Subtraction. To understand binary subtraction, two important terms must be understood: *one's complement* and *two's complement*.

One's complement For a given binary number, if all "0" bits are replaced with "1" and all "1" bits are replaced with "0", the transformed binary number is called the *one's complement* of the original binary number. For example:

Binary number:	01100101
Its one's complement:	10011010

Two's complement The *two's complement* of a given binary number is obtained by adding 1 to the one's complement of the binary number and neglecting the carry out (if any) of the most significant bit.

The most significant bit is the first bit from the left. In general, the most significant integer in any number is the integer on the left. For example, in the number 245, 2 is the most significant integer. It is clear that we can write 245 as 200 + 40 + 5. Thus, integer 2 represents the number 200. Accordingly, it is called the most significant integer. Equivalently, integer 5 in 245 is the least significant integer.

Now, consider the binary number 01100101:

Its one's complement is	10011010
Its two's complement is	10011011

EXAMPLE 10.6 Two's Complement

Find the two's complement of the binary number 0110101101.

SOLUTION

The solution has been summarized in Figure 10.17.

Therefore, the two's complement of the given binary number is 1001010100.

```
  0 1 1 0 1 0 1 1 0 0
          ↓
  1 0 0 1 0 1 0 0 1 1 (One's complement)
+                   1
  -------------------
  1 0 0 1 0 1 0 1 0 0 (Two's complement)
```

FIGURE 10.17 Finding the two's complement of 0110101101.

Subtraction of binary numbers utilizes the two's complement.

Subtraction using the two's complement The two's complement can be used to represent negative numbers and perform the subtraction. The first bit (i.e., the most significant bit) is taken as the sign bit (i.e., if the number is positive or negative) of the two's complement. The first bit is 0 for positive numbers and 1 for negative numbers. Negative numbers are represented as the two's complement of the corresponding positive number. For example, the decimal number +2 is equivalent to 00000010 in binary form (using 8 bits), and the decimal number −2 is 11111110 in binary form which is exactly the two's complement of number 2 in binary form. Therefore, subtraction can be performed by finding the two's complement of the subtrahend and then adding. For example:

$$\begin{aligned} 33 - 31 &= 33 + (-31) = 33 \text{ (binary)} + \text{Two's complement of } 31 \text{ (binary)} \\ &= 100001 + 100001 \text{ (Two's complement of 31 is 100001)} \\ &= 000010 \\ &= 2 \text{ (decimal)} \end{aligned}$$

Note that again the binary answer can be checked by converting the answer to decimal form.

EXAMPLE 10.7 Binary Subtraction

a. Find 2110 − 1250 using the binary form.
b. Find 1250 − 2110 using the binary form.

SOLUTION

a. First, convert 2110 and 1250 to binary form.

2110 (decimal) = 100000111110 (binary)

1250 (decimal) = 010011100010 (binary)

The two's complement of 1250 (decimal) is 101100011110. The operation is shown in Figure 10.18.

The operation 2110 − 1250 is equivalent to 2110 (binary) + the two's complement of 1250 (binary). This operation has been shown in Figure 10.19.

The answer in binary is positive because the leading binary is 0. Here, we ignore the leading 1 (from the last carry) in Figure 10.19.

Next, convert the result back to decimal to check the validity of our operation. This process has been shown in Figure 10.20.

b. The two's complement of 2110 (decimal) is 101100011110. The process of generating the two's complement is shown in Figure 10.21.

The operation 1250 − 2110 is equivalent to 1250 (binary) + the two's complement of 2110 (binary). This operation is shown in Figure 10.22.

The result is the signed two's complement representation for −860.

```
  010011100010                Complement means to flip all
       ↓                      the bits (0 to 1, 1 to 0)
  101100011101 ——→ 1's complement
+ 000000000001                1's complement + 1 = 2's complement
  ————————————
  101100011110 ——→ 2's complement
```

FIGURE 10.18 Finding the two's complement of 1250.

```
  2110 − 1250 =

  100000111110 ←—— 2110 in binary
+ 101100011110 ←—— 2's complement of
  ————————————     1250 in binary
 1001101011100
```

FIGURE 10.19 Binary subtraction using the two's complement.

```
Ignore it
  ↓
 (1) 0  0  1  1  0  1  0  1  1  1  0  0
```

$0 \times 2^{11} + 0 \times 2^{10} + 1 \times 2^{9} + 1 \times 2^{8} + 0 \times 2^{7} + 1 \times 2^{6} + 0 \times 2^{5} + 1 \times 2^{4} + 1 \times 2^{3} + 1 \times 2^{2} + 0 \times 2^{1} + 0 \times 2^{0} = 860$

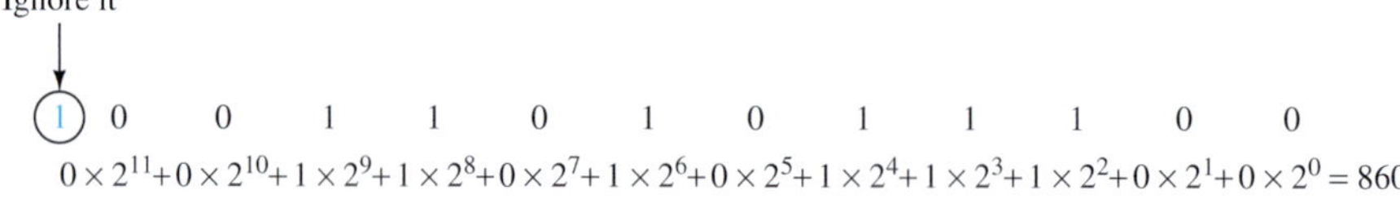

FIGURE 10.20 Binary subtraction result verification.

```
  100000111110 ——→ 2110 in binary form
       ↓
  011111000001 ——→ 1's complement
+ 000000000001
  ————————————
  011111000010 ——→ 2's complement
```

FIGURE 10.21 Finding the two's complement of 2110.

```
  1250 − 2110 =

  010011100010  ←—— 1250 in binary form
+ 011111000010  ←—— 2's complement of 1210
  ————————————
  110010100100  ←—— 2's complement of 860
```

FIGURE 10.22 Binary subtraction.

EXERCISE 10.2

Check the validity of the answer in Example 10.7, part b.

Note that if the result is beyond the maximal value that the limited number of bits can represent, *overflow* occurs. For example, 55 (decimal) + 96 (decimal) = 151 (decimal); the result 151 (decimal) is beyond the maximal value that can be represented by eight bits [+127 (decimal)].

55		00110111	(Positive Sign)	
+96	⇒	+01100000	(Positive Sign)	(Overflow occurs)
151		10010111	(Negative Sign)	

Similarly, if the result is less than the minimal value that the limited number of bits can represent, *underflow* occurs. For example, −100 (decimal) −32 (decimal) = −132 (decimal); the result −132 (decimal) is less than the minimal value that the limited number of bits can represent (−128 (decimal)).

−100		10011100	(two's complement, negative sign)	
−32	⇒	+ 11100000	(two's complement, negative sign)	
−132		01111100	(Positive sign)	(Underflow occurs)

In general, if the two numbers that are added have the same sign and the result has the opposite sign, *underflow or overflow* has occurred.

10.2.2 Hexadecimal Numbers

Hexadecimal numbers are converted from binary numbers, and can represent information more efficiently. The base is 16 for hexadecimal numbers; therefore, 16 symbols are required for a hexadecimal number (A through *F* are used to represent the digits for symbols 10 through 15). Table 10.2 shows the symbols used for hexadecimal numbers.

10.2.2.1 CONVERTING HEXADECIMAL NUMBERS TO BINARY

Based on Table 10.2, for a given hexadecimal number, replace each digit with its binary equivalent (4 bits). For example:

$$\begin{array}{c} \quad\quad\quad\; A \quad 1 \quad F\,.\,E \\ \text{A1F}\cdot E\text{ (hex)} = \underline{1010}\;\underline{0001}\;\underline{1111}.\underline{1110}\text{ (binary)} = 101000011111.1110\text{ (binary)}. \end{array}$$

10.2.2.2 CONVERTING BINARY NUMBERS TO HEXADECIMAL

To convert a given binary number to hexadecimal, first, group each four bits starting from the binary point and working outward. If the number of bits is less than four in the leading and/or tailing group, insert 0 bits in the leading group and append 0 bits in the tailing group. Next, replace each group of four bits with its hexadecimal equivalent symbol. For example, the process of converting the binary number 10011000101010.0010011 to hexadecimal is shown in Figure 10.23.

As shown, the hexadecimal form for the given binary number is 262A.26.

TABLE 10.2 Symbols for Hexadecimal Numbers

Hexadecimal	Binary (4 bits)	Decimal
0	0000	0
1	0001	1
2	0010	2
3	0011	3
4	0100	4
5	0101	5
6	0110	6
7	0111	7
8	1000	8
9	1001	9
A	1010	10
B	1011	11
C	1100	12
D	1101	13
E	1110	14
F	1111	15

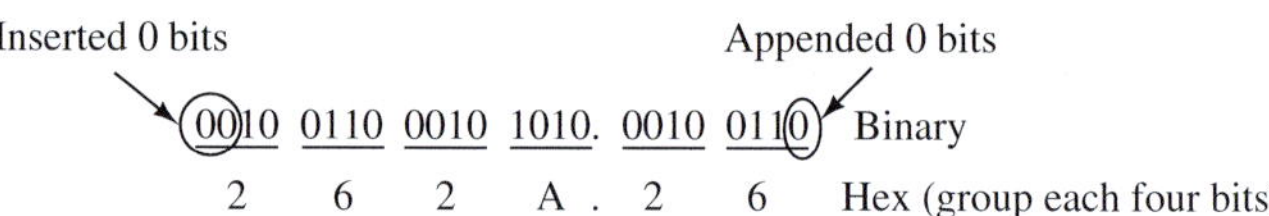

FIGURE 10.23 Converting a binary number to hexadecimal.

EXAMPLE 10.8 Conversion of Binary to Hexadecimal

Convert 100001110010010 and 0011111101110001000111110 to hexadecimal.

SOLUTION

$$100001110010010 \text{ (binary)} = 4392 \text{ (hex)}.$$

$$0011111101110001000111110 \text{ (binary)} = 3F711E \text{ (hex)}.$$

EXERCISE 10.3

Convert binary hexadecimal numbers 4392 and 3F711E in Example 10.8 back to binary.

10.2.3 Octal Numbers

The base of octal numbers is 8. Therefore, 8 symbols are required for these numbers. Table 10.3 shows the symbols used for the octal numbers.

10.2.3.1 CONVERTING OCTAL NUMBERS TO BINARY

Based on Table 10.3, for a given octal number, replace each digit with its binary equivalent (3 bits). For example:

$$\begin{matrix} & 2 & 7 & 1 & . & 5 & 3 \end{matrix}$$

$$271.53 \text{ (octal)} = \underline{010}\ \underline{111}\ \underline{001}.\underline{101}\ \underline{011} \text{ (binary)} = 010111001.101011 \text{ (binary)}.$$

TABLE 10.3 Symbols for Octal Numbers

Octal	Binary (3 bits)	Decimal
0	000	0
1	001	1
2	010	2
3	011	3
4	100	4
5	101	5
6	110	6
7	111	7

10.2.3.2 CONVERTING BINARY NUMBERS TO OCTAL

To convert a given binary number to octal, first, group each three bits starting from the binary point and working outward. If the number of bits is less than three in the leading and/or tailing groups, insert bits of 0 in the leading group and append bits of 0 in the tailing group. Next, replace each group of three bits with its octal equivalent symbol. For example, the process of converting binary number 11110010101.00101 to octal form is shown in Figure 10.24.

The octal form for the given binary number is 3625.12.

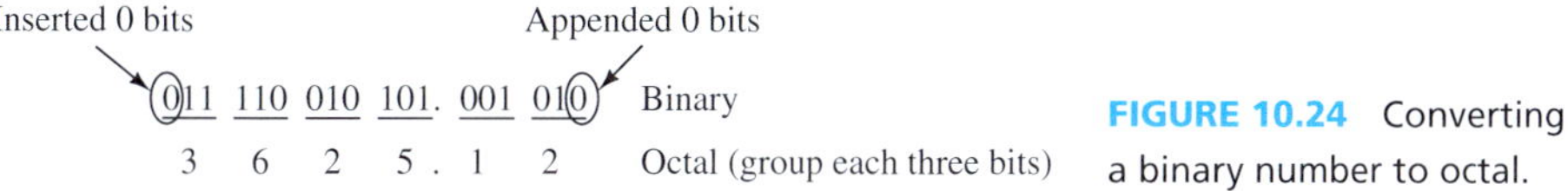

FIGURE 10.24 Converting a binary number to octal.

EXAMPLE 10.9 Conversion of Binary to Octal

Convert 111100101011001000 and 010010111011110111 to octal form.

SOLUTION

111 100 101 011 001 000 (binary) = 745310 (octal)
7 4 5 3 1 0

010 010 111 011 110 111 (binary) = 227367 (octal)
2 2 7 3 6 7

10.3 BOOLEAN ALGEBRA

Boolean algebra defines the arithmetic operations on the binary bits 0 and 1, which are sometimes referred to as true (1) and false (0).

10.3.1 Boolean Inversion

The inversion of 0 is 1 and the inversion of 1 is 0, that is,

$$\overline{1} = 0 \qquad \textbf{(10.1)}$$

$$\overline{0} = 1 \qquad \textbf{(10.2)}$$

10.3.2 Boolean AND Operation

The AND operation on two variables (A and B) is represented by AB or $A{\cdot}B$. It is also called logical multiplication. The AND operation can be denoted as $F = A \cdot B$. Here, F is equal to 1 only when A and B are both 1, otherwise, F is equal to 0. The truth table for AND operation is shown in Table 10.4.

TABLE 10.4 Truth Table for the AND Operation

A	B	$F = AB$
0	0	0
0	1	0
1	0	0
1	1	1

A truth table lists all possible combinations of the input variables and the corresponding output values. Logic circuits can be shown to be equivalent by using a truth table and/or through Boolean algebra.

10.3.3 Boolean OR Operation

The OR operation on two variables (A and B) is represented by $A + B$. It is also called logical addition. The OR operation can be denoted as $F = A + B$: F equals 0 only when A and B are both 0, otherwise, F equals 1. The truth table for the OR operation is shown in Table 10.5.

TABLE 10.5 Truth Table for the OR Operation

A	B	$F = A + B$
0	0	0
0	1	1
1	0	1
1	1	1

10.3.4 Boolean NAND Operation

The NAND operation on two variables (A and B) is the inversion of the result obtained by the Boolean AND operation. The NAND operation can be denoted as $F = \overline{A \cdot B}$, and F is equal to 0 only when A and B are both 1, otherwise, F is equal to 1. The truth table for the NAND operation is shown in Table 10.6.

TABLE 10.6 Truth Table for the NAND Operation

A	B	$F = \overline{A \cdot B}$
0	0	1
0	1	1
1	0	1
1	1	0

10.3.5 Boolean NOR Operation

The NOR operation on two variables (A and B) is the inversion of the result obtained by the Boolean OR operation. The NOR operation can be denoted as $F = \overline{A + B}$, and F is equal to 1 only when A and B are both 0, otherwise, F is equal to 0. The truth table for the NOR operation is shown in Table 10.7.

TABLE 10.7 Truth Table for the NOR Operation

A	B	$F = \overline{A + B}$
0	0	1
0	1	0
1	0	0
1	1	0

10.3.6 Boolean XOR Operation

The XOR operation on two variables (A and B) is also called modulo-two addition. If A and B are equal, the result of the XOR operation is 0; if A and B are different, the result of the XOR operation is 1. The XOR operation can be denoted as $F = A \oplus B$. The truth table for the XOR operation is shown in Table 10.8. $F = A \oplus B$ corresponds to $F = A\overline{B} + \overline{A}B$ as well. In other words:

$$A \oplus B = A\overline{B} + \overline{A}B \tag{10.3}$$

TABLE 10.8 Truth Table for the XOR Operation

A	B	$F = A \oplus B$
0	0	0
0	1	1
1	0	1
1	1	0

10.3.7 Summary of Boolean Operations

$$\overline{1} = 0 \tag{10.4}$$
$$\overline{0} = 1 \tag{10.5}$$
$$0 \cdot 0 = 0 \tag{10.6}$$
$$0 \cdot 1 = 0 \tag{10.7}$$
$$1 \cdot 0 = 0 \tag{10.8}$$
$$1 \cdot 1 = 1 \tag{10.9}$$
$$0 + 0 = 0 \tag{10.10}$$
$$0 + 1 = 1 \tag{10.11}$$
$$1 + 0 = 1 \tag{10.12}$$
$$1 + 1 = 1 \tag{10.13}$$
$$0 \oplus 0 = 0 \tag{10.14}$$
$$0 \oplus 1 = 1 \tag{10.15}$$
$$1 \oplus 0 = 1 \tag{10.16}$$
$$1 \oplus 1 = 0 \tag{10.17}$$

10.3.8 Rules Used in Boolean Algebra

$$\overline{\overline{A}} = A \tag{10.18}$$
$$A \cdot A = A \tag{10.19}$$

$$A \cdot \overline{A} = 0 \tag{10.20}$$

$$A \cdot 0 = 0 \tag{10.21}$$

$$A \cdot 1 = A \tag{10.22}$$

$$A + \overline{A} = 1 \tag{10.23}$$

$$A + 0 = A \tag{10.24}$$

$$A + 1 = 1 \tag{10.25}$$

$$A + A = A \tag{10.26}$$

10.3.9 De Morgan's Theorems

$$\overline{A \cdot B} = \overline{A} + \overline{B} \tag{10.27}$$

$$\overline{A + B} = \overline{A} \cdot \overline{B} \tag{10.28}$$

De Morgan's theorems also apply to three or more variables.

$$\overline{A \cdot B \cdot C} = \overline{A} + \overline{B} + \overline{C} \tag{10.29}$$

$$\overline{A + B + C} = \overline{A} \cdot \overline{B} \cdot \overline{C} \tag{10.30}$$

De Morgan's theorems can be stated as: if the variables in a logic expression are replaced by their inverters (e.g., A is replaced by $\overline{A}$, $\overline{B}$ is replaced by B), then the AND operation is replaced by OR, and the OR operation is replaced by AND. Here, any term consisting of two or more variables must be parenthesized before changing the operator (AND, OR).

A truth table can be used to prove the theorems for two variables (Tables 10.9 and 10.10). Sometimes in truth tables "1" is represented by "T" (true) and "0" is represented by "F" (false). Indeed, this is why traditionally these tables are called "truth" tables.

EXERCISE 10.4

Use a truth table to prove De Morgan's theorems for three variables.

TABLE 10.9 Truth Table for Proving De Morgan's Theorems (10.27)

A	B	$A \cdot B$	$\overline{A \cdot B}$	$\overline{A}$	$\overline{B}$	$\overline{A} + \overline{B}$
0	0	0	1	1	1	1
0	1	0	1	1	0	1
1	0	0	1	0	1	1
1	1	1	0	0	0	0

TABLE 10.10 Truth Table for Proving De Morgan's Theorems (10.28)

A	B	$A + B$	$\overline{A + B}$	$\overline{A}$	$\overline{B}$	$\overline{A} \cdot \overline{B}$
0	0	0	1	1	1	1
0	1	1	0	1	0	0
1	0	1	0	0	1	0
1	1	1	0	0	0	0

10.3.10 Commutativity Rule

The AND, OR, and XOR operations satisfy the commutativity rule.

$$A \cdot B = B \cdot A \quad \textbf{(10.31)}$$

$$A + B = B + A \quad \textbf{(10.32)}$$

$$A \oplus B = B \oplus A \quad \textbf{(10.33)}$$

A truth table can be used to prove the commutativity rule (see Tables 10.11, 10.12, and 10.13).

10.3.11 Associativity Rule

The AND and OR operations satisfy the associativity rule.

$$A \cdot (B \cdot C) = (A \cdot B) \cdot C = A \cdot B \cdot C \quad \textbf{(10.34)}$$

$$A + (B + C) = (A + B) + C = A + B + C \quad \textbf{(10.35)}$$

A truth table can be used to prove the associativity rule (see Tables 10.14 and 10.15).

10.3.12 Distributivity Rule

$$A + B \cdot C = (A + B) \cdot (A + C) \quad \textbf{(10.36)}$$

$$A \cdot (B + C) = A \cdot B + A \cdot C \quad \textbf{(10.37)}$$

A truth table can be used to prove the distributivity rule (see Tables 10.16 and 10.17).

We have seen that given the expression, we can generate the truth table. Inversely, given the truth table we can generate the expression. To do so, we simply add the expressions corresponding to the outcomes 1 of the table. For example, suppose we have the outcomes of the fifth

TABLE 10.11 A truth Table Used to Prove the Commutativity Rule for the AND Operation (10.31)

A	*B*	$A \cdot B$	$B \cdot A$
0	0	0	0
0	1	0	0
1	0	0	0
1	1	1	1

TABLE 10.12 A Truth Table Used to Prove the Commutativity Rule for the OR Operation (10.32)

A	*B*	$A + B$	$B + A$
0	0	0	0
0	1	1	1
1	0	1	1
1	1	1	1

TABLE 10.13 A Truth Table Used to Prove the Commutativity Rule for the XOR Operation (10.33)

A	*B*	$A \oplus B$	$B \oplus A$
0	0	0	0
0	1	1	1
1	0	1	1
1	1	0	0

TABLE 10.14 A Truth Table Used to Prove the Associativity Rule for the AND Operation (10.34)

A	B	C	$B \cdot C$	$A \cdot (B \cdot C)$	$A \cdot B$	$(A \cdot B) \cdot C$	$A \cdot B \cdot C$
0	0	0	0	0	0	0	0
0	0	1	0	0	0	0	0
0	1	0	0	0	0	0	0
0	1	1	1	0	0	0	0
1	0	0	0	0	0	0	0
1	0	1	0	0	0	0	0
1	1	0	0	0	1	0	0
1	1	1	1	1	1	1	1

TABLE 10.15 A Truth Table Used to Prove the Associativity Rule for the OR Operation (10.35)

A	B	C	$B + C$	$A + (B + C)$	$A + B$	$(A + B) + C$	$A + B + C$
0	0	0	0	0	0	0	0
0	0	1	1	1	0	1	1
0	1	0	1	1	1	1	1
0	1	1	1	1	1	1	1
1	0	0	0	1	1	1	1
1	0	1	1	1	1	1	1
1	1	0	1	1	1	1	1
1	1	1	1	1	1	1	1

TABLE 10.16 A Truth Table Used to Prove the Distributivity Rule for (10.36)

A	B	C	$B \cdot C$	$A + B \cdot C$	$A + B$	$A + C$	$(A + B) \cdot (A + C)$
0	0	0	0	0	0	0	0
0	0	1	0	0	0	1	0
0	1	0	0	0	1	0	0
0	1	1	1	1	1	1	1
1	0	0	0	1	1	1	1
1	0	1	0	1	1	1	1
1	1	0	0	1	1	1	1
1	1	1	1	1	1	1	1

TABLE 10.17 A Truth Table Used to Prove the Distributivity Rule for (10.37)

A	B	C	$B+C$	$F=A\cdot(B+C)$	$A\cdot B$	$A\cdot C$	$A\cdot B+A\cdot C$
0	0	0	0	0	0	0	0
0	0	1	1	0	0	0	0
0	1	0	1	0	0	0	0
0	1	1	1	0	0	0	0
1	0	0	0	0	0	0	0
1	0	1	1	1	0	1	1
1	1	0	1	1	1	0	1
1	1	1	1	1	1	1	1

column of Table 10.17, but we don't know that it corresponds to $F=A\cdot(B+C)$. Here, only the last three rows correspond to 1; thus, F is:

$$F = A\overline{B}C + AB\overline{C} + ABC$$

Now, factorizing AB from the last two terms, we have:

$$F = A\overline{B}C + AB(\overline{C} + C)$$

Now, $(\overline{C} + C) = 1$. Factorizing A we have:

$$F = A\overline{B}C + AB = A(\overline{B}C + B)$$

Using distributivity rule,

$$F = A(\overline{B} + B)(C + B)$$

Since $(\overline{B} + B) = 1$, we have:

$$F = A(B + C)$$

In general, it is the role of a logic engineer to simplify a logic such that a lower number of gates is needed to create that.

EXERCISE 10.5

Using the outcomes of the Truth Table 10.17, show that the last column of this table corresponds to $A\cdot B+A\cdot C$.

EXAMPLE 10.10 Simplifying Logic Functions Using De Morgan's Law

Use De Morgan's laws to change the expression of the following logic function.

$$F = \overline{A}\cdot B + A + \overline{B}$$

SOLUTION

Consider $\overline{A} \cdot B$ as the first term and $A + \overline{B}$ as the second term. Replace $\overline{A} \cdot B$ with its inversions; change "·" to "+". We have:

$$F = \overline{A+\overline{B}} + A+\overline{B}$$

Now, take $A + \overline{B} = D$,

$$F = \overline{D} + D = 1$$

EXAMPLE 10.11 Simplifying Logic Functions

Simplify the logic expression:

$$F = A\overline{B} + B\overline{C} + \overline{B}C + \overline{A}B$$

SOLUTION

Because $(A + \overline{A}) = 1$, and $(C + \overline{C}) = 1$:

$$\begin{aligned} F &= A \cdot \overline{B} + B \cdot \overline{C} + \overline{B} \cdot C + \overline{A} \cdot B \\ &= A \cdot \overline{B} + B \cdot \overline{C} + \overline{B} \cdot C \cdot (A + \overline{A}) + \overline{A} \cdot B \cdot (C + \overline{C}) \end{aligned}$$

Using the distributivity rule:

$$F = A \cdot \overline{B} + B \cdot \overline{C} + A \cdot \overline{B} \cdot C + \overline{A} \cdot \overline{B} \cdot C + \overline{A} \cdot B \cdot C + \overline{A} \cdot B \cdot \overline{C}$$

Rearrange the expression using the distributivity:

$$\begin{aligned} F &= A \cdot \overline{B} + A \cdot \overline{B} \cdot C + B \cdot \overline{C} + \overline{A} \cdot B \cdot \overline{C} + \overline{A} \cdot \overline{B} \cdot C + \overline{A} \cdot B \cdot C \\ &= A \cdot \overline{B} \cdot (1 + C) + (1 + \overline{A}) \cdot B \cdot \overline{C} + \overline{A} \cdot C \cdot (\overline{B} + B) \\ &= A \cdot \overline{B} + B \cdot \overline{C} + \overline{A} \cdot C \end{aligned}$$

APPLICATION EXAMPLE 10.12 Stair-Way Light System

A civil engineer designs the light system for a building as shown in Figure 10.25. Switch A is installed upstairs and switch B is installed downstairs. The initial position of each switch is shown in Figure 10.26(a). Switch A touches node *a*, and switch B touches node *d*. When the engineer is upstairs, he turns on the light shown in Figure 10.26(b), and when he goes downstairs, he can turn off the light as shown in Figure 10.26(c). The status of each switch is shown in Table 10.18. Find the truth table for the switch and logic expression for switches A and B and the lamp. Find alternative expressions for switches A and B and the lamp using AND, OR, and NOT operations.

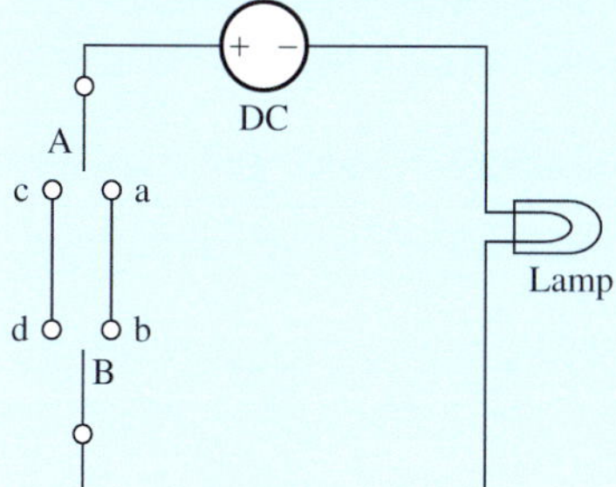

FIGURE 10.25 Light system for Example 10.12.

(*continued*)

APPLICATION EXAMPLE 10.12 Continued

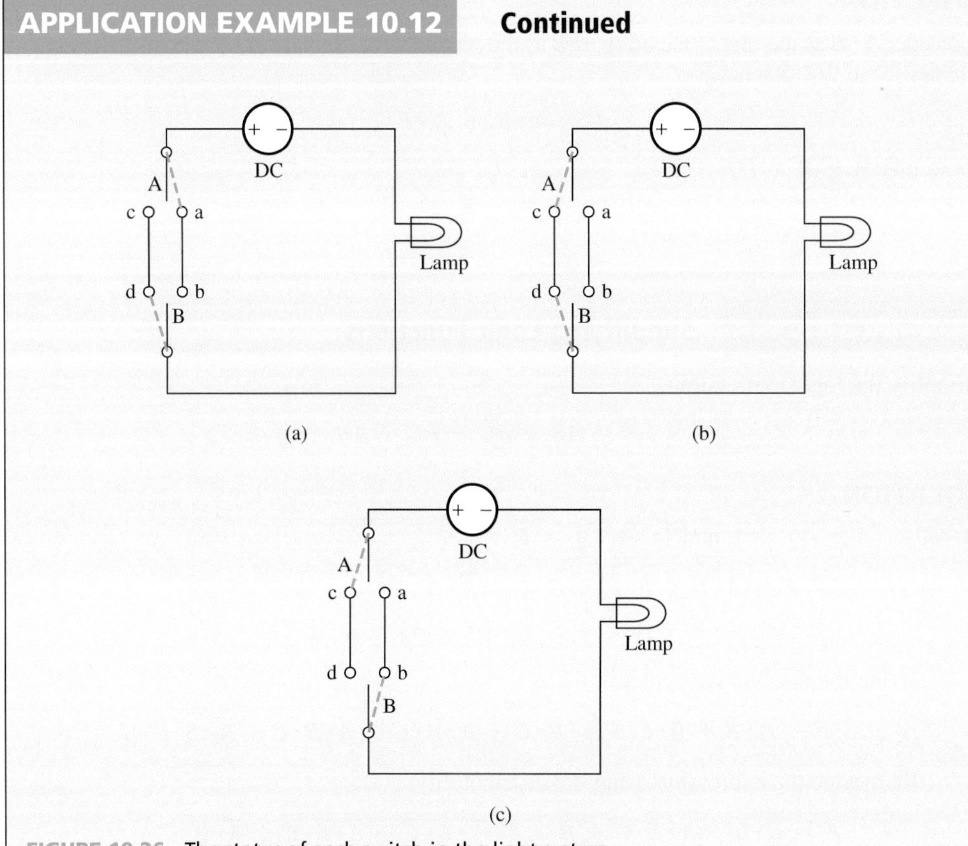

FIGURE 10.26 The status of each switch in the light system.

SOLUTION

For switch A:

Use 0 to denote the status when switch A touches node *c*.
Use 1 to denote the status when switch A touches node *a*.

For switch B:

Use 0 to denote the status when switch B touches node *d*.
Use 1 to denote the status when switch B touches node *b*.

Furthermore, use 1 to denote that the light is ON and 0 to denote that the light is OFF. Therefore, the truth table corresponds to Table 10.19.

TABLE 10.18 Switch Status

Switch A	Switch B	Lamp
c	d	ON
c	b	OFF
a	d	OFF
a	b	ON

TABLE 10.19 Switch Status

A	*B*	*F*
0	0	1
0	1	0
1	0	0
1	1	1

The expression to describe the relationships of A, B, and F corresponds to:

$$F = \overline{A \oplus B}$$
$$= \overline{A\overline{B} + \overline{A}B} \quad \text{(see Equation (10.3))}$$

Using De Morgan's theorem, an alternative expression is:

$$F = (\overline{A} + B)(A + \overline{B})$$
$$= \overline{A} \cdot \overline{B} + A \cdot B$$

10.4 BASIC LOGIC GATES

A logic gate is a physical device. Logic gates are the elementary building blocks of digital circuits. Any complex digital circuit can be built using basic logic gates. Most logic gates have two (or more) inputs and one output. The input or output terminal is in one of the two binary conditions *low* (0) *or high* (1), represented by different voltage levels. In most logic gates, the low state is approximately zero volts (0 V), while the high state is approximately five volts positive (+5 V).

This section discusses basic logic gates, including the AND gate, OR gate, XOR gate, NOT gate, NAND gate, and NOR gate. Table 10.20 summarizes the symbols used for these gates. These logic gates perform basic Boolean operations.

TABLE 10.20 Circuit Symbols for Gates

Gate	Circuit symbol
NOT	
AND	
OR	
NAND	
NOR	
XOR	
XNOR	

10.4.1 The NOT Gate

The NOT gate performs the NOT Boolean operation on a single input. Thus, it is also called an inverter. The NOT circuit symbol is shown in Figure 10.27, and its truth table is shown in Table 10.21.

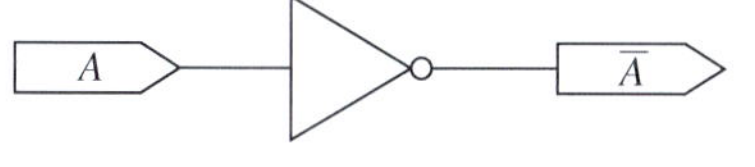

FIGURE 10.27 Circuit symbol for an inverter gate.

TABLE 10.21 Truth Table for the NOT Gate

A	$\overline{A}$
0	1
1	0

Note:

Complementing can be accomplished using a NOT gate. Here, "1" is changed to "0" and "0" is changed to "1".

10.4.2 The AND Gate

The AND gate performs the Boolean AND operation, which is applied to the input variables. Its circuit symbol is shown in Figure 10.28, and its truth table is shown in Table 10.22.

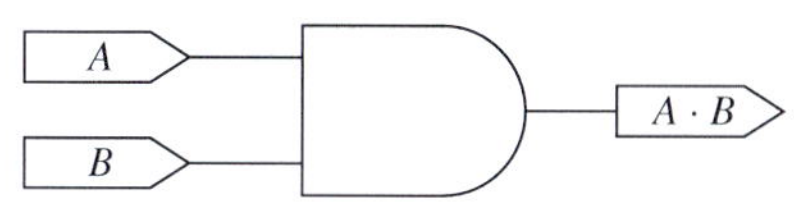

FIGURE 10.28 Circuit symbol for a two-input AND gate.

TABLE 10.22 Truth Table for an AND Gate

A	B	$A \cdot B$
0	0	0
0	1	0
1	0	0
1	1	1

Note:

The result is "1" if and only if both A and B are "1".

10.4.3 The OR Gate

The OR gate performs the Boolean OR operation, which is applied to the input variables. The OR circuit symbol is shown in Figure 10.29, and its truth table is shown in Table 10.23.

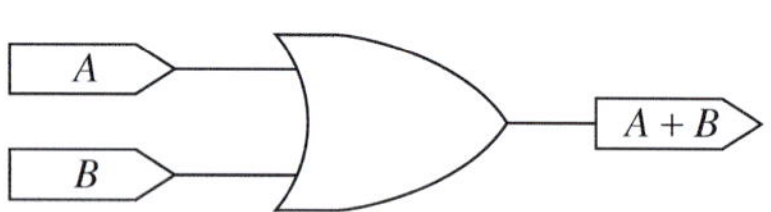

FIGURE 10.29 Circuit symbol for a two-input OR gate.

TABLE 10.23 Truth Table for an OR Gate

A	B	$A + B$
0	0	0
0	1	1
1	0	1
1	1	1

Note:

The result is "0" if and only if *A* and *B* are both "0".

10.4.4 The NAND Gate

The NAND gate performs the Boolean NAND operation, which is applied to the input variables. The NAND circuit symbol is shown in Figure 10.30, and its truth table is shown in Table 10.24.

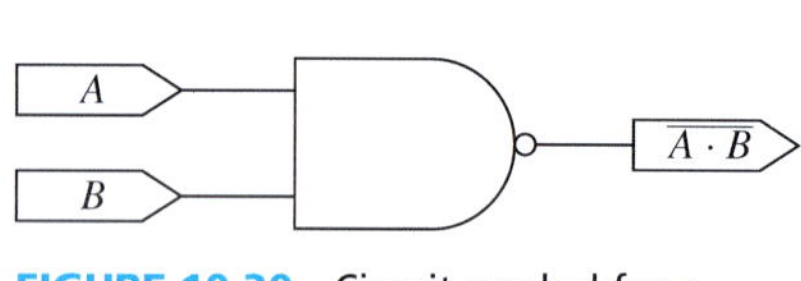

FIGURE 10.30 Circuit symbol for a two-input NAND gate.

TABLE 10.24 Truth Table for a NAND Gate

A	B	$A \cdot B$	$\overline{A \cdot B}$
0	0	0	1
0	1	0	1
1	0	0	1
1	1	1	0

Note:

The result is "0" if and only if *A* and *B* are both "1". The NAND gate is the reverse of the AND gate.

10.4.5 The NOR Gate

The NOR gate performs the Boolean NOR operation, which is applied to the input variables. The NOR circuit symbol is shown in Figure 10.31, and its truth table is shown in Table 10.25.

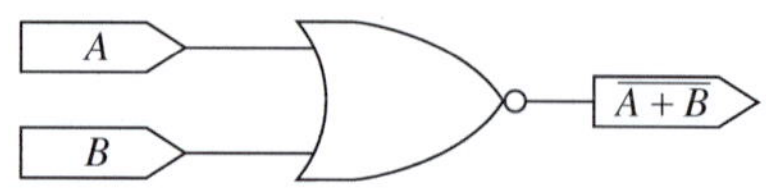

FIGURE 10.31 Circuit symbol for a two-input NOR gate.

TABLE 10.25 Truth Table for a NOR Gate

A	B	$A + B$	$\overline{A + B}$
0	0	0	1
0	1	1	0
1	0	1	0
1	1	1	0

Note:

The result is "1" if and only if *A* and *B* are both "0". The NOR gate is the reverse of the OR gate.

Important Notes:

(1) It is important to note that all Boolean operations can be represented by NAND and NOR operations. For example, we can represent $\overline{A}$ via $\overline{1 \cdot A}$ (NAND) or via $\overline{0 + A}$ (NOR). In addition, any expression can be solely represented by NAND or NOR gates. For example, using De Morgan's theorem:

$$A + B = \overline{\overline{A} \cdot \overline{B}}$$

Thus, $A + B$ can be represented by NAND operations, by creating $\overline{A}$ and $\overline{B}$ using $\overline{1.A}$ and $\overline{1.B}$ then applying another NAND to create $\overline{\overline{A} \cdot \overline{B}}$. Accordingly, the NAND and NOR operations are called *functionally complete*.

(2) As mentioned in Exercise 10.6 and Note (1), NAND and NOR are *functionally complete*. Thus, NAND and NOR gates can be used to represent any other gates. For example, we can represent a NOT gate using a NAND gate while one input of the gate is connected to a high voltage. As shown in Figure 10.32(a), we can represent XY using NAND gates. In addition, as shown in Figure 10.32(b), we can represent XY using NOR gates.

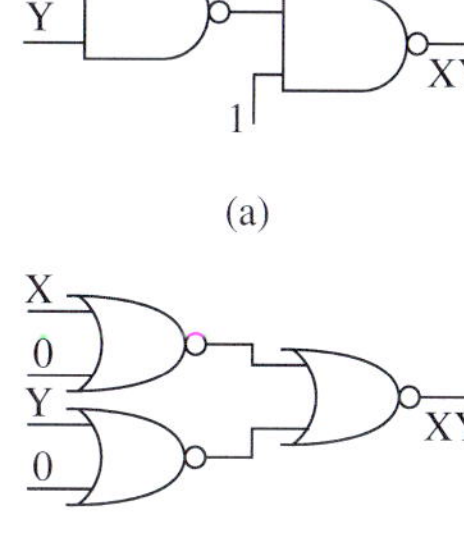

FIGURE 10.32
(a) Representing XY using (a) NAND gates and (b) NOR gates only.

EXERCISE 10.6

Represent the expression of Example 10.12 using (a) NAND gate only and (b) using NOR gate only.

EXERCISE 10.7

Show with details that in Figure 10.32, XY has been indeed produced via NAND and NOR gates. What theorems and operations are you using?

EXAMPLE 10.13 Representing Logic Functions Using NOR Gate

Use NOR gates only to represent $F = (A + \overline{B})(\overline{A} + C)$.

SOLUTION

First we represent *F* in terms of NOR operations only, and then we will use NOR gates to create *F*. To find *F* in terms of NOR gates, there are two approaches:

Approach 1: Convert *F* to:

$$F = (A + \overline{B})(\overline{A} + C) = AC + \overline{A}\,\overline{B} + \overline{B}C = F_1 + F_2 + F_3$$

(continued)

EXAMPLE 10.13 Continued

Using De Morgan's theorem,

$$F_1 = AC = \overline{(\overline{A} + \overline{C})} = \overline{(\overline{(0 + A)} + \overline{(0 + C)})}$$

Similarly, $F_2 = \overline{A}\,\overline{B} = \overline{(A + B)}$, and

$$F_3 = \overline{B}C = \overline{(B + \overline{(0 + C)})}$$

Up to this point, we have created F_1, F_2, and F_3 in terms of NOR gates. Now, we should create $F = F_1 + F_2 + F_3$ in terms of NOR gates. Because $\overline{\overline{(F_1 + F_2)}} = \overline{\overline{F_1}}\,\overline{\overline{F_2}} = F_1 + F_2$, it is easy to show:

$$F = \overline{\overline{(\overline{\overline{(F_1 + F_2)}} + F_3)}}$$

Figure 10.33(a) represents the creation of F using NOR gates only. As it is observed, this approach is very complex.

Approach 2: Directly use $F = (A + \overline{B})(\overline{A} + C)$ and write it as (use De Morgan's rule):

$$F = \overline{\overline{(A + \overline{B})} + \overline{(\overline{A} + C)}}$$

Figure 10.33(b) represents F using the second method. From this example, it is clear that if the approach is selected properly, lower number of gates can be used to create the same logic. It is the role of a good engineer to properly design the logic using lower number of gates.

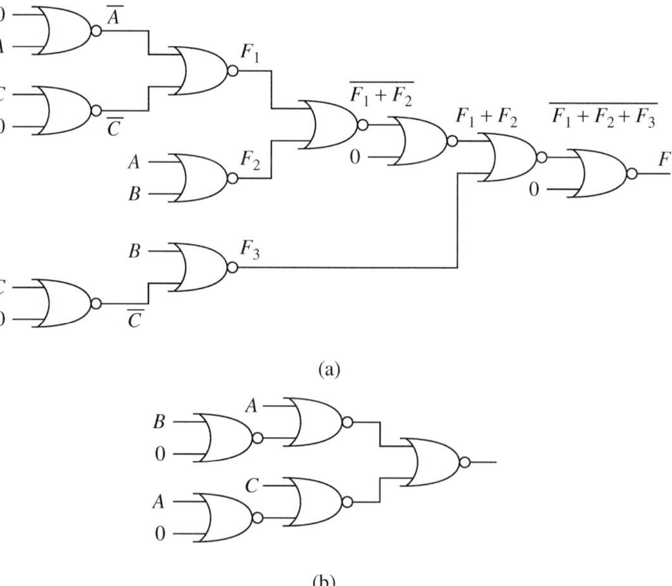

FIGURE 10.33 $F = (A + \overline{B})(\overline{A} + C)$ created using NOR gates only. (a) Approach 1 and (b) approach 2.

Conclusion: In order to implement a representation using NOR gates, it is better to write it first in terms of *multiplication of additions*. In order to implement a representation using NAND gates, it is better to first write it in terms of *addition of multiplications*.

EXERCISE 10.8

Repeat Example 10.13 using NAND only gates.

10.4.6 The XOR Gate

The XOR gate performs the Boolean XOR operation, which is applied to the input variables. The XOR circuit symbol is shown in Figure 10.34, and its truth table is shown in Table 10.26.

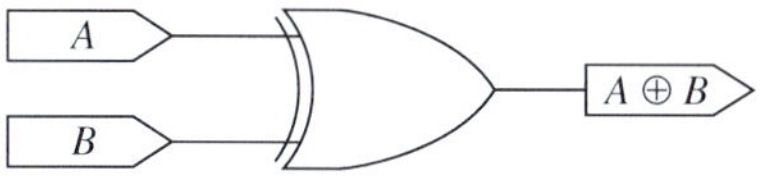

FIGURE 10.34 Circuit symbol for a two-input XOR gate.

TABLE 10.26 Truth Table for an XOR Gate

A	*B*	$A \oplus B$
0	0	0
0	1	1
1	0	1
1	1	0

Note:

The result is "1" if *A* is "0" and *B* is "1" or vice versa, that is, *A* and *B* are not equal.

10.4.7 The XNOR Gate

The XNOR gate inverts the result obtained by a Boolean XOR operation. The XNOR circuit symbol is shown in Figure 10.35, and its truth table is shown in Table 10.27.

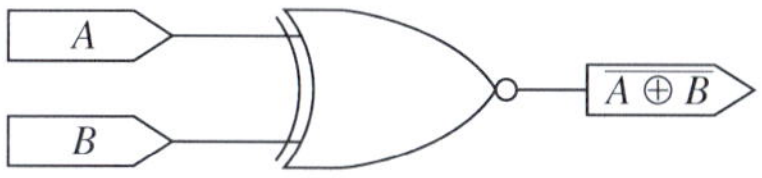

FIGURE 10.35 Circuit symbol for a two-input XNOR gate.

TABLE 10.27 Truth Table for an XNOR Gate

A	*B*	$A \oplus B$	$\overline{A \oplus B}$
0	0	0	1
0	1	1	0
1	0	1	0
1	1	0	1

Note:

The XNOR gate inverts the output of the XOR gate.

As mentioned before, any complex digital circuit can be built using these basic logic gates. From De Morgan's rule, any logic function can be implemented by AND gates and inverter gates. Similarly, any logic function can be implemented by OR gates and inverter gates. Karnaugh maps can be used to simplify a logic expression that includes multiple variables [1].

EXAMPLE 10.14 Using AND and OR Gates to Implement a Logic

Assume four inputs: *A*, *B*, *C*, and *D*. Use AND, OR, and inverter gates to implement the logic function:

$$F = A \cdot B + C \cdot \overline{D} + B \cdot C \cdot D$$

(continued)

EXAMPLE 10.14 Continued

SOLUTION

The process is very simple. Start with the inputs A, B, C, and D. Then, make $\overline{D}$, then, $A \cdot B$, $C \cdot \overline{D}$, and, $B \cdot C \cdot D$. Finally, make F the output. The logic circuit is shown in Figure 10.36.

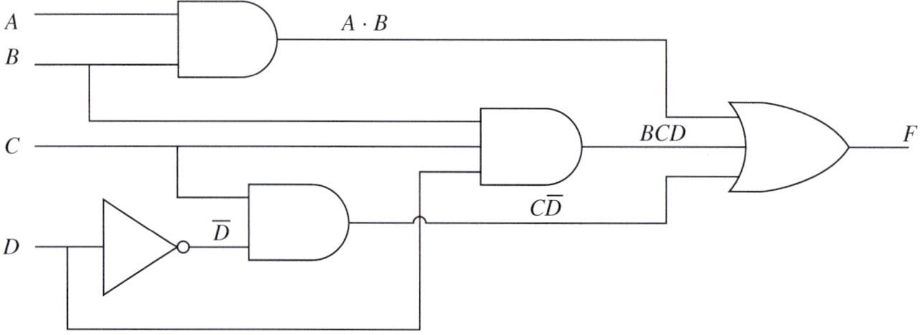

FIGURE 10.36 Circuit for Example 10.14.

EXERCISE 10.9

Use De Morgan's theorem to find an equivalent expression for F in Example 10.14. Then, sketch the equivalent logic circuit of that expression.

EXAMPLE 10.15 Using AND and OR Gates to Implement a Function

Assume four inputs: A, B, C, and D. Use AND, OR, and inverter gates to implement the logic function:

$$F = A \cdot B \cdot \overline{C} + C \cdot \overline{D} + (A + B + C) \cdot (\overline{C} + D)$$

SOLUTION

Sometimes, the logic function can be simplified to find an expression that is equivalent and easy to implement. In this case:

$$\begin{aligned} F &= A \cdot B \cdot \overline{C} + C \cdot \overline{D} + (A + B + C) \cdot (\overline{C} + D) \\ &= A \cdot B \cdot \overline{C} + C \cdot \overline{D} + A \cdot \overline{C} + B \cdot \overline{C} + C \cdot \overline{C} + A \cdot D + B \cdot D + C \cdot D \end{aligned}$$

The term $C \cdot \overline{C}$ is always 0. Thus, it can be simplified to:

$$\begin{aligned} F &= A \cdot B \cdot \overline{C} + C \cdot \overline{D} + A \cdot \overline{C} + B \cdot \overline{C} + A \cdot D + B \cdot D + C \cdot D \\ &= A \cdot B \cdot \overline{C} + C \cdot (\overline{D} + D) + A \cdot \overline{C} + B \cdot \overline{C} + A \cdot D + B \cdot D \end{aligned}$$

The term $\overline{D} + D$ is always 1. Further simplification yields:

$$\begin{aligned} F &= A \cdot B \cdot \overline{C} + C + A \cdot \overline{C} + B \cdot \overline{C} + A \cdot D + B \cdot D \\ &= (A \cdot B + A) \cdot \overline{C} + C + B \cdot \overline{C} + A \cdot D + B \cdot D \\ &= A \cdot (B + 1) \cdot \overline{C} + C + B \cdot \overline{C} + A \cdot D + B \cdot D \end{aligned}$$

The term $B + 1$ is always 1, thus:

$$\begin{aligned} F &= A \cdot \overline{C} + C + B \cdot \overline{C} + A \cdot D + B \cdot D \\ &= (A + B) \cdot \overline{C} + C + (A + B) \cdot D \\ &= (A + B) \cdot (\overline{C} + D) + C \end{aligned}$$

The corresponding logic circuit is shown in Figure 10.37.

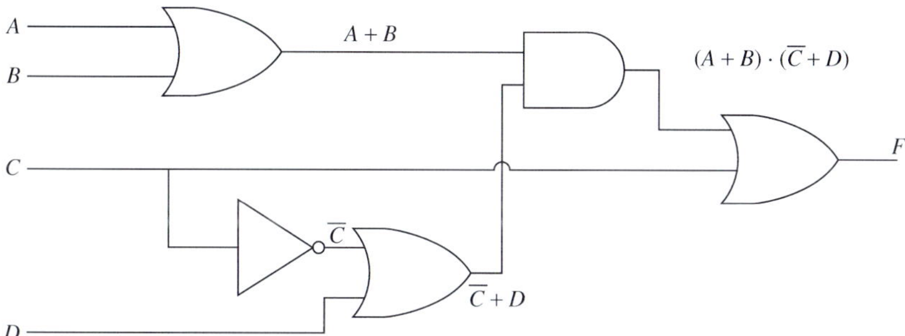

FIGURE 10.37 Circuit for Example 10.15.

APPLICATION EXAMPLE 10.16 Polling System

An association group has four members. They want to design a polling system to elect a leader among the four members. In the election meeting, each member holds an input terminal, and there is a button on the terminal device. If the button is pressed down, the terminal device output will be the logic 1 (e.g., a 5-V voltage). All the terminals are connected to a device equipped with a light. The light will be ON if it receives a 1. Therefore, whenever a member's name appears on the screen, if a member agrees to elect this member as their leader, each member can press the button on the terminal device. Design the circuit for the polling system. A member will be selected as leader if all the members simultaneously choose to elect him, thereby illuminating the light.

SOLUTION

Two-input AND gates can be used to design the circuit shown in Figure 10.38. When all members press down on their buttons, that is, $A = B = C = D = 1$, the output to the light is 1. In all other cases, the output to the light is 0.

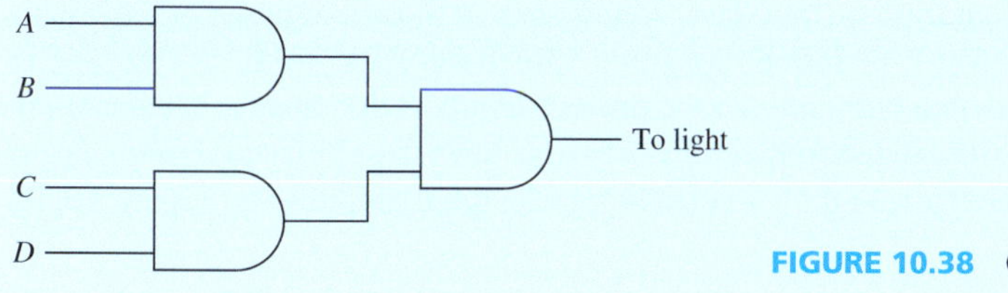

FIGURE 10.38 Circuit for Example 10.16.

APPLICATION EXAMPLE 10.17 Labtop Processor Design

A new labtop design has two processors P_1 and P_2 (core-duo) and four slots (S_1, S_2, S_3, S_4) for memory chips. When the processor P_1 is running, it uses the memory in slot S_1. When the processor P_2 is running, it uses the memory in slot S_4. When processors P_1 and P_2 are running simultaneously, they use the memory in slots S_2 and S_3. Design a circuit to connect the processors and the memory slots. If a processor is running, it produces an active, high output to activate the memory slots.

(*continued*)

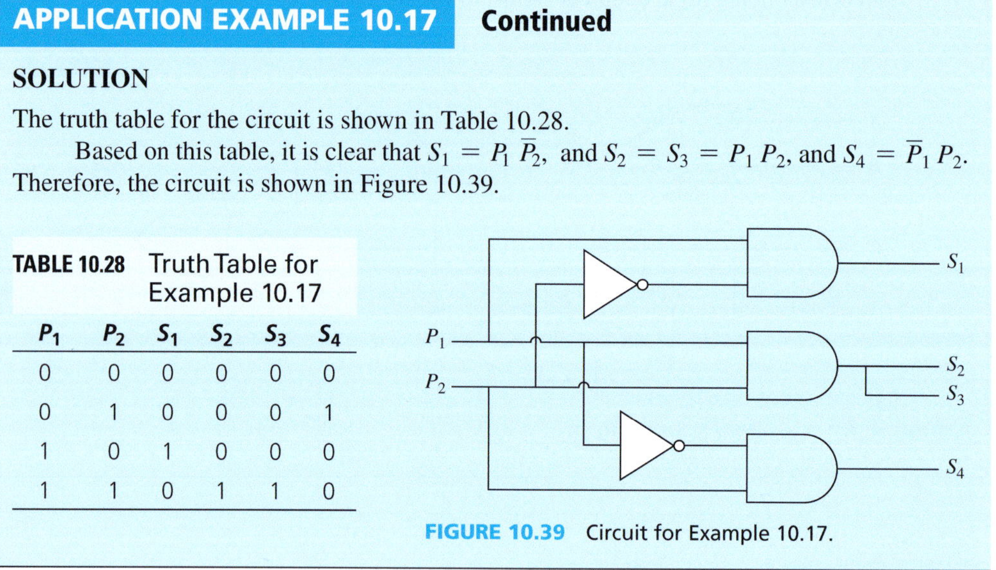

APPLICATION EXAMPLE 10.17 Continued

SOLUTION

The truth table for the circuit is shown in Table 10.28.

Based on this table, it is clear that $S_1 = P_1 \overline{P_2}$, and $S_2 = S_3 = P_1 P_2$, and $S_4 = \overline{P_1} P_2$. Therefore, the circuit is shown in Figure 10.39.

TABLE 10.28 Truth Table for Example 10.17

P_1	P_2	S_1	S_2	S_3	S_4
0	0	0	0	0	0
0	1	0	0	0	1
1	0	1	0	0	0
1	1	0	1	1	0

FIGURE 10.39 Circuit for Example 10.17.

Note:

As observed in these examples, the gates create the desired output as soon as an input is applied to them. Thus, their operation speed is very high. Today, field-programmable gated arrays (FPGA) are used for super high speed data/information processing. FPGA consists of a large number (e.g., hundreds) of gates implemented on an integrated circuit (IC). They are programmed to perform a digital task. Programming FPGA efficiently to use minimum number of gates requires specific skills. Many companies pay a good salary to the skilled FPGA programmers.

10.5 SEQUENTIAL LOGIC CIRCUITS

At any time instance, the output of combinational logic circuits, such as gates, depends only on the input values at that instant. Thus, these circuits are called *memoryless*. However, the output of sequential logic circuits depends on both past and present inputs. Sequential logic circuits do have memory. Thus, they are different from memoryless circuits such as those composed of basic gates.

A sequential circuit network is defined as a two-valued network. In these systems, the outputs at any time instance depend not only on the present inputs at that instant but also on the history of inputs. In other words, sequential circuit networks have the ability to remember past inputs. A sequential circuit is controlled by a periodic clock signal as shown in Figure 10.40.

10.5.1 Flip-flops

The core element of sequential networks is the flip-flop. The flip-flop is a device with two stable states. It remains in a stable state until its state is changed by new inputs. Different flip-flops respond differently to the clock signal or to input signals.

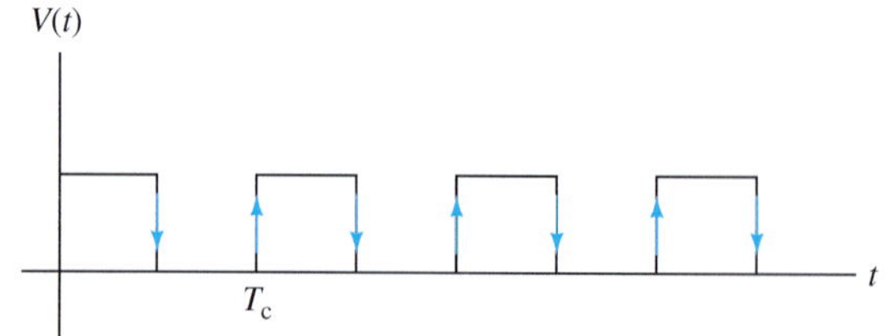

FIGURE 10.40 Clock signal used in the sequential circuit.

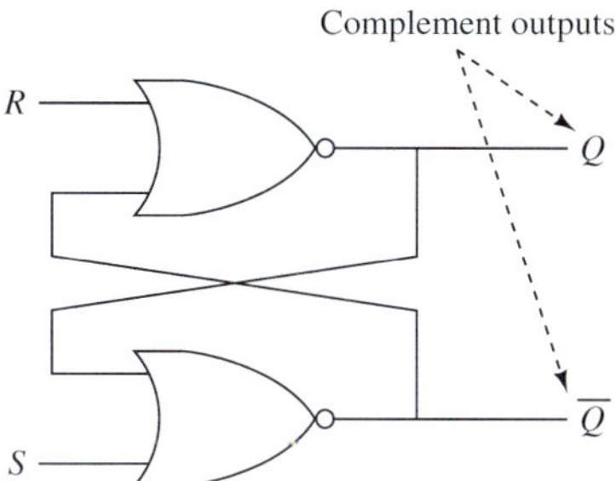

FIGURE 10.41 The SR latch.

10.5.1.1 THE LATCH

The latch is a basic flip-flop. A latch is a logic circuit with two outputs that are complements of each other. A gated latch or clocked latch is a latch which is synchronized by a control signal (usually a clock). The SR (set-reset) latch is used here as an example.

A SR latch consists of two NOR gates, and is shown in Figure 10.41. In this circuit:

1. Assume both inputs are initially 0; output Q is at 1 and $\overline{Q}$ is at 0; and that the circuit is stable with Q at 1 and $\overline{Q}$ at 0.
2. If R is set to 1, Q will be changed to 0, and $\overline{Q}$ changes to 1. If the R changes back to 0, the circuit remains in its stable state.
3. If S is set to 1, $\overline{Q}$ will change to 0 and Q will be 1.
4. S and R are not allowed to be equal to 1 simultaneously because in this case both Q and $\overline{Q}$ will be at 0 state. As a result, they will not be complemented.

   In addition, in this case, if S and R are changed to 0, the output will be ambiguous, that is, it is not clear whether they will be 0 or 1. For example, if Q is changed to 1 first (the two NORs have different delays), then $\overline{Q} = 0$; if $\overline{Q}$ is changed to 1 first, Q will be forced to 0. Thus, it is impossible to predict the output.

The logic symbol for the SR latch is shown in Figure 10.42. The truth table for the SR latch is sketched in Table 10.29.

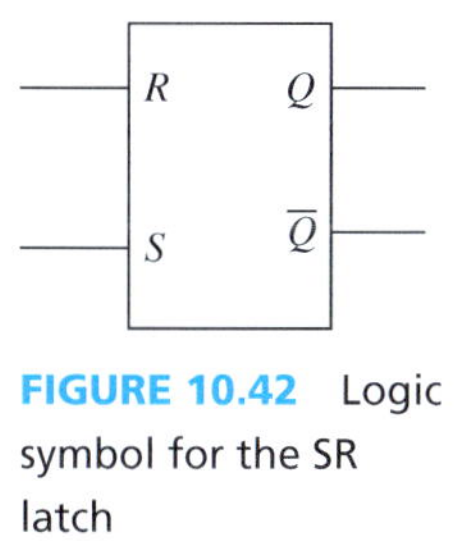

FIGURE 10.42 Logic symbol for the SR latch

TABLE 10.29 Truth Table for the SR Latch

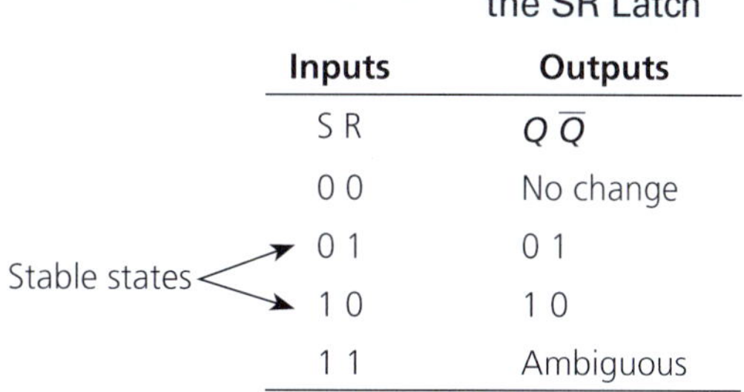

Inputs	Outputs
S R	$Q\ \overline{Q}$
0 0	No change
0 1	0 1
1 0	1 0
1 1	Ambiguous

Stable states (arrows to rows 0 1 and 1 0)

APPLICATION EXAMPLE 10.18 Switch De-Bouncing in a Circuit

Switch contacts are usually made of springy metals that are forced into contact by an actuator. When the contacts strike together, their momentum and elasticity act together to cause bounce. Therefore, when the switch is closed or opened, it may bounce for a period of time before settling down. This bouncing effect is harmful and should be prevented.

Switch de-bouncing has been discussed in Chapter 5 (see Example 5.2). Example 5.2 presented an RC circuit to prevent the harmful effects of switch bouncing. In addition, it will be discussed in Chapter 14 (Example 14.10) and Chapter 15 where the safety features of switches are discussed. Here, an RS flip-flop is used to avoid bouncing effects.

(continued)

APPLICATION EXAMPLE 10.18 Continued

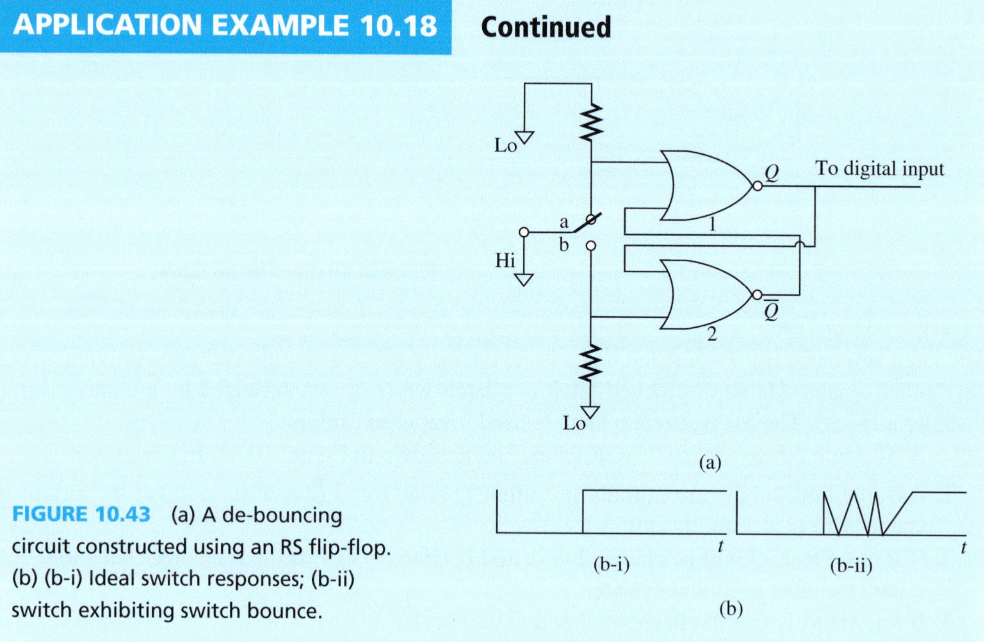

FIGURE 10.43 (a) A de-bouncing circuit constructed using an RS flip-flop. (b) (b-i) Ideal switch responses; (b-ii) switch exhibiting switch bounce.

The goal is to achieve a stable 1 at the input when a switch is turned on and to achieve a stable 0 at the input when the switch is turned off. A de-bouncing circuit constructed with an RS flip-flop for a digital system is shown in Figure 10.43(a). Figure 10.43(b) represents the bounce of switch. Explain the operation of this system.

SOLUTION

The operation of the de-bounce system is as follows. In general, when the switch is at position a, one input of the #1 NOR is 1, thus, the #1 NOR's output (Q) is 0. In this case, all inputs of the #2 NOR are 0 and its output ($\overline{Q}$) is 1. Therefore, the "digital input" will be stable on 0. When the switch is adjusted to position b, one of the #2 NOR's inputs is 1 and its output is 0. Both inputs of the #1 NOR are 0 and its output is 1. The "digital input" will be 1.

At the start, the switch is turned to position b and it bounces at position b. As a result, while the input of the #1 NOR becomes zero, the input of the #2 NOR that comes from the switch will change from 1 to 0, then 1, then 0, . . . , and finally it stabilizes on 1 after switch bouncing process is over [see Figure 10.43(b-ii)].

The first time the input b becomes 1, the output of #2 NOR gate becomes zero, and Q will be 1. When the input of the NOR gate becomes 0 due to bouncing, because the input of both gates will be zero, the flip-flop will stay in a no-change situation (see Table 10.29). Thus, Q will not bounce and instead will stay on 1. Thus, by applying an RS flip-flop, the switch bouncing effect is removed, resulting in a stable digital input.

10.5.1.2 SIMPLE JK FLIP-FLOP

A simple JK flip-flop is shown in Figure 10.44. The clock (CIk) is either "0" or "1" (see Figure 10.40).

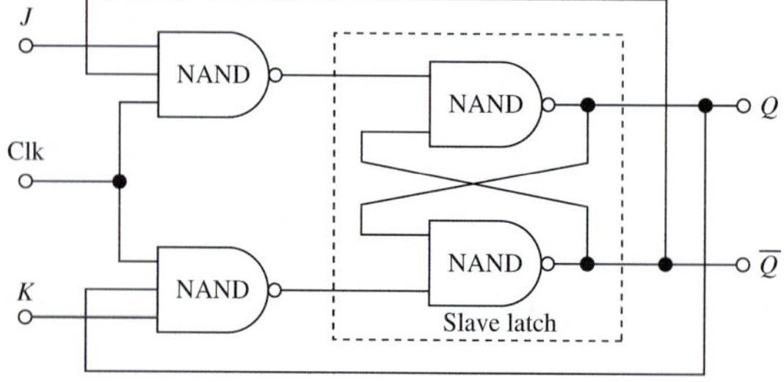

FIGURE 10.44 JK flip-flop.

J and K inputs control the mode in which the device operates.

1. When both J and K are 0, the flip-flop remains unchanged upon receipt of the next clock pulse;
2. When J is 1 and K is 0, upon receipt of the next clock pulse, the output of Q changes to 1, and $\overline{Q}$ changes to 0;
3. When J is 0 and K is 1, upon receipt of the next clock pulse, the output of Q changes to 0, and $\overline{Q}$ changes to 1;
4. When both J and K are 1, upon receipt of the next clock pulse, the output of Q and $\overline{Q}$ will flip to the opposite state.

The logic symbol for a JK flip-flop is shown in Figure 10.45. The truth table for a JK flip-flop is shown in Table 10.30.

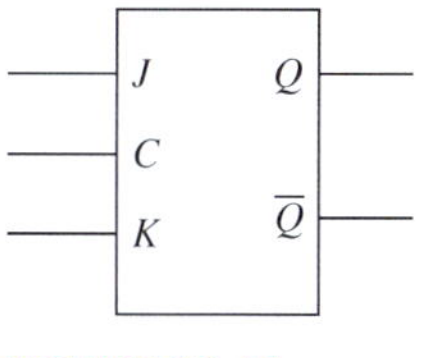

FIGURE 10.45 Logic symbol for a JK flip-flop.

TABLE 10.30 Truth Table for a JK Flip-Flop

Inputs		Outputs		
J	K	Q	$\overline{Q}$	
0	0	Q	$\overline{Q}$	← unchanged
0	1	0	1	
1	0	1	0	
1	1	$\overline{Q}$	Q	← flip

10.5.1.3 EDGE-TRIGGERED FLIP-FLOP

The inputs of the discussed JK flip-flop are enabled by the level of the clock signal. However, in some flip-flops, the changes of output are triggered by the leading or the trailing edge of the clock signal. Leading edge (rising edge) refers to the transition from low to high for the clock signal, and the trailing edge (falling edge) refers to the transition from high to low as shown in Figure 10.46. If the inputs are enabled when the clock signal is at the leading edge, the flip-flops are called *positive-edge-triggered flip-flops*. If the inputs are enabled when the clock signal is at the trailing edge, the flip-flops are called *negative-edge-triggered flip-flops*. For edge-triggered flip-flops, if the clock signal is steady (not at the leading edge or trailing edge), the inputs are disabled. Positive-edge-triggered D flip-flops and T flip-flops (see the next sections) are examples of edge-triggered flip-flops.

10.5.1.4 D FLIP-FLOP

A D flip-flop is also called a *delay flip-flop*. For the positive-edge-triggered D flip-flop, when the clock signal is at its leading edge, the D flip-flop output will be equal to the value of the input just prior to the triggering clock transition. The logic symbol for a positive-edge-triggered D flip-flop is shown in Figure 10.47.

In the logic symbol, if the clock signal is with the symbol ">", the flip-flop will be an edge-triggered flip-flop; otherwise, it will be a level-enabled flip-flop. The truth table for a positive-edge-triggered D flip-flop is shown in Table 10.31. Here, "↑" means the leading edge triggers the change of output of the flip-flop, and "↓" means the trailing edge triggers the change of output of the flip-flop. In addition, the sign X refers to 0 or 1.

10.5.1.5 T FLIP-FLOP

For the positive-edge-triggered T flip-flop, when $T = 0$, the flip-flop maintains its current state (holding), that is, Q is kept the same as it was before the clock edge. When $T = 1$, the output Q is

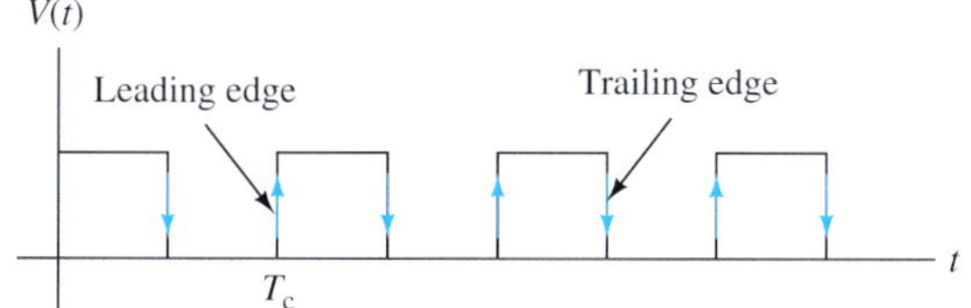

FIGURE 10.46 Diagram of a clock signal.

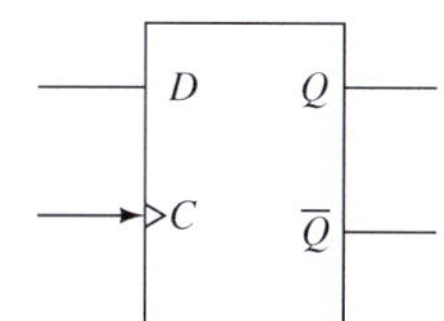

FIGURE 10.47 A positive-edge-triggered D flip-flop.

TABLE 10.31 Truth Table for a Positive-Edge-Triggered D Flip-Flop

Inputs		Outputs	
D	C	Q	$\overline{Q}$
0	↑	0	1
1	↑	1	0
x	0	Q	$\overline{Q}$
x	1	Q	$\overline{Q}$

TABLE 10.32 Truth Table for a Positive-Edge-Triggered T Flip-Flop

Inputs		Outputs	
T	C	Q	$\overline{Q}$
0	↑	Q	$\overline{Q}$
1	↑	$\overline{Q}$	Q ← Toggle
x	0	Q	$\overline{Q}$
x	1	Q	$\overline{Q}$

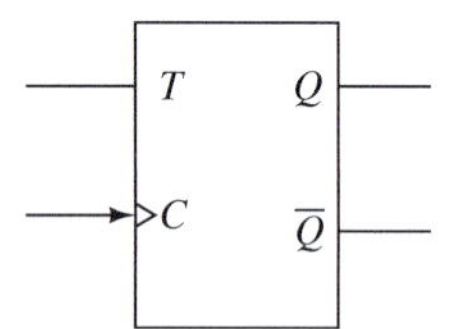

FIGURE 10.48 A positive-edge-triggered T flip-flop.

negated after the clock edge, compared to the value before the clock edge. Thus, the T flip-flop either maintains the current state's value for another cycle, or toggles the value (negates it) at the next clock edge. The logic symbol for a positive-edge-triggered T flip-flop is shown in Figure 10.48. The truth table for the positive-edge-triggered T flip-flop is shown in Table 10.32. The T and D flip-flop can be used to construct counters.

10.5.2 COUNTER

The edge of a clock signal can trigger the change of the output of a flip-flop. Therefore, a circuit consisting of flip-flops can be used to count the number of pulses of an input signal. The output of cascade flip-flops can produce a specified pattern. For example, for a binary counter, the output sequence of three flip-flops corresponds to the binary numbers (000, 001, 010, . . . , 111). A modulo-8 counter is shown in Figure 10.49.

The output sequence of this counter is shown in Figure 10.50. The original value of Q_0, Q_1, and Q_2 is 000. At the leading edge of the first pulse of the clock signal, the output of Q_0 changes from 0 to 1, and the output $\overline{Q}$ of the flip-flop *1* changes from 1 to 0. Therefore, the output of Q_1 does not change because the clock signal is not at the transition from 0 to 1.

Note that, the output $\overline{Q}$ of flip-flop *1* is the clock signal of flip-flop *2*. Consequently, the output of Q_2 does not change because the clock signal of flip-flop *2* does not change. At the leading edge of the second pulse of the clock signal, the output Q_0 changes back to 0, and the output of $\overline{Q}$ of flip-flop *1* changes from 0 to 1. In other words, the input *C* of the second flip-flop changes from 0 to 1. This makes the output of Q_1 change from 0 to 1. But the output of Q_2 does not change. Why?

To find the answer, look at the input *C* of the third flip-flop; does it change from 0 to 1? Note that, the output of the flip-flop changes if the *C* input changes from 0 to 1.

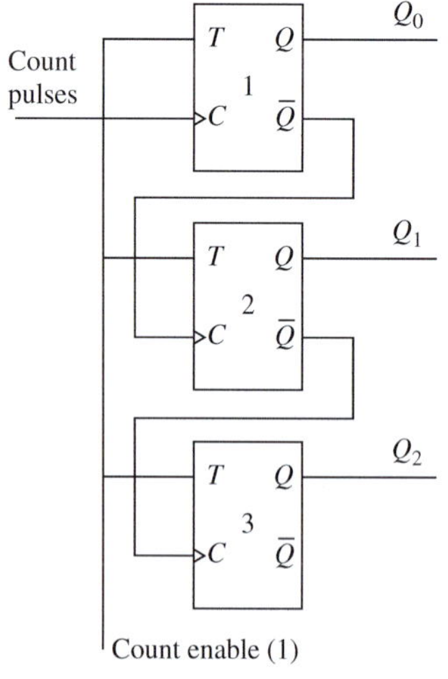

FIGURE 10.49 A modulo-8 counter constructed from T flip-flops.

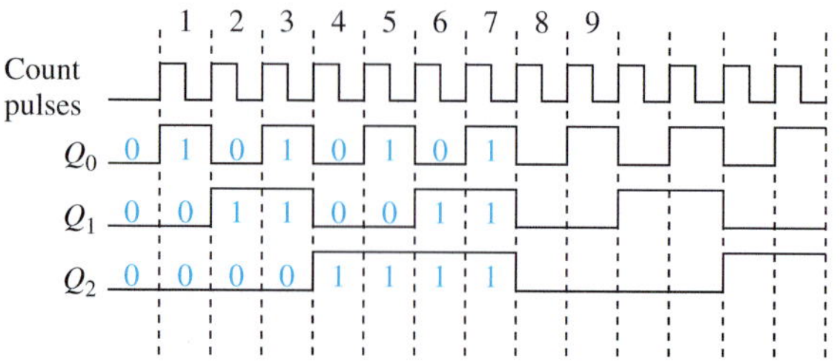

FIGURE 10.50 Output sequence of a modulo-8 counter.

At the leading edge of the third pulse, the output Q_0 changes to 1 and the output Q_1 stays at 1; therefore, the output of Q_2 remains unchanged. At the leading edge of the fourth pulse of the clock signal, the output Q_0 changes to 0, and the output of Q_1 changes from 1 to 0. The corresponding $\overline{Q}$ of the flip-flop changes from 0 to 1, and the output of Q_2 changes from 0 to 1. A thorough understanding of the details of this paragraph is important to understand the subsequent material.

Based on the analysis above, the change period for Q_0 is one period of the clock signal, the change period for Q_1 is two periods of the clock signal, and the change period for Q_2 is four periods of the clock signal. The simultaneous output of Q_2, Q_1, and Q_0 is shown in Figure 10.51.

Figure 10.51 shows that the output of the counter returns to a 000 state at the eigth clock pulse. In other words, the output of Q_2, Q_1, and Q_0 will repeat the pattern in Figure 10.51, therefore, the counter is called a modulo-8 counter.

Q_2	Q_1	Q_0
0	0	0
0	0	1
0	1	0
0	1	1
1	0	0
1	0	1
1	1	0
1	1	1

FIGURE 10.51 Output of Q_2, Q_1, and Q_0 in a modulo-8 counter.

D flip-flops can also be used to construct a 3-bit ring counter as shown in Figure 10.52. In a 3-bit ring counter, a single clock signal is applied to all three D flip-flops. Note that the Q output of each flip-flop is the D input (D_{in}) of the next flip-flop (the Q output of the flip-flop 1 is the D input of flip-flop 3).

A *reset* signal is connected to the reset inputs of flip-flops 1 and 2 to "0", and the *preset* input of flip-flop 3 to "1". An active *reset* signal can trigger the flip-flops to output logic "0" and an active *preset* signal can trigger the flip-flops to output logic "1." In Figure 10.52, the reset input of flip-flop 3 is high-active, that is, the reset signal is active when its values is logic "1"; the reset inputs of flip-flop 1, 2 and the preset input of flip-flop 3 are low-active, that is, those signals are active when their values are logic "0." Note that "reset" and "preset" can also be referred to as "clear" and "set."

When the reset signal is active, flip-flops 1 and 2 are cleared and flip-flop 3 is preset, that is, the output Q_1 and Q_2 is 0 and the output Q_3 is 1. When the reset signal changes to 1, the clock signal controls the outputs of the three flip-flops as follows:

1. Initial state: the output of Q_3 is 1, and the outputs of Q_2 and Q_1 are 0, that is:

$$Q_1\,Q_2\,Q_3 = 001$$

2. At the positive edge of the first pulse of the clock signal: the output of Q_1 in the last state is applied to flip-flop 3, and the output of Q_3 will be 0; the output of Q_3 in the last state is applied to flip-flop 2, therefore, the output of Q_2 will be 1; the output of Q_2 in the last state is applied to flip-flop 1 and the output of Q_1 is 0, that is:

$$Q_1\,Q_2\,Q_3 = 010$$

3. At the positive edge of the second pulse of the clock signal, the output of Q_1, Q_2, and Q_3 will be 100, that is:

$$Q_1\,Q_2\,Q_3 = 100$$

4. At the third pulse of the clock signal, the counter returns to the state 001, that is:

$$Q_1\,Q_2\,Q_3 = 001$$

The repeating pattern for the 3-bit ring counter is 001, 010, and 100, as shown in Figure 10.53. Because a 1 is moving periodically in a ring formed by the three flip-flops, this counter is called a *ring counter*.

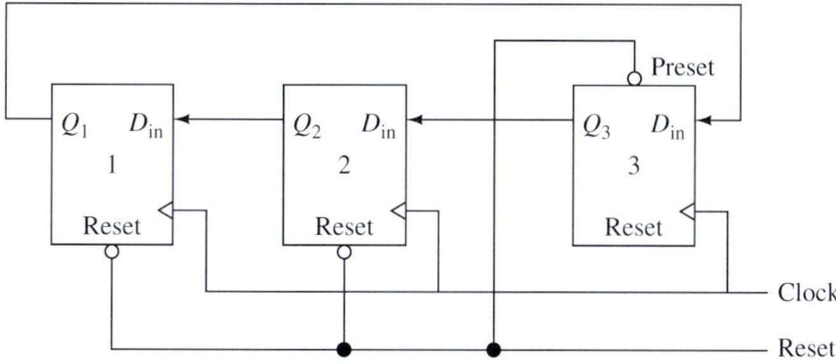

FIGURE 10.52 Circuit for a 3-bit ring counter.

FIGURE 10.53 Repeated pattern for a 3-bit ring counter.

EXAMPLE 10.19 D Flip Flop

For the negative-edge-triggered D flip-flop shown in Figure 10.54, the inputs are enabled when the clock signal is at the trailing edge (falling edge). The clock signal and the input data signal are shown in Figure 10.55. Sketch the output Q (the initial value of Q is zero).

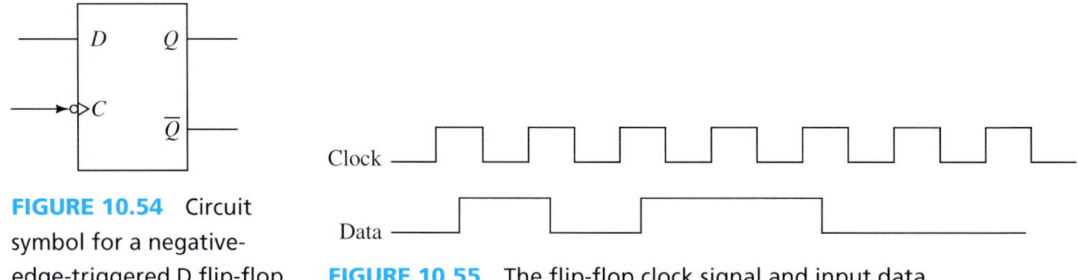

FIGURE 10.54 Circuit symbol for a negative-edge-triggered D flip-flop.

FIGURE 10.55 The flip-flop clock signal and input data

SOLUTION

The output, Q, is sketched in Figure 10.56.

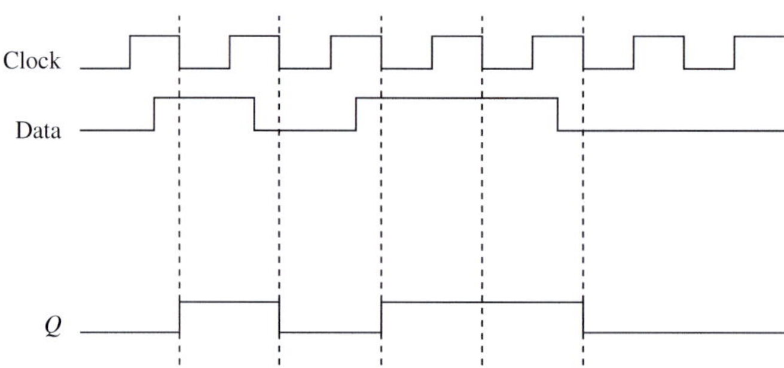

FIGURE 10.56 Output, Q, for Example 10.19.

APPLICATION EXAMPLE 10.20 Signal Light

A mechanical engineer intends to design a signal light for a small car model. He wants to use two bulbs for the signal light. If the bulbs blink as shown in Figure 10.57, the car will turn right. Design the circuit for the signal light system.

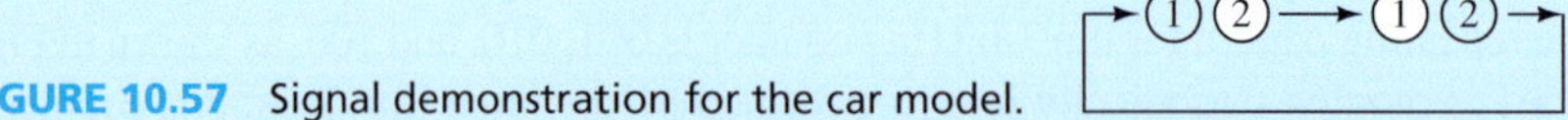

FIGURE 10.57 Signal demonstration for the car model.

SOLUTION

A 2-bit ring counter can be used to control the bulb. The circuit is shown in Figure 10.58.

Here, the output of Q_2 controls light bulb 2, and the output of Q_1 controls light bulb 1. The output of Q_1 and Q_2 will be zero if the car does not turn right (e.g., no power input to the device). When the car starts to turn right, the circuit is reset and the output of Q_2 will be 1 and the output of Q_1 will be 0, and therefore signal light bulb 2 will be ON, while light bulb 1 remains OFF. After the next clock signal, the output of Q_2 is 0 and the output of Q_1 is 1; therefore, signal light bulb 2 will be OFF and light bulb 1 will be ON. The process will repeat until the car finishes turning (power off for this device).

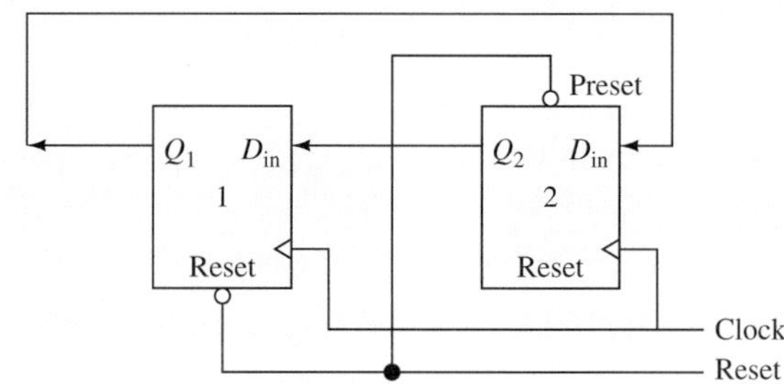

FIGURE 10.58 Circuit for a 2-bit ring counter.

APPLICATION EXAMPLE 10.21 Packing Machine

A packing factory uses a machine to fill a box with 12 bottles of soft drink. Sketch the simple logic circuit to demonstrate how this machine works.

SOLUTION

A counter flip-flop can be used for the logic circuit. The binary output for number 12 is 1100. Thus, given that the machine needs to fill the package with 12 bottles of the soft drink, the required binary output to trigger the machine is 1100. Therefore, given four flip-flops, with the outputs of Q_1, Q_2, Q_3, and Q_4, the instruction for packing is sent when Q_1 and Q_2 are 0, and Q_3 and Q_4 are 1. In other words, considering $A = \overline{Q}_1\overline{Q}_2Q_3Q_4$, A would be 1 if $Q_1 = Q_2 = 0$, and $Q_3 = Q_4 = 1$. Using De Morgan's theorem:

$$\begin{aligned} A &= \overline{Q}_1\overline{Q}_2 \cdot Q_3Q_4 \\ &= \overline{(Q_1 + Q_2) + (\overline{Q}_3 + \overline{Q}_4)} \\ &= \overline{Q_1 + Q_2} \cdot \overline{(\overline{Q}_3 + \overline{Q}_4)} \\ &= \overline{Q_1 + Q_2} \cdot (Q_3 \cdot Q_4) \end{aligned}$$

Therefore, the solution is shown in Figure 10.59. When Q_1 and Q_2 are 0, the output of the first NOR that is connected to Q_1 and Q_2 will be 1. In addition, when Q_3 and Q_4 are both 1, the

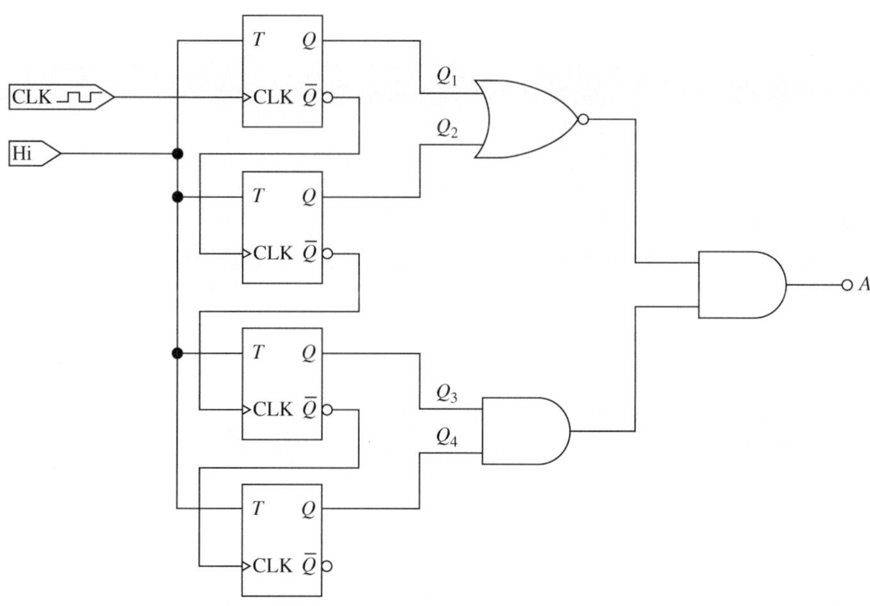

FIGURE 10.59 Logic circuit for application Example 10.21.

(continued)

APPLICATION EXAMPLE 10.21 Continued

output of the AND gate that is connected to Q_3 and Q_4 will be 1 as well. In this case, the inputs of the second AND logic circuit will be 1, and the resulting output of the AND will also be 1. This triggers the machine to close the box.

10.6 USING PSPICE TO ANALYZE DIGITAL LOGIC CIRCUITS

This section outlines how to use PSpice to set up a digital logic circuit. Unlike analog circuits, digital circuits do not need to be connected to a ground in order for the circuit to become a complete circuit. However, to study a bias point analysis of digital circuits, an analog circuit must be used in PSpice to force bias point to appear in the digital logic circuit. Because of this, the circuit must be connected to a ground at the end of the analog circuit (see Examples in this section). Note that use of an analog circuit is not required for time domain analysis in PSpice. This section examines two different kinds of simulation analysis in PSpice:

1. Bias point analysis: A set of fixed inputs is applied to study how the digital algorithm behaves.
2. Time domain analysis: The input varies over time to study the change of output with respect to the input.

At least one PSpice library file (*.lib and/or *.olb) is required to analyze a digital circuit using PSpice. However, it is advisable to have both of these library files installed. These files can be downloaded from the web. At least one of the following set of library files should be downloaded:

1. EVAL.lib and EVAL.olb (default in PSpice program)
2. 74LS.lib and 74LS.olb
3. DIG_PRIM.lib and DIG_PRIM.olb

To install these library files, choose "save" or "copy and paste" both the *.lib and *.olb files into the folder at file path C:\Program Files\OrCAD\Capture\Library\PSpice, or C:\Program Files\OrCAD_Demo\Capture\Library\PSpice for the demo version.

Table 10.33 summarizes some gates and their respective library that can be used for building digital logic circuits discussed in this chapter. There are other similar parts with different part names usable for digital logic circuits in the libraries. Students are encouraged to discover these unlisted parts and use them in the exercises.

EVAL, DIG_PRIM, and 74LS in Table 10.33 are the PSpice libraries that contain the required logic components for this chapter. Students can obtain the respective logic parts from PSpice by applying the part number/name shown in Table 10.33 into the "Place Part" menu. For example, to obtain an OR logic from the EVAL library, type "7432" in the Place Part menu (see Figure 2.63). Similarly, type "OR2" or "74LS32" to obtain the OR logic part from the DIG_PRIM and the 74LS library, respectively.

TABLE 10.33 PSpice Logic Part Names with Their Respective Library

	EVAL	DIG_PRIM	74LS
NOT	7404, 7405, 7406	INV	74LS04, 74LS05
AND	7408, 7409	AND2	74LS08, 74LS09
OR	7432	OR2	74LS32
NAND	7400, 7401, 7403	NAN2	74LS00, 74LS01
NOR	7402, 7428	NOR2	74LS02
XOR	74128, 74136	XOR	74LS136
XNOR	N/A	N/A	74LS266

EXAMPLE 10.22 PSpice Analysis

Let A and B represent the inputs for three digital logic circuits: AND, OR, and XOR. Use PSpice to sketch the digital circuits considering the following cases:

a. Bias point analysis for an AND logic circuit.
b. Time domain analysis for AND, OR, and XOR digital logic circuits.

SOLUTION

Part (a)

1. First, go to "Place Part" (see Figure 2.63) and use the reference in Table 10.33; pick an AND logic gate from one of the libraries and place it on the PSpice circuit board.
2. For bias point analysis, use the source that has only the output of 1 or 0 instead of a clock generator. Go to "Place Ground" and type "$D_HI" to choose the source that has the output of 1. Place and connect it to one of the inputs of AND, as shown in Figure 10.60.
3. Go to "Place Ground" again, and type "$D_LO" to choose the source with the output of 0. Place and connect it to another input of AND, as shown in Figure 10.60.
4. To force PSpice to show the bias point analysis results, an analog circuit part must be added to the digital logic circuit.
5. A logic part, "7407" in EVAL.lib library is used so that the PSpice circuit is forced to generate a digital output.
6. Go to "Place Part" and type "7407" under the "Part Name." Place it as shown in Figure 10.60, and also place the resistor, voltage source, and ground.

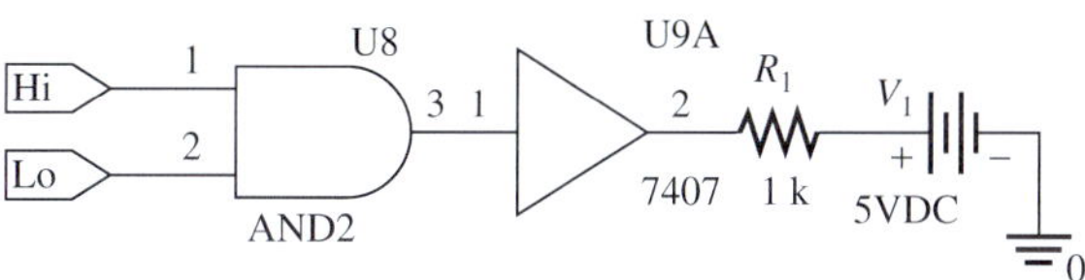

FIGURE 10.60 PSpice schematic circuit for Example 10.22 Part (a).

7. Set up a bias point simulation and press run (see Figure 2.71). The simulation results are shown in Figure 10.61.

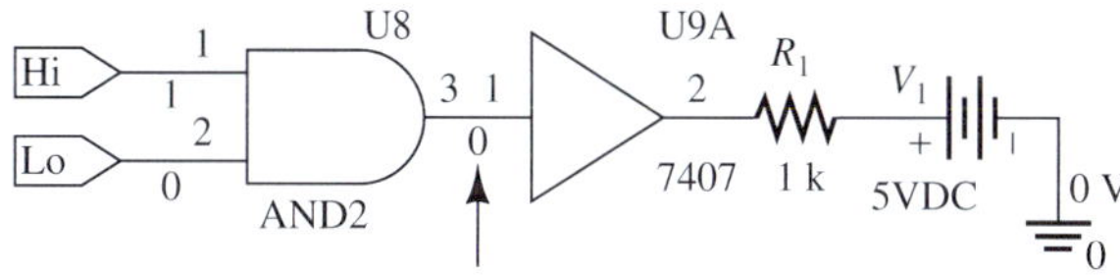

FIGURE 10.61 PSpice result for Example 10.22 Part (a).

8. The output of the AND logic is shown in the figure by the arrow; the output is 0 because one input is 1 and the other is 0.

Part (b)

1. First, go to "Place Part" (see Figure 2.63), and use Table 10.33 as reference. Place the logic parts AND, OR, and XOR onto the PSpice circuit board.
2. For time domain analysis, use the "STIM1" source. Go to "Place Part" and type "STIM1" under the "Part Name" and place it as shown in Figure 10.62.
3. Next, place the voltage level marker (shown by the arrow in Figure 10.63) as shown in Figure 10.62.

(*continued*)

EXAMPLE 10.22 **Continued**

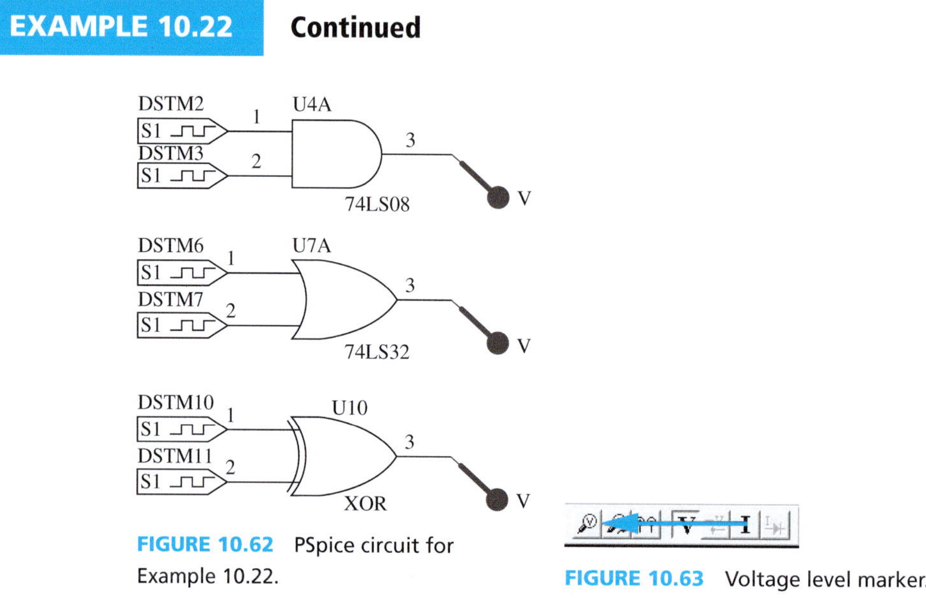

FIGURE 10.62 PSpice circuit for Example 10.22.

FIGURE 10.63 Voltage level marker.

4. Next, set up the output signal sequence of the DTSM (STIM1) source in the logic circuit. Click on the first STIM1 digital source of the AND logic part and go to its properties through "Edit Properties." Enter the values, "10 ms 0," "20 ms 1," and "30 ms 1" inside the Command 2, 3, and 4 columns (see Figure 10.64).

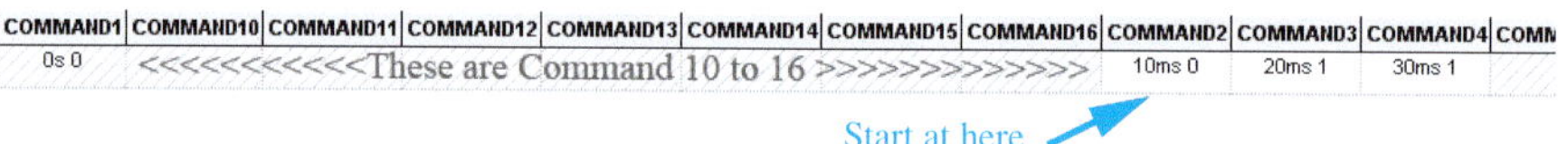

FIGURE 10.64 Command value of first digital source.

5. These value sets represent that at 10 ms, the digital source will generate the output of 0; at 20 ms the digital source will switch to output 1, etc.
6. Next, go to the properties of the second STIM1 digital source of the AND logic part. Enter "10 ms 1," "20 ms 0," and "30 ms 1" as shown in Figure 10.65. At 10 ms, the digital source switches to output 1; at 20 ms, the digital source switches the output back to 0. Finally, at 30 ms, the digital source switches to output 1 again.

FIGURE 10.65 Command value of second digital source.

7. Note that "Command 10 to 16" is beside the Command 1's column; while the Command 2's column is after the Command 16's column. PSpice simulates in the sequence of Command 1, 2, 3, and so on; therefore, the output of the digital source will not switch if the Command 2, 3, etc. are blank while there is value inside Command 10, 11, and so on.
8. Repeat steps 4 to 6 for the digital source for the OR and XOR logic parts.
9. Set the simulation to be time domain analysis (see Figure 2.75) with the simulation time at least 40 ms.
10. The simulation plot is shown in Figure 10.66. Notice that there are only two values for each output. The line with higher level is 1, while the lower level is 0.

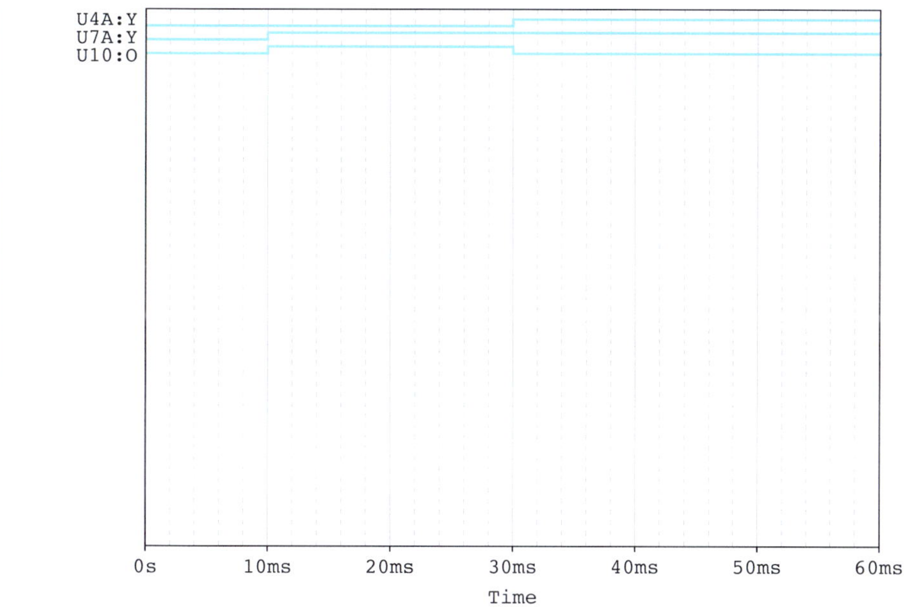

FIGURE 10.66 Digital output of Example 10.22 Part (b).

EXAMPLE 10.23 PSpice Analysis

An association group has four members, and they intend to design a polling system to elect a leader among the four members. In the election meeting, each member holds an input terminal, and there is a button on the terminal device. If the button is pressed down, the terminal device will output logic 1 (e.g., a 5-V voltage). Each terminal is connected to a device equipped with a light. The light will be ON, if it receives a 1 from all connected input terminals. Therefore, whenever a member's name appears on the screen, if all members agree to elect this member as their leader, they can press the button on the terminal device. Design the circuit for the polling system. The member whose name is listed on the screen will be the leader if all members elect him.

SOLUTION

The digital logic circuit for the question is given in Figure 10.67.

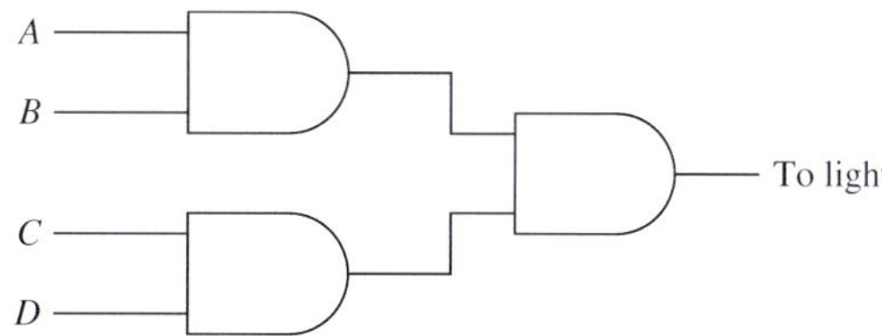

FIGURE 10.67 Circuit for Example 10.23.

1. Use the Part reference in Table 10.33 and follow the steps in Example 10.22 to set up a PSpice digital logic circuit, as shown in Figure 10.68.

(continued)

EXAMPLE 10.23 Continued

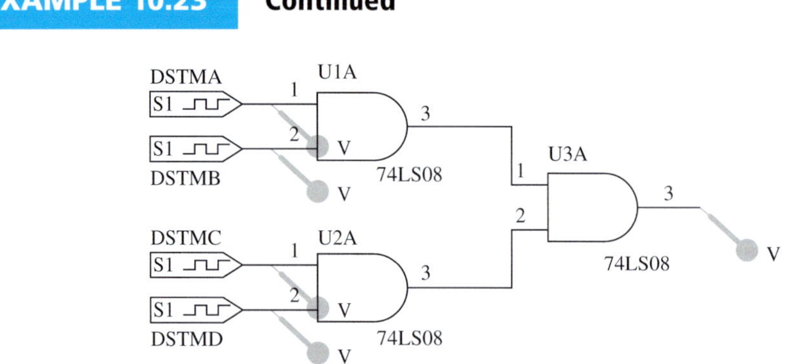

FIGURE 10.68 PSpice digital logic circuit for Example 10.23.

2. Using 10 ms as the time step, the output of each digital source is shown in Table 10.34, in order to study the output of the voting result. In each time step, a vote is allocated by the four members to each candidate. For example, at 0 s nobody votes for candidate 1 and at 10 ms two people are voting for candidate two, and so on.

TABLE 10.34 Setup of the Output Set of Each Digital Source

Time	0 s	10 ms	20 ms	30 ms	40 ms	50 ms	60 ms
DSTMA	0	0	1	1	1	0	1
DSTMB	0	1	1	0	1	1	0
DSTMC	0	1	1	1	0	0	1
DSTMD	0	0	1	1	0	0	1

3. Set the simulation analysis as time domain (see Figure 2.75), with a simulation time of at least 70 ms. The simulation output plot is shown in Figure 10.69.

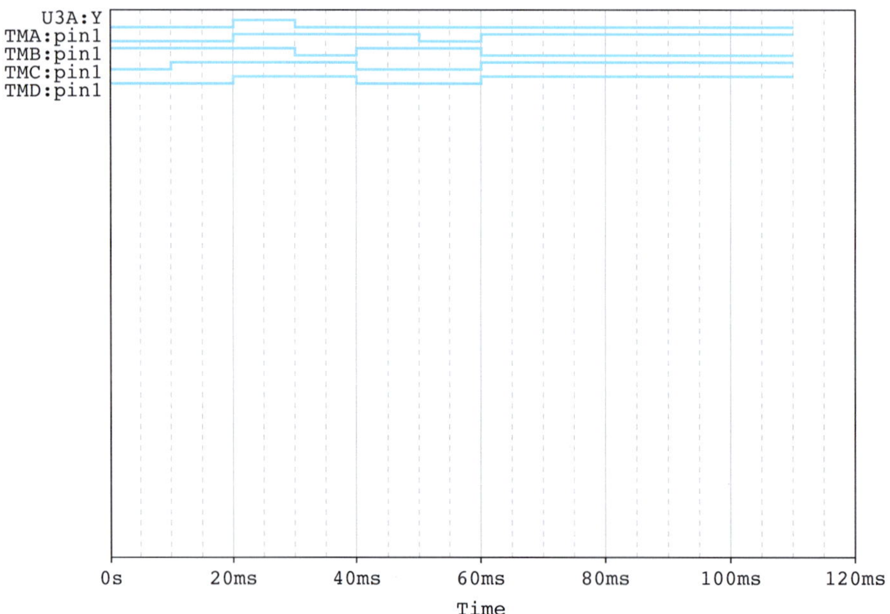

FIGURE 10.69 Digital output of Example 10.23.

4. TMA, TMB, TMC, and TMD are the input sources into the digital circuit. Use the reference in Table 10.34 and compare it to Figure 10.69. The result shows that there is only one member elected, which is the third vote.

EXAMPLE 10.24 PSpice Analysis

A new design of labtop has two processors P_1 and P_2 (core-duo) and four slots (S_1, S_2, S_3, and S_4) for the memory chips. When only processor P_1 is running, it uses only the memory in slot S_1; when only processor P_2 is running, it uses only the memory in slot S_4; when processors P_1 and P_2 are running simultaneously, they use the memory in slots S_2 and S_3. Design a circuit to connect the processors and the memory slots (if a processor is running, it produces an active, high output to activate the memory slots).

SOLUTION

The digital circuit of Example 10.17 is shown in Figure 10.70; the input values of P_1 and P_2 are shown in Table 10.35.

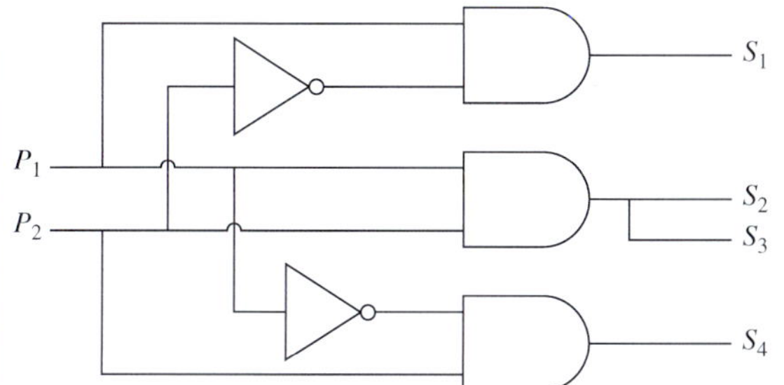

FIGURE 10.70 Circuit for Example 10.17.

TABLE 10.35 Input Value of P_1 and P_2 Sources

P_1	P_2
0	0
0	1
1	0
1	1

Use Table 10.33 as the Part reference and follow the steps in Example 10.22 to set up the PSpice digital logic circuit as shown in Figure 10.71. Edit the properties command for the DSTM1 and DSTM2 so that the input changes according to the values in Table 10.35. Select a constant time step, such as 10 ms. The output of S_1, S_2, S_3, and S_4 are shown in Figure 10.72. Figure 10.72 shows that both S_2 and S_3 share the same output, which is consistent with Figures 10.70 and 10.71.

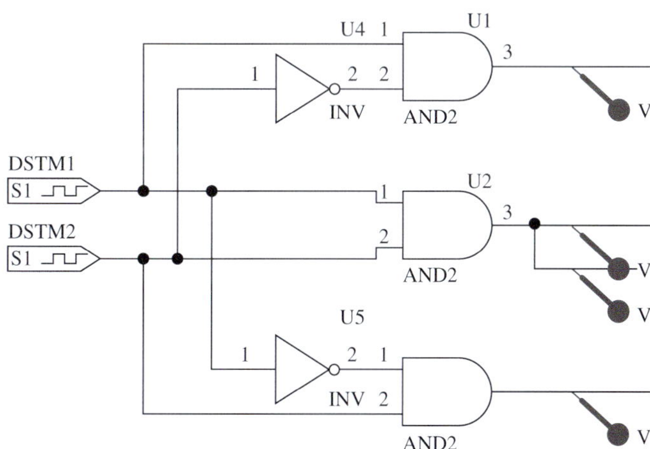

FIGURE 10.71 PSpice digital logic circuit for Example 10.24.

(*continued*)

EXAMPLE 10.24 Continued

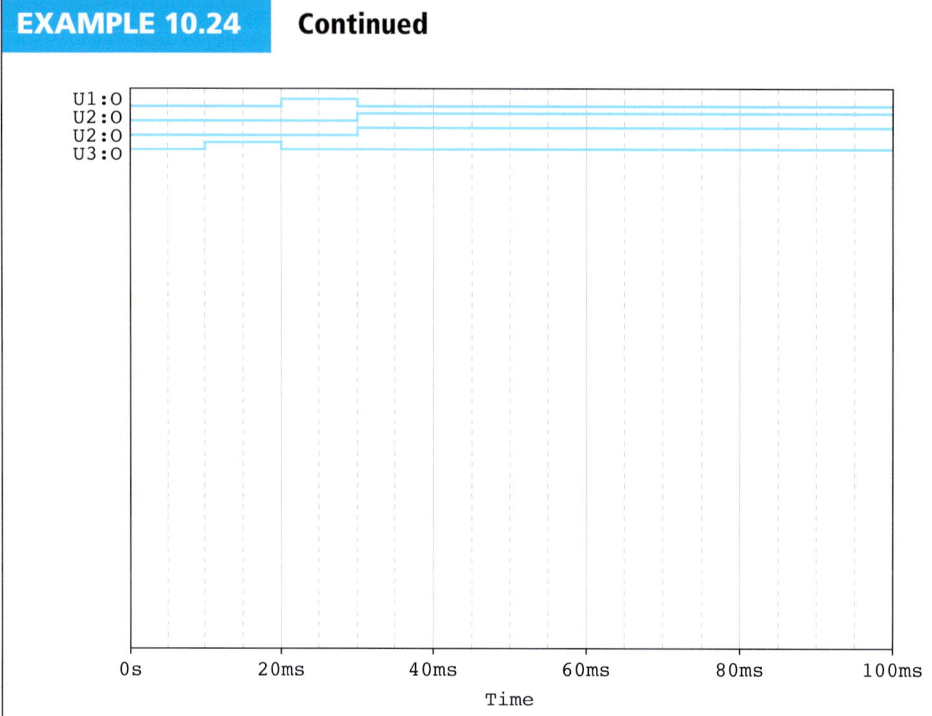

FIGURE 10.72 The output of Example 10.24.

EXAMPLE 10.25 PSpice Analysis

Use PSpice to set up an SR (NOR) latch as discussed in Section 10.5(c) and shown in Figure 10.73 with the input value given in Table 10.36.

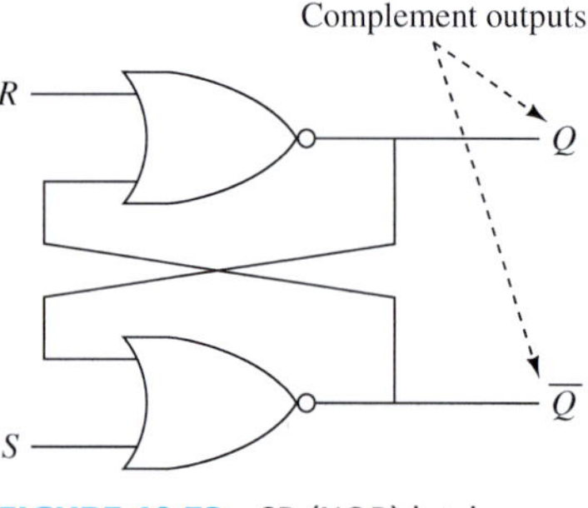

FIGURE 10.73 SR (NOR) latch.

TABLE 10.36 Input Value for *S* and *R*

R	*S*
1	0
0	0
0	1
1	1
1	0
0	0

SOLUTION

Use Table 10.33 and follow the steps in Example 10.22 to set up the PSpice logic circuit that is shown in Figure 10.17. Set the properties command for both input sources based on the values in Table 10.36 with a reasonable time step. Next, set the simulation type as time analysis (see Figure 2.75). Then, go to "Option," and choose "Gate-level Simulation" under the Category option as shown in Figure 10.74. Set the "Initialize all flip-flops to" as 0 for the initial condition. The setup, circuit, and simulation results are shown in Figures 10.75 and 10.76.

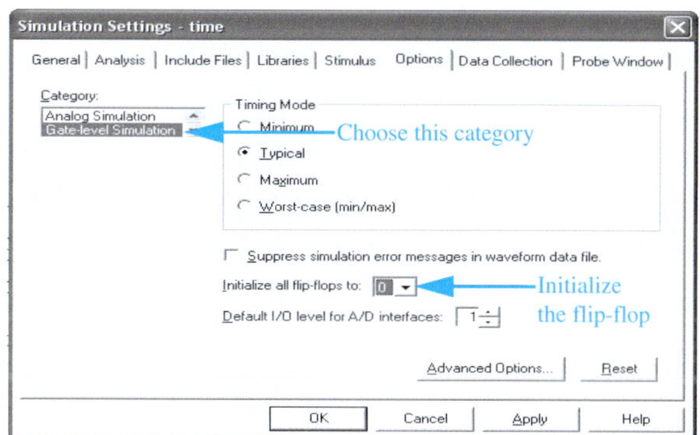

FIGURE 10.74 Setting the simulation option.

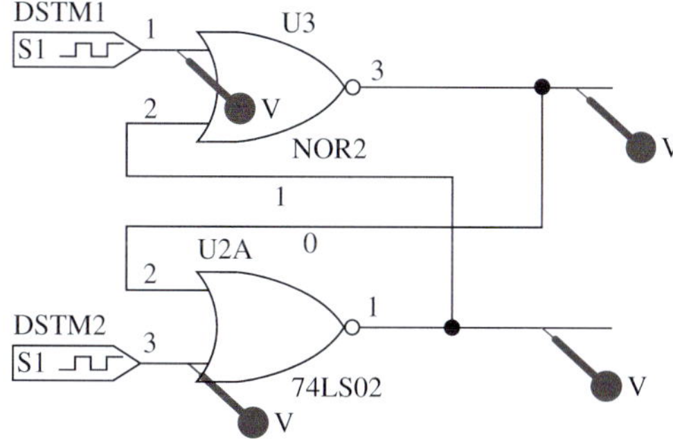

FIGURE 10.75 PSpice logic circuit for Example 10.25.

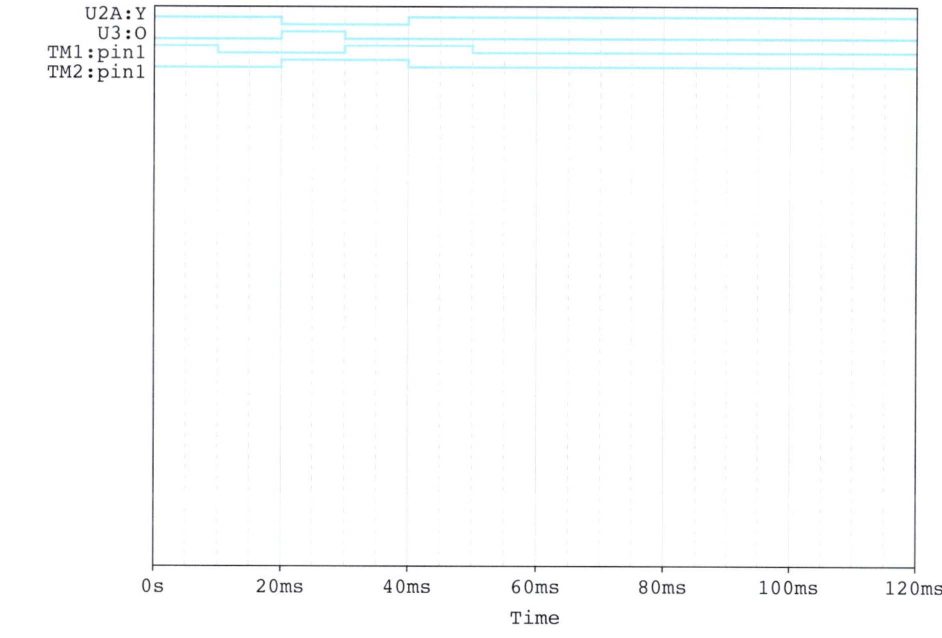

FIGURE 10.76 The output of Example 10.25.

10.7 WHAT DID YOU LEARN?

- In digital logic, the higher voltage represents logic 1 and the lower voltage represents logic 0.
- Numerical data can be represented in decimal, binary, octal, or hexadecimal forms.

- Subtraction of binary numbers are conducted by using the two's complement representation.
- A truth table lists all combinations of input variables and the corresponding output of each.
- Boolean operations include:

$$\overline{1} = 0$$
$$\overline{0} = 1$$
$$0 \cdot 0 = 0$$
$$0 \cdot 1 = 0$$
$$1 \cdot 0 = 0$$
$$1 \cdot 1 = 1$$
$$0 + 0 = 0$$
$$0 + 1 = 1$$
$$1 + 0 = 1$$
$$1 + 1 = 1$$
$$0 \oplus 0 = 0$$
$$0 \oplus 1 = 1$$
$$1 \oplus 0 = 1$$
$$1 \oplus 1 = 0$$

- Rules used in Boolean algebra are defined as:

$$\overline{\overline{A}} = A$$
$$A \cdot A = A$$
$$A \cdot \overline{A} = 0$$
$$A \cdot 0 = 0$$
$$A \cdot 1 = A$$
$$A + \overline{A} = 1$$
$$A + 0 = A$$
$$A + 1 = 1$$
$$A + A = A$$

- De Morgan's theorems are:

$$\overline{A \cdot B} = \overline{A} + \overline{B}$$
$$\overline{A + B} = \overline{A} \cdot \overline{B}$$

  De Morgan's theorems also apply to three or more variables, as shown here:

$$\overline{A \cdot B \cdot C} = \overline{A} + \overline{B} + \overline{C}$$
$$\overline{A + B + C} = \overline{A} \cdot \overline{B} \cdot \overline{C}$$

- Basic logic gates include the AND gate, OR gate, XOR gate, NOT gate, NAND gate, and NOR gate. These gates perform the corresponding Boolean operations.
- Sequential logic circuits have memory because the output of sequential logic circuits depends on both past and present inputs.
- Flip-flops include SR latch, JK flip-flop, D flip-flop, and T flip-flop.
- Flip-flops can be used to construct a counter.

References

1. Parag K. Lala. 1996. *Practical digital logic design and testing*. Englewood Cliffs, NJ: Prentice Hall.

Problems

*B refers to basic, A refers to average, H refers to hard, and * refers to problems with answers.*

SECTION 10.2 NUMBER SYSTEMS

10.1 (B) State the difference between an analog signal, a discrete signal, and a digital signal.

10.2 (A)
(a) Convert 1209 (decimal) to binary form.
(b) Convert 1225 (decimal) to binary form.

10.3 (B)* Convert 10001011(binary) to decimal form.

10.4 (A) Convert the decimal fraction number 0.68 into binary form (stop when the fraction part has 8 bits).

10.5 (A) Convert the decimal fraction number 0.141 into binary form and then convert it back into decimal fraction form. Stop when the fraction part has:
(a) 8 bits,
(b) 16 bits.
Compare both answers.
Express the decimal answer to the accuracy of 5 decimal points.

10.6 (B)* Convert the binary number 1101.1101 to a decimal number.

10.7 (A) Convert the decimal number 88.65 into binary form (stop when the fraction part has 4 bits).

10.8 (B) Use binary addition to calculate 11 (decimal) + 77 (decimal).

10.9 (B) Find both the one's complement and the two's complement of 120 (decimal).

10.10 (A) Find 168 − 66 using the binary form.

10.11 (A) Find 66 − 168 using the binary form.

10.12 (B) Convert the hexadecimal number 1FA3.0C42 to a binary number.

10.13 (B)* Convert the binary number 1010001101.11010011 to hexadecimal form.

10.14 (B) Convert the octal number 734.216 to a binary number.

10.15 (B) Convert the binary number 101110101.11010101 to octal form.

10.16 (A) Convert 1424 (decimal) to hexadecimal form.

10.17 (A) Convert 1211 (decimal) to octal form.

10.18 (B) Convert the octal number 272.61 to decimal form.

10.19 (B)* Convert the hexadecimal number 1FC2 to decimal form.

10.20 (H) Implement the conversion between a decimal number and a quaternary number (base 4): convert 123 (decimal) to quaternary form.

10.21 (B) Convert the octal number 742 to hexadecimal form.

10.22 (B) Convert the hexadecimal number FE021 to octal form.

10.23 (B)* Convert the quaternary number 3221.3 to hexadecimal form.

10.24 (B)* Convert the hexadecimal number 7F.32 to quaternary form.

10.25 (A) Calculate quaternary subtraction of 32 − 12 using binary form.

10.26 (A) Calculate octal subtraction of 45 − 32 using binary form.

10.27 (A) Calculate hexadecimal subtraction of 89 − 29 using binary form.

10.28 (A) Transform the quaternary number 20 (Q), octal number 20 (O), and hexadecimal number 20 (H) to decimal form and explain why they have the same form in quaternary, octal, and hexadecimal form but different values in decimal form.

10.29 (H) Provide an example to explain the limitations of a short-length word processor, for example an 8-bit word length processor. (Hint: consider the processing overflow.)

SECTION 10.3 BOOLEAN ALGEBRA

10.30 (B) State De Morgan's laws.

10.31 (B) Develop truth tables for the following Boolean expressions:

$$D = AB + B\overline{C}$$
$$F = BC + A\overline{D}$$

10.32 (B)* If $A = 1$, $B = 0$, and $C = 1$, find $F = A\cdot B + \overline{A}\cdot\overline{B} + C\cdot\overline{B} + ABC$.

10.33 (A) If $A = 0$, $B = 0$, and $C = 1$, find $F = (A + \overline{\overline{B} + C})\cdot(\overline{A\cdot B}\cdot C)$.

10.34 (H) Show that $A\cdot\left(\overline{\overline{B}\cdot(A + C)}\right) = AB$.

10.35 (H)* Simplify the logic circuit $F = \overline{A\cdot B + A\cdot\overline{C}}$.

10.36 (H) What is the output of $F = (\overline{C}\cdot B)\cdot\left[(A + C)\cdot(\overline{A + B})\right]$?

10.37 (A) Find the value of $E = \overline{A + \overline{\overline{C}}} + A\cdot(\overline{\overline{\overline{D}} + B}) + \overline{C} + B\cdot\overline{D}$ when $A = 1, B = 0, C = 0, D = 0$.

10.38 (A)* Find the value of $E = \overline{\overline{\overline{A + B} + \overline{\overline{B\cdot\overline{C}}}}} + \overline{\overline{C + B\overline{D}}}$ when $A = 0, B = 1, C = 0, D = 1$.

10.39 (H) From the truth table of Table 10.39, find the Boolean function $D = f(A, B, C)$.

TABLE P10.39 Truth Table

A	*B*	*C*	*D*
0	0	0	0
0	0	1	0
0	1	0	0
0	1	1	1
1	0	0	0
1	0	1	1
1	1	0	0
1	1	1	1

10.40 (H) Prove that $\overline{\overline{\overline{\overline{A} + \overline{B} + C}\cdot\overline{\overline{A + \overline{\overline{B}}\cdot\overline{\overline{C}}}}}} = 1$.

10.41 (H)* Simplify $D = \overline{\overline{A + \overline{\overline{B}}\cdot\overline{\overline{C}}\cdot A}} + \overline{A + B}$.

10.42 (H)* Simplify $C - B + \overline{\overline{\overline{A\cdot B}}\cdot\overline{A}}$.

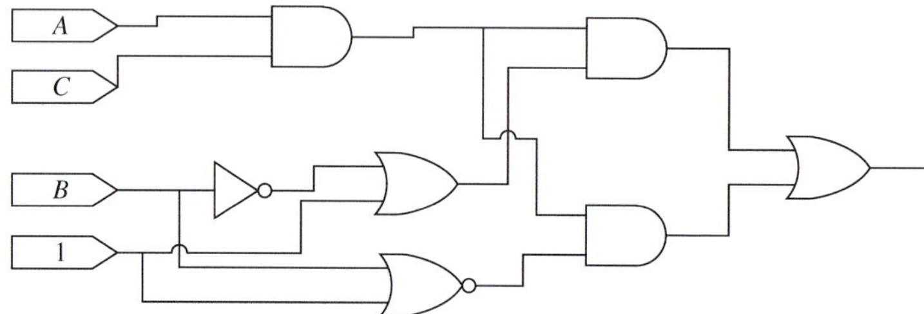

FIGURE P10.49 Logic circuit for Problem 10.49.

SECTION 10.4 BASIC LOGIC GATES

10.43 (A) Find the expression for the output of the logic circuits shown in Figure P10.43.

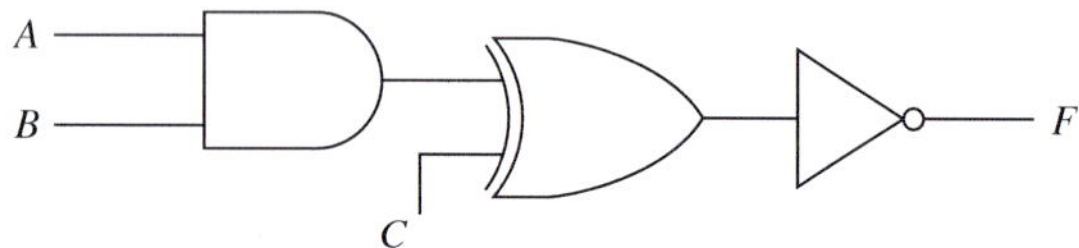

FIGURE P10.43 Circuit for Problem 10.43.

10.44 (A) Draw a circuit to realize the following expressions using AND gates, OR gates, and inverters:

$$F = A\overline{B}C + \overline{A}B$$

10.45 (A)* If $A = 0$, $B = 1$, which circuit as shown in Figure P10.45 will output 1?

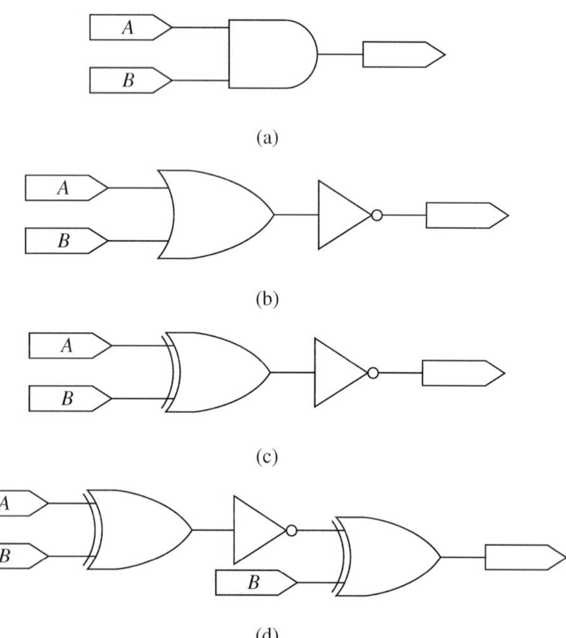

FIGURE P10.45 Circuit for Problem 10.45.

10.46 (A) Four inputs are A, B, C, and D. Use AND gates (two inputs), OR gates (two inputs), and inverter gates to implement the logic function:

$$F = A \cdot B \cdot D + B \cdot \overline{D} + C \cdot D$$

10.47 (H) Use only NAND gates to implement the logic function of Problem 10.46.

10.48 (A)* Simplify $(A + C) \cdot B + (\overline{B} + \overline{C}) \cdot C$ then draw the logic circuit for the solution.

10.49 (H) Write the logic expression for the circuit shown in Figure P10.49. Then, simplify the expression and draw the logic circuit that corresponds to the expression.

10.50 (H) A new design for an emergency demonstration system is composed of three bulbs installed on the top of a car, shown in Figure P10.50. A control console is installed inside the car. The console has two switches each with an output of 0 or 1. If the emergency level is 0 (no emergency), the green light is on. If the emergency level is 1, one red light is ON; if the emergency level is 2, both red lights are ON. Design the control system (assume that NOT gates, two-input AND gates, and OR gates are available).

FIGURE P10.50 Emergency demonstration car.

10.51 (B)* Use the smallest number of logic circuits to express $E = A\overline{C} + A\overline{D} + B\overline{C} + B\overline{D}$.

10.52 (H) Find the expression of the output of the given logic circuit, and then use the smallest number of logic circuits (a two-input AND operator, a two-input OR operator, and inversion operator) to express it.

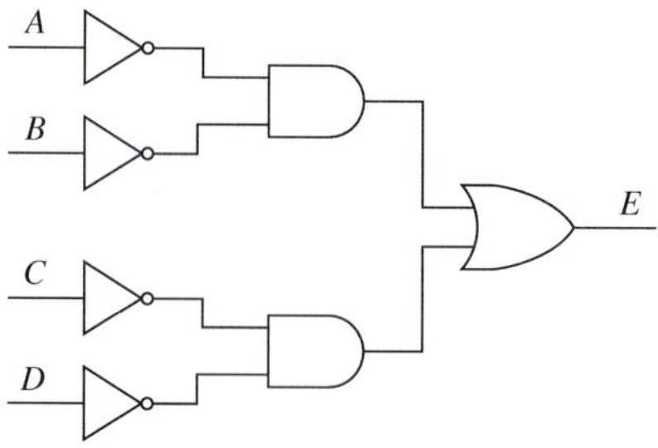

FIGURE P10.52 Logic circuit.

10.53 (A) Find the expression of the output of the logic circuit shown in Figure P10.53.

10.54 (H)* Simplify the result of Problem 10.53 to make it only include AND, OR, and inverse operators, and draw the corresponding logic circuit.

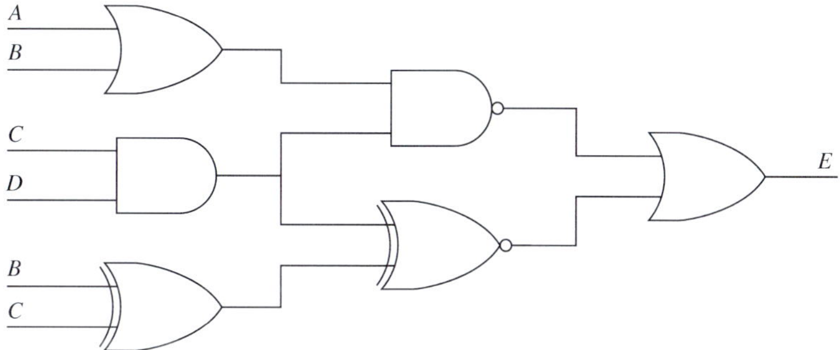

FIGURE P10.53 Logic circuit for P10.53.

10.55 (H) Using XOR, NAND, and AND operators, draw the logic circuit for the logic equation $E = A \cdot \overline{B} \cdot \overline{C} + A \cdot \overline{B} \cdot \overline{D} + \overline{A} \cdot B \cdot \overline{C} + \overline{A} \cdot B \cdot \overline{D}$.

10.56 (A) Simplify $E = A \oplus B + A \cdot B + \overline{A} \cdot \overline{B}$

10.57 (A) Find the truth table for the logic circuit shown in Figure P10.57.

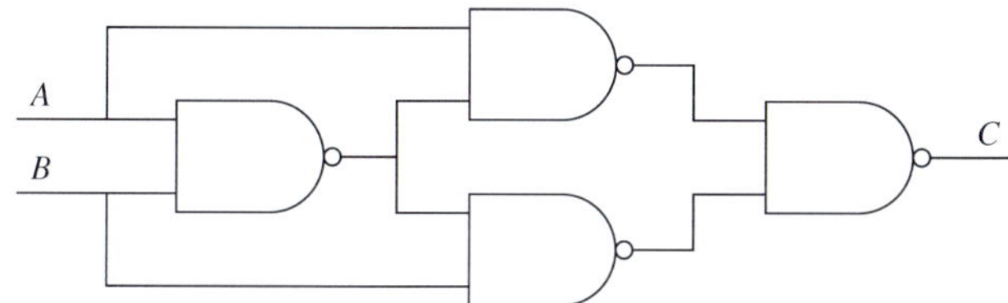

FIGURE P10.57 Logic circuit constructed with NAND and NOR.

10.58 (H)* Design a logic circuit $D = f(A, B, C)$ to satisfy the following truth table.

TABLE P10.58 Truth Table

A	*B*	*C*	*D*
0	0	0	0
0	0	1	1
0	1	0	0
0	1	1	1
1	0	0	0
1	0	1	1
1	1	0	1
1	1	1	1

SECTION 10.5 SEQUENTIAL LOGIC CIRCUITS

10.59 (B) For the JK flip-flop shown in Figure P10.59, the current values of J and K are $J = 0$, $K = 0$, and $Q = 0$. What is the value of Q when J changes to 1, and K changes to 0 on the arrival of another clock pulse?

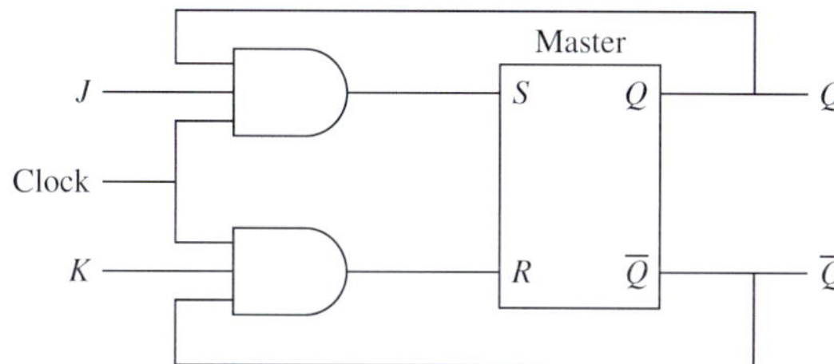

FIGURE P10.59 JK flip-flop for Problem 10.59.

10.60 (A)* How many D flip-flops are needed to represent a 4-bit ring counter?

10.61 (H) A student wants to use a simple logic circuit to design a three-second countdown for a car racing competition. The countdown is initiated by a red light. The initial light is followed by a yellow light when there is 1 s left. Finally, a green light represents the beginning of the race. Design a control system that meets the requirements.

10.62 (A) Determine the output, Q, of the JK flip-flop if the input is as shown in Table P10.62, given that the initial output, Q, of the flip-flop is 0.

TABLE P10.62 JK Flip-Flop Input for Problem 10.62

J	1	0	1	1	0	1	0	1	1	0
K	0	0	1	0	1	1	0	1	0	1

10.63 (H) Figure P10.63(a) shows the sequence of clock signal for both clock inputs, T_1 and T_2, while Figure P10.63(b) shows the JK flip-flop circuit. What is the output, Q_2 of JK flip-flop 2 immediately before the second trailing edge of the clock signal 2, T_2? Both the initial outputs Q_1 and Q_2 are zero. Assume that the input from the clock signal is 1 after the leading edge of the clock signal, and the input becomes 0 after the trailing edge of the clock signal. Also, assume a clock signal of 1, T_1 starts at the leading edge.

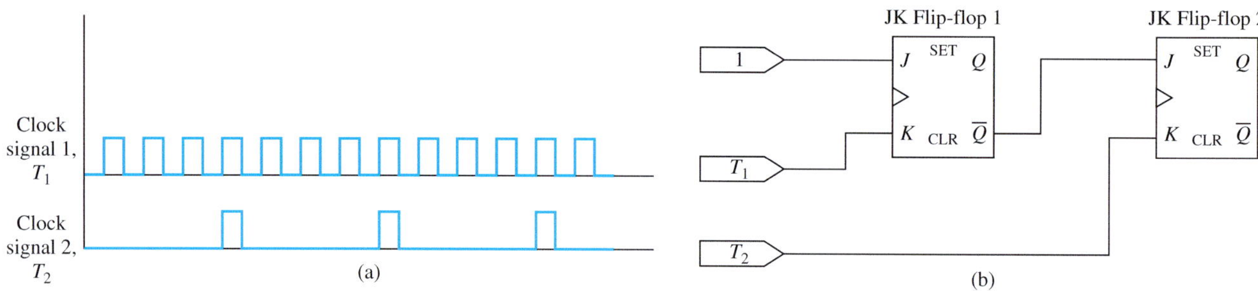

FIGURE P10.63 (a) Clock signal input for Problem 10.63; (b) logic circuit for Problem 10.63.

10.64 (A) A logic circuit and its input wave form are given in Figure P10.64. Assuming its initial output is 0, draw the output of the logic circuit with these input signals.

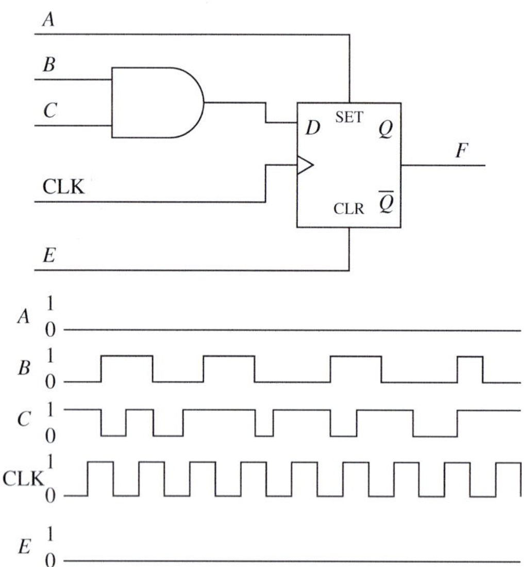

FIGURE P10.64 Logic circuit and its input.

10.65 (H) A frequency divider and the input clock are shown in Figure P10.65, and its initial outputs are all zeros. Draw the output waveform of A, B, and C.

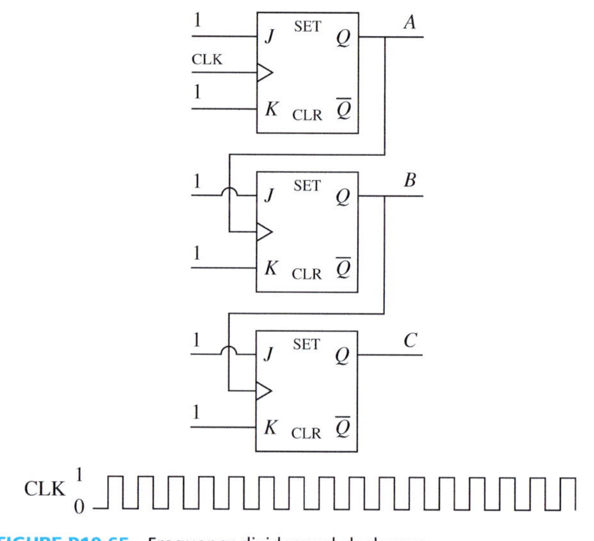

FIGURE P10.65 Frequency divider and clock wave.

10.66 (A) A 5-bit shift register constructed with D flip-flops is shown in Figure P10.66. At the first CLK rising edge, the output is as follows: $Q_1 = 0$, $Q_2 = 1$, $Q_3 = 1$, $Q_4 = 0$, $Q_5 = 0$. Provide the output at the second, third, fourth, and fifth CLK rising edges.

10.67 (H) A counter constructed with D flip-flops is shown in Figure P10.67. Assume that each D flip-flop's initial output is 0. Draw its output wave form.

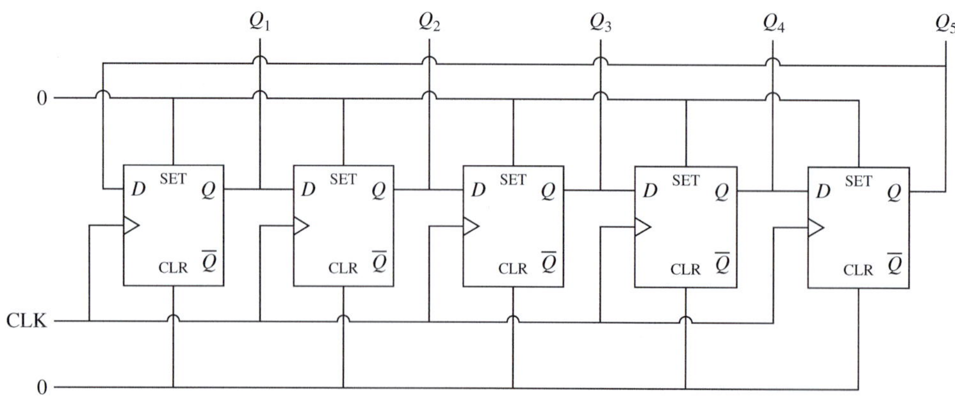

FIGURE P10.66 A 5-bit shift register.

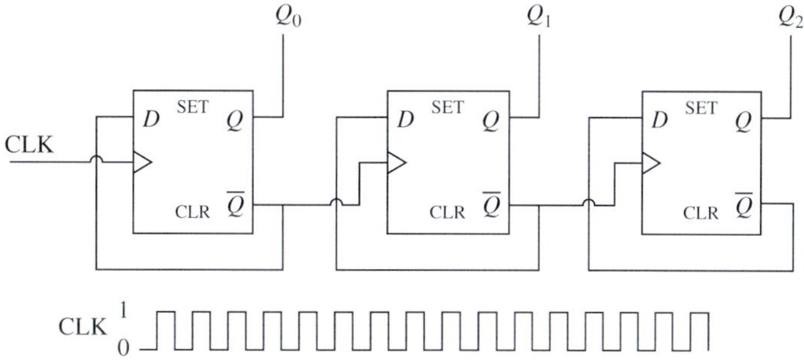

FIGURE P10.67 Counter and its input.

10.68 (H) Analyze the circuit shown in Figure P10.68 and provide the output waveform based on the input waveform.

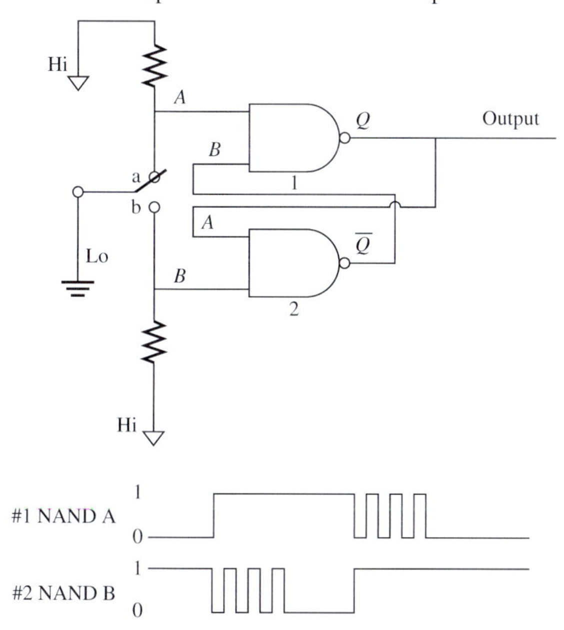

FIGURE P10.68 A logic circuit and its input.

SECTION 10.6 USING PSPICE TO ANALYZE DIGITAL LOGIC CIRCUITS

10.69 (A)* Use PSpice to solve the digital logic circuit, $F = \overline{A} \cdot \overline{B} + C \cdot \overline{B} + A \cdot B \cdot C$ if $A = 1, B = 0, C = 1$.

10.70 (H)* Use PSpice to solve the digital logic circuit, $F = \overline{(A + \overline{B} + C)} \cdot (\overline{A \cdot B \cdot C})$ if $A = 0, B = 0, C = 1$.

10.71 (B)* If $A = 0$ and $B = 1$, use PSpice to find the logic circuit which generates an output of 1 in Figure P10.45.

10.72(H) For the JK flip-flop shown in Figure P10.59, the current values of J and K are $J = 0$, $K = 0$, and $Q = 0$. Use PSpice to set up the digital logic circuit. The JK flip-flop circuit can be obtained from part "JKFF" in the DIG_PRIM library.Solution

TOPIC 6

Computer-Based Instrumentation Systems

The content of this topic is compiled from:

Chapter 11 (pp. 488–523), *Electrical Engineering: Concepts and Applications*

By S.A. Reza Zekavat

CHAPTER 11

Computer-Based Instrumentation Systems

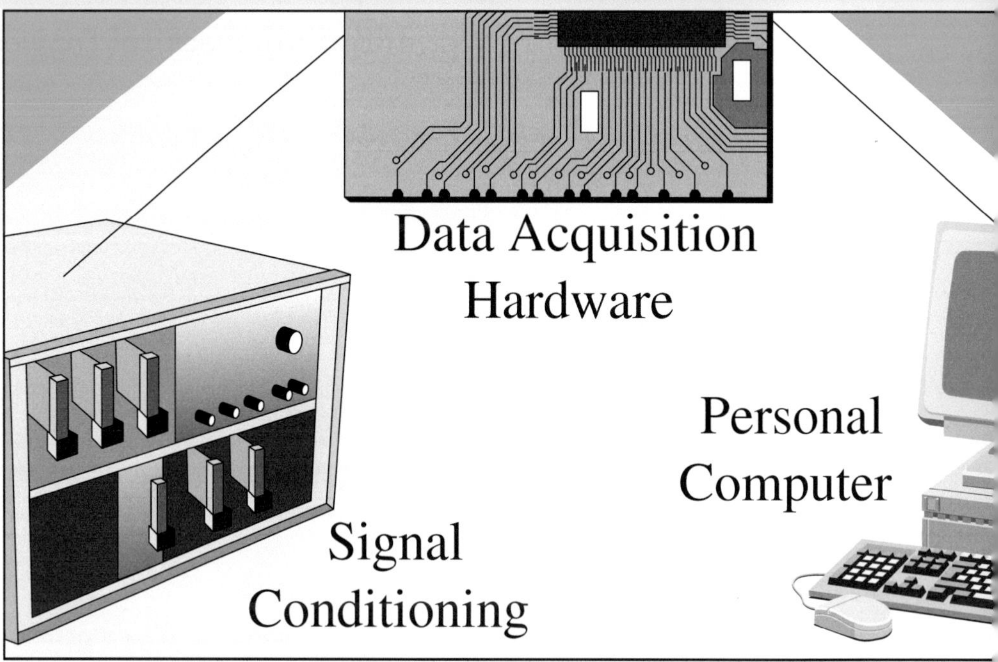

11.1 INTRODUCTION

In our daily lives, data are collected from many sources. Examples include measuring the tension of a bridge brace to gather data used to maintain the bridge (Figure 11.1) or noting the wind speed or humidity of the air to forecast the weather (Figure 11.2). The collected data are usually continuous in nature, and thus, it is represented by analog signals.

To obtain data about the tension on a bridge brace or wind speed, devices can be employed within the structure to measure the tension, or devices can be installed on poles in the air to measure the wind speed. Devices that collect data from the environment are called *sensors*. Many of these devices are capable of transmitting collected data wirelessly to a destination where the collected data are processed and deductions are made. Examples of real-world applications of sensors include temperature monitoring systems, smart buildings fire alarm systems, seismic pressure sensing in earthquake prediction, and crash-detecting systems in cars.

In general, due to the kind and amount of field data collected by sensors, advanced computer technology is incorporated to process the data (Figure 11.3). Statistical science techniques are used to summarize the collected data and make conclusions. For example, after analyzing

FIGURE 11.1 Bridge maintenance. (Used with permission from © Emily Lai/Alamy.)

FIGURE 11.2 Weather forecast. (Used with permission from © peter dazeley/Alamy.)

FIGURE 11.3 Data processing. (Courtesy Alamy Images.)

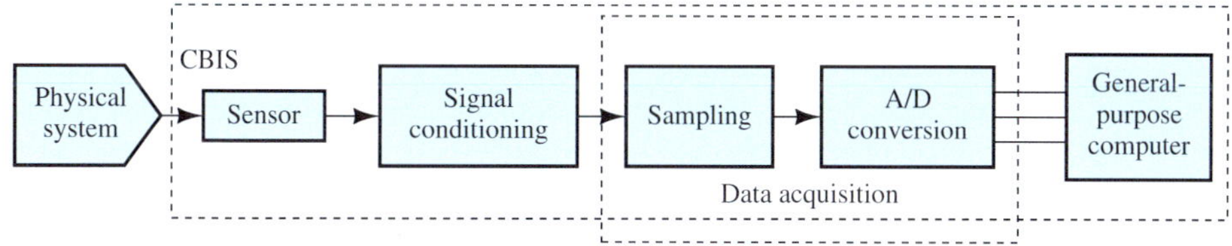

FIGURE 11.4 Computer-based instrumentation system (CBIS).

the data, the analyst may decide that a bridge brace needs replacement, a tornado is likely to form, a fire is ongoing, and so on.

Finally, assuming that the installed sensors have collected analog signals, the question is: How can digital signals be generated and processed when analog signals have been collected? The process of converting analog signals to digital (digitizing) includes sampling and coding. The coder is called an analog-to-digital converter (A/D converter; ADC). The signal generated by the ADC is applied to a computer or digital signal processor (DSP) for processing, making conclusions, and taking actions.

As summarized in Figure 11.4, the entire structure starting from the sensor through the computer (or DSP) forms a computer-based instrumentation system (CBIS).

Physical quantities such as pressure, temperature, stress, flow, angular speed, and displacement can be measured using sensors. These physical quantities produce changes in the voltages, currents, resistances, capacitances, and/or inductances of the sensors, which convert the change of the physical quantity into a corresponding change in an electrical quantity.

Signal conditioning transforms the changes in the electrical quantities into voltages, when the electrical quantity is not a voltage. Furthermore, signal conditioning alters the sensor output into an appropriate form through actions, such as amplifying, filtering, and so on. The conditioned signals are sent to the *data-acquisition system* that periodically samples the signal and converts the sampled value to digital "words" using an ADC. The digital words are then read by the computer for further processing. This chapter discusses the instrumentation system in detail and explains each of the components shown in Figure 11.4.

11.2 SENSORS

A sensor, also called a *transducer*, is defined as a device that responds to a physical stimulus (e.g., heat, light, sound, pressure, magnetism, or motion) and transmits a resulting impulse for measurement or control. Sensors/transducers detect and measure a signal or energy and convert it to a desirable output signal or energy. A microphone is one example of a transducer. Generally, sensors can be grouped according to their physical characteristics (e.g., electronic sensors or resistive sensors) or by the physical variable or quantity measured by the sensor (e.g., pressure, flow, or temperature). Sensors can also be grouped based on the domains to which they belong, such as thermal, mechanical, chemical, magnetic, radiant, or electrical. This section examines typical physical sensors.

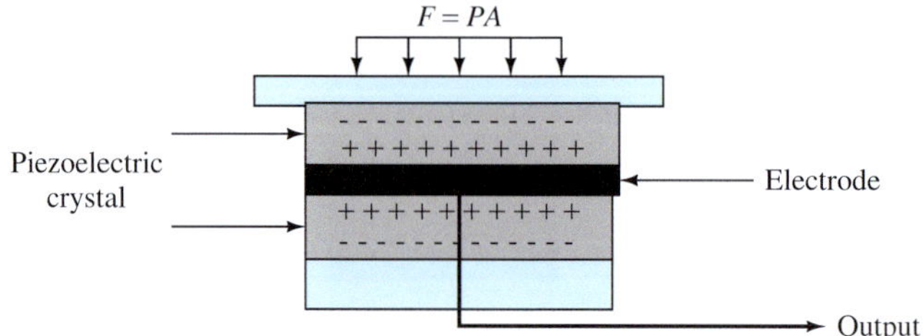

FIGURE 11.5 A piezoelectric pressure sensor.

11.2.1 Pressure Sensors

Pressure is a mechanical quantity defined as the normal force per unit area. Pressure can be sensed by elastic mechanical elements, such as plates, shells, or tubes, which are designed and constructed to deform when pressure is applied. This deformation is the basic mechanism of converting pressure to physical movement. The movement can then be transduced to obtain an electrical signal. Therefore, a pressure sensor is defined as a device that responds to the pressure applied to its sensing surface and converts the pressure to a measurable signal. The most common family of force and pressure sensors is made up of those based on strain gauges and piezoelectric sensors. This section examines how piezoelectric pressure sensors convert force or pressure to an electrical signal.

A representation of a typical piezoelectric pressure sensor is shown in Figure 11.5. Pressure and force sensors are nearly identical and rely on an external force to strain the crystals. A major difference is that pressure sensors use a diaphragm to collect pressure. Pressure is simply the force applied over a unit area, that is, force = pressure × area.

The piezoelectric pressure sensor contains a piezoelectric crystal that generates an electric charge in response to deformation. When a force or pressure is applied to the crystal, which produces a displacement, charges are generated within the crystal. If the external force generates a displacement, x_i, then the sensor will generate a charge, q, according to the expression:

$$q = K_p x_i \tag{11.1}$$

where K_p is the pressure sensitivity in coulomb/displacement unit. The magnitude of K_p depends on the piezoelectric material and the structure of the sensor. Figure 11.6 depicts a basic model of a piezoelectric sensor and the corresponding circuit. The model of a piezoelectric sensor consists of a piezoelectric crystal and two conducting electrodes, while the simple circuit contains a current source and a capacitor; V_o is the voltage output. In the circuit model, the current source represents the rate of change of the charge due to the motion of the crystal when an external force is applied.

In addition, the capacitance represents the effect of the sensor's structure responding to the external force. The sensor's output voltage, V_o, corresponds to:

$$V_o = \frac{1}{C}\int i\mathrm{d}t = \frac{1}{C}\int \frac{\mathrm{d}q}{\mathrm{d}t}\mathrm{d}t = \frac{q}{C} = \frac{K_p x_i}{C} \tag{11.2}$$

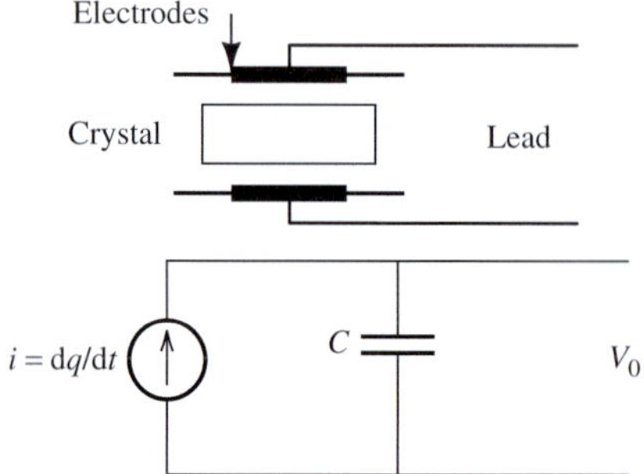

FIGURE 11.6 A piezoelectric pressure sensor and its equivalent circuit model.

In addition to piezoelectric sensors, potentiometric pressure sensors, inductive pressure sensors, capacitive pressure sensors, and strain-gauge pressure sensors are also widely used in practical applications.

EXAMPLE 11.1 Application of Pressure Sensors

Describe some applications of pressure sensors.

SOLUTION

Some applications of pressure sensors are listed below:

1. ***Fuel pressure:*** Pressure sensors are used to measure fuel pressure during fuel pumping.
2. ***Fluid flow:*** Pressure sensors are used to measure the differential pressure across an orifice. For example, intake airflow or engine coolant flow is measured in this way.
3. ***Air pressure:*** An aneroid barometer (a kind of pressure sensor) is used to measure the air pressure outside an aircraft.
4. ***Tire pressure:*** Pressure sensors are used to monitor the pressure in vehicle tires.
5. ***Blood pressure:*** A small sensor is penetrated into blood vessels to measure blood pressure.
6. ***Altitude sensing:*** The relationship between changes in pressure relative to altitude is used to measure the altitude in aircraft, rockets, satellites, and weather balloons.

EXAMPLE 11.2 Pressure Sensors

In piezoelectric sensors, the output voltage of the sensors can be measured to obtain the pressure from the object according to the equation:

$$V_o = K_E p x$$

where K_E is the voltage sensitivity of the sensor. For this example, assume $K_E = 0.055$ V m/N. In this equation, p is the pressure in N/m^2 and x is the displacement in m. If an object is placed on the sensor and measures $V_o = 3$ V, and $x = 1$ mm, determine the pressure from the object.

SOLUTION

Using the equation discussed in this section, and replacing for the voltage, displacement, and the parameter K_E, note that:

$$p = \frac{V_o}{K_E x} = \frac{3}{0.055 \times 0.001} = 5.45 \times 10^4 \text{ N/m}^2$$

11.2.2 Temperature Sensors

Temperature sensors measure the temperature, which is one of the most frequently measured physical quantities. Thermocouples, resistance temperature devices (RTDs), thermistors, and infrared thermometers represent different types of temperature sensors. The most common types of temperature sensors are thermocouples and RTDs. Thermistors are part of the RTD family.

11.2.2.1 THERMOCOUPLES

Thermocouples are the most common electrical output sensors used to that measure temperature. A thermocouple is formed by two dissimilar metals that are connected at one end (the measuring junction) and connected to a voltage-measuring instrument at the other end (the reference junction). This configuration is illustrated in Figure 11.7. Whenever the measuring junction is at a

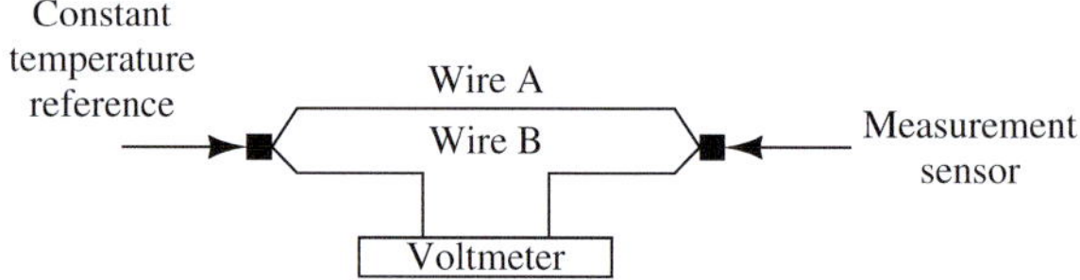

FIGURE 11.7 Thermocouple configuration.

temperature different from the reference junction, a voltage difference is produced across the two metals. Therefore, the temperature difference between the two junctions can be detected by measuring the change in the voltage across two dissimilar metals at the temperature measurement junction. This voltage varies with the temperature disparity of the junctions. If the temperature at one junction is known, the temperature at the other junction can be calculated.

Thermocouples generate an open-circuit voltage, called *Seebeck* voltage. This voltage is proportional to the temperature difference between the measuring and reference junctions.

Because thermocouple voltage is a function of the temperature difference between junctions, the voltage and reference junction temperature must be known to determine the temperature at the measuring junction. Consequently, a thermocouple measurement system must either measure the reference junction temperature or maintain it at a fixed, known temperature. Most industrial thermocouple measurement systems choose to measure, rather than to control, the reference junction temperature due to cost.

Thermocouples function based on the fact that the electromotive force (EMF) between two dissimilar metals is a function of their temperature difference. Three major effects are involved in a thermocouple circuit: the *Seebeck*, *Peltier*, and *Thomson effects*.

Seebeck effect: When two dissimilar conductors, A and B, comprise a circuit as shown in Figure 11.8, a current will flow in that circuit as long as the two junctions are at different temperatures, one junction at a temperature, T, and the other at a higher temperature $T + \Delta T$. The current will flow from A to B at the colder junction when conductor A has a higher potential with respect to B.

The Peltier, and Thomson effects, are related to the Seebeck effect.

Peltier effect: When current flows across a junction of two dissimilar conductors, heat is absorbed or created at the junction, depending upon the direction of current flow.

Thomson effect: When current flows through a conductor (generally in a long, thin bar to reduce thermal conductivity) along which a temperature gradient exists, heat is absorbed by or delivered from the wire.

The Seebeck effect describes the electromotive force (EMF) created across two dissimilar metallic materials. The change in material EMF with respect to a change in temperature is called the *Seebeck coefficient* or thermoelectric sensitivity. This coefficient is usually a nonlinear function of temperature.

A sample *thermocouple circuit* is illustrated in Figure 11.9. (Note: the ice bath shown is not a necessary component of every thermocouple, but is shown here for demonstration purposes.)

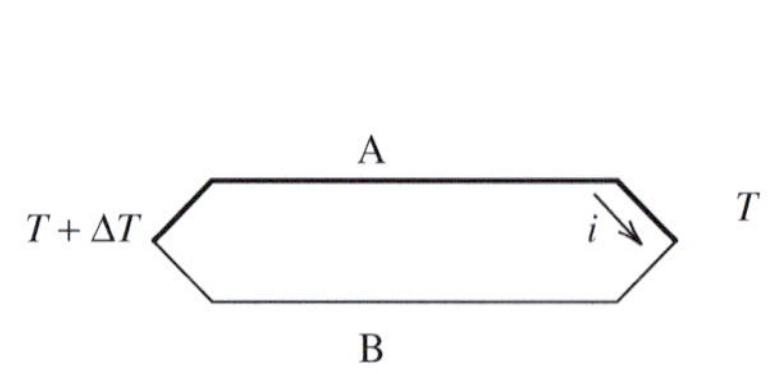

FIGURE 11.8 Seebeck effect.

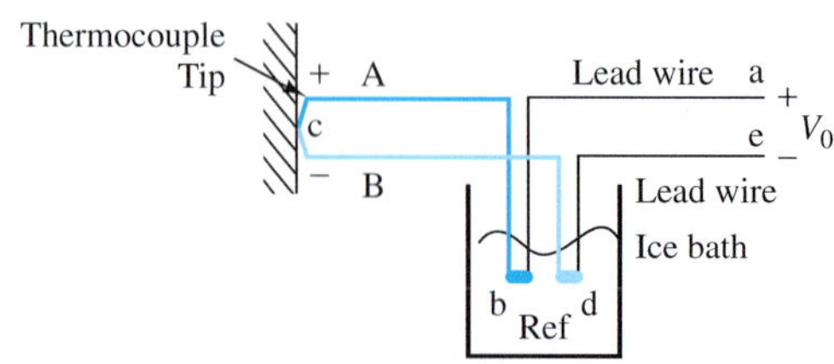

FIGURE 11.9 Example of a thermocouple circuit.

The voltage output, V_o, measured at the output of the thermocouple corresponds to:

$$V_o = \int_a^b S_{\text{Lead}}(T)\,dT + \int_b^c S_A(T)\,dT + \int_c^d S_B(T)\,dT + \int_d^e S_{\text{Lead}}(T)\,dT \qquad \textbf{(11.3)}$$

where b and d represent the reference points; c represents the probe tip; and a and e represent output terminals (see Figure 11.9). In addition, $S_A(t)$, $S_B(t)$, and $S_{\text{Lead}}(t)$ are the Seebeck coefficients (unit: volt/degree C) of two dissimilar metallic materials, metals A and B, and the lead wires, respectively. In general, all three Seebeck coefficients are nonlinear functions of temperature, T.

The voltage induced by the temperature of the lead wires cancels with each other, that is, $\int_a^b S_{\text{Lead}}(T)\,dT + \int_d^e S_{\text{Lead}}(T)\,dT = 0$. Thus:

$$V_o = \int_b^c S_A(T)\,dT + \int_c^d S_B(T)\,dT = \int_b^c [S_A(T) - S_B(T)]\,dT$$

The second equality is due to the fact that the temperature b equals the temperature d. If the Seebeck coefficient functions of the two thermocouple wire materials are pre-calibrated and the reference temperature is known (usually set by a 0°C ice bath), the temperature at the probe tip becomes the only unknown and can be directly related to the voltage readout.

If the Seebeck coefficients are nearly constant across the targeted temperature range, Equation (11.3) can be simplified to:

$$V_o = (S_A - S_B)(c - b) \Rightarrow c = b + \frac{V_o}{S_A - S_B} \qquad \textbf{(11.4)}$$

Here, C is the temperature at the tip and b is the reference temperature; thus,

$$T_{\text{Tip}} = T_{\text{Ref}} + \frac{V_o}{S_A - S_B} \qquad \textbf{(11.5)}$$

Equation (11.5) represents a linear relationship between the tip temperature and the output voltage. In practice, this linear relationship is not available and vendors provide calibration functions for their products. These functions are usually high-order polynomials and are calibrated with respect to a certain reference temperature, for example, 0°C (32°F). Assuming the coefficients of the calibration polynomials are $\alpha_0, \alpha_1, \alpha_2, \ldots, \alpha_n$, the relationship of the temperature at the probe tip and the output voltage corresponds to:

$$T_{\text{Tip}} = \alpha_0 + \alpha_1 V_o + \alpha_2 V_o^2 + \cdots + \alpha_n V_o^n \qquad \textbf{(11.6)}$$

where V_o is the thermocouple voltage in volts, T_{Tip} is the temperature in degree Celsius, and α_0 through α_n are coefficients that are specific to each thermocouple type. Note that Equation (11.6) is effective only if the reference temperature in the experiment is kept constant.

EXAMPLE 11.3 Thermocouple

The coefficients of calibration polynomials of a K-type thermocouple (commonly used due to their low cost) are shown in Table 11.1, and the voltage reading of the thermocouple is 3.47 mV. Find the room temperature computed by a K-type thermocouple, if the thermocouple considers the boiling water temperature as a reference temperature, while the other end is measuring the room temperature. Assuming the actual room temperature is 16°C, discuss the error in the thermocouple reading. Table 11.1 represents Equation (11.6) coefficients.

(continued)

EXAMPLE 11.3 **Continued**

TABLE 11.1 α_n Coefficient for a K-Type Thermocouple

α_n	Coefficient
0	0.226584602
1	24152.10900
2	67233.4248
3	2210340.682
4	–860963914.9
5	4.83506×10^{10}
6	-1.18452×10^{12}
7	1.38690×10^{13}
8	-6.33708×10^{13}

SOLUTION

Using Equation (11.6):

$$T = \sum_{n=0}^{\infty} a_n V^n$$

the thermocouple reference temperature is $T = 84.83°C$. Assuming that the water is boiled at 100°C, the room temperature reading with thermocouple reference is:

$$T = 100 - 84.83$$
$$T = 15.17°C$$

For this K-type thermocouple, the reading error is in the order of 0.83°C (that is less than 1°C). Therefore, the answer obtained from the thermocouple is reasonable.

11.2.2.2 THERMISTORS

A resistance temperature detector (RTD) is a variable-resistance device whose resistance is a function of temperature (Figure 11.10). RTDs are more accurate and stable than thermocouples.

Similar to an RTD, the **thermistor** (bulk semiconductor sensor) uses resistance to detect temperature. A thermistor is a resistance thermometer and is usually manufactured in the shape of beads, discs, or rods. It is made by combining two or more metal oxides. If oxides of cobalt, copper, iron, nickel, or tin are used, the resulting semiconductor will have a negative temperature coefficient (NTC) of resistance.

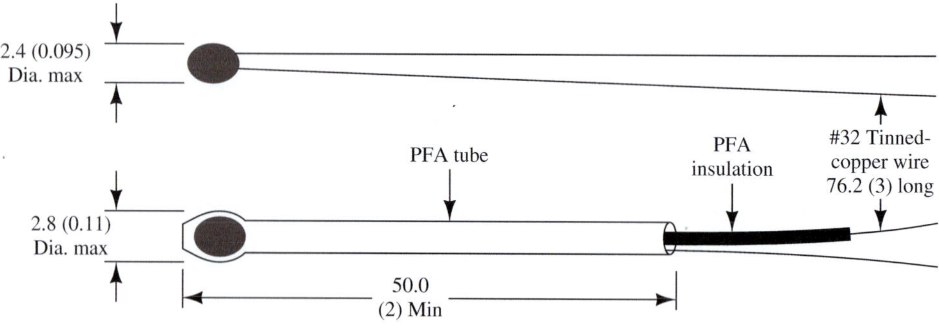

FIGURE 11.10 Precision thermistor elements, 44000 Series.

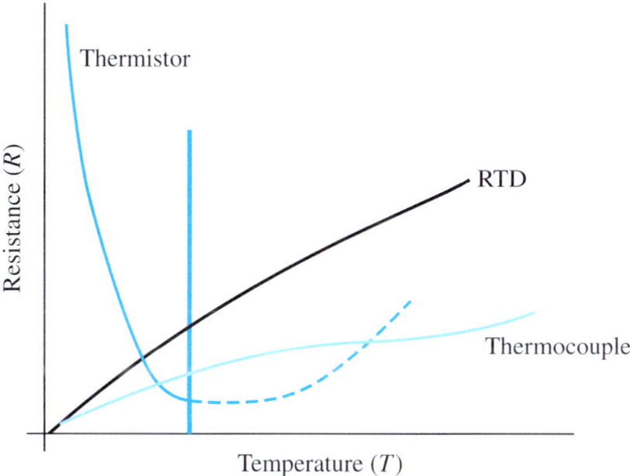

FIGURE 11.11 Characteristics of three temperature sensors.

Unlike an RTD's metal probe where the resistance increases with temperature, the thermistor responds inversely with temperature. As mentioned earlier, the relationship between resistance and temperature is nonlinear. In a thermistor, the resistance changes negatively and sharply with a positive change in temperature, as shown in Figure 11.11.

Note that in Figure 11.11, the resistance of the thermistor increases with temperature if the temperature is higher than a threshold. However, in this temperature range, the thermistor fails to give a correct temperature reading. Thermistors are fairly limited in their temperature range, working only over a nominal range of 0°C to 100°C.

The negative resistance–temperature relationship of a thermistor might be expressed by an exponential (i.e., nonlinear) function:

$$R = R_o \times e^{\beta\left(\frac{1}{T} - \frac{1}{T_o}\right)} \quad \textbf{(11.7)}$$

where

- T = absolute temperature (in kelvin)
- T_o = reference temperature, usually the room temperature (25°C; 77°F; 298.15 K)
- R = resistance at the temperature T (Ω)
- R_o = resistance at T_o
- β = calibration constant that varies with the thermistor material (usually between 3000 and 5000 K).

To solve for the temperature, T, take the logarithm (ln) of both sides of the Equation (11.7) because R and T are positive real numbers, and solve for $1/T$:

$$\frac{1}{T} = \frac{1}{T_o} + \frac{1}{\beta}[\ln(R) - \ln(R_o)] \quad \textbf{(11.8)}$$

Accordingly:

$$T = \frac{T_o \times \beta}{\beta + T_o[\ln(R) - \ln(R_o)]} \quad \textbf{(11.9)}$$

If Equation (11.7) is differentiated and rearranged, the sensitivity of the thermistor, α, at the temperature, T, corresponds to:

$$\alpha \approx \frac{1\,dR}{R\,dT} = -\frac{\beta}{T^2} \quad \textbf{(11.10)}$$

Because the constant β may vary slightly with temperature, several well-known temperature conditions may be used as check points, for example, ice water at 0°C (32°F) and boiling water at 100°C (212°F); other pre-calibrated thermometers may also be used to calibrate/curve-fit the value of β.

However, β may vary considerably across the temperature range of interest. In this case, a calibrated curve-fit of the R–T relationship may be used, rather than Equation (11.9). A suitable nonlinear function that can be generated via curve fitting and converts the measured resistance, R, of a thermistor to temperature, T, is:

$$\frac{1}{T} = A + B \ln R + C(\ln R)^3 \tag{11.11}$$

where R is measured in ohms, T in kelvin, and A, B, and C are curve-fitting constants. The A, B, and C values can be obtained by substituting three pairs of values of (R,T) within the operating range into Equation (11.11) to generate three equations. Then the A, B, and C constants can be acquired after solving the three generated equations.

EXERCISE 11.1

Assume the following measurement (R,T) pairs, (100,280), (95,290), (90,310). Use Equation (11.11) to find the constants, A, B, and C. (*Hint:* write three equations for constants A, B, and C; MATLAB can be used to solve the equations.)

EXAMPLE 11.4 Applications of Temperature Sensors

Describe some applications of temperature sensors.

SOLUTION

Some applications of temperature sensors include the following:

1. ***Air conditioning:*** Temperature sensors are used to measure the temperature of the air to control air conditioners.
2. ***Heating systems:*** Temperature sensors are used to control heating systems to maintain the temperature in a room.
3. ***Cooling systems in computers:*** When a personal computer is running for a long time, the motherboard generates a great deal of heat. Using the temperature sensor to sense the temperature inside the computer's case, the computer can automatically turn on the fan to reduce system temperature. It may also shut down the computer when high temperatures are detected.
4. ***Food transport and storage:*** Temperature sensors are used to make sure food does not exceed safe temperatures to minimize harmful bacterial growth.

APPLICATION EXAMPLE 11.5 RTD in Vehicles

The resistance of an RTD over a small temperature range can be expressed as:

$$R_{RTD} = R_o[1 + \alpha(T - T_0)]$$

The RTD has the resistance of 30 Ω at a temperature of 0°C. Assume the coefficient of the resistance, α, for this RTD is 0.004225°C^{-1}. The RTD is placed on a running car engine to measure the temperature. The measured resistance of the RTD is 36.78 Ω. Determine the car engine's temperature.

SOLUTION

Here, $R_o = 30\ \Omega$, and $\alpha = 0.004225°\text{C}^{-1}$. Using the equation introduced in the example:

$$T = \frac{\frac{R_{RTD}}{R_o} - 1}{\alpha} + T_0 = \frac{\frac{36.78}{30} - 1}{0.004225} + 0 = 53.49°\text{C}$$

11.2.3 Accelerometers

Accelerometers (acceleration sensors) are used to measure acceleration, vibration, and mechanical shock. Acceleration is the time rate of change of velocity with respect to a reference system. Acceleration is a vector quantity, and is closely related to velocity and displacement. The linear and angular relationships for movement correspond to:

$$\text{Linear:} \quad v = \frac{dx}{dt}, \quad a = \frac{d^2x}{dt^2} \tag{11.12}$$

$$\text{Angular:} \quad \omega = \frac{d\theta}{dt}, \quad \gamma = \frac{d^2\theta}{dt^2} \tag{11.13}$$

where

- x = linear displacement
- θ = angular displacement
- v = linear velocity
- ω = angular velocity
- α = linear accelerati on
- γ = angular acceleration
- t = the time.

In general, accelerometers sense acceleration by incorporating a seismic mass on which the external acceleration can be applied. They measure the displacement of the seismic mass, the force exerted by the seismic mass against the package, or the force required to keep them in place.

A typical accelerometer is illustrated in Figure 11.12. A seismic mass, m, is suspended by a spring, k, in a sensor package. When the sensor is moved with an acceleration, a, a relative displacement, x_{rel}, of the seismic mass is produced by the inertial force and is detected as an electrical signal. The motion equation corresponds to:

$$ma = m\frac{d^2x_{rel}}{dt^2} + \lambda\frac{dx_{rel}}{dt} + kx_{rel} \tag{11.14}$$

where

- k is the spring constant
- λ is the damping constant.

In the steady state, the relationship between the displacement, x_{rel}, and the acceleration, a, corresponds to:

$$F = ma = kx_{rel}$$

Accordingly:

$$\frac{x_{rel}}{a} = \frac{m}{k} \tag{11.15}$$

The displacement of the seismic mass, x_{rel}, can be detected by either monitoring the strain induced in the spring or directly measuring the position of the seismic mass.

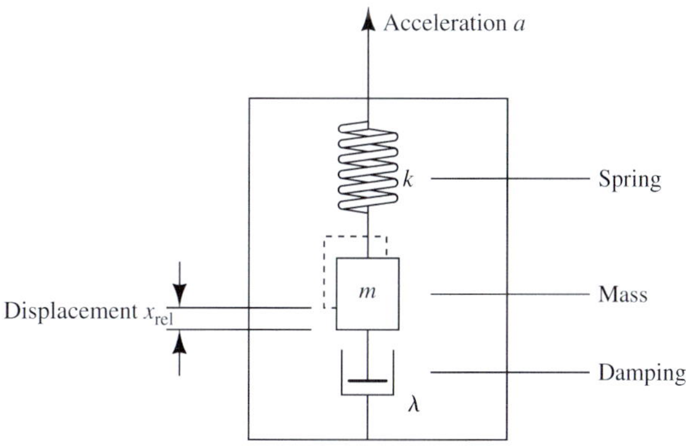

FIGURE 11.12 Accelerometer.

EXAMPLE 11.6 Applications of Accelerometers

Describe some applications of accelerometers.

SOLUTION

Some applications of accelerometers include the following:

1. ***Sport watches:*** When runners wear sport watches, their acceleration and speed can be determined by watches that contain accelerometers.
2. ***Digital cameras:*** Some digital cameras contain accelerometers to determine the orientation of the photo being taken.
3. ***Rockets:*** Accelerometers are used in rocket design to maintain the rocket's stability.
4. ***Vehicles:*** Accelerometers are used widely in vehicles to measure motion of the vehicle. A typical application is that accelerometers can help decide whether air bags should be deployed.

EXAMPLE 11.7 Accelerometer Measurement

An engineer intends to measure the acceleration of an elevator to make the elevator more comfortable for the passengers. The seismic mass, m, for the accelerometer is 50 kg and the spring constant for the accelerometer is 1.0×10^3 N/m. If the measured mass displacement of the accelerometer is 0.1 m, what is the acceleration of the elevator?

SOLUTION

According to the Equation (11.15):

$$\frac{x_{\text{rel}}}{a} = \frac{m}{k}$$

Using this equation:

$$a = \frac{k}{m}x_{\text{rel}} = \frac{1.0 \times 10^3}{50} \times 0.1 = 2 \text{ m/s}^2$$

11.2.4 Strain-Gauges/Load Cells

As discussed in Chapter 1, a load cell is a sensor that converts force into a measurable electrical output. Strain-gauge-based load cells are the most commonly used type of load cells. Strain is defined as the displacement and deformation when a force is applied to an object. Strain can be measured by a strain gauge. Strain-gauge load cells convert the load acting on them into electrical signals. The gauges used in the load cells are bonded onto a beam or structural member that deforms when weight is applied. When weight is applied, the strain changes the electrical resistance of the gauges in proportion to the load.

Load cell designs differ as the type of output signal generated (pneumatic, hydraulic, electric) changes. Load cells may also be categorized according to the technique they use to detect weight (bending, shear, compression, tension, etc.). This section examines the bonded semiconductor strain gauge as an example. The bonded semiconductor strain gauge depends on the piezoresistive effects of silicon and measures the change in the resistance with stress as opposed to strain. In order to measure strain with a bonded resistance strain gauge, it must be connected to an electric circuit that is capable of measuring small changes in resistance corresponding to strain.

A *Wheatstone bridge* is a circuit used to measure static or dynamic electrical resistance. The output voltage of the Wheatstone bridge is expressed in millivolts output per volt input. The Wheatstone circuit is also well suited for temperature compensation. Strain gauge transducers usually employ four strain-gauge elements electrically connected to form a Wheatstone bridge circuit as shown in Figure 11.13. Here, the output voltage corresponds to:

$$V_o = V_i\left[\frac{R_3}{R_3 + R_g} - \frac{R_2}{R_1 + R_2}\right] \quad \textbf{(11.16)}$$

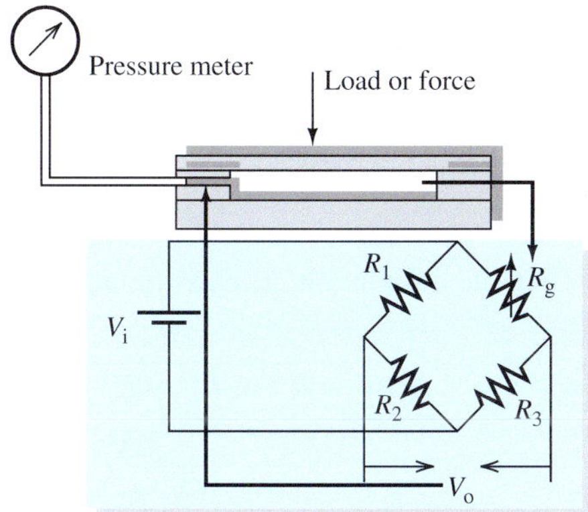

FIGURE 11.13 Strain-gauge load cell using a Wheatstone bridge.

If R_1, R_2, R_3, and R_g are equal, V_o will be zero. However, if R_g is changed to some value that does not equal R_1, R_2, and R_3, the bridge will become unbalanced and V_o will be greater than zero. The sensor may occupy one, two, or four arms of the bridge, depending on the application. The voltage, V_o, is equivalent to the difference between the voltage drop across R_1 and R_g.

In the field of measurement science, there are two ways to measure quantities: the "mass balance" (also called "null system") and the "bathroom or fish scale" methods. Like the mass balance scale, the Wheatstone bridge compares an unknown resistance against a known value until both are equal. On the other hand, a standard ohmmeter is an electronic device that converts resistance to current. The accuracy of a standard ohmmeter is dependent on a battery just like the accuracy of the bathroom or fish scale is dependent on the gravity constant.

The Wheatstone bridge circuit is almost universally used in load cells and other strain-gauge sensors because it facilitates the cancellation of unwanted temperature effects. Most strain-gauge materials are sensitive to temperature variations and their resistance tends to change resistance as they age.

Through proper flexure design and gauge placement, a linear relationship between the applied force and the sensed strain is achievable. Therefore, Equation (11.16) can be expressed by:

$$V_o = V_i \, K \, F \qquad \textbf{(11.17)}$$

where K is the calibration factor and F is the input force.

EXERCISE 11.2

Verify Equation (11.16) for the Wheatstone bridge of Figure 11.13. Assuming, $R_1 = 1\ \text{k}\Omega$, $R_2 = 2\ \text{k}\Omega$, $R_3 = 1.5\ \text{k}\Omega$, and $R_g = 2\ \text{k}\Omega$, $V_i = 5$ V, find the voltage, V_o.

EXAMPLE 11.8 Applications of Load Cells

Describe some applications of load cells.

SOLUTION

Some applications of load cells include the following:

1. ***Weighing of cargo:*** Load cells can be used to weigh trucks, tanks, or vessels.
2. ***Tension measurement:*** Load cells can be used to measure underwater tension.

EXAMPLE 11.9 Strain-Gauge Load Cell

For a strain-gauge load cell, the relationship between the output voltage and the mass weighed is:

$$M = 3.5V_o \text{ kg}$$

(continued)

EXAMPLE 11.9 Continued

The load cell has the same structure as shown in Figure 11.13. $V_i = 5$ V, and $R_1 = R_2 = R_3 = 100\ \Omega$. If $R_g = 80\ \Omega$, what is the mass?

SOLUTION

According to Equation (11.16), the output voltage corresponds to:

$$\begin{aligned} V_o &= V_i\left[\frac{R_3}{R_3 + R_g} - \frac{R_2}{R_1 + R_2}\right] \\ &= 5\left[\frac{100}{100 + 80} - \frac{100}{100 + 100}\right] \\ &= \frac{5}{18} \\ &\approx 0.278\ \text{V} \end{aligned}$$

According to the relationship between the output voltage and the mass, which is given in this example:

$$\begin{aligned} M &= 3.5\ V_o\ \text{kg} \\ &= 3.5 \times 0.278\ \text{kg} \\ &= 0.973\ \text{kg} \end{aligned}$$

11.2.5 Acoustic Sensors

A sound sensor is a device that converts acoustic energy into electrical energy. The frequency range of sound is extremely wide. Sound propagates in various media, such as solids, fluids, or gases. Accordingly, a wide range of sound sensors have been designed for different environments. This section concentrates on sound sensors, such as microphones, which are utilized for the detection of airborne sound in the audio-frequency range.

Generally, microphones consist of a diaphragm that vibrates by impinging waves of acoustic pressure. The motion of this diaphragm is converted into alternating electromotive forces or electric currents by a transducing mechanism. Piezoresistive microphones, capacitive microphones, and piezoelectric microphones are the most common types of microphones.

A typical capacitive microphone consists of a diaphragm, a back plate, a polarization voltage, resistors, and capacitors as depicted in Figure 11.14. The equivalent circuit of a capacitive microphone is shown in Figure 11.15. The diaphragm is placed close to the rigid metal back plate. The parallel diaphragm and back plate form a parallel plate capacitor. The sound pressure fluctuations alter the capacitor plate space and results in the capacitance changes [$\Delta C(t)$]. The output voltage across the capacitor fluctuates with change in capacitance. In Figure 11.14, the polarization voltage, E_0, is applied to generate an initial charge, Q_0, on the capacitors. If

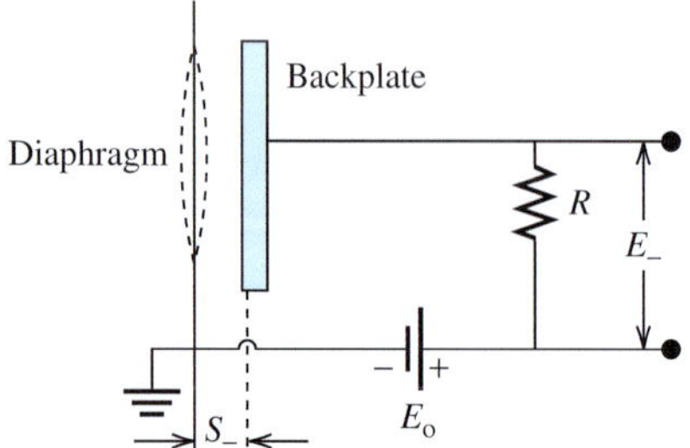

FIGURE 11.14 The basic principle of a capacitor microphone.

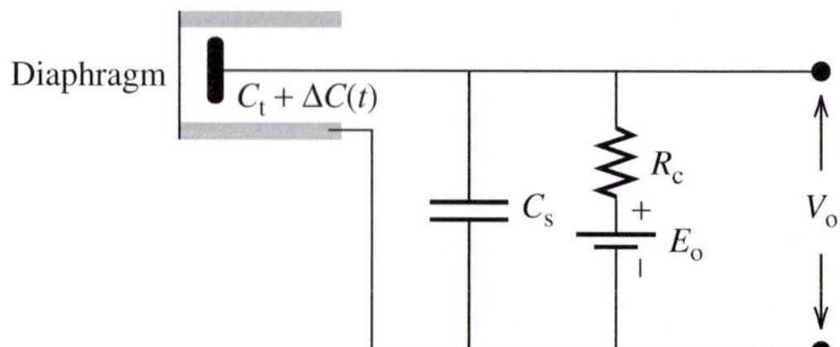

FIGURE 11.15 Equivalent circuit of a capacitive microphone.

charge is generated by the capacitor itself, sensors using such capacitors are often called electret microphones (a typical condenser microphone).

Now, considering Figure 11.15, the total capacitance, C_o, of the microphone corresponds to:

$$C_o = C_t + C_s \tag{11.18}$$

In Figure 11.15, C_s represents the stray capacitance in the sensor, and C_t represents the capacitance of the diaphragm. If C_t is constant, the capacitor will be equivalent to an open circuit. In this case, the voltage across the capacitor C_o is:

$$V_o = E_o \tag{11.19}$$

and the charge stored on the capacitor (C_o) is:

$$q_c = C_o E_o \tag{11.20}$$

The resistance, R_c, is large to ensure that the time constant, $R_c C_o$, is long compared to the lowest sound frequency to be measured. Accordingly, the charge stored in the capacitor will approximately be constant. If the charge is kept constant, a change in capacitance will lead to a change in voltage. When the capacitance varies, the following equation is used:

$$q_c = (C_o + \Delta C)V_c \tag{11.21}$$

Here, the voltage variation corresponds to:

$$\Delta V_o = V_o - V_c \tag{11.22}$$

Using Equations (11.19) to (11.22):

$$\begin{aligned} \Delta V_o &= V_o - V_c \\ &= E_o - \frac{C_o E_o}{C_o + \Delta C} \\ &= \frac{\Delta C}{C_o + \Delta C} E_o \end{aligned} \tag{11.23}$$

Now, because $C_o >> \Delta C$, Equation (11.23) is simplified to:

$$\Delta V_o \approx \frac{\Delta C}{C_o} E_o \tag{11.24}$$

Thus, as the capacitance of C_t changes, V_o will change. To ensure that the deflection of the diaphragm for a given sound pressure is independent of frequency and to achieve proportionality between the sound pressure and the output voltage, a stiffness-controlled operation of the membrane is required. Therefore, the fundamental resonance frequency of the membrane is placed at the upper limit of the operating frequency range.

EXAMPLE 11.10 Applications of Acoustic Sensors

Describe some examples of real-world applications of acoustic sensors.

SOLUTION

Some applications of acoustic sensors include the following:

1. ***Avalanche warning:*** By measuring the wind or air pressure, acoustic sensors can be used for avalanche warning and research.
2. ***Oceanographic data collection:*** Acoustic sensors are placed on the seafloor. Ranging measurements can be achieved by acoustic sensors that transmit and receive an acoustic signal from a near-surface projector and by noting timing and/or position measurements. These sensors help offshore exploration, reduce the impact of tsunami, aid in navigation, etc.

EXAMPLE 11.11 Equivalent Impedance and Resistance of Capacitive Microphone

Find the equivalent impedance of a capacitive microphone shown in Figure 11.15. Next, approximate the equivalent resistance of a capacitive microphone.

SOLUTION

The equivalent impedance is:

$$Z = \frac{\dfrac{R_c}{j\omega C_o}}{R_c + \dfrac{1}{j\omega C_o}}$$

$$= \frac{R_c}{1 + j\omega R_c C_o}$$

$$= \frac{R_c(1 - j\omega R_c C_o)}{1 + (\omega R_c C_o)^2}$$

$$= \frac{R_c}{1 + (\omega R_c C_o)^2} - j\frac{\omega R_c^2 C_o}{1 + (\omega R_c C_o)^2}$$

Because the time constant $R_c C_o$ is large we can take $\omega R_c C_o >> 1$:

$$Z \simeq \frac{1}{\omega^2 R_c C_o^2} + \frac{1}{j\omega C_o} = R_T + \frac{1}{j\omega C_o}$$

Thus, the equivalent resistance of a microphone is:

$$R_T = \frac{1}{\omega^2 R_c C_o^2}$$

EXERCISE 11.3

Replace the capacitance in Z in Figure 11.15 with a 10-mH inductance and repeat Example 11.11.

APPLICATION EXAMPLE 11.12 Microphone Impedance

Most microphones are designed while their impedance is not "matched" to the load to which they are connected. Impedance matching has been discussed in detail in Chapter 6. Impedance unmatch alters their frequency response and leads to a distortion effect (hum effect), especially at high sound pressure levels.

An engineer intends to test a new microphone and see whether it produces sound that is comfortable for listening. Based on the impedance calculated in Example 11.11, the circuit is shown in Figure 11.16. In this figure, 5 cos(3000t) is the output voltage of the microphone,

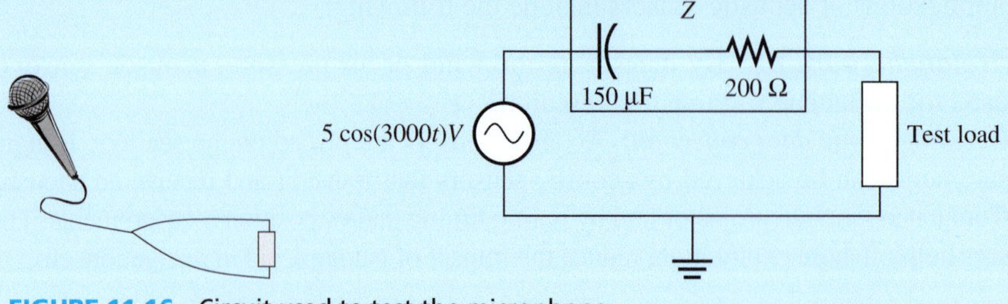

FIGURE 11.16 Circuit used to test the microphone.

and Z is its equivalent impedance. Determine the load at which the engineer needs to test the microphone to establish a "worst-case" sample.

SOLUTION

To test the relative listening comfort of the microphone's hum, the load of the microphone must match its impedance. Based on the concept of impedance matching studied in Chapter 6:

$$Z_L = Z^* = 200 - \frac{1}{j3000 \times 150 \times 10^{-6}} = 200 + j2.22\ \Omega$$

Because the imaginary part of Z_L is positive, it must be inductive. Therefore, Z_L consists of a resistor and an inductor in series, which is shown in Figure 11.17.

$$R_L = 200\ \Omega$$

$$L = \frac{2.22}{3000} = 0.74\ \text{mH}$$

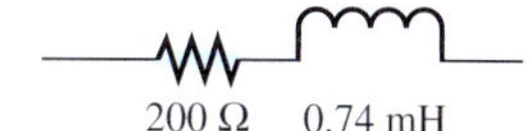

FIGURE 11.17 Equivalent circuit for the load.

11.2.6 Linear Variable Differential Transformers (LVDT)

The linear variable differential transformer (LVDT) is a displacement sensor, which consists of a transformer with a single primary coil and two identical secondary coils connected in a series-opposing manner, as depicted in Figure 11.18.

An LVDT is operated based on the mutual inductance concept that is discussed in more detail in Chapter 12. The primary coil is fed from an AC excitation supply and induces voltage across the two secondary coils. The output voltage is given by:

$$v_o = v_1 - v_2 \tag{11.25}$$

The iron core between the primary and secondary coils can be displaced by an external motion. This changes the magnetic coupling between the primary and secondary coils. The induced voltages have equal magnitudes when the iron core is in the central position, and the voltage output $v_o \sim 0$. An upward displacement of the iron core from the central position increases the coupling (mutual inductance) between the primary and the top secondary coil, while decreasing the coupling for the other secondary coil, and correspondingly induces a greater voltage in the top secondary coil.

As a result, $v_o > 0$ when the iron core has an upward displacement, and $v_o < 0$ when the iron core is displaced downward. If the primary coil has the resistance, R_p, and the self-inductance, L_p, then:

$$R_p \cdot i + L_p \cdot \frac{di}{dt} = v_s \tag{11.26}$$

Note that at the left side of Equation (11.26), the total voltage induced from secondary coils is zero. This is because the current directions in secondary coils are opposite. Thus, their

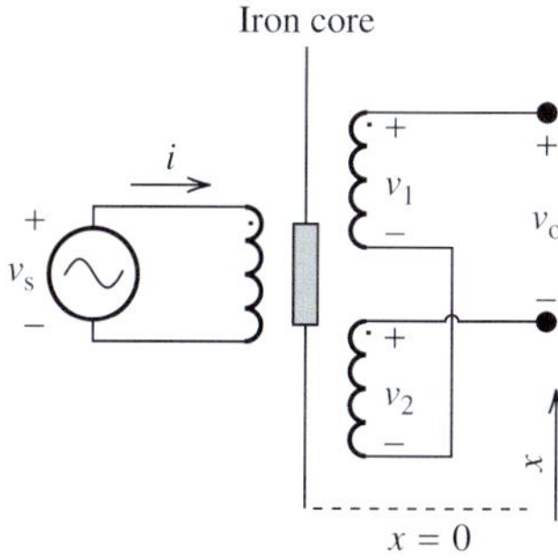

FIGURE 11.18 Linear variable differential transformer.

overall effect on the induced voltages at the primary coil would be zero. Then, the voltages induced in the secondary coils correspond to:

$$v_1 = M_1 \cdot \frac{di}{dt} \tag{11.27}$$

and

$$v_2 = M_2 \cdot \frac{di}{dt} \tag{11.28}$$

Thus:

$$v_o = (M_1 - M_2) \cdot \frac{di}{dt} \tag{11.29}$$

where M_1 and M_2 are the mutual inductances between the primary and the respective secondary coils. The position of the iron core determines M_1 and M_2. When the core is at the central position, $M_1 = M_2$, and $v_o = 0$; when the core is displaced upward away from the central position, $M_1 > M_2$, and $v_o > 0$; when the core is displaced downward away from the central position, $M_1 < M_2$, and $v_o < 0$.

Because an LVDT is usually excited by an AC signal, the current, i, can be represented by:

$$i = \cos(\omega t - \phi) \tag{11.30}$$

and v_1, v_2, and v_o correspond to:

$$v_1 = M_1 \sin(\omega t - \phi), \tag{11.31}$$

$$v_2 = M_2 \sin(\omega t - \phi), \tag{11.32}$$

$$v_o = v_1 - v_2 = (M_1 - M_2) \sin(\omega t - \phi) \tag{11.33}$$

The magnitude of the output voltage, v_o, will be the same if the magnitude of displacements $+x$ or $-x$ of the iron core away from the central position is equal. The directional information of the movement of the iron core is available in the phase of the output voltage: the movement in the two different directions differs by 180°. As a result, the amplitude and the phase of the output voltage, v_o, depends on the displacement, x.

In other words, the amplitude of v_o is modulated by the displacement. The relationship between the magnitude of the output voltage and the displacement, x, of the iron core is approximately linear over a reasonable range of movement of the iron core on either side of the central position. It is expressed by a constant of proportionality, G, as:

$$x = G \cdot v_o \tag{11.34}$$

where G is called the sensitivity (or gain) of the transformer.

EXAMPLE 11.13 Applications of LVDT

Describe some applications of LVDT.

SOLUTION

Some applications of LVDT include the following:

1. ***Weighing systems:*** LVDTs can be used to measure spring deformation in weighing systems. The measured displacement can be used to calculate the applied force based on the characteristics of the spring.

2. ***Displacement sensing:*** An LVDT may be mounted externally in parallel with the cylinder, thus making the core assembly free to move with the piston rod, and allowing measurement of the displacement of a piston.
3. ***Bill detector in ATMs:*** Using an LVDT, ATMs can detect the number of bills inserted. When bills pass, a motion signal is transferred to the LVDT core element and then a changing output signal is generated. As the bills pass between the rollers, the voltages vary according to the thickness of the bills.

EXAMPLE 11.14 Displacement in LVDT

For an LVDT shown in Figure 11.19, an engineer can obtain the displacement by measuring the voltage output according to Equation (11.34). Assume $G = 0.7282$ mm/V. Determine the displacement if the measured instant voltage $v_o = 5$ V.

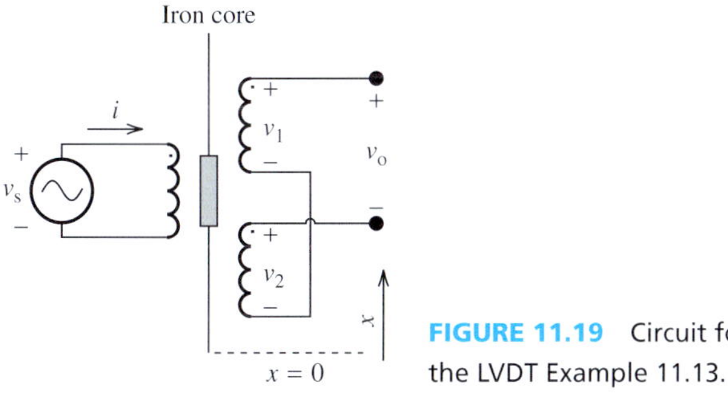

FIGURE 11.19 Circuit for the LVDT Example 11.13.

SOLUTION

Using Equation (11.34) and replacing for G and v_o:

$$x = G \times v_o = 0.7282 \times 5 = 3.641 \text{ mm}$$

11.3 SIGNAL CONDITIONING

The direct output of a sensor is generally not available in a form that computers can process. Signal conditioning is required to convert sensor output to an appropriate form. The two most important signal conditioning functions are *amplification* and *filtering*. If necessary, signal conditioning can also convert currents to voltages, or supply excitations to the sensors to convert a change in resistance, inductance, or capacitance to a change in voltage. This section describes amplifiers and filters.

11.3.1 Amplifiers

An amplifier is a device that increases the power contained within a signal. Often a low-level signal from a sensor needs to be amplified. Operational amplifier circuits have been discussed in Chapter 8.

11.3.2 Active Filters

Sensor signals may encounter interference from noise or undesired input. Therefore, filters are required to eliminate interference and noise effects. Active filters consist of resistors, capacitors, and amplifiers, whereas passive filters consist of resistors, inductors, and capacitors. Chapter 7

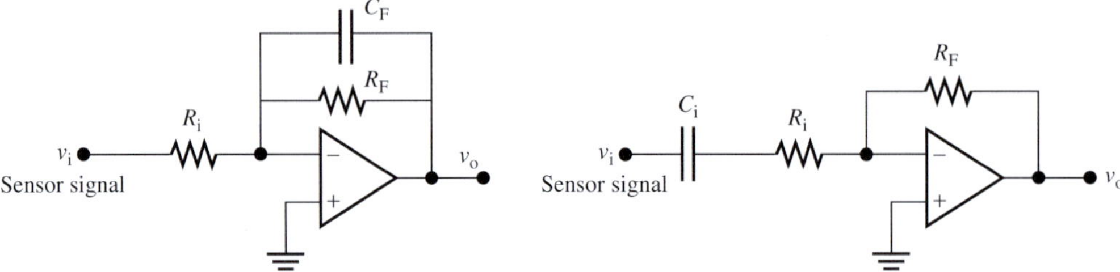

FIGURE 11.20 First-order, low-pass filter with an operational amplifier.

FIGURE 11.21 First-order, high-pass filter with an operational amplifier.

reviewed different types of passive filters in detail, including low-pass, high-pass, and band-pass filters.

Compared to active filters, passive filters have some disadvantages. First, inductors selected for the audio-frequency range, in which most mechanical measurements are made, are large and expensive. Second, passive filters restrict the sharpness of the cutoff frequency because they do not have power amplification, leading to loading in measurements. Finally, passive filters are limited to frequency selection.

Most active filters are built using operational amplifier circuits and different characteristics are defined by amplified input and feedback components that consist of resistors and capacitors. Some advanced active filters, such as Butterworth filters and Chebyshev filters, have widespread applications in instrumentation circuits.

Two examples of active filters are shown in Figures 11.20 and 11.21. A first-order, low-pass filter is shown in Figure 11.20.

According to Equation (8.54), the gain of an inverting amplifier corresponds to:

$$A_L(j\omega) = \frac{V_o}{V_i} \tag{11.35}$$

Based on Equation (8.55):

$$\frac{V_o}{V_i} = -\frac{Z_F}{Z_i} \tag{11.36}$$

where Z_F is the impedance of a paralleled capacitor, C_F, and a resistor, R_F, that is:

$$Z_F = \frac{1}{j\omega C_F} \parallel R_E = \frac{R_F}{1 + j\omega R_F C_F} \tag{11.37}$$

and

$$Z_i = R_i \tag{11.38}$$

Therefore, the gain of the low-pass, active filter is:

$$A_L(j\omega) = -\frac{\frac{1}{j\omega C_F} \parallel R_F}{R_i} = -\frac{\frac{R_F}{1 + j\omega R_F C_F}}{R_i} = -\frac{R_F}{R_i(1 + j\omega R_F C_F)} \tag{11.39}$$

A first-order, high-pass filter is shown in Figure 11.21.

According to Equation (8.55), the gain for the inverting amplifier is:

$$A_H(j\omega) = \frac{V_o}{V_i} = -\frac{Z_F}{Z_i} \tag{11.40}$$

where

$$Z_F = R_F \tag{11.41}$$

and

$$Z_i = R_i + \frac{1}{j\omega C_i} \tag{11.42}$$

Therefore, the gain of the high-pass, active filter is:

$$A_H(j\omega) = -\frac{R_F}{R_i + \dfrac{1}{j\omega C_i}} = -\frac{j\omega R_F C_i}{1 + j\omega R_i C_i} \tag{11.43}$$

The amplitude and phase of the frequency dependent gains in Equations (11.32) and (11.43) can be sketched in terms of frequency using the Bode plot or PSpice as discussed in Chapter 7.

EXERCISE 11.4

Refer to Chapter 7, sketch the Bode plot of the amplitude of the gain in Equation (11.43), assuming $C_i = 1\ \mu F$ and resistor $R_F = R_i = 1\ k\Omega$. Investigate the impact of changing R_i to $100\ \Omega$ and $10\ k\Omega$.

APPLICATION EXAMPLE 11.15 Robotic Arm

An operational amplifier is used to design controllers for mechanical applications. Assume that the transfer function of a robotic arm is $G(s) = 1/(s^2 + 1)$. The robotic arm's transfer function is an unstable system; therefore, a proportional-integral-derivative (PID) controller is added as depicted in Figure 11.22.

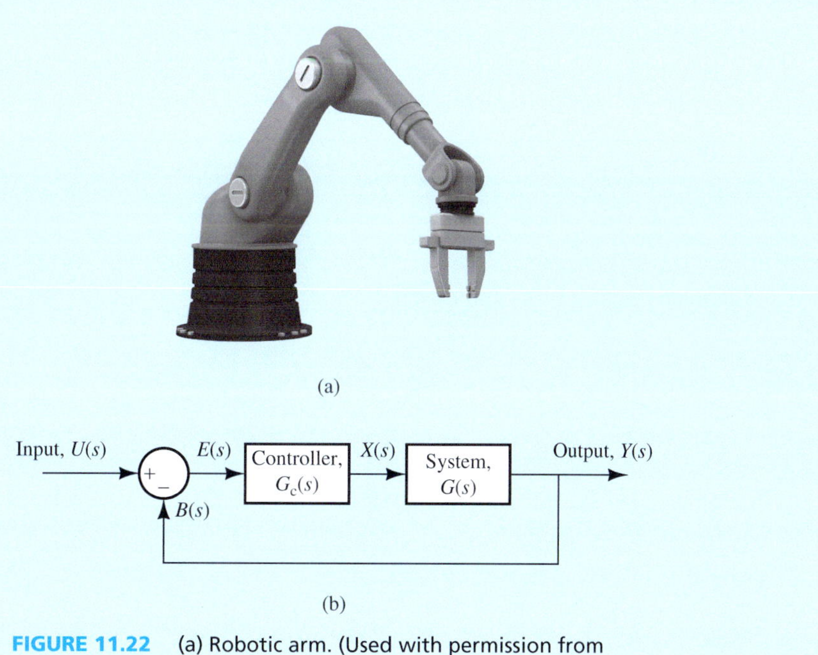

FIGURE 11.22 (a) Robotic arm. (Used with permission from Nuno Andre/Shutterstock.com.) (b) Block diagram of the robotic arm mechanical system.

(continued)

APPLICATION EXAMPLE 11.15 Continued

a. Given the circuit of the PID controller shown in Figure 11.23, determine the transfer function of the controller assuming $R_1 = 2000\ \Omega$, $C_1 = 875\ \mu F$, $R_2 = 2500\ \Omega$, $C_2 = 400\ \mu F$, $R_3 = 450\ \Omega$, $R_4 = 480\ \Omega$.

b. Plot the step response of a close-loop transfer function for the robotic arm:
   i. Without a PID controller
   ii. With a PID controller

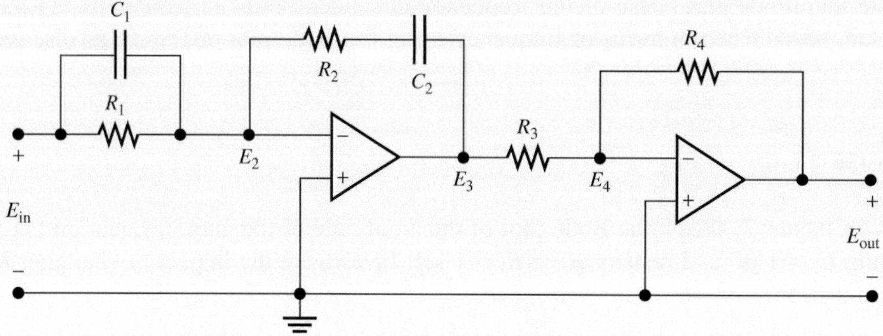

FIGURE 11.23 An operational amplifier circuit representing a PID controller.

SOLUTION

a. First, find the expression of the transfer function of the PID controller in terms of resistance and capacitance.
Using $s = j\omega$, the impedance of the capacitor can be expressed as:

$$Z_c = \frac{1}{j\omega c} = \frac{1}{CS}$$

The impedance Z_1 of R_1 and C_1 is:

$$Z_1 = \left(\frac{1}{R_1} + C_1 S\right)^{-1}$$

$$Z_1 = \frac{R_1}{R_1 C_1 s + 1}$$

The impedance Z_2 of R_2 and C_2 is:

$$Z_2 = R_2 + \frac{1}{C_2 s}$$

$$Z_2 = \frac{R_2 C_2 s + 1}{C_2 s}$$

Using the golden rules of op-amp, $E_2 = 0$ and:

$$\frac{E_{in} - E_2}{Z_1} = \frac{E_2 - E_3}{Z_2}$$

Thus,

$$\frac{E_{in}}{Z_1} = -\frac{E_3}{Z_2}$$

or

$$\frac{E_3}{E_{in}} = -\frac{Z_2}{Z_1}$$

Applying the same method to the second op-amp with $E_4 = 0$ shows:

$$\frac{E_3 - E_4}{R_3} = \frac{E_4 - E_{out}}{R_4}$$
$$\frac{E_3}{R_3} = -\frac{E_{out}}{R_4}$$
$$\frac{E_{out}}{E_3} = -\frac{R_4}{R_3}$$

Thus, the transfer function of the PID controller is:

$$\frac{E_{out}}{E_{in}} = \frac{E_{out}}{E_3} \times \frac{E_3}{E_{in}} = \frac{R_4}{R_3} \bullet \frac{Z_2}{Z_1}$$

or: $$\frac{E_{out}}{E_{in}} = \frac{R_4 R_2}{R_3 R_1} \frac{(R_1 C_1 s + 1)(R_2 C_2 s + 1)}{R_2 C_2 s}$$

Substituting the resistance and capacitance results in:

$$G_c = \frac{2.333(s + 1)(s + 0.5714)}{s}$$

b. Define the $E(s)$, $X(s)$, and $B(s)$ as shown in Figure 11.22 with $U(s)$ as the input and $Y(s)$ as the output of the system.
From Figure 11.22, $E(s) = U(s) - B(s)$. In addition:

$$\frac{X(s)}{E(s)} = G_c(s)$$
$$\frac{Y(s)}{X(s)} = G(s)$$
$$\frac{Y(s)}{E(s)} = \frac{Y(s)}{X(s)} \times \frac{X(s)}{E(s)} = G(s)G_c(s)$$

Using this equation and representing $E(s)$ in terms of $Y(s)$ in $E(s) = U(s) - B(s)$, and knowing that $B(s) = Y(s)$ shows:

$$\frac{Y(s)}{U(s)} = \frac{G(s)G_c(s)}{1 + G(s)G_c(s)}$$

i. Without a PID controller:

For a system without a PID controller, let the $G_c(s) = 1$. Then:

$$\frac{Y(s)}{U(s)} = \frac{G(s)}{1 + G(s)}$$

where $G(s) = 1/(s^2 + 1)$. To set up the transfer function, step response, or a Bode plot in MATLAB, first determine the vector for the numerator and denominator of the transfer function. The numerator of the $G(s)$ is 1; thus, write num = [1]. The denominator of $G(s) = s^2 + 1 = (1)s^2 + (0)s + 1$; thus, write den = [1 0 1].

Now, see that the numerator and denominator vectors can be defined in terms of the highest to the lowest order of the coefficient of "*s*." Use MATLAB to define the transfer function using "tf(num,den)." Next, plot the step response of the system using the function "step(G,T)." T is the total sampling time. The complete

(*continued*)

APPLICATION EXAMPLE 11.15 Continued

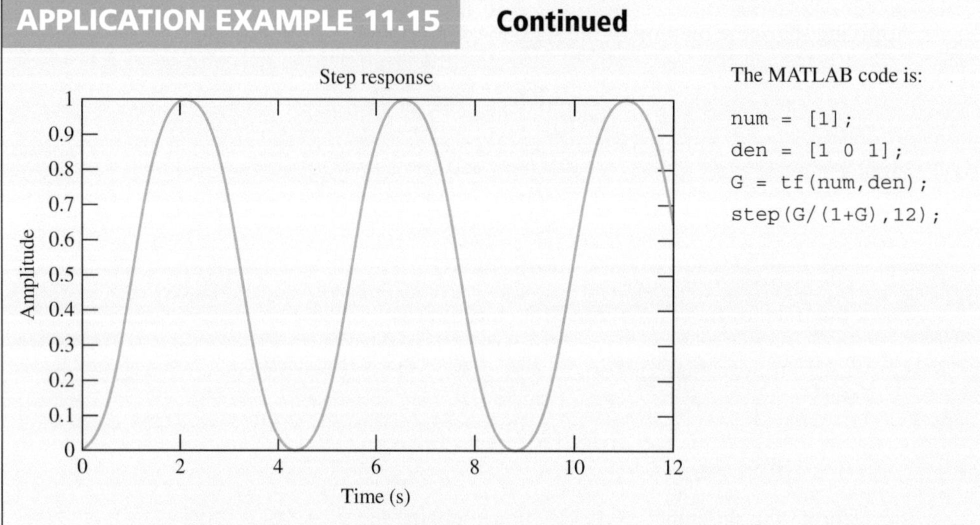

FIGURE 11.24 Step response for robotic arms.

MATLAB code is shown in Figure 11.24. Here, be aware that "step(G,T)" represents the open-loop step response. However, the system here is a closed-loop function, therefore it is "step(G/(1+G),T)."

ii. With a PID controller:

The closed-loop transfer function for the system with a PID controller is:

$$\frac{Y(s)}{U(s)} = \frac{G(s)G_c(s)}{1 + G(s)G_c(s)}$$

The given transfer function for the controller is:

$$G_c = \frac{2.333(s + 1)(s + 0.5714)}{s}$$

$$G_c = \frac{2.333(s^2 + 1.5714s + 0.5714)}{s}$$

Using this, define the transfer function for the controller in MATLAB, with the numerator and denominator of numc = 2.333*[1 1.5714 .5714], and denc = [1 0], respectively. The step response of the system with the PID controller is shown in Figure 11.25. Comparison of Figures 11.24 and 11.25 shows that a controller is required to let the robotic arm achieve stability during the operation.

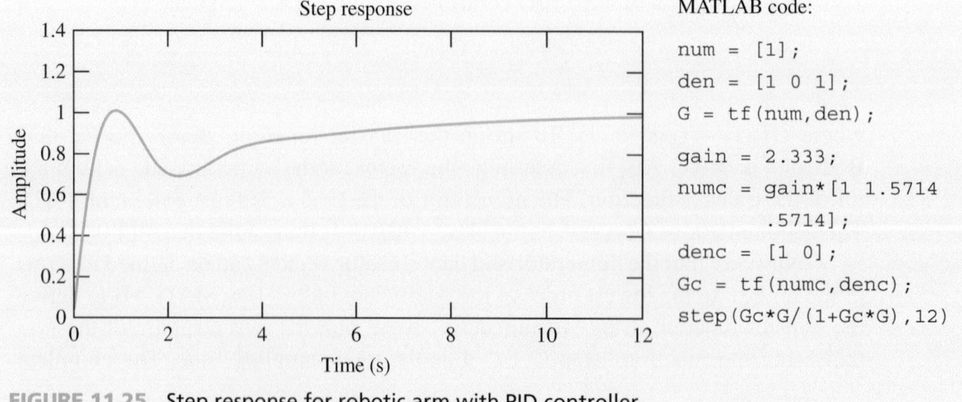

FIGURE 11.25 Step response for robotic arm with PID controller.

11.4 DATA ACQUISITION

A data acquisition system consists of an analog multiplexer (MUX), a buffer, a sample/hold amplifier, and an ADC as shown in Figure 11.26. As discussed in Chapter 8, a buffer made by operational amplifiers reduces the impact that two parts of a system may have on each other. A buffer has a very large (ideally infinity) input impedance and a small (ideally zero) output impedance.

11.4.1 Analog Multiplexer

A multiplexer (MUX) in a data acquisition system processes several different signals. In Figure 11.26, the channel to be monitored is selected and applied to a sampler. The sampled output is converted to digital using the multiplexer that is controlled by the control circuit, which simultaneously triggers the sampling circuit and the ADC. The multiplexer selects the next channel after the ADC completes the conversion and sends an appropriate signal to the control circuit.

11.4.2 Analog-to-Digital Conversion

Analog-to-digital converters (ADCs) are designed to transform continuous analog signals into a discrete binary code suitable for digital processing. There are two main steps in analog-to-digital conversion: *sampling* and *quantization*. First, the analog signal is sampled at regular time intervals to generate a discrete-time signal. Second, each sample of the analog voltage is converted into an equivalent digital value.

11.4.2.1 SAMPLING

The process of sampling is illustrated in Figure 11.27.

To be accurate, sampling must be conducted at a rate higher than twice the highest frequency of the analog signal. This rate is called the **Nyquist rate**. If the rate of sampling is less than the **Nyquist rate**, the reconstruction of the sample value cannot match the original analog signal. In this case, **aliasing** has occurred. Here, sampling may transmute a high-frequency signal into a lower frequency one, as illustrated in Figure 11.28. Thus, to avoid aliasing, the sampling rate, f_s, should be at least twice the highest frequency, f_H, in the sampled analog signal. That is, the Nyquist criterion corresponds to:

$$f_s \geq 2f_H \tag{11.44}$$

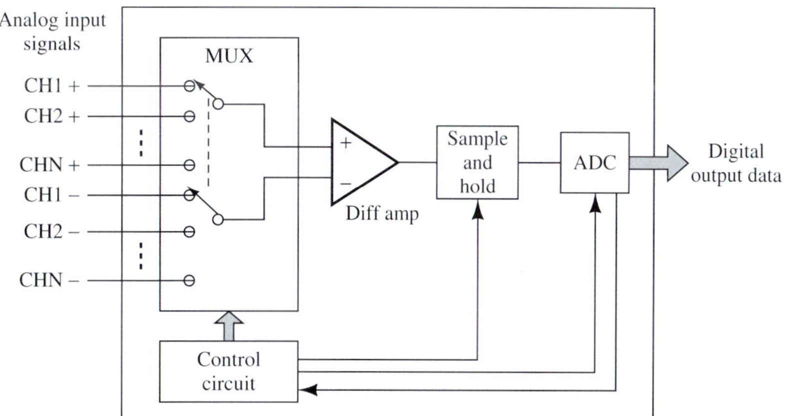

FIGURE 11.26 Data acquisition system.

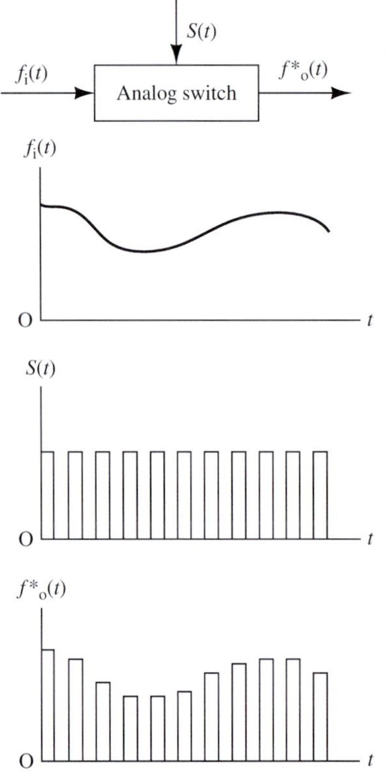

FIGURE 11.27 Analog signal sampling.

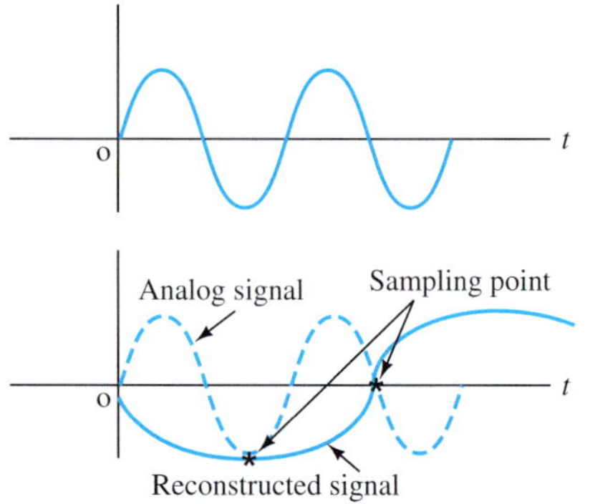

FIGURE 11.28 Conversion of an analog signal to a discrete signal if the sampling rate is not high enough.

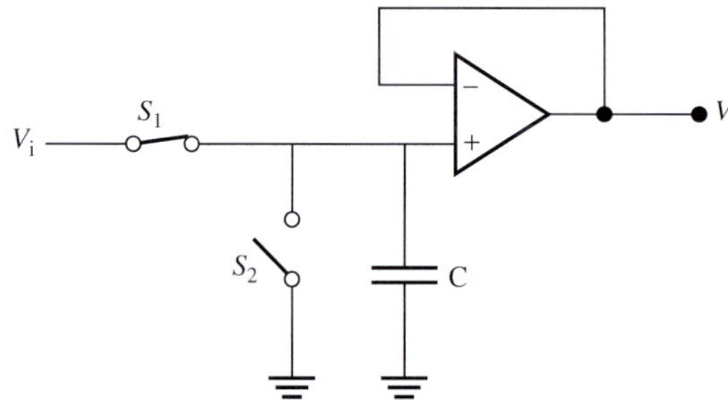

FIGURE 11.29 A simple sample and hold circuit.

EXAMPLE 11.16 **Sampling Rate**

A signal has frequency components up to 60 kHz. What is the minimum sampling rate that should be used to ensure that the signal is reconstructed properly?

SOLUTION

Based on the Nyquist rate criterion:

$$f_s = 2 \times f_H = 2 \times 60 \text{ kHz} = 120 \text{ kHz}$$

11.4.2.2 SAMPLE AND HOLD

The ADC requires a certain amount of time to perform the conversion, called the conversion time. If the analog input varies within the conversion time, a conversion error may occur. To hold the input signal at a constant level within the conversion time, a sample and hold circuit is required before applying the signal to the ADC.

A typical operational amplifier circuit that performs the sample and hold function is shown in Figure 11.29. When switch S_1 is closed (S_2 is open), the capacitor, C, is charged by the input sample signal. The output, V_o, changes with the voltage of capacitor, C. When the sampled signal pulse is over (the voltage is zero), the capacitor, C, holds the sampled signal voltage until the next sampled signal pulse, and the output, V_o, keeps the same voltage. Switch S_2 is used to short-circuit the capacitor, discharging it and setting the output to 0 V, if necessary.

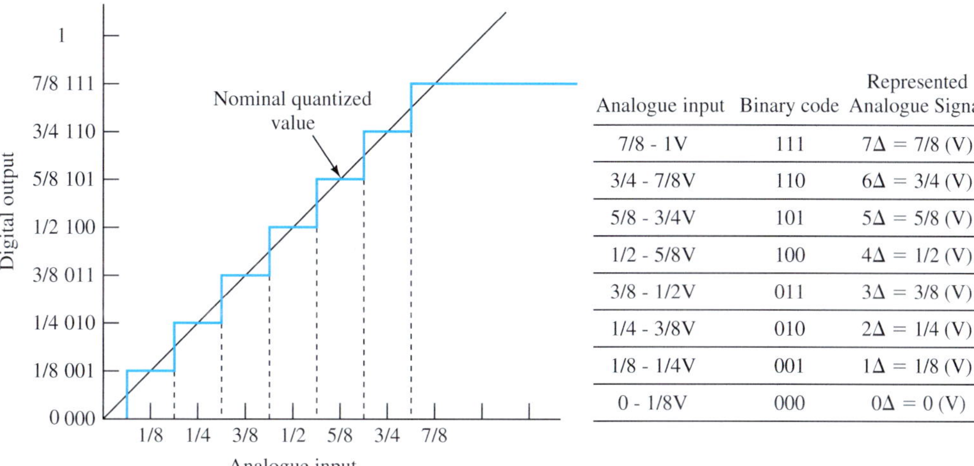

Analogue input	Binary code	Represented Analogue Signal
7/8 - 1V	111	7Δ = 7/8 (V)
3/4 - 7/8V	110	6Δ = 3/4 (V)
5/8 - 3/4V	101	5Δ = 5/8 (V)
1/2 - 5/8V	100	4Δ = 1/2 (V)
3/8 - 1/2V	011	3Δ = 3/8 (V)
1/4 - 3/8V	010	2Δ = 1/4 (V)
1/8 - 1/4V	001	1Δ = 1/8 (V)
0 - 1/8V	000	0Δ = 0 (V)

FIGURE 11.30 A/D conversion.

11.4.2.3 QUANTIZATION

Quantization describes the procedure whereby the continuous analog signal is quantized and encoded in the binary format. First, the range of the analog input signal is divided into a finite number $(2^N - 1)$ of intervals, where N is the number of bits available for the corresponding binary word. Second, a binary word is assigned to each interval; the binary word is the digital representation of any voltage (current) that falls within that interval. This is illustrated in Figure 11.30.

As shown in Figure 11.30, for eight levels of quantization, only three bits are needed. As the number of quantization levels increases, the number of bits required to represent those levels increases as well. For example, for a 16-level quantizer, four bits are needed to represent those levels and for a 32-level quantizer, five bits are needed to represent all levels. In general, for an L level quantizer, the number of required bits corresponds to:

$$N = \log_2 L \tag{11.45}$$

At any particular value of the analog signal, the digital representation is either the discrete level immediately above or below this value. If the difference between two successive discrete levels is represented by the parameter Δ, then, in Figure 11.30, the maximal quantization error is Δ, that is, $V/8$. This error is known as the quantization error and is proportional to the resolution of the analog-to-digital converter, that is, the number of bits used to represent the samples in digital form.

In the quantization process, all infinite amplitude levels of analog signals are mapped into certain predetermined levels. Obviously, the reconstructed signal will not be exactly the same as the original one, because the quantization will only represent certain levels of amplitude. As smaller number of intervals is selected, a larger length binary code is created at output and the digital code more accurately represents the original signal. In other words, quantization error decreases as the number of bits per sample increases. The number of bits controls the resolution of the ADC.

Resolution, Δ, is defined as the maximum analog voltage range divided by the number of quantization levels, $L = 2^N$, where N is the number of bits [see Equation (11.45)]. In Figure 11.30, the resolution is $V/8$. If the analog voltage range is ± 5 V, an 8-bit ADC's resolution would be $\Delta = 10/2^8 = 39.0625$ mV.

Therefore, the resolution of a computer-based measurement system is limited by the word length of the ADC. The effect of finite word length can be modeled by adding quantization noise to the reconstructed signal. It can be shown that the root mean square (rms) of the quantization noise is:

$$\sigma_{\text{qrms}} = \frac{\Delta}{2\sqrt{3}} \tag{11.46}$$

where Δ is the quantization zone or resolution.

EXAMPLE 11.17 Quantization Error

A 10-bit ADC is designed to accept signals ranging from −5 to 5 V. Determine the width of each quantization zone (the ADC's resolution), and the rms of quantization error.

SOLUTION

The converter has 2^{10} zones = 1024 zones. Total range covered by the ADC = 5 − (−5) = 10 V. Therefore, the width of each quantization (the ADC's resolution) is:

$$\Delta = \frac{10}{1024} = 9.8 \text{ mV}$$

The corresponding quantization rms error is:

$$\sigma_{\text{qrms}} = \frac{9.8}{2\sqrt{3}} = 2.83 \text{ mV}$$

EXAMPLE 11.18 Signal-to-Noise Ratio Created by Quantization

A 5-V_{peak} sinusoid is converted by a 10-bit ADC designed to accept a signal from −4 V to 4 V. Assuming quantization is the only source of error (noise), find the signal-to-noise ratio (SNR), expressed in decibels.

SOLUTION

The signal power corresponds to [see Equation (6.1)]:

$$P = \frac{V^2}{R} = \frac{\left(\frac{5}{\sqrt{2}}\right)^2}{R} = \frac{12.5}{R} \text{ W } (R \text{ is arbitrary})$$

The width of each quantization level is:

$$\Delta = \frac{8}{2^{10}} = 7.8125 \text{ mV}$$

Using Equation (11.46), quantization noise corresponds to:

$$\sigma_{\text{qrms}} = \frac{\Delta}{2\sqrt{3}} = \frac{7.8125}{2 \times 1.732} = 2.255 \text{ mV}$$

Accordingly, the quantization noise power is:

$$P_{\text{q}} = \frac{(2.255 \text{ mV})^2}{R} = \frac{5.085}{R} \,\mu\text{W}$$

Finally:

$$\text{SNR}_{\text{dB}} = 10 \log \left(\frac{\frac{12.5}{R} \text{ W}}{\frac{5.085}{R} \,\mu\text{W}} \right) = 10 \log(2.458 \times 10^6) = 63.91 \text{dB}$$

11.5 GROUNDING ISSUES

11.5.1 Ground Loops

A ground is a point where the voltage does not change regardless of the amount of current supplied to it or drawn from it. Ground specifies a reference voltage for a circuit or system.

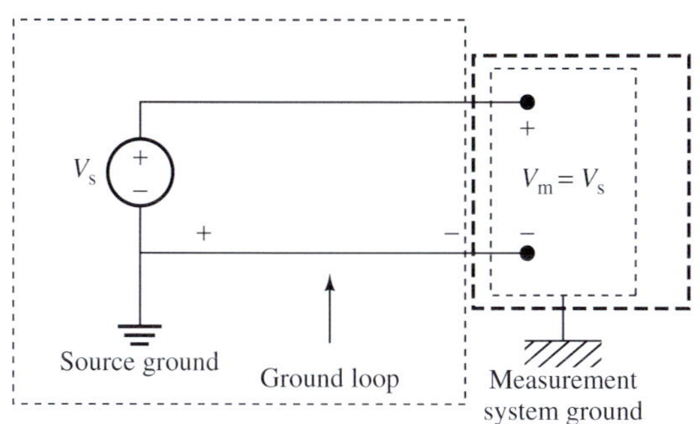

FIGURE 11.31 Ground loop.

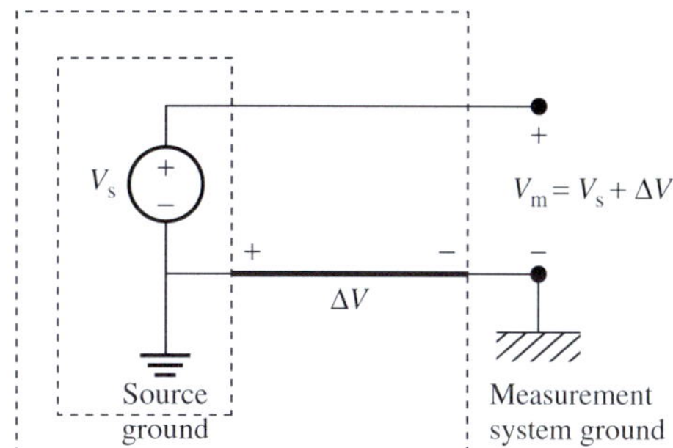

FIGURE 11.32 Differential measurement system.

For a measurement system, one important point in dealing with grounded signal sources is the *ground loop*. A ground loop is an undesired current path due to several reference voltages being connected together through different wires. A ground loop will not lead to any problem if these ground wires have zero impedance so that all of the ground points are at the same voltage. However, the resistance of some ground wires is nonzero, which changes the voltage between various ground points when current flows through them.

Figure 11.31 illustrates an example of a ground loop that is created across the measuring instrument and the instrument case grounds. Assuming that the resistance of the ground wires are not the same, a potential difference will be created across these two wires. When the measuring instrument and the instrument case grounds are connected together, a potential current will flow from one ground to the other through the small (but nonzero) resistance of the connecting wire. Thus, the voltage measured by the instrument will include the unpredictable voltage difference, ΔV. This leads to a measurement error.

As shown in Figure 11.32, to avoid ground loop in measurement instruments, the instrument case ground and the source ground are disconnected. Such a measurement system is called a differential measurement system. Proper grounding is also essential for safety, as discussed in Chapter 15.

APPLICATION EXAMPLE 11.19 Data Acquisition

Today, computers are widely used in research, control, testing, and measurement. Engineers can use computers to process the data collected from sensors. To obtain proper results from the rough data of sensors, using the knowledge gained through this chapter, construct a basic data acquisition system consisting of the system elements:

- Personal computer (software)
- Sensors
- Signal conditioning
- DAQ hardware

SOLUTION

Sensors are used to collect field data, and then data are applied to the *signal conditioning system* to filter or amplify the data, if necessary. To be readily accepted by a general-purpose computer, the data are then sent to the *data acquisition hardware* that periodically samples the signal and converts the sampled signals to digital words using an ADC. The digital words are then read by the *personal computer* for further processing. The system structure is shown in Figure 11.33.

(continued)

APPLICATION EXAMPLE 11.19 Continued

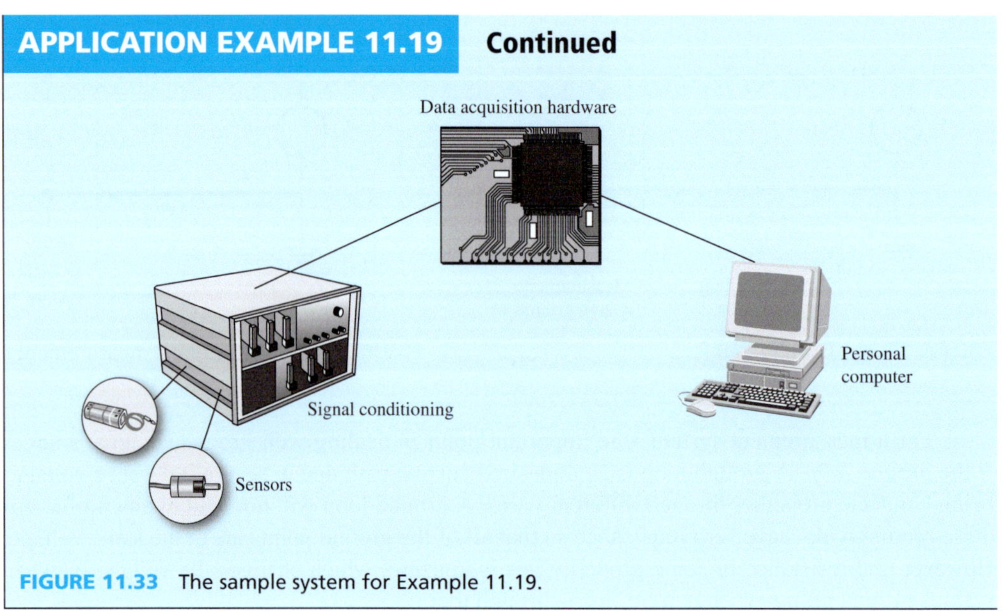

FIGURE 11.33 The sample system for Example 11.19.

11.6 USING PSPICE TO DEMONSTRATE A COMPUTER-BASED INSTRUMENT

This section introduces how to use PSpice for computer-based instrument analysis. Then, it addresses how to set up an op amp circuit, and run simulations to analyze relevant systems.

EXAMPLE 11.20 PSpice Analysis

For a strain-gauge load cell, the relationship between the output voltage and the mass weighed is $M = 3.5\ V_o$ kg. The load cell has the same structure as shown in Figure 11.13. $V_i = 5$ V, and $R_1 = R_2 = R_3 = 100\ \Omega$. Use PSpice to sketch the circuit and determine the mass if $R_g = 80\ \Omega$.

SOLUTION

The PSpice schematic solution is shown in Figure 11.34.

The output voltage, V_o, corresponds to:

$$V_o = 2.778 - 2.5$$
$$V_o = 0.278\ \text{V}$$

Thus, the mass of the load can be calculated, which is:

$$M = 3.5 \times V_o$$
$$M = 0.973\ \text{kg}$$

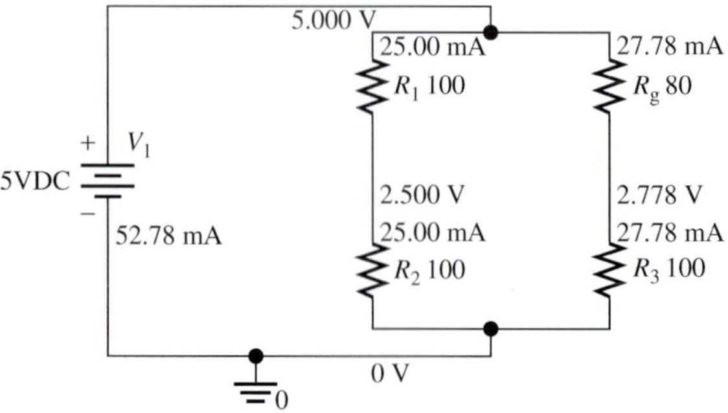

FIGURE 11.34 PSpice schematic solution for Example 11.20.

EXAMPLE 11.21 PSpice Analysis

Figure 11.35 shows an inverting operational amplifier. Consider the input voltage as 10 V, $R_i = 100\ \Omega$, $R_F = 200\ \Omega$, and the output load is 1000 Ω. Determine V_o using PSpice.

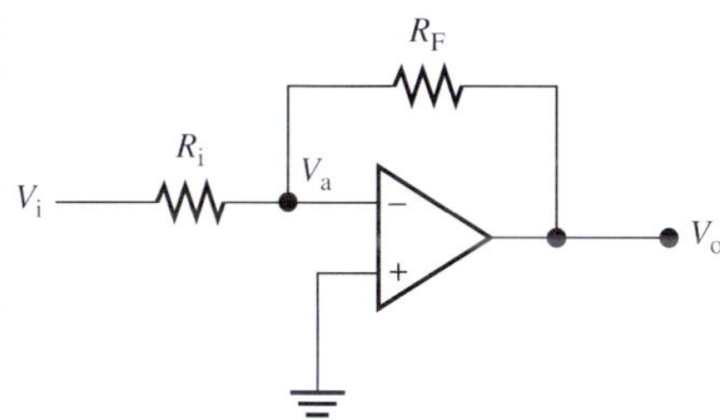

FIGURE 11.35 An inverting operational amplifier.

SOLUTION

1. First, go to "Place" > "Part" (see Figure 2.63) and type "uA741" to obtain the op amp.
2. Next, right click on the uA741 part and choose "Mirror Vertically" to invert the part upside down as shown in Figure 11.36.
3. Set up the PSpice schematic as shown in Figure 11.37, and run the simulation as a bias point analysis. The result is shown in Figure 11.37.

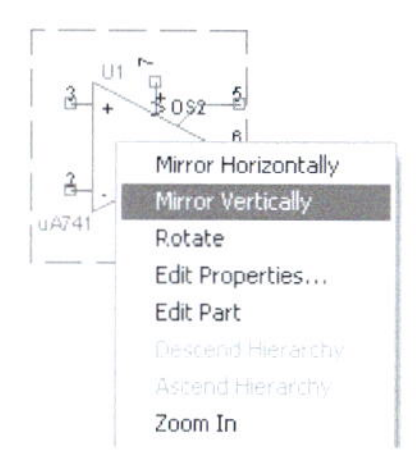

FIGURE 11.36 Mirror vertically.

The V_o of the circuit is −1.68 V. Notice that there is not any ideal op amp in PSpice, while μA741 is one of the op amp parts that is most commonly used. An ideal op amp model can be created through PSpice software. However, for the student/demo version, the PSpice model editing software is limited to diodes. Therefore, ideal op amps are not discussed in this section.

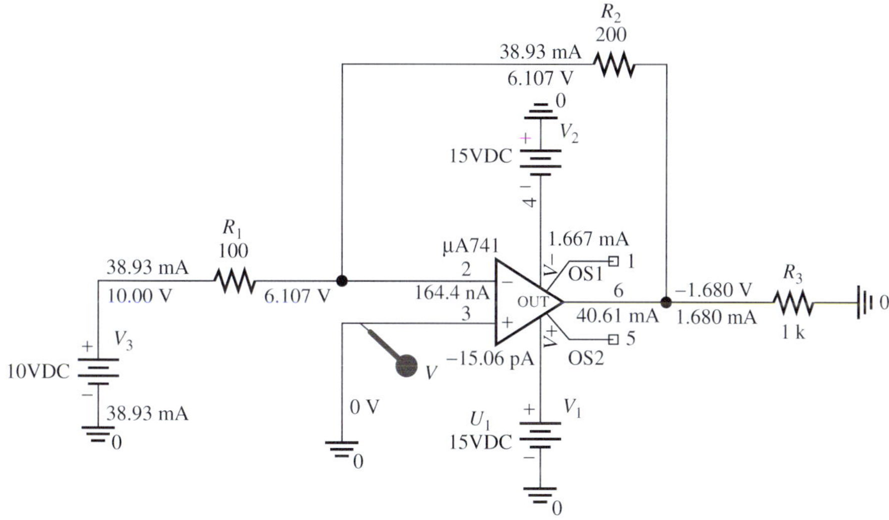

FIGURE 11.37 PSpice schematic solution for Example 11.21.

EXAMPLE 11.22 PSpice Analysis

Given the first-order, high-pass filter shown in Figure 11.38, plot the frequency response of the output with the load of 1 kΩ. Assume that the input voltage is 10 V 60 Hz, C is 10 μF, R_i is 100 Ω, and R_F is 200 Ω.

(*continued*)

EXAMPLE 11.22 Continued

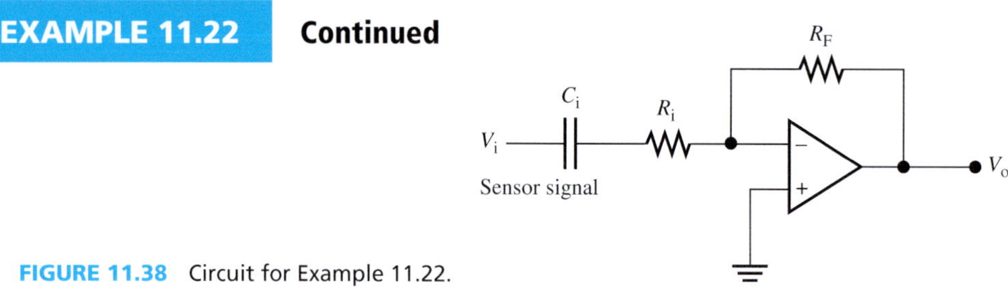

FIGURE 11.38 Circuit for Example 11.22.

SOLUTION

Follow the steps in Example 11.21 to set up an op amp in the PSpice circuit. The PSpice schematic solution is shown in Figure 11.39. Set the simulation analysis type to be AC sweep, with the logarithmic increment. The frequency plot and its phase diagram for the output are shown in Figure 11.40.

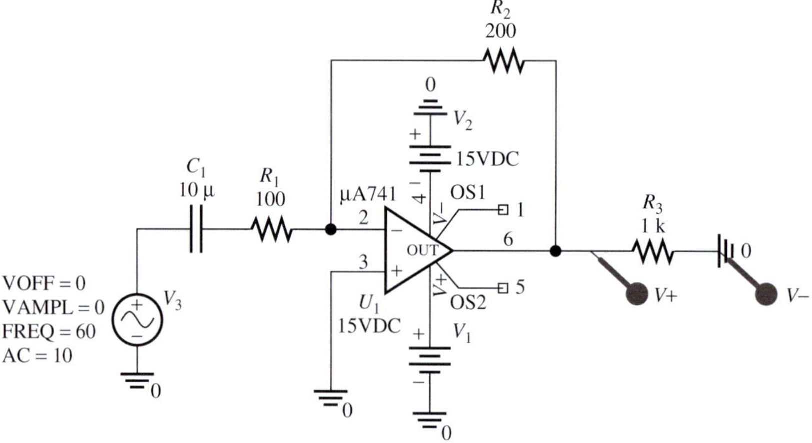

FIGURE 11.39 PSpice schematic solution for Example 11.22.

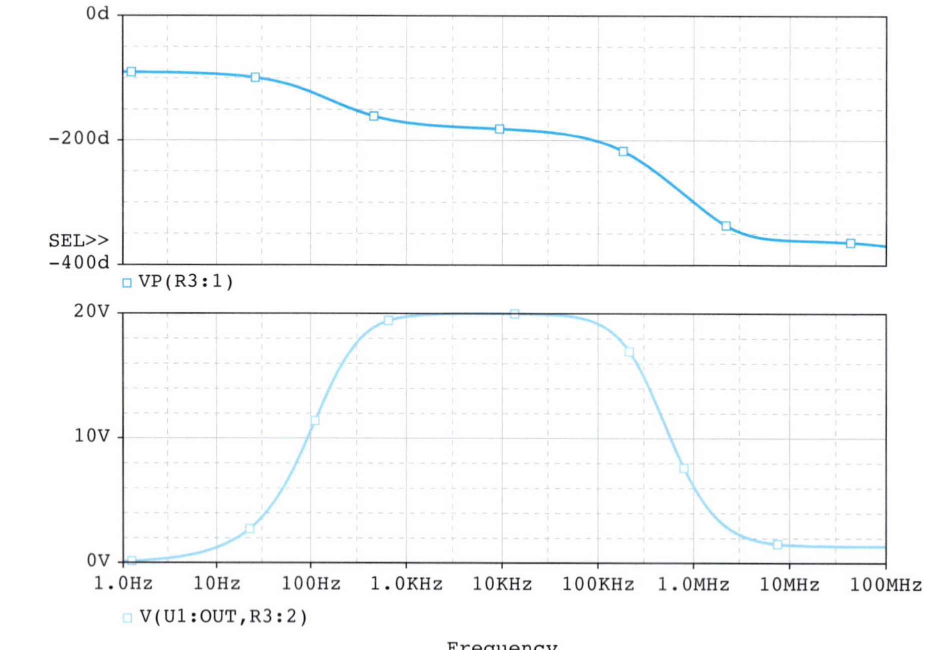

FIGURE 11.40 Bode plot for Example 11.22.

11.7 WHAT DID YOU LEARN?

- A computer-based instrumentation system consists of sensors, signal conditioning, a data-acquisition system, an analog-to-digital converter, and one or more general-purpose computer(s).
- A sensor, also called a transducer, is defined as a device that responds to a physical stimulus (e.g., heat, light, sound, pressure, magnetism, or motion) and transmits a resulting impulse for measurement or control.
- A pressure sensor is a device that responds to pressure applied to its sensing surface and converts the pressure to a measurable signal.
  Temperature sensors are used to measure temperature. For a thermocouple, the temperature difference between the two junctions of the thermocouple can be detected by measuring the change in the voltage across the dissimilar metals at the temperature measurement junction. A resistance temperature detector (RTD) is a variable-resistance device whose resistance is a function of temperature. RTDs are more accurate and stable than thermocouples.
- Accelerometers (acceleration sensors) are used to measure acceleration, vibration, and mechanical shock based on the effect of Newton's second law.
- A sound sensor is a device that converts acoustic energy into electrical energy. A microphone is one example of a typical sound sensor. The diaphragm of a microphone vibrates by impinging waves of acoustic pressure.
- A linear variable differential transformer (LVDT) is a displacement sensor that is operated based on the concept of mutual inductance.
- Signal conditioning is required to convert a sensor's output to an appropriate form. Two of the most important signal conditioning functions are amplification and filtering. An amplifier is a device that increases the power contained within a signal. A filter eliminates interference (noise or undesired input).
- A data acquisition system consists of an analog multiplexer, a buffer, a sample and hold amplifier, and an analog-to-digital converter.
- Analog-to-digital converters are designed to transform data in the form of continuous analog variables into a discrete binary code suitable for digital processing.
  There are two main steps in analog-to-digital conversion: sampling and quantization. First, the analog signal is sampled at regular time intervals. Next, quantization describes the procedure whereby the continuous analog signal is quantized and encoded in binary form.
- A ground specifies a reference voltage for a circuit or system.
- To avoid ground loop in measurement instruments, the instrument case ground and the source ground must be disconnected (see Figure 11.31).

Further Reading

Turner, J.D., and Prelove, A.J. 1991. *Acoustics for engineers*. New York: Macmillan.

Alan S. Morris. 2001. *Measurement and instrumentation principles*, 252–259. Oxford: Butterworth-Heinemann.

Chester L. Nachtigal. 1990. *Instrumentation and control: Fundamentals and applications*. New York: Wiley.

Turner, J., and Hill, M. 1999. *Instrumentation for engineers and scientists*. Oxford: Oxford University Press.

Gopel, W., Hesse, J., and Zemel, J.N. 1994. *Sensor: A comprehensive survey*. New York: VCH Verlagsgesellschaft, Weinheim (Federal Republic of Germany), VCH Publishers.

Francis, S. Tse, and Ivan E. Morse. 1989. *Measurement and instrumentation in Engineering*. New York: Marcel Dekker.

Problems

*B refers to basic, A refers to average, H refers to hard, and * refers to problems with answers.*

SECTION 11.2 SENSORS

11.1 (B) Explain the function of each element of a computer-based instrumentation system.

11.2 (B) Explain the basic concepts of sensors.

11.3 (B)* Select one: The relationship between pressure and force can be expressed as:
(a) Pressure = force × area
(b) Pressure = force/area

(c) Force = pressure/area
(d) None of above

11.4 (A) Piezoelectric sensors can measure the pressure of the object according to the equation:

$$p = \frac{V_0}{K_E x}$$

where, p is the pressure in N/m^2 and x is the displacement in meter. K_E is the voltage sensitivity of the sensor. Here, assume $K_E = 0.045$ V·m/N. If an object is placed on the sensor, and measures $V_0 = 2.75$ V, and $x = 0.85$ mm, then what is the pressure from the object?

11.5 (B) What is the Seebeck effect?

11.6 (A)* In Figure P11.6, which curve represents the characteristic of a thermistor?

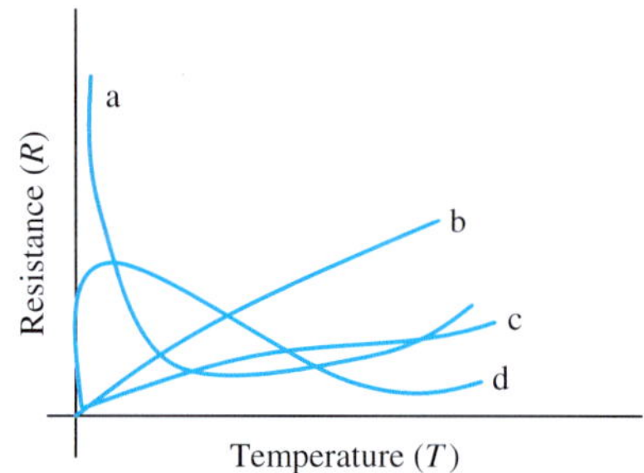

FIGURE P11.6 Resistance versus temperature curves for Problem 11.6.

11.7 (B) Which temperature sensors have a negative resistance–temperature relationship?
(a) Thermocouples
(b) All RTDs
(c) Thermistors
(d) Infrared thermometers

11.8 (H) A suitable curve fit that converts the measured resistance, R, of a thermistor to temperature, T, is given by:

$$\frac{1}{T} = A + B \ln R + C(\ln R)^3$$

where R is in ohms, T in kelvin, and A, B, and C are curve-fitting constants. An engineer has measured data pairs of resistance and temperature for the thermistor as follows:

$$(R_1, T_1),\ (R_2, T_2), \text{ and } (R_3, T_3).$$

Use these measured data to express the coefficients A, B, and C.

11.9 (A)* Which equation correctly expresses the acceleration? (x = linear displacement, θ = angular displacement, v = linear velocity, ω = angular velocity, a = linear acceleration, α = angular acceleration, t = time)

(a) $a = \frac{dx}{dt}$ (b) $\omega = \frac{d^2\theta}{dt^2}$ (c) $\alpha = \frac{d^2\theta}{dt^2}$ (d) $\omega = \frac{d^2x}{dt^2}$

11.10 (A) An engineer designs a toy car. To measure the acceleration of this kind of toy car, he also designs a simple accelerometer (a spring with spring constant 100 N/m and a mass $m = 2$ kg) and binds it to the car. When the car is accelerated, he measures the displacement of the mass which is 0.01 m. What is the acceleration of this toy car?

11.11 (A)* Accelerometers are used widely in vehicles to decide whether safety air bags should be deployed. Assume that the vehicle's air bag is deployed when the seismic mass m for the accelerometer is 0.2 kg and the vehicle's acceleration is 150 m/s^2. If at this moment, the mass displacement is measured as 0.02 m, then what is the spring constant for this accelerometer?

11.12 (A) How is a Wheatstone bridge used to construct a load cell?

11.13 (A)* A load cell is constructed as shown in Figure 11.13. There is a linear relationship between the output voltage, V_o, and the mass, M, weighted: $M = kV_o + b$, where k and b are constant. The steady input voltage to the cell is $V_i = 3$ V, and $R_1 = R_2 = R_3 = 200\ \Omega$. To calculate the value of k and b, two different masses ($M_1 = 1$ kg, $M_2 = 0.8$ kg) are loaded to the cell, respectively, and the R_g is measured as 100 Ω and 120 Ω, respectively. Find k and b.

11.14 (A) For an LVDT, the relationship between the magnitude of the output voltage and the displacement, x, of the iron core is approximately linear over a reasonable range of movement of the iron core and is expressed by a constant of proportionality, G, as: $x = G \cdot V_o$. An engineer uses this kind of LVDT (shown in Figure P11.14) to measure the displacement of a gear. The engineer measures the instant value of output voltage: $V_1 = 5$ V and $V_2 = 2$ V, and the sensitivity (or gain) G of the transformer is 0.6 mm/V. Determine the displacement of the gear.

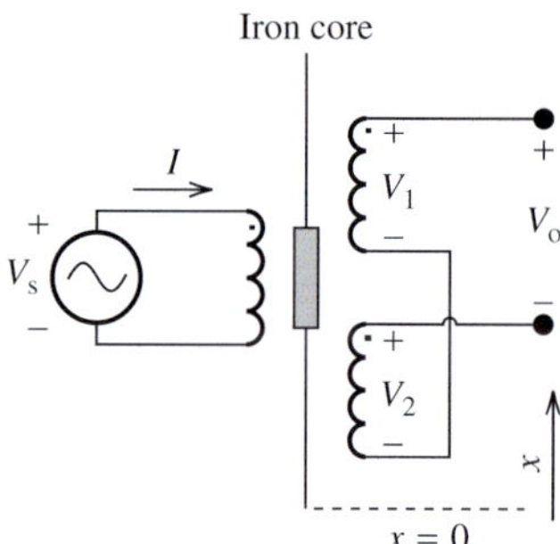

FIGURE P11.14 LVDT.

11.15 (H)* Figure P11.15(a) shows a capacitive microphone output. The voltage output of the circuit is shown in Figure P11.15(b). What is the change of capacitance due to the sound pressure on the diaphragm?

11.16 (B) Describe at least three sensors that are part of your body and explain their function.

11.17 (B) Describe at least three sensors applied in daily life and explain their function.

11.18 (A) The resistance of an RTD is calculated with $R_{RTD} = R_0[1 + \alpha(T - T_0)]$, and the coefficient $\alpha = 0.004°C^{-1}$. The room temperature is (25°C), $R_0 = 40\ \Omega$. An RTD is put on a running CPU and $R_{RTD} = 48\ \Omega$. What is the surface temperature of the CPU?

11.19 (A) Explain the basis of a thermoelectric cooler.

11.20 (H)* A set of accelerators are installed on an object as shown in Figure P11.20. Data are collected from each accelerator from time $T = 0$ s and $T = 3$ s, $a_x = 4$ m/s^2

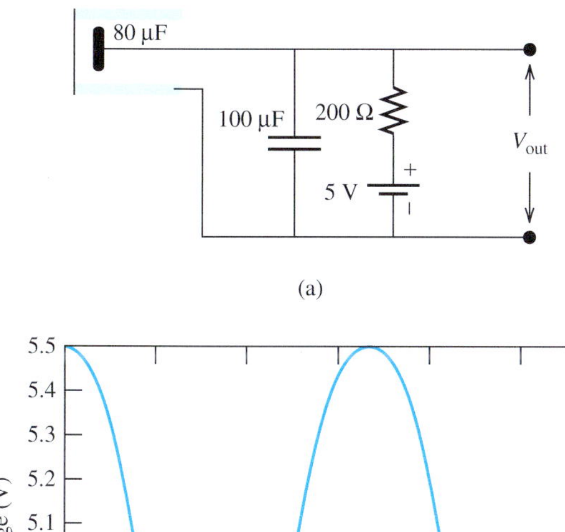

FIGURE P11.15 (a) Capacitive microphone circuit for Problem 11.15; (b) Voltage output plot for Problem 11.15.

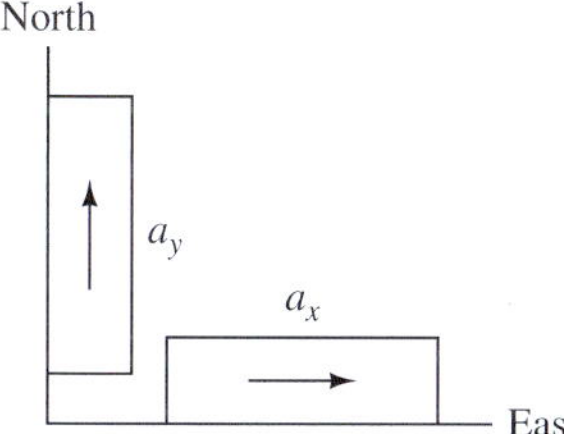

FIGURE P11.20 Accelerator set.

and $a_y = 3$ m/s^2. Assume the object is static at $T = 0$ s; calculate the object speed at $T = 3$ s.

11.21 (A) What is the advantage of applying a Wheatstone bridge in load cells compared to directly measuring the gauge resistance?

11.22 (H) If you connect a passive analog speaker to a microphone connector on a recorder, and you start to record when you speak to the passive speaker, can the recorder record your voice? If it can, what is the difference between the voices recorded using speaker and a microphone?

SECTION 11.3 SIGNAL CONDITIONING

11.23 (B) Explain why signal conditioning is mandatory in a computer-based instrumentation system.

11.24 (A)* Consider the high-pass, active filter shown in the Figure P11.24. If $v_i(t) = 10\cos(1000\pi \cdot t)$ V, $R_i = 150$ kΩ, $R_F = 200$ kΩ, and $C_i = 200$ pF, find $v_o(t)$.

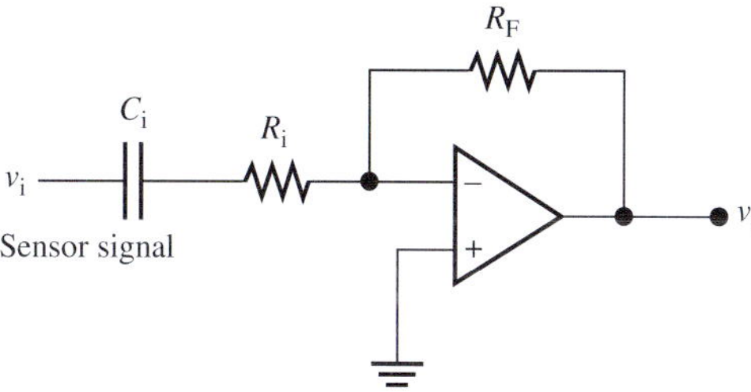

FIGURE P11.24 High-pass, active filter.

11.25 (A) Consider the low-pass, active filter shown in the Figure P11.25. If $v_i(t) = 4\cos(2000\pi t)$ V, $R_i = 10$ kΩ, $R_{F1} = 40$ kΩ, $R_{F1} = 60$ kΩ, and $C_F = 500$ pF, find $v_o(t)$.

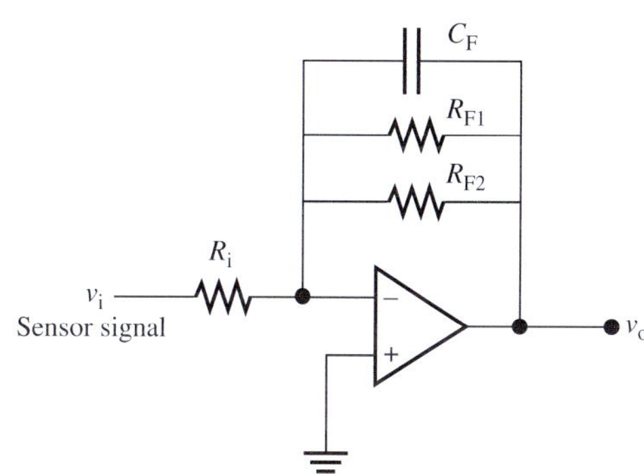

FIGURE P11.25 Low-pass, active filter.

11.26 (A) A simple band-pass filter with op amp is shown in the Figure P11.26. Derive the expression for the gain of this active filter.

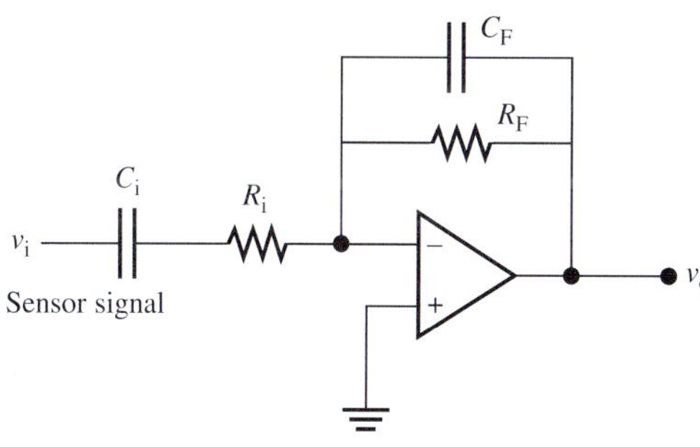

FIGURE P11.26 Band-pass, active filter.

11.27 (H) Given that both voltage sources, V_1 and V_2, are 120 V, 60 Hz what is the output voltage of the circuit shown in Figure P11.27?

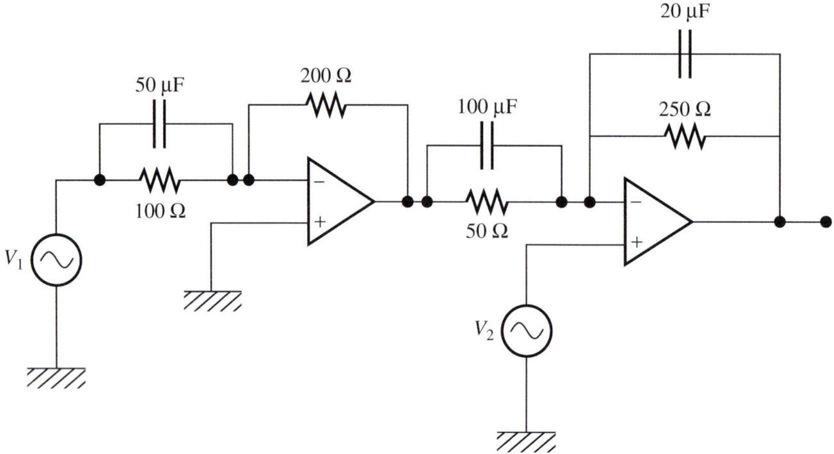

FIGURE P11.27 Circuit for Problem 11.27.

11.28 (H)* A data acquisition subsystem requires its input to be larger than 1 V and smaller than 10 V. The output of the sensor connected to the data acquisition sub-system is 1 to 5 mV. Calculate the expected voltage gain of the signal conditioner.

11.29 (A) Find the gain of the two low-pass filters shown in Figure P11.29, and analyze their pros and cons.

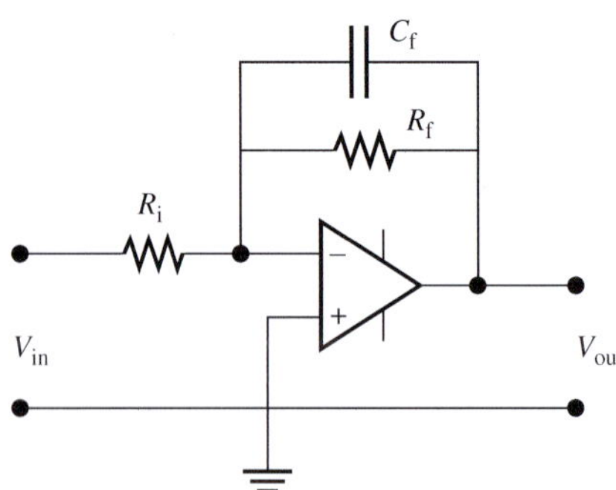

(a) Active, low-pass filter

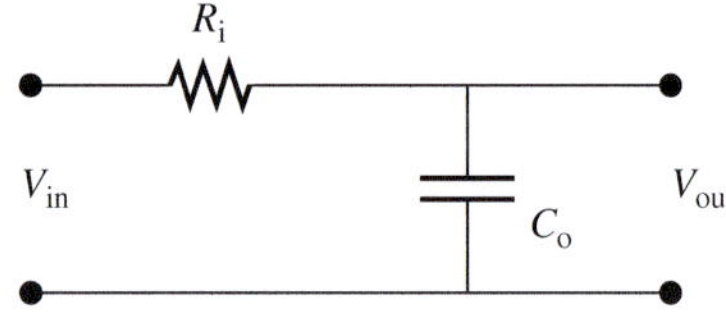

(b) Passive, low-pass filter

FIGURE P11.29 (a) Active and (b) passive low-pass filters.

11.30 (A)* An engineer wants to check a high power pulse transmitter's power supply wave form using an oscilloscope. The voltage range of the power supply is between 6 and 12 kV, and the oscilloscope's input range is 2 mV to 10 V. Which probe should the engineer select? (The parameter is the ratio of input to output).
(a) 10:1
(b) 100:1
(c) 1000:1
(d) 10,000:1

SECTION 11.4 DATA ACQUISITION

11.31 (B) What is signal sampling?

11.32 (B) A signal has frequency components up to 50 kHz. What is the minimum sampling rate that should be used to ensure that the signal is reconstructed properly?

11.33 (B) What is aliasing? Under what conditions does it occur?

11.34 (B) If the analog voltage range is ± 3 V, and a 16-bit ADC is used to convert the analog signal to digital signal, what is the resolution of this ADC?

11.35 (A)* A resolution of the sampled values of an ADC is desired to be 2%. What is the required number of bits for this ADC?

11.36 (A) Consider a voltage source, $V_S(t)$ of 5 V, 60 Hz. Use MATLAB to plot the output of sampling of this voltage source for 0.1 s if the sampling rate is:
(a) 60 Hz
(b) 120 Hz
(c) 10 times of voltage source frequency

11.37 (H) At any particular value of the analog signal, the digital representation is either the discrete level immediately above this value or the discrete level immediately below this value. For eight levels of quantization with three bits, if the difference between two successive discrete levels is represented by the parameter Δ, design a quantization method to make the maximal quantization error equal to $\Delta/2$.

11.38 (H)* A 3-V peak sinusoid is converted to a digital signal using an N-bit ADC which is designed to accept signals from −5 to 5 V. Find the smallest N to make the SNR (expressed in decibels) greater than 50 dB.

11.39 (B) Explain the necessity of ADC in a computer-based system.

11.40 (A) When the input is −5 ~ 5 V, compare the noise generated by an 8-bit and a 12-bit ADC. What are the advantages and disadvantages of using a 12-bit AD convertor compared to an 8-bit ADC?

11.41 (H) Prove that when an ADC output is increased by one bit, an extra 6 dB SNR is generated by the ADC.

11.42 (H)* Assume there is a signal between −6 and +6 V, and the required measurement RMS error is 1 mV. If a digital voltage meter is used, an ADC with how many bits is needed?

11.43 (A) The output SNR of an ADC is related to:
(a) Input SNR
(b) Input signal amplitude
(c) The number of ADC output bits
(d) The sampling rate

SECTION 11.5 GROUNDING ISSUES

11.44 (B) What is the ground loop problem?

11.45 (B) What is a data acquisition system?

11.46 (A)* Determine the voltage output, V_{out}, shown in Figure P11.46 if the ground of the voltage source and ground of the load are connected. The material of the wire that connects the devices in the circuit is copper with the resistance per length of 0.54 Ω/m.
If the connection between both grounds is cut, what is the voltage output, V_{out}?
(Provide the solution up to two decimal points.)

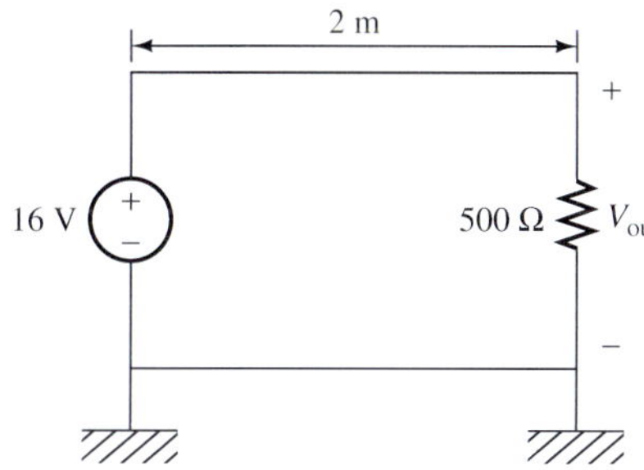

FIGURE P11.46 Circuit for Problem 11.46.

11.47 (B) How should the ground loop effect in a measurement system be minimized? Provide an example when the input is an AC signal.

11.48 (A)* A circuit provides 5 V power for an analog part and 3 V for a digital part shown in Figure P11.48. Calculate the voltage difference measured at V_{out} when $R_G = 0\ \Omega$ (no ground loop) and $R_G = 1\ \Omega$ (ground loop).

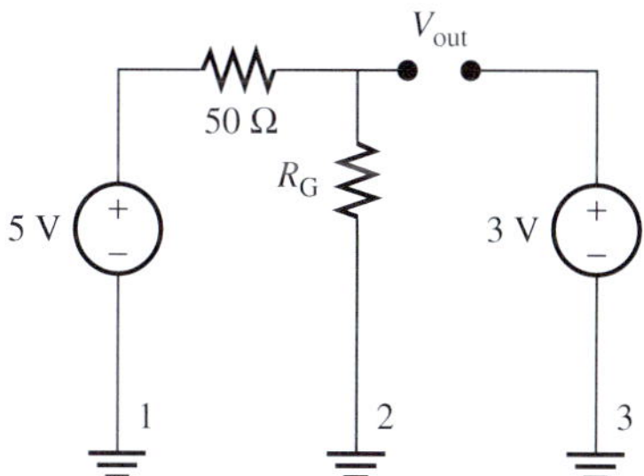

FIGURE P11.48 Ground loop.

SECTION 11.6 USING PSPICE TO DEMONSTRATE A COMPUTER-BASED INSTRUMENT

11.49 (A) Consider the high-pass, active filter shown in the Figure P11.4. If $v_i(t) = 10\cos(1000\ \pi t)$ V, $R_i = 150$ kΩ, $R_F = 200$ kΩ, and $C_i = 200$ pF. Use PSpice to plot the output of $v_o(t)$.

11.50 (A) Consider the low-pass, active filter shown in the Figure P11.5. If $v_i(t) = 4\cos(2000\ \pi t)$ V, $R_i = 10$ kΩ, $R_{F1} = 60$ kΩ, $R_{F2} = 40$ kΩ, and $C_F = 500$ pF. Use PSpice to plot the output of $v_o(t)$.

TOPIC 7

Electric Machines

The content of this topic is compiled from:

Chapter 13 (pp. 557–614), *Electrical Engineering: Concepts and Applications*

By S.A. Reza Zekavat

CHAPTER 13

Electric Machines

(Used with permission from © BL Images Ltd/Alamy.)

13.1 INTRODUCTION

In Chapter 12 you studied magnetic circuits, transformers, and the relationship between electricity and magnetism. In this chapter, you will learn about devices whose operation can be explained based on the relationship between electricity and magnetism. These devices are called *electric machines*. Do you know what central air conditioners, vacuum cleaners, and CD players all have in common? Each operates using an electric machine of one kind or another.

Electric machines are devices that convert energy from one form (e.g., potential energy of the stored water in the dam) to another, that is kinetic energy of the machine. Electric machines are classified into two types. Devices that generate electrical energy through the input of some other form of energy are called *generators*. Devices that transform electrical energy into mechanical energy are called *motors*. *Alternators* are generators that generate alternating current (AC). An example is a car alternator, which supplies the vehicle's equipment such as air conditioning, radio, lights, and speakers. You may be amazed at the wide range of applications in which electric motors are used. Electric motors are

TABLE 13.1 Examples of Various Applications of Electric Motors

Domestic and Office	Automobile	Medicine	Industry
Central air conditioners	Power windows	Dentist drills	Drilling machines
Food processors	Sunroof operation	Electric wheelchairs	Grinders
Vacuum cleaners	Radiator fans	Artificial hearts	Lathe machines
Refrigerators	Fuel pumps		Milling machines
Hair dryers	Starter motors		Compressors
Sewing machines	Windshield wipers		Industrial fans, blowers
CD players, cameras			Cranes, lifts
Ceiling fans			Elevators
Photocopiers			Motors in robots

found in applications as diverse as domestic appliances, automobiles, and industrial equipment. Table 13.1 lists various applications for electric motors. This leads us to an obvious question: Why have electric machines become so widespread across so many diverse fields?

13.1.1 Features of Electric Machines

- Wide range of torque–speed characteristics make them suitable for a variety of applications
- Smooth speed control can be attained via semiconductor devices, such as thyristors
- Compatible with different loads as operating characteristics can be easily modified
- High efficiency, high-load capacity, and long life
- Low noise and low maintenance
- Smooth acceleration and deceleration
- Can be operated in diverse environments (e.g., radioactive, submerged in liquids, etc.)
- Use a clean source of energy
- Fast response, that is, no warm-up time is required; thus, instant loading is achievable
- Compact, reliable, and economical
- Remote control is possible, if required

13.1.2 Classification of Motors

Now that you are familiar with various features of electric machines, we can investigate the governing principles of motors and generators. Motors are classified into two main categories: *alternating current (AC) motors* and *direct current (DC) motors*. AC and DC designations refer to the type of current supply required for the motor to operate. The tree diagram shown in Figure 13.1 depicts

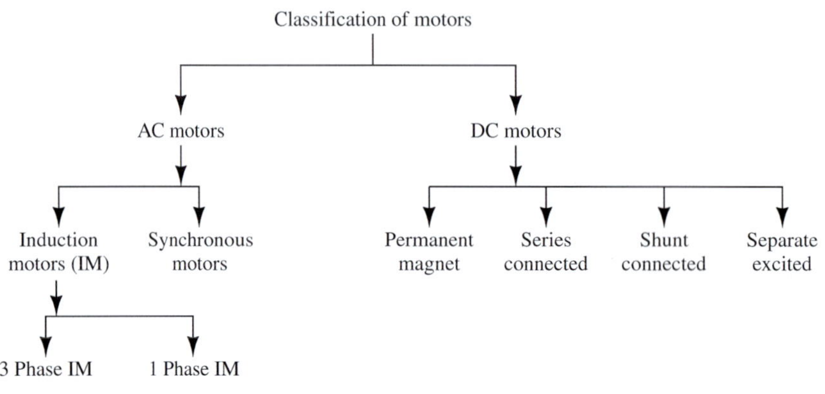

FIGURE 13.1 Classifications of motors.

the broad classifications of motors. The next sections of this chapter explain the different classes of motors revealed in Figure 13.1.

13.2 DC MOTORS

13.2.1 Principle of Operation

"When a current-carrying conductor is placed in a magnetic field, it experiences a mechanical force, the direction of which is given by the right-hand rule (or Fleming's left-hand rule)."

Based on this principle, the force applied to the current corresponds to:

$$\mathbf{F} = l\mathbf{I} \times \mathbf{B}$$

where **F** is the electromagnetic force vector, l is the length of the current-carrying wire, **I** is the current vector, and **B** is the magnetic field vector measured in tesla. Sometimes, the current, **I**, is replaced with the current density, **J**, measured in terms of amperes per square meter (A/m^2). In that case, the **F** would be the electromagnetic force density measured in terms of newton per square meter (N/m^2). The multiplication shown by "×" refers to the fact that the direction of the force is perpendicular to the plane formed by the direction of the current and the magnetic field vectors. The direction of this force can be determined by what is called the *right-hand rule*. This rule is shown visually in Figure 13.2(a). In this figure, you can see that the direction of all fingers (except the thumb) are first pointed toward the direction of current **I**. Then, when the vector formed by **I** is rotated toward vector **B**, the direction of the thumb indicates the direction of force applied to current **I**.

The direction of the force can be equivalently explained by Fleming's "left-hand rule" as shown in Figure 13.2(b). Here, the thumb, the first, and the second fingers on the left hand are held so that they are at right angles to each other. If the first finger points in the direction of the magnetic field and the second finger in the direction of the current in the wire, then the thumb will point toward the direction of the force on the conductor.

13.2.2 Assembly of a Typical DC Motor

To function, a DC motor needs a DC supply. A DC supply can be provided via rectifiers (AC to DC converters) as explained in Chapter 8. The DC supply can also be provided by batteries. The main components of a DC motor include the stator, rotor, windings, and commutator as explained below.

- ***Stator:*** A stator is usually the static component of the motor. Most electric motors employ a cylindrical stator that has an even number of magnetic poles. The stator can be formed either by permanent magnets (PMs) [as shown in Figure 13.3(a)] or by field windings [see Figure 13.3(b)]. Field windings are applied to the slots cut within the surface of the stator structure.

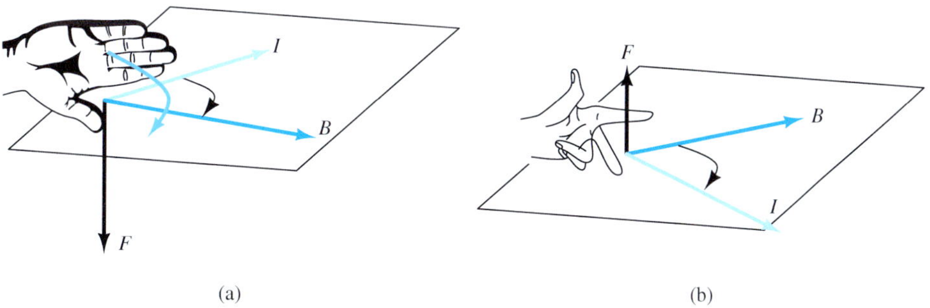

FIGURE 13.2 Evaluating the direction of the force applied to a current, **I**, using (a) right-hand rule; (b) Fleming's left-hand rule.

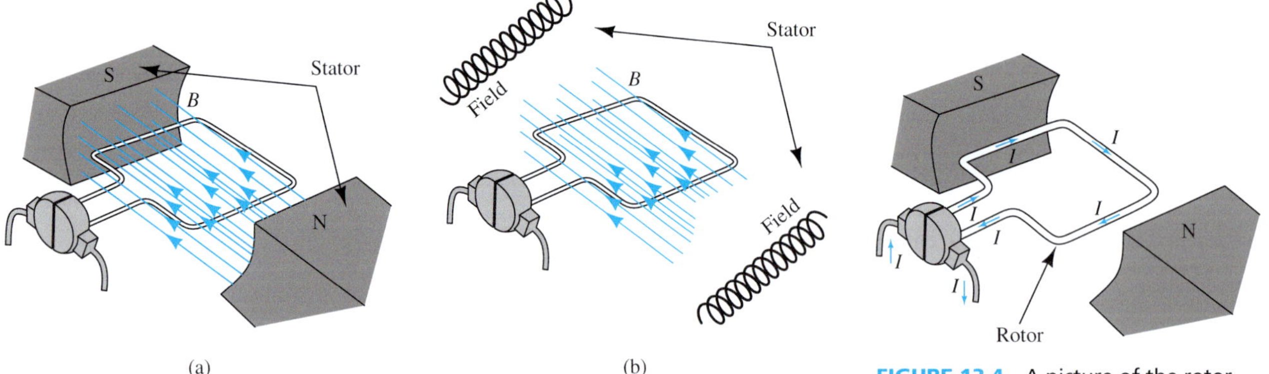

FIGURE 13.3 DC motor: (a) Permanent magnet stator; (b) field winding stator.

FIGURE 13.4 A picture of the rotor with slots.

- ***Rotor:*** A rotor is the rotating component of the motor. It is located inside the stator and is basically a laminated iron cylinder mounted on a shaft. The slots are cut lengthwise on the surface of the rotor to accommodate the armature conductors. Figure 13.4 represents one winding loop of the rotor.
- ***Windings:*** There are two types of windings: field and armature. Both types carry DCs. Field winding sets up the magnetic field of the stator, whereas armature winding does the same for the rotor. Figure 13.3 shows the two types of windings.
- ***Commutator:*** A commutator reverses the direction of the current in each conductor through the armature as it passes from one pole to another and helps to develop continuous unidirectional torque. Figure 13.5 is an actual photograph of a commutator.
- ***Brushes:*** The electrical connections to the commutator are made by brushes as illustrated in Figure 13.6. In contrast to the slip rings found in AC motors, brushes are more prone to wear.

13.2.3 Operation of a DC Motor

In many DC motors, field windings (in the stator) create the required magnetic field, while armature conductors (rotor conductors) carry the current. Thus, the force experienced by the current-carrying conductors placed in a magnetic field is transferred to the armature conductors. The direction of this force is given by the right-hand rule or Fleming's left-hand rule as shown in Figure 13.2. The armature conductors are placed in slots on the periphery of the rotor. The force applied to each conductor acts as a twisting force on the rotor. The cross product of twisting force and the radius vector of rotation is called *torque*. Figure 13.7 explains the operation of a DC motor.

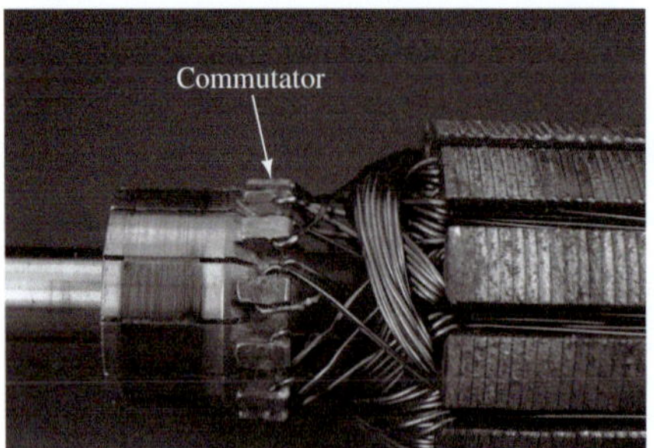

FIGURE 13.5 Commutator. (Used with permission from © sciencephotos/Alamy.)

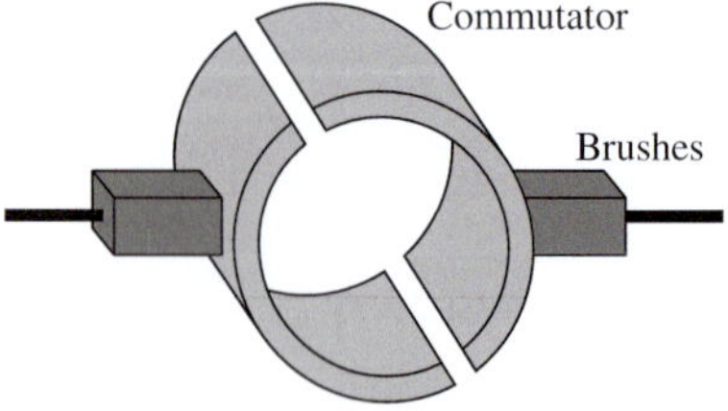

FIGURE 13.6 Brushes.

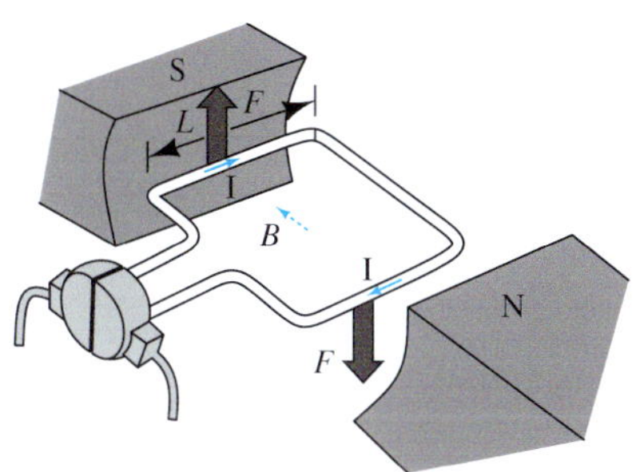

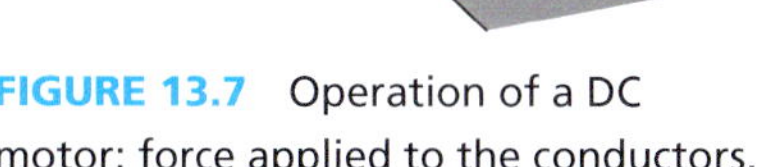
FIGURE 13.7 Operation of a DC motor: force applied to the conductors.

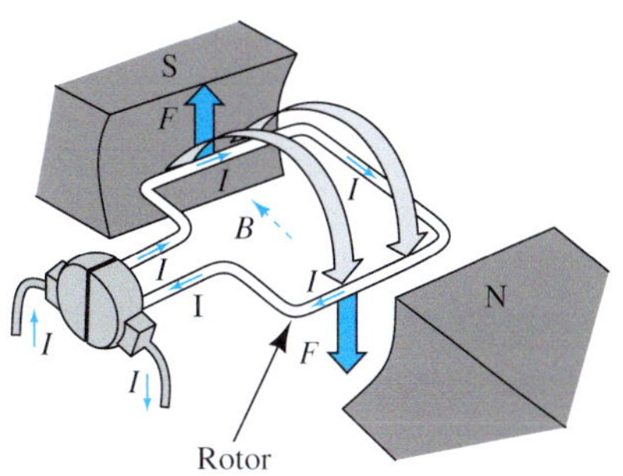

FIGURE 13.8 Rotation created by the torque produced.

In Figure 13.7, the direction of the magnetic field is depicted by the dashed arrow. The direction of current through the armature conductors is shown by the colored solid arrows. The direction of force experienced by each of the rotor conductors is denoted by the gray arrow. The curved gray arrows in Figure 13.8 depict the direction of the resultant torque produced by the force experienced by the individual rotor conductors.

EXERCISE 13.1

Apply the right-hand rule (or Fleming's left-hand rule) to Figures 13.7 and 13.8. Is the direction of the resultant force shown correctly?

EXAMPLE 13.1 **Force Direction**

In Figures 13.7 and 13.8, what happens when the direction of the field is reversed? What happens when the direction of the field and the direction in which the current is flowing reverse simultaneously?

SOLUTION

If the field direction reverses, the force will reverse. However, if the current and the field reverse their directions simultaneously, then the direction of the resultant force will remain unchanged. This can be easily verified using the right-hand rule.

13.2.4 Losses in DC Machines

In both generators and motors, a certain amount of energy is always dissipated during the energy conversion process. This dissipated energy appears in the form of heat and therefore increases the temperature of the machine. This phenomenon limits the performance and the lifetime of the components of the machine and reduces its power output. In addition, the dissipated energy is the detrimental factor in determining the kilowatt (kW) rating of the motor. Different losses that occur within a DC machine are classified into three main types:

i. *Copper* losses
ii. *Iron* or *core* losses
iii. *Mechanical* losses.

Each type of loss is explained in detail:

i. ***Copper losses:*** These losses are due to the current flowing through the resistance of various windings. These losses take place in armature as well as field circuits. Copper losses in armature are load dependent and are therefore named *variable losses.*

ii. ***Iron or core losses:***
Following are the sources of these losses:

- ***Eddy current losses:*** Eddy currents are created when there is a relative motion between a conductor and the magnetic field produced by the stationary stator. According to Lenz's law, the generated electric field is directed in such a way that it opposes the cause producing it. Thus, there is a loss of energy. Eddy currents also produce heat, which results in losses. These losses can be minimized by using materials with low electrical conductivity, such as ferrites, or by using laminations.
- ***Hysteresis losses:*** Hysteresis loss is a heat loss, which is caused by the magnetic properties of the armature. Armature rotation leads to the change of direction of magnetic flux, and thus continuous movement of magnetic particles. Magnetic particles are particles within a magnetic element. These particles create the magnetic property of the material. Indeed, a magnet is formed by the integration of many magnetic particles. Figure 13.9 represents the schematics of magnetic particles that form a magnet.

  A molecular friction is produced as magnetic particles try to align themselves with the changing direction of the magnetic field. This leads to an increase in temperature, which increases armature resistance. This loss can be minimized by proper selection of materials for armature. Examples of proper materials include steel or laminated iron.
- ***Mechanical losses:*** These losses are due to the rotation of the armature and include bearing friction loss, brush friction loss, and windage or air friction loss. These losses vary with speed and are considered constant for machines having fairly constant speed.

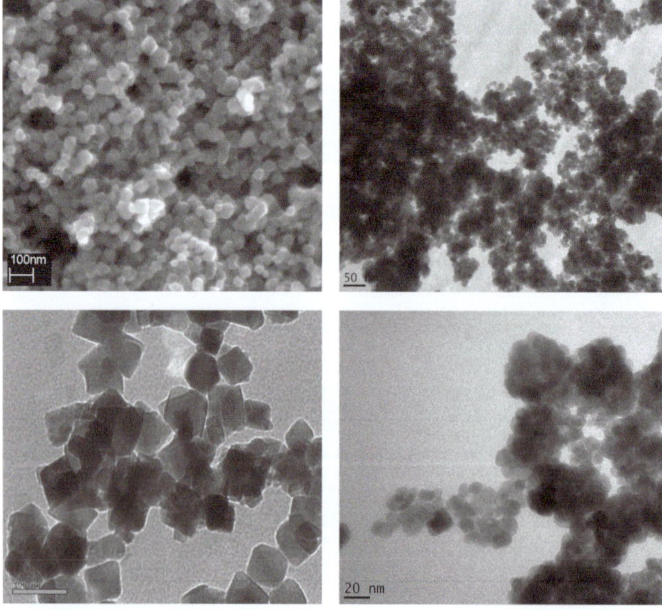

FIGURE 13.9 Magnetic particle under the microscope. (Photos courtesy of Silvio Dutz, IPHT Jena.)

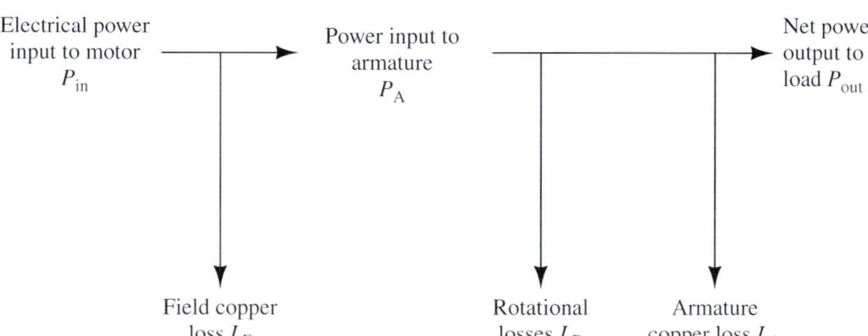

FIGURE 13.10 Power stages in a DC motor.

iii. *Iron and copper losses* are often grouped together and called *rotational losses*. The lost power is converted into heat and it is independent of the load; thus, it is called a *constant loss*.

Some losses in a machine are not exactly deterministic and are therefore called *stray losses*. These losses could be due to the magnetic flux leakages and short-circuit currents in the coils undergoing commutation. In small machines, these losses are negligible.

Figure 13.10 provides a summary of all losses. As depicted in this figure, the net output power is calculated after deducting the losses explained in this section.

EXAMPLE 13.2 **Drawbacks of DC Motors**

What is the most defining drawback of DC motors in general?

SOLUTION

The biggest drawback in any DC motor is the use of the commutator and brush arrangement and the relative motion between them. As a result of this arrangement, they:

- Require additional maintenance
- Have a shorter life span
- Possess a higher chance of sparking at the commutator and brush assembly

13.3 DIFFERENT TYPES OF DC MOTORS

As summarized in Figure 13.1, DC motors are divided into four main categories: (1) shunt-connected, (2) series-connected, (3) permanent magnet, and (4) separately excited motors. This section examines the equations that maintain the relationship of the voltage across armature, applied torque, generated power, and its angular speed (see Section 13.3.1). You will learn to analyze the characteristics of DC motors in other sections.

13.3.1 Analysis of a DC Motor

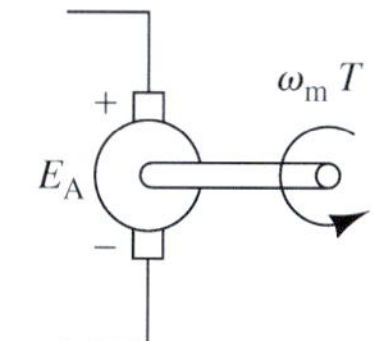

FIGURE 13.11 Armature back-emf and its relationship with torque and angular speed.

The emf, E_A, that is created across the rotor (see Figure 13.11) represents the average voltage induced in the armature due to the motion of conductors relative to the magnetic field. The concept of back-emf is discussed in greater detail later in the chapter. In general, in motors, E_A is referred to as back-emf because it opposes the externally applied voltage (i.e., the source that produces it). This is explained by Lenz's law.

Lenz's law states that an electromagnetic field interacting with a conductor generates electrical current which induces a magnetic field that opposes the magnetic field generating the current. In other words, the induced emf and the change in flux have opposite signs.

An expression for the induced emf is defined using the following parameters:

P = the number of stator poles

ϕ = flux per stator pole in weber

$\omega_m = 2\pi N$ is the angular velocity, with the unit of radians per minute

N = speed of armature in revolutions per minute (rpm)

Z = total number of armature conductors

According to Faraday's law of electromagnetic induction, the average emf induced in each conductor is proportional to the rate at which rotating parts cut the flux. It corresponds to:

$$e = \frac{d\phi}{dt} \tag{13.1}$$

Now, the total flux produced is given as:

$$d\phi = \phi \cdot P \tag{13.2a}$$

The time required for one revolution is:

$$dt = \frac{60}{N} \tag{13.2b}$$

Because the rate change of flux as described in Equation (13.1) is defined as total flux in one revolution, by replacing $d\phi$ and dt in Equation (13.1) with those in Equations (13.2a) and (13.2b), for the emf per conductor, we see that:

$$e = \frac{\phi \cdot P \cdot N}{60} \tag{13.3}$$

The total emf for Z conductors in the DC motor corresponds to:

$$E_A = \frac{\phi \cdot P \cdot N \cdot Z}{60} \tag{13.4}$$

Using Equation (13.5), we see that the induced voltage (emf) measured in volts in the armature is:

$$E_A = K \cdot \phi \cdot \omega \tag{13.5}$$

In Equation (13.5), $K = P \cdot Z/2\pi$ is called the machine constant and varies with the machine design. In addition,

$$\omega = \frac{2\pi N}{60} = \frac{\omega_m}{60} \tag{13.6}$$

is the angular velocity of the motor in radians per second. This equation represents that the induced voltage is directly proportional to the speed of the motor as well as the flux produced by each stator pole.

The torque (in newtons meter) produced by a conductor is:

$$T = B \cdot I_A \cdot l \cdot r \tag{13.7}$$

where:

- I_A = armature current in the conductor (in amperes)
- l = length of each conductor (in meters)
- r = radius at which the conductor is placed (in meters)

Because the motor contains Z conductors, the total torque corresponds to:

$$T = B \cdot I_A \cdot l \cdot r \cdot Z \tag{13.8}$$

We know that the relationship between the magnetic flux, magnetic field, and the cross-sectional area of the magnetic flux corresponds to:

$$B = \frac{\phi}{A} \tag{13.9}$$

This torque acting on the rotor is directly proportional to the magnetic flux and armature current. Replacing for B in Equation (13.8) using Equation (13.9), we see that:

$$T = \frac{\phi}{A} \cdot I_A \cdot l \cdot r \cdot Z \tag{13.10}$$

Thus, we can represent the torque as:

$$T = K_A \cdot \phi \cdot I_A \tag{13.11}$$

where:

$$K_A = \frac{lrZ}{A} \tag{13.12}$$

K_A is a constant that varies with the specifications of the motor such as its diameter, length, and the number of conductors. The generated power (mechanical power) by the motor corresponds to:

$$P = E_A \cdot I_A \tag{13.13}$$

Now, the relationship between generated power, torque, and its angular velocity corresponds to:

$$P = T \cdot \omega \tag{13.14}$$

Equating Equations (13.13) and (13.14) then replacing E_A with Equation (13.5) and T with Equation (13.11) results in $K = K_A$.

EXAMPLE 13.3 Induced emf

Consider a two-stator pole motor where each pole has a flux of 15 mWb. Assume the motor rotates at 600 rpm. Calculate the induced emf if the motor has two armature conductors.

SOLUTION

First, the angular velocity of the motor is:

$$\omega = 2\pi N/60$$
$$\omega = 20\pi$$

And the machine constant, K (defined in Equation (13.6)), is:

$$K = \frac{2 \times 2}{2\pi} = \frac{2}{n}$$

Therefore, the induced emf, E_A, is:

$$E_A = K \cdot \phi \cdot \omega$$
$$E_A = \frac{2}{\pi}(15 \times 10^{-3})(20\pi)$$
$$E_A = 0.6 \text{ V}$$

EXAMPLE 13.4 Induced emf

A DC motor that supplies a maximum torque of 50 N m rotates at the speed of 200 rpm. Assume that the induced current of the motor is 20 A. Determine the induced emf.

SOLUTION

Considering Equations (13.13) and (13.14), the induced emf is:

$$E_A = \frac{T\omega}{I_A}$$
$$E_A = \frac{50(200)(2\pi)/60}{20}$$
$$E_A = 52.3 \text{ V}$$

13.3.2 Shunt-Connected DC Motor

The shunt-connected DC motor derives its name from the fact that the armature and field windings are connected in parallel (shunt). The inductance effects of these windings are ignored because they behave as a short circuit for DC. Figure 13.12 represents an equivalent circuit of a DC shunt motor. The rotor and armature circuits are connected in parallel.

In Figure 13.12, R_A represents the effect of the resistance of the armature windings plus the brush resistance. The resistance, R_f, represents the field resistance, and I_f and V are the field current and voltage across the field resistance, respectively. Normally, in a DC motor, R_f is selected to be large; thus, the field current is small: it should be mainly enough to produce the required flux. Thus, almost the entire current from the supply is applied to the armature. The voltage across the armature is partially dissipated in the armature windings and partially converted to the required emf that leads to the rotation.

We can now analyze the equivalent circuit of a shunt DC motor (Figure 13.12). Considering steady-state status:

$$V = R_f \cdot I_F \qquad \textbf{(13.15)}$$

Now, examine the torque and speed relationship for a shunt motor. Applying Kirchhoff's voltage law (KVL) to the circuit of Figure 13.12 results in:

$$V = R_A I_A + E_A \qquad \textbf{(13.16)}$$

Next, replacing for I_A using Equation (13.11), and for E_A using Equation (13.5):

$$V = \frac{R_A T}{K\phi} + \frac{K\phi\omega_m}{60} \qquad \textbf{(13.17)}$$

Applying some mathematical manipulations to Equation (13.17), the torque can be computed as:

$$T = \frac{K\phi}{R_A}\left(V - \frac{K\phi\omega_m}{60}\right) \qquad \textbf{(13.18)}$$

The relationship of the torque and the angular frequency, ω_m, is derived using Equation (13.18). Observe that the speed at no load condition ($T = 0$) is maximum and the maximum torque would lead to zero speed ($\omega_m = 0$). Here, the load refers to mechanical load. Thus, when there is no load, the torque required to move the load is zero.

Equation (13.11) depicts that torque, T, is directly proportional to the armature current. At zero current, the torque is zero. This fact has been shown in Figure 13.13(a): torque increases when the armature current increases and vice versa. Now, consider the speed of a shunt motor when it is operating at a no load condition.

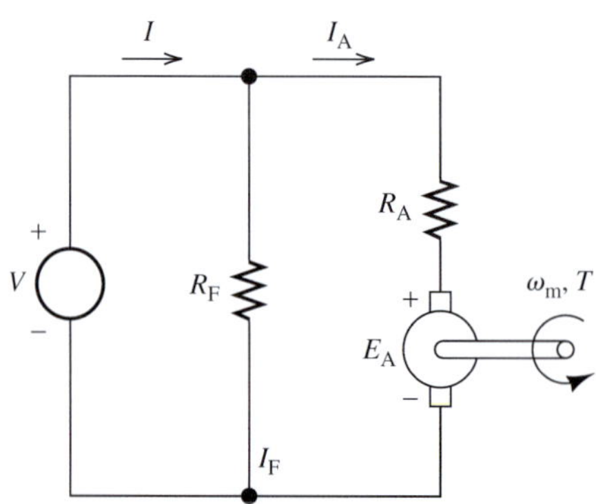

FIGURE 13.12 Equivalent circuit of a DC shunt motor.

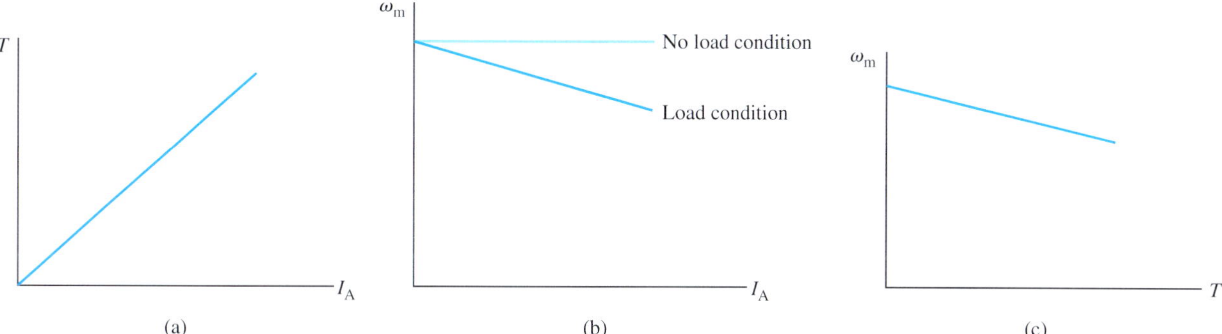

FIGURE 13.13 Torque and speed characteristics.

As shown in Figure 13.13(b), because flux and back-emf are constants, at a no load condition, using Equation (13.5), the speed would be constant and does not change with the armature current. However, under load conditions, with increase in armature current, using Equation (13.16), the emf decreases and thus using Equation (13.6) we can see that as emf decreases, the speed decreases as well.

Now consider Figure 13.13(c), which illustrates that when torque increases, the speed of the motor will decrease. This fact can be easily determined from Equation (13.17): For a constant V, as T increases, the motor speed decreases.

Applications: DC shunt motors are normally preferred for constant speed applications, such as automotive applications. However, because AC motors are simpler and less expensive, they are used in wide range of applications compared to DC motors. See Section 13.7 for more details.

EXAMPLE 13.5 Back-emf

A 220-V supply provides a current of 20 A to a shunt motor. The armature resistance is 0.3 Ω. Find back-emf and power created in the motor when the field resistance is 150 Ω.

SOLUTION

Using Figure 13.12 and Equation (13.15) the field current corresponds to:

$$I_F = \frac{\text{Supply voltage}}{R_F} = \frac{220}{150} = 1.46 \text{ A}$$

The value of the armature current is:

$$I_A = I_S - I_F = 20 - 1.46 = 18.53 \text{ A}$$

Thus:

$$E_A = V_S - I_A \, . \, R_A = 220 - 18.53 \times 0.3 = 214.44 \text{ V}$$

As a result, power corresponds to:

$$P = E_A I_A = 214.44 \times 18.53 = 3973.57 \text{ W}$$

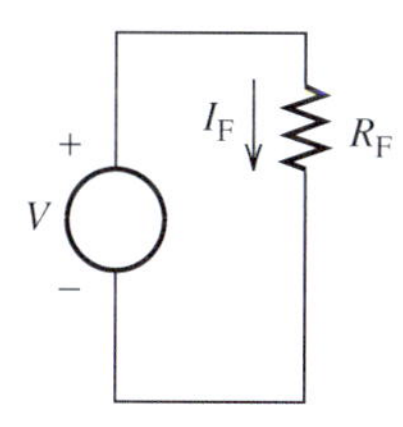

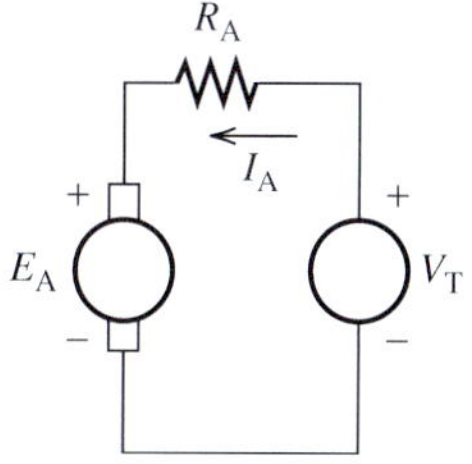

FIGURE 13.14 Equivalent circuit of a separately excited DC motor.

13.3.3 Separately Excited DC Motors

As shown in Figure 13.14, two sources of power are used in these motors: one for the stator windings and the other for rotor. This facilitates speed control by varying either of the two voltage sources. Characteristics of a separately excited DC motor are the same as those of the DC shunt motor. The main advantage of using two separate sources is to be able to control motor speed by varying either of the two sources. The disadvantage of this type is the use of a secondary DC voltage source.

13.3.4 Permanent Magnet (PM) DC Motor

The operation and characteristics of permanent magnet DC motors are the same as DC shunt motors with a wound stator. The only difference in architecture is that in a permanent magnet (PM) DC motor, the field is created by permanent magnets on the stator instead of a field winding. Figure 13.15 shows the components of a permanent magnet DC motor with all the components distinctly marked.

Figure 13.16 represents the equivalent circuit of a permanent magnet DC motor. Here, R represents the armature resistance. Remember that the field winding does not exist in a PM DC motor. Instead, a permanent magnet is used. As described in Section 13.3.1, the torque is related to the armature current by Equation (13.11). Now, because the flux is constant in PM DC motors, Equation (13.11) becomes:

$$T = K_T \cdot I_A \tag{13.19}$$

where $K_T = K_A\,\phi$ is the torque constant. Due to the relationship of the constant, K_A, and the motor dimensions and structure, [see Equation (13.12)] K_T is determined by the dimensions and structure of the motors. In addition, using Equation (13.5), the back-emf of PM motors corresponds to:

$$E_A = K_A \cdot \omega \tag{13.20}$$

where $K_A = K \,.\, \phi$ is a constant.

Now, using the circuit model of Figure 13.5, note that:

$$V_S = I_A R + E_A \tag{13.21}$$

Replacing for E_A and I_A using Equations (13.20) and (13.19), after some simple mathematical manipulations:

$$\omega = \frac{V_S}{K_A} - \frac{TR}{K_A K_T} \tag{13.22}$$

Equation (13.22) and Figure 13.17 illustrate the torque–speed characteristics of a permanent motor. From Figure 13.17, we can conclude that the torque developed by a PM DC motor is directly proportional to the applied voltage and inversely proportional to the motor speed.

FIGURE 13.15 Permanent magnet DC motor. (From www.cst.com. Reprinted with permission of Computer Simulation Technology.)

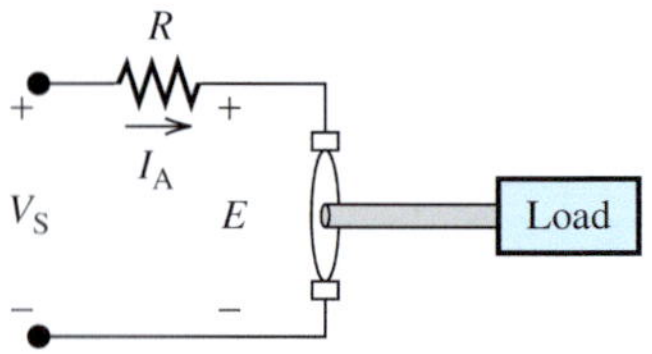

FIGURE 13.16 Equivalent circuit of a PM DC motor.

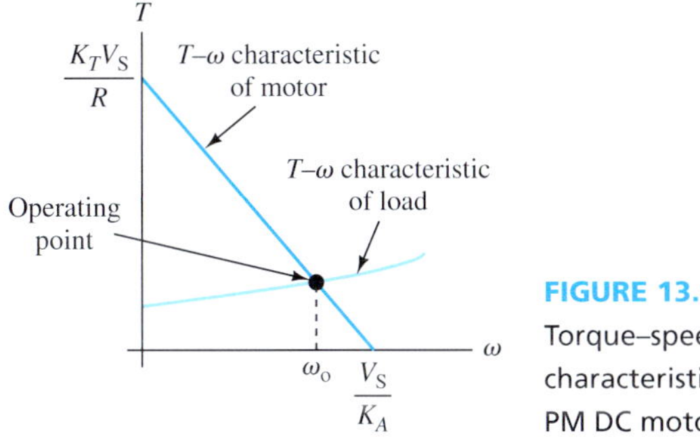

FIGURE 13.17 Torque–speed characteristics of the PM DC motor.

Figure 13.13(b) shows that when a load is applied to a motor, the torque and angular velocity of the motor decreases. In Figure 13.17, the T–ω characteristic of load represents the required torque with respect to the angular velocity to lift a load. The T–ω characteristic of motor and load intersect at the operating point, ω_o. At the operating point, the torque that is produced by the motor equals the torque required to move the load. If the torque supplied by the motor exceeds the torque required (i.e., in the left region of operating point), it will allow more load onto the motor until the motor speed is slowed down at the operating point. If the load increases after reaching the operating point, the motor will fail to function.

EXERCISE 13.2

What peculiar and important characteristic of DC series motors can we infer from Figure 13.17?

Advantages

- Higher efficiency as no power is needed to produce the stator flux.
- It is smaller in size and has a simpler architecture; thus, it can be used for simple applications such as toys.
- Motors can produce high torque with low heat loss.

Disadvantages

- Magnets can become demagnetized due to excessive use and overheating.
- The PM flux density is smaller compared to that of motors with stator winding. Thus, the torque produced is lower compared to a wound stator motor with the same power rating. This limits the use of PM DC motors to low-power applications such as motors used in toys, or low-power home applications, such as washing machines.
- PM motors have more variable characteristics than other types of motors. This can add to manufacturing cost and make PM motors less desirable if little variability is required.

EXAMPLE 13.6 Back-emf

A certain PM motor is used in a grass cutter (see Figure 13.18). The motor produces a back-emf of $E = 12$ V at the speed of 4000 rpm. Assuming the field current remains constant, find the back-emf for the speed of 1000 and 4500 rpm.

FIGURE 13.18 PM DC motor used in a grass cutter. (Used with permission from © David J. Green–technology/Alamy.)

SOLUTION

In PM DC motors $K_A = K \, . \, \phi$ is constant; therefore, using Equation (13.20):

$$\frac{E_1}{\omega_{m1}} = \frac{E_2}{\omega_{m2}}$$

(*continued*)

EXAMPLE 13.6 Continued

Replacing for E_1 and ω_{m1} and ω_{m2}:

$$\frac{12}{2\pi \times 4000} = \frac{E_2}{2\pi \times 1000}$$

or

$$E_2 = 3 \text{ V}$$

Similarly, for $N = 4500$:

$$E_2 = 13.5 \text{ V}$$

EXAMPLE 13.7 Armature Current and Voltage

Find the armature current and armature voltage (without load) of a PM DC motor when the source current is 20 A, field current is 1.2 A, source voltage is 225 V, and the armature resistance is 0.3 Ω.

SOLUTION

Using Figure 13.12, the armature current corresponds to:

$$I_A = I_S - I_F = 20 - 1.2 = 18.8 \text{ A}$$

In addition, the voltage can be found using:

$$E_A = V_S - I_A . R_A = 225 - 18.8 \times 0.3 = 219.36 \text{ V}$$

EXAMPLE 13.8 DC Motor Torque and Speed

Using the result of Example 13.7, find the speed and the torque of a DC motor with the power of 2300 W, $K = 1$, and $\phi = 3.2$ Wb.

SOLUTION

The speed of a motor is given by Equation (13.5):

$$\omega = \frac{E_A}{K \cdot \phi}$$

$$\omega = \frac{219.36}{3.2} = 68.55 \text{ rad/s}$$

In order to find the torque, use Equation (13.14):

$$T = \frac{P}{\omega}$$

$$T = \frac{2300}{68.55} = 33.55 \text{ N m}$$

EXAMPLE 13.9 DC Shunt versus Permanent Magnet Motor

Considering the similarity in the construction of the DC shunt motor and the permanent magnet DC motor, what are the differences between the two in terms of characteristics and applications?

SOLUTION

Permanent magnet (PM) DC motors support lower torques compared to the DC shunt motor because of their use of permanent magnets. PM DC motors have limited power range up to 10 hp while DC shunt motors offer power ranges up to 200 hp.

13.3.5 Series-Connected DC Motor

DC motors exhibit considerable differences in characteristics when the connection between the armature and field is changed from parallel to series. Series DC motors have higher starting torque compared to parallel motors; however, their speed may have variations, in other words their speed stability (or speed regulation) is poor. Speed regulation is defined as the difference between full-load and no-load speed. Series-connected motors are used for applications such as cranes, where a high starting load is required.

The equivalent circuit diagram of a series-connected DC motor is shown in Figure 13.19. The inductance of the field winding is neglected because inductance behaves like a short circuit for DC under steady-state conditions.

Let's analyze the equivalent circuit of the DC series motor. As shown in Figure 13.19, in this case, the armature and field windings will be in series and thus their currents will be equal. Accordingly, the relationship between the magnetic flux and the armature current corresponds to:

$$\phi = K_1 \cdot I_A \qquad \textbf{(13.23)}$$

Here, K_1 is a constant that varies with the physical parameters of the field winding. The relationship between force and current was discussed in Equation [12.11(a)] (see Chapter 12). This relationship corresponds to magnetic motive force (mmf).

Replacing for the flux in Equations (13.5) and (13.11) with the flux in Equation (13.23), the created emf, E_A, and torque, T, correspond to:

$$E_A = K \cdot K_1 \cdot \omega I_A \qquad \textbf{(13.24)}$$

and

$$T = K \cdot K_1 \cdot I_A^2 \qquad \textbf{(13.25)}$$

Here, I_A is armature current which is proportional to field current in series motors, and ω_m is the angular speed. Applying Kirchhoff's voltage law to the equivalent circuit of Figure 13.19:

$$V = (R_F + R_A) \cdot I_A + E_A \qquad \textbf{(13.26)}$$

Replacing Equation (13.5) for E_A in Equation (13.26), and substituting for the flux using Equation (13.23), the expression for I_A can then be shown as:

$$I_A = \frac{V}{R_A + R_F + K \cdot K_1 \cdot \omega} \qquad \textbf{(13.27)}$$

Now, squaring both sides of Equation (13.27) and using Equation (13.25) results in:

$$T = \frac{K \cdot K_1 \cdot V^2}{(R_A + R_F + K \cdot K_1 \cdot \omega)^2} \qquad \textbf{(13.28)}$$

Equation (13.28) shows that the torque is inversely related to the speed: As the speed increases, the torque decreases. This fact is depicted in Figure 13.20.

Applications: DC series motors can develop very high torque at low speed and therefore are preferred for traction type loads. This is due to the high armature current in Equation (13.27). Thus, they are used in cranes, electric locomotives, conveyors, and so on.

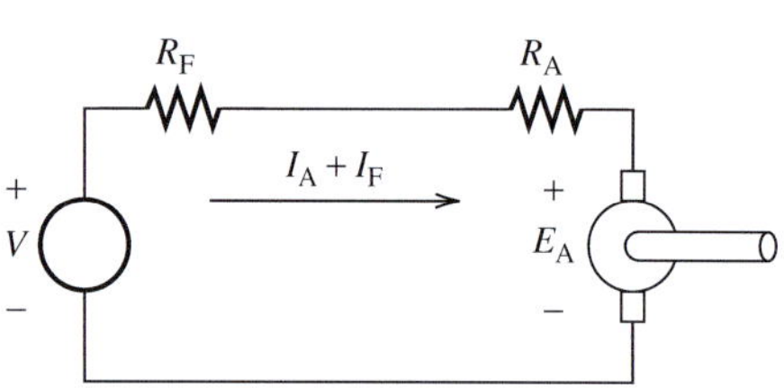

FIGURE 13.19 Equivalent circuit of the DC series motor.

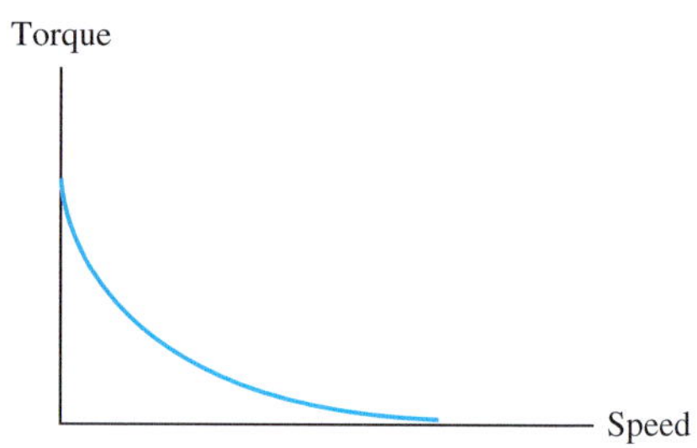

FIGURE 13.20 Torque and speed characteristics of a DC series motor.

EXAMPLE 13.10 DC Motor Speed

A DC series motor running at the speed of 800 rpm draws 18 A from the supply. If the load is changed such that the current drawn by the motor is 55 A, calculate the speed of the motor on the new load. Armature and field winding resistances are 0.2 and 0.4 Ω, respectively. Assume a supply voltage of 230 V and also that the flux produced is proportional to the current.

SOLUTION

Using Equation (13.27), calculate the constant $K \cdot K_1$:

$$K \cdot K_1 = \frac{V - I_A(R_A + R_F)}{I_A \cdot \omega}$$
$$= 0.1454$$

Next, calculate the speed of the motor at the current of $I_A = 55$ A. Using Equation (13.27):

$$\omega = \frac{230 - 33}{55 \cdot 0.1454} = 24.6 \frac{\text{rad}}{\text{s}}.$$

EXAMPLE 13.11 DC Motor Voltage and Torque

A DC motor is used in a conveyor belt. The belt has been wrapped around the rollers, and one of the rollers is connected to the motor. The series-connected motor rotates at 1200 rpm with the torque of 12 N m and the field current of 40 A. Find (a) output voltage, and (b) torque, assuming $R_A = R_F = 0$.

SOLUTION

a. In order to find the output voltage, equate Equations (13.13) and (13.14), which leads to:

$$E_A = \frac{T \cdot \omega}{I_A}$$

Before solving, first find ω. We know (Equation (13.6)):

$$\omega = \frac{2\pi N}{60} = \frac{2\pi \cdot 1200}{60} = 1215.66 \text{ rad/s}.$$

Thus, we can solve for the voltage:

$$V = \frac{12 \times 125.66}{40} = 37.69 \text{ V}$$

b. The field current is given and we know that we are using a series DC motor; thus, $I_A = I_F$ and $I_A = 40$ A. Using Equation (13.25):

$$K \cdot K_1 = \frac{T}{I_A^2} = \frac{12}{1600}$$

Now, to find the torque when R_A and $R_F = 0$, use Equation (13.28):

$$T = \frac{K \cdot K_1 \cdot V^2}{(R_A + R_F + K \cdot K_1 \cdot \omega)^2}$$

or:

$$T = \frac{37.69}{125.66} = 0.3 \text{ N m}$$

13.3.6 Summary of DC Motors

Table 13.2 summarizes the different types of DC motors, their torque characteristics, power, and applications.

13.4 SPEED CONTROL METHODS

Speed control in motors is required in many applications, such as conveyors, extruders, surface winders, machine tool spindles of lathe machines, pumps, some fans, and so on. In these applications, the motor speed must be able to be adjusted either to very high or low values. Several methods can be used to control the speed the of DC motors.

13.4.1 Speed Control by Varying the Field Current

Consider a shunt-connected DC motor. Equation (13.5) shows that keeping E_A constant, the motor speed and the flux created by the field winding are inversely proportional. E_A can be kept constant if the armature resistance, R_A, is low and thus the voltage drop across this resistance is negligible. In this case, E_A will be equal to the supply voltage which is assumed constant. Therefore, a resistance in series with the field winding (see Figure 13.21) can be used to adjust the field and, accordingly, the motor speed. As the resistance increases, the field current and, therefore, the flux decreases. This leads to an increase in the speed of the motor.

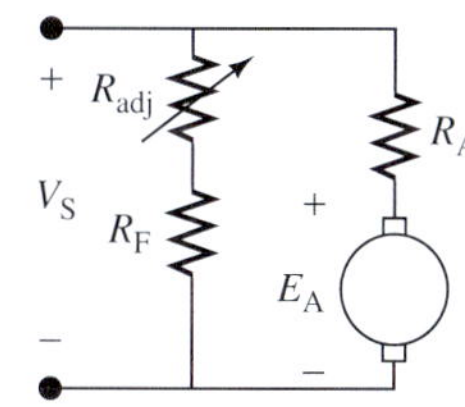

FIGURE 13.21 Speed control by varying the field current.

Accordingly, this method of speed control is only applicable to shunt-connected DC motors. It should be noted that in the case of a PM DC motor, field windings are not available. In

TABLE 13.2 Summary of all DC Motors Studied

Motor Types	Characteristics	Power Range	Applications
DC shunt motor	Constant flux constant speed motor creates moderate torque at start	Up to 200 hp	Machine tools like lathes, milling machines, grinding machines, centrifugal and reciprocating pumps, blowers, and fans
DC series motor	Creates dangerously high torques at low speeds and should be always connected to the load; speed can be varied	Up to 200 hp	Preferred for traction-type loads; employed in electric locomotives, conveyors, cranes, elevators, trolleys
Permanent magnet DC motor	Higher efficiency, smaller size and simpler architecture; magnets can become demagnetized due to excessive use and overheating; produces lower torque	Up to 10 hp	Power windows in automobiles, computer peripherals
Separately excited motors	Can be controlled either by varying the voltage applied to the field winding or by varying the voltage applied to the armature; can produce high torques	Up to 100 hp	Traction applications, to control the speed and torque of the motor by changing both armature voltage and stator current

addition, in series-connected motors, the current through the armature is equal to the field current [see Equation (13.23)]. Thus, it is not possible to control the speed using field current in either PM DC motors or series-connected motors.

Characteristics of the field current speed control system:

- Speed control is smooth and easy.
- Speed control above the rated value is also possible by making I_A very large (rated value is the speed that the motor rotates without causing overheating).
- The method is efficient as the power loss in the field circuit is minimal due to the lower value of the field current.
- As a disadvantage: if the speed increases to very high values, the commutation operation might be affected—an undesirable result.

EXAMPLE 13.12 Supply Voltage and Field Impact on Shunt-Connected Motor

What would happen if supply voltage varies in a shunt-connected motor? What would happen if the field circuit gets opened?

SOLUTION

I_A and I_F depend on supply voltage. Thus, if supply voltage decreases, both currents will decrease and, accordingly, the produced flux will decrease. Because the voltage drop across armature resistance is very small, $V_S \approx E_A = K\,\phi\,\omega$. Therefore, reducing V_S decreases the flux. In this case, there will only be a very small reduction in speed.

If the field circuit is open-circuited, the field current and therefore electromagnetic flux will be zero. Now assuming the motor has an initial speed, its speed increases infinitely while its torque decreases to zero. Note that as a practical matter, the speed cannot increase infinitely. Mechanical losses would overheat the motor and cause damage that would ultimately stop its rotation.

EXAMPLE 13.13 DC Shunt Motor Current and Speed

A 250-V DC shunt motor has a shunt field resistance of 150 Ω and an armature resistance of 0.3 Ω. For a given load, the motor runs at 1400 rpm, and drawing 20 A current. If a resistance of 75 Ω is added in series with the field, find the new armature current and speed. Assume load torque and flux are constant.

SOLUTION

Figure 13.12 represents the equivalent circuit of a DC shunt motor. The field current corresponds to:

$$I_f = \frac{V_f}{R_f} = \frac{250}{150} = 1.66 \text{ A}$$

Now:

$$I_A = I_L - I_f = 20 - 1.66 = 18.34 \text{ Amps}$$

Observe that the current of the field circuit is much smaller than the armature current. Thus, adding the speed control resistance leads to minor power dissipation. Now, applying KVL to the loop of Figure 13.12:

$$E_A = V_S - I_A R_A = 250 - 5.502 = 244.5 \text{ V}$$

Using Equation (13.5):

$$E_A = K\phi\omega$$

$$K\phi = \frac{E_A}{\omega} = \frac{244.5}{2 \times \pi \times 1400} = 0.0278$$

Considering Figure 13.12, note that:

$$I_f = \frac{V_S}{R_F + R_{adj}} = \frac{250}{225} = 1.11 \text{ Amps}$$

The new armature current corresponds to:

$$I_A = I_L - I_f = 20 - 1.11 = 18.89 \text{ Amps}$$

Thus, the new E_A is:

$$E_A = V_S - I_A R_A = 250 - 5.667 = 244.33 \text{ V}$$

Again, using Equation (13.6):

$$\omega = \frac{244.33}{0.0278} = 8789 \text{ rad/s}$$

13.4.2 Speed Control by Varying the Armature Current

This method is applicable to all types of DC motors. The voltage across the armature is controlled when a resistance is inserted in series with the armature. Speed has a direct relationship with the voltage applied across the armature. For a certain load, the value of the armature current remains constant. Thus, when a resistance is added in series with the armature, the new resistance is called R_A as illustrated in Figure 13.22. In this case, the same relationship as Equation (13.18) can be calculated for the torque, that is:

$$T = \frac{K\phi}{R_A}(V_T - K\phi\omega)$$

The torque–speed relationship for different armature resistances is sketched in Figure 13.23.

The main disadvantage of this method of speed control is the fact that it needs the addition of an extra resistor as shown in Figure 13.22. This resistor consumes energy. Specifically, the speed can be adjusted to lower values if the series resistance is adjusted to higher values. This leads to the waste of considerable energy in the form of the heat dissipated in the series resistance. It should be noted that this problem is not too severe when the speed control resistance is located in the shunt resistor. As discussed earlier, the current through the shunt resistor is very minimal and so is the dissipated power.

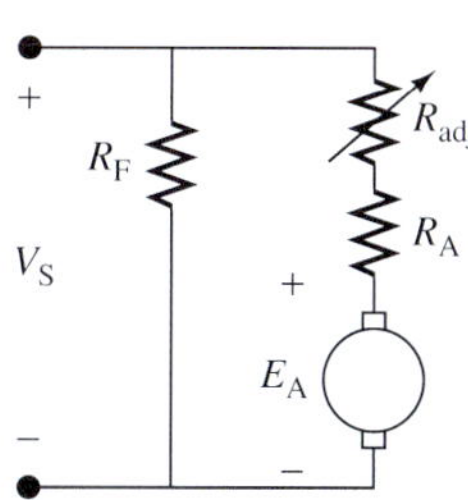

FIGURE 13.22 Speed control by the armature current.

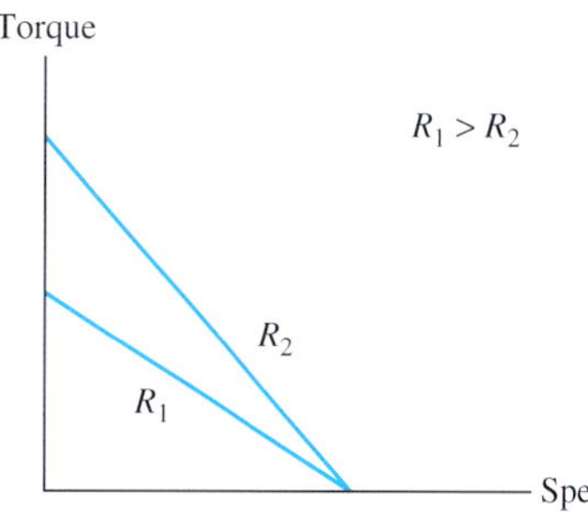

FIGURE 13.23 Torque–speed characteristics.

13.5 DC GENERATORS

A DC generator converts mechanical energy into electrical energy. Examples of the sources of mechanical energy include a steam turbine, diesel engine, or wind power. DC generators have applications in the production of industrial materials, vehicle battery charging, and street lights. Small generators produce power in the range of 1 to 10 kVA. Large industrial generators produce power in the range of 8 to 30 kVA for homes or small shops.

Generators have a standard *rating*. Rating defines the values for voltage and current of a machine at which the machine works with minimal copper losses. These standard rating definitions are designed to allow correct selection of machines. This section details the principles of DC generators and describes different types of DC generators, their features, and characteristics.

13.5.1 The Architecture and Principle of Operation of a DC Generator

The architecture of DC generators is very similar to that of DC motors, which have already been discussed. A generator consists of a stationary stator. The rotor receives the input from the prime mover. An example of a prime mover is a water turbine. The main function of the stator is to create the required flux via windings. The DC generator also consists of a commutator, whose purpose is to convert the alternating emf generated into unidirectional emf. It is made up of copper segments insulated from each other by a layer of mica (Figure 13.24).

FIGURE 13.24 Commutator in a DC generator. (Photo courtesy P. D. Simpson & Company.)

Accordingly, the fundamental principle underlying the operation of a DC generator is Faraday's law of electromagnetic induction: when the flux linkage of a coil or conductor changes, electromagnetic force (emf) is created across the coil. Changes in flux linkage occur only when there is a relative motion between the flux and the coil. The direction of this emf can be found via Fleming's left-hand rule or the right-hand rule explained in Figure 13.2.

A relative motion between flux and coil is created by rotating the conductors—with respect to the flux—using a prime mover. To achieve higher voltages, a large number of conductors needs to be connected together in a specific manner to form a winding. This winding is placed on a rotor and is called an armature winding. There are two types of armature windings: *lap winding* and *wave winding*, which are shown in Figures 13.25 and 13.26.

Lap winding: Here, the connections start from the conductor in slot 1 and subsequently overlap each other as the winding proceeds until the starting point is reached. As illustrated

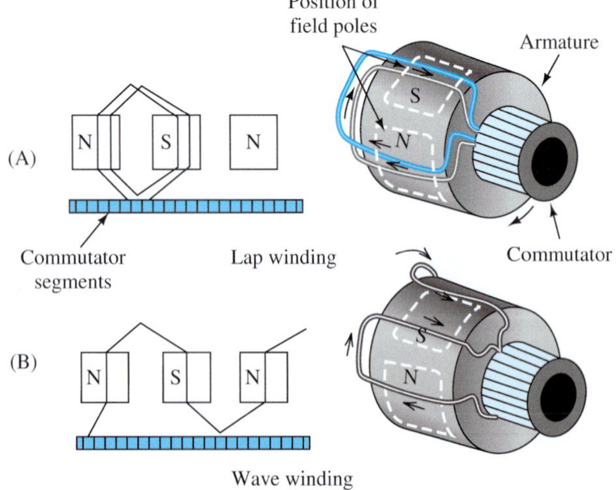

FIGURE 13.25 (a) Lap and (b) wave winding.

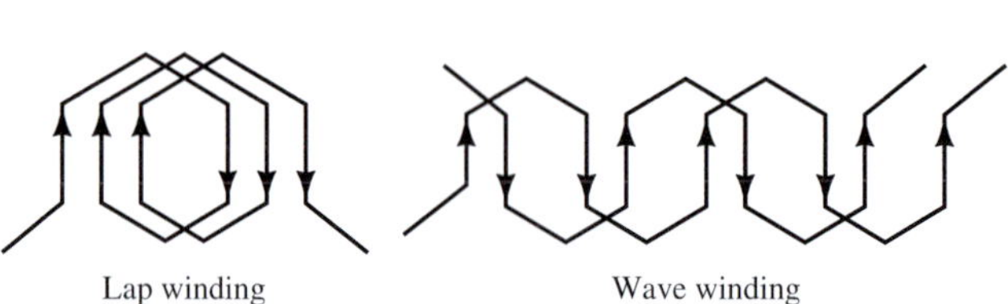

FIGURE 13.26 Types of armature winding.

in Figure 13.25, the end of one coil is connected to a commutator and the end of other coil is placed just under the same pole and in this way all coils are connected.

Accordingly, the total number of conductors is divided into a number of parallel paths, which is equal to the number of magnetic poles in the machine. The current-carrying capacity of this type of winding is higher because of the presence of several parallel paths. This type of winding (lap winding) is used in DC generators designed for high-current applications. DC generators need many pairs of poles and brushes as shown in Figure 13.25.

Wave winding: In this type, windings are organized in a way to avoid overlapping. They are organized like a progressive wave and hence the name. The winding is divided into two parallel paths irrespective of the number of poles on the machine. The lower number of parallel paths results in lower carrying current capacity. This type of winding is used in high-voltage applications. As shown in Figure 13.26, the two ends of each coil are connected to a commutator separated by the distance between poles. This configuration makes the series addition of the voltages in all the windings between brushes. This type of winding only requires one pair of brushes to provide only two paths regardless of the number of poles.

13.5.2 emf Equation

Let us now find an expression for the induced emf in terms of the parameters of DC generators. The expression is similar to Equation (13.5). The only difference is that here there are R parallel paths within which the armature conductors are divided. R equals the number of poles for lap winding, and $R = 2$ for wave winding.

There are Z conductors and R parallel paths; therefore, Z/R series-connected conductors will be available and the same emf is induced in each of them. The total induced emf defined in Equation (13.5) for motors can then be replaced by the following for DC generators:

$$E_A = e \cdot \frac{Z}{R} = \frac{\phi \cdot P \cdot N \cdot Z}{60 \cdot R} \qquad \textbf{(13.29)}$$

Here, $N/60$ can be replaced by $\omega/2\pi$ and $K = (P \cdot N \cdot Z)/(2\pi R)$. Thus, the result is an equation similar to Equation (13.6) in Section 13.2.

$$E_A = K \cdot \phi \cdot \omega \qquad \textbf{(13.30)}$$

Equation (13.30) is known as the emf equation for a DC generator. Equation (13.6) for the emf of DC motors represents the average voltage induced in the armature due to the motion of conductors relative to the magnetic field. In motors, E_A is sometimes referred to as back-emf because it opposes the externally applied voltage (the cause producing it). This voltage is exactly the same as what is found using Equation (13.30).

EXAMPLE 13.14 DC Shunt Generator

A four-pole DC shunt generator has the speed of 1200 rpm. The armature has 700 conductors and the flux per pole is 30 mWb. Find the emf voltage. Assume the number of parallel paths equals the number of poles.

SOLUTION

Using Equation (13.29):

$$E = \frac{\phi \cdot P \cdot N \cdot Z}{60R} = \frac{30 \times 10^{-3} \times 4 \times 1200 \times 700}{60 \times 4} = 420 \text{ V}$$

APPLICATION EXAMPLE 13.15 **Vehicle Battery Charging**

Explain the source of energy for charging the battery of an automobile.

SOLUTION

The battery is charged by the car itself. In the engine, the chemical energy of the fuel is converted into mechanical energy (rotary motion) available at the crankshaft. This is delivered to the input of the alternator using a fan belt. The shaft of the alternator has a cylindrical structure and has electric conductors on its periphery attached to it. Thus, it forms the rotor of the car's generator. The rotary movement at the input causes the rotor to move, and therefore, the conductors cut the magnetic flux present inside the alternator, and electricity is generated. This electrical energy is fed to rectifiers that convert alternating (time varying) voltage at the alternator to constant (DC) voltage required to charge the battery. Rectifiers were studied in Chapter 8. Figure 13.27 represents the car generator structure and the circuit for charging the battery.

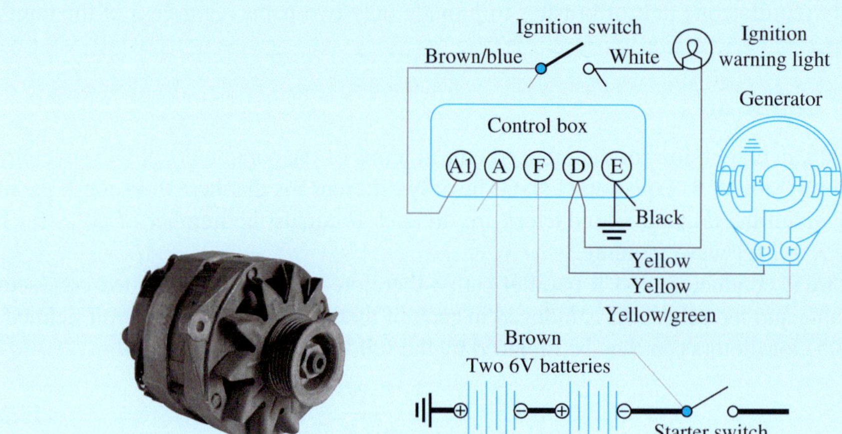

FIGURE 13.27 Car generator (Used with permission from © INSADCO Photography/Alamy.) and battery charging circuit.

13.6 DIFFERENT TYPES OF DC GENERATORS

This section examines different types of DC generators, including separately excited DC generators and shunt-connected DC generators. The section also investigates load regulation characteristics of DC generators.

A DC generator is basically an energy converter. In DC generators, the field windings must be excited with DC. The two main types of DC generator are called *shunt* and *series*. When a shunt field configuration is used, it is called a *shunt generator*. As the name indicates, the shunt field is connected in parallel with the armature. When separate voltage sources are used, the generator is called a *separately excited generator*. These types of generators are examined further in the rest of this section.

13.6.1 Load Regulation Characteristics of DC Generators

In any generator, the induced emf generates current that flows through the load. As shown in the equivalent circuits in Figure 13.28, as the load increases from zero, the current increases and the voltage drop across the armature resistance, R_A, increases. Therefore, the terminal voltage (load voltage) of any generator reduces as the load increases. As shown in Figure 13.28, the armature current increases if the load increases. This in turn decreases the back-emf.

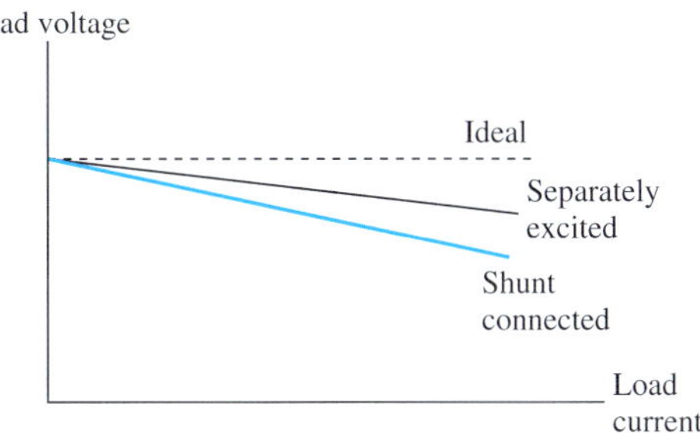

FIGURE 13.28 Load regulation characteristics for different types of DC generators.

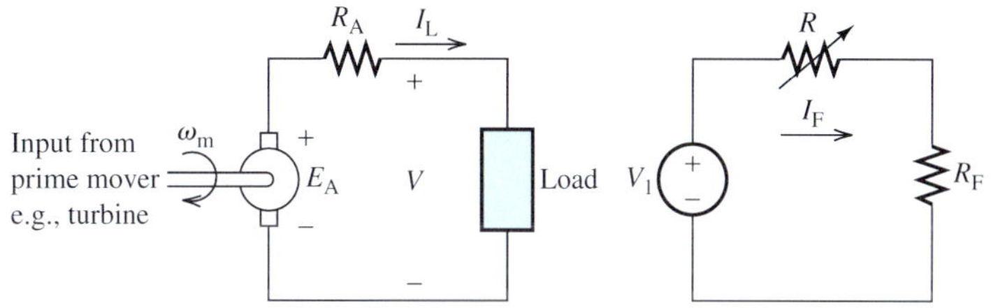

FIGURE 13.29 Equivalent circuit for a separately excited DC generator.

Voltage regulation of a generator is defined as the change in its output voltage from no load to full load. It is usually expressed as a percentage.

$$\%\ \text{Voltage regulation} = \frac{(V_{NL} - V_{FL})}{V_{FL}} \times 100 \tag{13.31}$$

where:

- V_{NL} = terminal voltage of the generator at no load ($I_L = 0$)
- V_{FL} = terminal voltage of the generator at full load

Figure 13.28 compares the load (voltage) regulation characteristics of different types of generators. For a shunt-connected DC generator, as the load on the generator increases (load current increases), the load voltage drops considerably. On the other hand, the load regulation characteristics for the separately excited DC generator show a much lower drop in load voltage with increasing load current. Thus, the separately excited DC generator has better performance than its shunt counterpart.

13.6.2 Separately Excited DC Generator

As its name suggests, with this generator there are two separate sources of voltage, one for the armature circuit and the other for the field circuit. As shown in Figure 13.29, in a separately excited DC generator, a prime mover creates the rotational movement of the generator at an angular velocity of ω_m. Induced armature voltage (i.e., the back-emf E_A) generates current flow through the load. Here, I_L and I_F are the load and field currents, respectively. Figure 13.29 (right side) shows the field circuit in which V_1 is the field voltage, and R_A, and R_F are armature and field resistances, respectively. The voltage across load is represented as V. E_A is the emf generated by the prime mover and R is the variable resistance inserted to control the flux generated, and therefore the load voltage.

EXAMPLE 13.16 Voltage Regulation of Generator

A separately excited generator is rated for a load voltage of 100 V with a full-load current of 23 A at 1800 rpm. The no-load voltage is 120 V. Find:

a. The armature resistance and the developed torque at full load
b. The voltage regulation

SOLUTION

a. Considering Figure 13.29:

$$R_A = (V_{\text{noloud}} - V_{\text{fullload}})/I_L = (120 - 100)/23 = 0.87\ \text{ohms}$$

(continued)

EXAMPLE 13.16 Continued

Using Equation (13.30):

$$E_A = K \cdot \phi \cdot \omega$$

$$K.\phi = \frac{E_A}{\omega} = \frac{120}{2\pi \times 1800} = 0.0106$$

Using Equation (13.11):

$$T = K \cdot \phi \cdot I_A = 0.0106 \times 23 = 0.2438 \text{ N m}$$

b. Voltage regulation is given by Equation (13.31):

$$\text{Voltage regulation} = \frac{V_{\text{noLoad}} - V_{\text{fullLoad}}}{V_{\text{fullLoad}}} \times 100 = \frac{120 - 100}{120} \times 100 = 16.67\%$$

Similar to the case of DC motors, the disadvantage of a separately connected machine is that two power sources are required. In the case of a shunt-connected DC generator, only one power source is required as discussed in the next section.

13.6.3 Shunt-Connected DC Generator

In these generators, the field is in shunt with the armature. Here, the load voltage can be controlled by changing the resistor, R. The load regulation of a shunt-connected DC generator is not as good as its separately excited counterpart because as the load current increases, the voltage drop across the armature resistance increases and as a result the field current reduces as described in Figure 13.30. Here, the armature current, I_A, is a combination of field and load currents, I_L and I_F. Thus, the voltage drop across R_A will be larger due to the high resistance of R_A. Shunt-connected generators have large load regulation as compared to separately excited generators because the field current falls as the load current increases due to the drop across the armature resistance.

13.7 AC MOTORS

This section outlines AC machines such as AC motors and generators. Here, AC stands for *alternating current*. The section covers three-phase as well as single-phase motors.

- An AC motor takes in AC supply as an input and converts it to rotational motion: conversion of electrical energy to mechanical energy.
- An AC generator accepts mechanical (rotational motion) as input and produces alternating supply at its output: conversion of mechanical energy to electrical energy.

AC motors are discussed first. According to the classification of motors discussed earlier, AC motors are basically divided into two types:

1. Synchronous motors
2. Induction motors

It should be noted that when comparing an AC and DC motor, the speed of a DC motor is controlled by its voltage, whereas an AC motor speed is a function of supply frequency. That

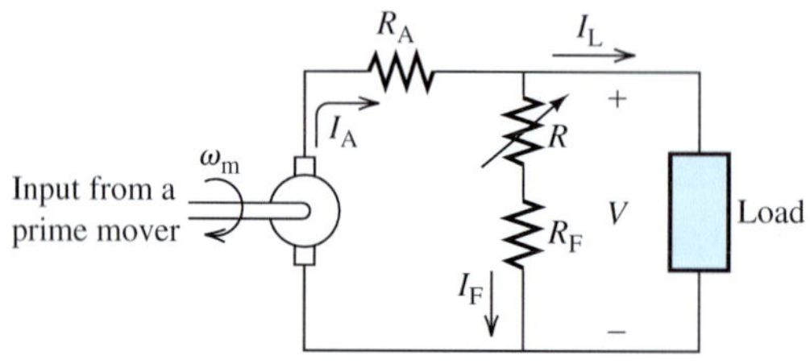

FIGURE 13.30 Equivalent circuit for a shunt-connected DC generator.

is why conventional analog electric clocks with motors are so accurate. In addition, in an AC motor, the commutator that is required in DC motors no longer needed because the AC itself causes a reversal of the magnetic force. Therefore, no change of brushes needed and no cleaning is required. In other words, AC motors are mechanically simpler. Thus they are less expensive. Finally, in many AC motors, the power consumption is fixed and is not purely a function of voltage, lowering the voltage, the current increases.

13.7.1 Three-Phase Synchronous Motors

These motors operate at a constant speed and hence the name *synchronous speed*. Section 13.7.1.3 details the principles of synchronous speed. These motors need a rotating magnetic field (RMF) for their operation. Let's see how that can be created.

13.7.1.1 CREATING A ROTATING MAGNETIC FIELD (RMF)

The generation of synchronous speed motors involves the use of polyphase (multiple phase) electric supply and stationary windings in the stator (and rotor) placed physically apart in space. Each of these stationary windings is called a *phase*. The physical separation between these phases should be equal to the electrical phase difference between the currents from the polyphase supply. Typically, a three-phase supply is used for these motors.

As discussed in Chapter 9, the phase difference between the output currents from a three-phase supply is 120°. Hence, it is also mandatory that the physical separation between the stationary windings (phases) is 120°. The number of stationary windings (phases) has to be equal to the number of supply phases.

When all of these conditions are satisfied, a rotating magnetic field is generated. Because three-phase supply is applied to the three stationary windings, three fluxes are produced. The resultant interaction of these fluxes is a flux of constant magnitude with its axis rotating in space. This corresponds to the fact that windings don't need to be physically rotated, which is definitely an advantage as it avoids any possible mechanical damage in the windings.

13.7.1.2 RMF IN A SYNCHRONOUS MOTOR

A three-phase synchronous motor consists of three-phase windings, placed in lengthwise slots cut in the stator. The three phases (stationary windings) are separated from each other by 120° and connected in either a star (having a common point called neutral) or delta (windings connected end-to-end to form a triangle) arrangement and powered by a balanced, three-phase electric supply. The same concept has been discussed in Chapter 9 for power systems.

Figure 13.31 shows a star-connected, balanced, three-phase winding and the direction of rotation of RMF. In reality, each winding consists of a large number of turns that spread over the entire circumference of the stator. A balanced supply corresponds to the production of equal magnitude of sinusoidal flux across all windings.

Assume the phase sequence of the windings is denoted as R–Y–B. The fluxes generated in these three phases are depicted by ϕ_R, ϕ_Y, ϕ_B. These are sinusoidal as shown:

$$\phi_R = k\, i_R(t)\cos\theta \quad \textbf{(13.32)}$$

$$\phi_Y = k\, i_Y(t)\cos(\theta - 120^\circ) \quad \textbf{(13.33)}$$

$$\phi_B = k\, i_B(t)\cos(\theta - 240^\circ) \quad \textbf{(13.34)}$$

where k is a constant that depends on the geometry and materials of the stator and rotor, and $i_R(t)$, $i_Y(t)$, and $i_B(t)$ are currents of windings. As we have learned, applying a balanced three-phase source to the windings results in currents with the same magnitude, which are 120° apart in phase. This fact is shown in Figure 13.32.

The currents are given by:

$$i_R(t) = I_m\cos(\omega t) \quad \textbf{(13.35)}$$

$$i_Y(t) = I_m\cos(\omega t - 120^\circ) \quad \textbf{(13.36)}$$

$$i_B(t) = I_m\cos(\omega t - 240^\circ) \quad \textbf{(13.37)}$$

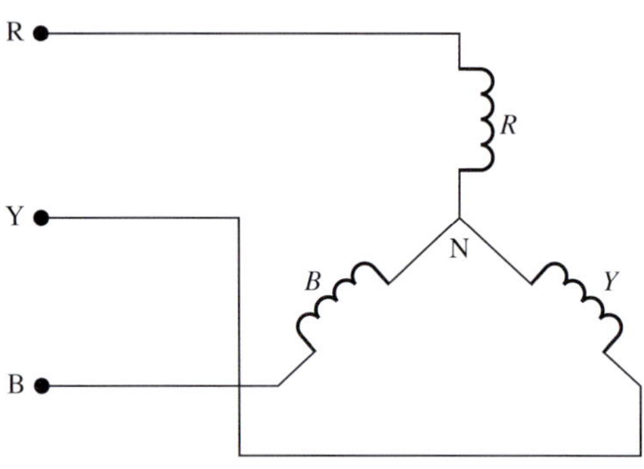

FIGURE 13.31 Star-connected three-phase winding.

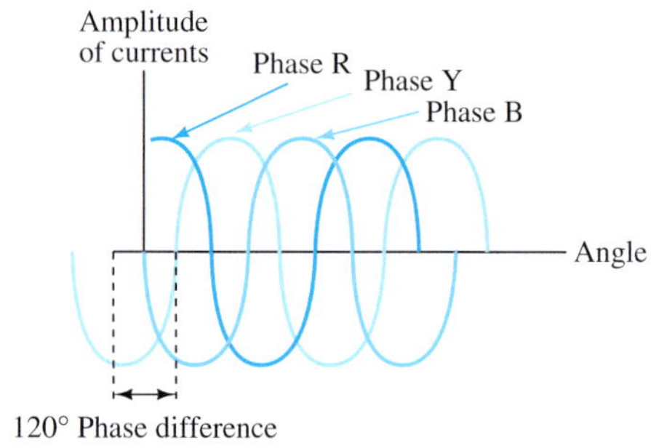

Amplitude of currents
Phase R
Phase Y
Phase B
Angle
120° Phase difference

FIGURE 13.32 Waveform of the three currents.

Here, I_m represents the maximum current. From Equations (13.32) to (13.34), the magnitude of the flux in each phase can be inferred to be the same.

Figure 13.33 represents the phasor diagram for the three fluxes and that results for the condition of $\theta = 0$. In addition, $\phi_T = \phi_R + \phi_Y + \phi_B$ is the total or the resultant of the three fluxes obtained by phasor addition.

In the phasor diagram of Figure 13.33, the magnitudes of ϕ_R, ϕ_Y, ϕ_B can be found using Equations (13.32) to (13.34). Assuming $\theta = 0$, $\phi_Y = 0.25\,\phi_R$, $\phi_B = 0.25\,\phi_R$. Using these values, the total amplitude of $\phi_T = 1.5\phi_R$. This can be easily proven by applying some basic geometric principles.

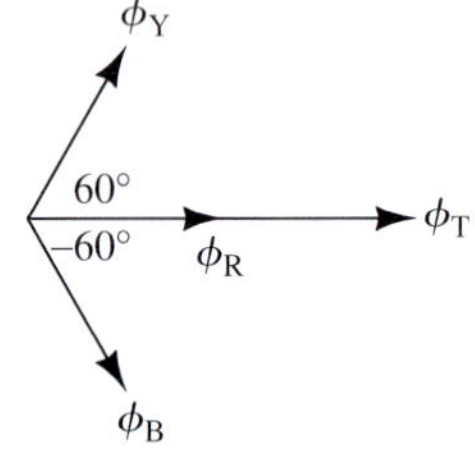

FIGURE 13.33 Directions of fluxes and the calculation of resultant flux when $\theta = 0$.

EXERCISE 13.3

Calculate the total flux when $\theta = 90$. $\phi_Y = 0.25\ \phi_R$, and $\phi_B = 0.25\ \phi_R$

13.7.1.3 SYNCHRONOUS SPEED AND TORQUE

The speed of the rotation of the resultant flux, that is, the rotating magnetic field in rpm is N rpm. Accordingly, the revolutions per second correspond to $N/60$. P refers to the number of poles. The speed of a synchronous motor is determined by the number of poles and the frequency of signal, not line voltage. In a three-phase system $P = 6$, and for a two-phase system $P = 2$. As illustrated in Figure 13.34, a three-phase induction motor has six poles.

Equation (13.6) was presented for the angular speed of a two phase system. In general, for a P pole system the frequency of signal at the output of a generator (or the frequency of signal applied to an AC motor) corresponds to the equation:

$$f = \frac{N}{60} \cdot \frac{P}{2} \tag{13.38}$$

Accordingly:

$$N = \frac{120 \cdot f}{P} \tag{13.39}$$

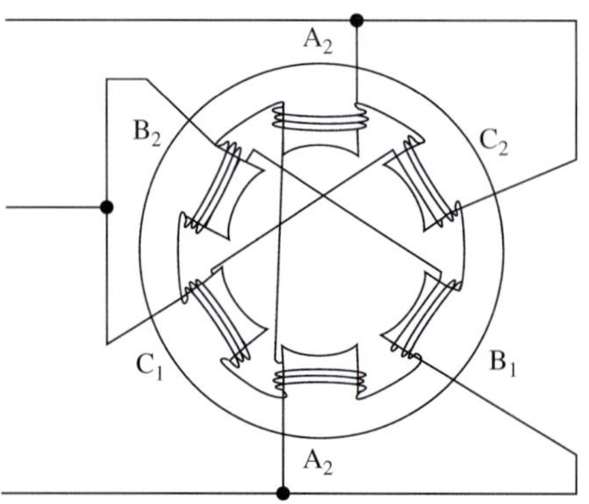

FIGURE 13.34 A three-phase synchronous motor.

Thus, the rotation speed, N, varies with the supply frequency as well as the number of poles. The torque developed by a synchronous motor can be written in terms of stator and rotor currents. It corresponds to:

$$T = \phi_s \cdot \phi_r \cdot \sin(\gamma) = K \cdot I_s \cdot I_r \sin(\gamma) \quad \textbf{(13.40)}$$

where γ is the angle between the stator and rotor fields, and $I_s(\phi_s)$ and $I_r(\phi_r)$ are the stator and rotor currents (fields). In addition, similar to K_1 in Equation (13.23), K is a constant that varies with the physical parameters of the field winding.

EXAMPLE 13.17 Synchronous Speed

A six-pole, three-phase synchronous motor is supplied by a 60-Hz supply. Determine its synchronous speed.

SOLUTION

Using Equation (13.39), the synchronous speed can be found to be:

$$N = \frac{120 \cdot f}{P} = \frac{120 \times 60}{6} = 1200 \text{ rpm}$$

13.7.1.4 STRUCTURE OF SYNCHRONOUS MOTORS

The earlier sections of this chapter examined the structure of DC machines. There is a minor constructional difference between AC and DC machines. The synchronous motor basically consists of a stator and a rotor. The stator is powered by a three-phase AC supply while a rotor might be powered by a DC or an AC supply to create poles and support the process of rotation. Figure 13.35 shows the architecture of a synchronous motor with a salient pole rotor.

Stator: The stator contains a set of windings (also called an *armature*) that creates the stator rotating magnetic field (RMF). These windings might be star- or delta-connected (see Chapter 9). These fields consist of P number of magnetic poles. The number P is an even number, as for every north pole there is a south pole.

Rotor: The rotor of a synchronous machine can be either a salient pole (also called a projected pole) or non-salient pole (also called a cylindrical pole) as shown in Figure 13.35(a) and (b), respectively. Many manufacturers use a salient pole type of construction because it has a large starting torque. Salient poles are divided into two types: permanent magnet or electromagnet. Non-salient pole rotors are also called *drum* or *wound* rotors. Salient rotors are designed for lower speeds (less than 1500 rpm) while non-salient rotors are designed for higher speeds (higher than 1500 rpm).

The field current can be supplied by an external DC source; however, in most cases an AC generator output is rectified and applied to the field. This AC generator is called an *exciter*. The input energy to this generator is created by the synchronous motor rotation. The exciter avoids the requirement of using a DC source and its related maintenance issues.

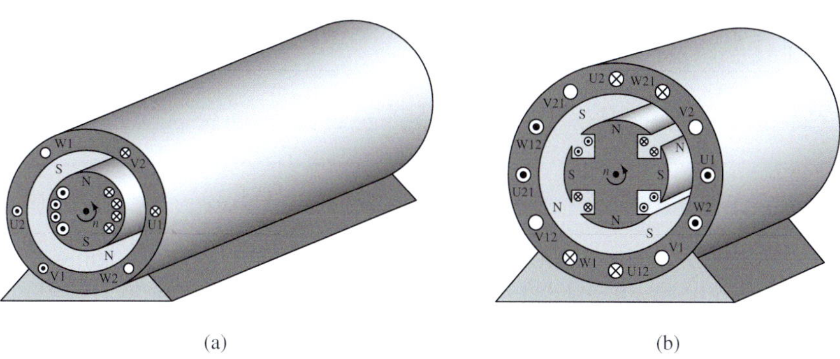

FIGURE 13.35 The architecture of a synchronous motor. Adapted with permission from Hermann Merz, Electrical Machines and Drives: Fundamentals and Calculation Examples for Beginners. Berlin: VDE Verlag, 2002.

13.7.1.5 OPERATION

The excitation provided to the stator generates a magnetic field (RMF) in the air gap between the stator and rotor. The principle of operation can be summarized as follows: "When a current-carrying conductor (rotor) is placed in a magnetic field (RMF) it experiences a force."

The rotating magnetic field is equivalent to a physically rotating magnet. This rotation occurs at the synchronous speed of N_s. For simplicity, let us consider that the stator consists of two poles N_1 and S_1 rotating at a synchronous speed. The rotor field current also creates two rotor poles N_2 and S_2. Consider that at a given instant, the stator and rotor poles are aligned such that like poles are near each other. Like poles repel each other and so if the stator poles are rotating clockwise, then the rotor poles will experience a torque in the anticlockwise direction. But due to inertia, it takes some time for the rotor to start rotating. This time period might be in the order of one half-cycle of the electric supply of the stator.

At the start of the next half-cycle, the positions of the two poles of the stator would be exactly opposite to the start of first half-cycle. Thus, now the unlike poles are near each other and as a result the rotor poles will experience torque in the clockwise direction. Hence, in one cycle of electric supply to the stator, the rotor experiences zero net torque. Even if it is assumed that at the start, the unlike stator and rotor poles are facing each other, still—because of the inertia of the rotor—the net torque experienced by the rotor would be zero. Accordingly, a synchronous motor is not self-starting.

This leads to the question: How can a synchronous motor be started?

The answer lies in rotating the rotor of the synchronous motor using an external drive at a speed almost equal to the synchronous speed. At the beginning of this process, the rotor does not have any excitation. After the rotor starts rotating at speeds almost equal to N_s, it is then supplied with electric current to generate poles. At a certain instant, unlike stator and rotor poles may face each other with magnetic axes almost aligned toward each other. In this case, they experience a force of attraction between the two and it is as if the stator and rotor are locked together magnetically and from now on they continue to occupy same relative positions.

The rotor now experiences a unidirectional torque in the direction of stator field and is said to be *synchronized* with the stator. Now, the external drive can be removed and the synchronous motor is now up and running at the speed of N_s and will continue to run. As a result, synchronous motors are inherently constant speed motors. Even though mechanical load on the motor may be increased (up to a certain limit which is determined by other parameters of the motor) it continues to rotate at a constant speed. This is an advantage of synchronous motors. The speed may change if there are fluctuations in the supply frequency.

Synchronous motors have the following drawbacks:

- Because three-phase constant voltage and constant frequency are supplied, speed variation/control is not possible unless a supply with variable frequency is provided.
- They require an external drive to start and bring them up to their synchronous speed.
- Two separate drives, one AC and one DC are needed.

However, synchronous motors still find applications in many devices. Examples include fans and blowers, rolling mills, cement mills, textile mills, and motor generator sets.

EXERCISE 13.4

Can the impedance of the rotor conductors be assumed to be purely resistive?

13.7.2 Three-Phase Induction Motor

The three-phase induction motor is used in a wide range of applications. More than 80% of motors used today are induction motors. In this motor, torque is developed in the rotor due to induced currents which react with the stator flux.

The *three-phase induction* motor derives its name from the fact that currents are "induced" in the rotor due to the RMF produced in the stator. Thus, no separate excitation is required for the rotor.

13.7.2.1 STRUCTURE

The generation of RMF and the architecture of the stator are exactly the same as the synchronous motor discussed in the previous section. The difference between the two motors from an architectural point of view lies in the rotor. Figure 13.36 shows a photograph of an induction motor, normally used in a home for water pumping purposes.

FIGURE 13.36 Three-phase induction motor. (Used with permission from © Scott Bowman/ Alamy.)

Rotors. Two kinds of rotor architectures are used in induction motors:

1. Squirrel cage or short-circuited rotors [Figure 13.37(a)]
2. Slip ring or wound rotors [Figure 13.37(b)].

Squirrel Cage Rotors. This rotor is characterized by simplicity and ruggedness as shown in Figure 13.37. It consists of aluminum bars called rotor conductors placed in lengthwise slots cut along the circumference of a cylindrical rotor made up of iron. These aluminum bars are shorted at the end by rings and the entire structure looks like a cage, and hence the name (see Figure 13.38). A slip ring and brush arrangement is not needed which helps to reduce the required maintenance and keeps the architecture simple. This type of rotor is most commonly used. When rotor and stator fields are stationary with respect to each other, starting torque is generated. If this starting torque is sufficient to spin the rotor, then it will move up to its operating speed. Moderate starting torque is achievable using this rotor design. Starting torque is an important feature of an electric machine. It is the torque provided by the motor at the zero speed.

Slip Ring Rotors. In this case, the rotor is exactly identical to the stator. It contains a set of three-phase coils placed in slots and configured in such a way that the number of poles on the stator and rotor are equal. Three terminals of the windings are brought out to the external terminals through the use of a slip ring and brush arrangement. Advantages of slip ring rotors include: (1) good speed and torque control can be achieved with variable resistances connected to its terminals and (2) supporting very high-starting torques.

Disadvantages of this type of rotor include:

- High cost of construction.
- Less rugged than alternatives.
- Maintenance is required due to the brush and slip ring arrangement.

13.7.2.2 PRINCIPLE OF OPERATION

An earlier section outlined how a rotating magnetic field produces the same effect as rotating magnets. Suppose that the direction of rotation is clockwise. Now, because the rotor is stationary, there is a relative motion between the RMF and the rotor. The RMF that is cut by the rotor conductors leads to the induction of electromotive force (emf) in the rotor conductors. This makes the current flow within the rotor.

(a)

(b)

FIGURE 13.37 (a) Squirrel cage (Used with permission from © David J. Green–electrical/Alamy.) (b) wound rotor. (Used with permission from © Lyroky/Alamy.)

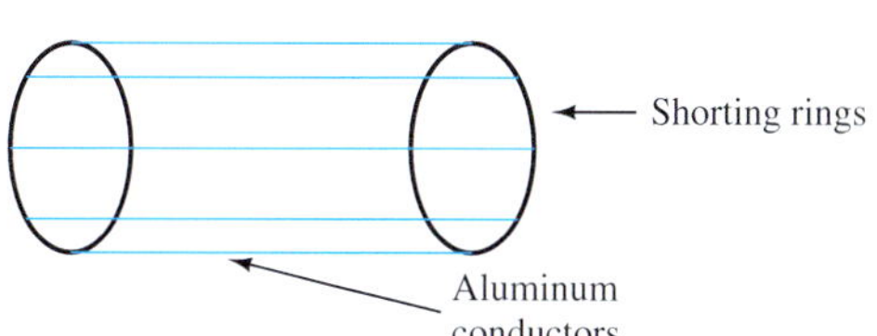

FIGURE 13.38 Diagram of a squirrel cage rotor.

According to Lenz's law, this current should be in a direction such that it opposes the cause producing it. The cause producing the current is the induced emf, and an emf exists in the first place because of the relative motion between the stator and the rotor conductors. Therefore, to oppose or reduce the relative motion, the rotor starts rotating in the same direction as the stator and also tries to catch up with the stator in terms of speed.

EXAMPLE 13.18 Synchronous Speed

Does the rotor ever catch up with the synchronous speed of the stator rotating field? What would happen in such a scenario?

SOLUTION

If the rotor does manage to rotate at the same speed as that of the stator rotating field then there would be no relative motion between those two and consequently there would be no motoring action. In this case, the motor speed would reduce to zero, but as soon as the motor speed began to decline immediately a relative motion would be created again and the motor would start. In practice, however, this does not happen owing to the large inertia of the motor. As a result: *in the steady state, the rotor always rotates at a speed less than the synchronous speed.*

13.7.2.3 THE CONCEPT OF SLIP

The rotor's induced voltage depends on (1) the relative speed of the stator field with respect to the rotor and (2) on the number of poles. N denotes the rotor speed, and N_s denotes the synchronous speed. The rotor speed ranges from zero to synchronous speed. The difference between these two speeds is called the *slip speed*. The slip speed is defined as a fraction of N_s and corresponds to:

$$s = \frac{N_s - N}{N_s} \tag{13.41}$$

The maximum value of slip is $s = 1$, which occurs when the motor starts as $N = 0$ or when the rotor is stationary. The $s = 0$ condition occurs when rotor runs at synchronous speed as can be verified using Equation (13.41).

EXAMPLE 13.19 Induction Motor Speed

A six-pole, three-phase induction motor has a full-load slip of 2% with the frequency of 60 Hz. Calculate the full-load speed of the motor.

SOLUTION

Using Equation (13.39), the synchronous speed of motor N_s is:

$$N_s = \frac{120 \times \text{frequency}}{\text{No. of poles}} = \frac{60 \text{ s/min} \times 60 \text{ r/s}}{6/2} = 1200 \text{ rpm}$$

Using Equation (13.41):

$$N = N_s(1 - s) = 0.98 \times 1200 = 1176\ r/\text{min}$$

In radians per second (see Equation (13.6)):

$$\omega = 123.15 \text{ rad/s.}$$

Full-load speed is 1176 rpm.

13.7.2.4 TORQUE EQUATION

The torque developed inside the induction motor depends on the following factors:

- The amount of RMF, which interacts with rotor conductors and the induced emf
- The magnitude of the rotor current when the rotor runs
- The power factor of the rotor when it runs

Stator and rotor circuits of an induction motor are shown in Figure 13.39. Important elements include:

I_{rr} = magnitude of the rotor current when it runs
ϕ = flux that induces emf in the rotor
ϕ_{rr} = power factor of the rotor when it runs
E_r = rotor's voltage
E_s = stator's voltage
R_c = core loss resistance
X_m = mutual reactance
X_r, X_s = reactance of the stator and rotor, respectively
R_r, R_s = resistance of the stator and rotor, respectively

Based on the definition of slip speed in Equation (13.41) and the corresponding discussion, when a load is applied to a motor, the actual speed, N, is different from the synchronous speed, N_s. The slip denoted by s is defined as a fraction of N_s: $s = (N_s - N)/N_s$. Using the relationship between the rotor voltage and its angular speed in Equation (13.30), the rotor-induced voltage at the slip, s, E_{rr} can be related to the rotor voltage at synchronous speed, E_r, by:

$$E_{rr} = E_r \cdot s \tag{13.42}$$

The reactance, X_r, is computed using the synchronous speed which is proportional to the AC signal frequency. An applied load decreases that frequency by ratio s. Therefore, the actual reactance is $s \cdot X_r$. Accordingly, applying KVL to the circuit of Figure 13.39(b), and assuming that an applied load creates a slip, the rotor current with slip, s, corresponds to:

$$I_{rr} = \frac{E_{rr}}{\sqrt{R_r^2 + (s \cdot X_r)^2}} \tag{13.43}$$

$$I_{rr} = \frac{sE_r}{\sqrt{R_r^2 + (s \cdot X_r)^2}} \tag{13.44}$$

The parameter s that appears in the denominator of Equation (13.44) represents the change in speed when a load is applied.

Equations (13.13) and (13.14) show that power can be expressed in terms of torque, angular velocity, voltage, and current. Assuming resistor R_r represents the applied load, the power output at R_r can be expressed as:

$$P_{out} = I_{rr}^2 R_r \tag{13.45}$$

Combining Equations (13.14) and (13.45) and knowing that the angular frequency is reduced to $S\omega$:

$$ST\omega = I_{rr}^2 R_r \tag{13.46}$$

Substituting Equation (13.44) into Equation (13.46), the torque can be expressed in terms of voltage, resistance, and angular velocity.

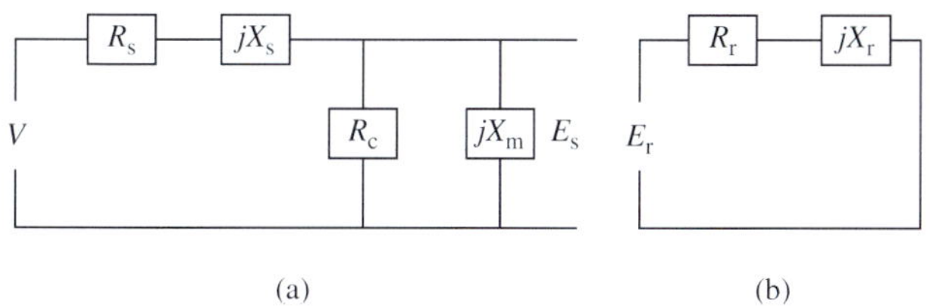

FIGURE 13.39 The (a) stator and (b) rotor circuits of an induction motor.

$$T = \frac{E_r^2}{\omega} \frac{sR_r}{R_r^2 + (sX_r)^2} \tag{13.47}$$

The maximum torque that corresponds to the slip s_{max} can be obtained by taking its derivative with respect to slip and setting it to zero, which corresponds to:

$$\frac{dT}{ds} = \left(\frac{E_r^2}{\omega}\right)\left[\frac{R_r}{R_r^2 + (sX_r)^2} - \frac{2s^2X_r^2R_r}{(R_r^2 + (sX_r)^2)^2}\right] \tag{13.48}$$

By taking $dT/ds = 0$, the slip value that corresponds to the maximum torque, s_{max} is:

$$s_{max} = \frac{R_r}{X_r} \tag{13.49}$$

By substituting the s_{max} obtained in Equation (13.49) into Equation (13.47), the maximum torque (T_m) is represented as:

$$T_m = \frac{E_r^2}{2\omega X_r} \tag{13.50}$$

The maximum torque is also known as *breakdown torque* and the speed at this torque is called *breakdown speed.*

13.7.2.5 TORQUE–SPEED CHARACTERISTICS

The region near the synchronous speed is called the *stable region* and usually represents the operating range of the induction motor. In this region, the speed ranges between 90% and 95% of the synchronous speed. In this region, up to the point of maximum torque, the slip is directly proportional to the torque (considering small slip) as illustrated in Figure 13.40. When the torque of the motor decreases, the magnitude of inductive reactance in Figure 13.39 will increase and the magnitude of the current will become almost independent of slip.

As the slip increases beyond the point of maximum torque, the inductive component in Equation (13.44) dominates the resistive one and the equation reduces to satisfy the condition in which the torque varies inversely with slip. In any machine, when load increases the speed reduces. In this case, the reduction in speed refers to an increase in slip, which in turn reduces the torque. This is undesirable as one would want the torque to increase as a response to higher loads. Accordingly, higher loads lead to further decrease in speed and torque and eventually the motor will stop. As depicted in Equation (13.48), at any speed, the developed torque is proportional to the square of the applied voltage. Figure 13.40 also illustrates that motor speed does not change significantly around the no-load condition. Thus, practically, the induction motor is called a*constant speed motor.*

13.7.2.6 ROTOR INPUT POWER VERSUS ROTOR COPPER LOSS AND MECHANICAL POWER

According to Equation (13.14), the mechanical power of a motor is related to the torque by $P = T \cdot \omega$, in which $\omega = 2 \cdot \pi \cdot N/60$, and N is the speed in rpm. P_s is defined as the total input power applied to the induction motor at the stator. In addition, assume the percentage of this

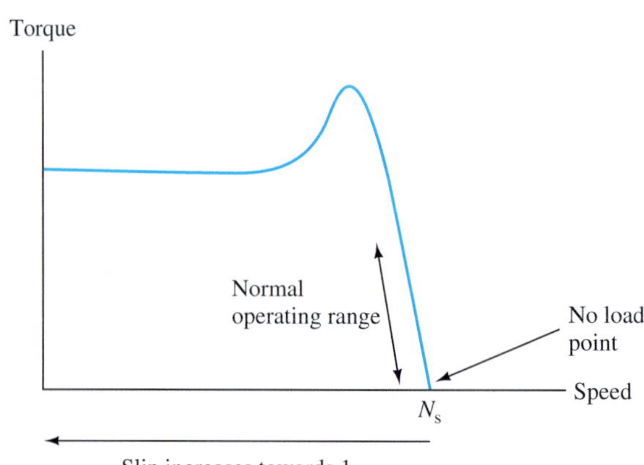

FIGURE 13.40 Torque–speed characteristics of an induction motor.

power that is the power P_r is applied to the rotor. P_r is coming from the stator side through the rotating magnetic field that has the synchronous speed, N_s. Assuming that the rotor rotates with synchronous speed, the power then corresponds to:

$$P_r = T \times \frac{2 \cdot \pi \cdot N_s}{60} \quad \textbf{(13.51)}$$

Now, the rotor tries to apply all of this power to the mechanical load on the motor. But the rotor has the speed of N, which is slightly lower than the synchronous speed. If the synchronous speed N_s is replaced with the rotor speed N, the equation of P_m is obtained—which is the power provided by the rotor to the mechanical load—and is shown as:

$$P_m = T \times \frac{2 \cdot \pi \cdot N}{60} \quad \textbf{(13.52)}$$

The difference between Equations (13.51) and (13.52) is the value of the rotor copper losses which corresponds to:

$$P_C = P_r - P_m = T \times \frac{2 \cdot \pi}{60} \times (N_s - N) \quad \textbf{(13.53)}$$

Dividing Equation (13.53) by Equation (13.51) results in:

$$\frac{P_C}{P_r} = \frac{N_s - N}{N_s} \quad \textbf{(13.54)}$$

However, based on the definition in Equation (13.41):

$$s = \frac{N_s - N}{N_s}$$

Therefore:

$$\frac{P_C}{P_r} = s \quad \textbf{(13.55)}$$

Using a similar approach, the ratio is:

$$P_r : P_C : P_m \text{ is equal to } 1 : s : 1 - s \quad \textbf{(13.56)}$$

This demonstrates a very important identity relating the power at various stages inside the motor to slip s. This equation shows the role of slip, s, in getting good performance in an induction motor.

EXERCISE 13.5

Verify Equation (13.56).

EXAMPLE 13.20 **Slip and Speed of Induction Motor**

An eight-pole, three-phase induction motor is supplied by a 60-Hz source. At full load, the frequency of the emf induced in the rotor is 5 Hz. Find the full-load slip and the full-load speed.

SOLUTION

First, calculate the synchronous speed using Equation (13.39):

$$N_s = \frac{120 \times f}{P} = \frac{120 \times 60}{8} = 900 \text{ rpm}$$

The full-load slip corresponds to:

$$s = \frac{f_R}{f} = \frac{5}{60} = 0.083$$

Using Equation (13.41) to compute the full-load speed:

$$N = (1 - s)\, N_s = 825.3 \text{ rpm}$$

EXAMPLE 13.21 Total Power Transferred to the Rotor from Stator

The torque at the load of a three-phase, 60-Hz, six-pole induction motor is 150 N m. The frequency of the emf induced in the rotor is 4 Hz. Mechanical losses are 500 W. Calculate the total power available to the rotor from the stator.

SOLUTION

To calculate the synchronous speed, use Equation (13.39):

$$N_s = \frac{120 \times f}{P} = \frac{120 \times 60}{6} = 1200 \text{ rpm}$$

Next, the full-load slip corresponds to:

$$s = \frac{f_R}{f} = \frac{4}{60} = 0.067$$

Thus:

$$N = (1 - s)\, N_s = 1119.6 \text{ rpm}$$

Now, using Equation (13.52):

$$P_m = T \times \frac{2\pi N}{60} = 150 \times 117.71 = 17{,}656 \text{ W}$$

Therefore, the total power available to the rotor is:

$$P_r = P_m + P_c = 17{,}656 + 500 = 18.16 \text{ KkW}$$

13.7.2.7 EFFECT OF EXTERNAL RESISTANCE ON TORQUE

An external resistance might be applied to a motor to control the speed of the motor. Adding an external resistance is only possible for slip ring or wound rotor induction motors.

Recall the analysis of Section 13.7.2.4 to compute the equation of the torque of induction motors. From Equation (13.49), observe that the maximum value of the torque is independent of the rotor resistance at standstill. Therefore, adding an external resistance does not alter the maximum torque value but can definitely impact the value of the slip at which that maximum value occurs. As observed in Section 13.7.2.3, s_{max} is the value of slip at which the maximum torque is attainable. Based on Equation (13.49), s_{max} corresponds to:

$$s_{max} = \frac{R_r}{X_r}$$

Therefore, the addition of external resistance to the rotor changes the value of s_{max}. Accordingly, the maximum torque will occur at higher values of slip. Now, by referring to Equation (13.48), the expression for the torque corresponds to:

$$T = \frac{s \cdot E_r^2 \cdot R_r/\omega}{R_r^2 + (s \cdot X_r)^2} \tag{13.57}$$

When the motor starts, the value of slip is 1, that is, $s = 1$ in Equation (13.41). Thus, the starting torque can be controlled by adding an external resistance to the increasing rotor resistance.

If the maximum torque is desired at the motor's start, then the value of slip should be equal to 1 at maximum torque. Equation (13.49) shows that this is only possible when $R_r = X_r$. However, such a high value of resistance—if kept permanently in the circuit—could lead to very high copper losses. Therefore, this level of resistance is undesirable. In practice, once the motor starts, this high resistance is gradually reduced to zero. Reducing the resistance after start ensures good performance at the time of start and during running conditions.

13.7.2.8 APPLICATIONS OF THE INDUCTION MOTOR

Squirrel cage induction motors are used in applications that need moderate starting torque, such as lathe machines, water pumps, grinders, and printing machines.

Slip ring induction motors are used in applications that need higher starting torque such as lifts, cranes, elevators, and compressors.

Induction motors are considered more reliable than synchronous or DC machines because they do not require slip rings or brushes.

13.7.3 Losses in AC Machines

The losses in AC machines (motors as well as generators) are almost the same as those discussed in Section 13.2.5 for DC machines. However, in the case of AC machines, iron losses or core losses that occur in the stator and rotor core are frequency dependent. Stator iron losses constitute the major part of these losses because stator frequency is the supply frequency. In the case of the rotor (for induction motors) the rotor frequency is directly proportional to the slip. Therefore, it is much smaller compared to supply frequency. Accordingly, rotor iron losses are quite minimal.

When a conductor is rotated in a magnetic field, currents are induced in it. These induced currents are called *eddy currents*. These currents create some power dissipation (loss) in the form of heat. *Hysteresis loss* is a heat loss caused by the magnetic properties of the armature. *Mechanical losses* are due to the rotation of the armature and include bearing friction loss, brush friction loss, and windage or air friction loss.

13.7.4 Power Flow Diagram for an AC Motor

Figure 13.41 represents the power flow diagram for an AC motor. In summary, there are basically three stages of losses during power flow of an AC motor. The first stage is stator losses that occur at the stator part of motor when input power is applied to the stator. The second stage is rotor losses that occur at the rotor part of the motor. The third stage of losses is mechanical losses. Power is reduced because of these losses.

Besides the machine type, that is, motor or generator, an electric machine nameplate usually includes some information that represents its nominal characteristics such as its rated voltage, its rated current, and its voltage or current regulation. The rated voltage is the terminal root mean square (rms) voltage for which the machine is designed and the rated current is the terminal rms current that does not cause overheating. Voltage or current regulation is the maximum voltage or current that maintains a constant speed or torque when the load varies.

Table 13.3 summarizes different types of AC motors and compares their characteristics.

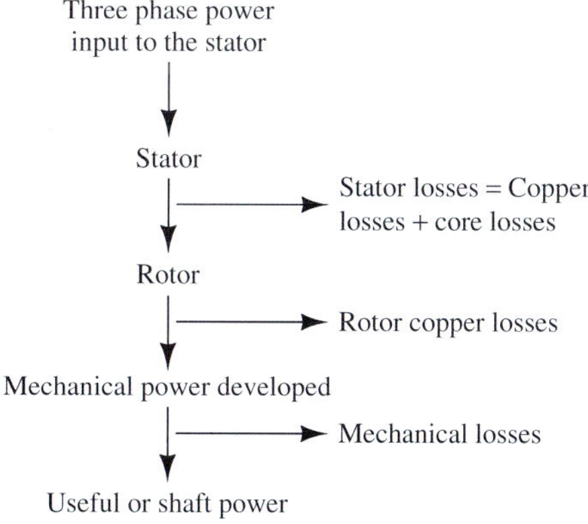

FIGURE 13.41 Power flow diagram for AC motors.

TABLE 13.3 Summary of All AC Motors Studied

Motor Types	Characteristics	Power Range	Applications
Three-phase induction motor (Squirrel cage rotor)	Moderate starting torque. Simple robust and maintenance free construction. Speed control by addition of rotor resistance is not possible	1–5000 hp	Lathe machines, water pumps, grinders, printing machines, large refrigeration and air-conditioning units, small compressors
Three-phase induction motor (Slip ring rotor)	Very high starting torque. Speed control by rotor resistance possible. High cost. Complicated construction which requires maintenance due to use of slip rings and brush	1–5000 hp	Cranes, hoists, elevators, large compressors, industrial fans and blowers
Three-phase synchronous motor	Constant speed irrespective of load, need of a starting device, need for two different excitation sources, variable frequency drives required for speed control	Up to 50,000 hp	Motor generator sets, timing devices, centrifugal pumps, textile mills, cement mills, rolling mills

13.8 AC GENERATORS

This section examines synchronous generators. Recall that synchronous machines refer to those that operate at their synchronous speed, N_S. Close to 98% of the world's power generators are synchronous generators. Induction generators are not often employed because of their lower performance. The principle of operation of synchronous generators is similar to that of DC generators.

Section 13.5 on DC generators explained that the electromotive force (emf) generated in the armature winding of DC generators has an alternating nature. The commutator is a device, which converts the alternating emf into unidirectional emf. As a result, theoretically, replacing the commutator with a slip ring and brush assembly (to tap the generated AC voltages) and making the armature rotate should convert a DC generator to an AC generator (see Figure 13.42). However, this concept cannot be implemented practically. Why not?

As we discussed previously, the relative motion between a conductor and the magnetic flux is the basis of producing emf in the conductor. Thus, both the rotation of the conductor in a stationary magnetic field and the rotation of the magnetic field with stationary conductors lead to generation of emf. In DC generators, the preferred architecture involves the conductors rotating in a stationary magnetic field. In AC generators, it is more common for the conductors to be stationary in a rotating magnetic field. Thus, the stator, also called the armature (stationary part), carries the conductors while the rotor is the rotating part, which carries the field winding (the source of flux).

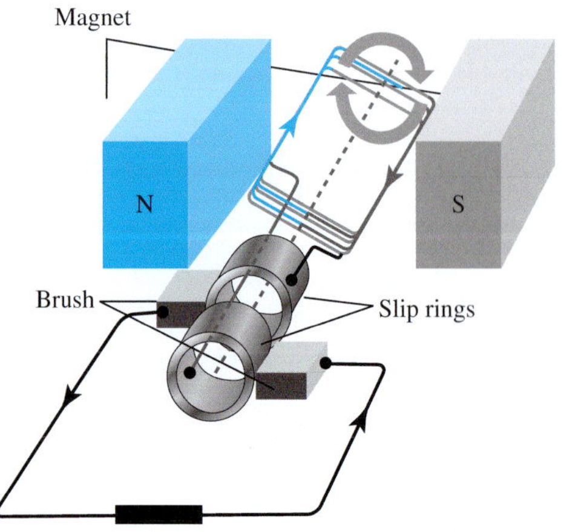

FIGURE 13.42 The structure of AC generators.

AC generators are configured differently than DC generators because of the following reasons:

- AC generators are usually of much higher capacity than their DC counterparts. Higher demands for power require the current-carrying conductor to be much thicker. Thicker conductors require deeper slots to house them. This results in an increase in the overall size of the stator. Rotating a large stator calls for more output from the prime mover. In addition, it is not easy or recommended to tap high voltages from rotating armatures in spite of the slip ring and brush arrangement.
- We can completely eliminate the necessity of a slip ring and brush assembly for the stator if it is kept stationary. This is because an emf is generated in the armature conductors, which can be directly connected to the transmission line.
- A cooling arrangement is an important consideration for the stator of an AC generator because very high voltage levels are generated. Efficient cooling is possible when the component to be cooled is stationary.
- A stationary stator winding avoids mechanical damage due to rotation and protects the armature from the centrifugal forces due to rotation.

13.8.1 Construction and Working

The construction of the synchronous generator is not very different from the synchronous motor studied earlier. A stator consists of slots to hold windings. Windings are separated from each other by 120° in space. The voltages generated by the windings have 120° phase difference with respect to each other. Steel is used as the material for construction to minimize hysteresis losses.

The rotor can be either salient pole or cylindrical in type, as shown in Figure 13.43:

- The *salient pole* has poles projecting out and field windings are located on them.
- The *cylindrical* style houses the winding in slots and the remaining un-slotted portion forms the magnetic poles. The cylindrical style is preferred for many applications that demand high speeds. Higher speeds are achievable because of the mechanical strength of the cylindrical style, which creates the ability to bear the heavy centrifugal forces due to the rotation at high speeds. A rotor has to be driven by a prime mover like an internal combustion (IC) engine (an engine that converts chemical energy into useful mechanical energy by burning fuel) or a steam turbine.

The rotor is supplied by a DC excitation and also it is rotated via a prime mover. This creates the rotation of the magnetic field. Now, the armature conductors (stator windings) cut this flux and the emf is induced in them. The synchronous speed for a rotor is given in Equation (13.38).

13.8.2 Winding Terminologies for the Alternator

Before deriving the emf equation of the alternator, a few important winding terms must be defined. A familiarity with the terminology of winding is needed because the winding impacts the amount of emf generated.

In the stator of synchronous generators, there are six terminals (ports) for the three windings (two per winding). Three out of these six terminals are connected in star or delta and the

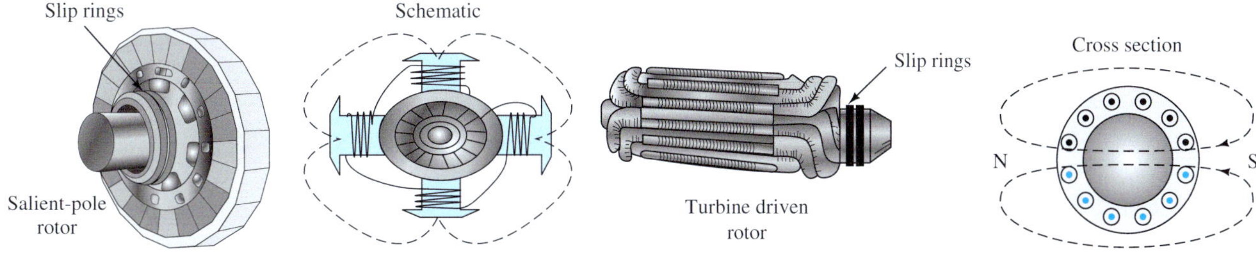

FIGURE 13.43 Salient pole and cylindrical rotor. (a) Salient pole rotor and its cross section; (b) an example of a cylindrical rotor— turbine driven rotor, and its cross section.

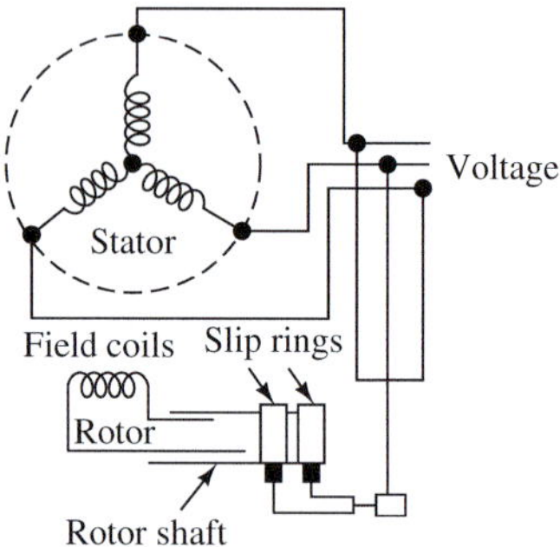

FIGURE 13.44 Circuit of an AC generator.

remaining three are used as outputs. Output terminal voltage is tapped as shown in Figure 13.44. Each of the three windings are called a *phase* and the emf generated in each is called *emf per phase*, E_{ph}. All the emfs generated should be added together.

- A *conductor* is a metallic wire under the influence of the magnetic field.
- A *turn* is formed when two conductors from two different slots are connected.
- When a number of turns are grouped together, a *coil* is formed (e.g., the field coil shown in Figure 13.44).
- The number of slots per pole or the distance between two adjacent poles is called *pole pitch*.

Accordingly, one pole corresponds to 180° of emf. This 180° is called *one pole pitch*. The number of slots per pole is referred to as *n*; therefore, these *n* slots are responsible for producing 180° phase difference.

- *Slot angle* is the phase difference created as a result of each slot on the armature of the alternator and it is denoted by β which corresponds to:

$$\beta = \frac{180}{n} \tag{13.58}$$

- *Single and double-layered winding* is distinguished by the number of coil sides (one or two) in each slot. Double-layered winding is preferred to save space.
- When the coil side in one slot is connected to another coil side, that is, one pole pitch away the winding is said to be *full-pitched winding*. Anything shorter is called *short-pitched winding*.

Figure 13.45(a) and (b) show a one-slot-per-pole alternator. Each coil in the figures is aligned with each slot. The coil group in Figure 13.45(a) is also called half-coil winding, because only one side of the slot has a coil. In Figure 13.45(b), both sides of the slot contain a coil. Therefore, the arrangement in Figure 13.45(b) is known as a whole-coil winding. Figure 13.45(c)

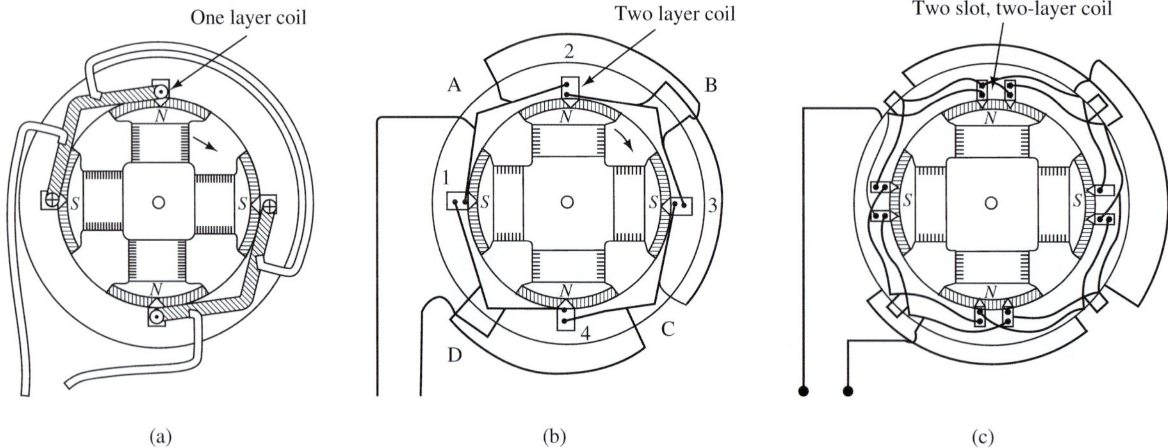

FIGURE 13.45 Alternator: (a) One-layer, one slot per pole; (b) double-layer, one slot per pole; (c) double-layer, two slots per pole. (From "A Course in Electrical Engineering, Volume 2, from ALTERNATING CURRENTS" by Chester Laurens Dawes. Copyright © 1952 The McGraw-Hill Companies, Inc.)

shows a two-slot-per-pole, two-layer alternator. Two whole-coil winding slots are aligned with each pole in the alternator.

For example, suppose an alternator has eight slots per pole. If the coil side in slot 1 is connected to the coil side in slot 9, such that the two slots are one pole pitch apart, then the winding is called full-pitched winding. If the coil side in slot 1 is connected to the coil side in slots 7 or 8, then it is called short-pitched winding.

In practice, short-pitched windings are employed more frequently because:

1. shorter copper conductors may be used for end connections;
2. high frequency harmonics are eliminated, which in turn results in reduced eddy current and hysteresis losses.

In Figure 13.46, α denotes the angle by which the coils are short pitched. The factor by which the generated emf is used is called the **pitch factor** and it corresponds to:

$$K_{\text{coil}} = \cos\left(\frac{\alpha}{2}\right) \quad \textbf{(13.59)}$$

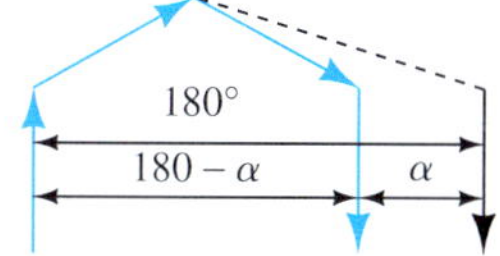

FIGURE 13.46 Short-pitched winding.

- In a *concentrated winding*, conductors that belong to a particular phase are placed in a single slot whereas in a *distributed winding* the conductors belonging to a particular phase are distributed across all slots. Distributed windings reduce harmonics heat dissipation and perform better than concentrated windings.

In practice, double-layered, short-pitched, and distributed winding is used for alternators.

In concentrated windings, because the entire winding is kept in one slot, the induced emf in each coil has the same phase. In contrast, in distributed windings, each slot adds a phase difference of β degrees to the generated emf. Consequently, there is a reduction in the emf. The factor by which the emf is reduced is called the *distribution factor* and corresponds to:

$$K_{\text{dist}} = \frac{\sin\left(\frac{m\beta}{2}\right)}{m \cdot \sin\left(\frac{\beta}{2}\right)} \quad \textbf{(13.60)}$$

where m is the number of slots per pole, per phase. For a three-phase machine, $m = n/3$ (n is the number of slots per pole).

13.8.3 The emf Equation of an Alternator

For this example, assume a full-pitch concentrated winding. The average induced emf for a single conductor is $d\phi/dt$, which is the ratio of the change of flux per unit time. For one revolution of the rotor, the emf induced in a single conductor is the ratio of the total cut flux and the time required for each revolution. If ϕ is the flux per pole, and P is the number of poles:

$$d\phi = \phi \cdot P \quad \textbf{(13.61)}$$

$$dt = \frac{60}{N_s} \quad \textbf{(13.62)}$$

Using Equations (13.61) and (13.62):

$$E_{av} = \frac{\phi \cdot P}{\frac{60}{N_s}} \quad \textbf{(13.63)}$$

In the Equation (13.63), $60/N_s$ is the time of one revolution. Here, refer to Equation (13.39), N_s is the synchronous speed and corresponds to:

$$N_s = \frac{120 \cdot f}{P} \quad \textbf{(13.64)}$$

Substituting Equation (13.64) into (13.63):

$$E_{av} = \phi \cdot 2f \quad \textbf{(13.65)}$$

Equation (13.65) represents the average value of emf induced in a single conductor. For the induced emf per turn (a turn consists of two conductors as defined previously), Equation (13.65) should be revised to:

$$E_{av} = 2 \cdot 2f\phi \qquad \textbf{(13.66)}$$

A factor of 2 in Equation (13.66) can only come into the picture when the winding is full pitch and the emfs produced by the two conductors (separated by 180° for full pitch) assist each other.

Assume there are T turns per phase and these turns are connected in series. Considering a concentrated winding, and that all these turns are within the same slot and they are in phase, then the average value of the induced emf per phase corresponds to:

$$E_{phase} = T \cdot E_{av} = T \cdot 4f\phi \qquad \textbf{(13.67)}$$

APPLICATION EXAMPLE 13.22 Hydroelectric Power Generation

Potential energy of stored water in dams is converted to kinetic energy when it flows through pipes onto the blades of a turbine. The turbine blades are connected to the rotor inside the generator by a common shaft. The rotation of turbine blades creates the rotation of the rotor conductors. Accordingly, the magnetic flux is cut inside the generator, and an emf and, finally, electricity is generated. This method is called hydroelectric power generation.

Other techniques like nuclear power generation, electricity from wind energy, and electricity from solar energy are slowly gaining momentum and their generation capacities are increasing manifold every year. For example, in 2003, 85% of the total power generation in France happened via nuclear power generation.

EXAMPLE 13.23 AC Generator Speed

Find the speed of a six-pole AC generator when its frequency is 60 Hz.

SOLUTION

To find the speed of a generator, the frequency and the number of poles must be known; in this case, 60 Hz and six (6), respectively. Using Equation (13.64):

$$N_s = \frac{120 \cdot f}{P} = \frac{120 \cdot 60}{6} = 1200 \text{ rpm}$$

EXAMPLE 13.24 Induced emf of AC Generator

Calculate the induced emf per phase of a four-pole AC generator which has a torque of 6 N·m. The speed of the AC generator is 1500 rpm and the flux of the AC generator is 0.6 Wb. The number of turns per phase is six.

SOLUTION

In order to calculate the induced emf per phase of an AC generator, first find the frequency of the generator. Because the number of poles and the speed of the generator are given, using Equation (13.64):

$$f = \frac{N_s \cdot P}{120} = \frac{1500 \cdot 4}{120} = \frac{150}{3} = 50 \text{ Hz}$$

Now, calculate the induced emf per phase. Using Equation (13.67):

$$E_{phase} = T \cdot 4 \cdot f \cdot \phi = 6 \cdot 4 \cdot 50 \cdot 0.6 = 720 \text{ V}$$

13.9 SPECIAL TYPES OF MOTORS

13.9.1 Single-Phase Induction Motors

The first single-phase induction motor was built by Nikola Tesla in the USA in 1943. He was a Yugoslav–American scientist. The SI unit of magnetic flux density is named after Tesla. The stator of a single-phase induction motor has a winding called the *main winding* and the rotor is squirrel cage just like the three-phase induction motor discussed earlier. As described for the three-phase motor, the flux (RMF) rotates in a given direction and its direction can be reversed by interchanging the Y and B phases.

However, in single-phase motors, the total flux can be expressed as the sum of two equal flux components rotating in opposite directions. One of these components rotates in the same direction as the rotor and is called the forward component, while the other is called the reverse component. Thus, it can be inferred that because the torques produced by these two flux components are equal and opposite they will cancel each other and, as a result, net starting torque will be zero. Therefore, this motor cannot deliver power to a load from a standing start.

The mechanism of a single-phase induction motor is shown in Figure 13.47. Figure 13.47(a) shows the rotor at its initial stationary position. Here, the forces in each pole are in equilibrium states. In Figure 13.47(b), when an initial torque, T_o, is applied, it makes the rotor begin to rotate. At this point, the forces are not in equilibrium anymore, and the south pole of the rotor, S_R, is pulled to the left, toward the north pole of the stator, N_S. Then, the north pole of the rotor, N_R, will be pulled toward to the right, toward the south pole of the stator, S_S. The nonequilibrium force causes the rotor to rotate after the initial torque has been applied. In order to allow the rotor to rotate continuously, the stator is connected to an AC source. The AC source switches the pole of the stator continuously, which prevents the equilibrium state from being achieved.

13.9.2 Stepper Motors

A stepper motor is a motor that maps electrical pulses into a series of discrete angular movements called steps. There is a one-to-one relationship between the input pulse and the step movement of the shaft. These motors are normally divided into two types, *variable reluctance* and *permanent magnet* stepper motors, as explained below.

Applications of these motors include the following:

- Precise position control such as controlling the position of the magnetic head for reading or writing information onto computer discs
- Odometers in automobiles
- Controlling fuel-to-air ratio before injection into a cylinder in a car
- Sound and vision equipment
- Driving the arms of a watch

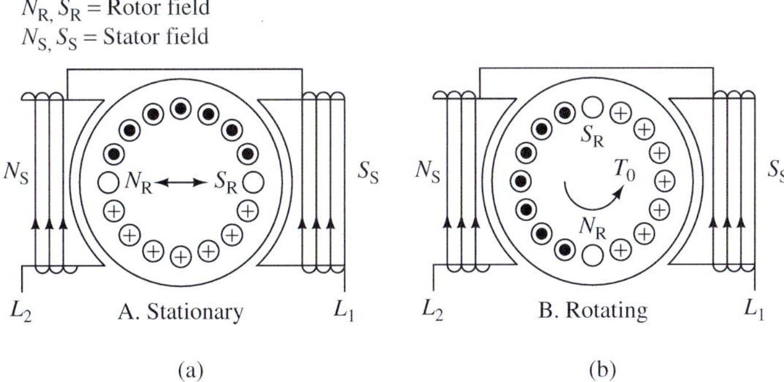

FIGURE 13.47 A single-phase induction motor. (a) At initial stationary status; (b) an initial torque is applied to begin the rotation.

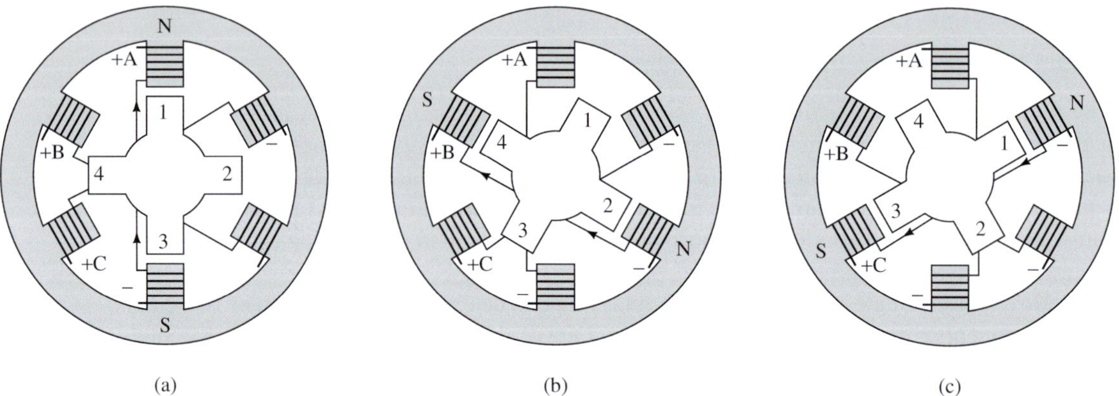

FIGURE 13.48 A variable reluctance stepper motor (From INDUSTRIAL ELECTRONICS, 1st Edition by Colin D. Simpson. Copyright © 1996 by Colin D. Simpson. Printed and Electronically reproduced by permission of Pearson Education, Inc., Upper Saddle River, New Jersey.)

13.9.2.1 VARIABLE RELUCTANCE STEPPER MOTORS

In a variable reluctance stepper motor, the stator usually consists of six poles and phase winding as shown in Figure 13.47. One stator phase includes windings on two diametrically opposite poles that are serially connected. The rotor has poles with no excitation winding.

Figure 13.48 shows how a variable reluctance stepper motor works. Three poles in the stator are labeled as A, B, and C, and the poles in the rotor are labeled as 1, 2, 3, and 4 [see Figure 13.48(a)]. The magnetic field is activated across A, B, and C sequentially. Based on the status of the poles shown in Figure 13.48(a), when current passes through A (ON), while it does not pass through B and C (OFF), the rotor will be in stationary. Next, poles A and C are turned OFF and pole B is turned ON. Then the rotor is turned and its poles (2 and 4) are aligned with pole B in the stator that is shown in Figure 13.48(b). Again, when poles B and A are turned OFF, with pole C turned ON, the rotor will rotate and its poles (1 and 3) will be aligned with pole C in the stator. At this status, the rotor rotates in the clockwise direction. However, it can be carefully designed to ensure its rotation in the desired direction.

13.9.2.2 PERMANENT MAGNET STEPPER MOTOR

In a permanent magnet stepper motor, the stator has four poles around which excitation coils are wound. These are spaced 90° apart as shown in Figure 13.49(a). These four excitation coils form the four phases of the stator.

The rotor is cylindrical in nature and permanently magnetized (made of materials such as ferrite). Now, if the phases are excited in a particular sequence there will be an interaction across the two fluxes: (a) the stator flux due to the current passing through the poles, and (b) the permanent rotor flux. Due to this interaction, a torque (angular movement or step) will be produced.

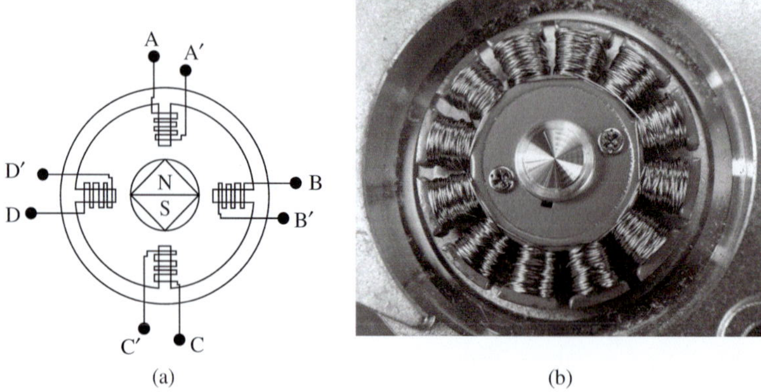

FIGURE 13.49 (a) Permanent magnet stepper motor. (b) Brushless DC motor. (Used with permission from © Lyroky/Alamy.)

The rotor position is stepped (angular movement) by applying a sequence of pulses to the stator windings. The step size will be 90°.

An increase in the number of stator poles guarantees a smaller step size but increases the difficulties in construction. As a result, hybrid stepper motors have been developed to combine the properties of variable reluctance and permanent magnet stepper motors.

13.9.3 Brushless DC Motors

Do you know which motor is used in the fan present in your computers for ventilation?
Do you know which motor is used in the hard drive in your computer?

The answer to both questions is the *brushless DC motor* as shown in Figure 13.49(b).

Learning the operation of a brushless DC motor is a logical extension of the principles of the previously studies permanent magnet stepper motor. The only difference in the architecture of brushless DC motors is the use of position sensors and control devices such as a thyristor that is an electronic gate. The position sensors can be based on Hall effect principles or can be based on optical position sensors. The Hall effect was discovered by Edwin Hall in 1879. Based on the *Hall effect*, a potential difference (*Hall voltage*) is created across an electric conductor located in a magnetic field when an electric current flows through the conductor.

This motor is equipped with a sensor that senses the magnetic field which is called Hall IC. The magnets of the motor are labeled 1, 2, 3, and 4. As one magnet gets closer to the Hall IC in Figure 13.50(a), the magnetic field of magnet 1 is detected by the Hall IC sensor. Then, the Hall IC sensor sends a signal to the transistor which produces a current in the electromagnet that creates a force capable of rotating the motor. Figure 13.50(b) shows that when a motor rotates and there is no magnet close to the Hall IC, there will be no signal in the transistor and as a result no force is created by the electromagnet. However, the inertia created by the initial force maintains the rotation. Thus, magnet 2 in Figure 13.50(c) will go into a position similar to the magnet 1 in Figure 13.50(a). In this case, the magnetic field is detected by the Hall IC sensor and a signal is sent to rotate the motor. As this process continues, it maintains the rotation of the rotor.

Figure 13.51 represents a more complex brushless motor. Here, H_1, H_2, and H_3 are sensors. The most important element in the operation of these motors is the timing of the switching of the power between the three stator windings which is determined by the rotor position. The rotor position is an input to the position sensor whose output is a digital signal.

The speed of the brushless DC motor is a function of the amplitude and the duration of the firing pulses applied to each of the stator windings. The brushless DC motor is gaining considerable popularity in industry due to the following characteristics:

- The absence of brushes and the commutator means no maintenance is required.
- Longer life.
- Greater reliability.

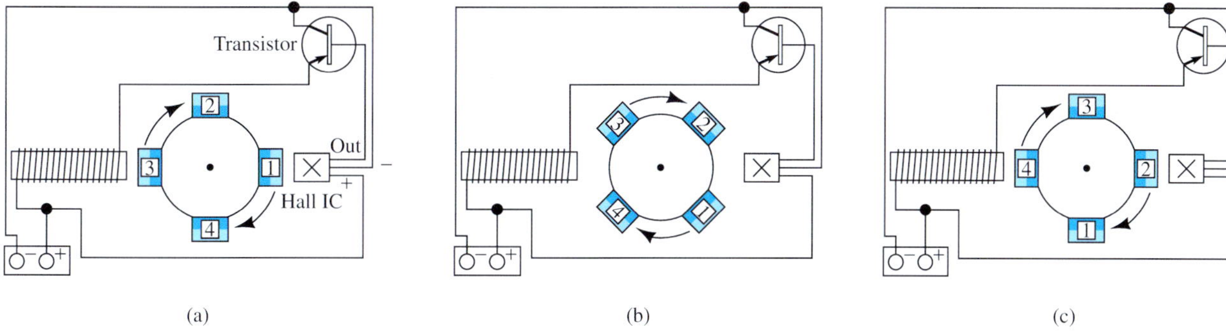

FIGURE 13.50 A simple illustration of the Hall effect switch in a motor. (Reprinted with permission. http://www.simplemotor.com.)

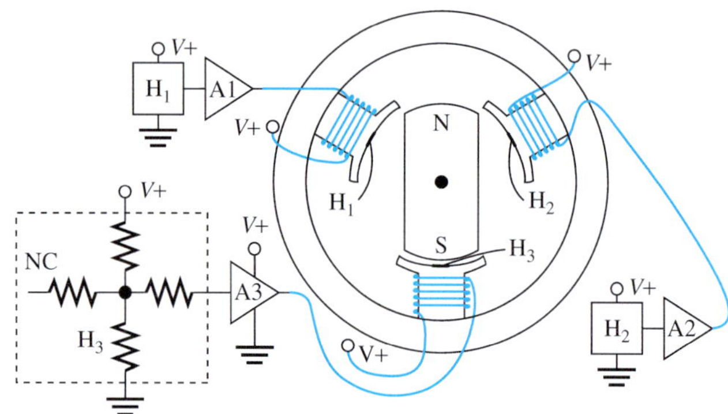

FIGURE 13.51 A sensor-based brushless DC motor.

- Greater efficiency.
- Lower inertia and friction which means they can accelerate faster and move to higher speeds quicker.
- The armature winding is located on the stator which supports effective cooling.
- They have low starting torque and high cost; however, their cost reduces as the cost of thyristorized circuits decreases.

Applications

- Video recorders
- Computer peripherals
- Biomedical instruments
- Commercial ovens
- Film processing equipment
- Printing technology
- Material handling equipment
- Centrifuges, grinders, and fans
- Hybrid vehicles

13.9.4 Universal Motors

DC series motors operated using single-phase AC sources are called *universal motors*. Universal motors are possible only because DC series motors can be operated using either AC or DC sources. This flexibility occurs because the commutator arrangement permits the field and armature currents to remain in phase. An AC motor with a brush and commutator arrangement is classified as a universal motor. Other names for the universal motor include *AC commutator motor* or *AC series motor*. The name AC series motor is derived from the fact that the stator and rotor are in series and are designed such that they operate on a 50 or 60 Hz supply. Different components of a universal motor are shown in Figure 13.52.

FIGURE 13.52 Universal motor. (Reproduced with permission from Blackstone Industries.)

From the discussion on DC series motors, you learned that the torque produced is proportional to the square of the applied voltage [see Equation (13.28)]. This leads to the conclusion that the torque produced is unidirectional and it is independent of the polarity of the applied voltage. The only restriction in powering a DC series motor using a single-phase AC source is that the stator must use laminated construction in order to reduce the eddy current losses. This is a constructional difference between DC series motors and universal motors.

In the equivalent circuit of a DC series motor, we ignored the effect of the inductance of the armature and field windings. As we explained in Chapter 4, an inductance is equivalent to a short circuit when a DC source is applied. However, when the same motor is connected to a single-phase AC source, the inductance cannot be ignored. If it were ignored, there would be a finite voltage drop across this inductance. This voltage drop would then reduce the current in the circuit. To cope with this problem, the inductance of the windings must be reduced. This is accomplished by using fewer turns on the winding.

However, fewer turns corresponds to lower flux. This can be addressed by incorporating a larger number of armature conductors. A larger number of conductors increases the armature reaction which can be offset using a compensatory winding. This is defined as a winding connected in series with armature winding to produce mmf.

Characteristics of the universal motor include the following:

- The ratio of the developed power to the motor weight is very high.
- The value of starting torque is very high without causing the current to rise too high.
- The universal motor suits applications in which load torque varies over a wide range.
- The universal motor can be operated at very high speeds.
- One disadvantage is the short lifespan of brushes and commutators used.
- As a result of the short lifespan, in applications which require the motor to run for a long time, induction motors could replace universal motors.

Applications

- Small appliances like mixers, blenders, drills, saws
- Vacuum cleaners
- Sewing machines

Table 13.4 represents a summary of all motors discussed in this section.

TABLE 13.4 Summary of all Special Types of Motors Studied

Motor Types	Characteristics	Power Range	Applications
Single-phase induction motor	Simple and rugged construction	Up to 5 hp	Fans, water pumps, and refrigerators
Stepper motor	Smooth rotation, linear relation between electrical input and step rotation	Sub-fractional horsepower rating	Used for precise position control in printers, floppy disk drives, robotics, process control, machine tools
Brushless DC motor	No maintenance required, long life, high reliability and efficiency, low inertia and friction, low starting torque and high costs, ability to run at speeds as high as 50,000 rpm	Up to 300 hp	Video recorders, computer peripherals, biomedical instruments, commercial ovens, film processing equipment, printing technology, and material handling equipment
Universal motor	High-power-to-weight ratio, can produce high starting torque, can operate at very high speeds, suits wide variety of applications demanding different torques, shorter lifespan	Used exclusively in fractional horsepower range	Small appliances like mixers, blenders, drills, saws, vacuum cleaners, sewing machines

13.10 HOW IS THE MOST SUITABLE MOTOR SELECTED?

The beginning of the chapter outlined how each motor suits a wide variety of applications. Because of this variety, more than one motor or more than one design of the same motor could be used for a specific application. However, engineers are often called upon to select the best possible design for each situation. Proper motor selection can help achieve the greatest possible efficiency and productivity while putting the lowest load on resources like power and cost.

This section covers a few points that should be taken into account before finalizing the selection of a motor for a specific application:

- Starting as well as running characteristics (operating voltages and currents)
- Life span
- Possible speed control and maximum operating speed
- Size and weight considerations
- Level of noise present in the operating environment
- Suitability of the motor for constant, intermittent, and frequently variable loads
- Capacity in case of overloads and torque characteristics
- Available power supply
- Installation as well as running cost
- Frictional characteristics

To explain these points more clearly, consider an application-based problem and design modifications that can be applied in induction motors to make them suitable for different torque requirements and therefore different applications.

EXAMPLE 13.25 Motor Design

What are possible modifications in the design of a squirrel cage induction motor so that it can serve different applications with a variety of torque–speed characteristics?

SOLUTION

Here, to understand the different torques present in the motor, it is helpful to review the torque–speed characteristics of a squirrel cage induction motor as shown in Figure 13.53.

Starting torque, also called locked-rotor torque, is the torque produced by the induction motor at start-up of the motor. Therefore, *starting current* of the motor is also called the *locked-rotor current.*

Pull-up torque is also called the minimum torque developed by the induction motor.

Breakdown torque is the maximum torque generated by the induction motor without stalling. Sometimes breakdown torque is called *pull-out torque.*

Full-load torque is the torque which the motor produces at rated load.

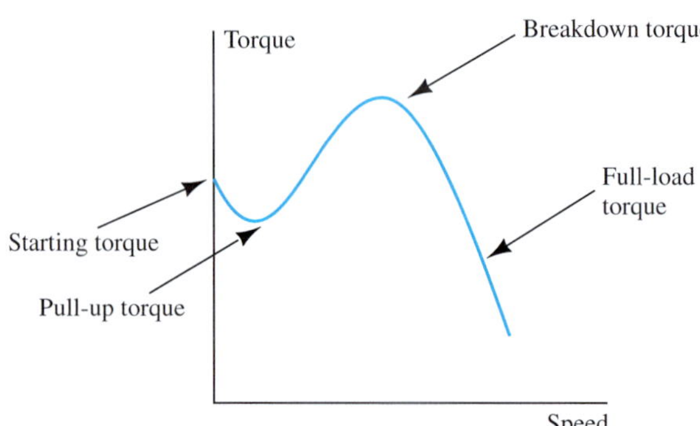

FIGURE 13.53 Torque–speed characteristics.

Torque–speed characteristics of a squirrel cage induction motor can be altered to suit different applications by modifying the construction and design. Different torque–speed characteristics can be generated to suit different applications as illustrated in Figure 13.54.

The National Electrical Manufacturers Association (NEMA) has identified four such designs as part of their standard. Each is considered, along with its corresponding performance curve.

Design A This design is characterized by a starting current close to 10 times that of the full-load current, high breakdown torque, and low slip. Starting torque is close to 150% of full-load torque and breakdown torque is close to 200% of full-load torque. Motors having this design are characterized by high efficiency.

Design B This design has a high reactance obtained by narrower rotor bars, which limits the starting current to five times the full-load current. Starting torque and slip are the same as in the design A motor. Breakdown torque is lower than design A motors, which is a drawback of this design and consequently the design B motor cannot be used for loads which have high peak values. Motors of this design are used in applications with relatively low starting torque requirements like fans, blowers, centrifugal pumps, and so on.

Design C This motor has the highest starting torque of any of the designs discussed so far, with values going up to 200% of full-load torque. However, the breakdown torque is lower than either design A or B. Full-load torque is, however, the same as in designs A and B. At rated load, the values of slip are higher and, as a result, these motors are less efficient at rated load. This motor is particularly used for accelerating heavy loads like conveyors, agitators, reciprocating pumps, and so on.

Design D This design has the highest starting torque of any of the designs discussed with values approximately 280% to 300% of full-load torque. However, in this design, torque is inversely proportional to speed. Starting current is the same as in the motor of design B and slip is high. This design has the highest slip among any of the designs and is primarily used where the starting torque requirement is very high but the load is not continuous, such as in hoists and elevators.

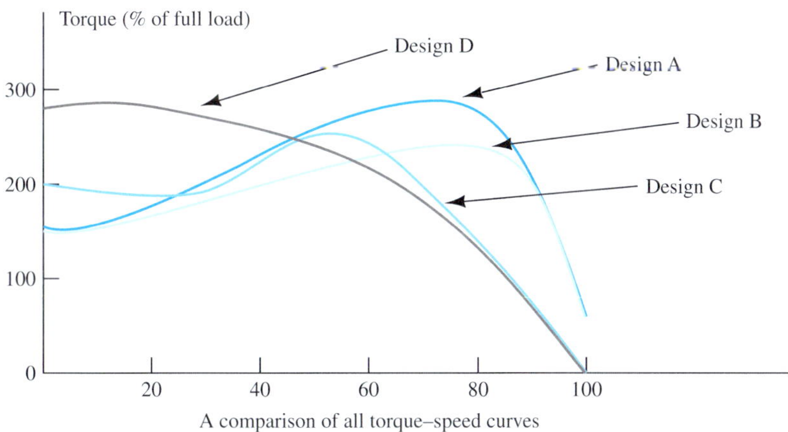

FIGURE 13.54 Different torque–speed characteristics.

13.11 SETUP OF A SIMPLE DC MOTOR CIRCUIT USING PSPICE

In this section, a few methods to simulate a DC motor circuit are introduced using a simple PSpice schematic. In addition, you will learn how to set up a global parameter sweep simulation. In some cases during PSpice simulation setup, you may require multiple electronic parts of the same kind, for example, impedance with the same value. Instead of changing the value of each part one by one, a part can be defined called "PARAMETERS" for all electronic parts. Then, the value assigned to PARAMETERS can be changed and the value of all associated components will change.

The part defined by PARAMETERS has a few functions in PSpice. First, it serves as a model in PSpice to connect the electronic parts you may wish to simultaneously control to the global parameter sweep option in the simulation setting. In addition, it works as a global parameter where it can assign a single value to multiple parts at once. For example, if there are 10 resistors and you wish to divide them into two sets, each consisting of five resistors, and allocate the same magnitude to each five resistors, you can use PARAMETERS to configure such that R_1 and R_2 represent the value for each set of five resistors. You only need to change the value of R_1 and R_2 in the PARAMETERS part, instead of changing the resistance of each resistor one by one.

It should be mentioned that for many electronic parts such as variable resistors, "R_var" cannot be referred directly by a parameter model in the simulation. Thus, parameter sweep is not available for them. The solution to this problem is to use the part, PARAMETERS, which serves as an intermediate part that connects the electronic part to the parameter model in the parameter sweep simulation. The impedance/parameters of the electronic parts are connected and assigned through the PARAMETERS. Then, the PARAMETERS passes the impedance/parameters that are assigned into the simulation to run.

EXAMPLE 13.26 PSpice Analysis

A DC shunt motor is designed to drive a load in the range of 20 to 50 N. The armature resistance, R_A, and the field resistance, R_F, are 0.5 and 80 Ω, respectively. In addition, assume the torque–emf relationship is $T = K_A E_A$ where $K_A = 10$; note that this relationship can be maintained using Equations (13.24) and (13.25), in which $K_A = I_A / \omega_m$. Use PSpice to simulate and plot the power consumed by the motor assuming the voltage supply is 10 V.

SOLUTION

First, determine the induced emf by the motor, which is:

$$E_A = \frac{T}{K_A}$$

Using this equation and given the range of torque is 20 to 50 N, the range of induced emf is calculated as 2 to 5 V. Assume that the induced emf is a voltage source. The PSpice circuit is shown in Figure 13.55.

Here, the label of voltage source, "VM" (in the colored box) represents the induced emf produced by the motor. The label of any components in PSpice can be renamed by double-clicking on the label (such as the "VM" inside the colored box).

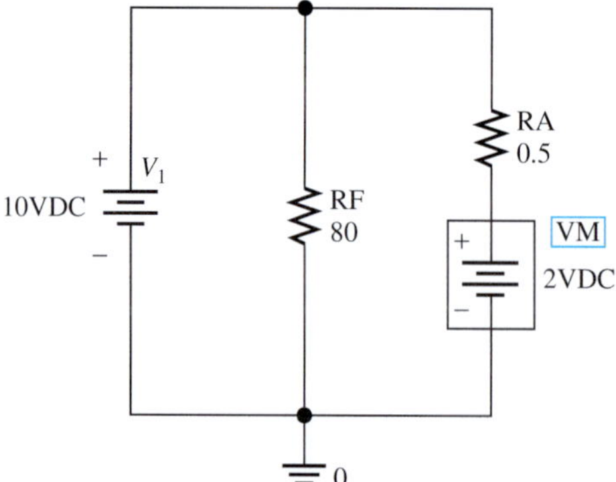

FIGURE 13.55 PSpice schematic for Example 13.26.

Next, simulate the circuit for a given range of induced emf. Set the analysis type of the simulation to "DC sweep" (colored solid arrow in Figure 13.56). Then, determine which voltage source needs to be swept in the simulation, that is, the induced emf, "VM" (dashed arrow).

The minimum and maximum induced emf is set to 2 and 5 V, respectively. Using an appropriate increment, the sweep type is selected to be linear (black arrow), with the minimum and maximum induced emf obtained from the calculation (dotted arrow).

The PSpice Add Trace menu allows us to plot various expressions for the circuit using the basic expression of mathematic functions. Then, to plot the power consumed by motor, first, go to the Add Trace menu. The power consumed by an electronic device corresponds to $P = V \cdot I$. Here, V is the voltage across the motor (also known as induced emf), and I is the current flow through the motor. Then, inside the Trace Expression column (see the arrow in Figure 13.57), type input "V(VM:+,VM,−)*I(VM)." Here, "V(VM:+,VM,−)" is the voltage across the motor, and I(VM) is the current flow through the motor and the symbol "*" represents the multiplication between two variables. The simulation plot is shown in Figure 13.58.

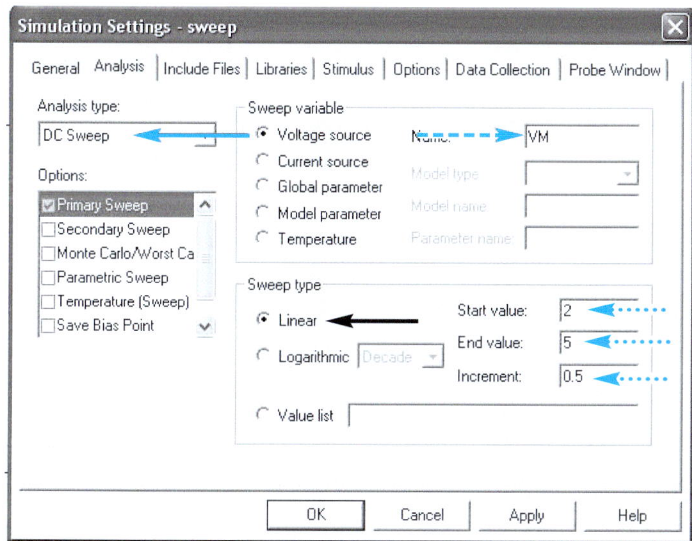

FIGURE 13.56 Simulation setting for Example 13.26.

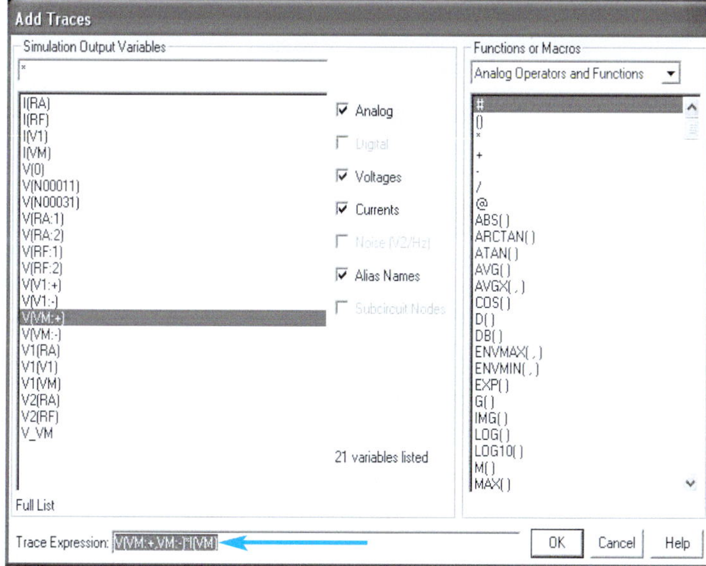

FIGURE 13.57 Add trace.

(continued)

EXAMPLE 13.26 Continued

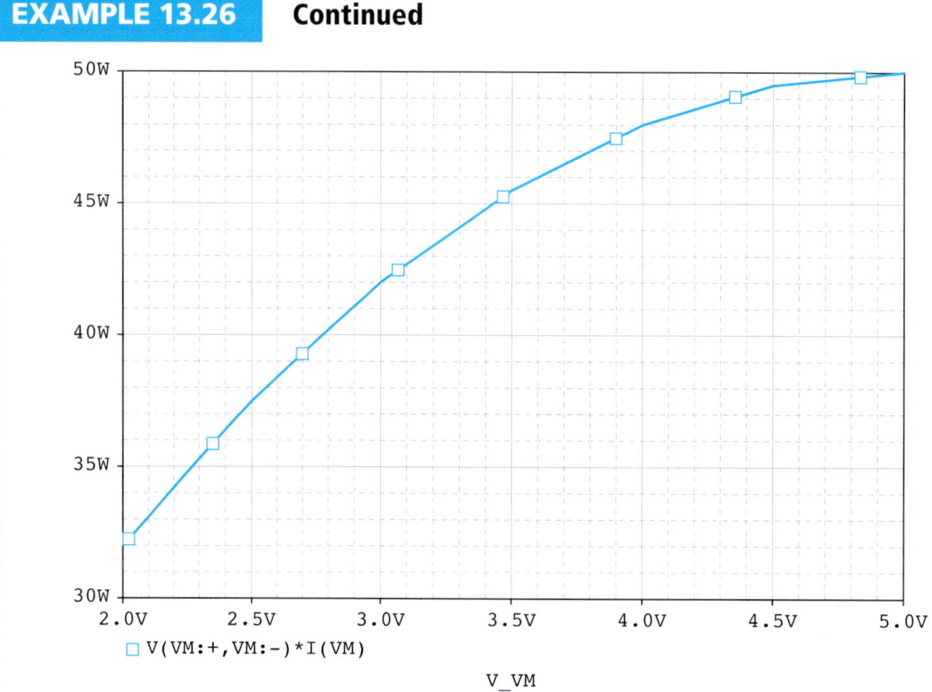

FIGURE 13.58 Power consumed by the motor with respect to different torque.

EXAMPLE 13.27 PSpice Analysis

A DC series motor with constant $K = 0.008$ and $K_1 = 2.5$ is connected to a voltage source of 40 V. Given that the motor's minimum and maximum speeds are 500 and 800 rpm, use PSpice to simulate and plot the power consumed by the motor if the motor's field resistance, R_F, is 0.4 Ω and the armature resistance, R_A, is 0.2 Ω.

SOLUTION

Based on Equation (13.24):

$$E_A = KK_1 I_A \omega$$

The resistance of the motor, R_M, can be expressed as:

$$\frac{E_A}{I_A} = KK_1\omega = R_m$$

Then, based on the minimum and maximum speed of 500 and 800 rpm, the minimum and maximum resistance of the motor corresponds to:

$$R_{m,\,min} = 10\ \Omega$$

$$R_{m,max} = 16\ \Omega$$

Next, set up the PSpice schematic as shown in Figure 13.59.

To obtain the variable resistance in the PSpice circuit, go to "Part" and type "R_var." To simulate the motor with different values of resistance in "R_var," go to the "R_var" properties by double clicking on it. Under the value column (arrow in Figure 13.60), type "RESISTANCE" instead of inserting any integer number.

Next, another new part is introduced, which is the PARAMETERS. Here, the PARAMETERS work as global parameters which can assign a single value to multiple parts at once.

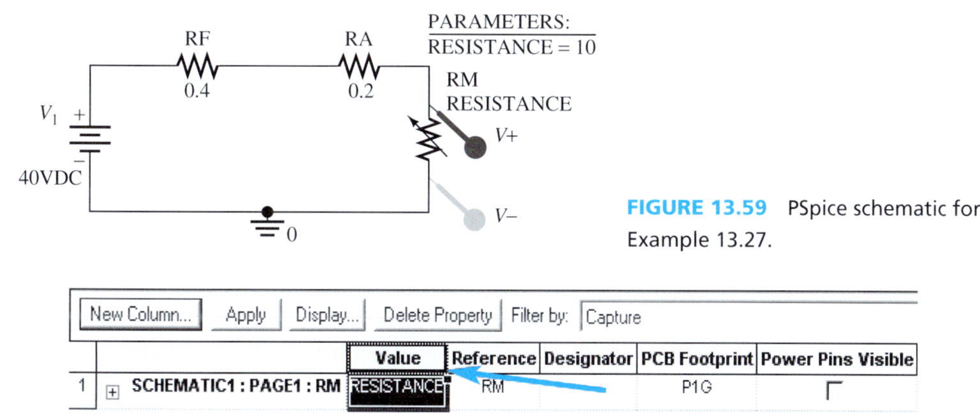

FIGURE 13.59 PSpice schematic for Example 13.27.

FIGURE 13.60 R_VAR's properties menu.

To place the PARAMETERS into the PSpice circuit, go to "Parts" and type "PARAM." It can be found under the "Special" library. Next, go into the PARAMETERS properties menu, and set the RESISTANCE value to minimum resistance, that is 10 Ω (arrow in Figure 13.61). If the "RESISTANCE" does not appear in the properties menu, click the "New Column" (solid arrow in Figure 13.62), then type "RESISTANCE" under the NAME column. Here, the number 10 represents 10 Ω under the VALUE column (dashed arrow in Figure 13.62).

There are three types of simulations that provide the parameter sweep option, which are DC Sweep, AC Sweep, and Time Transient. Here, the circuit is DC, thus, the DC option is used. Note that AC Sweep is used for AC circuits. In addition, the Time Transient option is used if you desire to study transient analysis of the circuits (e.g., RC, RL, and RLC circuit) with different sets of parameters.

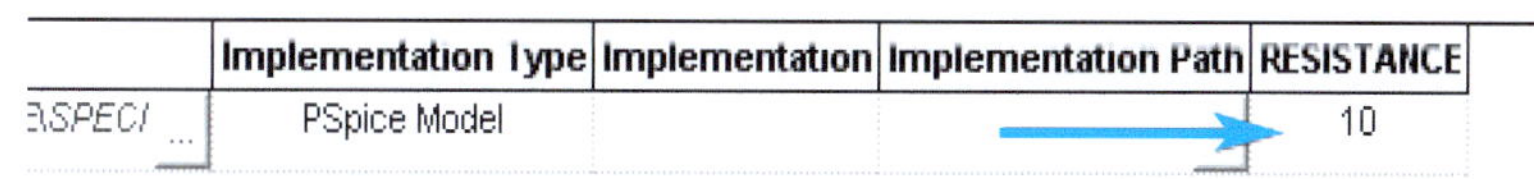

	Implementation Type	Implementation	Implementation Path	RESISTANCE
3.SPECI	PSpice Model			10

FIGURE 13.61 Properties menu of PARAMETERS.

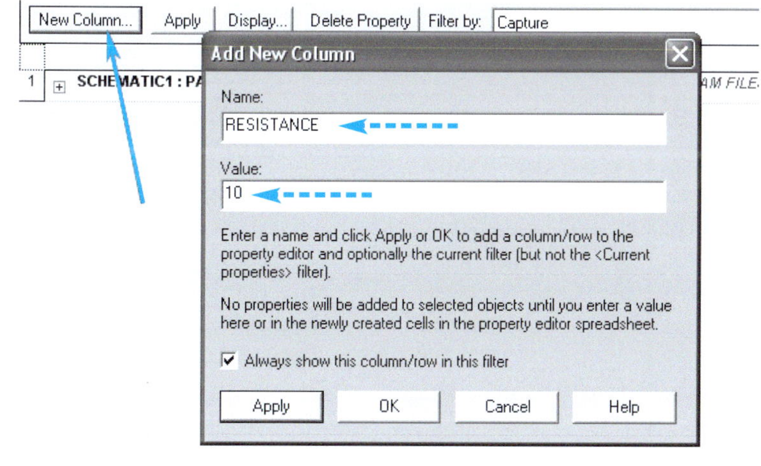

FIGURE 13.62 Insert a new column in the properties menu.

(continued)

EXAMPLE 13.27 Continued

Next, set the simulation type to be DC Sweep. Under the "Primary Sweep" (solid arrow in Figure 13.63) options, set the voltage source's name (dashed arrow) as the voltage supply in the PSpice circuit in Figure 13.63 (i.e., V1). Because the magnitude of voltage source is constant, then the start and end values of Voltage Sweep are the same (dotted arrows). Be aware that PSpice cannot run the Voltage Sweep simulation with zero increment; therefore, you must enter a number inside the increment column, such as 1.

Then, click and tick on the "Parametric Sweep" option (solid arrow in Figure 13.64). Select the "Global parameter" as the sweep variable (dashed arrow). Then, insert the parameter name that you wish to control to be swept, which is "RESISTANCE." Enter the start and end value of the sweep as the minimum and maximum resistance obtained from calculation, with an increment of 1 (dotted arrow). Then press "Apply" and "OK."

Run the simulation, and follow the steps as shown in Figure 13.57 to add the trace that represents the power consumed by the motor. The simulation plot is shown in Figure 13.65.

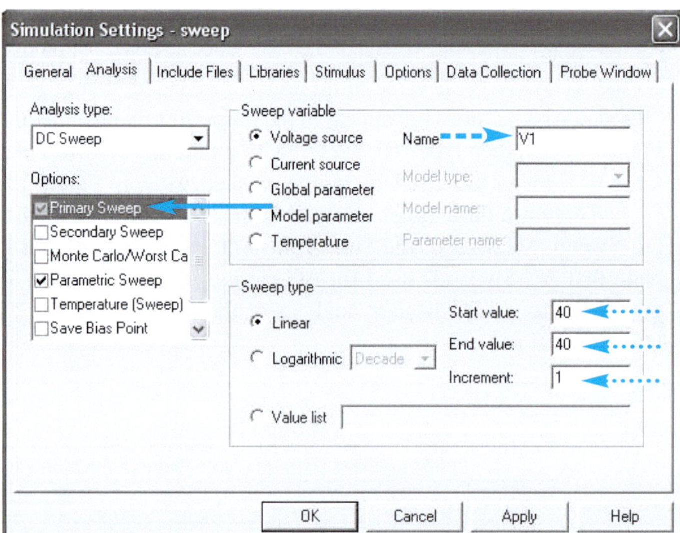

FIGURE 13.63 Primary sweep configuration.

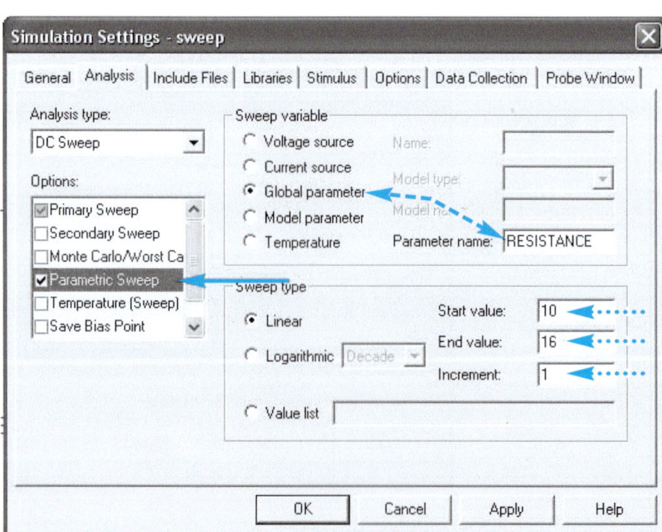

FIGURE 13.64 Parametric sweep configuration.

Note that the resistance of the motor is directly proportional to the angular velocity [see Equation (13.6)]. Then, you can conclude that the plot of Figure 13.65 also represents the power consumed by the motor with respect to the angular velocity of the motor.

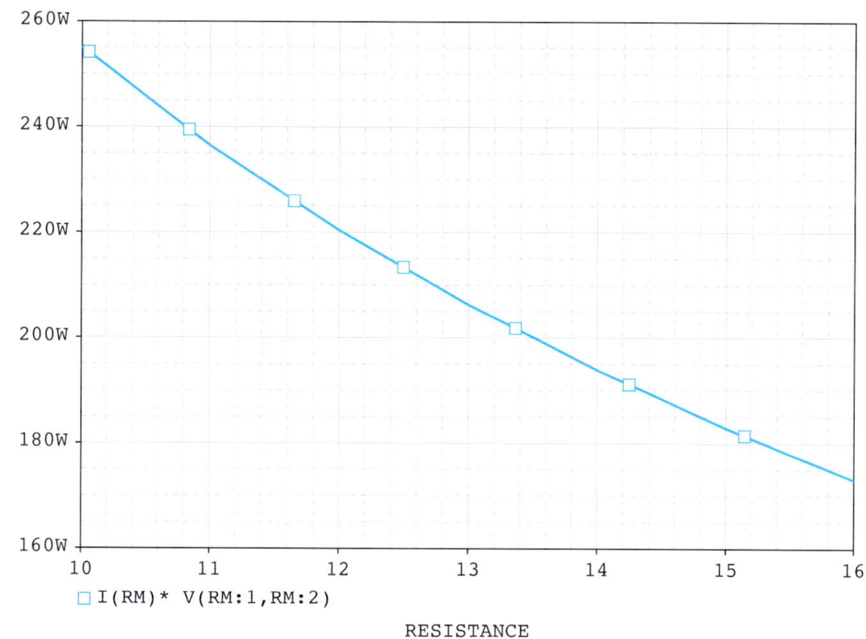

FIGURE 13.65 Power consumed by the motor with respect to resistance (or angular velocity).

EXAMPLE 13.28 PSpice Analysis

Consider a speed control motor in Figure 13.66 that has an armature resistance, R_A, of 0.4 Ω and a field resistance, R_F, of 100 Ω. Assume that the motor has an impedance of 80 Ω and the circuit is connected to a voltage source of 20 V. Use PSpice to simulate the induced emf produced by the motor if the variable resistor has a resistance that varies within 5 to 50 Ω.

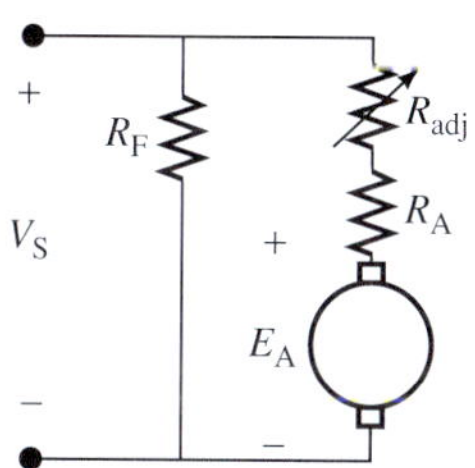

FIGURE 13.66 Speed control motor with varying current.

SOLUTION

Following the same setup in Example 13.26, the PSpice schematic is given in Figure 13.67.

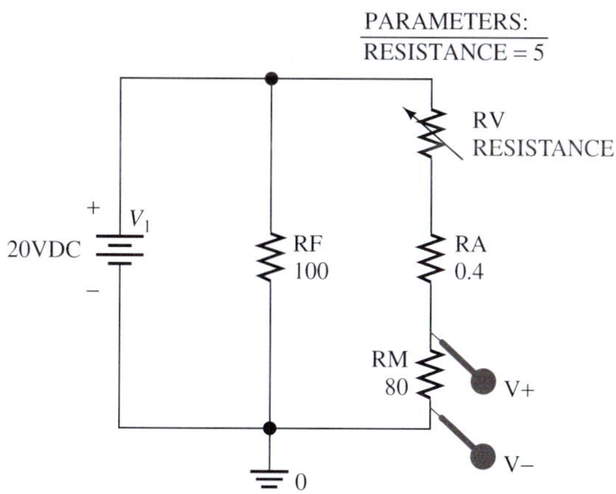

FIGURE 13.67 PSpice schematic for Example 13.28.

(continued)

EXAMPLE 13.28 Continued

Here, RM represents the motor's impedance, 80 Ω. Then, set up the PARAMETERS part for the circuit in PSpice by following the procedure in Figures 13.60 to 13.62. Next, refer to Figures 13.63 and 13.64. The simulation setup is configured as shown in Figure 13.68.

Then, run the simulation. The output of induced emf is given in Figure 13.69.

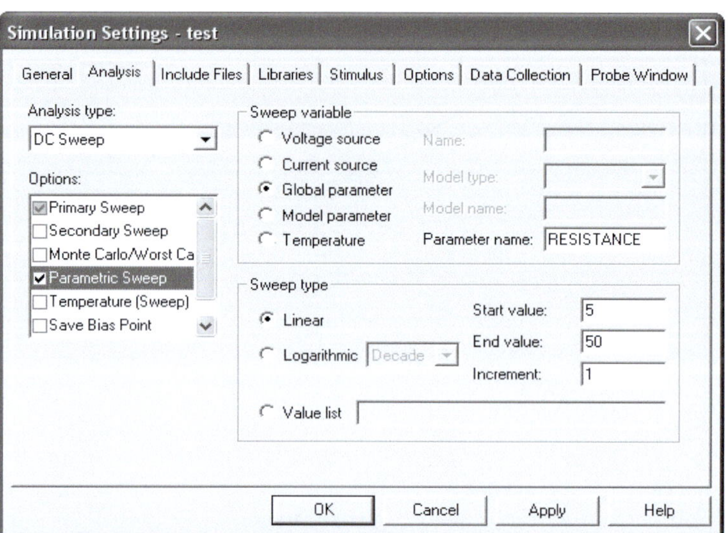

FIGURE 13.68 Global parameter configuration for Example 13.28.

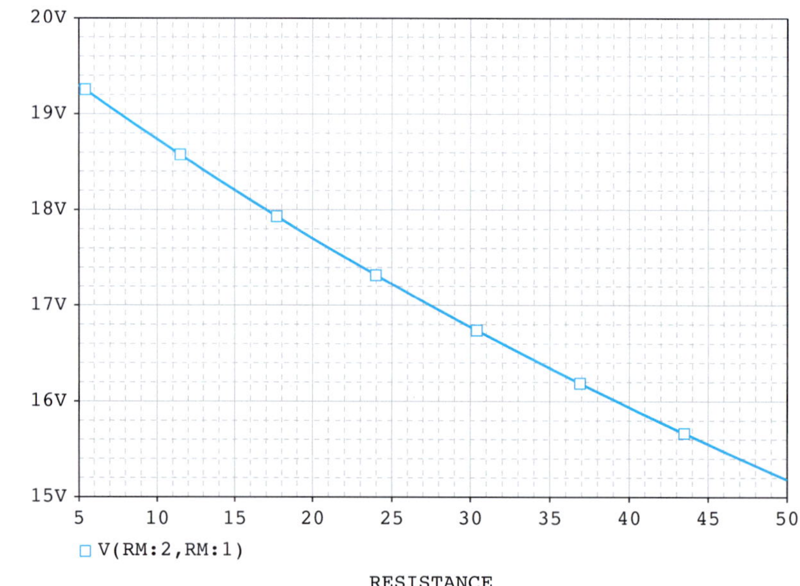

FIGURE 13.69 Output of induced emf by the motor.

13.12 WHAT DID YOU LEARN?

- Motors are classified into two main categories: AC motors and DC motors.
- DC motors have the following components: rotor, stator, windings, commutators, and brushes.
- Losses in a DC machine include: copper losses, iron or core losses, and mechanical losses.

- DC motors are divided into three main categories: (1) shunt-connected, (2) series-connected, (3) permanent magnet, and (4) separately excited motors.
- The following equations provide the basis for analyzing DC machines [Equations (13.6), (13.11), and (13.14)]:

$$E_A = K \cdot \phi \cdot \omega$$
$$T = K_A \cdot \phi \cdot I_A$$
$$P = T \cdot \omega$$

- DC generators include separately excited DC generators and shunt-connected DC generators.
- AC motors include synchronous motors and induction motors.
- Special types of motors include single-phase induction motors, stepper motors, brushless motors, and universal motors. A summary of the characteristics of each is given in Table 13.4.
- Factors to consider when selecting an electrical motor for a specific application include the following:
  - Starting as well as running characteristics (operating voltages and currents)
  - Life span
  - Speed control characteristics and maximum operating speed
  - Size and weight
  - Level of noise present in the operating environment
  - Suitability of the motor for constant, intermittent, frequently variable loads
  - Capacity in case of overloads
  - Torque characteristics
  - Available power supply
  - Installation and operation costs
  - Frictional characteristics

Further Reading

Hambley, A. 2005. *Electrical Engineering Principles and Applications.* Upper Saddle River, NJ: Prentice Hall.

Rizzoni, G. 2004. *Principles and Applications of Electrical Engineering.* New York: McGraw Hill.

Zorbas, D. 1989. *Electric Machines Principles, Applications, and Control Schematics.* St. Paul, MN: West Publishing Company.

Problems

*B refers to basic, A refers to average, H refers to hard, and * refers to problems with answers.*

SECTION 13.2 DC MOTORS

13.1 (B) If the current always flows from A to B (see Figure P13.1), briefly explain the mechanism of this motor.

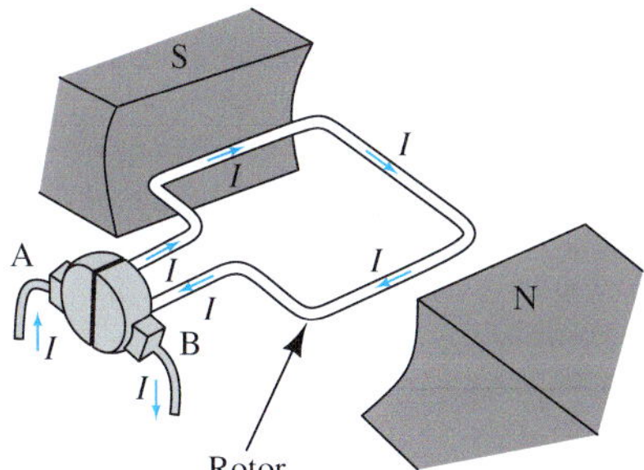

FIGURE P13.1 Rotor for Problem 13.1.

13.2 (B) Given the specfic properties in each case below, determine the result from the setup in Figure P13.2.

(a) Determine the moving direction of the electric rod if the current flows from A to B; C is the north pole and D is the south pole.

(b) What is the pole of the magnetic bars if current flows from B to A, and the electric rod moves in the *X*-direction?

(c) Determine the current flow direction if C is the south pole and D is the north pole, but the electric rod does not move.

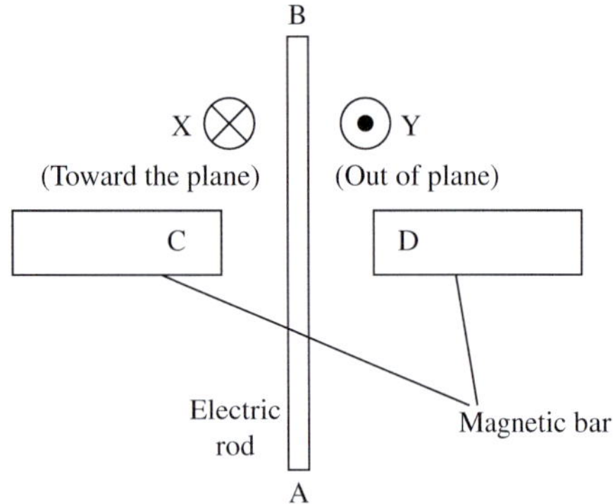

FIGURE P13.2 Diagrammatic representation for Problem 13.2.

13.3 (B) For a simple DC motor setup (see Figure P13.1), explain one way to configure the motor such that the rotor will always rotate in the same direction.

SECTION 13.3 DIFFERENT TYPES OF DC MOTORS

13.4 (B)* Consider a series-connected motor where the field and armature resistance are connected in series. The series current is 60 A, the rotating speed is 1800 rpm, and the power is 8000 W. Find the torque and the armature emf.

13.5 (B) Consider a DC motor that has 540 conductors, four poles, and the flux of 5 mWb per pole. What is the emf if its rotational speed is 1000 rpm?

13.6 (B) Equation (13.11) shows the relationship between the torque, flux, and the armature current. In many cases, the magnetic flux is considered to be constant. What will happen to the armature current if the torque is reduced by half the original torque? Assume that the number of conductors and poles remain the same.

13.7 (B)* A PM motor that drives a load at 500 N m with motor speed of 1200 rpm is connected to a voltage source of 50 V at 10 A. What is the torque constant, k_T, and armature constant, k_A?

13.8 (B) A motor is connected to a 110-V voltage source running at the speed of 500 rpm. The motor has a magnetic flux of 0.5 Wb and the constant K_1 is 4. Determine the current flow through the circuit.

13.9 (A)* Find the speed for a shunt DC motor that takes 35 A current with a supply of 240 V and a torque of 72 N m. The armature and field resistances are 0.30 and 200 Ω, respectively.

13.10 (A) A series-connected DC motor has 400 conductors and six poles. It produces a 120-N m torque when the armature current is 50 A. What is the magnetic flux per pole for the motor?

13.11 (A)* A separately excited DC motor has a supply voltage of 220 V to the field resistance. If the supply voltage to the armature is 75% of the supply voltage to field resistance, what is the back-emf of a motor with armature resistance of 0.6 Ω and armature current of 40 A?

13.12 (A) A 230-V DC shunt motor supplies the current at 93.5 A, with field resistance at 120 Ω and armature resistance at 0.3 Ω. What is the field current if the back-emf of the motor is 203 V?

13.13 (A)* The T–ω characteristics of a load is represented by $\omega = 0.02T + 900$ Write the expression of the speed at the operating point of a PM motor.

13.14 (A) What is the resistance required for a PM motor to drive a load at a torque of 2000 N m given that the T–ω relationship is $\omega = 0.04T + 120$ rad/s, the voltage source, V_S, is 100 V, k_a is 0.02, and k_T is 100.

13.15 (A)* A 5-A series motor that consumes 400 W is connected to a 110-V voltage source. What is the total resistance of R_A and R_F required if the motor is running at 400 rad/s? Also, what is the maximum torque that the motor can drive?

13.16 (A) A series-connected motor in Figure P13.16 is driving a load of 2.5 N m at 100 rad/s. The constant K is 2 and K_1 is 0.05. Determine the minimum voltage of the source and the armature current.

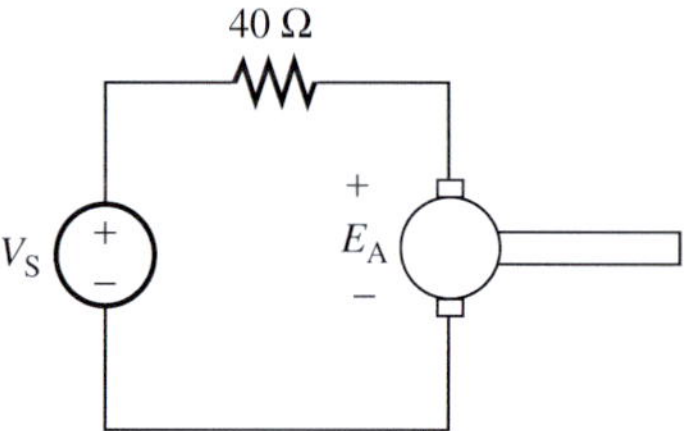

FIGURE P13.16 Circuit for Problem 13.16.

13.17 (H)* The DC shunt motor is running at full load with a supply voltage of 440 V and a supply current of 70 A. The rotation speed in full load is 1000 rpm, the field resistance is 200 Ω and the armature resistance is 0.2 Ω. Calculate its speed if the load torque is reduced to three-fourth of the full-load torque, and the armature resistance is increased by 0.5 Ω. Neglect the voltage drop due to the brushes and assume constant flux.

13.18 (H) A 230-V DC shunt motor runs at 1200 rpm on full load and the output power is 18.5 kW. Its efficiency is 85%, the armature resistance is 0.5 Ω, and field resistance is 50 Ω. If load torque is reduced to 100 N m and an additional 1.5 Ω resistance is connected in series with the armature, what will be the back-emf generated in the machine? Assume load torque to be constant.

13.19 (H) A 230-V DC shunt motor has an armature resistance of 0.05 Ω. When connected to a 230 V direct current supply it develops a back-emf of 225 V at 1200 rpm. Calculate armature current, armature current at start, and speed of the machine if operated as a generator in order to deliver an armature current of 80 A at 230 V. Assume a constant flux.

13.20 (H)* A resistor is connected in a circuit to control the current flow through a PM motor. The motor has an armature resistance of 0.5 Ω and is driving a load of 700 N m at 4 A with a 200 V supply. What is the maximum load that the motor is able to drive if the resistance is changed to 80 Ω?

13.21 (H) Given a PM motor with resistance, R, of 10 Ω and k_a of 4, what is the required source voltage, if the operating point of the PM motor is 40 rad/s and it drives a load of 25 N m?

SECTION 13.4 SPEED CONTROL METHODS

13.22 (A) For a speed control motor (see circuit in Figure P13.22), given $E_A = K\,\phi\,\omega_m$, express the speed of the motor in terms of K, ϕ, I_A, I_F, R_{adj}, R_F, and R_A.

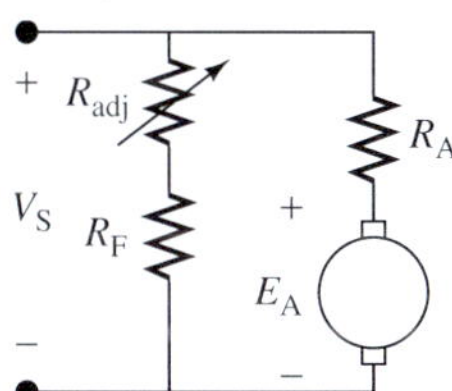

FIGURE P13.22 Circuit for Problem 13.22.

13.23 (A) Explain the mechanism of a speed control motor during the change of resistance of an adjustable resistor, for the cases that the adjustable resistor is connected in series with
(a) Armature resistor
(b) Field resistor

13.24 (A)* A speed control motor with flux constant, K_1 of 0.04, K of 0.125, is running at 1000 rpm 10 A when the resistance is set to 50 Ω. The R_A is 0.5 Ω and R_F is 150 Ω, respectively, and the adjustable resistor has a range of 50 to 100 Ω and is connected in series with the field resistor. What is the current flow through the field resistor?

13.25 (H) A speed control motor is input with a voltage source of 200 V at 25 A. The armature resistance, R_A, and field resistance, R_F, are 0.5 and 20 Ω, respectively. The adjustable resistor has a range of 20 to 180 Ω, which is connected in series with the field resistor. If the motor constant K is 50 and flux constant K_1 is 0.5, write the function of the torque produced by motor T in terms of R_{adj}, and sketch the T–R_{adj} plot.

13.26 (H) A speed control motor is input with a voltage source of 120 V at 30 A, with R_A and R_F are 0.2 and 40 Ω, respectively. The adjustable resistor has a range of 80 to 120 Ω and it is connected in series with the field resistor. If the motor constant, K, is 30, and flux constant, K_1, is 0.04, is it possible to drive a load that requires 1525 N m?

13.27 (H)* A student bought a 24 V, 4 A speed control DC motor, and he wants to know an estimate of the field and armature resistance of this motor. He connects the motor to a power source and runs two different speed and torque tests, which are (1) lowest resistance, and (2) highest resistance. The results he attained are:
1. $\omega = 32.80$ rad/s, $T = 2.312$ N m
2. $\omega = 29.93$ rad/s, $T = 2.738$ N m
Assume that the adjustable resistor is connected in series with the field resistor, which is stated in the manual. Also, assume that the lowest resistance of the adjustable resistor is 0 Ω. Find the field and armature resistance of this motor.

SECTION 13.5 DC GENERATORS

13.28 (B) A lap-winding generator has 40 slots and eight conductors per slot with four poles. The flux is 0.02 Wb per pole. Find the emf of the generator with a speed 1200 rpm.

13.29 (B) Consider a wave-winding generator with 540 conductors with eight poles. If the flux is 5 mWb per pole and the rotation speed is 1000 rpm, calculate the emf of the generator.

13.30 (B)* The terminal voltage of a DC generator at no load is 400 V. When full load is applied to the generator, the terminal voltage is read as 325 V. What is the voltage regulation of this DC generator?

13.31 (A) Calculate the output voltage when the supply voltage is 220 V and the current is 150 A. The shunt resistance is 50 Ω and the armature resistance is 0.02 Ω.

13.32 (A) A DC generator outputs a voltage of 220 V and a current of 150 A. The shunt field resistance is 50 Ω and armature resistance is 0.02 Ω. Find the power generated through the armature current.

13.33 (A)* A six-pole DC generator with a wave-wound armature has 50 slots and 25 conductors per slot. Find the generated emf if it is driven at 30 revolutions per second and the total useful flux in the machine is 0.05 Wb.

13.34 (A) A four-pole, wave-winding DC generator has 400 conductors on its armature with flux of 0.02 Wb per pole. The speed of the generator is 1000 rpm and the current is 100A. Calculate the output voltage and torque of the generator.

13.35 (A) A four-pole, wave-winding DC generator has 400 conductors. The input voltage is 220 V and the power is 12 kW. The generator runs at the speed of 1100 rpm. Calculate the flux and armature resistance.

13.36 (H)* A four-pole, wave-winding DC generator has 540 conductors. Its speed is 1000 rpm and flux per pole is 25 mWb. The armature resistance is 0.8 Ω and field resistance is 50 Ω. At no load, the terminal voltage is measured as 376 V. Calculate the load current for full load, when the voltage regulation is 22% and the power generated across the field resistance is 924.5 W.

13.37 (H) The armature of a four-pole, lap-wound DC shunt generator has 150 conductors. The resistance of the shunt field is 85 Ω and armature resistance is 0.02 Ω. If the flux per pole is 0.07 Wb, find the speed of the machine when supplying 100 kW at a terminal voltage of 220 V.

13.38 (H) A four-pole, lap-wound 800 rpm DC shunt generator has an armature resistance of 0.5 Ω and field resistance of 250 Ω. The armature has 720 conductors and flux per pole is 30 mWb. If the load resistance is 25 Ω, determine its terminal voltage.

SECTION 13.7 AC MOTORS

13.39 (B) For a star-connected, three-phase winding AC motor, what can you conclude from the resultant flux?

13.40 (B)* An AC induction motor has a synchronous speed of 2000 rpm. A load has been applied to the motor and the rotor speed is 1900 rpm. What is the slip of the induction motor?

13.41 (B) An AC motor outputs the power to the rotor at 13 kW. If the slip of the motor is 0.2 when a load is applied to it, what is the mechanical power and copper loss power?

13.42 (B) Determine the number of poles of an induction motor with a synchronous speed of 1800 rpm and frequency of 30 Hz.

13.43 (A) A two-pole induction motor supplies 12 kW power at 50 Hz to a load. Find the motor slip and torque when the rotor speed is 2800 rpm.

13.44 (A) Consider a six-pole induction motor rotating at synchronous speed of 1700 rpm. It produces the torque at 100 N m with frequency of 13 Hz. Find the power loss due to the copper loss effect.

13.45 (A) An eight-pole, 8 hp three-phase induction motor of 50 Hz runs at speed of 1200 rpm. Calculate the slip and frequency of the rotor currents.

13.46 (A)* Consider a two-pole 400 V induction motor with 60 Hz and power of 50 hp. It is operating at a speed of 1700 rpm. Calculate the slip rotor and frequency of the rotor current.

13.47 (A) A four-pole induction motor is operating at 80 Hz. The rotor resistance is 0.02 Ω and rotor reactance is 0.9 Ω. Find the rotor speed at the maximum slip.

13.48 (H) The torque expression of a six-pole induction motor is given as:

$$T = \frac{3sE_r^2R_1}{\omega\left[(R_1 + R_2)^2 + (sX)^2\right]}$$

(a) Find the expression of the breakdown torque.
(b) Given that $R_1 = 0.8$ Ω, $R_2 = 0.3$ Ω, $E_r = 550$ V, $X = 0.15$ Ω, $N_s = 1800$ rpm, and $N = 1700$ rpm. What is the torque produced by the motor?

13.49 (H) A mechanical lifting arm with length of 5 m is designed to lift a load of 200 kg. It is equipped with a four-pole, 60 Hz AC induction motor with 1% copper losses. Determine the required power output for this mechanical load. (Assume gravitation acceleration is 9.8 m/s^2.)

13.50 (H)* A six-pole, 60 Hz AC induction motor in Section 13.7.2.4 has the rotor resistance and reactance at 0.3 and 1.3 Ω, respectively. The rotor-induced voltage at breakdown torque is measured as 237 V. Determine the breakdown torque and the percentage of power losses due to copper losses.

SECTION 13.8 AC GENERATORS

13.51 (B) Why is a short-pitched winding alternator preferred in many applications as compared to one that has full-pitched winding?

13.52 (B) Determine the slot angles for the following configuration:
(a) Single phase, three slots per pole
(b) Single phase, eight slots, two poles
(c) Three phases, six slots per pole

13.53 (B) What is the distribution factor for a three-phase machine that has 12 slots and three poles?

13.54 (A) Does an eight-pole alternator generate more average induced emf than a four-pole alternator? Why? Assume both alternators have the same synchronous speed and magnetic flux.

13.55 (A) An AC generator is designed to generate an average 100 V voltage using an eight-pole, two-conductor configuration. What is the synchronous speed the generator should run at if the flux of the AC generator is 0.5 Wb?

13.56 (A)* Calculate the flux of a six-pole AC generator that generates 120 V per phase. The synchronous speed of the AC generator is 1200 rpm and there are six turns per phase.

13.57 (H) Using a simple two pole (a north pole and a south pole) AC generator, sketch the configuration diagram and plot to explain how the voltage/current is generated for one cycle.

SECTION 13.11 SETUP OF A SIMPLE DC MOTOR CIRCUIT USING PSPICE

13.58 (B) A DC shunt motor is designed to be such that its minimum and maximum induced emf are 2 and 5 V, respectively. Given that the voltage source is 12 V, the armature resistance, R_A, and field resistance, R_F, are 0.2 and 75 Ω, respectively. Use PSpice to simulate and plot the power consumed by the motor.

13.59 (H) Consider a DC motor that varies the armature current to control its speed has an armature resistance, R_A, of 2 Ω and a field resistance, R_F, of 80 Ω, respectively. Assume that the motor has an impedance of 50 Ω with the coefficient $K\phi$ of 0.04. Let the circuit connect to a voltage source of 20 V. Use PSpice to simulate and plot the angular velocity of the motor if the resistor controller has a resistance range from 0 to 40 Ω. Then, determine the approximate minimum and maximum angular velocity of the motor from the simulation plot.

13.60 (H) An electronic device is connected to a shunt DC generator where its armature resistance, R_A, and field resistance, R_F, are 0.5 and 100 Ω, respectively. The DC generator has a range of angular velocity between 1000 and 1800 rpm with the coefficient $K\phi$ of 0.02. Assume that the electronic device has an impedance of 25 Ω and requires a minimum current of 1 A. Use PSpice to determine if the DC generator is capable of supplying the power to the electronic device or not.

TOPIC 8

Lecturer's Notes

TOPIC 9

Capacitance and Inductance

The content of this topic is compiled from:

Chapter 4 (pp. 135–163), *Electrical Engineering: Concepts and Applications*

By S.A. Reza Zekavat

CHAPTER 4

Capacitance and Inductance

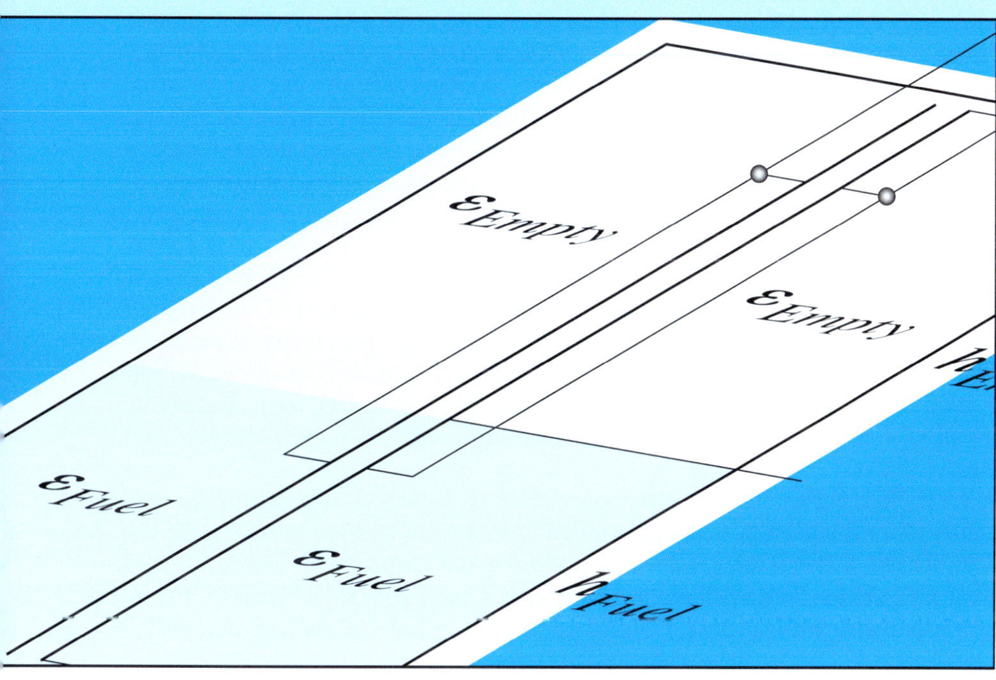

4.1 INTRODUCTION

In Chapter 3, we examined resistors and circuits. Resistors are elements that resist the flow of current. Resistance creates a voltage drop across the resistor which in turn leads to power consumption. Resistors are energy dissipaters. In this chapter, we consider energy storage elements. Resistors are a good model for many circuits but are not the ideal model for all circuits. Modeling of some circuits can be made more realistic by adding in other components. When investigating a system, it is important that the system is replaced with the best model.

For example, a light can be modeled by a resistor; however, the best model for a light is a combination of a resistor and an inductor. This combined model is more realistic because the inductor creates sparks when closing or opening a switch. These sparks are hazardous and should be prevented. (Note: Chapter 15 explains the details of preventing sparks and maintaining a safe environment.) Thus, a resistor alone is not the best circuit model for a light, because it does not account for the potential sparks caused by the switch.

Other circuits also benefit from models with additional elements. For example, a computer can be modeled using a combination of resistors and capacitors. Likewise, transmission lines can be modeled as a combination of capacitors, resistors, and inductors. Chapter 9 studies circuits made by transmission lines.

This chapter focuses on two popular circuit components, that is, capacitors and inductors, which store power and energy. Similar to resistors, these elements are passive; that is, they do not supply power. This chapter describes the relationship between voltage, current, power, and energy for inductors and capacitors. In addition, it explains the effects of connecting these components in series and in parallel.

A *capacitor* stores energy in its electric field. It consists of an insulating material, called dielectric, that is sandwiched between two conductors. An *inductor* stores energy in its magnetic field. Any piece of wire can create a magnetic field; thus, it can form an inductor. However, wires can be twisted to make inductances with higher values. Thus, an inductor is often in a spiral shape. It is interesting to note that because inductors are usually made using a twisted piece of wire, they also offer resistance. Thus, an inductor is most accurately modeled using a combination of inductance and resistance. This chapter primarily examines circuits that can be made using only capacitors and/or inductors. Circuits that are made using a combination of resistors and capacitors or inductors are studied in Chapter 5. This chapter also discusses the steps used to analyze the equivalent capacitor and inductor of capacitive or inductive circuits. Similar to previous chapters, PSpice will also be used to analyze these circuits.

Learning to analyze capacitors and capacitive circuits is useful in a variety of engineering disciplines. These skills help mechanical engineers to understand the operation of capacitive vibration or motion sensors. In addition, because coiled wires are critical in the design of electric motors and solenoids, an understanding of inductive circuits helps to understand the operation of these devices. Civil engineers see the application of capacitors in capacitive sensors that are often used to measure displacement, for example, beam deflection. Capacitors and inductors are also an integral part of tuning circuits for applications in radio, TV, and cellular phones.

Section 4.6 of this chapter outlines many applications of capacitors and inductors in sensors. This section also explains how a capacitor or an inductor can be used to build different types of sensors for use as measurement devices, for example, to measure the pressure, sense vibration, and/or measure the level of liquid in a container. A more detailed review of different types of sensors is provided in Chapter 11.

4.2 CAPACITORS

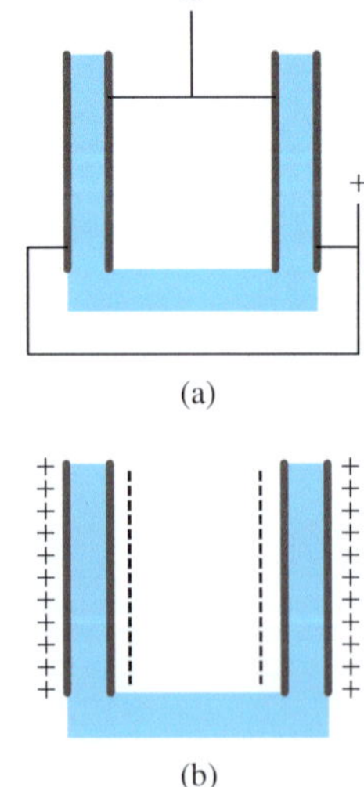

FIGURE 4.1 Leyden Jar (a) charging and (b) charged.

A capacitor stores energy in its electric field. It consists of an insulating material, the dielectric, sandwiched between two conductors. The basic function of a capacitor was famously illustrated by the Leyden jar in 1746. To demonstrate a capacitor using the Leyden jar approach, the interior and exterior of a glass jar are lined with a thin metal conductor [See Figure 4.1(a)]. The inner conductor is negatively charged, while the outer conductor is connected to a ground (zero charge), as shown in Figure 4.1(a). The glass material is between the two conductors. As a result of the glass material, no charge is able to reach the outer conductor. The negative charge on the inner conductor repels the negative charge on the outer to the ground, resulting in a net positive charge on the outer conductor. When the electrical connections are removed, the charges remain on the surface of the conductor. The glass and air are insulators, and they shield the charges. The net charges are equal in magnitude, and opposite in sign. The result is a stored electric charge on the jar, as shown in Figure 4.1(b).

Capacitors are used for storing electric charge in applications that require a quick discharge, for example, a camera flash or a defibrillator. Capacitors also play a major role in the tuning circuits of TVs and radios.

A basic capacitor is simply a dielectric placed between two conducting plates, as shown in Figure 4.2. A dielectric is an insulator. Examples include ceramic, vacuum, air, electrolyte, oil, paper, plastic, and mica. If the spacing between the two plates is filled with

TABLE 4.1 Permittivity of Some Typical Materials

Material	Permittivity
Polyethylene	2.1
Paper	3
Salt	3–15
Methanol	30
Sulfuric acid	83.6
Silicon	11.68
Rubber	7

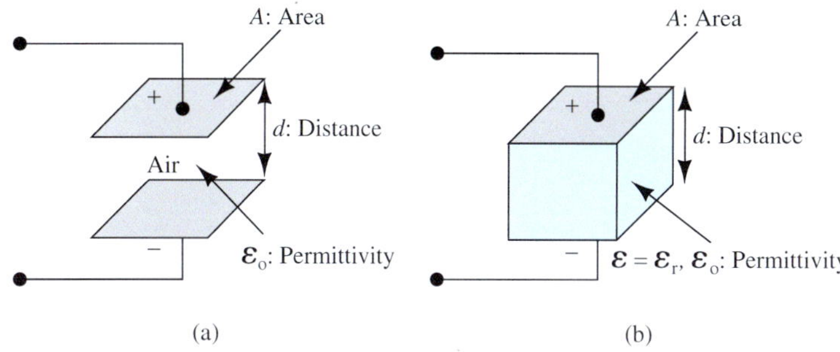

FIGURE 4.2 Parallel-plate capacitor with (a) an air dielectric and (b) a solid dielectric.

electrolyte, the capacitor will be a polarized one. It is very critical to connect polarized capacitors in the right order. If they are not connected in the right order, the capacitor could be exploded. Other dielectrics do not polarize the capacitor, and the capacitor can be connected in any direction. Electrical characteristic of dielectrics are usually determined by their permittivity.

The capacitance of a parallel-plate capacitor corresponds to:

$$C = \frac{\varepsilon_o \varepsilon_r \times A}{d} \tag{4.1}$$

where

ε_o = permittivity of free space ≈ 8.85×10^{-12} F/m

ε_r = relative permittivity of the dielectric (that is relative to air) (see Table 4.1 for examples of relative permittivity)

A = area of the plates (in square meters, m^2)

C = capacitance, in farads (F)

d = distance between the plates.

Capacitance is measured in *farads* (F), where 1 F = 1 coulomb/volt (C/V). The unit farad is named in honor of Michael Faraday. Faraday was the first to show the existence of electric and magnetic fields via experiments.

EXAMPLE 4.1 Capacitance

A parallel-plate capacitor has dimensions as shown in Figure 4.3. The dielectric is air, that is, $\varepsilon_r = 1$. Find the total capacitance.

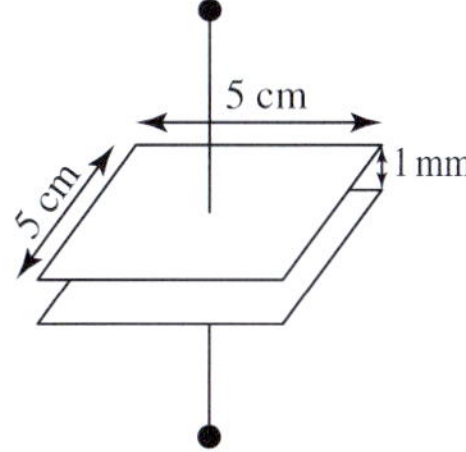

FIGURE 4.3 Capacitor for Example 4.1.

(continued)

EXAMPLE 4.1 Continued

SOLUTION

Using Equation (4.1):

$$C = \frac{\varepsilon_0 \varepsilon_r \times A}{d}$$

Replacing for all parameters,

$$C = \frac{8.85 \times 10^{-12} \times 1 \times (0.05)^2}{0.001}$$

Therefore,

$$C = 22.13 \times 10^{-12}\ \text{F}$$

4.2.1 The Relationship Between Charge, Voltage, and Current

The symbols used to designate a capacitor are shown in Figure 4.4. The symbol in the left might be used for polarized capacitors. As mentioned, the spacing between the plates of these capacitors is filled with electrolyte.

The charge on a capacitor, $q(t)$, is the product of the capacitance and voltage:

$$q(t) = C \cdot v(t) \tag{4.2}$$

+
$v(t)$
–
$i(t)$
+
$q(t)$
–
or
+
$v(t)$
–
$i(t)$
+
$q(t)$
–

FIGURE 4.4 Capacitor symbols.

EXAMPLE 4.2 Stored Charge

Find the stored charge in a 10-μF capacitor that is charged to 6 V.

SOLUTION

$$q = Cv$$
$$q = 10 \times 10^{-6} \times 6$$
$$q = 60\ \mu\text{C}$$

By definition, current is the derivative of charge with respect to time.

$$i(t) = \frac{\mathrm{d}q(t)}{\mathrm{d}t} \tag{4.3}$$

Substituting for $q(t)$ from Equation (4.2), the current through a capacitor corresponds to:

$$i(t) = C\frac{\mathrm{d}v(t)}{\mathrm{d}t} \tag{4.4}$$

EXAMPLE 4.3 Current of Capacitor

In Figure 4.5(a), calculate the current through the capacitor assuming:

a. $v(t) = 5\,\text{V}$
b. $v(t) = 5\cos(2\pi \times 10^3 t)\ \text{V}$

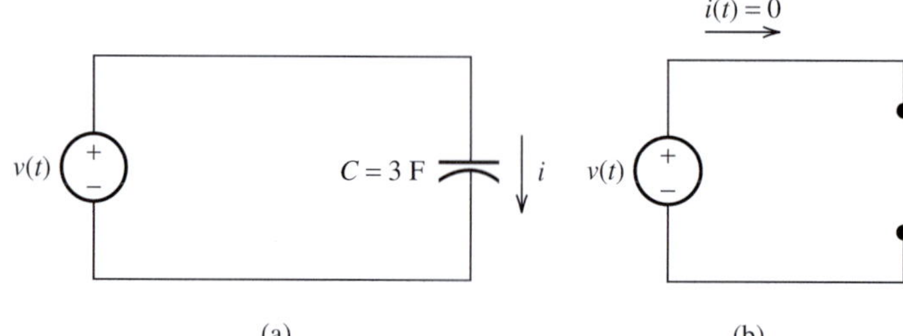

FIGURE 4.5 (a) The circuit for Example 4.3; (b) equivalent circuit of a capacitor when a constant current is applied to.

SOLUTION

Using Equation (4.4):

$$i(t) = 3 \times \frac{\mathrm{d}v(t)}{\mathrm{d}t}$$

a. For $v(t) = 5\,\mathrm{V}$, $\frac{\mathrm{d}v(t)}{\mathrm{d}t} = 0$; therefore $i(t) = 0$. Thus, the capacitor would be equivalent to an open circuit as shown in Figure 4.3 (b).

b. For $v(t) = 5\cos(2\pi \times 10^3 t)\,\mathrm{V}$, $\frac{\mathrm{d}v(t)}{\mathrm{d}t} = -10\pi \times 10^3 \sin(2\pi \times 10^3 t)$; therefore:

$$i(t) = -30\pi \times 10^3 \sin(2\pi \times 10^3 t)\ \mathrm{A}$$

Note: If a *constant* voltage is applied directly to a capacitor, the capacitor will ultimately (i.e., when it is fully charged) act as an *open circuit*, because the current through it will be zero. The equivalent circuit for this situation is shown in Figure 4.5(b). Note that the derivative of a constant voltage (see Equation (4.4)) is zero.

Based on Equation (4.3), the charge–current relationship is as follows:

$$q(t) = \int_{t_0}^{t} i(t)\mathrm{d}t + q(t_0) \tag{4.5}$$

Here, $q(t_0)$ represent the initial charge of capacitor at $t = t_0$. Setting the right-hand side of Equations (4.2) and (4.5) equal, and solving for $v(t)$:

$$v(t) = \frac{1}{C}\int_{t_0}^{t} i(t)\mathrm{d}t + \frac{q(t_0)}{C} \tag{4.6}$$

Because the initial voltage across the capacitor corresponds to:

$$v(t_0) = \frac{q(t_0)}{C} \tag{4.7}$$

Equation (4.6) may be rewritten as:

$$v(t) = \frac{1}{C}\int_{t_0}^{t} i(t)\mathrm{d}t + v(t_0) \tag{4.8}$$

If the capacitor has zero charge at $t = t_0 = 0$, that is $q(t_0) = 0$, or a zero initial condition, then based on Equation (4.7), Equation (4.8) simplifies to:

$$v(t) = \frac{1}{C}\int_{0}^{t} i(t)\mathrm{d}t \tag{4.9}$$

Comparing Equations (4.4) and (4.8), it is observed that current is a linear function of voltage [see Equation (4.4)]. Recall that linearity consists of two properties: additivity and homogeneity. Therefore, a function $f(x)$ is considered linear if it meets both conditions:

Additivity: $f(x_1 + x_2) = f(x_1) + f(x_2)$

Homogeneity: $f(\alpha x) = \alpha \cdot f(x)$

Additivity and homogeneity are summarized in the following equation:

$$f(\alpha_1 x_1 + \alpha_2 x_2) = \alpha_1 f(x_1) + \alpha_2 f(x_2) \tag{4.10}$$

EXERCISE 4.1

Show that Equation (4.8) represents a linear function with respect to $i(t)$ and $v(t_0)$. Show that Equation (4.9) is only linear with respect to $i(t)$.

4.2.2 Power

As mentioned in Chapter 2, instantaneous power equals voltage times current.

$$p(t) = v(t)i(t) \tag{4.11}$$

Using Equations (4.3) and (4.4), the stored delivered power by a capacitor corresponds to:

$$p(t) = v(t)\frac{dq(t)}{dt} = Cv(t)\frac{dv(t)}{dt} \tag{4.12}$$

EXAMPLE 4.4 **Power Delivered to a Capacitor**

A 1.5-μF capacitor shown in Figure 4.6 has been previously charged to 5 V. It is then discharged within 50 ms. What is the instantaneous power delivered?

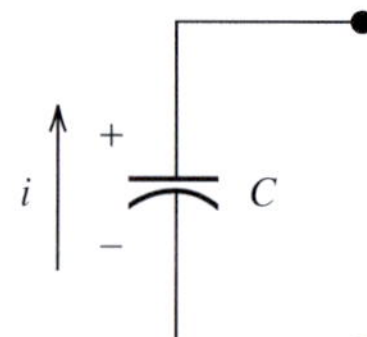

FIGURE 4.6 Circuit for Example 4.3.

SOLUTION

Using Equation (4.2):

$$q = Cv \rightarrow q = 1.5 \times 10^{-6} \times 5 = 7.5 \times 10^{-6}$$

$$\text{The rate of discharge} = \frac{7.5 \times 10^{-6}\text{C}}{50 \times 10^{-3}\text{s}} = 150\frac{\mu\text{C}}{\text{s}} = 150\ \mu\text{A} = i_{\text{capacitor}}$$

Referring to the sign of the voltage and current (as discussed in Chapter 2) as indicated in Figure 4.6:

$$p = -vi = 5 \times 150 \times 10^{-6} = -750\ \mu\text{W}$$

The minus sign indicates that the power is delivered by the capacitor.

4.2.3 Energy

The power–energy relationship corresponds to:

$$p(t) = \frac{dE(t)}{dt} \tag{4.13}$$

Therefore:

$$E(t) = \int_{t_0}^{t} p(t)\,\mathrm{d}t \tag{4.14}$$

Using Equation (4.12):

$$p(t) = Cv(t)\frac{\mathrm{d}v(t)}{\mathrm{d}t} = \frac{1}{2}C\frac{\mathrm{d}v^2(t)}{\mathrm{d}t} \tag{4.15}$$

The last equality holds since $v(t)\frac{\mathrm{d}v(t)}{\mathrm{d}t} = \frac{1}{2}\frac{\mathrm{d}v^2(t)}{\mathrm{d}t}$.

Comparing Equations (4.12) and (4.15), energy stored in or delivered by a capacitor is given by:

$$E(t) = \frac{1}{2}Cv^2(t) \tag{4.16}$$

EXAMPLE 4.5 Voltage and Energy of Capacitor

At time $t = 0$, a 2-mA current source is applied to an uncharged 1.2-μF capacitor, as shown in Figure 4.7 below. Determine the voltage, v_C, and the energy stored in the capacitor, E_C, at $t = 20$ ms assuming zero initial voltage across the capacitor prior to the closing of the switch, that is, a zero initial condition.

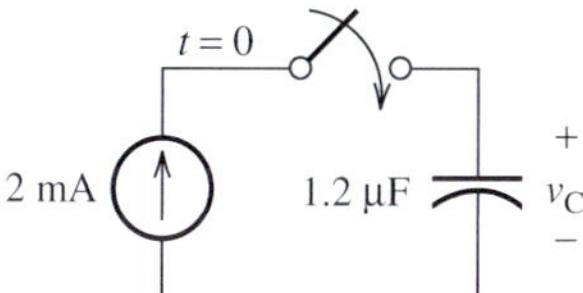

FIGURE 4.7 Circuit for Example 4.5.

SOLUTION

Given a constant current and zero initial charge, using Equation (4.5), the charge corresponds to:

$$q = i \cdot t = 2\text{ mA} \cdot 20\text{ ms} = 40\ \mu\text{C}$$

Now, using Equation (4.2):

$$q = Cv_C \rightarrow v_C = \frac{q}{C} = \frac{40\ \mu\text{C}}{1.2\ \mu\text{F}} = 33.3\frac{C}{C/v} = 33.3\text{ V}$$

Finally, applying Equation (4.16):

$$E_C = \frac{1}{2}Cv^2 = \frac{1}{2} \times 1.2 \times 10^{-6} \times 33.3^2 = 667\ \mu\text{J}$$

4.3 CAPACITORS IN SERIES AND PARALLEL

Capacitors, like resistors, can be arranged into series and parallel connections. In this section, the equivalent capacitances of simple series and parallel combinations are derived. The formulas derived can in turn be used to calculate the capacitance of more complex capacitor networks.

4.3.1 Series Capacitors

Figure 4.8(a) represents three capacitors, C_1, C_2, and C_3, connected in series.

Applying KVL to the structure of Figure 4.8(a):

$$v_C = v_1 + v_2 + v_3 \tag{4.17}$$

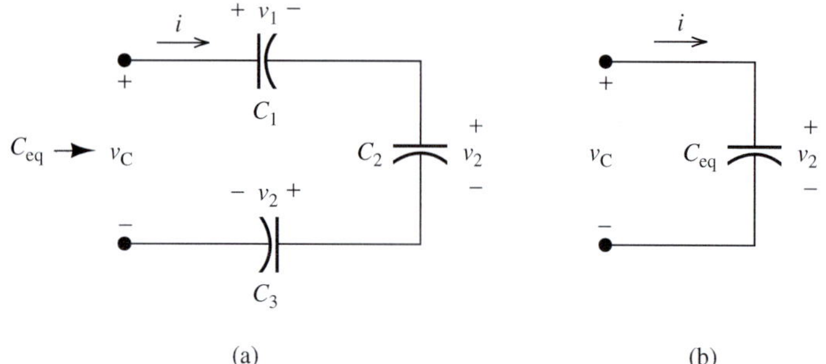

FIGURE 4.8 (a) Series capacitors; (b) an equivalent circuit.

Because the current $i(t)$ flowing through all capacitors is equivalent, the voltage across each capacitor corresponds to:

$$v_k = \frac{1}{C_k}\int i(t)\mathrm{d}t + v_k(0), \quad k \in \{1,2,3\} \tag{4.18}$$

Assuming $v_k(0) = 0$ for all values of k (i.e., the initial voltage of all capacitors is zero), and substituting Equation (4.18) into Equation (4.17), the total voltage is:

$$v_C(t) = \frac{1}{C_1}\int i(t)\mathrm{d}t + \frac{1}{C_2}\int i(t)\mathrm{d}t + \frac{1}{C_3}\int i(t)\mathrm{d}t \tag{4.19}$$

Factorizing the integral component that is the same across all terms above:

$$v_C(t) = \left(\frac{1}{C_1} + \frac{1}{C_2} + \frac{1}{C_3}\right)\cdot \int i(t)\mathrm{d}t \tag{4.20}$$

Also, the relationship between voltage, v_C, and current, i, for the equivalent circuit of Figure 4.8(b) corresponds to:

$$v_C(t) = \left(\frac{1}{C_{eq}}\right)\cdot \int i(t)\mathrm{d}t \tag{4.21}$$

Comparing Equations (4.20) and (4.21), the equivalent capacitance for three capacitors in series corresponds to:

$$C_{eq} = \left(\frac{1}{C_1} + \frac{1}{C_2} + \frac{1}{C_3}\right)^{-1} \tag{4.22}$$

For a general case with N capacitors in series:

$$C_{eq} = \left(\frac{1}{C_1} + \frac{1}{C_2} + \cdots + \frac{1}{C_N}\right)^{-1} \tag{4.23}$$

As a result, the method for calculating the equivalent capacitance of capacitors in series is equivalent to that of parallel resistors.

EXERCISE 4.2

Explain how Equation (4.22) can be generalized to Equation (4.23).

4.3.2 Parallel Capacitance

Next, consider the equivalent capacitance of parallel capacitors. As an example, consider the three capacitors connected in parallel in Figure 4.9(a).

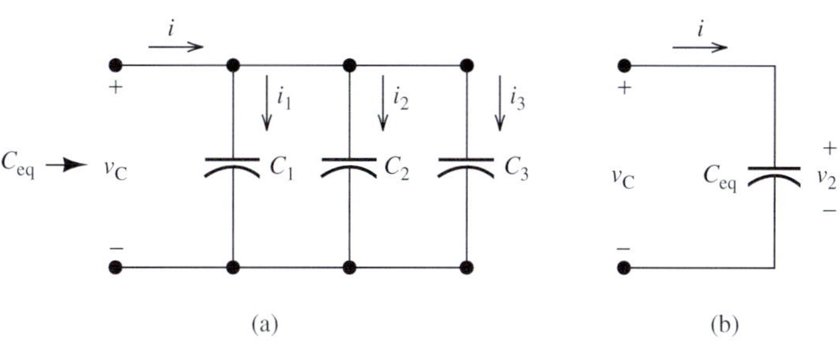

FIGURE 4.9 (a) Parallel capacitors; (b) their equivalent circuit.

Because the voltage across all capacitors is equivalent, the current–voltage relationship for each capacitor $R \in \{1,2,3\}$ corresponds to:

$$i_k(t) = C_k \frac{\mathrm{d}v_C(t)}{\mathrm{d}t},\ k \in \{1,2,3\} \quad \textbf{(4.24)}$$

Applying KCL to Figure 4.9(a), the total current is:

$$i(t) = i_1(t) + i_2(t) + i_3(t) \quad \textbf{(4.25)}$$

Substituting Equation (4.24) into (4.25):

$$i(t) = C_1 \frac{\mathrm{d}v_C(t)}{\mathrm{d}t} + C_2 \frac{\mathrm{d}v_C(t)}{\mathrm{d}t} + C_3 \frac{\mathrm{d}v_C(t)}{\mathrm{d}t} \quad \textbf{(4.26)}$$

Rearranging the terms:

$$i(t) = (C_1 + C_2 + C_3) \frac{\mathrm{d}v_C(t)}{\mathrm{d}t} = (C_{eq}) \frac{\mathrm{d}v_C(t)}{\mathrm{d}t} \quad \textbf{(4.27)}$$

Also, the current–voltage relationship for the equivalent circuit of Figure 4.8(b) corresponds to:

$$i(t) = C_{eq} \frac{\mathrm{d}v_C(t)}{\mathrm{d}t} \quad \textbf{(4.28)}$$

Comparing Equations (4.27) and (4.28), the equivalent capacitance of the parallel capacitors in Figure 4.9 is:

$$C_{eq} = C_1 + C_2 + C_3 \quad \textbf{(4.29)}$$

For a general case with N capacitors:

$$C_{eq} = C_1 + C_2 + \cdots + C_N \quad \textbf{(4.30)}$$

Thus, the equivalent capacitance of *parallel* capacitors is computed in a way similar to the method used for *series* resistors. To demonstrate the calculation of the equivalent circuit of capacitor networks connected in more complex configurations, the following examples are presented.

EXAMPLE 4.6 Equivalent Capacitance

Find the equivalent capacitance in the circuit shown in Figure 4.10(a).

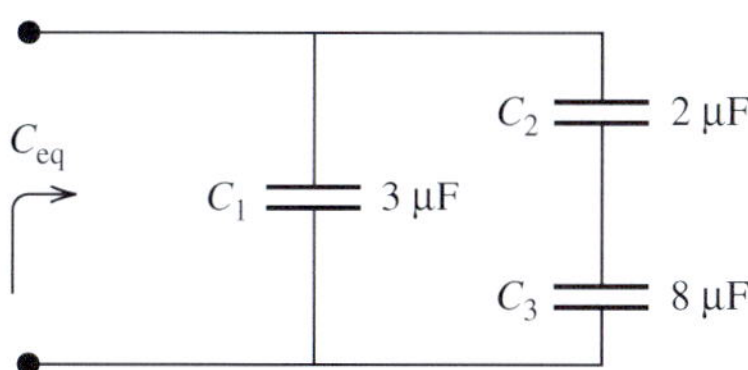

FIGURE 4.10(a) The circuit for Example 4.6.

(continued)

EXAMPLE 4.6 Continued

SOLUTION

The total capacitance is found in a way that is similar to finding the total resistance of a resistive network. Here, there are two parallel branches. One branch includes two capacitors, C_2 and C_3, that are in series. Equation (4.21) is used to find the series capacitance.

$$C_{23} = \left(\frac{1}{2 \times 10^{-6}} + \frac{1}{8 \times 10^{-6}}\right)^{-1} = 1.6\ \mu\text{F}$$

Now, in Figure 4.10(a), replace the series capacitors, C_2 and C_3, with the equivalent capacitor C_{23}, as shown in Figure 4.10(b).

Figure 4.10(b) represents two parallel capacitors, C_1 and C_{23}. As shown by Equation (4.30), the total equivalent capacitance corresponds to:

$$C_{eqv} = 3\ \mu\text{F} + 1.6\ \mu\text{F} = 4.6\ \mu\text{F}$$

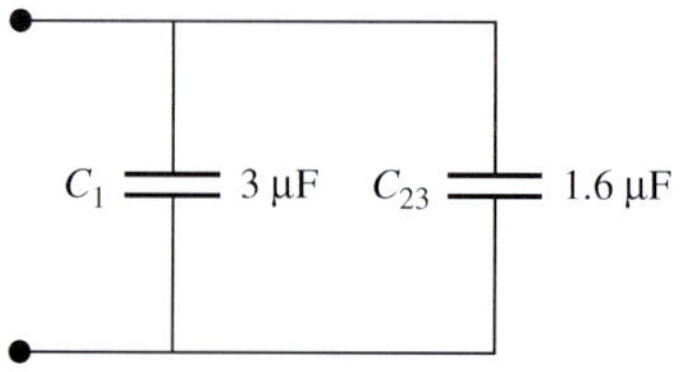

C_1 3 μF C_{23} 1.6 μF

FIGURE 4.10(b) Modification of Figure 4.10(a).

EXAMPLE 4.7 Equivalent Capacitance

Find the equivalent capacitance of the capacitor network shown in Figure 4.11(a) below.

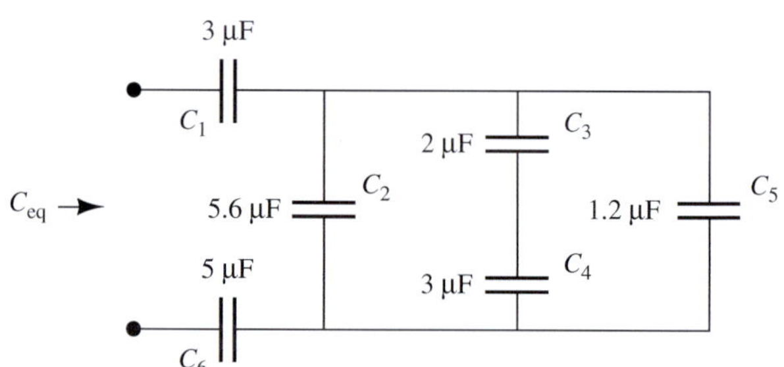

FIGURE 4.11(a) The circuit for Example 4.7.

SOLUTION

Note that C_3 and C_4 are in series. Combine these capacitors using Equation (4.23).

$$C_{34} = \left(\frac{1}{2 \times 10^{-6}} + \frac{1}{3 \times 10^{-6}}\right)^{-1} = 1.2\ \mu\text{F}$$

Next, replace series C_3 and C_4 capacitors with C_{34}, as shown in Figure 4.11(b).

In Figure 4.11(b), note that C_2, C_{34}, and C_5 are in parallel, and their equivalent capacitance corresponds to:

$$C_{2345} = C_2 + C_{34} + C_5 = 8\ \mu\text{F}$$

Replacing these three capacitors with C_{2345}, results in the circuit shown in Figure 4.11(c).

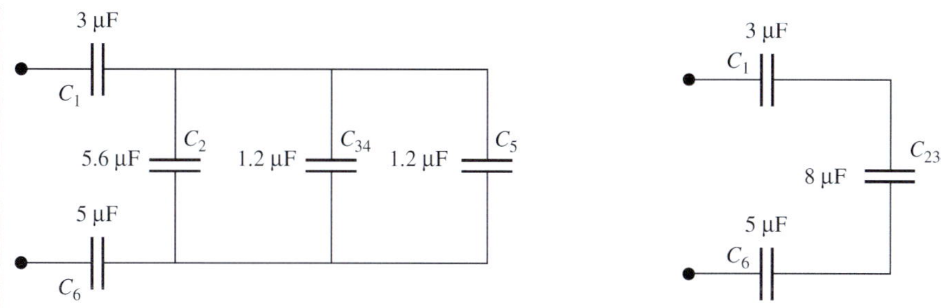

FIGURE 4.11(b) Modification of Figure 4.11(a). **FIGURE 4.11(c)** Modification of Figure 4.11(b).

Finally, find the series combination of the three capacitors, C_1, C_{2345}, and C_6:

$$C_{eq} = \left(\frac{1}{3 \times 10^{-6}} + \frac{1}{8 \times 10^{-6}} + \frac{1}{5 \times 10^{-6}}\right)^{-1} = 1.52\ \mu\text{F}$$

APPLICATION EXAMPLE 4.8 Tuning Circuits

The tuning circuit in a radio consists of several capacitors. The tuning knob of the radio adjusts the value of a variable capacitor. Figure 4.12 represents the capacitive portion of a radio tuning circuit. Find the minimum and maximum values for C_{eqv}.

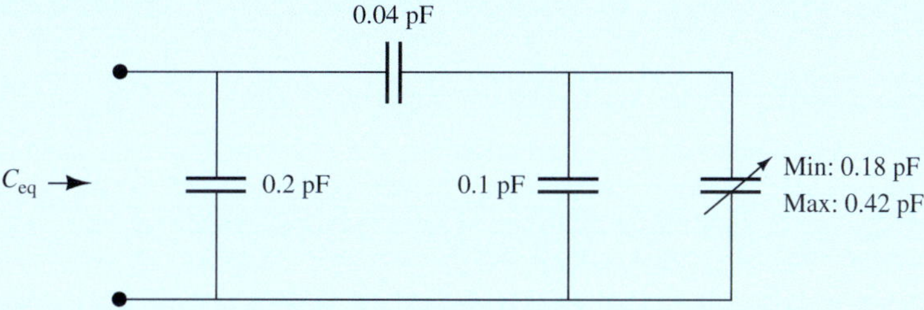

FIGURE 4.12 A radio tuner's capacitor network.

SOLUTION

To find the minimum and maximum, consider two scenarios. Considering the minimum capacitance for the tuning capacitor:

$$C_{eq} = \left(\frac{1}{0.18\ \text{pF} + 0.1\ \text{pF}} + \frac{1}{0.04\ \text{pF}}\right)^{-1} + 0.2\ \text{pF} = 0.2350\ \text{pF}$$

Now, considering the maximum capacitance for the tuning capacitor:

$$C_{eq} = \left(\frac{1}{0.42\ \text{pF} + 0.1\ \text{pF}} + \frac{1}{0.04\ \text{pF}}\right)^{-1} + 0.2\ \text{pF} = 0.2371\ \text{pF}$$

Therefore, the range of values for the circuit capacitance becomes:

$$C_{eq} \in [0.2350, 0.2371]\ \text{pF}$$

APPLICATION EXAMPLE 4.9 A Capacitive Displacement Sensor

In civil engineering, it is important to know how far a beam or other structure made of a certain material deflects with a given force. Study of deflections and other force-related deformities is known as *solid mechanics*. A simple parallel-plate capacitor sensor can be used to determine this displacement.

(*continued*)

APPLICATION EXAMPLE 4.9 Continued

This capacitive sensor is shown in Figure 4.13(a). The sensor has three plates, two fixed and one movable, using air as the dielectric. Moving the center plate changes the capacitance across the terminals. The movable plate is connected to the deflecting beam and the fixed plate is connected to a fixed structure. Note that $C_2 = C_3$.

Consider a sample device where all the plates have dimensions of 4 cm $\times$ 4 cm. Spacing $S_1 = 2.5$ mm and $S_2 = 1$ mm. Air has a relative permittivity, $\varepsilon_r = 1$ and $\varepsilon_0 = 8.85 \times 10^{-12}$.

a. Find the total capacitance as a function of the displacement, d.
b. Find the displacement if the total capacitance is 6.181×10^{-12} F.

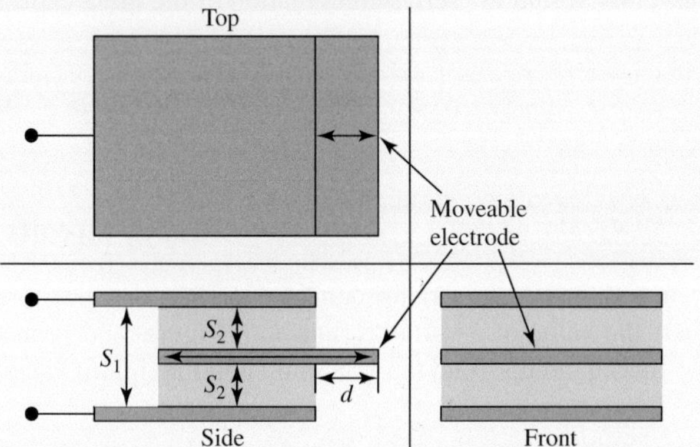

FIGURE 4.13(a) Figure for Example 4.7.

SOLUTION

a. To find the total capacitance as a function of the displacement, d, use Equation (4.1).

$$C_1 = 8.85 \times 10^{-12}\frac{0.04 \cdot d}{0.0025} = 141.6 \times 10^{-12} \cdot d$$

$$C_2 = C_3 = 8.85 \times 10^{-12} \times \frac{0.04 \cdot (0.04 - d)}{0.001}$$

$$= 8.85 \times 10^{-9} \cdot (0.0016 - 0.04d)$$

$$= -354 \times 10^{-12} \cdot d + 14.16 \times 10^{-12}$$

Now, using the sensor network model of Figure 4.13(b) and knowing that $C_3 = C_2$:

$$C_{\text{total}} = C_1 + \left(\frac{1}{C_2} + \frac{1}{C_3}\right)^{-1} = C_1 + \left(\frac{2}{C_2}\right)^{-1} = C_1 + \frac{C_2}{2}$$

$$= 141.6 \times 10^{-12} \cdot d + \frac{-354 \times 10^{-12} \cdot d + 14.16 \times 10^{-12}}{2}$$

$$= 141.6 \times 10^{-12} \cdot d - 177 \times 10^{-12} \cdot d + 7.08 \times 10^{-12}$$

$$C_{\text{total}} = -35.4 \times 10^{-12} d + 7.08 \times 10^{-12}\,\text{F}$$

b. To find the displacement if the total capacitance is 6.181×10^{-12} F, solving for d in the equation above produces:

$$d = \frac{C_{\text{total}} - 7.08 \times 10^{-12}}{-35.4 \times 10^{-12}}$$

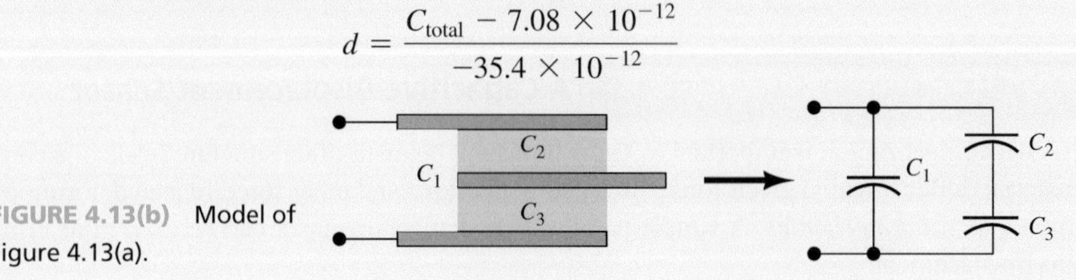

FIGURE 4.13(b) Model of Figure 4.13(a).

$$d = \frac{6.181 \times 10^{-12} - 97.08 \times 10^{-12}}{-35.4 \times 10^{-12}}$$

$$d = 25.4 \times 10^{-3}\ \text{m} = 2.54\ \text{cm}$$

4.4 INDUCTORS

An inductor is a passive element that stores energy in its magnetic field. It consists of a *coil* of wire that is wrapped around a *core*. Current flowing through this coil produces a magnetic flux (see Figure 4.14). This flux, $\phi(t)$, is a function of the current, $i(t)$. Current flowing through any wire produces a slight magnetic field. Coiling the wire intensifies both this field and its related flux. A magnetic flux is the integral of the magnetic field. The concepts of magnetic field and flux will be introduced in Chapter 12.

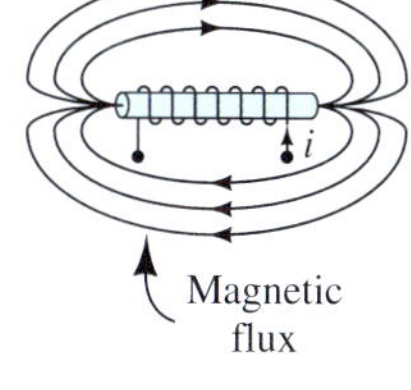

FIGURE 4.14 A basic inductor.

A core of magnetic material is often used to direct the magnetic flux for a given current. Increasing the flux for a given current increases the inductance. Inductors, like capacitors, play an important role in tuning circuits (e.g., radios). Different types of inductors are shown in Figure 4.15.

The circuit schematic symbol for an inductor is shown in Figure 4.16.

Inductance, L, is measured in *henrys* (H):

$$1\ \text{H} = 1\ \text{V} \cdot \text{s/A} \tag{4.31}$$

The induced voltage across an inductor is the rate of change of the flux with time. Therefore, if the flux does not vary with time, the voltage will be zero. As a result, the induced voltage corresponds to:

$$v(t) = \frac{d\varphi}{dt} \frac{\text{weber}}{\text{s}} \tag{4.32}$$

4.4.1 The Relationship Between Voltage and Current

In an inductor, flux is proportional to current:

$$\phi(t) = Li(t) \tag{4.33}$$

Here, L is the inductance, and is defined as:

$$L = \frac{\phi}{i}\left(\frac{\text{weber}}{\text{ampere}} = \text{henry}\right) \tag{4.34}$$

Therefore, the induced voltage, $v(t)$, corresponds to:

$$v = \frac{d\varphi}{dt} = L\frac{di}{dt} \tag{4.35}$$

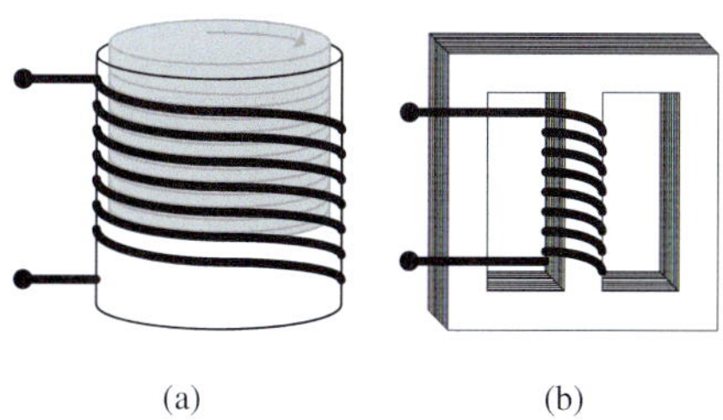

FIGURE 4.15 Different kinds of inductors. (a) Adjustable inductor via movable magnetic core; (b) laminated core transformer.

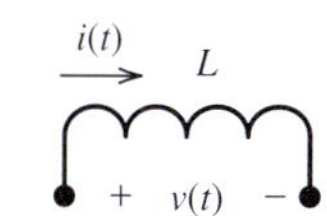

FIGURE 4.16 The symbol for an inductor.

Thus, for an inductor, voltage is a linear function of current. Rearranging the terms results in:

$$\mathrm{d}i = \frac{1}{L}v(t)\mathrm{d}t \tag{4.36}$$

Integrating both sides to find the current:

$$i(t) = \frac{1}{L}\int_{t_0}^{t} v(t)\mathrm{d}t + i(t_0) \tag{4.37}$$

Here, $i(t_0)$ is the initial current at time t_0.

EXAMPLE 4.10 Inductor Voltage

Calculate the voltage across the inductor in Figure 4.17(a), assuming:

a. $i(t) = 5\text{A}$
b. $i(t) = 5\cos(2\pi \times 10^3 t)\text{A}$

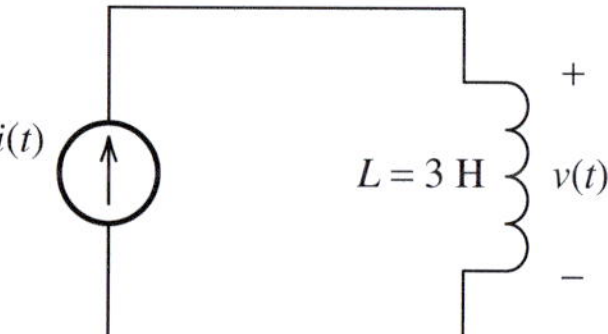

FIGURE 4.17(a) Circuit for Example 4.10.

SOLUTION

Using Equation (4.35):

$$v(t) = 3 \times \frac{\mathrm{d}i(t)}{\mathrm{d}t}$$

a. For $i(t) = 5$, $\frac{\mathrm{d}i(t)}{\mathrm{d}t} = 0$, therefore, $v(t) = 0\text{ V}$

b. For $i(t) = 5\cos(2\pi \times 10^3 t)$, $\frac{\mathrm{d}i(t)}{\mathrm{d}t} = -10\pi \times 10^3 \sin(2\pi \times 10^3 t)$, therefore,

$$v(t) = -30\pi \times 10^3 \sin(2\pi \times 10^3 t)\text{ V.}$$

Note: If a constant current is applied directly to an inductor, the inductor will ultimately i.e., after a long time duration, act as a *closed circuit*. This is because the voltage across the inductor is zero. The equivalent circuit of the inductor is shown in Figure 4.17(b).

FIGURE 4.17(b) Equivalent circuit of an inductor when constant current is applied to it.

4.4.2 Power and Stored Energy

The power delivered to a circuit element is the product of the voltage and current.

$$p(t) = v(t) \times i(t) \tag{4.38}$$

Using Equation (4.35) to substitute for the voltage:

$$p(t) = Li(t) \times \frac{\mathrm{d}i}{\mathrm{d}t} \tag{4.39}$$

The energy delivered to the inductor within the time period t_0 to t corresponds to the integration of the power from t_0 to t.

$$E(t) = L\int_{t_0}^{t} i(t) \times \frac{di}{dt}dt \tag{4.40}$$

Canceling differential time and changing the limits of integration to the corresponding instantaneous currents result in:

$$E(t) = L\int_{i(t_0)}^{i(t)} i(t)\ di \tag{4.41}$$

Integrating (4.41), assuming the initial current $i(t_0)$ is zero, leads to:

$$E(t) = \frac{1}{2}Li^2(t) \tag{4.42}$$

EXAMPLE 4.11 Inductor Current

Given the circuit in Figure 4.18, find the current, i, as a function of time.

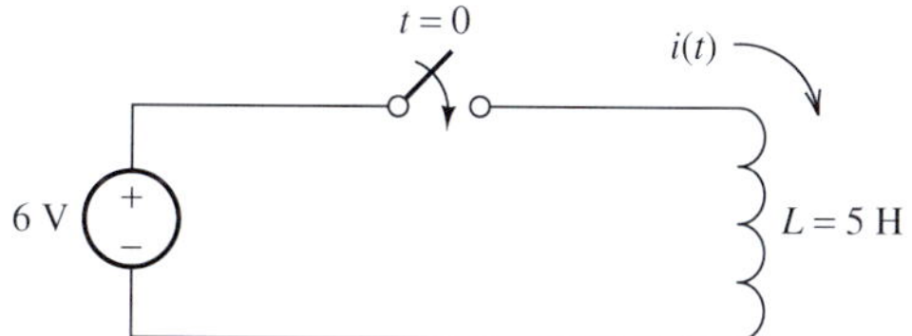

FIGURE 4.18 Circuit for Example 4.11.

SOLUTION

Find the current using Equation (4.37).

$$i(t) = \frac{1}{L}\int_{t_0}^{t} v(t)\ dt + i(t_0)$$

The switch closes at $t = 0$, and $t_0 = 0$. No current flows prior to closing the switch, that is, $i(t_0) = 0$. Therefore, the current after closing the switch is:

$$i(t) = \frac{1}{5}\int_{0}^{t} 6\ dt$$

Integrating and evaluating results in:

$$i(t) = \frac{6}{5}t \text{ for } t > 0$$

Thus, applying a DC voltage to an inductor, the current linearly increases to infinity with time ($i(t) = (6/5)t$). As current increases beyond a limit, the connecting wires and the inductor will not be able to tolerate the heat created due to high values of current. As a result, the wires and/or inductor may burn, therefore opening the circuit. To avoid this situation, a resistor, limiter, or fuse can be used to limit the current in the circuit.

4.5 INDUCTORS IN SERIES AND PARALLEL

Inductors, like resistors and capacitors, can be arranged into networks. The equivalent inductance of an inductor network can be calculated by the same method used to calculate the equivalent resistance of a resistor network. **Series inductor combinations are simply added together.**

Parallel combinations are calculated by taking the reciprocal of the sum of the reciprocals of the inductance values.

4.5.1 Inductors in Series

Applying Kirchhoff's voltage law (KVL) around the loop in Figure 4.19, results in:

$$v = v_1 + v_2 + v_3 \tag{4.43}$$

Substituting Equation (4.35) into Equation (4.43) for the voltages yields:

$$v = L_1 \times \frac{di}{dt} + L_2 \times \frac{di}{dt} + L_3 \times \frac{di}{dt} \tag{4.44}$$

Separating the terms produces a result comparable to Equation (4.35), that is:

$$v = (L_1 + L_2 + L_3) \times \frac{di}{dt} = (L_{eq}) \times \frac{di}{dt} \tag{4.45}$$

Therefore, it can be seen that the equivalent inductance of the three series inductors is:

$$L_{eq} = L_1 + L_2 + L_3 \tag{4.46}$$

Similarly, for N inductors in series:

$$L_{eq} = L_1 + L_2 + \cdots + L_N \tag{4.47}$$

4.5.2 Inductors in Parallel

Applying Kirchhoff's current law (KCL) to the upper node of Figure 4.20 results in:

$$i = i_1 + i_2 + i_3 \tag{4.48}$$

Assuming that the initial current for all inductors is zero, Equation (4.37) can be substituted for i_1, i_2, i_3 in Equation (4.48), which leads to:

$$i = \frac{1}{L_1}\int_{t_0}^{t} v(t)dt + \frac{1}{L_2}\int_{t_0}^{t} v(t)\,dt + \frac{1}{L_3}\int_{t_0}^{t} v(t)\,dt \tag{4.49}$$

Equation (4.49) can be rewritten as:

$$i = \left(\frac{1}{L_1} + \frac{1}{L_2} + \frac{1}{L_3}\right)\int_{t_0}^{t} v(t)\,dt = \frac{1}{L_{eq}}\int_{t_0}^{t} v(t)\,dt \tag{4.50}$$

Therefore, the equivalent inductance corresponds to:

$$L_{eq} = \left(\frac{1}{L_1} + \frac{1}{L_2} + \frac{1}{L_3}\right)^{-1} \tag{4.51}$$

Therefore, the equivalent inductance for N inductors in parallel is:

$$L_{eq} = \left(\frac{1}{L_1} + \frac{1}{L_2} + \cdots + \frac{1}{L_N}\right)^{-1} \tag{4.52}$$

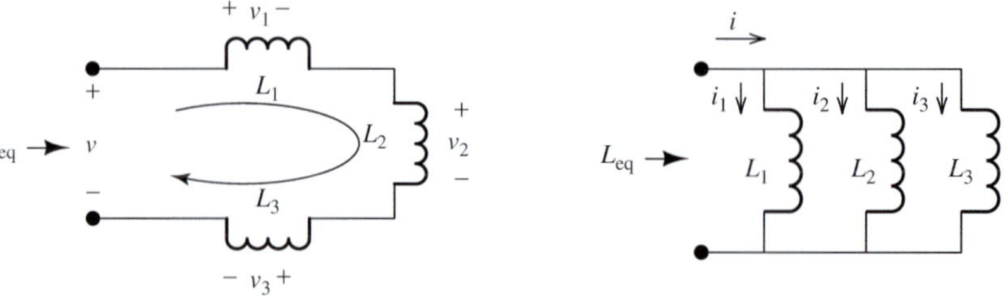

FIGURE 4.19 Inductors in series.

FIGURE 4.20 Inductors in parallel.

EXAMPLE 4.12 Equivalent Inductance

Find the equivalent inductance in Figure 4.21(a).

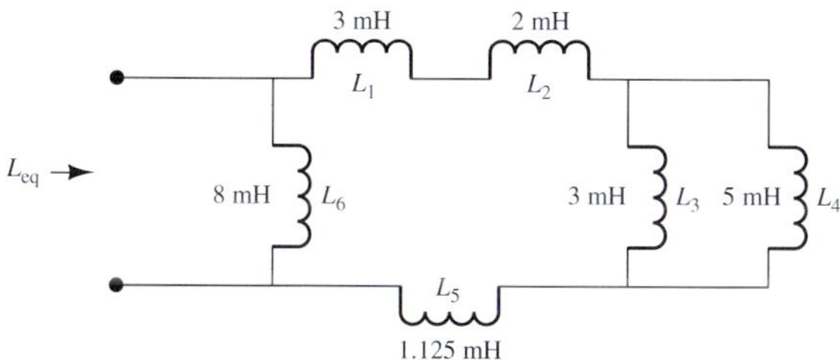

FIGURE 4.21(a) Circuit for Example 4.12.

SOLUTION

L_3 and L_4 are in parallel. Thus, the circuit can be simplified to that shown in Figure 4.21(b).

$$L_{34} = \left(\frac{1}{L_3} + \frac{1}{L_4}\right)^{-1}$$

Because L_1, L_2, L_{34}, and L_5 are in series, the inductor network simplifies to Figure 4.21(c).

$$L_{12345} = L_1 + L_2 + L_{34} + L_5$$

Now L_6 and L_{12345} are in parallel. As a result, the equivalent inductance corresponds to:

$$L_{eq} = \left(\frac{1}{8\text{ mH}} + \frac{1}{8\text{ mH}}\right)^{-1} = 4\text{ mH}$$

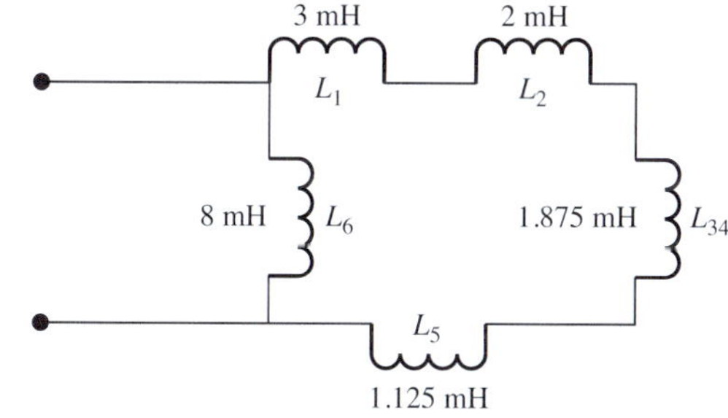

FIGURE 4.21(b) Simplified circuit of Figure 4.21(a).

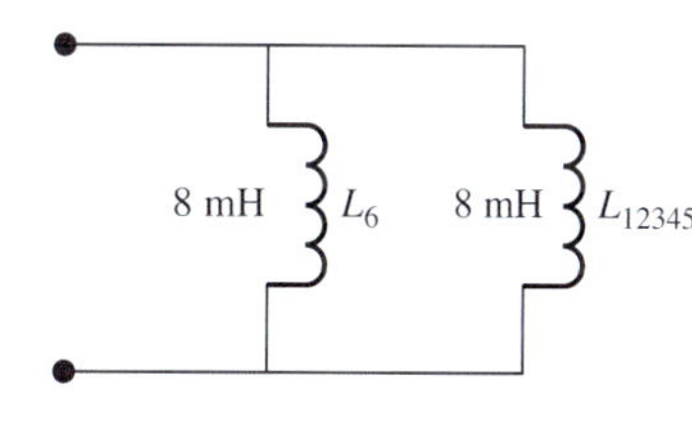

FIGURE 4.21(c) Simplified circuit of Figure 4.21(b).

EXAMPLE 4.13 Equivalent Inductance

Find the equivalent inductance of the circuit shown in Figure 4.22(a).

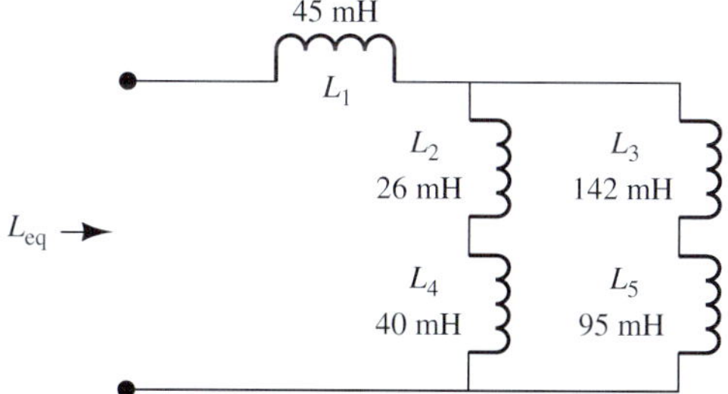

FIGURE 4.22(a) Circuit of Example 4.13.

SOLUTION

Note that in Figure 4.22(a), there are two parallel branches, each consisting of two inductances in series: L_2 is in series with L_4, and L_3 is in series with L_5. Thus, the equivalent inductances of $L_{24} = L_2 + L_4$ and $L_{35} = L_3 + L_5$ and simplify Figure 4.22(a) to Figure 4.22(b).

(*continued*)

EXAMPLE 4.13 Continued

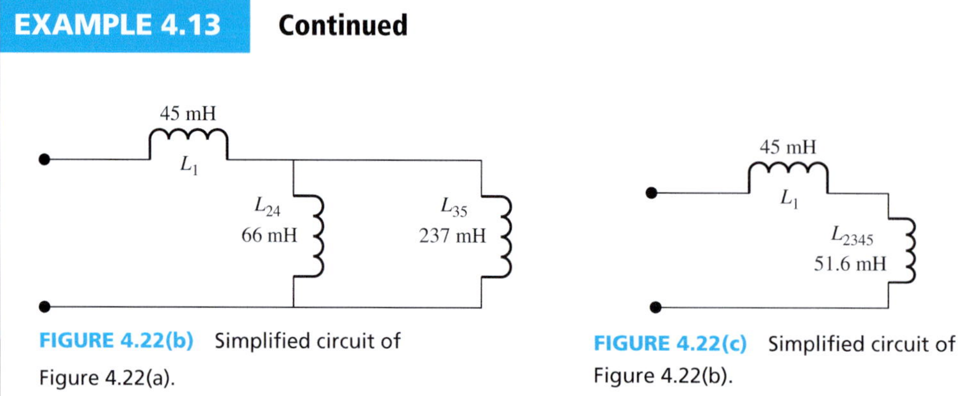

FIGURE 4.22(b) Simplified circuit of Figure 4.22(a).

FIGURE 4.22(c) Simplified circuit of Figure 4.22(b).

The parallel connection of L_{24} with L_{35} can be combined as shown in Figure 4.22(c).

$$L_{2345} = \left(\frac{1}{L_{24}} + \frac{1}{L_{35}}\right)^{-1}$$

Now, the network of Figure 4.22(c) is a simple series connection; therefore, the equivalent inductance will correspond to:

$$L_{eq} = 96.62 \text{ mH}$$

4.6 APPLICATIONS OF CAPACITORS AND INDUCTORS

Capacitors and inductors have many applications to different systems and technologies. This section explains some examples of these practical applications.

4.6.1 Fuel Sensors

Capacitors can be used as fuel gauges, for example, the fuel gauges used in airplanes. Automotive fuel gauges use a resistive sensor instead of capacitors. For a capacitive fuel gauge, low-voltage capacitors are stacked vertically in the fuel tank, as shown in Figure 4.23(a). In addition, one large parallel-plate capacitor is used, as shown in Figure 4.23(b).

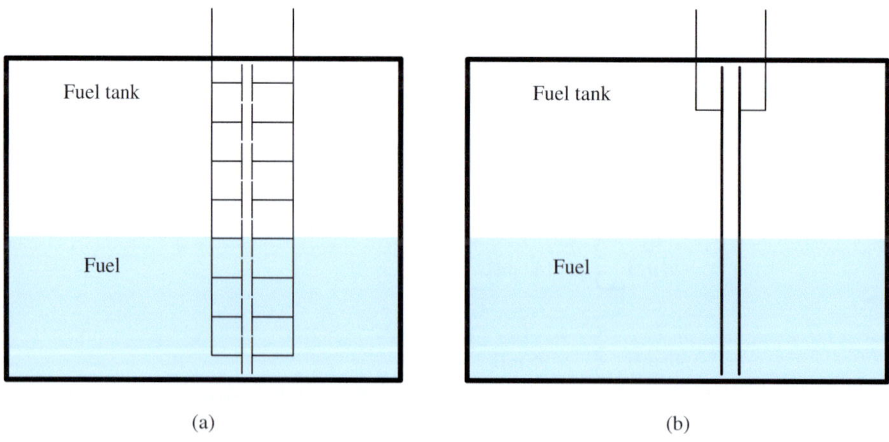

FIGURE 4.23 Capacitive fuel sensors: (a) series of capacitors; (b) large parallel-plate capacitor.

These capacitors are designed to allow the fuel to penetrate the area between the plates to contact the conductors. As the fuel level changes, the dielectric between the parallel plates changes and as a result its capacitance also changes. The magnitude of the capacitance is used to determine the fuel level.

The capacitance can be modeled as two capacitors in parallel as shown in Figure 4.24. The permittivity of the empty section of the tank (air) differs from the permittivity of the fuel. When the fuel level decreases, the capacitance associated with the air, C_{empty}, increases, while the capacitance associated with the fuel, C_{fuel}, decreases.

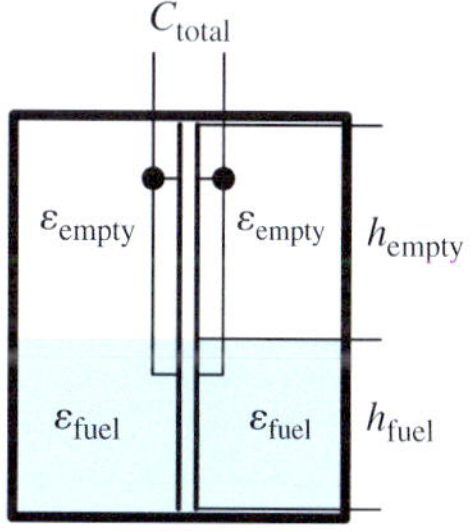

FIGURE 4.24 Capacitance model for Figure 4.23.

The total capacitance corresponds to:

$$C_{total} = C_{empty} + C_{fuel} \tag{4.53}$$

Applying the formula for the capacitance of a parallel-plate capacitor:

$$C = \varepsilon_o \varepsilon_r \times \frac{A}{d} \tag{4.54}$$

The total capacitance is:

$$C_{total} = \frac{w(\varepsilon_{empty} h_{empty} + \varepsilon_{full} h_{full})}{d} \tag{4.55}$$

where:

w = width of the capacitor plates
h_{empty} = height of the empty section
h_{full} = height of the full section
ε_{empty} = permittivity of the empty section
ε_{full} = permittivity of the full section
d = displacement of the capacitor plates.

Equation (4.55) maintains a relationship (scale) between the capacitor and the fuel level. Accordingly, this design can function as an indicator for the fuel level.

4.6.2 Vibration Sensors

Capacitors and inductors can each be used to sense vibration and motion. These sensors have two components: a fixed component and a moving component. The moving component is attached to the moving object, while the fixed component is anchored to a fixed structure (see Figure 4.25). Vibrations in the moving object create pulsation in the moving component. This changes the inductance or the capacitance of the sensor. As shown in Figure 4.25, the sensor can be connected between two structures where the vibration needs to be measured.

Figure 4.25 details two capacitive vibration sensors. In Figure 4.25(a), the capacitance that changes the moving conducting plate changes position between the plates of a parallel-plate

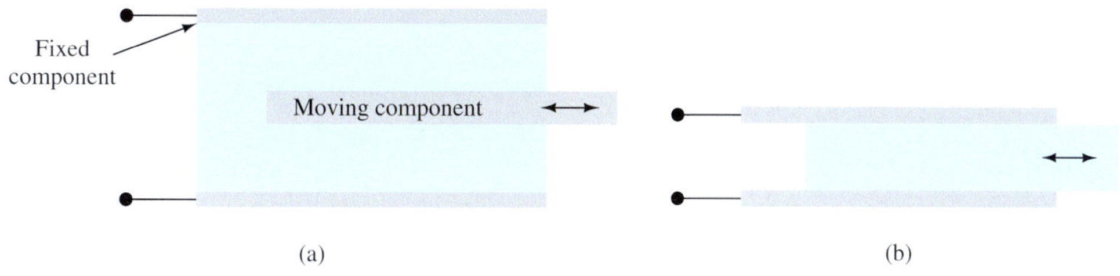

FIGURE 4.25 Capacitive vibration sensors: (a) Moving plate. (b) Moving dielectric.

capacitor. The dielectric can be a fluid or air. In the sensor shown in Figure 4.25(b), the capacitance varies by moving a solid dielectric in and out between the conducting plates. Capacitive displacement sensors have been detailed in Example 4.9 of Section 4.3.

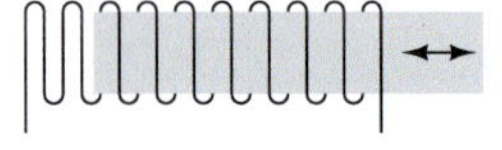

FIGURE 4.26 Inductive vibration sensor.

Figure 4.26 represents a simple inductive vibration sensor. Here, the inductance varies by moving a magnetic core in and out of the inductance coil.

This sensor can be modeled using two series-variable inductors as shown in Figure 4.27. As the core moves out of the coil, L_{core} decreases and L_{nocore} increases. Conversely, moving the core into the coil increases L_{core} and decreases L_{nocore}. As explained in Section 4.4, a magnetic core increases the overall inductance. Therefore, when L_{core} increases, the total inductance increases, and vice versa.

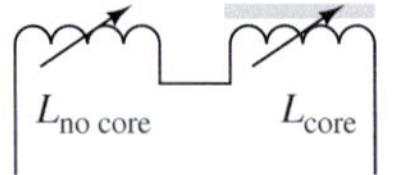

FIGURE 4.27 Model of Figure 4.26.

A linear variable differential transformer (LVDT) is a commonly used inductive sensor. A LVDT, as shown in Figure 4.28, is very similar to the sensor shown in Figure 4.26.

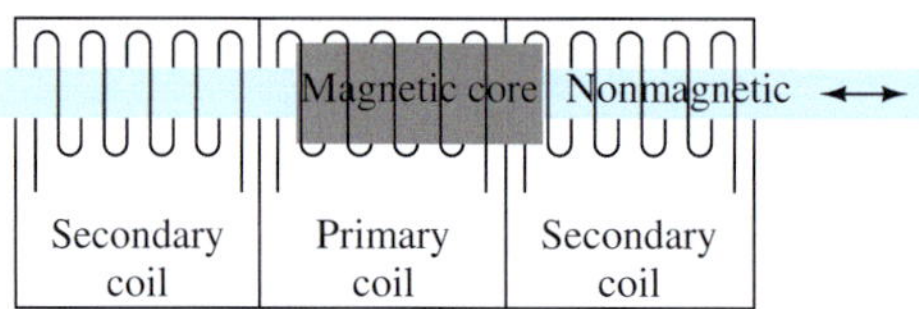

FIGURE 4.28 Linear variable differential transformer.

In a zero position, the magnetic core rests completely within the primary coil. When movement is applied to the core, it moves partially into one of the secondary coils. LVDT will be discussed in detail in Chapter 11.

APPLICATION EXAMPLE 4.14 Pressure Sensors

Capacitive sensors can also be used to measure the change of pressure. A change in pressure leads to a change of the distance between two capacitor plates. This changes the capacitance of the capacitor (Figure 4.29). If a reference pressure is given, with a known coefficient that represents pressure changes as a change in capacitance, the capacitance can be computed for any given pressure.

Assume that the change of capacitance due to the change of pressure is $\Delta C = 0.0085 \times \Delta P$, where P is the pressure measured in Pascal, Pa. If the pressure changes from 1000 Pa to 1050 Pa, with an initial capacitance of 3 F, what is the capacitance after the pressure changes?

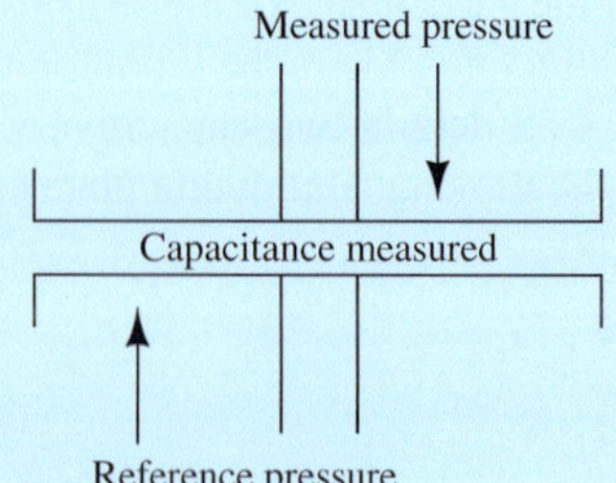

FIGURE 4.29 Capacitive pressure sensor.

SOLUTION

Change of pressure = 1050 − 1000 = 50 Pa, thus, the change in capacitance is:

$$\Delta C = 0.0085 \times \Delta P$$
$$\Delta C = 0.0085 \times 50 = 0.425 \text{ F}$$

The capacitance after the pressure change is:

$$C_{\text{final}} = C_{\text{initial}} + \Delta C$$
$$C_{\text{final}} = 3 + 0.425 = 3.425 \text{ F}$$

APPLICATION EXAMPLE 4.15 Traffic Light Detectors

In an automated traffic light system, like the one shown in Figure 4.30, the red traffic light stays ON longer if no vehicle stops at it; but it turns to green earlier when at least one vehicle is stopped at the light. This system operates as follows.

First, an inductor is placed underneath the ground in the lane where a vehicle will stop in the event of the red light. When there is a vehicle at the stop light, the size of the vehicle changes the inductance. A change in the inductance is detected by a specific device, which makes the traffic light to turn to green. Given the voltage source of 12 V, the inductance when there is not any vehicle is 2 H, and the sensitivity of the device to the change of inductance is 0.005 H.

Calculate the change of inductance if:

a. A car is stopped at the light, with the dimensions (mm) of: 4450 × 1750 × 1420
b. A motorcycle is stopped at the light, with the dimensions (mm) of: 1945 × 785 × 1055.

Assume the inductance changes with the respect to the size of vehicle as shown here:

$$L = 0.002 \times \text{volume}$$

Note:

$$\text{Volume} = \text{height} \times \text{width} \times \text{length, in m}^3.$$

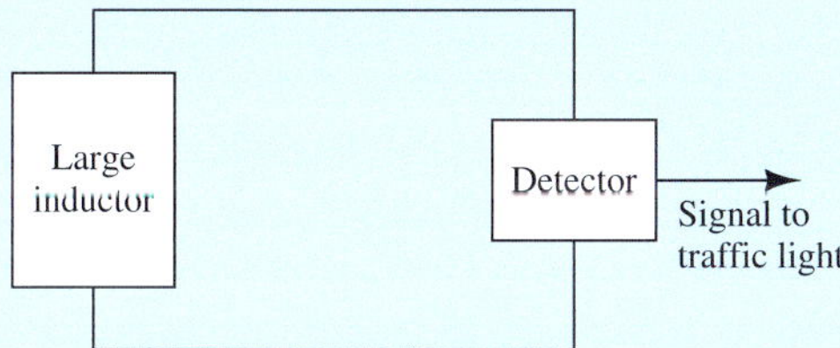

FIGURE 4.30 A traffic light detector.

SOLUTION

For part a:

The volume of the car is:

$$v = 4.45 \times 1.75 \times 1.42$$
$$v = 11.058 \text{ m}^3$$

The change in inductance is:

$$L = 0.002 \times 11.058$$
$$L = 0.0221 \text{ H} > 0.005 \text{ H (the device sensitivity)}$$

Thus, the device responds to the presence of the car.

For part b:

The volume of the motorcycle is:

$$v = 1.945 \times 0.785 \times 1.055$$
$$v = 1.611 \text{ m}^3$$

(*continued*)

APPLICATION EXAMPLE 4.15 **Continued**

The change of inductance is:

$$L = 0.002 \times 1.611$$
$$L = 0.0032\,\text{H} < 0.005\text{ H (the device sensitivity)}$$

Thus, the detector is not able to detect the change of inductance when there is a motorcycle stopped at the light.

4.7 ANALYSIS OF CAPACITIVE AND INDUCTIVE CIRCUITS USING PSPICE

This section describes the process of analyzing capacitive and inductive circuits using PSpice. Here, we use examples to explain the approach.

EXAMPLE 4.16 **PSpice Analysis**

In the Figure used in Example 4.3, repeated as Figure 4.31, use PSpice to simulate the current through the capacitor, assuming $v(t) = 5\cos(2\pi \times 10^3 t)$ V.

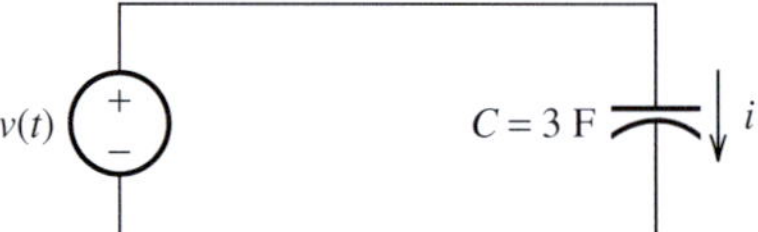

FIGURE 4.31 The circuit of Example 4.3.

SOLUTION

First, determine the frequency of the sinusoidal voltage source.

$$f = \frac{\omega}{2\pi}$$
$$f = \frac{2\pi \times 10^3}{2\pi}$$
$$f = 1000\text{ Hz}$$

1. Choose VSIN as the voltage source, following the process shown in Example 2.24 in Chapter 2 to calculate and set up the parameters for VSIN.
2. To choose the capacitor, go to "Part" and type "C" in part name. Instead of using "C," choose "C_elect." If "C" is unavailable in "Part," add "Analog" into the library (see Figure 2.63 in Chapter 2).
3. To set the initial condition for the capacitor, right-click the capacitor and choose "Edit properties." Under the "IC" tab (solid arrow), insert the desired initial condition, "0" because $v_C(t = 0)$ is 0 V. The initial condition can optionally be displayed in the PSpice circuit by highlighting the "IC" tab then choosing "Display…" (dashed arrow). Choose "Name and Value" under the Display format.

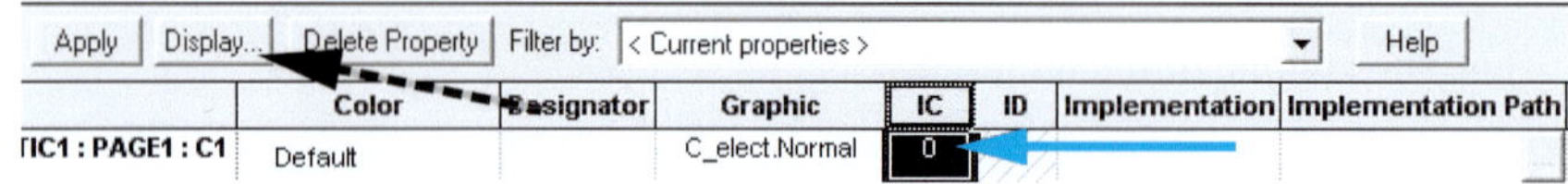

FIGURE 4.32 Setting the capacitor's initial conditions.

4. Set up the circuit as shown in Figure 4.33.
5. To run the simulation for the capacitor and inductor circuit, choose the "Analysis Type" to be time domain (see Figure 2.71 in Chapter 2). Set the "Run to time" as 10 ms. The result is shown in Figure 4.34. Use "Add Trace" in the Trace menu (Figure 2.77 in Chapter 2), to add the current flow through the capacitor into the plot.

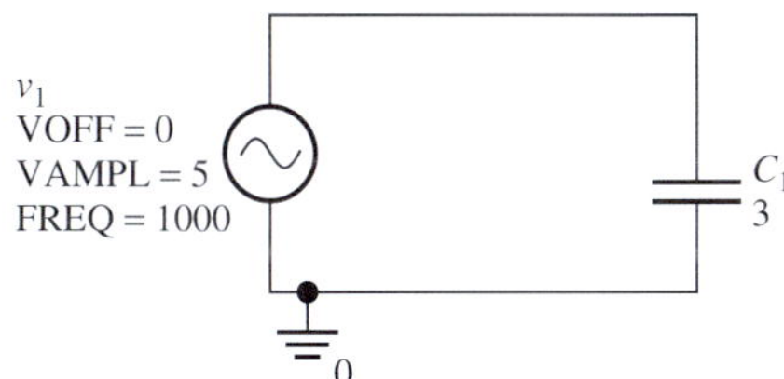

FIGURE 4.33 PSpice solution for Example 4.3.

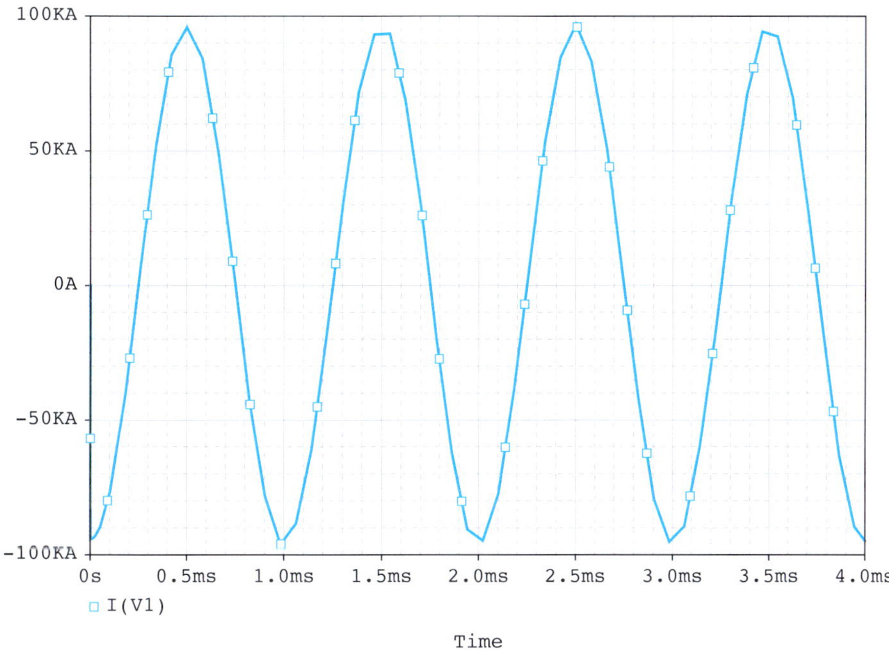

FIGURE 4.34 Results for Example 4.3.

EXAMPLE 4.17 PSpice Analysis

In Figure 4.17(a), repeated in Figure 4.35, use PSpice to simulate the voltage across the inductor, assuming $i(t) = 5\cos(2\pi \times 10^3 t)$ A.

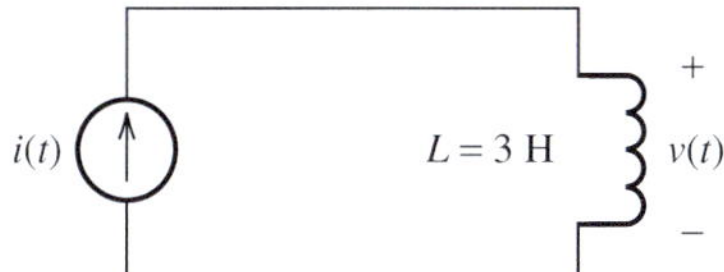

FIGURE 4.35 Circuit for Example 4.10.

SOLUTION

First, determine the frequency of the sinusoidal current source.

$$f = \frac{\omega}{2\pi}\, f = \frac{2\pi \times 10^3}{2\pi}\, f = 1000\,\text{Hz}$$

(continued)

EXAMPLE 4.17 Continued

1. An "ISIN" sinusoidal current source has been chosen for this example. Go to "Place" > "Part" and type "ISIN" to obtain the sinusoidal current source part (see Figure 2.63 in Chapter 2). Follow the steps in Example 2.24 in Chapter 2 to set up the sinusoidal current source parameters.
2. To add an Inductor into the PSpice circuit, go to "Part" and type "L."
3. Define the initial condition, as shown in Example 4.16. Set IC to 0 for the case of I_L $(t = 0) = 0$ A.
4. Wire the circuit as shown in Figure 4.36 and run the simulation for 5 ms. The result is shown in Figure 4.37.

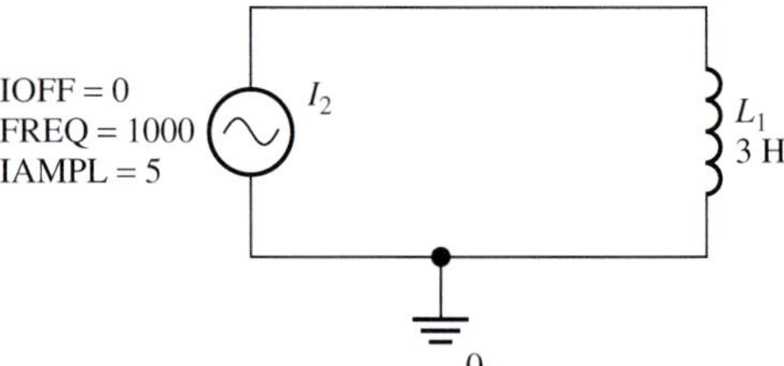

FIGURE 4.36 PSpice solution for Example 4.10.

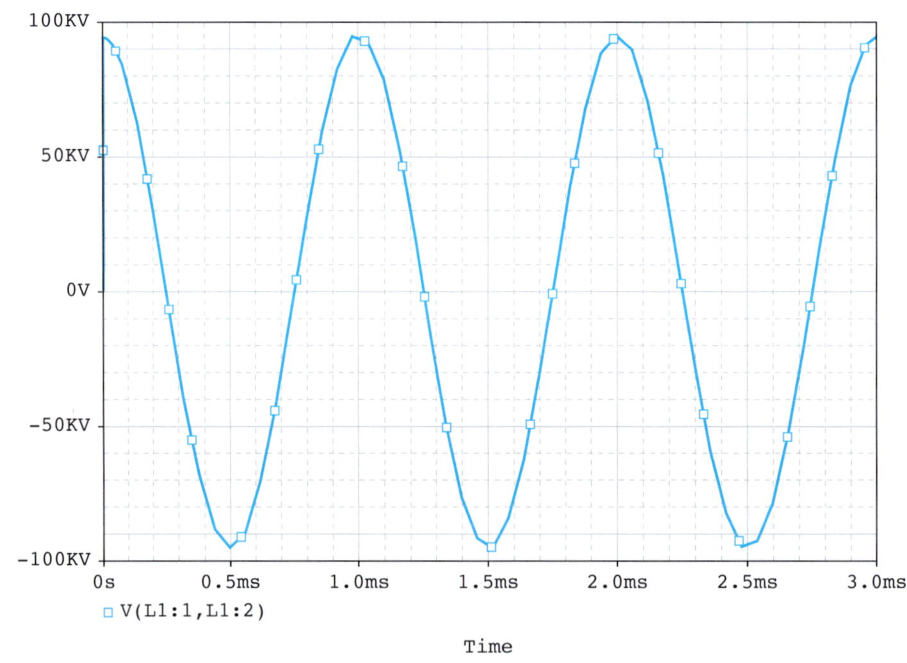

FIGURE 4.37 Results for Example 4.10.

4.8 WHAT DID YOU LEARN?

- The capacitance of a parallel-plate capacitor is proportional to the area of the plates and is inversely proportional to the distance between the plates [Equation (4.1)]:

$$C = \frac{\varepsilon_0 \varepsilon_r \times A}{d}$$

- The relationship between capacitors and inductors is expressed through the following equations:

Capacitor	Inductor
$q(t) = C \cdot v(t)$	$L = \frac{\phi}{i}$
$i(t) = C\frac{dv(t)}{dt}$	$di = \frac{1}{L}v(t)dt$
$v(t) = \frac{1}{C}\int_{t_0}^{t} i(t)\, dt + v(t_0)$	$i(t) = \frac{1}{L}\int_{t_0}^{t} v(t)\, dt + i(t_0)$
$P(t) = Cv(t)\frac{dv(t)}{dt}$	$P(t) = Li(t) \times \frac{di}{dt}$
$E(t) = \frac{1}{2}Cv^2(t)$	$E(t) = \frac{1}{2}Li^2(t)$

- Capacitors in series and parallel:
  Capacitors *in series* can be combined like resistors in parallel [Equation (4.23)]:

$$C_{eq} = \left(\frac{1}{C_1} + \frac{1}{C_2} + \cdots + \frac{1}{C_N}\right)^{-1}$$

  Capacitors *in parallel* can be combined like resistors in series [Equation (4.30)]:

$$C_{eq} = C_1 + C_2 + \cdots + C_N$$

- Inductors in series and parallel:
  Inductors *in series* can be combined like resistors in series [Equation (4.47)]:

$$L_{eq} = L_1 + L_2 + \cdots + L_N$$

  Inductors *in parallel* can be combined like resistors in parallel [Equation (4.52)]:

$$L_{eq} = \left(\frac{1}{L_1} + \frac{1}{L_2} + \cdots + \frac{1}{L_N}\right)^{-1}$$

Problems

*B refers to Basic, A refers to Average, H refers to Hard, and * refers to problems with answers.*

SECTION 4.1 INTRODUCTION

4.1 (B)* Which of the following devices is not passive?
- **a.** Resistor
- **b.** Capacitor
- **c.** Inductor
- **d.** Current-controlled current source.

4.2 (B) Which of the following devices store energy?
- **a.** Resistor
- **b.** Capacitor
- **c.** Voltage source
- **d.** Inductor.

SECTION 4.2 CAPACITORS

4.3 (B) A parallel plate capacitor has gasoline as the dielectric. The area of the plate is 0.25 m^2 and the distance between two plates is 5 cm. Find the capacitance given that the relative permittivity of gasoline is 2.

4.4 (B) In problem 4.3, how does the capacitance change if:
- **a.** the plate area is doubled
- **b.** the distance between plates is doubled
- **c.** the dielectric becomes air.

4.5 (B)* Find the stored charge in a 10-μF capacitor that is charged to 6 V.

4.6 (H) Application: How can one electron be detected?
- **a.** In the market, we can find many capacitors in the range of 1 μF. How many electrons should be available in a 1-μF capacitor to create a total voltage of 1 V?
- **b.** Suppose we want to see a change of 1 V in a capacitor before and after one electron is stored in it. What should the capacitance be?
- **c.** We can detect whether one electron is being stored in the capacitor by observing the change of the capacitor voltage. Why can we not detect one electron being stored in a 1-μF capacitor? (*Hint:* compute the capacitor voltage change while one electron being stored)

4.7 (H)* In Problem 4.6, if the smallest observable voltage is 0.1 V, compute the side-length of a square plate capacitor that can detect one electron, assuming the dielectric is air with relative permittivity of 1 and the distance between plates is 10 nm (10×10^{-9} m).

4.8 (B) If a 1.5-μF capacitor, charged to 5 V, is discharged in a time of 50 ms, what is the average power provided?

4.9 (A) A 5-μF capacitor is charged with a steady current of 50 μA. How long does it take to charge to 12 V?

4.10 (B) Find the voltage, v, in Figure P4.10.

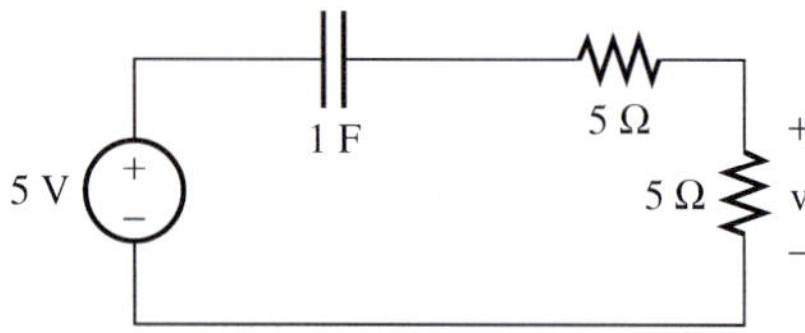

FIGURE P4.10 Circuit for Problem 4.10.

4.11 (A) The voltage across a 2.5-μF capacitor is 10 sin (200 πt) V as shown in Figure P4.11. Find an expression for (a) i_C through the capacitor and (b) the stored energy.

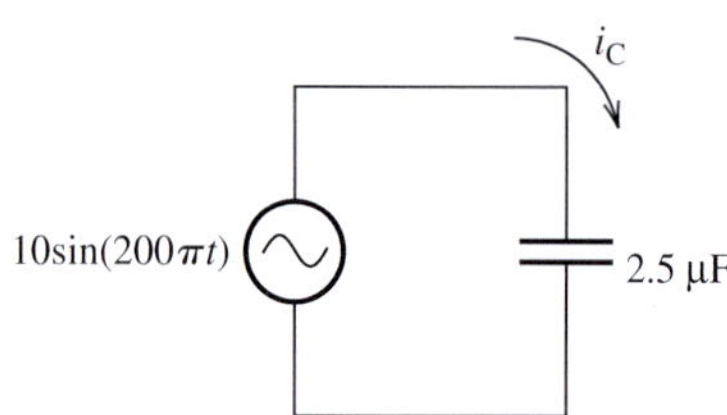

FIGURE P4.11 Circuit for Problem 4.11.

4.12 (H) Suppose $C = 2$ F in the circuit shown in Figure P4.12(a) and the voltage, $v(t)$, is shown in Figure P4.12(b).

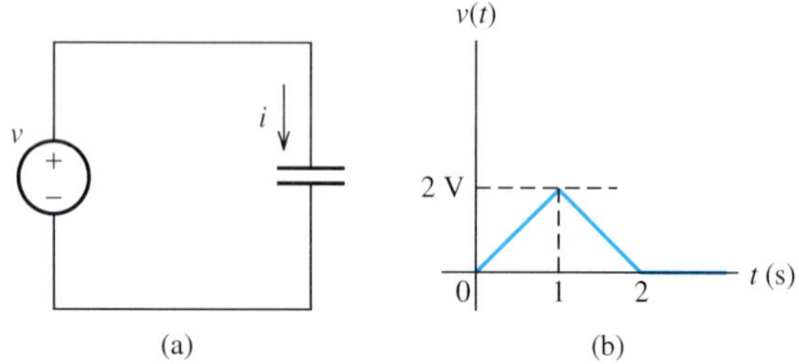

FIGURE P4.12 (a) A capacitor circuit; (b) waveform of $V(t)$.

Compute the current, $i(t)$, through the capacitor.

4.13 (A) In Problem 4.12:

a. Compute the power absorbed by the capacitor

b. Compute the energy stored in the capacitor.

4.14 (A)* At $t = 0$, a 2-mA source is applied to an uncharged 1.2-μF capacitor as shown in Figure P4.14. Determine (a) the capacitor voltage, v_C. at $t = 20$ ms and (b) the energy, w_C, stored in the capacitor at $t = 20$ ms.

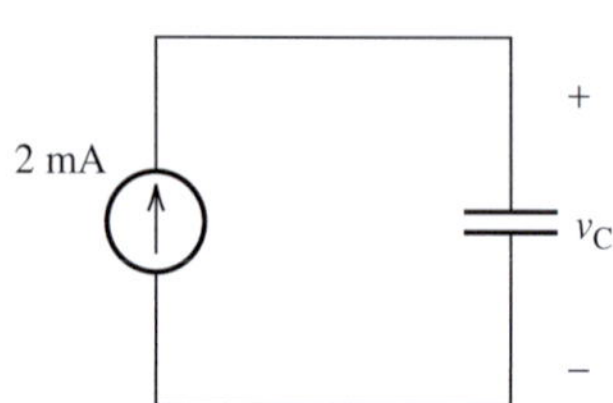

FIGURE P4.14 Circuit for Problem 4.14.

4.15 (H) At $t = 0$, a 1-mA source is connected to a 2-μF capacitor which has been previously charged to 3 V. Give (a) the expression for the capacitor voltage, v_C. as a function of time, and (b) v_C at $t = 10$ ms.

4.16 (H) For the circuit shown in Figure P4.16(a), $i(t)$ is given in Figure P4.16(b). Assume the initial value of v, $v(0) = 0$.

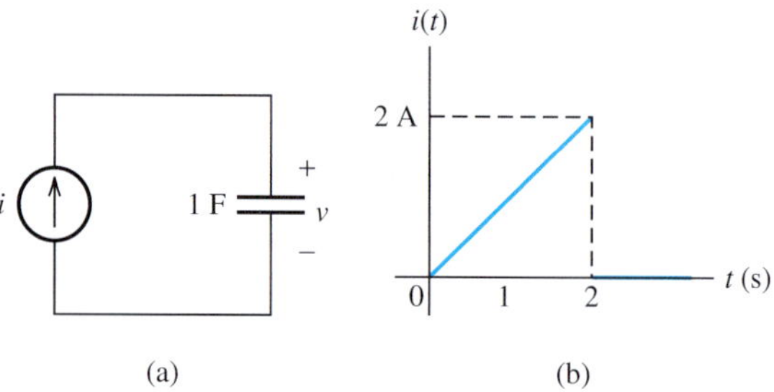

FIGURE P4.16 (a) Capacitor circuit; (b) waveform of $i(t)$.

Compute the voltage $v(t)$ across the capacitor.

4.17 (A) In Problem 4.16:

a. Compute the power absorbed by the capacitor

b. Compute the energy stored in the capacitor.

4.18 (H) Prove that the voltage across a capacitor cannot change instantaneously. (*Hint:* if the voltage at time a is $v(a) = 1/C\int_0^a i(t)\,dt$, compute the voltage at time $(a + \varepsilon)$, where ε is a positive number. When ε gets arbitrarily small, what will be the behavior of $v(a + \varepsilon)$?)

4.19 * Find ε_4 and the current through each resistor shown in Figure P4.19. $\varepsilon_1 = 4$ V, $\varepsilon_2 = 8$ V, $\varepsilon_3 = 12$ V, $C_1 = C_2 = C_3 = C_4 = 1$ μF. All resistors have the same resistance of 1Ω.

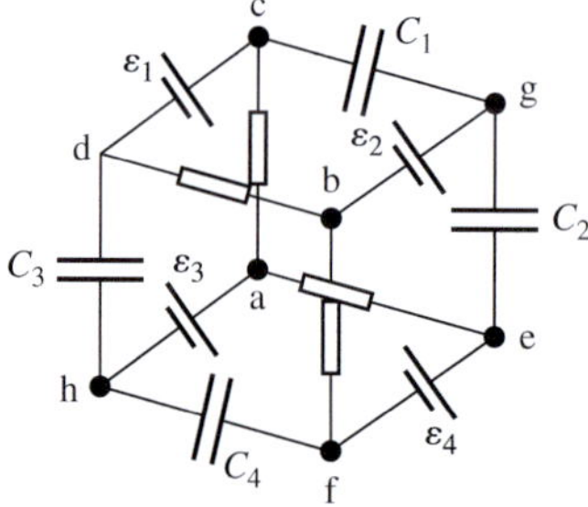

FIGURE P4.19 Circuit for Problem 4.19.

4.20 (A) Find the charge on each capacitor shown in Figure P4.19. Use the parameters in Problem 4.19.

4.21 (A) Find the total energy stored in capacitors shown in Figure P4.19. Use the parameters in Problem 4.19.

4.22 (H)* Connect nodes a and b in Figure P4.19, repeat Problem 4.19.

4.23 (A) Connect nodes a and b in Figure P4.19, repeat Problem 4.19.

4.24 (A)* Connect nodes a and b in Figure P4.19, repeat Problem 4.19.

SECTION 4.3 CAPACITORS IN SERIES AND PARALLEL

4.25 (B) Find the equivalent capacitance between points A and B in Figure P4.25.

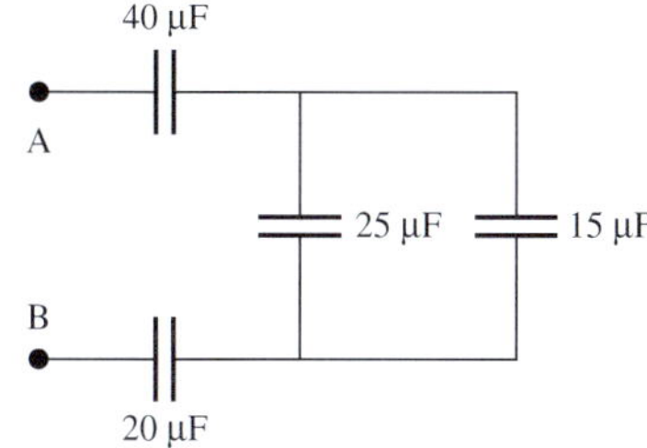

FIGURE P4.25 Capacitor network.

4.26 (B) For the capacitor network in Figure P4.26, what is the equivalent capacitance, C_{eq}?

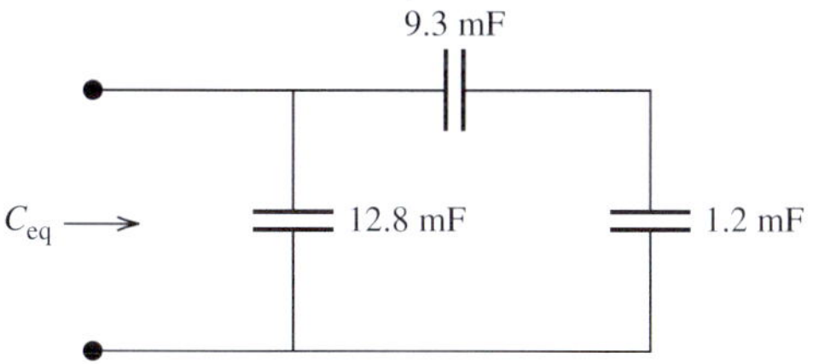

FIGURE P4.26 Circuit for Problem 4.26.

4.27 (A) Find the capacitance, C_A, needed to produce a network capacitance of $C_{eq} = 11.545$ μF.

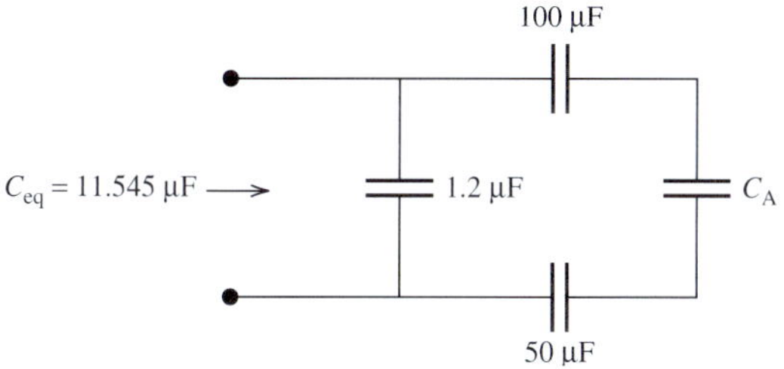

FIGURE P4.27 Capacitor network.

4.28 (A)* For the capacitor network in Figure P4.28, what is the equivalent capacitance, C_{eq}?

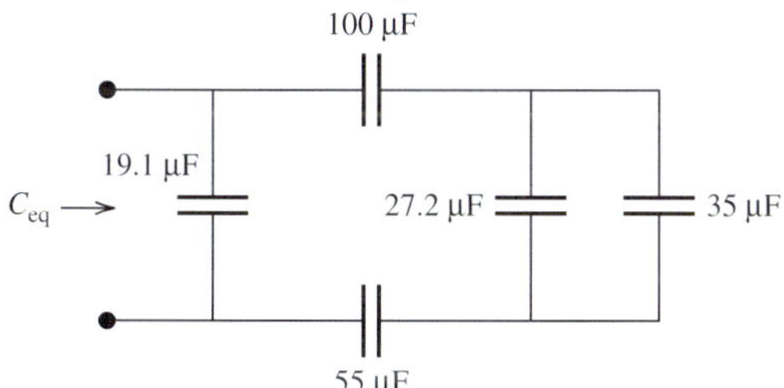

FIGURE P4.28 Circuit for Problem 4.28.

4.29 (A) Find the equivalent capacitance between points A and B in Figure P4.29.

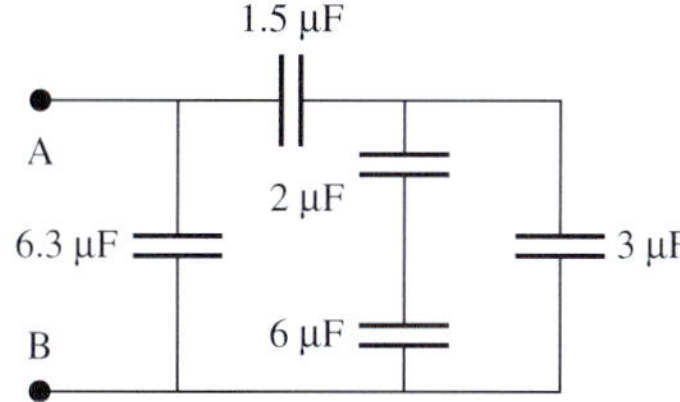

FIGURE P4.29 Capacitor network.

4.30 (A) Find the minimum and maximum values for C_{eq} as shown in Figure P4.30.

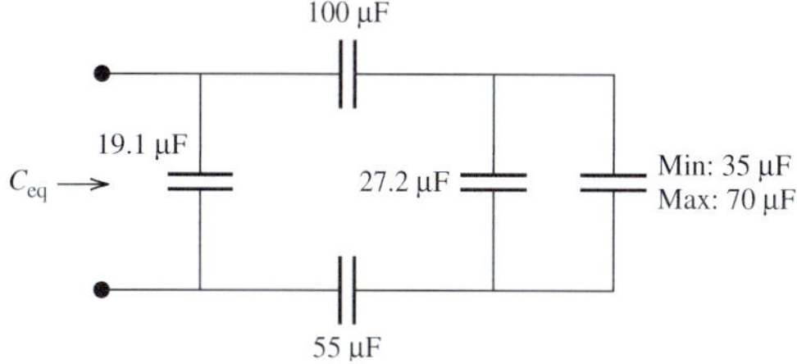

FIGURE P4.30 Capacitor network.

4.31 (H) Find the equivalent capacitance between points A and B in Figure P4.31.

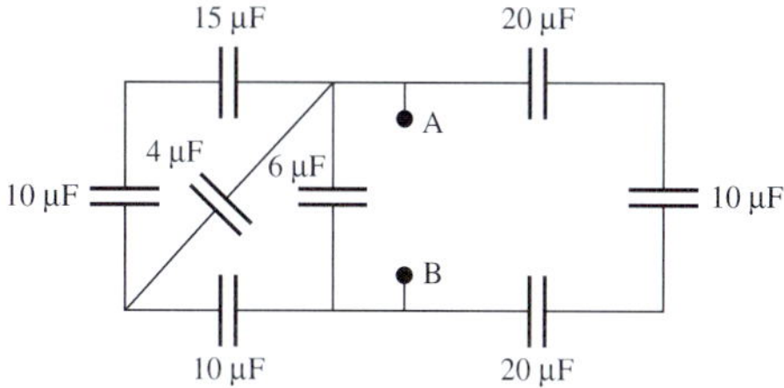

FIGURE P4.31 Capacitor network.

4.32 (H)* Look at the fuel sensor example in Section 4.6.1. The permittivity of air is 1 and the permittivity of gasoline is 2. Let the height of the gasoline level be h.

a. Find the expression of the total capacitance C_{total} in terms of h, width of the capacitor w, distance between plates d, and the free space permittivity ε_0.

b. When the tank is full, C_{total} is 12×10^{-12} F; when the tank is empty, C_{total} is 6×10^{-12} F. Suppose the capability of the tank is 15 gallon. How much gasoline in gallon is in the tank when C_{total} is read as 8×10^{-12} F?

SECTION 4.4 INDUCTORS

4.33 (B) Given the circuit in Figure P4.33, find the current, i, as a function of time. (*Note:* v_s is a sinusoid.)

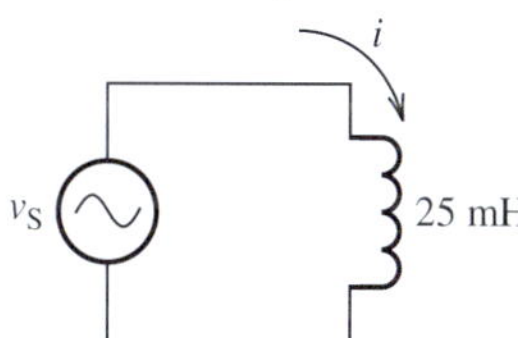

FIGURE P4.33 Inductor circuit.

4.34 (B) Find i_1 in the circuit as shown in Figure P4.34.

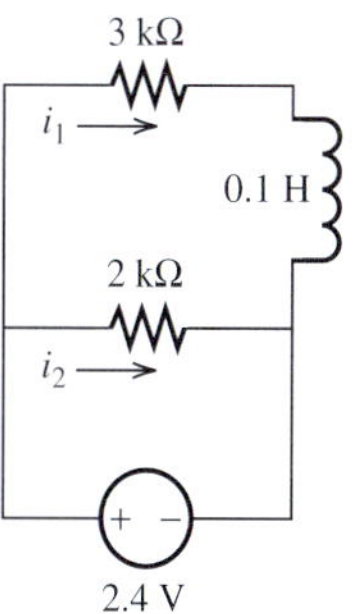

FIGURE P4.34 Circuit with an inductor.

4.35 (A)* Find v_o in the circuit given in Figure P4.35.

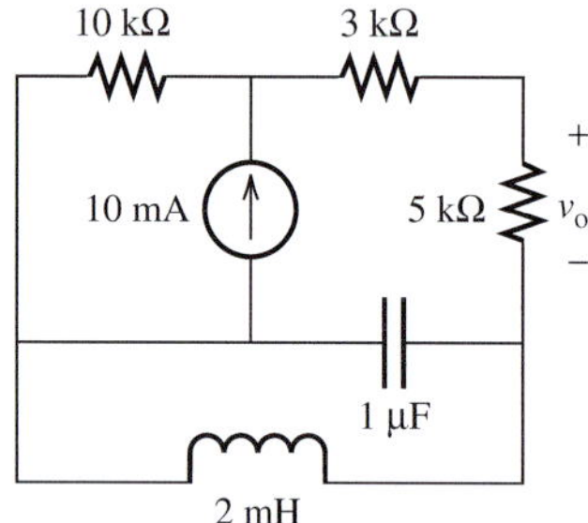

FIGURE P4.35 Circuit with capacitor and inductor.

4.36 (B) If a current of $i(t) = 20\cos(100\pi t)$ mA flows through an inductor of $L = 45$ mH, find the voltage induced across the inductor.

4.37 (A) Current through an inductor is turned on at time $t = 0$, as shown in Figure P4.37. $v_s = \cos(200\pi t)$. Calculate the energy delivered to the inductor at $t = 21$ ms.

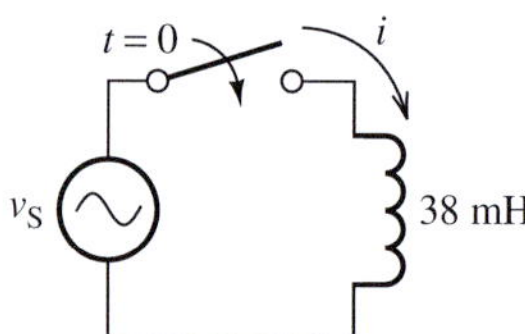

FIGURE P4.37 Circuit for Problem 4.37.

4.38 (H) For the circuit shown in Figure P4.38(a), $v(t)$ is given in Figure P4.38(b). Assume the initial value of i, $i(0) = 0$.

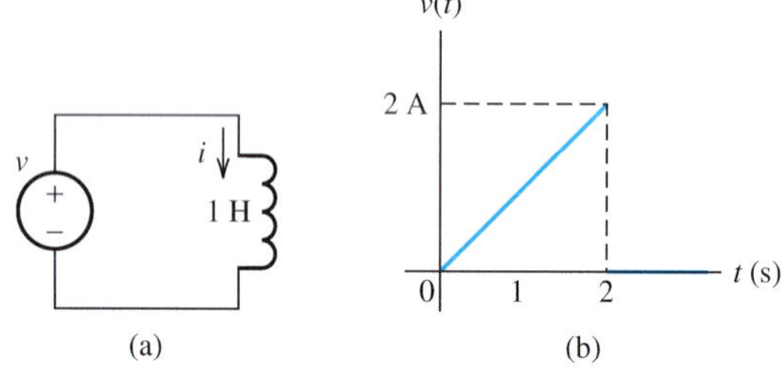

FIGURE P4.38 (a) Inductor circuit; (b) waveform of $v(t)$.

Compute the current, $i(t)$, through the inductor.

4.39 (A) In Problem 4.38:

a. Compute the power absorbed by the inductor

b. Compute the energy stored in the inductor.

4.40 (H) For the circuit shown in Figure P4.40(a), the current $i(t)$ is shown in Figure P4.40(b).

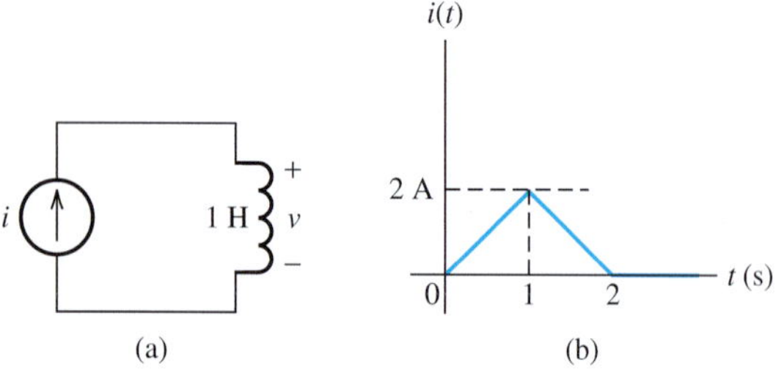

FIGURE P4.40 (a) Inductor circuit; (b) waveform of $i(t)$.

Compute the voltage, $v(t)$, across the inductor.

4.41 (A) In Problem 4.40:

a. Compute the power absorbed by the inductor.

b. Compute the energy stored in the inductor.

4.42 (H) Prove that the current through an inductor cannot change instantaneously. (*Hint:* refer to the hint of Problem 4.18.)

SECTION 4.5 INDUCTORS IN SERIES AND PARALLEL

4.43 (B)* For the inductor network in Figure P4.43, what is the equivalent inductance, L_{eq}?

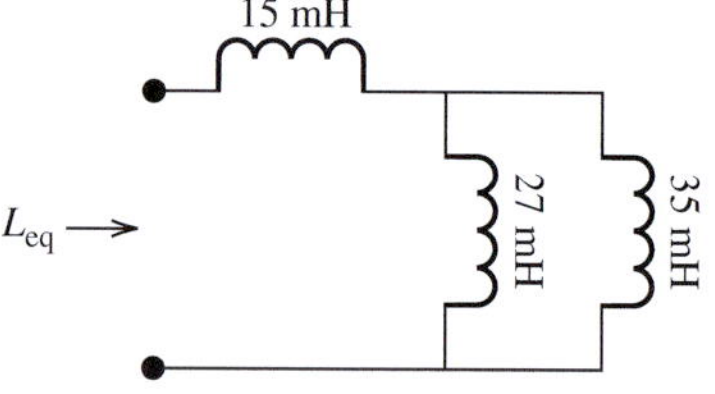

FIGURE P4.43 Circuit for Problem 4.43.

4.44 (B) For the inductor network in Figure P4.44, what is the equivalent inductance, L_{eq}?

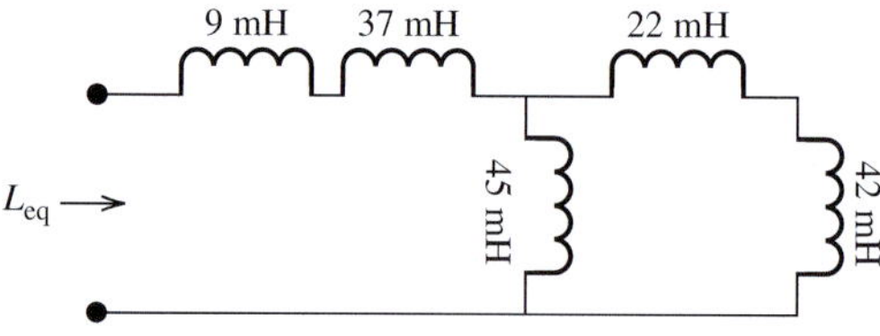

FIGURE P4.44 Circuit for Problem 4.44.

4.45 (A)* Find the equivalent inductance of the inductor network of Figure P4.45.

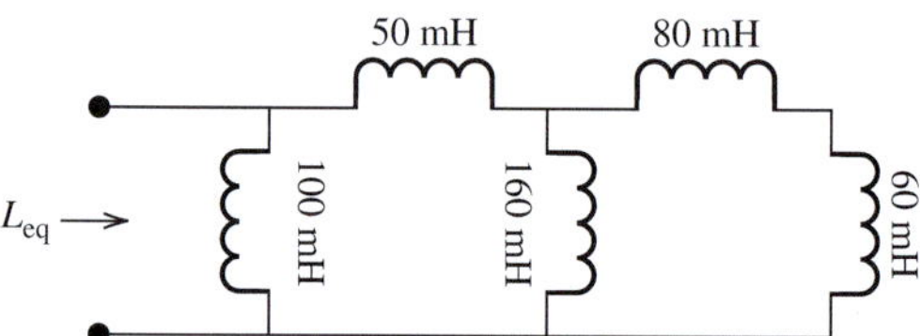

FIGURE P4.45 Inductor network.

4.46 (H) Find the equivalent inductance across points A and B in Figure P4.46.

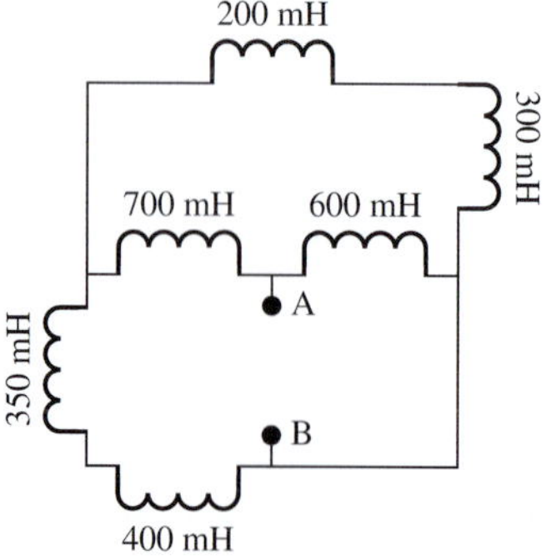

FIGURE P4.46 Inductor network.

4.47 (A) Find the equivalent inductance across points A and B in Figure P4.47.

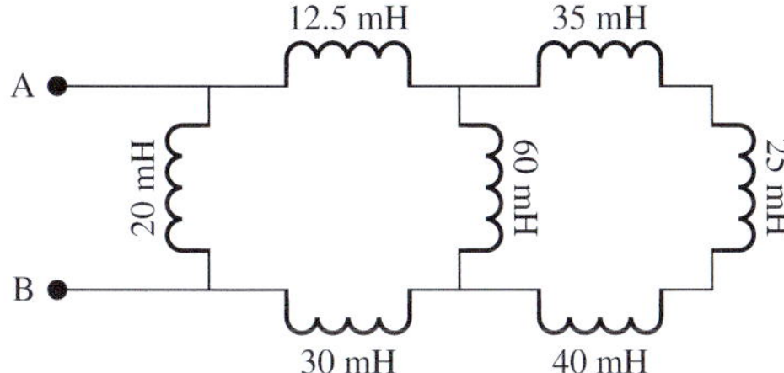

FIGURE P4.47 Inductor network.

4.48 (A)* (Application) oil pipeline, running from Mosul, Iraq to Beirut, Lebanon is being restored so it can be placed back into service. A new pump is needed to replace a broken one. A large electric motor is needed to drive this pump. The motor has the equivalent circuit as shown in Figure P4.48. Solve for the equivalent inductance.

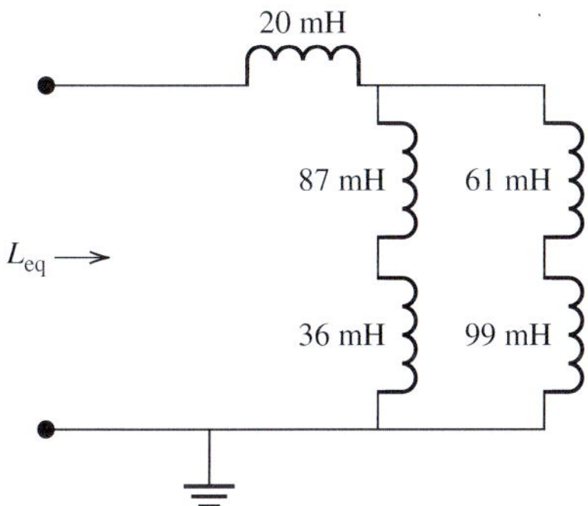

FIGURE P4.48 Circuit for Problem 4.48.

4.49 (H) Find the expression for the equivalent inductance across the points A and B in Figure P4.49.

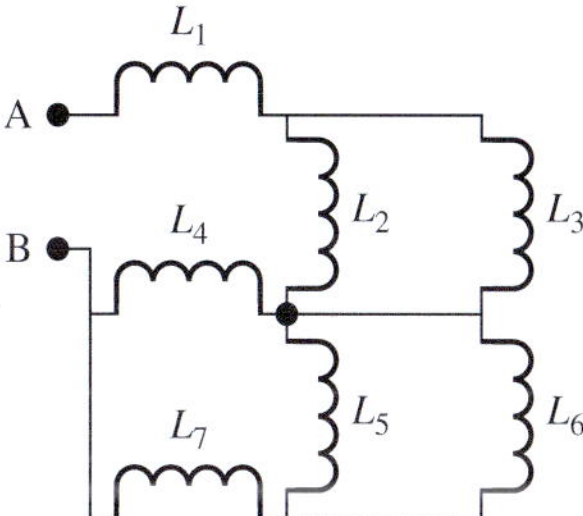

FIGURE P4.49 Inductor network.

SECTION 4.6 ANALYSIS OF THE CIRCUITS USING PSPICE

4.50 (B) The voltage across a 2.5-μF capacitor is 10 sin(200 πt) V, as shown in Figure P4.11. Use PSpice and plot the current, I_C, through the capacitor.

4.51 (A) Consider a circuit with three capacitor C_1, C_2, and C_3 connected in parallel to a voltage source of 10 $\sin(120\pi t)$ V. Use PSpice to plot the current flow through C_3 if C_1 and C_2 is 100 μF and C_3 is 200 μF.

4.52 (H) Given the voltage source is $5\sin(\pi t)$, current through an inductor is turned on at time, t = 0, as shown in Figure P4.37. Use PSpice to plot the voltage and current flow through the inductor for 25 ms.

TOPIC 10
Steady-State AC Analysis

The content of this topic is compiled from:

Chapter 6 (pp. 215–273), *Electrical Engineering: Concepts and Applications*

By S.A. Reza Zekavat

CHAPTER 6

Steady-State AC Analysis

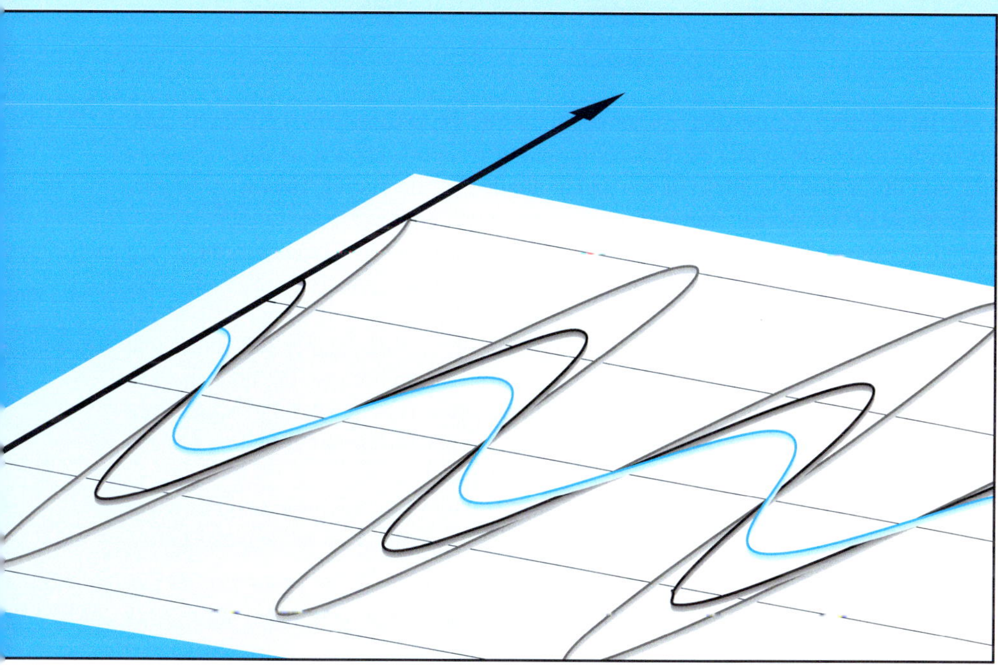

6.1 INTRODUCTION: SINUSOIDAL VOLTAGES AND CURRENTS

Sinusoidal signals (also called sinusoids) play an important role in science and engineering. Figure 6.1 shows an example of a sinusoid. Sinusoidal signals describe the characteristics of many physical processes; examples include (1) ocean waves; (2) radio or television carrier signals; (3) acoustic pressure variations of a single musical tone; and (4) voltages and currents generated by alternating-current (AC) sources. Figure 6.2 shows examples of signals formed by sinusoids.

Chapter 5 outlined the transient behavior of circuits when time-varying inputs are applied. An example of a time-varying input is when a source is applied to a circuit by turning a switch ON or OFF. Other examples are sinusoidal signals, square wave signals, and triangular signals. Chapter 5 explained that when a time-varying source is applied to a circuit that includes capacitors and/or inductors in addition to resistors, the response to the circuit will include two main components: the transient response and the steady-state response. Recall that a response can be a current that flows through any circuit component or the voltage at

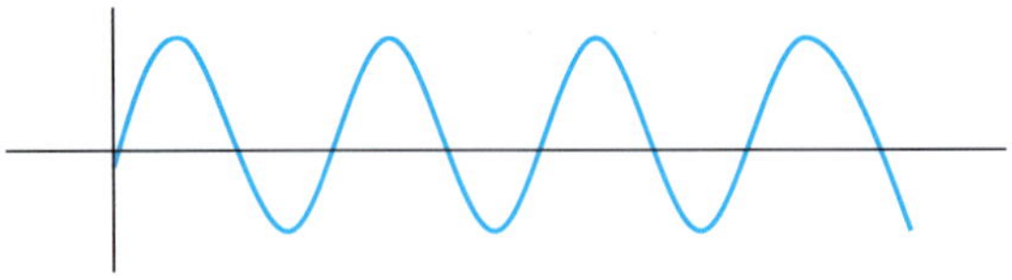

FIGURE 6.1 Sinusoidal signals.

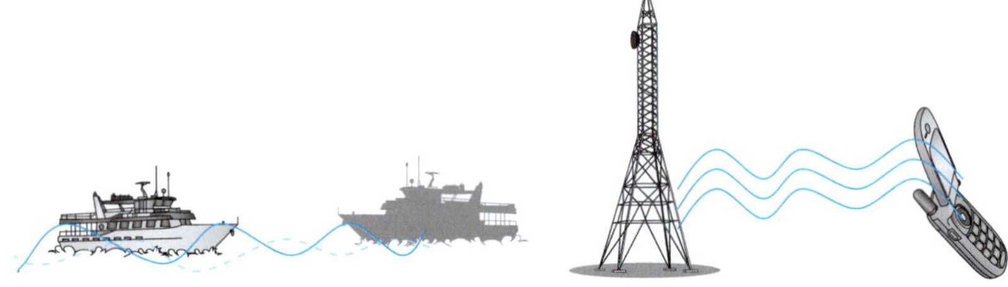

FIGURE 6.2 Examples of sinusoidal signals.

any circuit node. In addition, steady-state response represents the response to a circuit as time goes to infinity. It should be noted that circuits composed of resistors, capacitors, and inductors impact both the amplitude and the phase of the signal applied to the circuit. However, circuits that include only resistors change only the amplitude of the signal applied to the circuit.

Chapter 5 also explained that when a sinusoidal signal is applied to a linear circuit (i.e., a circuit that is composed of linear resistive, capacitive, and inductive elements), the steady-state response only includes a sinusoidal signal. The frequency of the sinusoidal signal will be the same as the frequency of the applied signal. Thus, the exercises in this chapter will show that the sinusoidal signal frequency does not include any new information because the response has the same frequency. This means that only the amplitude and phase (i.e., phasors) of signals are needed to study the steady-state response of linear circuits. Nonlinear circuits, such as circuits that include diodes and transistors, will be investigated in Chapter 8.

This chapter studies the steady-state response to circuits with sinusoidal sources. The study of steady-state AC is helpful in analyzing loudspeakers, electric heaters (such as a hot plate), motors that operate elevators, or pumps in a heart–lung machine.

Definition 1: Sinusoid A **sinusoid** is a signal that has the form of the sine and cosine function.

Definition 2: Alternating current (AC) circuits Circuits driven by sinusoidal current or voltage sources are called **AC circuits**.

A sinusoidal voltage is shown in Figure 6.3 and the following equation:

$$v(t) = V_{\text{m}} \cos(\omega t + \theta) \tag{6.1}$$

where

V_{m} = peak value of the voltage, also called amplitude
ω = angular frequency in rad/s.

Note that $\omega = 2\pi f = 2\pi/T$, and f is the frequency measured in hertz and T is the period measured in seconds, and θ is the phase angle.

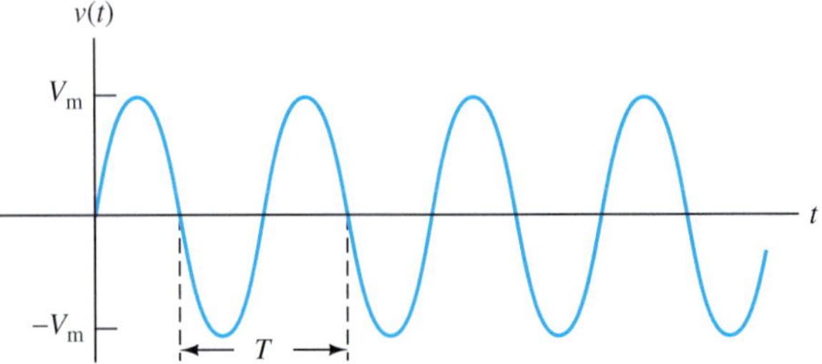

FIGURE 6.3 A sinusoidal voltage.

A sinusoid can be expressed in either sine or cosine form. Using trigonometric identities:

$$\sin\left(\alpha - \frac{\pi}{2}\right) = -\cos\alpha \tag{6.2a}$$

$$\cos\left(\alpha - \frac{\pi}{2}\right) = \sin\alpha \tag{6.2b}$$

$$\sin\left(\alpha + \frac{\pi}{2}\right) = \cos\alpha \tag{6.2c}$$

$$\cos\left(\alpha + \frac{\pi}{2}\right) = -\sin\alpha \tag{6.2d}$$

Therefore, Equation (6.1) can be expressed as:

$$v(t) = V_{\mathrm{m}} \sin\left(\frac{\pi}{2} + \omega t + \theta\right) = V_{\mathrm{m}} \sin(\omega t + \theta') \tag{6.3}$$

where $\theta' = \left(\frac{\pi}{2}\right) + \theta$.

EXAMPLE 6.1 Cosine-Sine Conversion

$$\begin{aligned}
\cos 35^\circ &= \sin(90^\circ - 35^\circ) = \sin 55^\circ \\
-\cos 35^\circ &= \sin(-90^\circ + 35^\circ) = \sin -55^\circ \\
\sin 10^\circ &= \cos(90^\circ - 10^\circ) = \cos 80^\circ \\
-\sin 10^\circ &= \cos(90^\circ + 10^\circ) = \cos 100^\circ
\end{aligned}$$

EXERCISE 6.1

Find θ in the following:

$$\begin{aligned}
\cos 5^\circ &= \sin\theta \\
-\sin 15^\circ &= \cos\theta
\end{aligned}$$

A **periodic signal** is one that repeats itself every T seconds, that is, $f(t) = f(t + nT)$, where T is the fundamental period and n is any integer. T is called the fundamental period or simply the period of the signal. Sinusoidal signals are periodic. In Figure 6.3, T represents the signal period. The square wave shown in Figure 6.4 is an example of a periodic signal.

Note:

In general, based on the principles of Fourier series, any periodic signal such as the square wave shown in Figure 6.4 can be represented via a combination of multiple sinusoids. Each of these sinusoids have different amplitudes and angular frequencies. That is, if $v(t)$ is periodic, then we can represent $v(t)$ by:

$$v(t) = \sum_{i=-\infty}^{+\infty} V_i \cos(\omega_i t + \theta_i)$$

In this chapter, we consider the impact of a linear time invariant (LTI) circuit on a sinusoidal signal that has certain frequency component. Thus, the response of a circuit would be explicitly calculated for a certain frequency component. When, the frequency component changes, the response would also change.

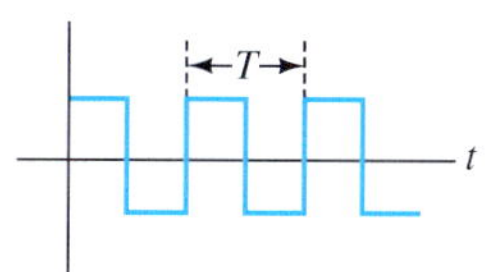

FIGURE 6.4 An example of a periodic signal: square waves.

Now, if we intend to consider the effect of all frequency components available in a periodic signal, superposition principles can be applied. For example, if $i_j(t)$ is the response of the circuit to one of the components of $v(t)$, such as $V_j \cos(\omega_j t + \theta_j)$, then the response of the circuit to all components of $v(t)$ will be $\sum_{i=-\infty}^{+\infty} i_i(t)$.

Recall that an LTI circuit is composed of components whose input–output relationships are expressed by a linear relationship as explained in Chapter 4 (Section 4.2.1). In addition, the values of the elements of an LTI circuit do not change with time.

EXAMPLE 6.2 The Period of a Sinusoid

Find the period of $x(t) = \cos(\pi t + 60)$.

SOLUTION

We know:

$$\cos(\pi t + 60) = \cos(\pi t + 60 + 2k\pi), \qquad k = 1, 2, 3, \ldots$$

Thus, it is a periodic signal. In addition:

$$\cos(\pi t + 60 + 2k\pi) = \cos(\pi(t + 2k) + 60)$$

Thus:

$$\cos(\pi t + 60) = \cos(\pi(t + 2k) + 60)$$

or

$$x(t) = x(t + 2k)$$

Based on the definition, the fundamental period (or, simply, the period) is 2.

EXERCISE 6.2

Determine if $x(t) = \cos(10\pi t + 60)$ and $y(t) = \cos(10\pi t + 60) + \sin(20\pi t + 90)$ are periodic. What is the period of these two signals?

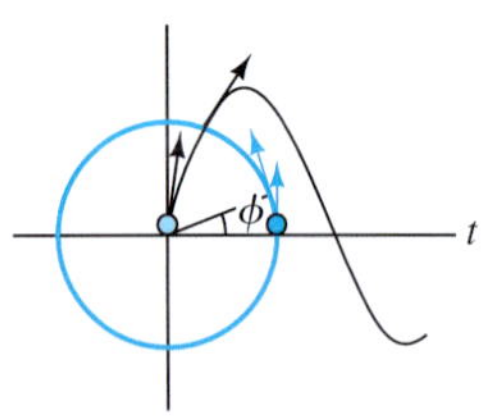

FIGURE 6.5 Creation of a sine curve.

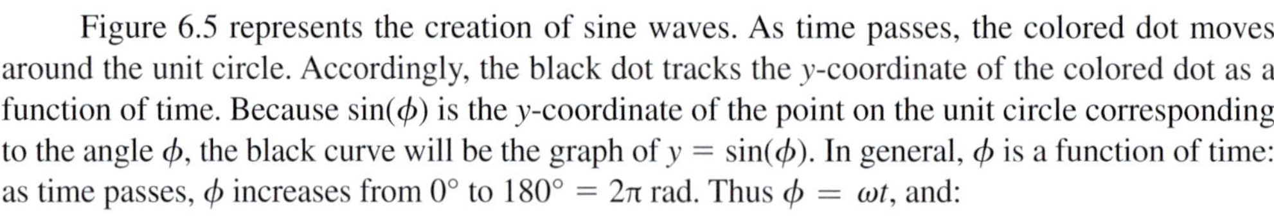

Figure 6.5 represents the creation of sine waves. As time passes, the colored dot moves around the unit circle. Accordingly, the black dot tracks the y-coordinate of the colored dot as a function of time. Because $\sin(\phi)$ is the y-coordinate of the point on the unit circle corresponding to the angle ϕ, the black curve will be the graph of $y = \sin(\phi)$. In general, ϕ is a function of time: as time passes, ϕ increases from 0° to 180° = 2π rad. Thus $\phi = \omega t$, and:

$$y = \sin \omega t \tag{6.4}$$

where ω represents the angular speed of the colored dot. Considering T as the time over that the colored dot completes a turn around the circle of Figure 6.5:

$$\omega T = 2\pi \tag{6.5}$$

$$T = \frac{2\pi}{\omega} \tag{6.6}$$

Defining f as the frequency of rotation of the colored dot:

$$T = \frac{1}{f} \tag{6.7}$$

Substituting Equation (6.7) into (6.6), results in:

$$\omega = 2\pi f \tag{6.8}$$

Although a sinusoidal function can be expressed in either sine or cosine form, for uniformity, sinusoidal functions are expressed here only by the cosine function.

In general, an angle can be expressed in degrees or radians. The relationship between radians and degrees is:

$$\frac{\theta_{\text{rad}}}{\theta_{\text{deg}}} = \frac{\pi}{180} \tag{6.9}$$

EXAMPLE 6.3 Components of a Sinusoid

The purpose of this example is to understand different components of a sinusoidal signal, that is, amplitude, phase, and frequency. It should be noted that sinusoids can be fully represented by these components.

a. Find the amplitude, phase, period, and frequency of the sinusoidal signal's voltage if:

$$v(t) = 5\cos(120\pi t + 30°)$$

b. Compute the voltage at $t = 5$.

c. Sketch $v(t)$.

SOLUTION

a. The amplitude is the component that is multiplied in the cosine term:

$$V_{\text{m}} = 5 \text{ V}$$

The phase is the component that is added to the ωt term in the cosine term:

$$\theta = 30°, \quad 30° = 30 \times \frac{\pi}{180}\text{ rad} = \frac{\pi}{6}\text{ rad}$$

The angular frequency is:

$$\omega = 120\pi \text{ rad/s}$$

The period is:

$$T = \frac{2\pi}{\omega} = \frac{2\pi}{120\pi} = \frac{1}{60} = 0.01667 \text{ s}$$

The frequency is:

$$f = \frac{1}{T} = 60 \text{ Hz}$$

b. The voltage at $t = 5$ corresponds to:

$$v(5) = 5\cos(120\pi \times 5 + 30°) = 5\cos\left(120\pi \times 5 + \frac{\pi}{6}\right) = 5\cos\left(\frac{\pi}{6}\right) = 4.3301$$

Note that the units of all angles must be the same before adding; i.e., we cannot add an angle in radians and another one in degrees.

c. $v(t)$ is sketched in Figure 6.6. Note that the point that the signal crosses the x-axis is the point at which the voltage is zero. Finding these points helps to sketch the sinusoid. Equating the voltage to zero:

$$v(t) = 5\cos(120\pi t + 30°) = 0$$

(continued)

EXAMPLE 6.3 Continued

Now, recall that $\cos\phi = 0$ if $\phi = \frac{(2k+1)\pi}{2}$, $k = \ldots, -2, -1, 0, 1, 2, \ldots$

For this example $\phi = 120\pi t + 30°$. Thus:

$$(120\pi t + 30°) = \frac{(2k + 1)\pi}{2}$$

This equation can be used to find the zero (x-axis) crossing points. For example, for $k = -1$, $t = -1/180$ as shown in the figure. Note that $\pi = 180°$, which can be used to simplify the calculations. In addition, for $k = 0$, $t = 1/360$. The cross section with the y-axis also helps to sketch Figure 6.6. This point occurs when time is set $t = 0$. Note that $\cos 30° = \sqrt{3}/2$; thus, at this point, the amplitude is $5\sqrt{3}/2$, as shown in Figure 6.6 using a dashed line.

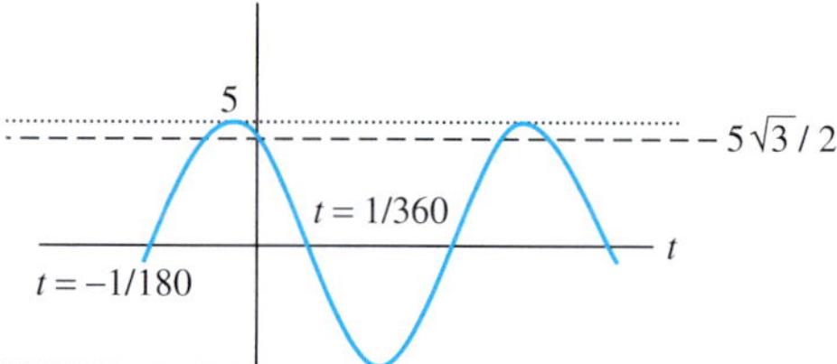

FIGURE 6.6 Figure for Example 6.3, part (c).

Sinusoids are time-varying signals and their value changes with time. Thus, sinusoid voltage, current, and power all vary with time. A time-varying power is called instantaneous power. However, instantaneous power might not be the best measure to represent the signal power. For example, to represent the voltage or power of an instrument, it is useful to have a number and not a time-varying function. This section investigates different measures of expressing a sinusoid's power, such as *root-mean-square* and *average power*.

6.1.1 Root-Mean-Square (rms) Values (Effective Values)

The **root-mean-square (rms)** for the periodic voltage, $v(t)$, is defined as:

$$V_{\mathrm{rms}} = \sqrt{\frac{1}{T}\int_0^T v^2(t)\,\mathrm{d}t} \tag{6.10}$$

In the United States, AC power in residential wiring is distributed as a 60-Hz, 115-V sinusoid. Note that, here, 115 V is the rms value, and not the peak value. The rms value is also called the **effective value**. Substituting $v(t) = V_{\mathrm{m}}\cos(\omega t + \theta)$ in (6.10):

$$V_{\mathrm{rms}} = \sqrt{\frac{1}{T}\int_0^T V_{\mathrm{m}}^2\cos^2(\omega t + \theta)\,\mathrm{d}t} \tag{6.11}$$

Note that $\cos^2\phi = (1/2)(1 + \cos 2\phi)$; therefore:

$$V_{\mathrm{rms}} = \sqrt{\frac{V_{\mathrm{m}}^2}{2T}\int_0^T (1 + \cos(2\omega t + 2\theta))\,\mathrm{d}t} \tag{6.12}$$

The integral of $\cos(2\omega t + 2\theta)$ over one period is zero; thus, what is left is the integral over $\mathrm{d}t$ from zero to T, which is T. This leads to the final result of:

$$V_{\mathrm{rms}} = \frac{V_{\mathrm{m}}}{\sqrt{2}} \tag{6.13}$$

Similarly, the rms value of the periodic current, $i(t)$, is defined as:

$$I_{\text{rms}} = \sqrt{\frac{1}{T}\int_0^T i^2(t)\text{d}t} \tag{6.14}$$

Defining $i(t) = I_{\text{m}} \cos(\omega t + \theta)$:

$$I_{\text{rms}} = \frac{I_{\text{m}}}{\sqrt{2}} \tag{6.15}$$

Thus, given 115 V as the rms value, the peak value of the voltage of AC power in the United States is $V_{\text{m}} = V_{\text{rms}} \times \sqrt{2} = 115 \times \sqrt{2} \approx 163$ V. Again, it must be noted that usually in electric product descriptions, the voltage and current range refer to the rms value, and not the peak value. For example, the voltage for a laptop is from 100 to ~120 V, which is the rms value variation range.

EXAMPLE 6.4 Peak and rms Value of a Sinusoid

A sinusoidal voltage is given by:

$$v(t) = 311 \cos(120\pi t + 30°)$$

Find the peak value and rms value of the sinusoidal voltage.

SOLUTION

The peak value that the amplitude corresponds to is:

$$V_{\text{m}} = 311 \text{ V}$$

Using Equation (6.13), the rms value is:

$$V_{\text{rms}} = \frac{311}{\sqrt{2}} = 220.6 \text{ V}$$

Note: In many countries such as England and France, the rms voltage of electric devices is 220 V.

6.1.2 Instantaneous and Average Power

The instantaneous power delivered by the voltage source in Figure 6.7 to the resistor, R, corresponds to:

$$p(t) = v(t) \cdot i(t) \tag{6.16}$$

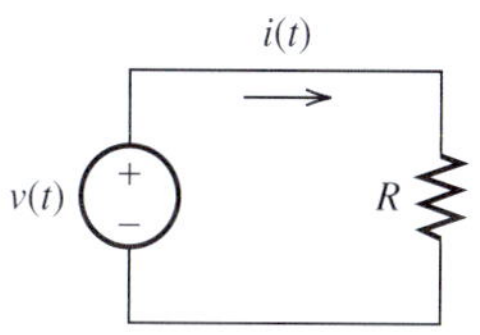

FIGURE 6.7 General circuit for a resistive load.

where

$$i(t) = \frac{v(t)}{R} \tag{6.17}$$

Substituting Equation (6.17) into (6.16) results in:

$$p(t) = \frac{v^2(t)}{R} \tag{6.18}$$

Average power corresponds to:

$$P_{\text{av}} = \frac{1}{T}\int_0^T p(t)\text{d}t \tag{6.19}$$

Using the same integration approach as in Equation (6.11):

$$P_{av} = \frac{V_m^2}{2R} \tag{6.20}$$

EXERCISE 6.3

Show the mathematical details of the derivation of Equation (6.20).

Note that $V_{rms}^2 = V_m^2/2$, thus:

$$P_{av} = \frac{V_{rms}^2}{R} \tag{6.21}$$

Equivalently, $v(t)$ in Equation (6.16) can be replaced by $v(t) = R \cdot i(t)$. In this case:

$$p(t) = R \times i^2(t) \tag{6.22}$$

Similarly, it can be depicted that:

$$P_{av} = R \times I_{rms}^2 \tag{6.23}$$

EXERCISE 6.4

Use the definition of Equation (6.19) to prove Equation (6.23).

EXAMPLE 6.5 Average Delivered Power

A sinusoidal voltage is given by:

$$v(t) = 311 \cos(120\pi t + 30°)$$

Find the average power delivered to a 100-Ω resistance.

SOLUTION

The peak value is:

$$V_m = 311 \text{ V}$$

The rms value is:

$$V_{rms} = \frac{311}{\sqrt{2}} = 220.6 \text{ V}$$

Using Equation (6.21):

$$P_{av} = \frac{V_{rms}^2}{R} = \frac{(220.6)^2}{100} = 486.6 \text{ W}$$

6.2 PHASORS

Phasors are complex numbers represented in the polar domain. Phasors are used to characterize sinusoidal voltages or currents. Considering a sinusoidal voltage:

$$v(t) = V_m \cos(\omega t + \theta) \tag{6.24}$$

its phasor is defined as:

$$V = V_m\angle\theta = V_m e^{j\theta} \tag{6.25}$$

Based on Euler's formula:

$$e^{j\theta} = \cos\theta + j\sin\theta \tag{6.26}$$

Thus (see Appendix C for a review of complex numbers):

$$\cos\theta = \mathrm{Re}\{e^{j\theta}\}, \quad \sin\theta = \mathrm{Im}\{e^{j\theta}\} \tag{6.27}$$

where $Re\{\cdot\}$ and $Im\{\cdot\}$stand for the real and imaginary components, respectively. Equivalently:

$$\begin{aligned} v(t) &= V_m\cos(\omega t + \theta) \\ &= V_m\,\mathrm{Re}\{e^{j(\omega t+\theta)}\} \\ &= \mathrm{Re}\{V_m e^{j(\omega t+\theta)}\} \\ &= \mathrm{Re}\{(V_m e^{j\theta})\,e^{j\omega t}\} \end{aligned} \tag{6.28}$$

Thus:

$$v(t) = \mathrm{Re}\{Ve^{j\omega t}\} \tag{6.29}$$

where V is the phasor of $v(t)$. Note that sometimes phasors are represented by $\overline{V}$, i.e., a bar added on the top of the parameter. The magnitude of a phasor equals the peak value and its angle equals the phase of the sinusoid in cosine form. If a sinusoidal signal is represented in sine form, it can be converted into cosine form using the method explained in Equation (6.2).

It is important to understand why a sinusoidal voltage or current is converted to a phasor. This chapter focuses on the study of linear circuits. In the steady-state analysis of linear circuits, the frequency of the voltage across and current flow through any given element does not change. As a result, the information of amplitude and phase of a voltage or current, that is, its phasor, are sufficient to characterize that voltage or current. Note that linear circuits consist of linear elements, which maintain a linear relationship between current and voltage. Recall that a linear function is that which fulfills the properties of homogeneity $[f(\alpha x) = \alpha \cdot f(x)]$ and additivity $[f(x_1 + x_2) = f(x_1) + f(x_2)]$.

EXAMPLE 6.6 Specifying a Sinusoid from its Phasor

A voltage $v(t) = V_m\cos(2\pi \times 10^3 t)$ is applied to a linear circuit. The current phasor flowing through one of the elements is determined to be $2e^{j\pi/6}$ A. Specify the sinusoidal current flowing through the element. Can the circuit be a resistive circuit?

SOLUTION

The frequency of the current flowing through the element is the same as the applied voltage and corresponds to $f = 10^3$ Hz $= 1$ kHz, therefore:

$$i(t) = 2\cos\left(2\pi \times 10^3 t + \frac{\pi}{6}\right)$$

The circuit cannot be resistive as the current and voltage phases are not the same. Later, it will be shown that the voltage and current phase in resistive circuits are equivalent.

EXAMPLE 6.7 Phasor or a Sinusoid

a. The phasor for the sinusoidal voltage, $v_1(t) = 150\cos(120\pi t + 30^o)$, is $V_1 = 150\angle 30°$.
b. The phasor for the sinusoidal voltage, $v_2(t) = 312\sin(120\pi t + 45^o)$, is:

$$V_2 = 312\angle(45° - 90°) = 312\angle -45°.$$

(continued)

EXAMPLE 6.7 Continued

Note that in this case, the sine function must first be converted to cosine to find its phase. The techniques of converting a sine function to cosine and vice versa are discussed in Equation (6.2).

The phasors for sinusoidal currents have the same form. For the current:

$$i(t) = I_{\mathrm{m}} \cos(\omega t + \theta) \tag{6.30}$$

and the phasor is:

$$I = I_{\mathrm{m}} \angle \theta = I_{\mathrm{m}} e^{j\theta} \tag{6.31}$$

Correspondingly:

$$i(t) = \mathrm{Re}\{Ie^{j\omega t}\} \tag{6.32}$$

6.2.1 Phasors in Additive or (Subtractive) Sinusoids

This section discusses how sinusoids can be added by adding their phasors. As an example, assume the following:

$$v_1(t) = V_{m1} \cos(\omega t + \theta_1)$$
$$v_2(t) = V_{m2} \cos(\omega t + \theta_2)$$
$$v_3(t) = V_{m3} \cos(\omega t + \theta_3)$$

Next, calculate $v(t) = v_1(t) + v_2(t) + v_3(t)$. The phasors for $v_1(t)$, $v_2(t)$, and $v_3(t)$, respectively, correspond to:

$$V_1 = V_{m1} \angle \theta_1, \quad V_2 = V_{m2} \angle \theta_2, \quad V_3 = V_{m3} \angle \theta_3$$

According to Equation (6.29):

$$v_1(t) = \mathrm{Re}\{V_1 e^{j\omega t}\}$$
$$v_2(t) = \mathrm{Re}\{V_2 e^{j\omega t}\}$$
$$v_3(t) = \mathrm{Re}\{V_3 e^{j\omega t}\}$$

Therefore:

$$v(t) = v_1(t) + v_2(t) + v_3(t) = \mathrm{Re}(V_1 e^{j\omega t}) + \mathrm{Re}(V_2 e^{j\omega t}) + \mathrm{Re}(V_3 e^{j\omega t})$$
$$= \mathrm{Re}(V_1 e^{j\omega t} + V_2 e^{j\omega t} + V_3 e^{j\omega t})$$
$$= \mathrm{Re}[(V_1 + V_2 + V_3) e^{j\omega t}]$$

Accordingly, the addition of sinusoids is converted into the addition of complex numbers.

EXAMPLE 6.8 Addition of Two Sinusoids Using Their Phasors

Assume:

$$v_1(t) = 5 \cos(\omega t + 45°)$$
$$v_2(t) = 10 \cos(\omega t + 30°)$$

Calculate:

$$v(t) = v_1(t) + v_2(t)$$

SOLUTION

The phasors of $v_1(t)$ and $v_2(t)$, respectively, are:

$$V_1 = 5\angle 45^\circ$$
$$V_2 = 10\angle 30^\circ$$

Thus, their addition corresponds to:

$$\begin{aligned} V_1 + V_2 &= 5\angle 45^\circ + 10\angle 30^\circ \\ &= 5 \times \cos(45^\circ) + j5 \times \sin(45^\circ) + 10 \times \cos(30^\circ) + j10 \times \sin(30^\circ) \\ &= 3.54 + j3.54 + 8.66 + j5 \\ &= 12.2 + j8.54 \\ &= \sqrt{(12.2)^2 + (8.54)^2}\angle \tan^{-1}\left(\frac{8.54}{12.2}\right) \\ &= 14.89\angle 34.99^\circ \end{aligned}$$

Then, the amplitude of the signal is 14.89 and its phase is 34.99°, or:

$$v(t) = 14.89 \cos(\omega t + 34.99^\circ)$$

6.3 COMPLEX IMPEDANCES

In the context of steady-state AC circuits, resistors, capacitors, and inductors are described using a parameter called **impedance** that is viewed as a complex resistance that changes with frequency. Note that phasors and complex impedances are applicable to sinusoidal steady-state conditions.

6.3.1 The Impedance of a Resistor

Consider that the source $v_R(t) = A \cos\omega t$ is applied to a resistor. According to Ohm's law, the current flowing through resistor R is:

$$i_R(t) = \frac{v_R(t)}{R} = \frac{A}{R}\cos\omega t \tag{6.33}$$

Using phasors to express voltage and current:

$$V_R = A\angle 0^\circ \tag{6.34}$$

$$I_R = \frac{A}{R}\angle 0^\circ \tag{6.35}$$

The impedance of the resistor (Z_R) is defined as the ratio of the phasor voltage across the resistor to the phasor current:

$$Z_R = \frac{\mathrm{V_R}}{\mathrm{I_R}} = \frac{A\angle 0^\circ}{\frac{A}{R}\angle 0^\circ} = R \tag{6.36}$$

As a result, for a resistor, $Z_R = R$. In addition, according to Equations (6.34) and (6.35), the current flowing through and the voltage across resistors are in phase.

6.3.2 The Impedance of an Inductor

As discussed in Chapter 4, the relationship of voltage and current for an ideal inductor corresponds to:

$$v_L(t) = L\frac{di_L(t)}{dt} \tag{6.37}$$

Let $i_L(t) = A\cos\omega t$, then:

$$\begin{aligned} v_L(t) &= L\frac{di_L(t)}{dt} \\ &= -A\omega L\sin\omega t \\ &= A\omega L\cos\left(\omega t + \frac{\pi}{2}\right) \end{aligned} \tag{6.38}$$

In other words, the voltage has a 90° ($\pi/2$ radian) lead with respect to the current flowing through it. Accordingly, voltage and current phasors are:

$$I_L = A\angle 0 \tag{6.39}$$

$$V_L = A\omega L\angle\frac{\pi}{2} \tag{6.40}$$

Equivalently, it can be shown that:

$$V_L = A\angle 0 \tag{6.41}$$

$$I_L = A\omega L\angle -\frac{\pi}{2} \tag{6.42}$$

In other words, if the voltage across an inductor is $v_L(t) = A\cos\omega t$, then the current will be $i_L(t) = (A/\omega L)\cos(\omega t - \pi/2)$. This can be verified through the following exercise.

EXERCISE 6.5

Start with the following voltage–current relationship in an inductor, that is:

$$i_L(t) = \frac{1}{L}\int v_L(t)dt + I_0$$

Take $I_0 = 0$, $v_L(t) = A\cos\omega t$ and validate the voltage and current relationship presented in Equations (6.41) and (6.42).

Based on Equations (6.39) and (6.40), or (6.41) and (6.42), the impedance of the inductor (Z_L) is defined as:

$$Z_L = \frac{V_L}{I_L} = \frac{A\angle 0}{\frac{A}{\omega L}\angle -\frac{\pi}{2}} = \omega L\angle\frac{\pi}{2} = \omega L\left(\cos\frac{\pi}{2} + j\sin\frac{\pi}{2}\right) = j\omega L \tag{6.43}$$

Equation (6.43) uses the fact that $\cos(\pi/2) = 0$, and $\sin(\pi/2) = 1$.

Note:

In Equation (6.43), $j\omega L$ is also called the **reactance** of an inductor.

6.3.3 The Impedance of a Capacitor

As discussed in Chapter 4, the relationship between voltage and current in an ideal capacitor is:

$$i_C(t) = C\frac{dv_C(t)}{dt} \tag{6.44}$$

Let $v_C(t) = A\cos\omega t$, then:

$$\begin{aligned} i_C(t) &= C\frac{dv_C(t)}{dt} \\ &= C\frac{d(A\cos\omega t)}{dt} \\ &= -C(A\omega\sin\omega t) \\ &= \omega CA\cos\left(\omega t + \frac{\pi}{2}\right) \end{aligned} \tag{6.45}$$

The phasors of $v_C(t)$ and $i_C(t)$ are:

$$V_C = A\angle 0 \tag{6.46}$$

$$I_C = \omega CA\angle\frac{\pi}{2} \tag{6.47}$$

In other words, in capacitors, the voltage across the capacitor lags the current through it. The same voltage–current relationship can be established using:

$$v_C(t) = \frac{1}{C}\int i_C(t)dt + V_0 \tag{6.48}$$

EXERCISE 6.6

Use Equation (6.48), set $V_0 = 0$, and take $i_C(t) = A\cos\omega t$. Show that:

$$I_C = A\angle 0 \tag{6.49}$$

$$V_C = \frac{A}{\omega C}\angle -\frac{\pi}{2} \tag{6.50}$$

The impedance of the capacitor (Z_C) is defined as:

$$Z_C = \frac{V_C}{I_C}$$

Using Equations (6.46) and (6.47) to replace for the current and voltage phasors results in:

$$Z_C = \frac{A\angle 0}{\omega CA\angle\frac{\pi}{2}} = \frac{1}{\omega C}\angle -\frac{\pi}{2}$$

Now, using Euler's formula in Equation (6.26):

$$Z_C = \frac{1}{\omega C}\left[\cos\left(-\frac{\pi}{2}\right) - j\sin\left(\frac{\pi}{2}\right)\right]$$

Using the fact that $\cos(-\pi/2) = 0$, and $\sin(\pi/2) = 1$:

$$Z_C = -j\frac{1}{\omega C} = \frac{1}{j\omega C} \tag{6.51}$$

Note:

In Equation (6.51), $1/j\omega C$ is also called the *reactance* of a capacitor.

Based on Equations (6.36), (6.43), and (6.51), for the impedance of resistors, inductors, and capacitors, it can be seen that the relationship between the voltage and current phasors for each of these impedances corresponds to:

$$V = Z \times I \tag{6.52}$$

In Equation (6.52), $Z = Z_R$ [Equation (6.36)] if the element is a resistor, $Z = Z_L$ [Equation (6.43)] if the element is an inductor, and $Z = Z_C$ [Equation (6.51)], if the element is a capacitor. However, in general, Z can be a combination of resistors, capacitors, and/or inductors.

Equation (6.52) shows that the voltage–current phasor relationship is similar to the instantaneous voltage–current relationship in the resistors $[v(t) = R \times i(t)]$. As a result, the impedance of series and parallel impedances are computed similar to series and parallel resistances. Note that according to Equation (6.52), the unit of impedance is ohm (Ω), which is the same as in resistors. However, impedance is not only a real resistor, but also a complex resistor, which consists of a real part and an imaginary part.

EXAMPLE 6.9 Impedance of a Capacitance

A voltage $v_L(t) = 110\cos(120t + 45°)$ is applied to a 47-μF capacitance. Find the impedance of the capacitance.

SOLUTION

From the equation of the voltage, it is clear that $\omega = 120$, thus:

$$Z_C = \frac{1}{j\omega C} = \frac{1}{j120 \times 47 \times 10^{-6}} = j177.305\ \Omega$$

Note:

In general, an impedance can be represented by a complex number. The real part of this complex number represents the resistance and the imaginary part of that represents the reactance.

6.3.4 Series Connection of Impedances

The total impedance of series-connected impedances (independent of its nature, whether it comes from a resistor, capacitor, inductor or any combination of these components) shown in Figure 6.8 can be calculated using a method similar to that used for series-connected resistors.

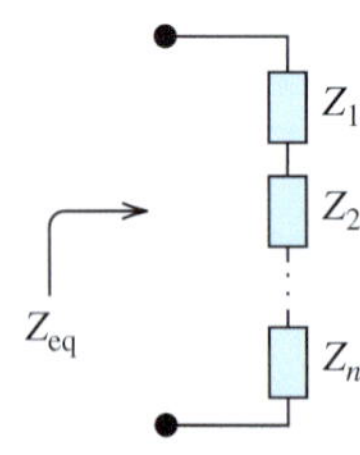

FIGURE 6.8 Series connection of impedance.

$$Z_{eq} = Z_1 + Z_2 + \cdots + Z_n = \sum_{k=1}^{n} Z_k \tag{6.53}$$

where Z_i, $i \in \{1, 2, \ldots, n\}$, can be Z_R, Z_L, Z_C, or a combination of them.

The proof of Equation (6.53) is very simple. Because the current flowing through all impedances is the same, for any impedance Z_i, $i \in \{1, 2, \ldots, n\}$, $V_i = Z_i I$. Based on Kirchhoff's voltage law:

$$V = V_1 + V_2 + \cdots + V_n = \sum_{k=1}^{n} V_k = \sum_{k=1}^{n} Z_k I$$

As a result:

$$Z_{eq} = \frac{V}{I} = \sum_{k=1}^{n} Z_k$$

6.3.5 Parallel Connection of Impedances

Simultaneously, writing the phasor voltage–current relationship in Figure 6.9 for each parallel impedance, Z_i, results in:

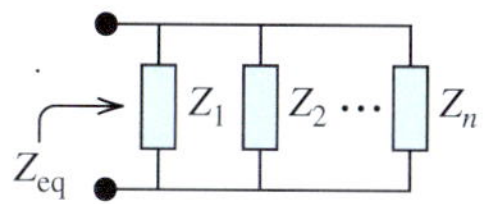

FIGURE 6.9 Parallel connection of impedance.

$$I_i = \frac{V}{Z_i}$$

Because, the phasor voltage, V, is constant across all impedances and the total current $\vec{I} = \sum \vec{I_i}$, it can be shown that:

$$Z_{eq} = \frac{1}{\sum_{i=1}^{n} \frac{1}{Z_i}}$$

In other words:

$$\frac{1}{Z_{eq}} = \frac{1}{Z_1} + \frac{1}{Z_2} + \cdots \frac{1}{Z_n} \qquad \textbf{(6.54)}$$

where Z_i, $i \in \{1, 2, \ldots, n\}$, can be a resistor, a capacitor, or an inductor.

EXERCISE 6.7

Write the details of the proof of Equation (6.54).

EXAMPLE 6.10 **Total Impedance**

A voltage, $v(t) = 220\cos(360t + 30°)$, is applied to the circuit shown in Figure 6.10. Find the total impedance of the series inductance and the resistor.

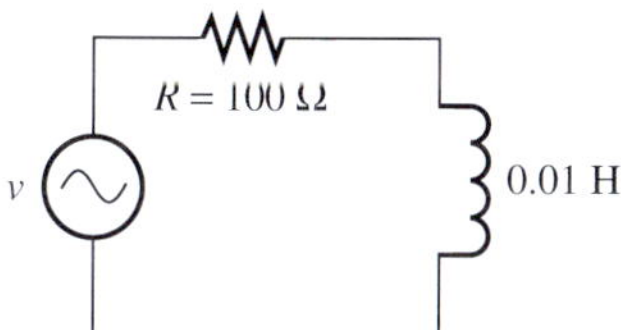

FIGURE 6.10 Circuit for Example 6.10.

SOLUTION

Here, $\omega = 360$, thus:

$$Z = Z_R + Z_L = 100 + j360 \times 0.01 = 100 + j3.6\ \Omega$$

EXAMPLE 6.11 **Total Impedance**

The voltage, $v(t) = 10\cos(500t + 60°)$, is applied to the circuit shown in Figure 6.11. Find the total impedance seen through the terminals of the voltage source.

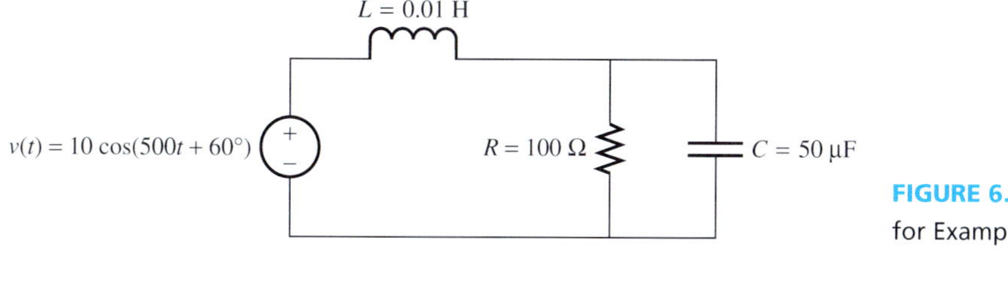

FIGURE 6.11 Circuit for Example 6.11.

(continued)

EXAMPLE 6.11 Continued

SOLUTION

The equivalent phasor voltage is $V = 10\angle 60°$. Thus, the phasor format is shown in Figure 6.12.

For the impedance of Z_L and Z_C:

$$Z_L = j\omega L = j0.01 \times 500 = j5\ \Omega$$

$$Z_C = \frac{1}{j\omega C} = \frac{1}{j500 \times 50 \times 10^{-6}} = -j40\ \Omega$$

The total parallel impedance, $Z_{RC} = Z_R \| Z_C$, in polar form is:

$$Z_{RC} = \frac{1}{\dfrac{1}{Z_R} + \dfrac{1}{Z_C}} = \frac{1}{0.01 + j0.025}$$

$$= \frac{1\angle 0°}{0.0269\angle 68.2°}$$

$$= 37.17\angle -68.2\ \Omega$$

The rectangular form for Z_{RC} corresponds to:

$$Z_{RC} = 13.80 - j34.51\ \Omega$$

Therefore, the total impedance corresponds to:

$$Z_T = Z_L + Z_{RC}$$

$$= j5 + 13.80 - j34.51$$

$$= 13.80 - j29.51$$

$$= 32.58\angle -64.94\ \Omega$$

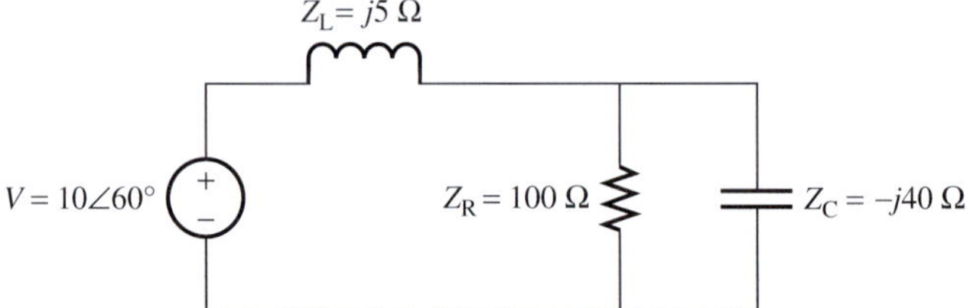

FIGURE 6.12 Transformed circuit of Figure 6.11.

APPLICATION EXAMPLE 6.12 A Loudspeaker

An audiophile wants to ensure her new speakers can shake the house, but do not harm the windows. She places a pressure sensor 0.5 m from the center of one of her new loudspeakers (see Figure 6.13). She connects a test signal of $v_{sig}(t) = 3.5\cos(400t - 45°)$ to the amplifier. The gains of the amplifier and pressure shown in Figure 6.13 are specific for this frequency. The pressure gain (pascal per volt) of the speaker (as seen by the sensor) is available. Find the pressure seen by the pressure sensor as a function of time.

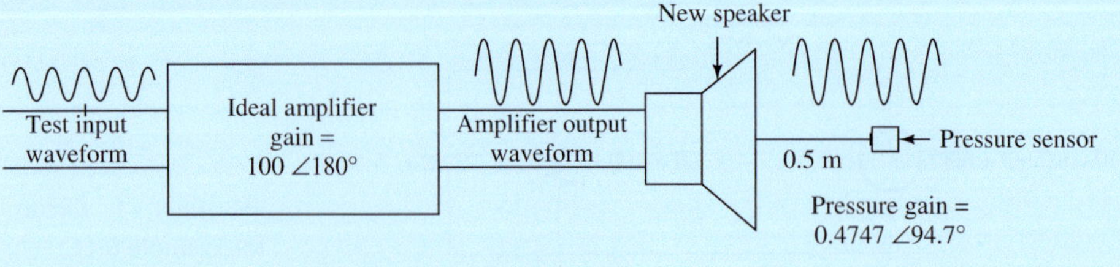

FIGURE 6.13 Circuit for Example 6.12 (loudspeaker).

SOLUTION

The test signal, $v_{sig}(t) = 3.5\cos(400t - 45°)$, can be expressed in phasor form as $3.5\angle -45°$. Applying the amplifier and pressure gain shown in Figure 6.13, the pressure, P, corresponds to:

$$P = 3.5\angle{-45°} \times 100\angle 180° \times 0.4747\angle 94.7°$$
$$= 166.145\angle 229.7°$$

In the time domain, the pressure is:

$$P(t) = 166.145\cos(400t + 229.7°)\text{ Pa (pascals or n/m}^2)$$

6.4 STEADY-STATE CIRCUIT ANALYSIS USING PHASORS

This section examines the steady-state circuit response to a sinusoidal source using phasors and impedance. If the circuit is in the steady state, the output will be sinusoidal with the same frequency as the input, as illustrated in Figure 6.14. In this figure, $v_i(t)$ is the input and $v_o(t)$ is the output. Chapter 5 explained that for a sinusoidal input, the output voltage includes two components: the transient and the steady state. As stated before, the steady-state response to a circuit is explicitly computed for a given frequency. As the frequency changes, the response would also change.

In the steady state, all voltages and currents are sinusoids. Assuming a linear circuit, the frequency of these sinusoids will be the same as the input signal frequency. Thus, they can be represented by their corresponding phasors. In addition, circuit elements, such as resistors, inductors, and capacitors, are represented by impedances, as illustrated in Figure 6.15. KVL and KCL can be applied to the voltage and current in phasor form, respectively, that is, the summation of the phasor voltages for any closed path in an electrical network is equal to zero, that is $\sum V_i = 0$.

In addition, the summation of the phasor currents $I_k^{(\text{enter})}$ entering a node is equal to the summation of the phasor currents leaving $I_m^{(\text{leave})}$:

$$\sum_k I_k^{(\text{enter})} = \sum_m I_m^{(\text{leave})}$$

Therefore, the same principles discussed in Chapters 2 and 3 to conduct steady-state analysis for sinusoidal signals can be used here. Here, all elements are in phasor form. Analyzing

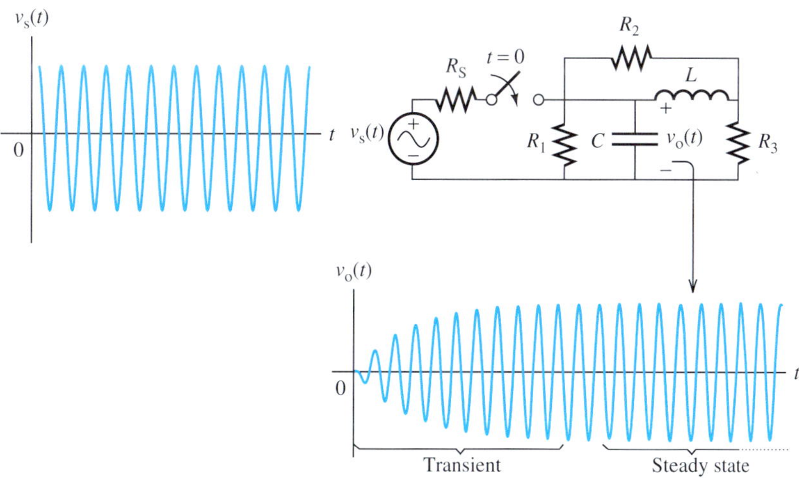

FIGURE 6.14 A steady-state circuit.

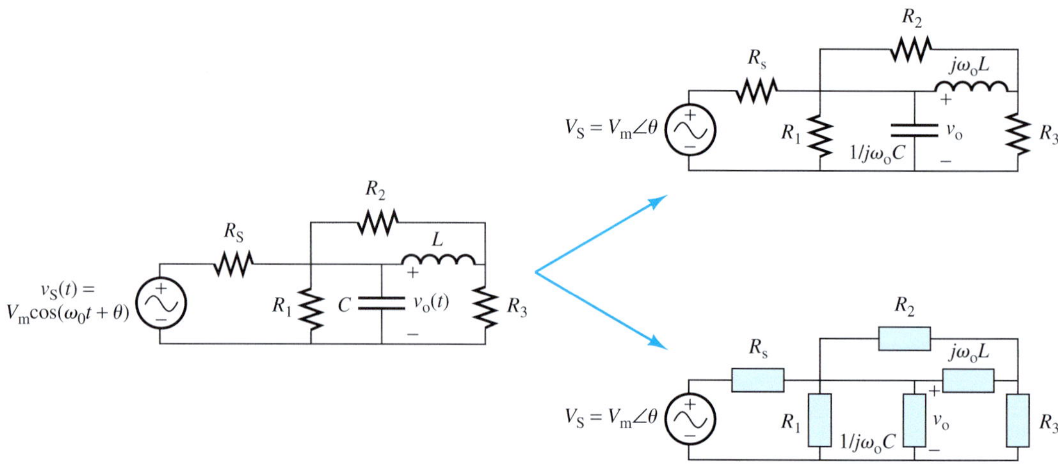

FIGURE 6.15 Steady-state circuit analysis.

circuits with impedances is similar to analyzing resistive circuits. In this analysis, real currents and voltages are replaced with complex (phasor) currents and voltages.

The procedure for steady-state analysis of circuits with sinusoidal sources is as follows:

1. Replace the instantaneous voltage and current source (s) (i.e., $v(t)$ and $i(t)$) with the corresponding phasors (i.e., V and I).
2. Replace inductance, L, with the complex impedance $Z_L = j\omega L = \omega L\angle 90°$, and capacitance, C, with the complex impedance $Z_C = 1/j\omega C = -j(1/\omega C) = (1/\omega C)\angle -90°$. Resistances have impedances equal to their resistances.
3. Analyze the circuit using techniques similar to those used for resistive circuits, and perform the calculation using complex arithmetic.

EXAMPLE 6.13 Steady-State Current

Find the steady-state current $i(t)$ and $v_L(t)$ shown in Figure 6.16.

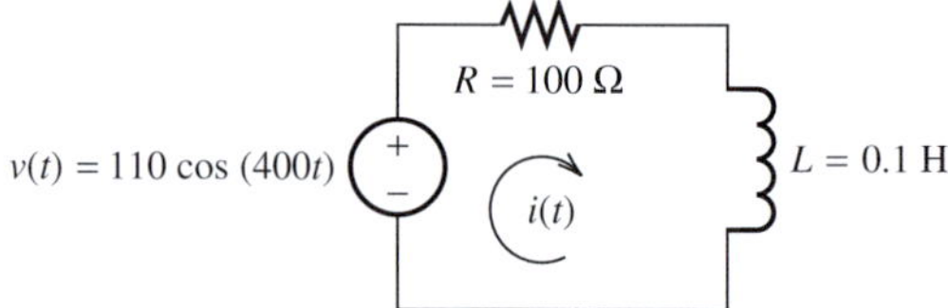

FIGURE 6.16 Circuit for Example 6.13.

SOLUTION

The voltage phasor of the source is $V = 110\angle 0°$ and the impedance of the resistor is $Z_R = 100\ \Omega$. In addition:

$$Z_L = j\omega L = j400 \times 0.1 = j40 = 40\angle 90°$$

The total impedance of the resistor and inductor is:

$$Z_{total} = Z_R + Z_L = 100 + j40 = 107.7\angle 21.8°$$

As a result, the current phasor corresponds to:

$$I = \frac{V}{Z_{total}} = \frac{110\angle 0^\circ}{107.7\angle 21.8^\circ} = 1.02\angle -21.8^\circ$$

The voltage phasor of the inductor is:

$$V_L = j\omega L \times I = 40\angle 90^\circ \times 1.02\angle -21.8^\circ = 40.8\angle 68.2^\circ$$

The corresponding instantaneous current and voltage are:

$$i(t) = 1.02\cos(400t - 21.8^\circ), \quad \text{and} \quad v_L(t) = 40.8\cos(400t + 68.2^\circ)$$

EXAMPLE 6.14 Steady-State Voltage

Find the steady-state voltage, $v_A(t)$, in Figure 6.17.

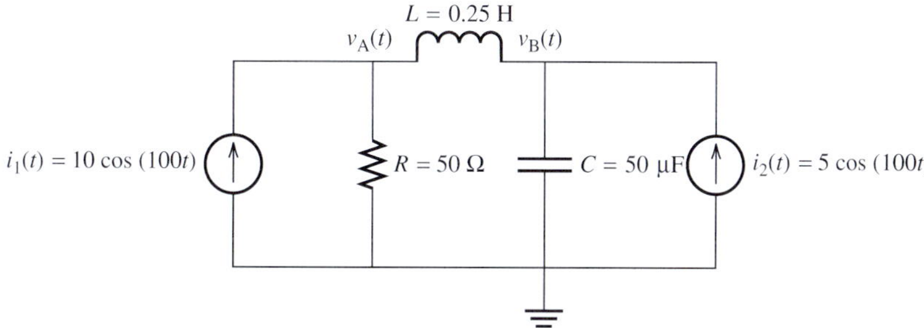

FIGURE 6.17 Circuit for Example 6.14.

SOLUTION

The equivalent phasor domain circuit of Figure 6.17 is presented in Figure 6.18. All impedances are shown in Figure 6.18.

In Figure 6.18, $Z_R = 50\ \Omega$, $Z_L = j25\ \Omega$, and $Z_C = -j200\ \Omega$. Writing KCL for nodes A and B:

$$I_A + I_{AB} = 10\angle 0^\circ$$
$$I_B + (-I_{AB}) = 5\angle 0^\circ$$

Replacing the currents with voltage divided by the impedance results in:

$$\frac{V_A}{50} + \frac{V_A - V_B}{j25} = 10\angle 0^\circ$$
$$\frac{V_B}{-j200} + \frac{V_B - V_A}{j25} = 5\angle 0^\circ$$

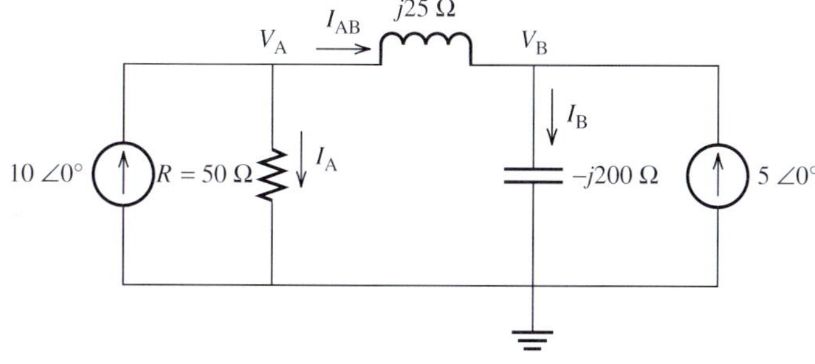

FIGURE 6.18 Equivalent circuit for Example 6.14.

(continued)

EXAMPLE 6.14 Continued

Converting the polar phasor to rectangular form and factorizing:

$$(0.02 - j0.04)V_A + j0.04V_B = 10$$
$$j0.04V_A + j0.035V_B = 5$$

Solving this system of equations using Cramer's rule (see Appendix A):

$$V_A = \frac{\begin{vmatrix} 10 & j0.04 \\ 5 & j0.035 \end{vmatrix}}{\begin{vmatrix} 0.02 - j0.04 & j0.04 \\ j0.04 & j0.035 \end{vmatrix}} = 48.69\angle 76.86°$$

In the time domain:

$$v_A(t) = 48.69\cos(100t + 76.86°)$$

EXAMPLE 6.15 Steady-State Voltage

Considering the circuit shown in Figure 6.19, find the steady-state voltage across the capacitor, $v_C(t)$, and the phasor current through both the inductor and the capacitor.

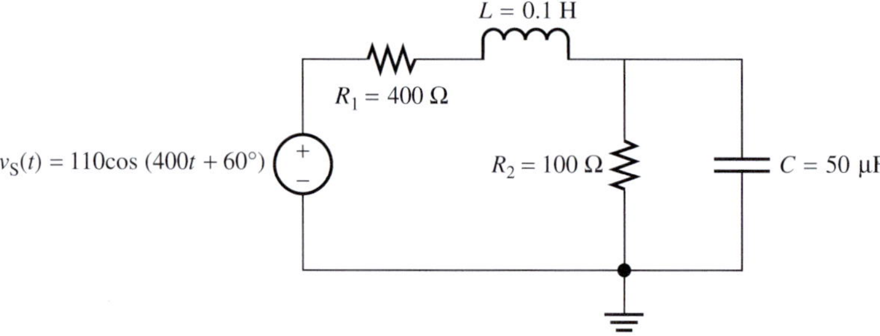

FIGURE 6.19 Circuit for Example 6.15.

SOLUTION

The equivalent circuit of Figure 6.19 in the phasor domain is shown in Figure 6.20.

In Figure 6.20:

$$Z_L = j\omega L = j0.1 \times 400 = j40$$

The equivalent series impedance of R and L is:

$$Z_{RL} = Z_{R_1} + Z_L = 400 + j40$$

$$Z_C = \frac{1}{j\omega C} = \frac{1}{j400 \times 50 \times 10^{-6}} = -j50\ \Omega$$

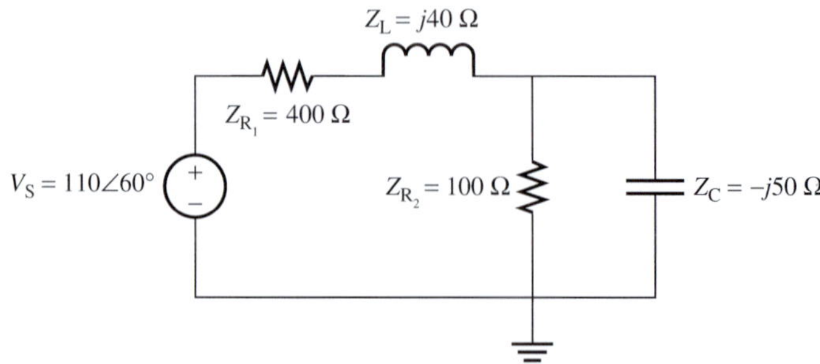

FIGURE 6.20 Equivalent circuit for Example 6.15.

The equivalent impedance of the parallel *RC* corresponds to:

$$Z_{R_2C} = \frac{1}{\frac{1}{Z_{R_2}} + \frac{1}{Z_C}} = \frac{1}{0.01 + j0.02} = \frac{1\angle 0^\circ}{0.02236\angle 63.43^\circ} = 44.6429\angle -63.43^\circ$$

The rectangular form for Z_{RC} can be given by:

$$Z_{R_2C} = 19.968 - j39.928$$

Therefore, the circuit in Figure 6.21 will be the equivalent circuit for Figure 6.20.

According to the voltage division principle:

$$\begin{aligned} V_C &= \frac{Z_{R_2C}}{Z_{R_1L} + Z_{R_2C}} V_S \\ &= \frac{44.6429\angle -63.43^\circ}{419.968 + j0.072} \cdot 110\angle 60^\circ \\ &= \frac{4910.719\angle -3.43^\circ}{419.968\angle 0.01} \\ &= 11.693\angle -3.44 \end{aligned}$$

Therefore:

$$v_C(t) = 11.693 \cos(400t - 3.44^\circ)$$

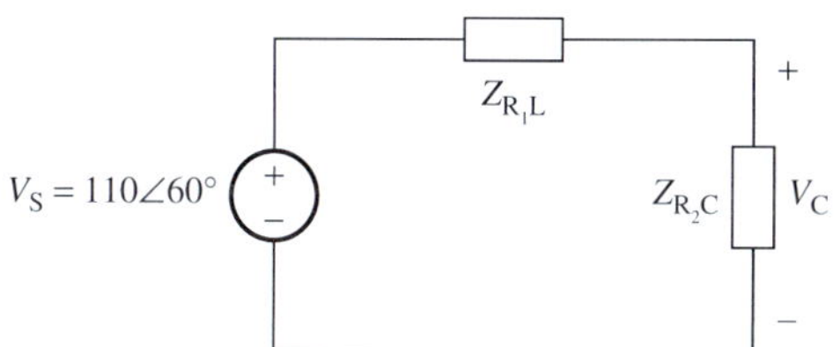

FIGURE 6.21 Equivalent circuit for Example 6.13.

APPLICATION EXAMPLE 6.16 Elevator

Figure 6.22 shows a simple block diagram of an elevator. The passenger selects a destination floor by pressing the floor button on the Input/Control. The Input/Control processes the message and "orders" the motor to "move" the elevator to the particular floor. The sinusoidal voltage source, $v_S(t) = 160 \cos(120t)$, the input control has a resistance of 100 Ω, and the motor has an impedance of 20 + *j*10 Ω. The equivalent circuit is shown in Figure 6.23. Determine the expression for the current in the circuit.

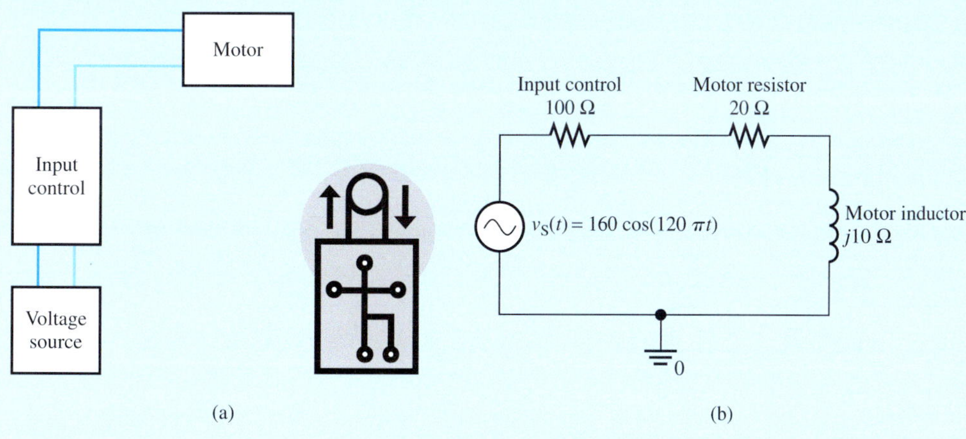

FIGURE 6.22 (a) An elevator's block diagram; (b) its equivalent circuit.

(continued)

APPLICATION EXAMPLE 6.16 Continued

SOLUTION

Based on the circuit shown in Figure 6.22(b), the total impedance seen through the voltage source terminals is:

$$Z_{\text{total}} = Z_{\text{control}} + Z_{\text{resistor}} + Z_{\text{inductor}}$$
$$Z_{\text{total}} = 100 + 20 + j10$$

$$Z_{\text{total}} = 120 + j10\ \Omega$$
$$Z_{\text{total}} = 120.416\angle 4.76°\ \Omega$$

The sinusoidal voltage source is, $V_S(t) = 160\cos(120\pi t)$, which has the phasor of:

$$V_S = 160\angle 0°$$

Therefore, the current phasor in the circuit is:

$$I = \frac{V_S}{Z_{\text{total}}}$$
$$= \frac{160\angle 0°}{120.416\angle 4.76°}$$
$$= 1.33\angle -4.76°$$

Accordingly, the expression of the current in the circuit corresponds to:

$$i(t) = 1.33\cos(120\pi t - 4.76°)$$

APPLICATION EXAMPLE 6.17 Automated Traffic Light

Figure 6.23(a) represents the block diagram of an automated traffic light sensor and controller. In an automated traffic light system, the red traffic light stays ON longer if there is not any vehicle stopped at the light. However, it turns to green earlier when there is at least one vehicle waiting at the light. This system operates as follows: There is an inductor that is underneath the ground. When there is a vehicle at the stop light, the vehicle size changes the inductance. A change in the inductance is detected by a specific device, which makes the traffic light turn green.

Assume that the sinusoidal voltage source, $v_S(t) = 110\cos(120\pi t)$, the traffic light has the resistance of 300 Ω, the resistance of the detector is 20 Ω, and the inductor has an initial inductance of 100 mH. Find the expression for the voltage across the inductor.

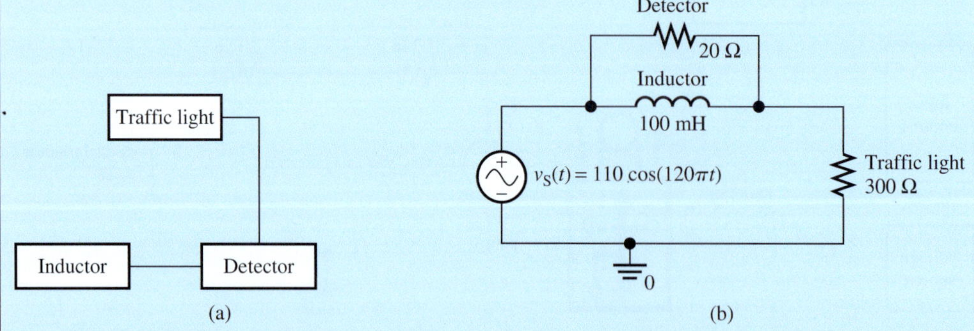

FIGURE 6.23 (a) Block diagram of an automated traffic light sensor and controller; (b) its equivalent circuit.

SOLUTION

The circuit of the automated traffic light and detector is shown in Figure 6.23 (b). The impedance of the traffic light is:

$$Z_{\text{light}} = 300\ \Omega$$

The inductor is 100 mH or 0.1 H; thus, impedance of the inductor is:

$$Z_{\text{I}} = j0.1 \times 120\pi \rightarrow Z_{\text{I}} = j12\pi\ \Omega$$

The impedance of the parallel circuit of the detector and inductor is calculated using:

$$\frac{1}{Z_{\text{par}}} = \frac{1}{20} + \frac{1}{(j12\pi)}$$

Thus:

$$Z_{\text{par}} = \frac{j240\pi}{20 + j12\pi} \rightarrow Z_{\text{par}} = 17.667\angle 27.946°\ \Omega$$

To calculate the total impedance of the circuit, both impedances must be expressed in rectangular form. Now:

$$Z_{\text{par}} = 15.61 + j8.28\ \Omega$$

Thus, the total impedance is:

$$Z_{\text{total}} = 300 + 15.61 + j8.28 \rightarrow Z_{\text{total}} = 315.61 + j8.28 \rightarrow Z_{\text{total}} = 315.72\angle 1.5°\ \Omega$$

Because the detector and the inductor are in parallel, the voltages across them are the same. Therefore, using voltage division principle results in:

$$V_{\text{L}} = \frac{Z_{\text{par}}}{Z_{\text{total}}} \times V_{\text{s}} \rightarrow V_{\text{L}} = \frac{17.667\angle 27.946°}{315.72\angle 1.5°} \times 110\angle 0° \rightarrow V_{\text{L}} = 6.155\angle 26.446°$$

In the time domain:

$$v_{\text{L}}(t) = 6.155\cos(120\pi t + 26.446°)$$

In a circuit network, if there is more than one source, and those sources have different frequencies, the circuit can be analyzed using the superposition principle.

EXAMPLE 6.18 Output Voltage Computation

For the circuit shown in Figure 6.24, find $v_o(t)$.

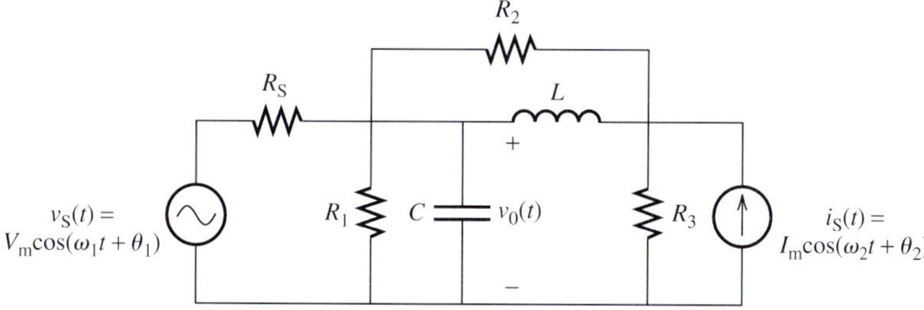

FIGURE 6.24 Circuit for Example 6.18.

(continued)

EXAMPLE 6.18 Continued

SOLUTION

Here, apply the superposition principle. Note that the frequencies of the two sources are different (one is ω_1 and the other is ω_2). Therefore, the superposition cannot be applied to the phasors (see Exercise 6.9 for the case in which the sources have the same frequency. Thus, two sets of analysis (one for each frequency) are needed.

1. Set $i_S(t)$ to zero and obtain the response $v_1(t)$ to $v_S(t)$. The corresponding circuit is illustrated in Figure 6.25.
2. Set $v_S(t)$ to zero to obtain the response $v_2(t)$ to $v_S(t)$. The corresponding circuit is illustrated in Figure 6.26.
   Finally, $v_o(t) = v_1(t) + v_2(t) = \text{Re}\{V_1 e^{j\omega_1 t}\} + \text{Re}\{V_2 e^{j\omega_2 t}\}$.

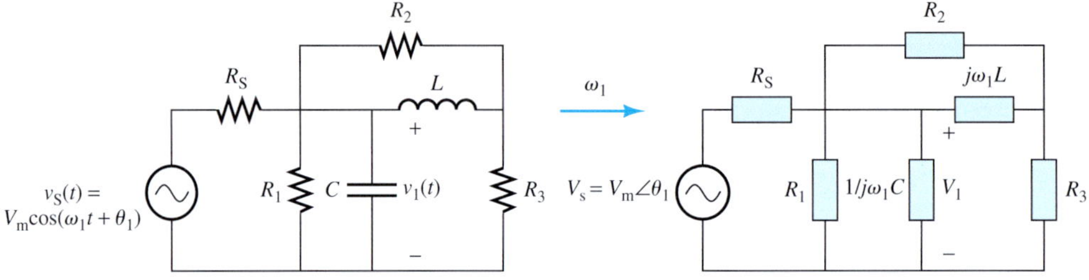

FIGURE 6.25 Circuit for Example 6.18.

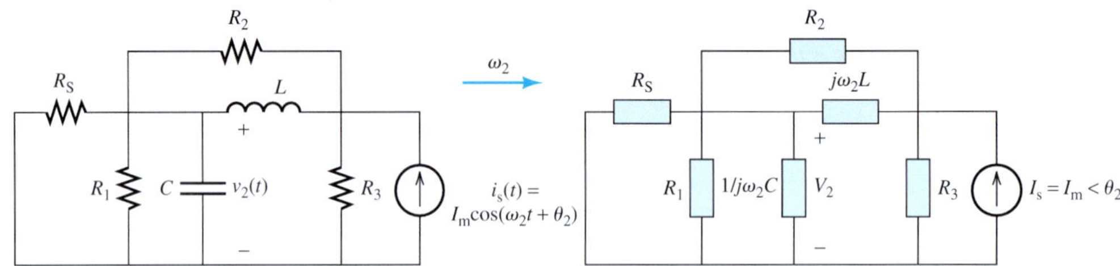

FIGURE 6.26 Circuit for Example 6.18.

EXERCISE 6.8

In Example 6.18, calculate the voltage phasors, V_1 and V_2, and the total voltage, $v_o(t)$, if $R_1 = R_2 = R_3 = 1\ \text{k}\Omega$, $L = 3$ H, and $C = 2$ F.

EXERCISE 6.9

In Example 6.18, calculate the output voltage if both sources have the same frequency (e.g., ω_1). Note that in this case the sources can be replaced with their phasor equivalents and the circuit can be analyzed considering both sources are available simultaneously. This analysis is possible using nodal analysis as discussed in Chapter 3. However, in Example 6.18, this process cannot be used because the frequencies of the two sources are different. Thus, in Example 6.18, the response to each source must be calculated separately, and the voltages in time domain are added.

6.5 THÉVENIN AND NORTON EQUIVALENT CIRCUITS WITH PHASORS

The Thévenin and Norton theorems are applied for AC steady-state analysis of circuits that include resistors, capacitors, and inductors (i.e., impedances) in the same way as they are applied to the analysis of DC circuits that include resistors as discussed in Chapter 3. A Thévenin equivalent circuit consists of a voltage source in series with a resistance. This chapter examines steady-state conditions for circuits composed of sinusoidal sources, resistances, inductances, and capacitances. In this case, the Thévenin equivalent circuit consists of a phasor voltage source in series with complex impedances.

6.5.1 Thévenin Equivalent Circuits with Phasors

6.5.1.1 THÉVENIN VOLTAGE

From the perspective of any load, the Thévenin voltage, V_{th}, equals the open-circuit voltage, V_{AB}, of the circuit, as shown in Figure 6.27.

Calculating the Thévenin voltage (see Figure 6.28):

1. Remove the load, leaving the load terminals open-circuited.
2. Define the open-circuit voltage V_{oc} ($V_{\text{oc}} = V_{\text{AB}}$, in Figure 6.28).
3. Use the corresponding phasors to describe the voltage and current source.
4. Replace inductance, capacitance, and resistance with the corresponding impedance (Z_{L}, Z_{C}, and Z_{R}).
5. Use any method (e.g., nodal analysis, voltage division, etc.) to calculate V_{oc}.
6. The resulting voltage is the Thévenin equivalent voltage, that is, $V_{\text{th}} = V_{\text{oc}}$.

6.5.1.2 THÉVENIN IMPEDANCE

The Thévenin impedance, Z_{th}, seen through two terminals is found by equating the independent sources to zero. Recall that a zero-independent voltage source is equivalent to a short circuit. In addition, a zero-independent current source is equivalent to an open circuit. As discussed in Chapter 3, dependent sources cannot be set to zero.

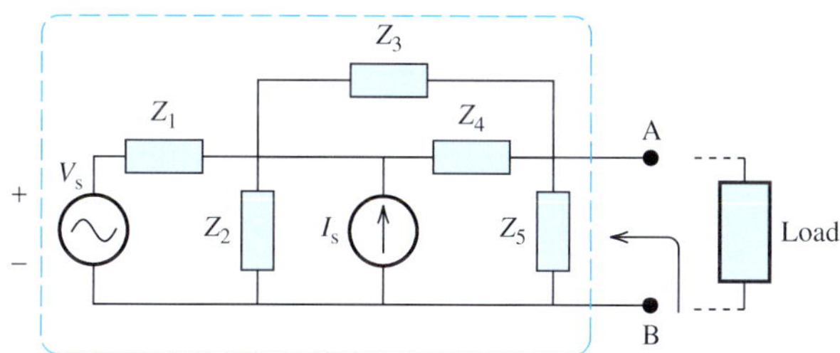

FIGURE 6.27 A Thévenin equivalent circuit.

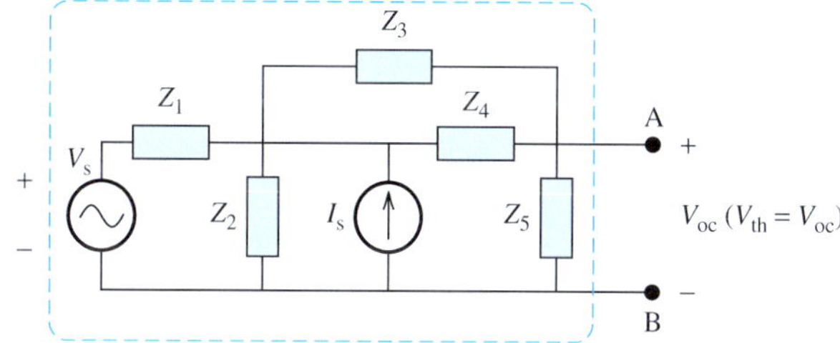

FIGURE 6.28 Thévenin equivalent voltage.

Calculating the Thévenin impedance (see Figure 6.28):

1. Remove the load
2. Equate all independent voltage and current sources to zero
3. Replace all inductance, capacitance, and resistance with the corresponding impedance (Z_{L}, Z_{C}, and Z_{R})

4. Compute the total impedance, Z_{AB}, seen across load terminals (*with the load removed*)
5. The Thévenin impedance $Z_{th} = Z_{AB}$

The Thévenin equivalent circuit is shown in Figure 6.29.

6.5.2 Norton Equivalent Circuits with Phasors

For steady-state AC circuits, another equivalent circuit is the Norton equivalent circuit, which consists of a phasor current source, I_n, in parallel with the Thévenin impedance.

6.5.2.1 NORTON CURRENT

The Norton current can be found by equating the load to zero. Thus, the load is replaced by a short circuit, and the current flowing through the short circuit can be computed.

Calculating the Norton current (see Figure 6.30):

1. Replace the load with a short circuit
2. Define the short-circuit current as I_{sc}
3. Use any method (e.g., nodal or mesh analysis) to calculate I_{sc}
4. The Norton equivalent current $I_n = I_{sc}$

6.5.2.2 NORTON IMPEDANCE

The Norton impedance, Z_n, is the same as the Thévenin impedance, Z_{th}, and can be computed similarly. Another approach is to use the following equation:

$$Z_{th} = Z_n = \frac{V_{oc}}{I_{sc}} \tag{6.55}$$

Note that in Figure 6.29, $I_{sc} = V_{th}/Z_{th}$, $V_{th} = V_{oc}$. Figure 6.31 can be used to prove Equation (6.55). The Norton equivalent circuit is shown in Figure 6.30.

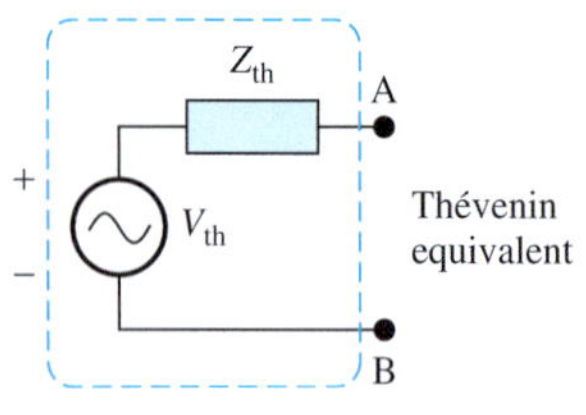

FIGURE 6.29 Thévenin equivalent impedance.

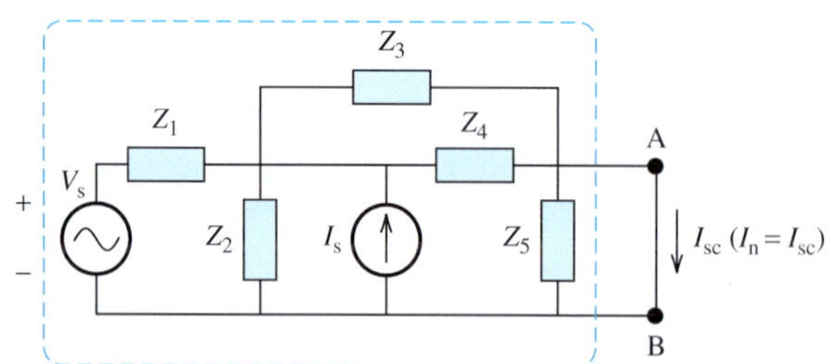

FIGURE 6.30 Norton equivalent current.

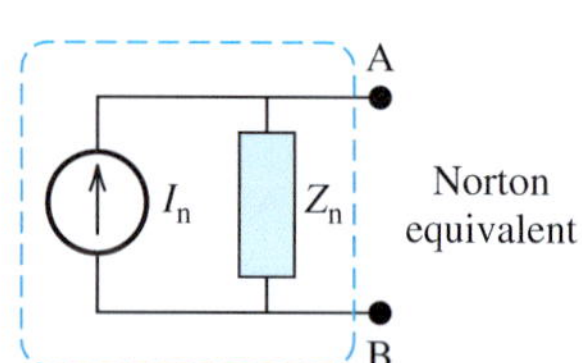

FIGURE 6.31 Norton equivalent impedance.

EXERCISE 6.10

Use Figure 6.31 to prove Equation (6.55).

EXAMPLE 6.19 Thévenin and Norton Equivalent Circuit

Find the Thévenin and Norton equivalent circuits for the circuit shown in Figure 6.32, as seen by the load resistor, R_L.

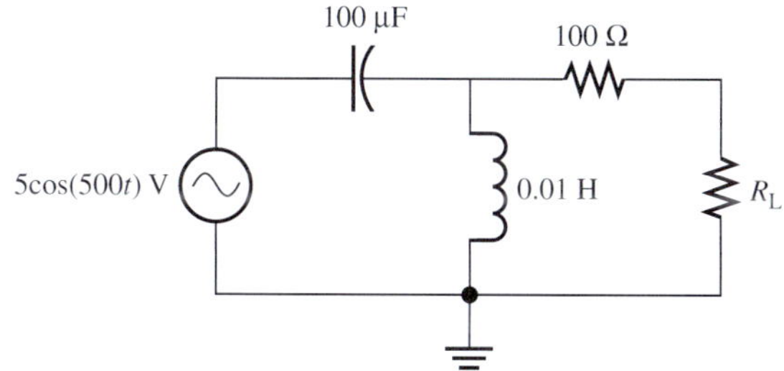

FIGURE 6.32 Circuit for Example 6.19.

SOLUTION

First, find the Thévenin impedance. The equivalent phasor circuit is shown in Figure 6.33.

Next, remove R_L and set the source equal to zero. The corresponding figure is shown in Figure 6.34.

The Thévenin impedance seen through A–B terminals is:

$$Z_{th} = 100 + \frac{1}{\dfrac{1}{-j20} + \dfrac{1}{j5}}$$

$$= 100 + j\frac{20}{3}$$

$$= 100.222\angle 3.8141°$$

The Norton impedance is then:

$$Z_n = Z_{th} = 100 + j\frac{20}{3} = 100.222\angle 3.8141°$$

Because the current through the 100-Ω resistor is zero, the voltage across the terminals of A and B is equal to the voltage across the inductor. Using a simple voltage dividing technique, the Thévenin voltage can be obtained:

$$V_{th} = 5\angle 0° \frac{j5}{-j20 + j5}$$

$$= \frac{5\angle 0°}{-3}$$

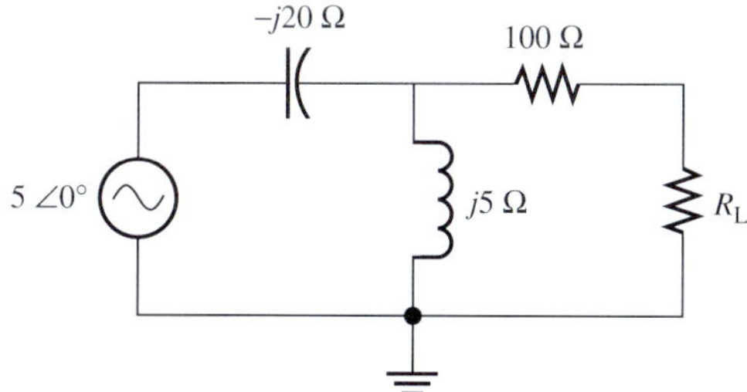

FIGURE 6.33 Equivalent circuit for Example 6.19.

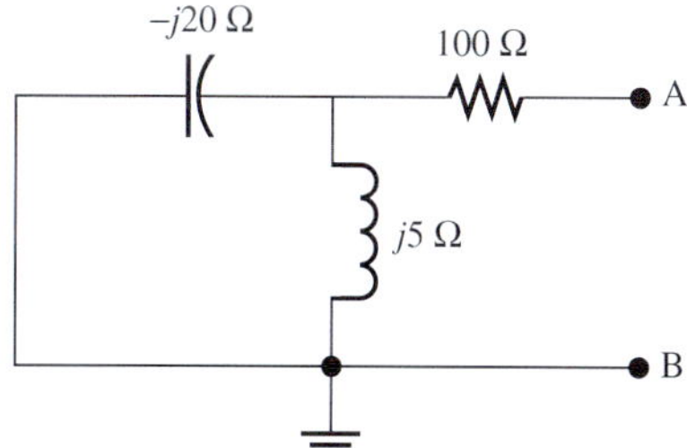

FIGURE 6.34 Equivalent circuit for Example 6.19.

(continued)

EXAMPLE 6.19 Continued

$$= \frac{5\angle 0^\circ}{3\angle 180^\circ}$$
$$= \frac{5}{3}\angle -180^\circ$$
$$= 1.667\angle -180^\circ$$

Finally, the Norton current is:

$$I_n = \frac{V_{th}}{Z_{th}} = \frac{1.667\angle -180^\circ}{100.222\angle 3.8141^\circ}$$
$$= 0.0166\angle -183.814^\circ$$
$$= 0.0166\angle +176.186$$

The Thévenin and Norton equivalent circuits are shown in Figure 6.35.

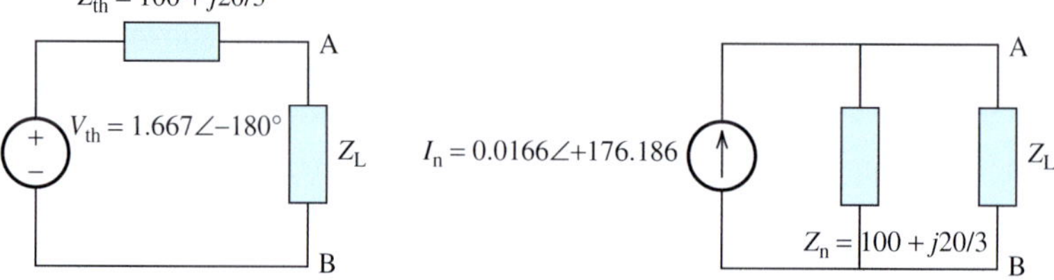

FIGURE 6.35 Thévenin and Norton equivalent circuits for Example 6.19.

APPLICATION EXAMPLE 6.20 Controlling Building Sway

An accelerometer is connected to a skyscraper to monitor building sway. It is wired to a computer, as shown in Figure 6.36. The parasitic capacitance and resistance of the cable connecting the sensor to the computer are also included. Find the Thévenin voltage as seen by the load.

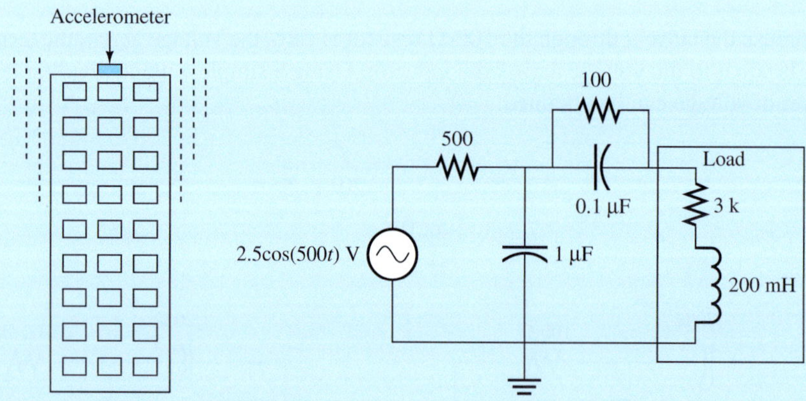

FIGURE 6.36 Controlling building sway.

SOLUTION

The equivalent circuit is shown in Figure 6.37. First, find V_{oc}.

Based on the circuit shown in Figure 6.38, V_{oc} corresponds to the voltage across the $-j2000$ impedance. Now, using the voltage division rule:

$$V_{oc} = 2.5\angle 0^\circ \times \frac{-j2000}{500 - j2000}$$

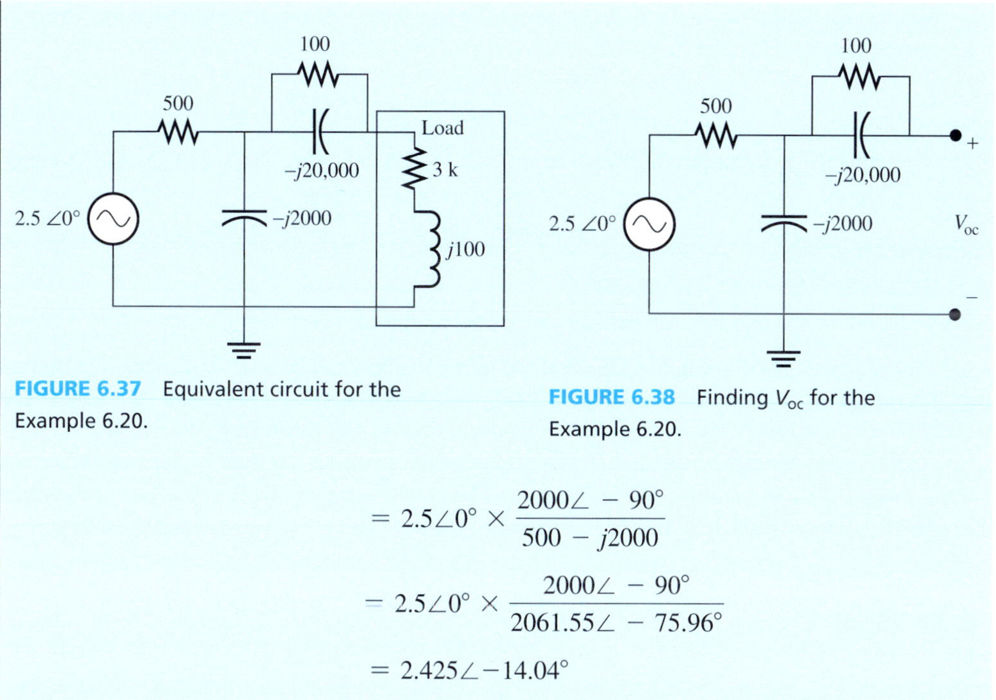

FIGURE 6.37 Equivalent circuit for the Example 6.20.

FIGURE 6.38 Finding V_{oc} for the Example 6.20.

$$= 2.5\angle 0^\circ \times \frac{2000\angle -90^\circ}{500 - j2000}$$

$$= 2.5\angle 0^\circ \times \frac{2000\angle -90^\circ}{2061.55\angle -75.96^\circ}$$

$$= 2.425\angle -14.04^\circ$$

6.6 AC STEADY-STATE POWER

This section discusses the power delivered by the source to a general load that can be any RLC network as shown in Figure 6.39.

The first section examines the power delivered to a pure resistive load, a pure inductive load, and a pure capacitive load.

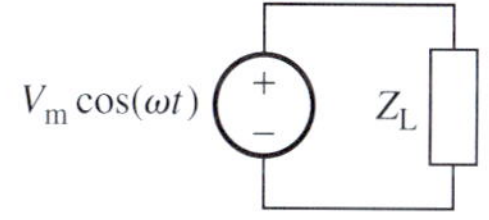

FIGURE 6.39 An RLC network.

1. ***A resistive load:*** **voltage, current, and power**

   Figure 6.40 is an example of a resistive load. In this case, the voltage is:

$$v(t) = V_m \cos(\omega t) \tag{6.56}$$

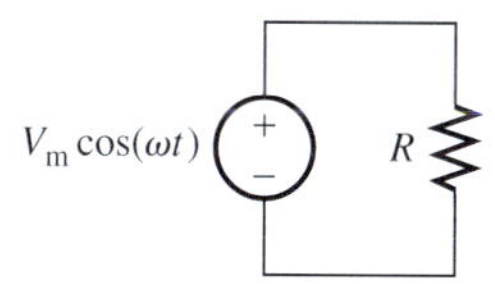

FIGURE 6.40 A resistive load.

   and the current is:

$$i(t) = \frac{V_m}{R}\cos(\omega t) = I_m \cos(\omega t), \quad \text{where } I_m = \frac{V_m}{R} \tag{6.57}$$

   Thus, the instantaneous power is:

$$p(t) = v(t)i(t) = V_m I_m \cos^2(\omega t) \tag{6.58}$$

   For resistive loads, $p(t)$ is always positive. Therefore, the energy flows continually from the source to the resistance and is converted into heat.

2. ***An inductive load:*** **voltage, current, and power**

   Figure 6.41 is an example of a circuit with an inductive load. In this case, if the voltage is $v(t) = V_m \cos(\omega t)$, then the voltage phasor is:

$$V = V_m \angle 0^\circ \tag{6.59}$$

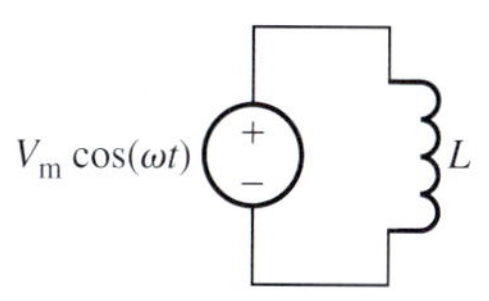

FIGURE 6.41 An inductive load.

   and the current phasor is:

$$I = \frac{V}{Z} = \frac{V_m\angle 0^\circ}{j\omega L} = \frac{V_m\angle 0^\circ}{\omega L\angle 90^\circ} = I_m\angle -90^\circ \tag{6.60}$$

where $I_m = V_m/\omega L = V_m/|Z|$, $Z = j\omega L = \omega L\angle 90°$. Accordingly, the time domain current is:

$$i(t) = I_m \cos(\omega t - 90°) \quad \textbf{(6.61)}$$

Note that using Equation (6.2), $\cos(\omega t - 90°) = \sin(\omega t)$. Thus, the instantaneous power is:

$$p(t) = v(t)i(t) = V_m I_m \cos(\omega t)\sin(\omega t) = \frac{V_m I_m}{2}\sin(2\omega t) \quad \textbf{(6.62)}$$

Equation (6.62) uses the equation:

$$\sin(2\omega t) = 2\cos(\omega t)\sin(\omega t)$$

The phase of the current is $-90°$, and the phase of the voltage is $0°$. Therefore, the current lags the voltage by 90°. For the sinusoidal signal, the value is positive in one-half of a period and negative in the other half. Therefore, intuitively, it can be said that the average current will be zero.

In addition, based on Equation (6.62), the power is positive one-half of the time when the energy flows from the source to the inductance. The power is negative the other half of the time, and the energy returns back to source from the inductance. Thus, the average power is zero.

EXERCISE 6.11

Sketch $v(t)$, $i(t)$, and $p(t)$ in Equation (6.62) and compare these three functions. What is the period of $p(t)$ compared to $v(t)$ and $i(t)$? Theoretically calculate and compare the period of these three functions.

3. *A capacitance load:* voltage, current, and power

Next, consider a circuit with the capacitive load shown in Figure 6.42.

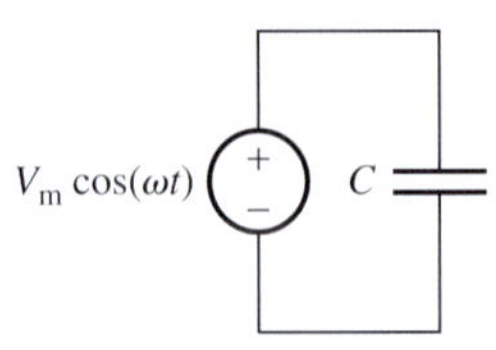

FIGURE 6.42 A capacitance load.

In this case, if the voltage is $v(t) = V_m \cos(\omega t)$, then the current phasor will be:

$$I = \frac{V}{Z} = \frac{V_m\angle 0°}{\dfrac{1}{j\omega C}} = \frac{V_m\angle 0°}{\dfrac{1}{\omega C}\angle -90°} = I_m\angle 90° \quad \textbf{(6.63)}$$

where

$$I_m = \frac{V_m}{\dfrac{1}{\omega C}} = \frac{V_m}{|Z|}, \quad Z = \frac{1}{j\omega C} = \frac{1}{\omega C}\angle -90°$$

Accordingly:

$$i(t) = I_m \cos(\omega t + 90°) = -I_m \sin(\omega t) \quad \textbf{(6.64)}$$

Thus, the instantaneous power corresponds to:

$$p(t) = v(t)i(t) = -V_m I_m \cos(\omega t)\sin(\omega t) = -\frac{V_m I_m}{2}\sin(2\omega t) \quad \textbf{(6.65)}$$

In this case, the phase of the current is 90°, and the phase of the voltage is 0°. Therefore, the current leads the voltage by 90°. The instantaneous power corresponds to Equation (6.65). Therefore, the power is negative in one-half of the cycle—when the energy flows from the source to the capacitance. It is then positive in the next half cycle—when the energy returns back to the source from the capacitance. Accordingly, the average power will be zero.

4. *RLC load:* voltage, current, and power

A load that is a combination of resistance, inductance, and capacitance is called a RLC load as shown in Figure 6.43. Here, the RLC load impedance can be represented by two components, the real (resistive) and the imaginary (inductive/capacitive) components. In this case, if the voltage is $v(t) = V_m cos(\omega t + \theta_v)$, then voltage phasor will be:

$$V = V_m \angle \theta_v \tag{6.66}$$

In addition, the current phasor is:

$$I = \frac{V}{Z} = \frac{V_m \angle \theta_v}{|Z| \angle \theta_z} = I_m \angle \theta_i \tag{6.67}$$

where $I_m = V_m/|Z|$, $\theta_i = \theta_v - \theta_z$, and $Z = R + jX$. Accordingly:

$$i(t) = I_m \cos(\omega t + \theta_i) \tag{6.68}$$

Then, the instantaneous power is:

$$p(t) = v(t)i(t) = V_m I_m \cos(\omega t + \theta_v) \cos(\omega t + \theta_i) \tag{6.69}$$

Using the trigonometric equation:

$$\cos(\omega t + \theta_v)\cos(\omega t + \theta_i) = \frac{1}{2}\left[\cos(2\omega t + \theta_v + \theta_i) + \cos(\theta_v - \theta_i)\right] \tag{6.70}$$

The power is calculated to be:

$$p(t) = \frac{V_m I_m}{2}\cos(2\omega t + \theta_v + \theta_i) + \frac{V_m I_m}{2}\cos(\theta_v - \theta_i) \tag{6.71}$$

6.6.1 Average Power

The average power is calculated by averaging the instantaneous power over one period of the input signal. In Equation (6.71), the average value of the first term is zero. Considering the period of sinusoid voltage is T, $\omega = 2\pi/T$.

$$\begin{aligned}\frac{1}{T}\int_0^T \cos(2\omega t + \theta_v + \theta_i)dt &= \frac{1}{T}\left[\frac{1}{2\omega}\sin(2\omega t + \theta_v + \theta_i)\right]_0^T \\ &= \frac{1}{2\omega T}(\sin(2\omega T + \theta_v + \theta_i) - \sin(\theta_v + \theta_i))\end{aligned}$$

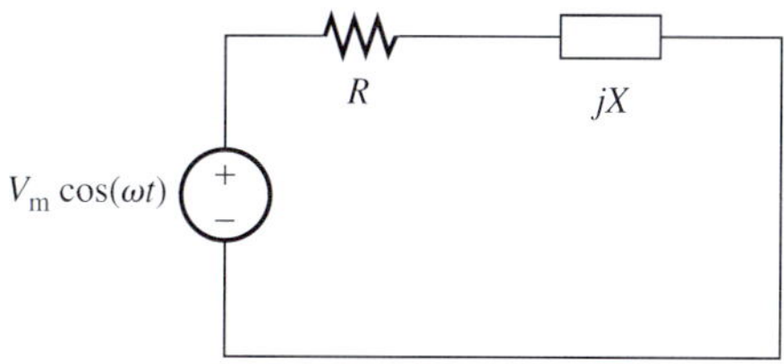

FIGURE 6.43 A general (RLC) load.

$$\left(\text{Note: } \omega = \tfrac{2\pi}{T}\right) \quad = \frac{1}{2\omega T}\left(\sin\left(2T\frac{2\pi}{T} + \theta_v + \theta_i\right) - \sin(\theta_v + \theta_i)\right)$$

$$= \frac{1}{2\omega T}\left[\sin(2\pi + \theta_v + \theta_i) - \sin(\theta_v + \theta_i)\right]$$

$$= \frac{1}{2\omega T}(\sin(\theta_v + \theta_i) - \sin(\theta_v + \theta_i))$$

$$= 0$$

This calculation uses the fact that sin $(2\pi + \theta) = \sin(\theta)$. As a result, the average power, P, [the unit is watt (W)] is calculated based on the second term of Equation (6.71) and corresponds to:

$$P = \frac{1}{T}\int_0^T p(t)\mathrm{d}t = \frac{V_m I_m}{2}\cos(\theta_v - \theta_i) \qquad \textbf{(6.72)}$$

According to Equations (6.13) and (6.15):

$$V_{rms} = \frac{V_m}{\sqrt{2}}, \quad I_{rms} = \frac{I_m}{\sqrt{2}}$$

Using $\theta_z = \theta_v - \theta_i$, defined in Equation (6.67), results in:

$$P = V_{rms} I_{rms} \cos(\theta_v - \theta_i) = V_{rms} I_{rms} \cos(\theta_z) \qquad \textbf{(6.73)}$$

6.6.2 Power Factor

Power factor (PF) is defined as:

$$\text{PF} = \cos\theta_z = \cos(\theta_v - \theta_i) \qquad \textbf{(6.74)}$$

where θ_z is called the **power angle**. Because $\cos(\theta_v - \theta_i) \leq 1$, power factor is expressed as a percentage. This power is indeed the ratio of the power delivered to the load (customer) and the apparent power that is the multiplication of rms voltage and current in the circuit (see Section 6.6.5). The power factor will be **inductive** or **lagging** if current lags voltage (when θ_z is positive because θ_i is less than θ_v); the power factor will be **capacitive** or **leading** if current leads voltage (when θ_z is negative as θ_i is greater than θ_v).

6.6.3 Reactive Power

In an AC circuit, the **peak instantaneous power** (its unit is VAR, volt–ampere–reactive) associated with the energy storage elements (inductance and capacitance) for an RLC load is called **reactive power**, and corresponds to:

$$Q = V_{rms} I_{rms} \sin(\theta_z) \qquad \textbf{(6.75)}$$

where θ_z is the power angle introduced in Equation (6.73), and V_{rms} and I_{rms} are the effective voltage and current across the load. Note that:

- For a pure resistive load, $\theta_z = 0$, and $Q = 0$
- For a pure inductive load, $\theta_z = 90°$, and $Q = V_{rms} I_{rms}$
- For a pure capacitive load, $\theta_z = -90°$, and $Q = -V_{rms} I_{rms}$

Accordingly, for a resistive load, the reactive power is zero, while it is maximum (minimum) for an inductive (capacitive) load. Therefore, it is used to express the peak instantaneous power that is associated with the inductance or capacitance. Note that the instantaneous power across a load that contains energy storage elements (inductance and capacitances) varies as voltage and current change.

6.6.4 Complex Power

Complex power is represented by:

$$P_{\text{comp}} = \frac{1}{2}VI^{*} \tag{6.76}$$

In Equation (6.76), the * refers to the complex conjugate, that is, if $I = I_m e^{j\theta_i}$, then $I^* = I_m e^{-j\theta_i}$. Replacing V with $V_m\angle\theta_v$, and I with $I_m\angle\theta_i$ results in:

$$\begin{aligned} P_{\text{comp}} &= \frac{1}{2}V_m\angle\theta_v \cdot (I_m\angle\theta_i)^{*} \\ &= \frac{1}{2}V_m I_m \angle\theta_v - \theta_i \end{aligned}$$

Converting the terms from polar to rectangular coordinates:

$$P_{\text{comp}} = \frac{1}{2}V_m I_m \cos(\theta_v - \theta_i) + j\frac{1}{2}V_m I_m \sin(\theta_v - \theta_i)$$

Replacing the maximum voltage and current (V_m and I_m) with their rms values:

$$P_{\text{comp}} = V_{\text{rms}} I_{\text{rms}} \cos(\theta_v - \theta_i) + jV_{\text{rms}} I_{\text{rms}} \sin(\theta_v - \theta_i)$$

The real part of the P_{comp} corresponds to the average power introduced in Equation (6.73), that is:

$$\begin{aligned} P = \text{Re}\{P_{\text{comp}}\} &= \frac{1}{2}V_m I_m \cos(\theta_v - \theta_i) \\ &= V_{\text{rms}} I_{\text{rms}} \cos(\theta_v - \theta_i) \\ &= V_{\text{rms}} I_{\text{rms}} \cos(\theta_z) \end{aligned} \tag{6.77}$$

where Re{·} denotes the real part. The imaginary part of the P_{comp} corresponds to the reactive power as introduced in Equation (6.75), that is,

$$\begin{aligned} Q = \text{Im}\{P_{\text{comp}}\} &= \frac{1}{2}V_m I_m \sin(\theta_v - \theta_i) \\ &= V_{\text{rms}} I_{\text{rms}} \sin(\theta_v - \theta_i) \\ &= V_{\text{rms}} I_{\text{rms}} \sin(\theta_z) \end{aligned} \tag{6.78}$$

where Im{·} denotes the imaginary part. As a result:

$$P_{\text{comp}} = P + jQ \tag{6.79}$$

Based on Equations (6.76) and (6.79), for complex power:

$$P_{\text{comp}} = \frac{1}{2}V \cdot I^{*} = V_{\text{ms}} I_{\text{ms}} (\cos\theta_z + j\sin\theta_z) = P + jQ \tag{6.80}$$

Now, for an RLC circuit, voltage phasor, V, and current phasor, I, have the following relationship:

$$V = Z \times I \tag{6.81}$$

In Equation (6.81):

$$Z = |Z|\angle\theta_z = |Z|e^{j\theta_z} = R + jX \tag{6.82}$$

Substituting Equation (6.82) into Equation (6.81):

$$V = |Z| \times I \times e^{j\theta_Z} \quad \textbf{(6.83)}$$

Substituting Equation (6.83) into Equation (6.80):

$$P_{\text{comp}} = \frac{|Z|}{2} \times e^{j\theta_Z} \times I \times I^* \quad \textbf{(6.84)}$$

Now, $I \times I^* = |I|^2 = I_m^2$, thus:

$$P_{\text{comp}} = \frac{|Z|}{2} \times I_{\text{m}}^2 \angle\theta_z \quad \textbf{(6.85)}$$

Replacing $|Z|\angle\theta_z$ with its rectangular equivalent $R + jX$:

$$P_{\text{comp}} = \frac{1}{2}RI_{\text{m}}^2 + j\frac{1}{2}XI_{\text{m}}^2 \quad \textbf{(6.86)}$$

In other words, the average (active), P, and reactive power, Q, respectively, correspond to:

$$P = \frac{1}{2}RI_{\text{m}}^2 \quad \textbf{(6.87)}$$

and:

$$Q = \frac{1}{2}XI_{\text{m}}^2 \quad \textbf{(6.88)}$$

From Equation (6.87) it is clear that the active power, P, is a function of resistive loads in the circuit while the reactive power, Q, is a function of capacitive and inductive loads in the circuit.

EXERCISE 6.12

If in Equation (6.80), V is replaced with:

$$V = I|Z| \times e^{j\theta_z} = I|Z| \times e^{j\theta_z} = I(R + jX) \quad \textbf{(6.89)}$$

and prove Equation (6.86).

EXERCISE 6.13

Show that Equation (6.86) can be rewritten as

$$P_{\text{comp}} = RI_{\text{rms}}^2 + jXI_{\text{rms}}^2 \quad \textbf{(6.90)}$$

and specify the active and reactive components of the complex power.

6.6.5 Apparent Power

Apparent power (its unit is VA, volt–ampere) is defined as the product of the effective voltage and the effective current:

$$\text{Apparent power} = V_{\text{rms}} I_{\text{rms}} \quad \textbf{(6.91)}$$

Actually, apparent power is the **maximal average power** that is delivered to the load. This maximum power is delivered when the load is purely resistive.

Based on the definitions in Equations (6.73) and (6.75):

$$P^2 + Q^2 = (V_{\text{rms}} I_{\text{rms}})^2 \cos^2(\theta_z) + (V_{\text{rms}} I_{\text{rms}})^2 \sin^2(\theta_z) = (V_{\text{rms}} I_{\text{rms}})^2 \quad \textbf{(6.92)}$$

In addition, apparent power represents the amplitude of the complex power. Thus, according to Equation (6.80), a triangle can be used to express the relationship between average power, P, reactive power, Q, apparent power, $V_{\text{rms}} I_{\text{rms}}$, and the power angle, θ_z, as shown in Figure 6.44. Table 6.1 summarizes all discussed equations.

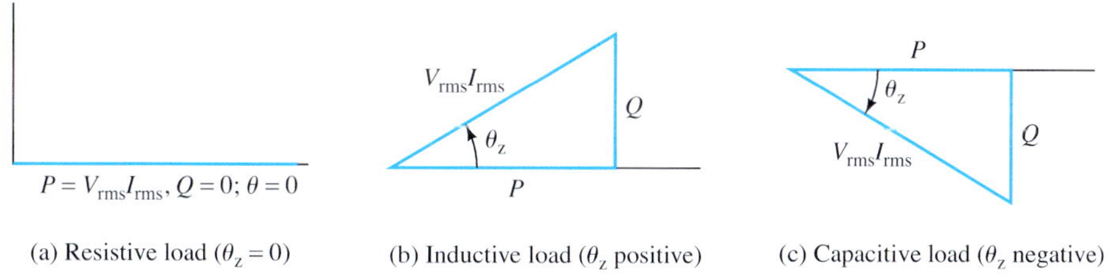

(a) Resistive load ($\theta_z = 0$) (b) Inductive load (θ_z positive) (c) Capacitive load (θ_z negative)

FIGURE 6.44 Power triangle.

TABLE 6.1 Summary of Power Definitions

	Formula	Unit
Instantaneous power	$p(t) = v(t)i(t) = V_m I_m \cos(\omega t + \theta_v)\cos(\omega t + \theta_i)$	watt (W)
Average power	$P = V_{\text{rms}} I_{\text{rms}} \cos(\theta_v - \theta_i) = V_{\text{rms}} I_{\text{rms}} \cos(\theta_z)$	watt (W)
Reactive power	$Q = V_{\text{rms}} I_{\text{rms}} \sin(\theta_z)$	VAR
Apparent power	Apparent power $= V_{\text{rms}} I_{\text{rms}}$	VA
Complex power	$P_{\text{comp}} = \frac{1}{2} V I^* = V_{\text{rms}} I_{\text{rms}} (\cos\theta_z + j\sin\theta_z) = P + jQ$	VA

EXAMPLE 6.21 Average, and Reactive Power and Power Factor

Consider the circuit shown in Figure 6.45. Compute the average power and reactive power taken from the source and the power factor.

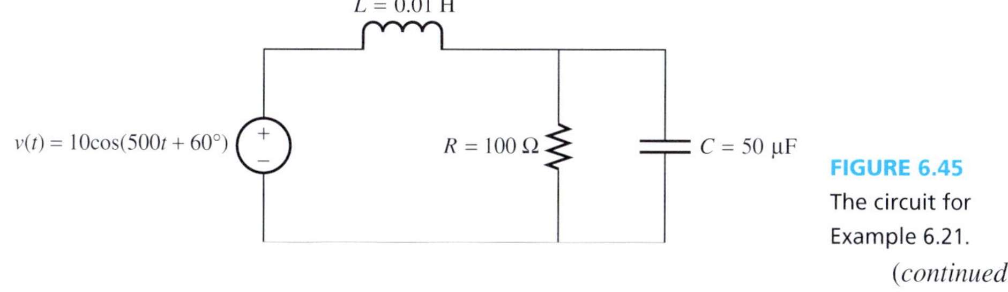

FIGURE 6.45 The circuit for Example 6.21.

(continued)

EXAMPLE 6.21 Continued

SOLUTION

The equivalent circuit for the phasor domain is shown in Figure 6.46.

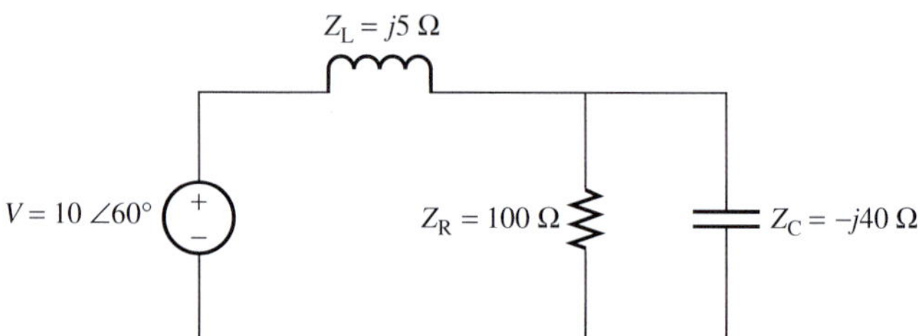

FIGURE 6.46 Equivalent circuit for Example 6.21.

where

$$Z_L = j\omega L = j0.01 \times 500 = j5\ \Omega$$

$$Z_C = \frac{1}{j\omega C} = \frac{1}{j500 \times 50 \times 10^{-6}} = -j40\ \Omega$$

The equivalent impedance of the parallel resistor and capacitor is:

$$Z_{RC} = \frac{1}{\dfrac{1}{Z_R} + \dfrac{1}{Z_C}} = \frac{1}{0.01 + j0.025} = \frac{1\angle 0°}{0.0269\angle 68.2°} = 37.17\angle -68.2°\ \Omega$$

In rectangular form:

$$Z_{RC} = 13.80 - j34.51\ \Omega$$

The total impedance seen by the voltage source is $Z_L + Z_{RC}$. In addition, the current through the circuit is:

$$I = \frac{V}{Z_L + Z_{RC}} = \frac{10\angle 60°}{j5 + 13.80 - j34.51}$$

By simplifying:

$$I = \frac{10\angle 60°}{13.80 - j29.51}$$

$$= \frac{10\angle 60°}{32.58\angle -64.94°}$$

$$= 0.3069\angle 124.94°$$

Therefore, the power angle is:

$$\theta_z = \theta_v - \theta_i = 60° - 124.94° = -64.94°$$

and the power factor is:

$$PF = \cos\theta_z = 0.4236$$

The effective value of voltage and current are:

$$V_{rms} = \frac{V_m}{\sqrt{2}} = \frac{10}{\sqrt{2}} = 7.071\ \text{V}$$

$$I_{rms} = \frac{I_m}{\sqrt{2}} = \frac{0.3069}{\sqrt{2}} = 0.217\ \text{A}$$

Using Equations (6.73) and (6.75), the average power and reactive power correspond to:

$$P = V_{rms} I_{rms} \cos\theta_z = 7.071 \times 0.217 \times \cos(-64.94°) = 0.6499 \text{ W}$$

$$Q = V_{rms} I_{rms} \sin\theta_z = 7.071 \times 0.217 \times \sin(-64.94°) = -1.39 \text{ VAR}$$

APPLICATION EXAMPLE 6.22 Chemical Heater

An electric heater is used to control the temperature of a chemical solution. The heater is required to heat evenly with precise temperature control. The heater element is a coil of resistive material, similar to that in an electric toaster. If the circuit in Figure 6.47 is the equivalent of the heater element, find the average power that is consumed by the heater.

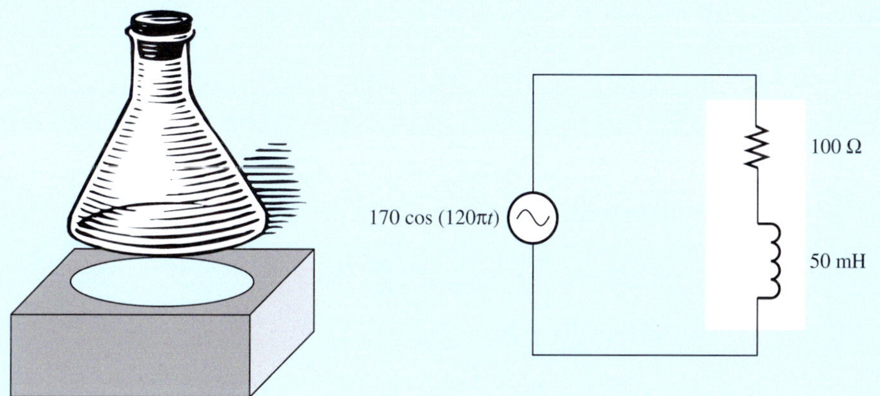

FIGURE 6.47 Chemical heater.

SOLUTION

The equivalent circuit of Figure 6.47 is shown in Figure 6.48.

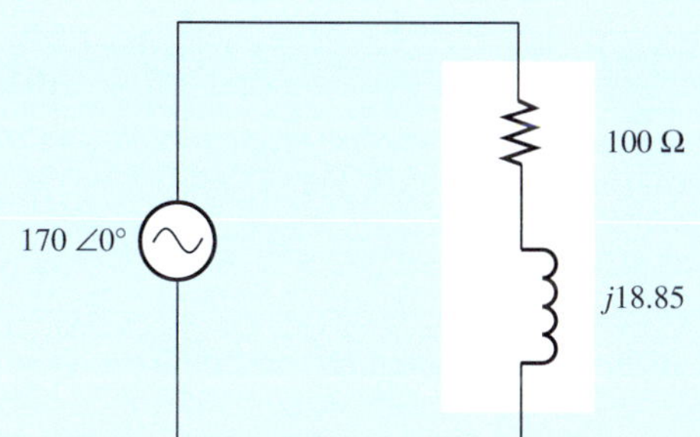

FIGURE 6.48 Equivalent circuit for Example 6.22.

The complex power corresponds to:

$$P_{comp} = \frac{1}{2} V I^* = \frac{1}{2} V \frac{V^*}{Z^*} = \frac{1}{2} \frac{V_m^2}{Z^*}$$

$$= \frac{1}{2} \frac{170^2}{100 - j18.85}$$

(continued)

APPLICATION EXAMPLE 6.22 Continued

$$= \frac{14{,}450}{101.76\angle -10.68^\circ}$$

$$= 142.0\angle 10.68^\circ = 139.54 + j26.3$$

Therefore, the average power consumed by the heater is 139.54 W, which is the real part of the complex power.

6.6.6 Maximum Average Power Transfer

Given a circuit network, the maximal average power delivered to the load by the source is a function of the load impedance. The circuit network can be represented by its Thévenin equivalent, as shown in Figure 6.49. Observe that in general, the Thévenin equivalent circuit includes a voltage source and an impedance.

The Thévenin impedance is $Z_{th} = R_{th} + jX_{th}$, and the impedance of the load is $Z_L = R_L + jX_L$. Therefore, the total impedance is $Z_{tot} = R_{th} + R_L + j(X_{th} + X_L)$ and its magnitude is:

$$|Z_{tot}| = \sqrt{(R_{th} + R_L)^2 + (X_{th} + X_L)^2}$$

The current flowing through the load is:

$$I_{out} = \frac{V_{out}}{Z_L} \quad \textbf{(6.93)}$$

Using the voltage division formula:

$$V_{out} = \frac{Z_L}{Z_{th} + Z_L} \cdot V_s = \frac{Z_L}{Z_{tot}} \cdot V_s \quad \textbf{(6.94)}$$

The complex power delivered to the load is (see Appendix C for additional information about the calculations in this section):

$$P_{comp} = \frac{1}{2} V_{out} I^*_{out}$$

Replacing $\boldsymbol{I}_{out}$ using Equation (6.93) results in:

$$P_{comp} = \frac{1}{2} \frac{|V_{out}|^2}{Z^*_L} \quad \textbf{(6.95)}$$

Replacing V_{out} using Equation (6.94):

$$P_{comp} = \frac{1}{2} \left| \frac{Z_L}{Z_{tot}} \right|^2 \cdot \frac{|V_s|^2}{Z^*_L} \quad \textbf{(6.96)}$$

Note that $|V_s| = V_m$. Multiplying by Z_L in both the numerator and denominator of Equation (6.96) yields:

$$P_{comp} = \frac{1}{2} Z_L \frac{V_m^2}{|Z_{tot}|^2}$$

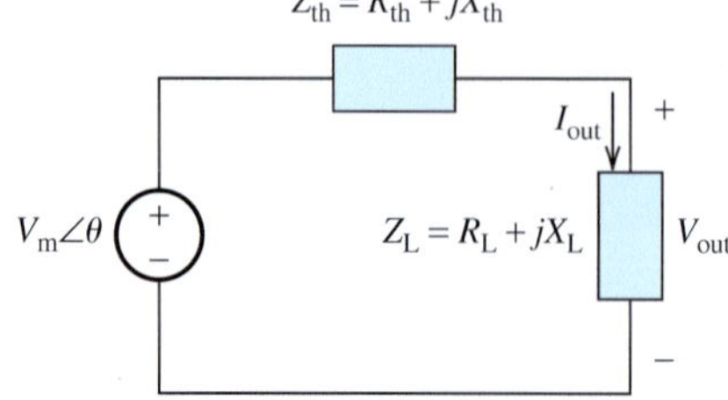

FIGURE 6.49 Computation of the maximal average power transfer to a load.

The average delivered power is the real part of P_{comp}, which is:

$$P_{\text{av}} = \frac{R_{\text{L}}}{2\left((R_{\text{th}} + R_{\text{L}})^2 + (X_{\text{th}} + X_{\text{L}})^2\right)} V_{\text{m}}^2 \tag{6.97}$$

Now, to find the Z_{L} that maximizes the transferred power, find the derivative of P_{av} with respect to R_{L} and X_{L}, and equate it to zero. The Z_{L} that maximizes the average power is:

$$Z_{\text{L}} = Z_{\text{th}}^* = R_{\text{th}} - jX_{\text{th}} \tag{6.98}$$

In other words, $R_{\text{L}} = R_{\text{th}}$, and $X_{\text{L}} = -X_{\text{th}}$. Accordingly, the maximal average power is:

$$P_{\text{max}} = \frac{V_{\text{m}}^2}{8R_{\text{th}}} \tag{6.99}$$

EXERCISE 6.14

Take the derivative of Equation (6.97) with respect to R_{L} and X_{L} and verify Equation (6.98). Then, use Equation (6.98) to verify Equation (6.99).

If the load is purely resistive, that is, $X_{\text{L}} = 0$, using Equation (6.99):

$$P = \frac{V_{\text{m}}^2 R_{\text{L}}}{2(R_{\text{th}}^2 + 2R_{\text{th}}R_{\text{L}} + R_{\text{L}}^2 + X_{\text{th}}^2)} \tag{6.100}$$

Dividing the numerator and denominator of Equation (6.100) by R_{L}:

$$P = \frac{V_{\text{m}}^2}{2\left(\dfrac{R_{\text{th}}^2 + X_{\text{th}}^2}{R_{\text{L}}} + R_{\text{L}} + 2R_{\text{th}}\right)}$$

Now, to find Z_{L} that maximizes the transferred power, find R_L that maximizes the power. Alternatively, find R_L that minimizes the denominator. Taking the derivative of the denominator with respect to R_L:

$$R_{\text{L}}^2 = R_{\text{th}}^2 + X_{\text{th}}^2$$

Therefore, for the pure resistance load, taking:

$$R_{\text{L}} = \sqrt{R_{\text{th}}^2 + X_{\text{th}}^2} = |Z_{\text{th}}| \tag{6.101}$$

the load can attain the maximal average power:

$$P_{\text{max}} = \frac{V_{\text{m}}^2}{4\left(\sqrt{R_{\text{th}}^2 + X_{\text{th}}^2} + R_{\text{th}}\right)} \tag{6.102}$$

Comparing Equation (6.99) to Equation (6.102), it is clear that the power delivered to a pure resistance load is less than that delivered to a complex one because:

$$\frac{V_{\text{m}}^2}{8R_{\text{th}}} \geq \frac{V_{\text{m}}^2}{4\left(\sqrt{R_{\text{th}}^2 + X_{\text{th}}^2} + R_{\text{th}}\right)} \tag{6.103}$$

In Equation (6.103), equality holds when $X_{\text{th}} = 0$, that is, the Thévenin impedance is also a pure resistance.

Note: If the load impedance is given and the equivalent R_{th} needs to be adjusted, taking R_{th} as small as possible, maximizes the maximum power transfer (see Example 6.25).

EXERCISE 6.15

Verify the inequality of Equation (6.103).

EXAMPLE 6.23 Maximum Delivered Power

For the circuit shown in Figure 6.50, determine the maximum average power that can be delivered to the load. Also, determine the maximum average power that can be delivered to a purely resistive load.

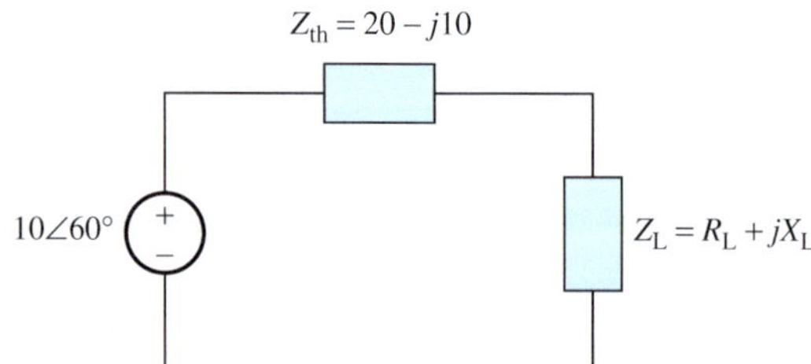

FIGURE 6.50 Circuit for Example 6.23.

SOLUTION

If the load is not purely resistive, using Equations (6.98) and (6.99):

$$Z_L = Z_{th}^* = 20 + j10$$

$$P_{max} = \frac{V_m^2}{8R_{th}} = \frac{100}{8 \times 20} = 0.625 \text{ W}$$

If the load is purely resistive, using Equations (6.101) and (6.102)

$$R_L = |Z_{th}| = \sqrt{400 + 100} = 22.3607,\ X_L = 0$$

$$P_{max} = \frac{V_m^2}{4(\sqrt{R_{th}^2 + X_{th}^2} + R_{th})} = \frac{100}{4(\sqrt{400 + 100} + 20)} = 0.5902 \text{ W}$$

6.6.7 Power Factor Correction

Equation (6.73) represents the total power consumed by the consumer and apparent power delivered to the consumer. The consumer is indeed charged for the power that it consumes. However, the power plant generates the power that is consistent with the apparent power in Equation (6.91). Power factor ($\cos\theta_z$) maintains the relationship of Equations (6.91) and (6.73).

Now, if the power factor is small (i.e., the case for inductive loads such as motors), the power company creates more power than consumed by the customer. To avoid this problem, a power factor correction is needed. That is we should increase the power factor to 1. For example, let's say a customer has a load of 740 W, but the apparent power is 1000 VA. The power company can only bill the customer for the 740 W used even though they have to have an infrastructure to supply 1000 VA to the customer. Applying power factor correction, then the power company only has to supply 740 VA to the customer which they can fully bill since the customer is consuming 740 W.

In industrial plants, usually, a heavy inductive load (e.g., motors) increases the reactive power. When reactive power, Q, increases, the current flowing from the source to the load increases according to Equation (6.88) (see the circuit example shown in Figure 6.51). Consequently, the power factor ($\cos\theta_z$) decreases because θ_z increases when reactive power

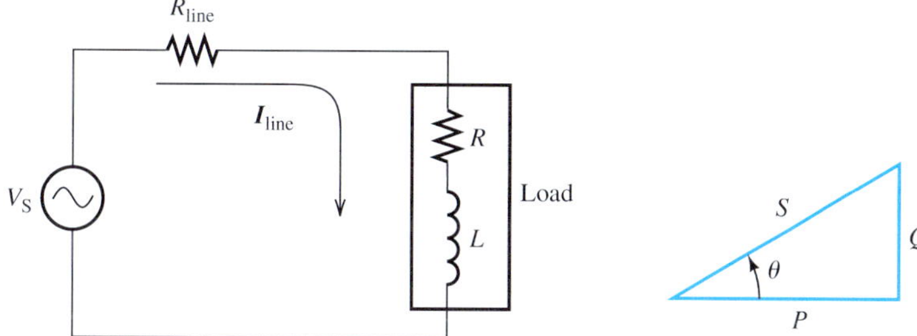

FIGURE 6.51 Circuit example.

FIGURE 6.52 Power triangle.

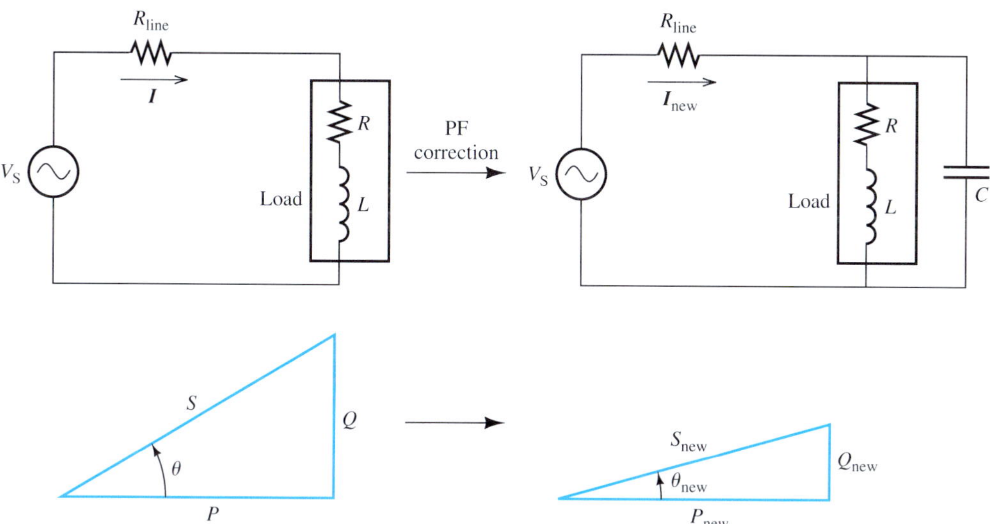

FIGURE 6.53 Power factor correction.

($\sin\theta_z$) increases. According to the power triangle shown in Figure 6.52, to increase the power factor, the reactive power should be reduced.

This can be easily achieved using the fact that this problem is mainly due to the inductance of motor windings. In addition, it is known that the effect of the positive impedance of an inductive load ($X_L = j\omega L$) can be compensated using the negative impedance of a capacitive load ($X_C = 1/j\omega C = -j/\omega C$). In other words, if a load contains inductance and capacitance with reactive powers of equal magnitude, their reactive powers will cancel. Therefore, a general approach to reduce the reactive power is to add capacitors in parallel with an inductance. This increases the power factor. This fact is illustrated in Figure 6.53. In Figure 6.53, $P = P_{new}$, thus, when Q is reduced to Q_{new}, the θ_{new} decreases. Accordingly, the power factor increases.

EXAMPLE 6.24 Power Factor Correction Via Capacitor

A 60-W load powered by a 60-Hz, 240-V line with a power factor of 0.7071 lag is shown in Figure 6.54. Calculate the parallel capacitance required to correct the power factor to 0.9 lag.

(*continued*)

EXAMPLE 6.24 Continued

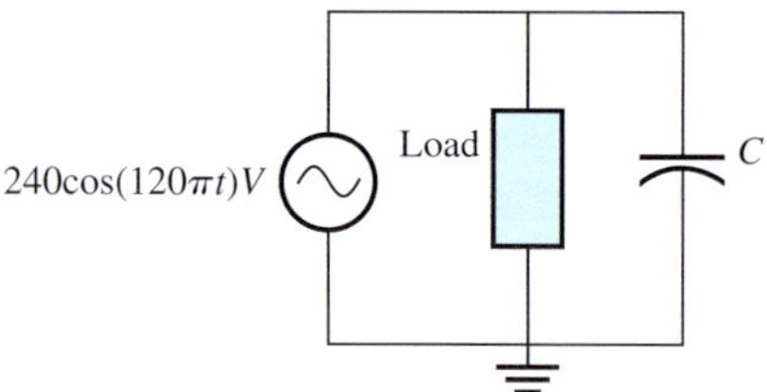

FIGURE 6.54 Circuit for Example 6.24.

SOLUTION

In order to solve this problem, find the load power angle:

$$\theta_L = \arccos(0.7071) = 45°$$

Use the power triangle in Figure 6.52 to find the reactive power Q_L.

$$Q_L = P_L \tan(\theta_L) = 60 \times \tan(45°) = 60 \text{ VAR}$$

After adding the parallel capacitor, the new power angle is:

$$\theta_{new} = \arccos(0.9) = 25.84°$$

The new value of the reactive power is:

$$Q_{new} = P_L \tan(\theta_{new}) = 29.06 \text{ VAR}$$

Note that $P_L = 60$ W is a function of the resistive elements in the circuit and does not change after adding the capacitor.

The reactive power due to the addition of the capacitance is:

$$Q_C = Q_{new} - Q_L = -30.94 \text{ VAR}$$

The reactance of the capacitor is:

$$X_C = \frac{V_{rms}^2}{Q_C} = \frac{(240/\sqrt{2})^2}{-30.94} = -930.83 \ \Omega$$

The angular frequency is:

$$\omega = 2\pi f = 2\pi \times 60 = 376.99$$

The required capacitance is then found to be:

$$C = \frac{1}{\omega |X_C|} = \frac{1}{376.99 \times 930.83} = 2.85 \ \mu\text{F}$$

APPLICATION EXAMPLE 6.25 Heart–Lung Machine

A heart–lung machine has an electric pump that takes over for a biological heart. Suppose a pump for a heart–lung machine is connected as shown in Figure 6.55. Determine:

a. The impedance Z needed for maximum power transfer to the pump
b. The circuit component values needed for maximum power transfer

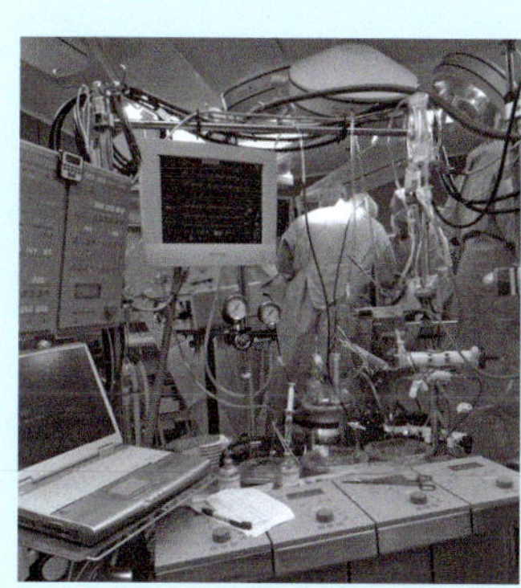

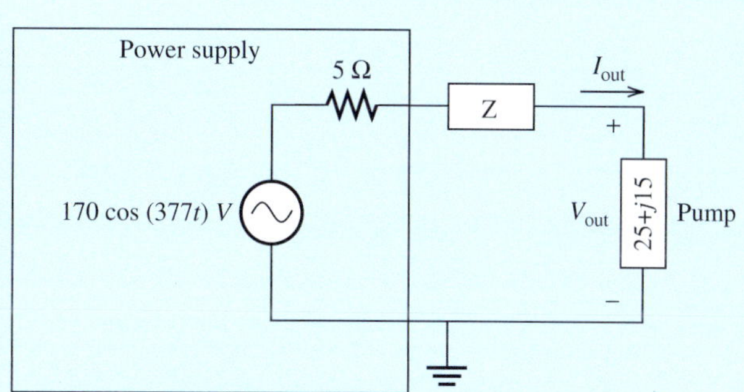

FIGURE 6.55 Heart–lung machine. (Used with permission from © Black Star/Alamy.)

SOLUTION

a. The complex power consumed by the pump can be written as:

$$P_{comp} = \frac{1}{2} V_{out} I^*_{out}$$

By Ohm's law:

$$I_{out} = \frac{V_{out}}{25 + j15}$$

Therefore, using Equation (6.95) P_{comp} becomes:

$$P_{comp} = \frac{1}{2}\frac{|V_{out}|^2}{25 - j15} = \frac{1}{2}\frac{|V_{out}|^2(25 + j15)}{(25 - j15)(25 + j15)} = \frac{1}{2}\frac{|V_{out}|^2(25 + j15)}{|25 + j15|^2}$$

The average delivered power to the pump is the real part of P_{comp}, which is:

$$P_{av} = \frac{1}{2}\frac{|V_{out}|^2\, 25}{|25 + j15|^2}$$

Observing the expression of P_{av}, the power delivered to the pump is maximized when $|V_{out}|^2$ is maximized. By using the voltage division rule, we find:

$$V_{out} = \frac{25 + j15}{5 + Z + 25 + j15} 170\angle 0°$$

Therefore, $|V_{out}|^2$ is maximized when $|5 + Z + 25 + j15|^2$ is minimized. Note that the impedance, Z, always has a positive (or zero) real part. Any positive value for Z makes $|V_{out}|$ smaller. In order to maximize $|V_{out}|^2$ and P_{av}, we select $Z = -j15$.

b. In order to determine the components that make up Z, look at the value of Z. Because the imaginary part of Z is negative, it must be capacitive. Therefore, Z is made by a capacitor. The impedance of the capacitor is $-j15$ and $Z = 1/(j\omega C_Z)$, and $\omega = 377$. Thus:

$$C_z = \frac{1}{j\omega Z} = \frac{1}{j377 \times (-j15)} = 176.83\ \mu F$$

The process of conversion and the final circuit are shown in Figure 6.56.

(continued)

APPLICATION EXAMPLE 6.25 Continued

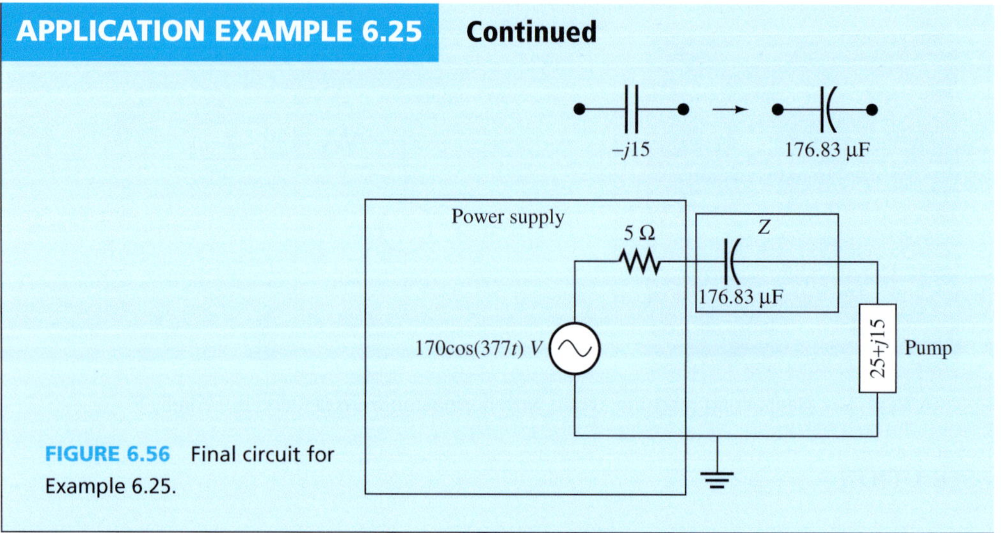

FIGURE 6.56 Final circuit for Example 6.25.

EXERCISE 6.16

Find an equivalent circuit for the following impedances:

a. $12 - j15$, $\omega = 10$
b. $20 + j\,25$, $\omega = 25$
c. $+j110$, $\omega = 220\pi$

APPLICATION EXAMPLE 6.26 Elevator

A large AC motor is used to operate an elevator as shown in Figure 6.57. It is desired that the motor use the least amount of current possible. A capacitor C_{PF} is included for power factor correction due to the motor inductance. If $\omega = 377$ rad/s, determine the current I, if $C_{PF} = 150$ μF.

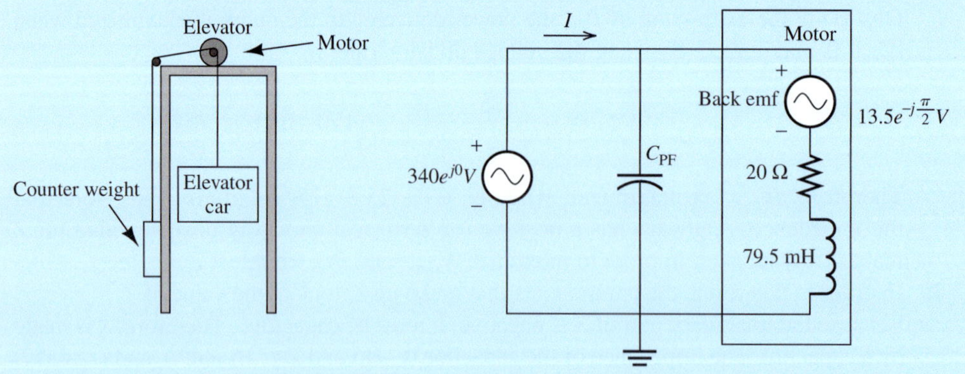

FIGURE 6.57 An elevator and its circuit.

SOLUTION

The equivalent circuit elements are shown in Figure 6.58. The impedance of the capacitor is calculated using $Z = 1/(j\omega C_{PF}) = -j17.68$. Note that the current, I, is the sum of the

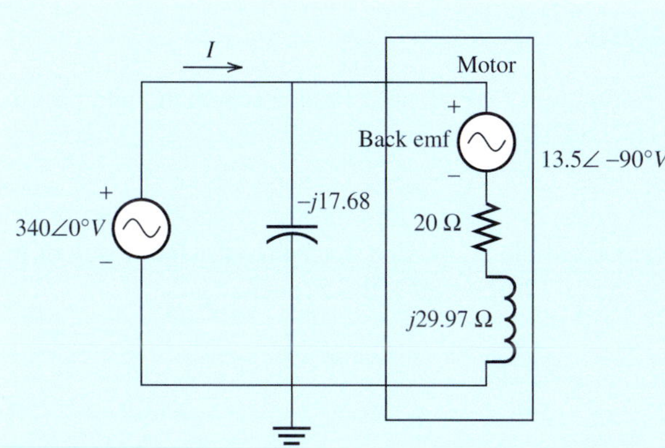

FIGURE 6.58 Equivalent circuit elements for Example 6.26.

two currents flowing through the two parallel branches. The current through the capacitor is $\frac{340}{-j17.68}$, and the current through the motor is:

$$I_{\text{motor}} = \frac{340 - 13.5\angle -90^\circ}{20 + j29.97}$$

Therefore, the total current is:

$$I = \frac{340 - 13.5\angle -90^\circ}{20 + j29.97} + \frac{340}{-j17.68}$$

Some simplifications (detailed below) arrive at the final answer:

$$\begin{aligned} I &= \frac{340 - 13.5\angle -90^\circ}{20 + j29.97} + \frac{340}{-j17.68} \\ &= \frac{340 + j13.5}{20 + j29.97} + j19.23 \\ &= \frac{340.27\angle 2.27^\circ}{36.03\angle 56.28^\circ} + j19.23 \\ &= 9.44\angle -54.01^\circ + j19.23 \\ &= 5.55 - j7.64 + j19.23 \\ &= 5.55 + j11.59 \\ &= 12.85\angle 64.4^\circ \end{aligned}$$

In the time domain:

$$i(t) = 12.85 \cos(377t + 64.41^\circ)$$

6.7 STEADY-STATE CIRCUIT ANALYSIS USING PSPICE

This section outlines the use of PSpice for steady-state circuit analysis. The tutorial and examples in Chapter 2 as well as examples in other chapters also help to develop this analysis.

EXAMPLE 6.27 PSpice Analysis

Use the values obtained in part (a) of Example 6.3 to set up a PSpice schematic and plot the voltage across a resistor, with the resistance of 100 Ω.

SOLUTION

To set up a sinusoidal voltage source (see Example 2.24), first determine the frequency of the source. Given the voltage source, $v(t) = 5\cos(120\pi + 30°)$, the frequency is:

$$\begin{aligned} f &= \frac{120\pi}{2\pi} \\ &= 60 \text{ Hz} \end{aligned}$$

Next, determine the phase angle of the sinusoidal voltage source. The sinusoidal voltage source function in PSpice is "sinus." Therefore, a default phase angle of +90° is needed if the given voltage source is a "cosine" function. Then, the phase angle is:

$$\begin{aligned} \theta &= 90° + 30° \\ &= 120° \end{aligned}$$

Next, the PSpice schematic can be set up using the following steps:

1. Example 2.24 shows the process of adding a sinusoidal voltage source, "VSIN" into a circuit. Set VOFF = 0, VAMPL = 5, FREQ = 60. VOFF is the offset voltage, VAMPL is the amplitude or the peak of the voltage, and FREQ is the frequency of the voltage.
2. To set the phase angle of the voltage source, go to "edit properties" of voltage source by right-clicking on it, or double-click on the voltage source. A window similar to Figure 6.59 will appear.
3. Set the value under "PHASE" (see the black arrow) to 120. Also, go to "Display…" (see the colored arrow) and choose "Display both Name and Value."
4. Add the resistor and ground (GND) into the circuit.
5. Create a simulation profile, with the analysis type set as "Time Domain (Transient)." Set the simulation in between 50 and 100 ms and run the simulation (see Figure 2.75).
6. The PSpice schematic is shown in Figure 6.60.
7. The voltage across the resistor plot is shown in Figure 6.61.

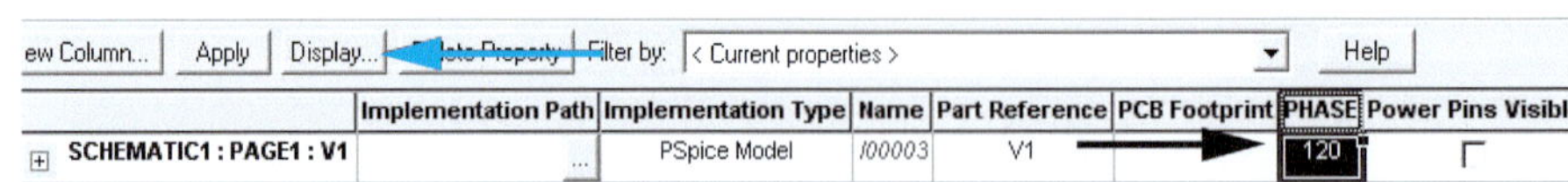

FIGURE 6.59 Set the phase angle.

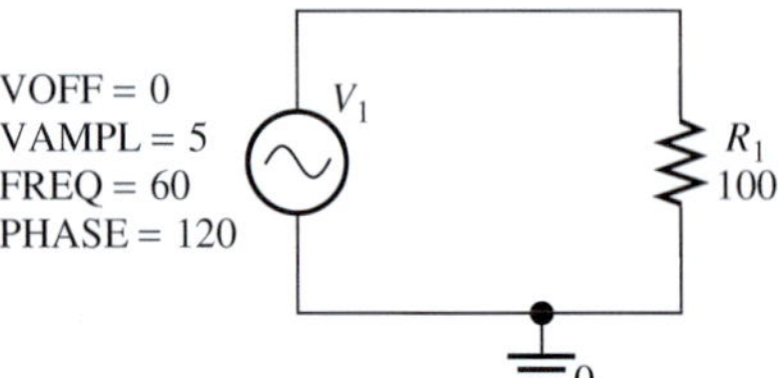

FIGURE 6.60 PSpice schematic solution for Example 6.3.

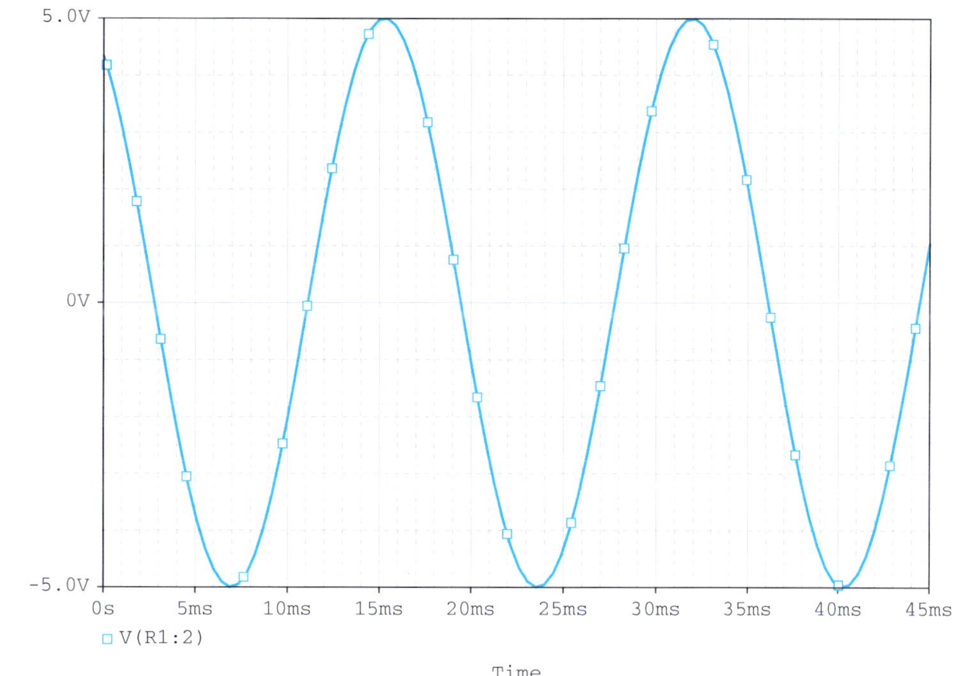

FIGURE 6.61 Sinusoidal voltage source across the resistor plot.

EXAMPLE 6.28 PSpice Analysis

Use PSpice to find the steady-state current $i(t)$ and $v_L(t)$ shown in the figure for Example 6.13.

SOLUTION

First, determine the frequency and the phase angle of the voltage source. The voltage source is: $v(t) = 110\cos(400t)$. The frequency is then:

$$f = \frac{400\pi}{2\pi}$$
$$= 63.662 \text{ Hz}$$

The sinusoidal voltage source is a cosine function; therefore, the phase angle is not zero for the PSpice sinusoidal voltage source, which is:

$$\theta = 90°$$

Add the sinusoidal voltage source, VSIN, into the PSpice schematic, with VOFF = 0, VAMPL set to 110, and FREQ set to 63.662. The phase angle can be set by going to the "edit properties" of VSIN (see Figure 6.59).

Create a new simulation profile and set the analysis type to be "Time Domain (Transient)." Set the simulation time between 50 and 100 ms and run the simulation (see Figure 2.71). The PSpice schematic solution is shown in Figure 6.62, and the plots of $i(t)$ and $v_L(t)$ are shown in Figures 6.63 and 6.64, respectively.

(continued)

EXAMPLE 6.28 Continued

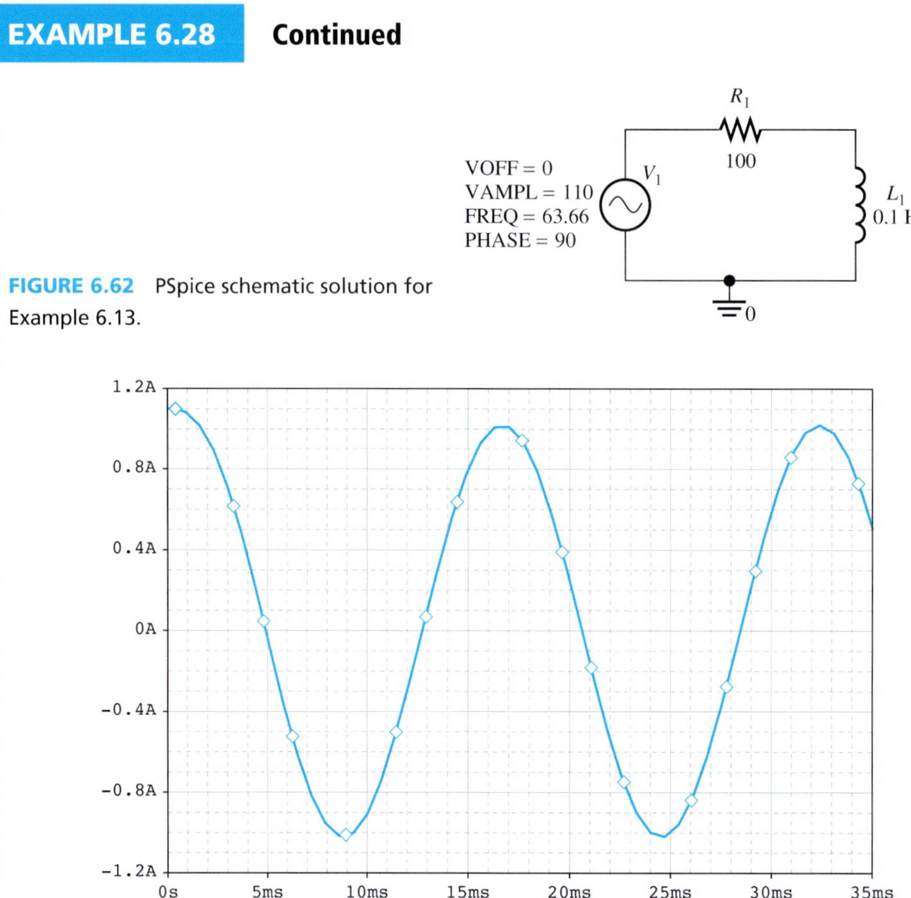

FIGURE 6.62 PSpice schematic solution for Example 6.13.

FIGURE 6.63 Current across the circuit in Example 6.13.

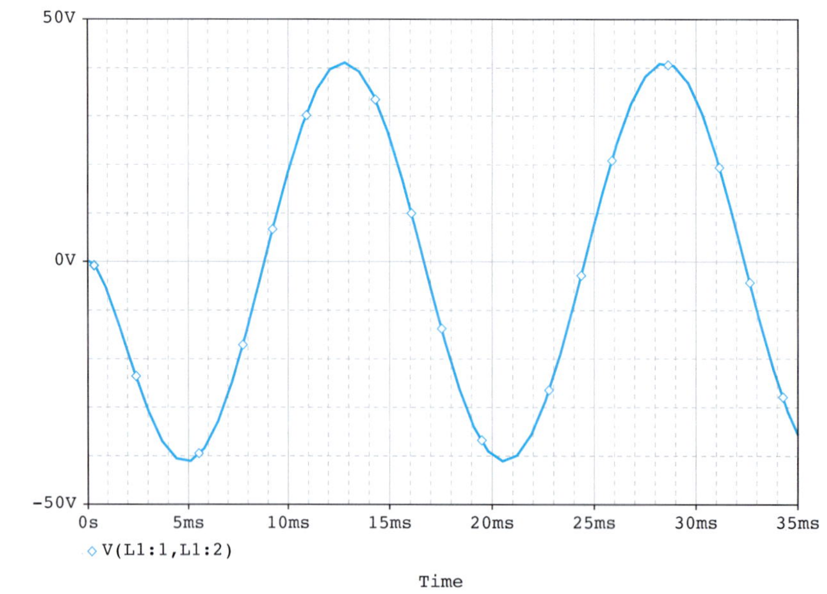

FIGURE 6.64 Voltage across the inductor in Example 6.13.

EXAMPLE 6.29 PSpice Analysis

Considering the circuit shown in Figure 6.64 for Example 6.15, set up a PSpice schematic plot for the $v_C(t)$ in steady state.

SOLUTION

First, determine the frequency and phase angle of the voltage source. Given that the voltage source is $v(t) = 110\cos(400t + 60°)$, the frequency is:

$$f = \frac{400\pi}{2\pi} = 63.662 \text{ Hz}$$

Because the sinusoidal voltage source is a cosine function, the phase angle is:

$$\theta = 60° + 90° = 150°$$

Add the sinusoidal voltage source, VSIN, into the PSpice schematic, with VOFF = 0, VAMPL set to 110 and FREQ set to 63.662. The phase angle can be set by going to the "edit properties" of VSIN (see Figure 6.59).

Add the resistor, inductor, and capacitor into the circuit as shown in Figure 6.65. Set the initial condition for both the inductor and the capacitor to zero (see Example 4.16).

Set the simulation time to run for 100 ms (see Figure 2.75). The voltage across capacitor, C1, is shown in Figure 6.66.

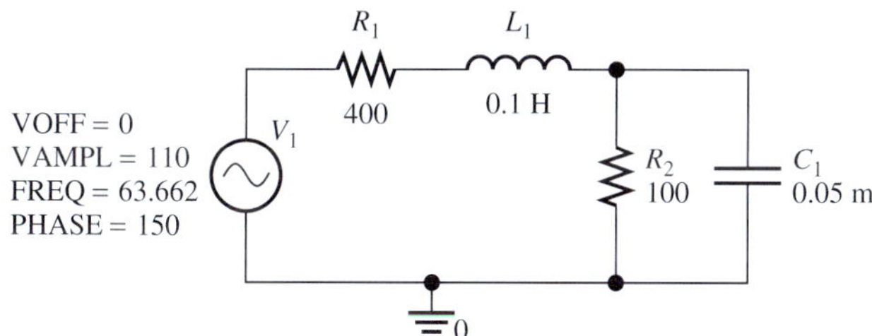

FIGURE 6.65 PSpice schematic solution for Example 6.15.

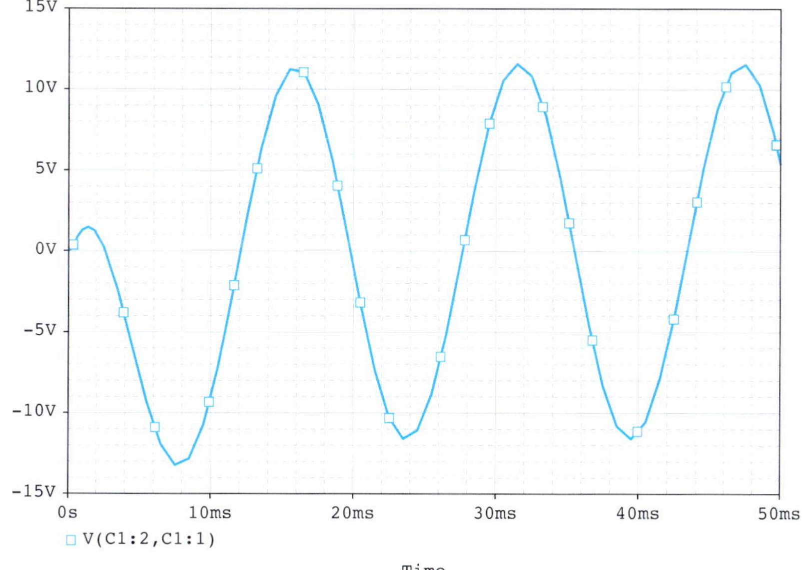

FIGURE 6.66 Voltage across the capacitor, C1, for Example 6.15.

APPLICATION EXAMPLE 6.30 Controlling Building Sway

In Example 6.20, find the Thévenin voltage as seen by the load using PSpice.

SOLUTION

First, determine the frequency and phase angle of the voltage source. Given that the voltage source, $v(t) = 2.5\cos(500t)$, the frequency is:

$$f = \frac{500\pi}{2\pi}$$
$$= 79.577 \text{ Hz}$$

Because the sinusoidal voltage source is a cosine function, the phase angle is:

$$\theta = 90°$$

Add the sinusoidal voltage source, VSIN, into the PSpice schematic, with VOFF set to 0, VAMPL set to 2.5 and FREQ set to 79.577. The phase angle can be set by going to the "edit properties" of VSIN (see Example 6.29).

The PSpice schematic solution is shown in Figure 6.67. Notice that there is no inductor in Figure 6.67. Because the goal is to observe the Norton equivalent voltage across the series resistor and inductor, both the resistor and the inductor are replaced by a large value of resistance (100 GΩ). Also, a "Voltage Differential Marker(s)" (see Figure 6.68, noted by the arrow) is placed around the Norton resistance, so that PSpice will plot the voltage across that resistor automatically.

Set the initial condition of the capacitors to 0 V (see Example 4.16). Set the simulation time to 100 ms and press the run button (see Figure 2.71). The Norton equivalent voltage is plotted in Figure 6.69.

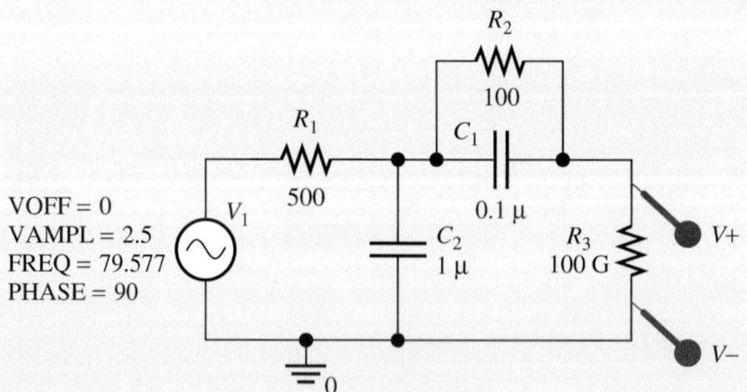

FIGURE 6.67 PSpice schematic solution for Example 6.20.

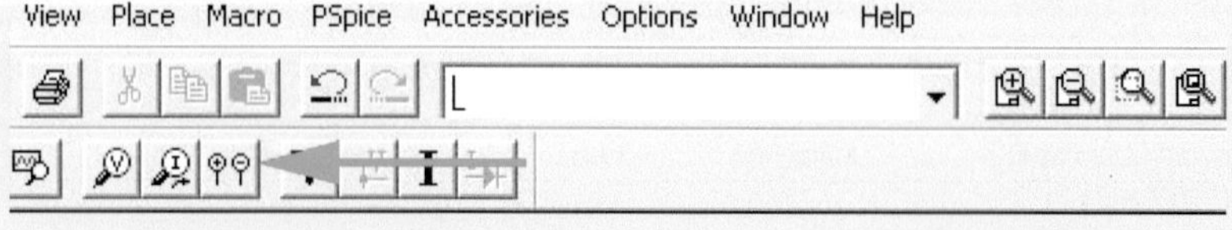

FIGURE 6.68 Voltage differential marker(s).

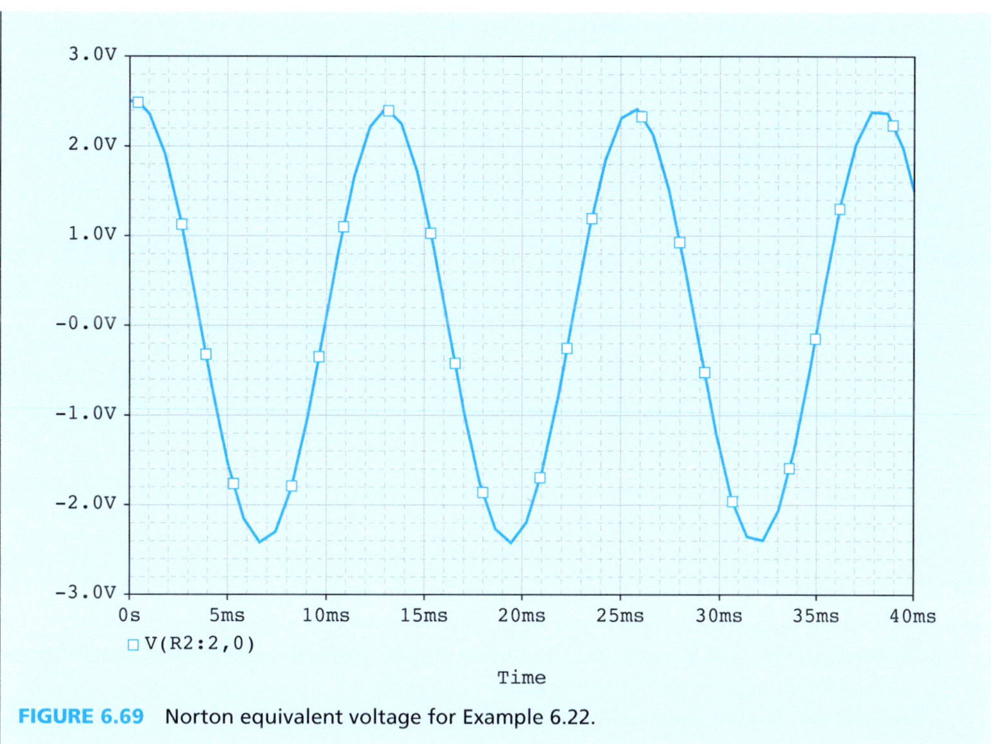

FIGURE 6.69 Norton equivalent voltage for Example 6.22.

6.8 WHAT DID YOU LEARN?

- The sinusoidal function $v(t) = V_m \cos(\omega t + \theta)$ has an amplitude of V_m, a radian frequency of ω, a period of $2\pi/\omega$, and a phase angle of θ.
- The *effective value* of the sinusoidal voltage $v(t) = V_m \cos(\omega t + \theta)$ as shown in Equation (6.13):

$$V_{rms} = \frac{V_m}{\sqrt{2}}$$

- The *instantaneous power* delivered by the voltage source to the resistor, R, corresponds to Equation (6.16):

$$p(t) = v(t) \cdot i(t)$$

  The *average power* corresponds to Equation (6.11):

$$P_{av} = \frac{V_m^2}{2R}$$

- The *sinusoidal voltage*, $v(t) = V_m \cos(\omega t + \theta)$, can be written in phasor form as shown in Equation (6.25):

$$V = V_m\angle\theta = V_m e^{j\theta}$$

- The impedance of R, L, and C are explained as follows: The impedance of the resistor is [Equation (6.36)]:

$$Z_R = R$$

  The impedance of the inductor is [Equation (6.43)]:

$$Z_L = j\omega L$$

The impedance of the capacitor is [Equation (6.51)]:

$$Z_C = \frac{1}{j\omega C}$$

- Impedances in series and parallel:

  Impedances in *series* can be combined like resistors in series [Equation (6.53)]:

$$Z_{eq} = \sum_{k=1}^{n} Z_k$$

  Impedances in *parallel* can be combined like resistors in parallel [Equation (6.54)]:

$$\frac{1}{Z_{eq}} = \frac{1}{Z_1} + \frac{1}{Z_2} + \cdots \frac{1}{Z_n}$$

- To perform circuit analysis with phasors:
  1. Replace the instantaneous voltage and current source(s) [i.e., $v(t)$ and $i(t)$] with the corresponding phasors (i.e., V and I)
  2. Replace inductance, L, with the complex impedance $Z_L = j\omega L = \omega L\angle 90°$, and capacitance, C, with the complex impedance $Z_C = 1/j\omega C = -j1/\omega C = 1/\omega C\angle -90°$. Resistances have impedances equal to their resistances
  3. Analyze the circuit using the techniques used for resistive circuits and perform the calculation using complex arithmetic
- Thévenin and Norton equivalent circuits with phasors:

  Thévenin voltage and impedance

  From the perspective of any load, the Thévenin voltage, V_{th}, equals the open-circuit voltage, V_{AB}, of the circuit. The Thévenin impedance, Z_{th}, seen through two terminals is found by setting the independent sources to zero.

  Norton current and impedance

  The Norton current can be found by equating the load to zero. The Norton impedance, Z_n, is the same as the Thévenin impedance, Z_{th}, and can be computed in the same fashion.
- Power definitions (see Table 6.1).
- Maximum average power transfer:

  Given a circuit network, if the Thévenin impedance is $Z_{th} = R_{th} + jX_{th}$, and the impedance of the load is $Z_L = R_L + jX_L$, then the Z_L that maximizes the average power is [Equation (6.97)]:

$$Z_L = Z_{th}^* = R_{th} - jX_{th}$$

  Accordingly, the maximal average power is [Equation (6.98)]:

$$P_{max} = \frac{V_m^2}{8R_{th}}$$

  If the load is pure resistance, $X_L = 0$, then the R_L that maximizes the average power is:

$$R_L = \sqrt{R_{th}^2 + X_{th}^2} = |Z_{th}|$$

  The maximal corresponding average power is then:

$$P_{max} = \frac{V_m^2}{4(\sqrt{R_{th}^2 + X_{th}^2} + R_{th})}$$

- The power factor (PF) is defined as [Equation (6.74)]:

$$\text{PF} = \cos\theta_z = \cos(\theta_v - \theta_i)$$

The power factor will be *inductive* or *lagging* if current lags voltage (when θ_z is positive); the power factor will be *capacitive* or *leading* if current leads voltage.

- *Power factor correction*
  In order to increase the power factor and reduce the reactive power, a capacitor can be added in parallel with the inductor.

Problems

*B refers to Basic, A refers to Average, H refers to Hard, and * refers to problems with answers.*

SECTION 6.1 SINUSOIDAL VOLTAGES AND CURRENTS

6.1 (B)* Determine e^{2+3j}. Provide the answer to two decimal places.

6.2 (B) Determine $e^{3j} \cdot e^{1-j}$. Provide the answer to two decimal places.

6.3 (B)* A sinusoidal voltage is given by:

$$v(t) = 100 \sin(100\pi t + 30°)$$

Find the peak value of it and the voltage at $t = 10$.

6.4 (B) Find the amplitude, phase, period, and frequency of the sinusoid voltage:

$$v(t) = 10 \cos(110\pi t + 60°)$$

6.5 (B) Find the period of the following sinusoids:

a. $v(t) = \cos(2\omega t)$
b. $v(t) = \sin(3t)$
c. $v(t) = \sin(t + \theta)$
d. $v(t) = \cos(2\omega t + \theta) + \sin(2\omega t + \phi)$
e. $v(t) = \cos(t/2 + \theta) + \sin(t/3 + \phi)$

6.6 (B) Write the following sine functions in terms of cosine:

a. $v(t) = \sin(100t + 30°)$
b. $v(t) = \sin(50°)$
c. $v(t) = \sin(115°)$
d. $v(t) = \sin(\omega t + \theta)$
e. $v(t) = \sin(\omega t + \pi - \theta)$

6.7 (B) Write the following cosine functions in terms of sine:

a. $v(t) = \cos(100\pi t + 10°)$
b. $v(t) = \cos(t + 150°)$
c. $v(t) = \cos(195°)$
d. $v(t) = \cos(\omega t + \theta)$
e. $v(t) = \cos(\omega t - \theta)$

6.8 (B)* A sinusoidal voltage is given by:

$$v(t) = 156 \cos(120\pi t + 45°)$$

Find the peak value and rms value of the sinusoid voltage.

6.9 (B) A sinusoidal current $i(t) = 2\cos(120\pi t + 60°)$ passes a 100-Ω resistance; find the rms value of the sinusoid current.

6.10 (B) A sinusoidal voltage is given by:

$$v(t) = 200 \cos(120\pi t + 45°)$$

Find the average power delivered to a 1.5-kΩ resistance.

6.11 (A)* Given a sinusoidal voltage source that is $v_s(t) = 6\cos(20t) + 2$ V, what is the rms voltage of this voltage source in a period cycle?

6.12 (A) Sketch the following sinusoidal voltage source in a period cycle with appropriate label on the axes and their cross sections, $v_s(t) = 4\cos((\pi/5)t + 30)$.

6.13 (H)* A voltage source plot is shown in Figure P6.13. Determine the period and function of this sinusoidal voltage.

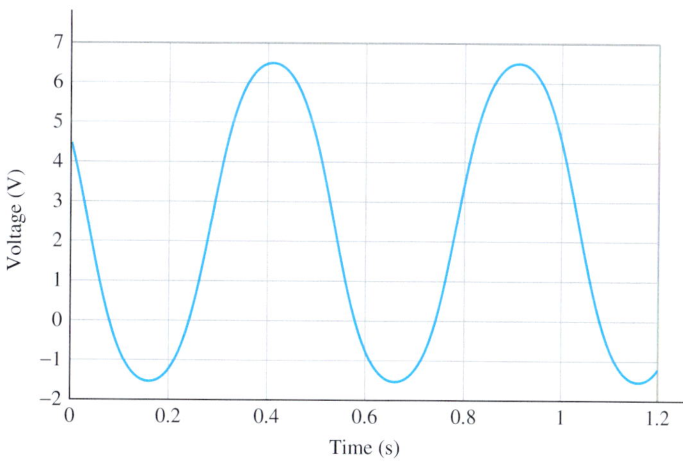

FIGURE P6.13 Voltage plot for Problem 6.13.

SECTION 6.2 PHASORS

6.14 (B) Find the phasor for the sinusoid voltages $v_1(t) = 156\cos(110\pi t + 60°)$ and $v_2(t) = 220\sin(120\pi t - 45°)$.

6.15 (B) Find the phasor for the sinusoid current $i(t) = 3\cos(120\pi t + 20°)$.

6.16 (A) Comment on the phasor between these two voltage sources, $v_1(t) = 110\cos(120\pi t + 60°)$ and $v_2(t) = 110\cos(100\pi t + 60°)$.

6.17 (A)* Two sinusoid voltage signals are given by $v_1(t) = 100\cos(120\pi t + 30°)$, and $v_2(t) = 250\cos(120\pi t + 60°)$. Use phasors to compute $v_a = v_1(t) + v_2(t)$.

6.18 (A) Suppose that $v_1(t) = 100\cos(120\pi t + 30°)$ and $v_2(t) = 250\cos(120\pi t + 60°)$, use phasors to find $v_d = v_1(t) - v_2(t)$.

6.19 (A) Given that two phasor voltage source are $V_1 = 20\angle 15°\ V$ and $V_2 = 10\angle 45°\ V$, find the sinusoidal voltage expression of $V_1 + V_2$. Express the solution to two decimal places.

SECTION 6.3 COMPLEX IMPEDANCES

6.20 (B)* What is the total impedance of the circuit in the Figure P6.20, if the frequency of the circuit is $200/\pi$ Hz? Provide the answer to two decimal places.

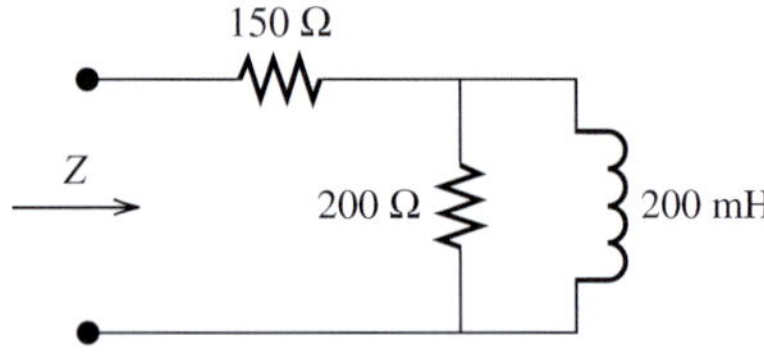

FIGURE P6.20 Circuit for Problem 6.20.

6.21 (B) Find Z in Figure P6.21 for $\omega = 200$, $\omega = 500$, and $\omega = 1000$.

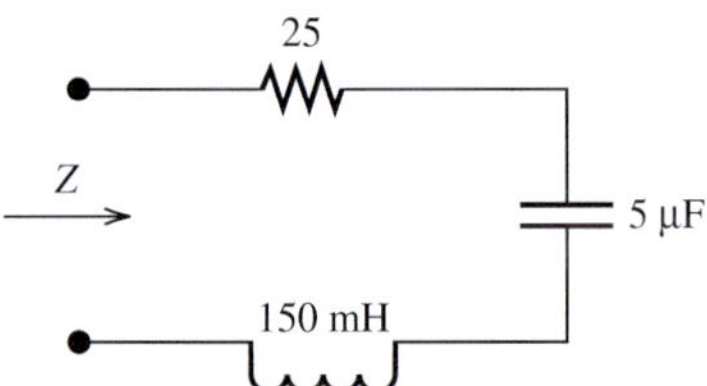

FIGURE P6.21 Circuit for Problem 6.21.

6.22 (B) Find Z in Figure P6.22 for $\omega = 200$, $\omega = 800$, and $\omega = 1500$.

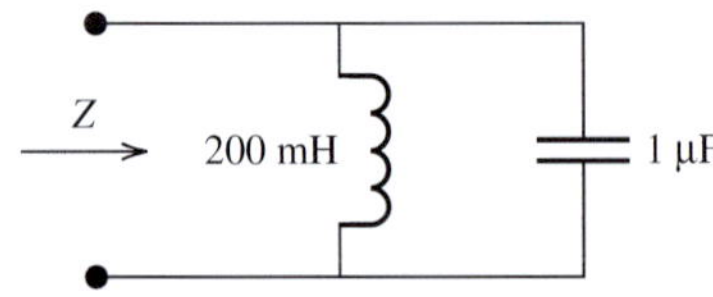

FIGURE P6.22 Circuit for Problem 6.22.

6.23 (A) If $\omega = 600$, find Z in Figure P6.23.

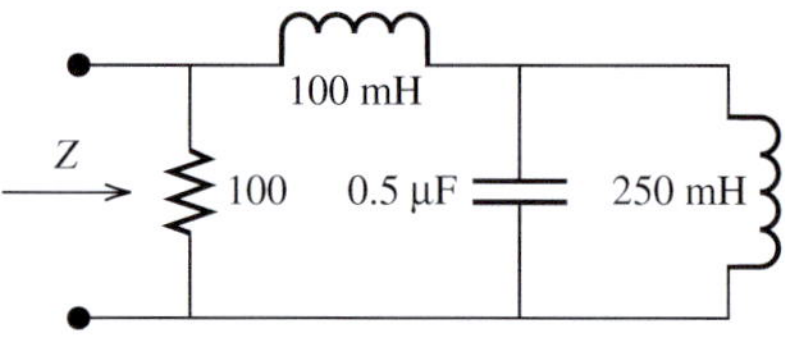

FIGURE P6.23 Circuit for Problem 6.23.

6.24 (A)* If $\omega = 250$, find Z in Figure P6.24.

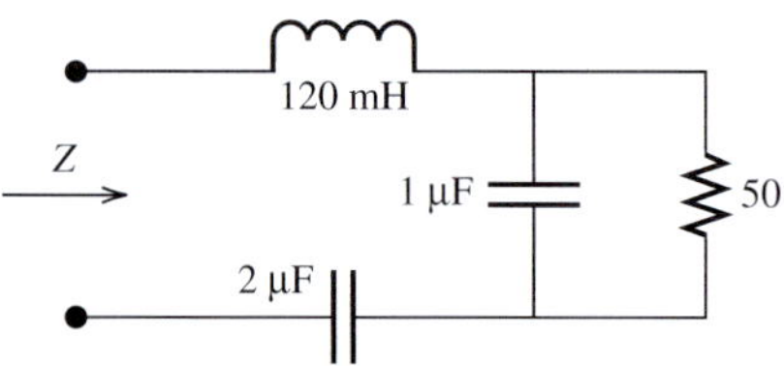

FIGURE P6.24 Circuit for Problem 6.24.

6.25 (A) Find Z in Figure P6.25.

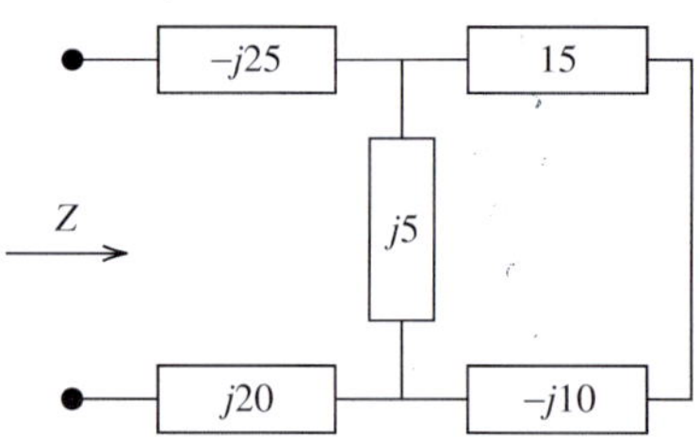

FIGURE P6.25 Circuit for Problem 6.25.

6.26 (A) Find Z in Figure P6.26.

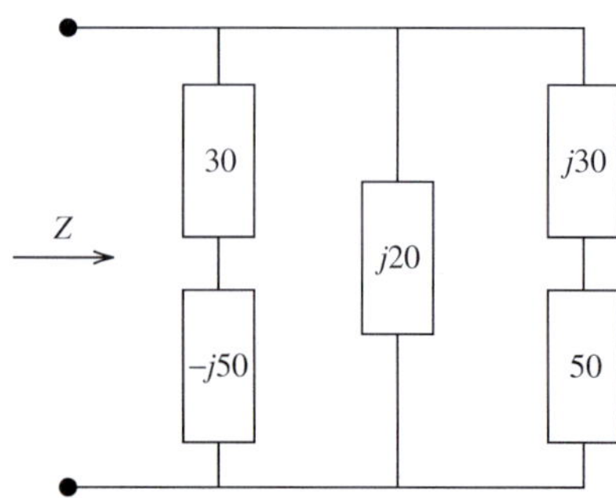

FIGURE P6.26 Circuit for Problem 6.26.

6.27 (H) Determine the power consumed by the inductor for the circuit in Figure P6.27 if the voltage source is $15\cos(200t)$. Provide the solution up to two decimal places.

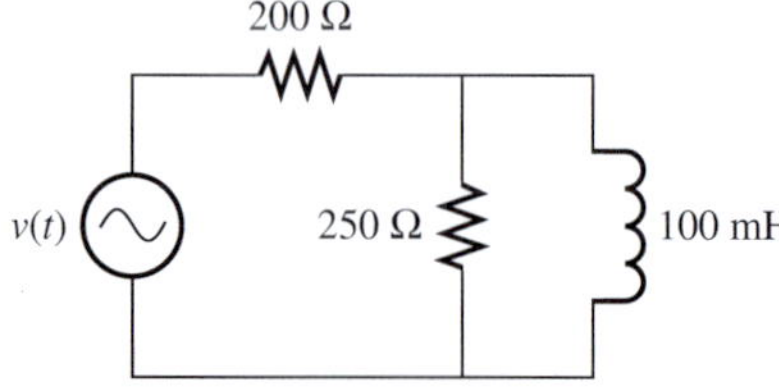

FIGURE P6.27 Circuit for Problem 6.27.

6.28 (A)* Sketch a simplified circuit for the circuit in Figure 6.28. Then, calculate the total impedance for the circuit, if the voltage source is $v_S(t) = V\cos(200t + 30°)$.

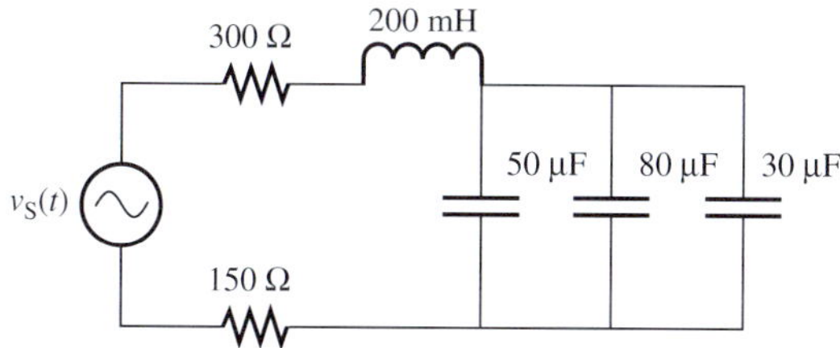

FIGURE P6.28 Circuit for Problem 6.28.

6.29 (H) What is the angular rate, ω, for the circuit in Figure P6.29, if the total impedance is $48.4125 - j137.5\ \Omega$, the resistance R is $77.46\ \Omega$, and capacitance C is $50\ \mu F$?

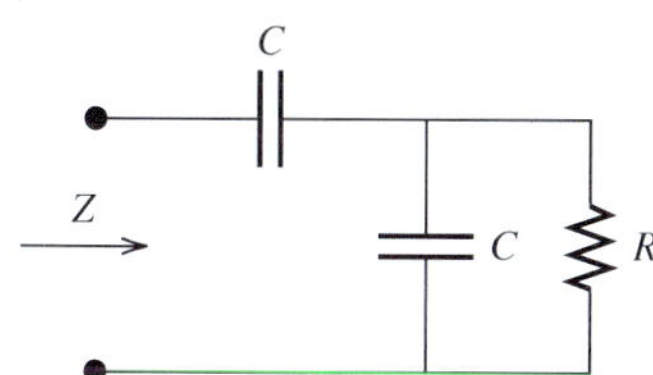

FIGURE P6.29 Circuit for Problem 6.29.

6.30 (H)* Application: voltage divider
In the circuit shown in Figure 6.30, the output and input voltage ratio is always

$$\frac{V_2}{V_1} = \frac{R_2}{R_1 + R_2}.$$

Find the relationship of R_1, R_2, C_1, and C_2.

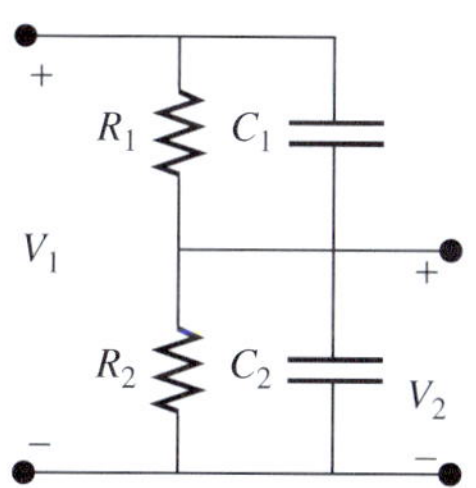

FIGURE P6.30 Circuit for Problem 6.30.

6.31 (H) Find the impedance, Z_{AB}, as seen from terminals A and B in the circuit shown in Figure P6.31.

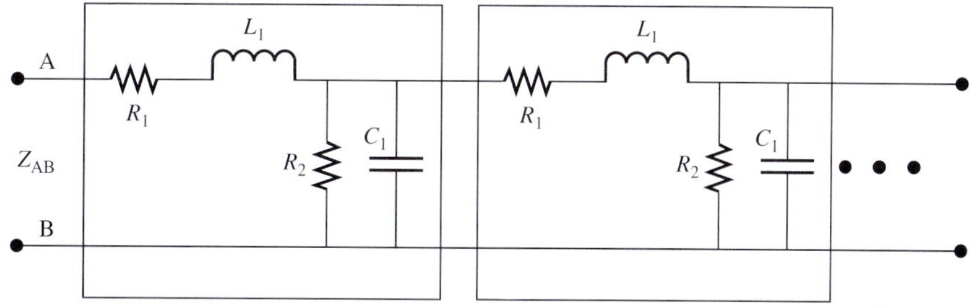

FIGURE P6.31 Circuit for Problem 6.31.

SECTION 6.4 STEADY-STATE CIRCUIT ANALYSIS USING PHASORS

6.32 (B) Find the steady-state expression of the current for the circuit in Figure P6.32 if the voltage source is $v(t) = 12\cos(120t)$.

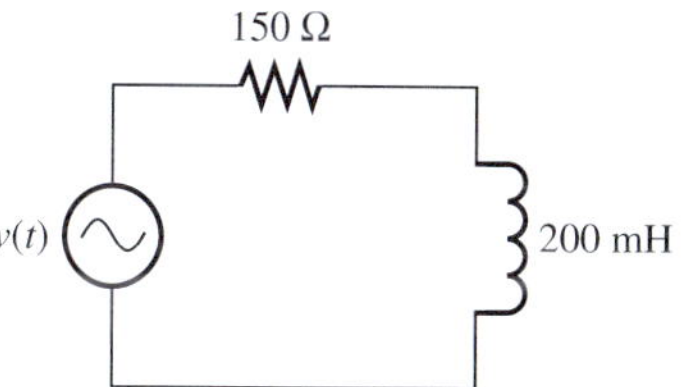

FIGURE P6.32 Circuit for Problem 6.32.

6.33 (B)* Find the steady-state expression of current for the circuit in Figure P6.33 if the voltage source is $v(t) = 10\cos(200t + 30°)$.

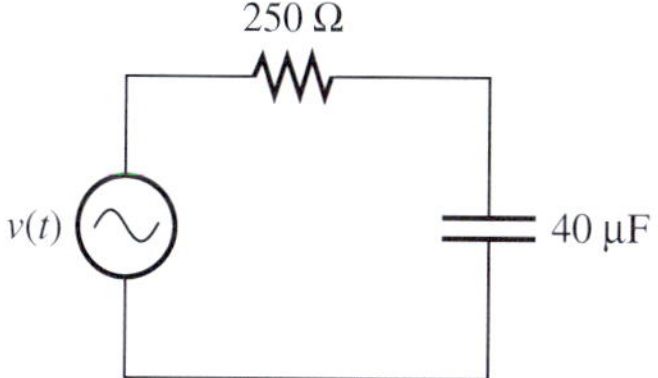

FIGURE P6.33 Circuit for Problem 6.33.

6.34 (A) Find V_1 and V_2 in Figure P6.34 using nodal analysis.

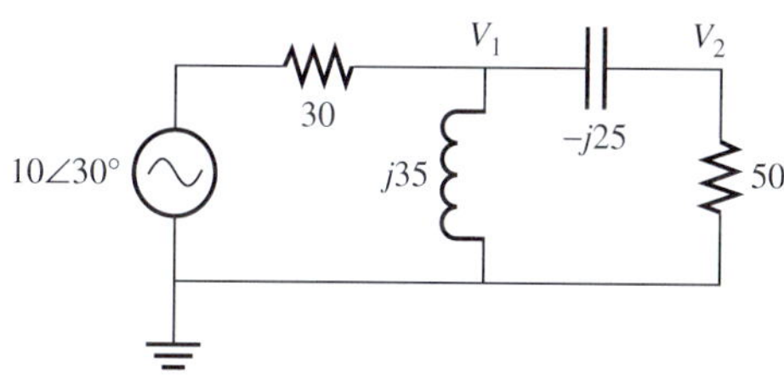

FIGURE P6.34 Circuit for Problem 6.34.

6.35 (A) Find node voltage V_1 and V_2 in Figure P6.35.

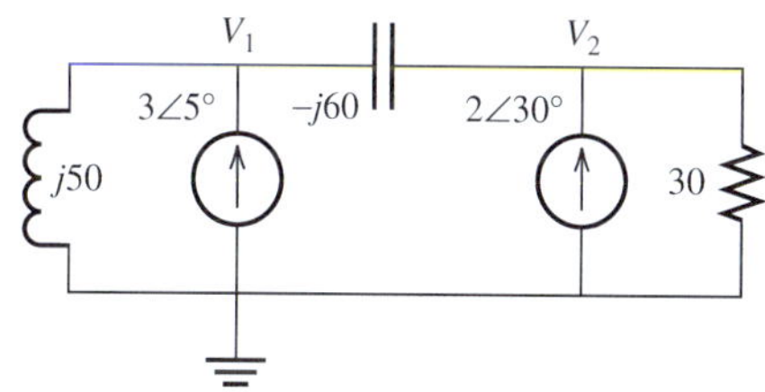

FIGURE P6.35 Circuit for Problem 6.35.

6.36 (A)* Find V_2 in Figure P6.36 using nodal analysis.

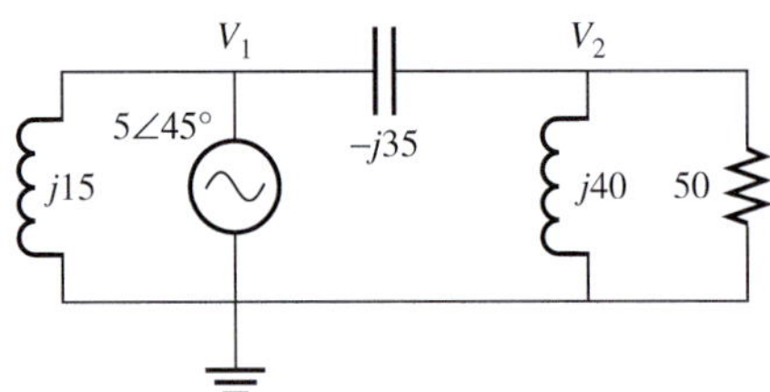

FIGURE P6.36 Circuit for Problem 6.36.

6.37 (A) Determine the rms current that flows across the resistor in Figure P6.37 if the voltage source is $v(t) = 24\cos(500t + 60°)$.

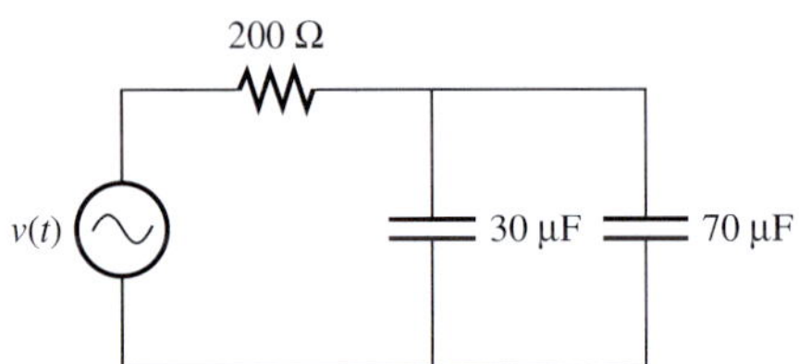

FIGURE P6.37 Circuit for Problem 6.37.

6.38 (A) Find the node voltages V_1 and V_2 in Figure P6.38.

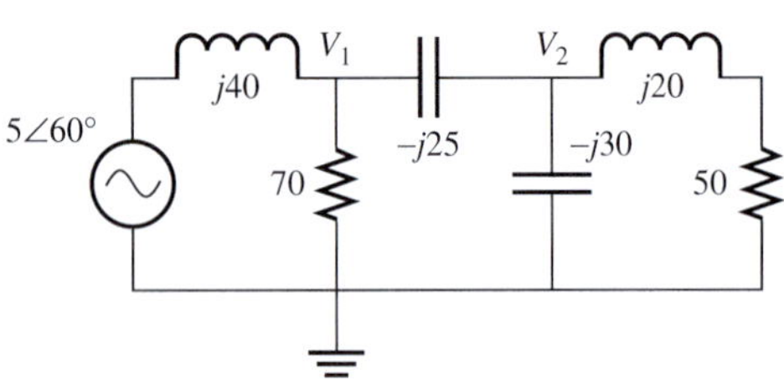

FIGURE P6.38 Circuit for Problem 6.38.

6.39 (A)* If $i_s(t) = 2\cos(200t)A$ in Figure P6.39, find the steady-state expressions for I_R and I_C.

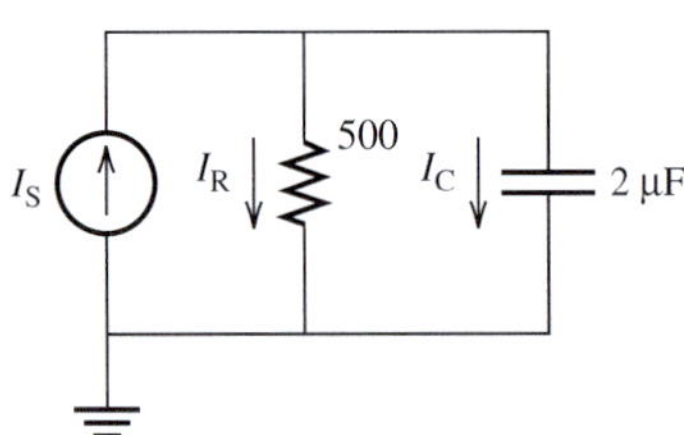

FIGURE P6.39 Circuit for Problem 6.39.

6.40 (A) An op amp is used as an amplifier to increase the magnitude of voltage output given the input voltage source. An **inverting amplifier** (see Figure P6.40) is one popular type of op amp. Given that the inverting amplifier has a gain of $G = -(Z_{out}/Z_{in})$, what is the output voltage if the voltage source is $v_S(t) = 50\cos(120\pi t + 30°)$ and $Z_{in} = 35\ \Omega$, $Z_{out} = 105\ \Omega$?

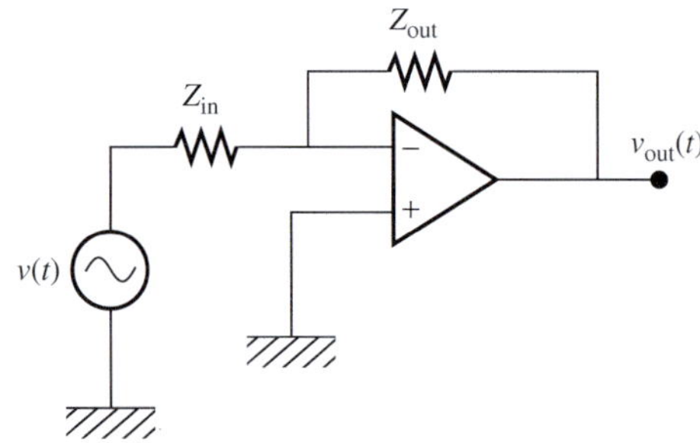

FIGURE P6.40 An inverting amplifier.

6.41 (A) Find the steady-state expressions for V_1 and V_2 in Figure P6.41.

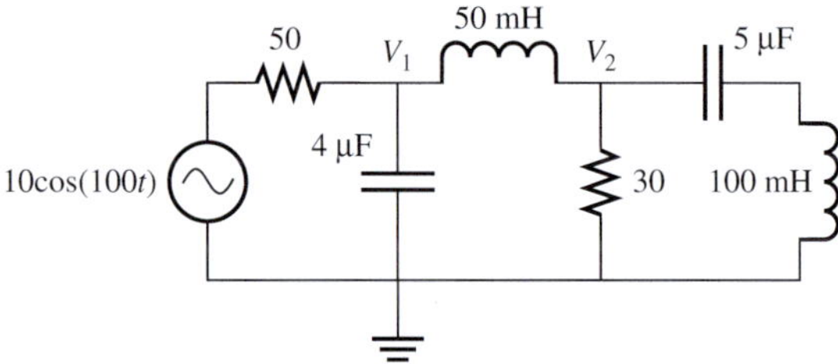

FIGURE P6.41 Circuit for Problem 6.41.

6.42 (A) Find the steady-state expressions for V_1 and V_2 in Figure P6.42.

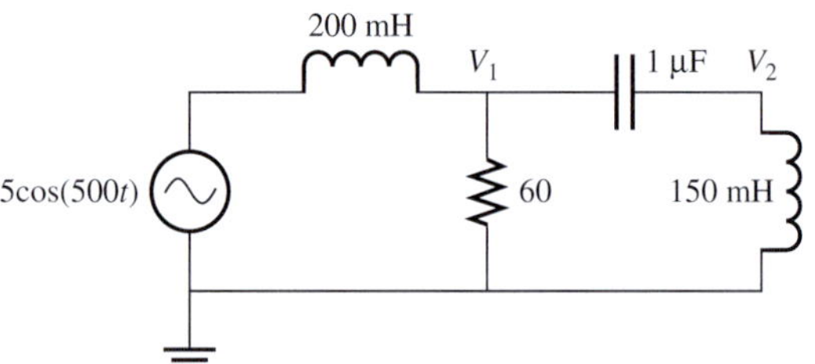

FIGURE P6.42 Circuit for Problem 6.42.

6.43 (H)* The electric fan circuit shown in Figure P6.43 is connected to US standard AC power (assume no phase angle). What is the cost of electric utility if the fan rotates at a constant speed for 12 h? Assume that the cost of electricity is 23 cents per kilowatt hour. (*Hint*: Assume the average power is $P_{av} = V_{rms}I_{rms}$.)

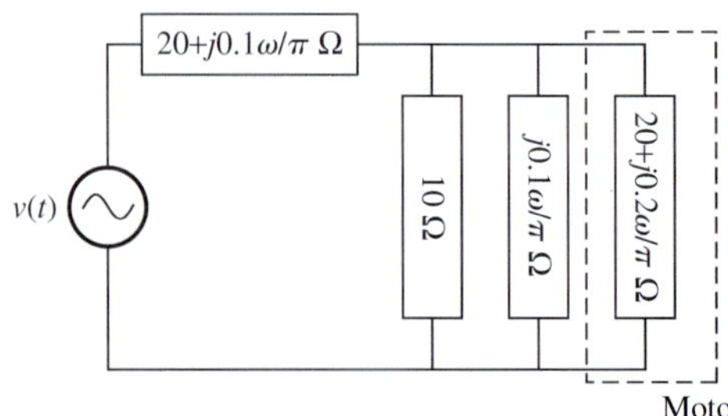

FIGURE P6.43 Circuit for Problem 6.43.

6.44 (H) The voltage source has a voltage of $v(t) = 4\cos^3 \omega t$. The radius frequency, ω, and the inductance, L, satisfies $\omega L = R$.

Find the voltage across the inductor at $t = 0$. (*Hint:* $4\cos^3 \omega t = 3\cos\omega t + \cos 3\omega t$.)

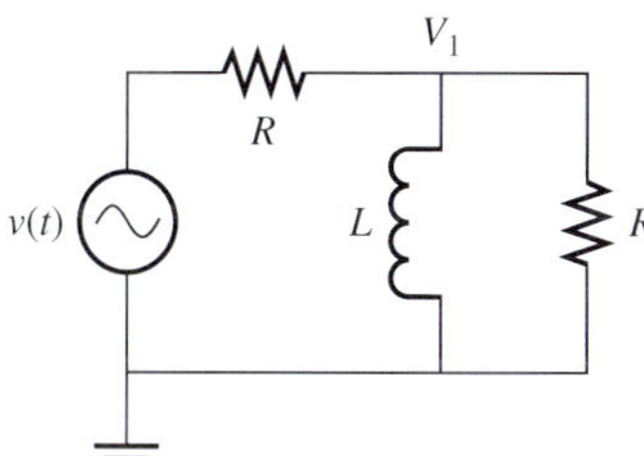

FIGURE P6.44 Circuit for Problem 6.44.

SECTION 6.5 THÉVENIN AND NORTON EQUIVALENT CIRCUITS WITH PHASORS

6.45 (A) Find the Thévenin and Norton equivalent circuits for the circuit shown in Figure P6.45, as seen by the load resistor, R_L.

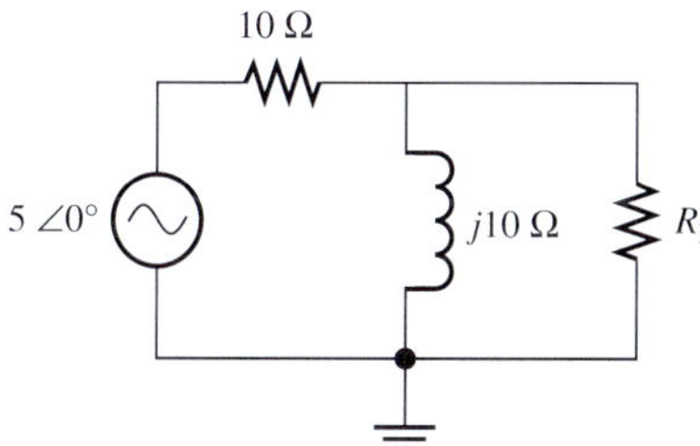

FIGURE P6.45 Circuit for Problem 6.45.

6.46 (B)* Determine the Norton equivalent current and equivalent impedance observed by R_L for the circuit in Figure P6.46 with the current source, $4\cos(200t + 30°)$.

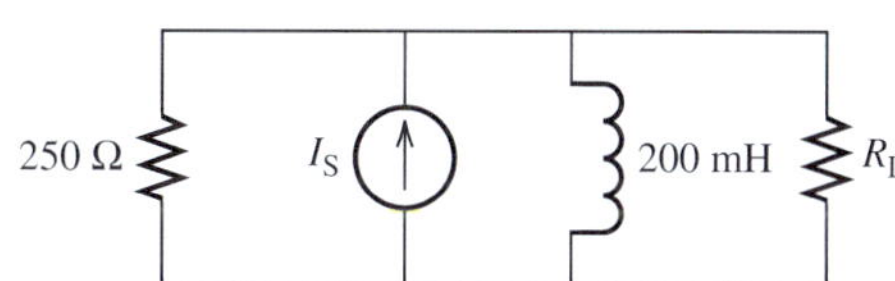

FIGURE P6.46 Circuit for Problem 6.46.

6.47 (A) Find the Norton equivalent circuits for the circuit shown in Figure P6.47.

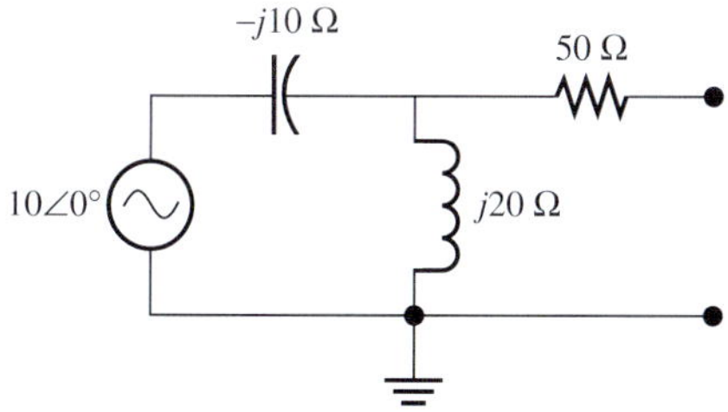

FIGURE P6.47 Circuit for Problem 6.47.

6.48 (A) Sketch the Thévenin and Norton equivalent circuit for the circuit shown in Figure P6.48, where $V_S = 20\angle 30°$.

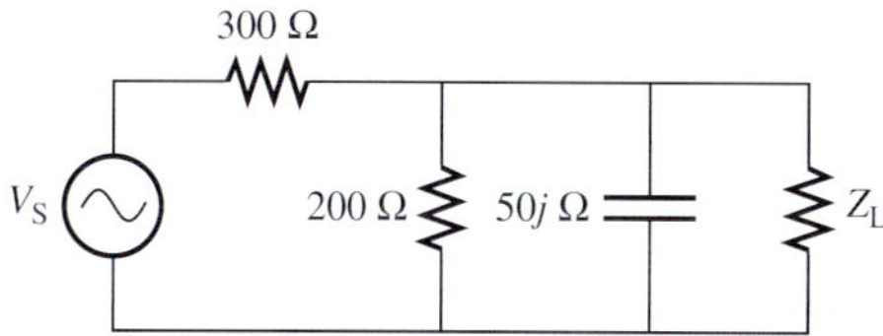

FIGURE P6.48 Circuit for Problem 6.48.

6.49 (H) Given that the Thévenin equivalent voltage, $V_{th} = 12 + j4\sqrt{6}$ V, what are the values of Z_R and Z_C if the voltage source, $V_S = 20\angle 0°$ V and $Z_C/R = -\sqrt{6/5}$?

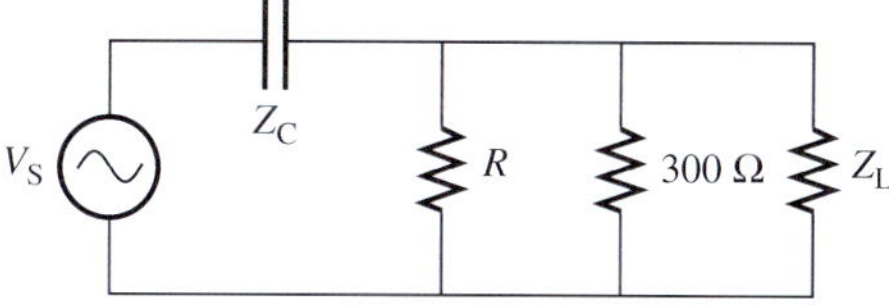

FIGURE P6.49 Circuit for Problem 6.49.

6.50 (B)* What are the Thévenin/Norton equivalent resistance observed by R_L for the circuit shown in Figure P6.50? Provide the answer to two decimal places.

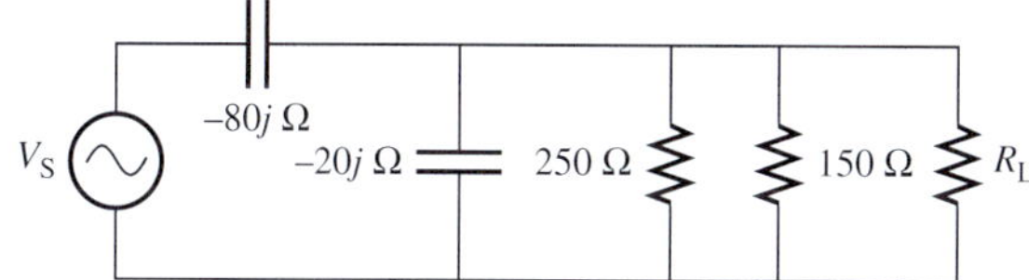

FIGURE P6.50 Circuit for Problem 6.50.

6.51 (A) Find the Thévenin equivalent circuits for the circuit shown in Figure P6.51.

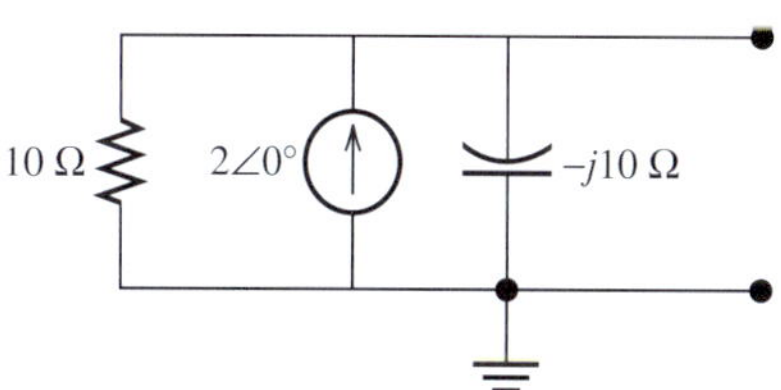

FIGURE P6.51 Circuit for Problem 6.51.

6.52 (H) Find the Thévenin and Norton equivalent circuits for the circuit shown in Figure P6.52, as seen by the load resistor, R_L.

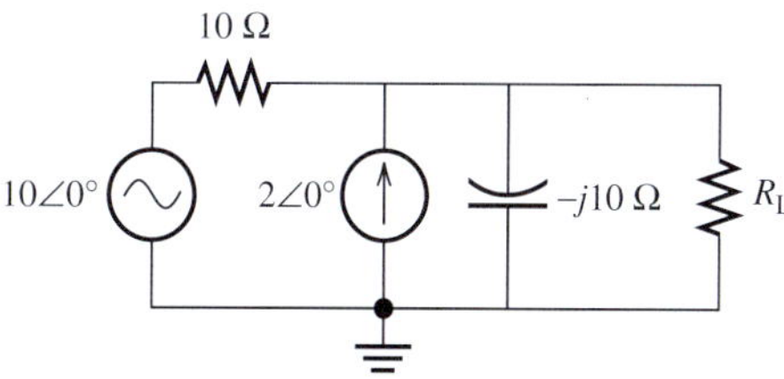

FIGURE P6.52 Circuit for Problem 6.52.

6.53 (B)* Express the Thévenin/Norton equivalent resistance, R_L, in terms of the resistance, R, and impedance of inductor Z_L for the circuit in Figure P6.53.

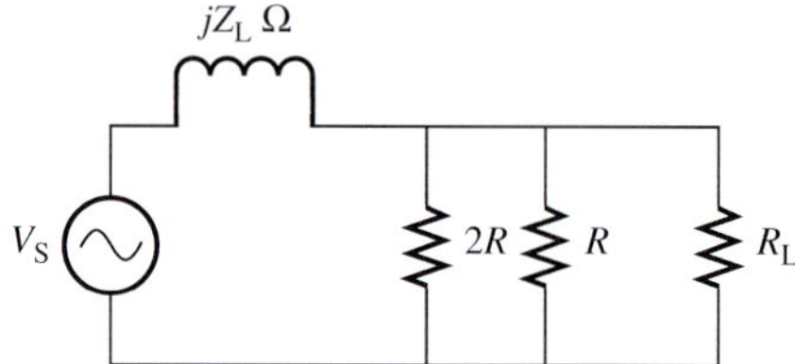

FIGURE P6.53 Circuit for Problems 6.53 and 6.54.

6.54 (H) In Figure P6.53, the voltage source is $V_{th} = 50\angle 0°$ the Thévenin voltage and Norton current are $V_{th} = 20\angle -60°$ and $I_{th} = 2\angle -90°$. Determine the value of R and Z_L.

6.55 (H) Use Thévenin equivalent circuit to solve Problem 6.44.

SECTION 6.6 AC STEADY-STATE POWER

6.56 (B) Find the equivalent circuit of the following impedances:

a. $-j25, \omega = 110\pi$
b. $12 + j110, \omega = 220\pi$
c. $20 - j220, \omega = 110\pi$
d. $+j110, \omega = 220\pi$

6.57 (B) Find the equivalent circuit of the following admittances:

a. $j250, \omega = 110\pi$
b. $1 + j10, \omega = 220\pi$
c. $2 - j20, \omega = 110\pi$
d. $1 - j, \omega = 1$

6.58 (B)* Determine the average power, reactive power, and apparent power generated by the resistor for the circuit in Figure P6.58 with the voltage source $160\angle 30°$.

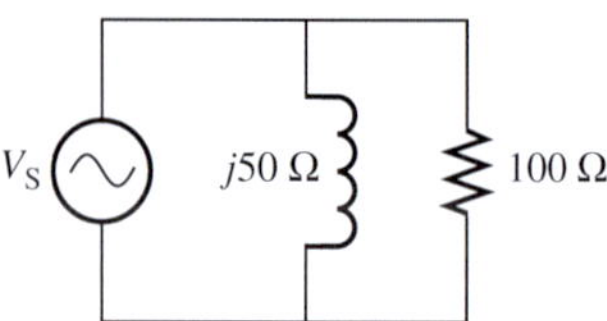

FIGURE P6.58 Circuit for Problem 6.58.

6.59 (A) Find the real and reactive power consumed by the load in Figure P6.59.

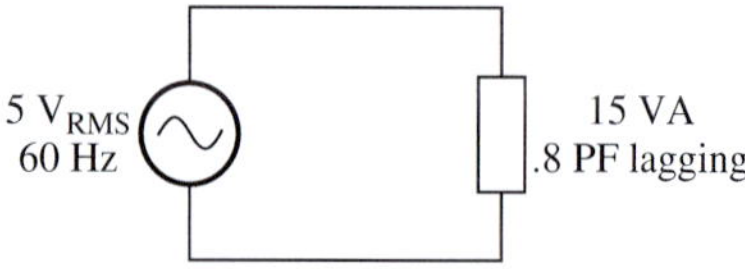

FIGURE P6.59 Circuit for Problem 6.59.

6.60 (A) Find the current, I, in Figure P6.60:

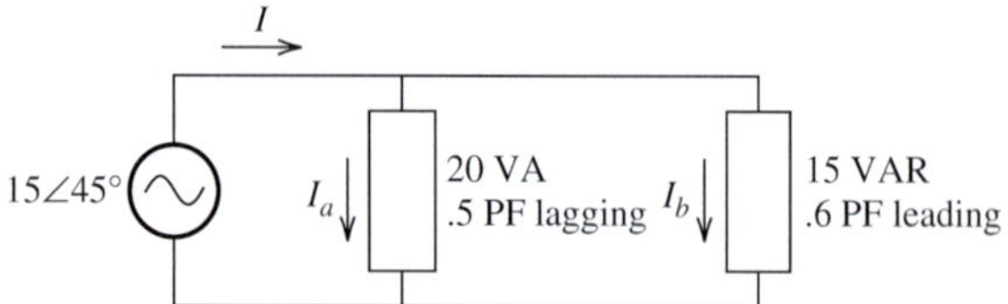

FIGURE P6.60 Circuit for Problem 6.60.

6.61 (A) Find the real, reactive, and apparent power drawn by the circuit, and the power factor for Figure P6.61.

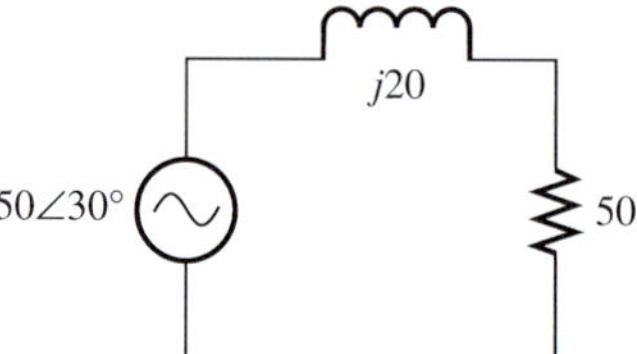

FIGURE P6.61 Circuit for Problem 6.61.

6.62 (A)* Find I, and the apparent power consumed by the circuit for Figure P6.62.

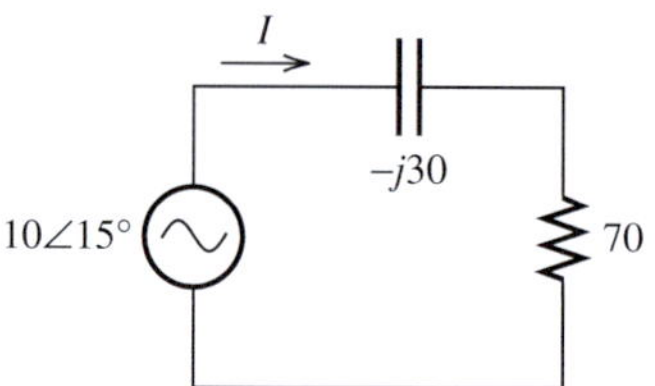

FIGURE P6.62 Circuit for Problem 6.62.

6.63 (H) Find the complex power taken from the source for the circuit shown in Figure P6.63.

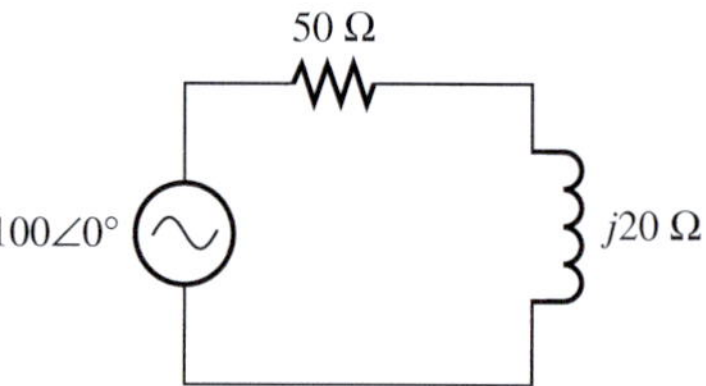

FIGURE P6.63 Circuit for Problem 6.63.

6.64 (A)* Consider the circuit shown in Figure P6.64. Determine the maximum average power that can be delivered to the load if (1) the load is pure resistance and (2) the load is the combination of resistance, inductance, and capacitance.

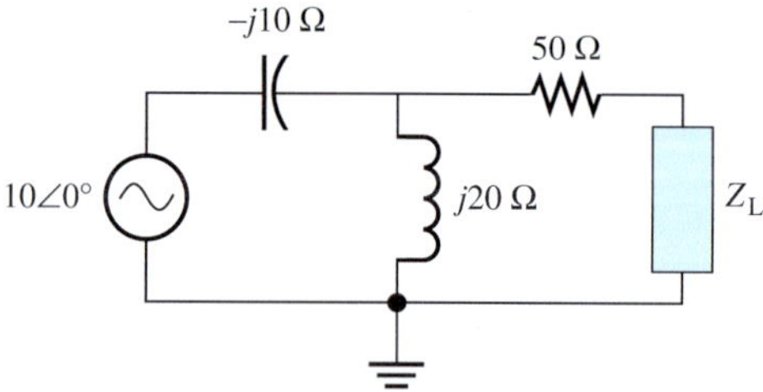

FIGURE P6.64 Circuit for Problem 6.64.

6.65 (B) What kind of load is shown in Figure P6.65?

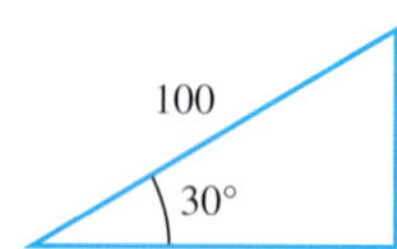

FIGURE P6.65 Triangle relationship for Problems 6.65 and 6.66.

6.66 (H) Determine θ_v and θ_i using the value shown in Figures P6.65 and P6.66.

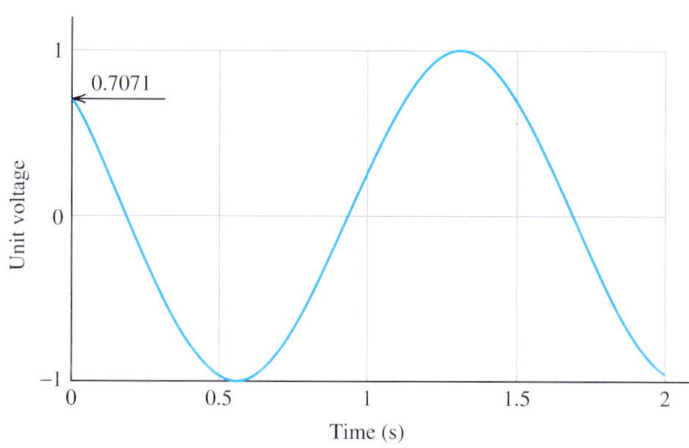

FIGURE P6.66 Unit voltage output plot for Problem 6.66.

6.67 (A)* A 30-kW load is powered by a 60-Hz, 480-V_{RMS} line with a power factor of 0.8 lag. Calculate the parallel capacitance required to correct the power factor to 0.9 lag.

6.68 (A) A 100-kW load is powered by a 60-Hz, 1-kV_{rms} line, power factor of 0.5 lag. To correct the power factor, a 100-μF capacitor bank is placed in parallel with the load. What is the new power factor?

6.69 (H)* An 8-kW load is powered by a 240-V_{peak}, PF = 1.00 source.

a. Calculate the current drawn by the load.

b. Calculate the current if power factor is 0.4

6.70 (H) Consider the circuit shown in Figure P6.70, if the load, Z_L, is the combination of resistance, inductance, and capacitance, how can R be adjusted to make the maximum average power delivered to the load less than 0.01 W?

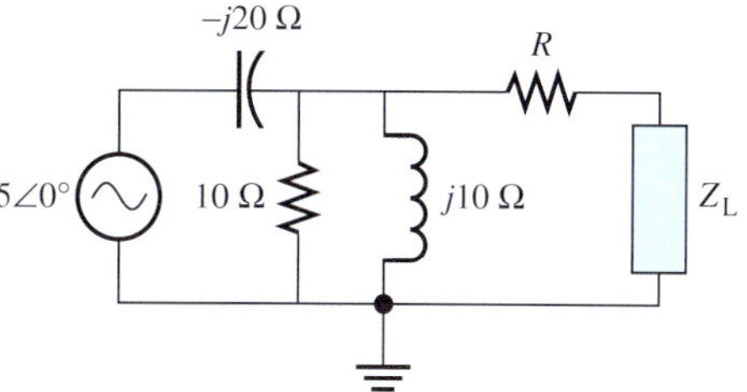

FIGURE P6.70 Circuit for Problem 6.70.

6.71 (H)* A lamp is rated "110 V, 15 W" and its working current is 0.7 A.

a. Find the power factor of the lamp.

b. What can be done to increase the lamp's power factor to 1? The voltage source has a frequency of w = 60 Hz. (*Hint:* Connect a capacitor in parallel with the lamp.)

SECTION 6.7 CIRCUIT STEADY-STATE ANALYSIS USING PSPICE

6.72 (A) A sinusoidal current $i(t) = 2\cos(120\pi t + 60°)$ flows through a 100-Ω resistance. Set up a PSpice schematic, plot the voltage across the resistance and determine the *rms* value of sinusoid voltage from the plot.

6.73 (B) If $i_S(t) = 2\cos(200t)$ in Figure P6.39, use PSpice to plot the current, I_R and I_C.

6.74 (H) In Figure P6.41, use PSpice to plot V_1 and V_2.

6.75 (A) Replace the sinusoidal voltage source to the sinusoidal current that is $i_S(t) = 5\cos(100t)$ in Figure P6.45, use PSpice to plot the Norton equivalent current flows through R_L.

TOPIC 11

Electrical Safety

The content of this topic is compiled from:

Chapter 15 (pp. 646–669), *Electrical Engineering: Concepts and Applications*

By S.A. Reza Zekavat

CHAPTER 15

Electrical Safety

(Used with permission from Jhaz Photography/Shutterstock.com.)

15.1 INTRODUCTION

The preceding chapters covered the various applications of electricity. This chapter discusses safety issues regarding electricity. In addition to electrical shock, this chapter addresses hazards relating to electromagnetic (radio) waves, electrical arcs, and explosive atmospheres. Moreover, the chapter presents the National Electric Code (NEC) and defines its role and regulations that promote electrical safety. The entire chapter can be considered as an application chapter for many concepts presented in this book.

15.2 ELECTRIC SHOCK

Electrical shock is caused by bodily connection to two points that have different electrical potentials. In other words, electric shock occurs when a person simultaneously touches two points with a voltage drop across them. The *ground* or the *Earth* can be one of the contacts. As a result, touching a live wire and the ground at the same time can deliver an electrical shock. Electric shock can be lethal when it passes through the heart or the head. Electric current passing through the head affects the brain, while current passing through the chest can stop the heart. For these reasons, if a worker needs to test a live circuit, one hand can be placed in the worker's pocket so that if something goes wrong, there is less likelihood the current will pass through the worker's heart.

15.2.1 Shock Effects

Alternating current (AC) is the main source of most electrocutions. AC current can interfere with the natural electrical pulse of the heart. This interference happens even at low voltage levels. The AC causes ventricular fibrillation, which can quickly lead to death. However, direct current can also be fatal at high current levels. With either AC or DC, the electrical *current* passing through the body is the main cause of death. Whether a shock victim lives or dies depends on how much current flows through his or her body. It only takes 100 mA or 0.1 A to disrupt a victim's heart. Low currents can send the victim's heart into ventricular fibrillation.

Higher currents can stop a victim's heart. However, it is more difficult to treat ventricular fibrillation than restarting a stopped heart. Therefore, low-current shocks can be even more dangerous than high-current shocks. Table 15.1 summarizes the effects of various current levels on the human body.

Electric shock is not the only hazard posed by electrical equipment. Electric energy generates electromagnetic radiation proportional to the magnitude of the current. This radiation can be hazardous, as covered in Section 15.3.2.

According to Ohm's law, the amount of current flowing through the body depends on the body's resistance. A low body resistance and a moderate voltage will produce a dangerously high current through the body. Factors that can lower bodily resistance include wetness or moisture on the skin, bare feet on bare earth (dirt), and broken or bleeding skin. Normal hand-to-hand resistance is about 300 kΩ. But the resistance with wet skin is less than 20 kΩ, which means the current is 15 times greater. Having bare feet on bare ground also gives a better electrical contact with ground, lowering resistance even further.

However, electrical contact to broken or bleeding skin yields the highest, most dangerous current levels. Most of the body's resistance to electricity is found in the skin. Due to the highly conductive blood, water, and nervous system, the internal resistance behind the skin is very low. As a result, the body's resistance is lower at any point of broken skin. Touching electrical contacts with bleeding wounds or stabbing electrified leads into the skin can cause death, even when the contact is with low-voltage DC because the resulting current is so high.

TABLE 15.1 Electric Shock Effects on the Human Body
Cadick, J. 1994. *Electrical Safety Handbook*, 1.4. New York: McGraw-Hill.

Current	Effect on Body	Potential for Fatality
<1 mA	None	None
1 mA	Perception threshold, mild sensation	None
1–10 mA	Mild to painful sensation	None
10 mA	Paralysis threshold, cannot release hand grip	Minimum level that has a slight risk of fatality
30 mA	Respiratory paralysis (cannot breathe)	Frequently fatal
75 mA	Fibrillation threshold 0.5%	Heart action disrupted, possibly fatal
250 mA	Fibrillation threshold 99.5%	Usually fatal (very high risk)
4 A	Heart paralysis threshold (heart stops)	Possible to revive victim; fatal if not intervened quickly
5 A	Tissue burning	Yes, if vital organs are burned

EXAMPLE 15.1 Fatal Voltage

If a person's body resistance is 100 kΩ, what voltage can potentially be fatal?

SOLUTION

The minimal current for a fatality is in the order of 0.1 A. According to Ohm's law:

$$V = 100\ \text{k}\Omega \times 0.1\ \text{A} = 10\ \text{kV}$$

EXAMPLE 15.2 Electric Current in a Swimming Pool

An electrical power cord falls into a swimming pool, as shown in Figure 15.1. If the pool water has the resistivity of 175 kΩ m, find the current, I, that flows through the water. Assume that each longitude cross section of the pool acts as an insulator. In this case, we can consider the simplified model shown in Figure 15.2 for the problem. Here, the two dashed planes represent the two sides of the water resistor. The separation of these two planes is assumed to be 2 cm.

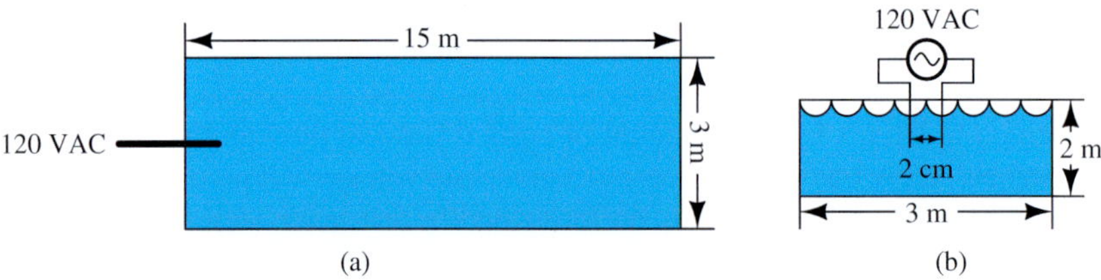

FIGURE 15.1 Swimming pool configuration for Example 15.2. (a) Overhead view; (b) cross-sectional view.

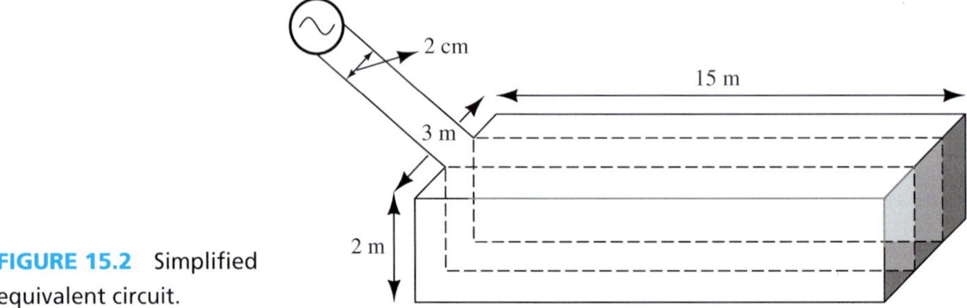

FIGURE 15.2 Simplified equivalent circuit.

SOLUTION

The electrical current flows through a conductor consisting of water. This conductor is 2 cm long and has a cross-sectional area of $15 \times 2 = 30\ \text{m}^2$.

$$R = \frac{175{,}000 \times 0.02}{30} = 116.7\ \Omega$$

Therefore,

$$I = \frac{120}{116.7} = 1.029\ \text{A}$$

Note that this current is 10 times the lethal limit that is in the order of 0.1 A (see Table 15.1). Also, note that the pool wall can act as a ground, causing current to flow through the entire body of water, and not just the "conductor" section.

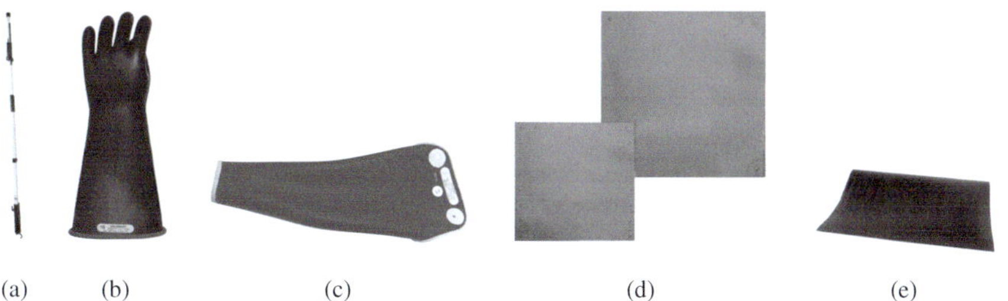

(a) (b) (c) (d) (e)

FIGURE 15.3 Safety tools: (a) hot stick; (b) insulating gloves; (c) insulating sleeve; (d) insulating blanket; and (e) insulating mat. (Photos courtesy of Salisbury by Honeywell.)

15.2.2 Shock Prevention

To prevent shock, prior to working on equipment, all power should be turned off if at all possible. If de-energizing the circuit is not practical, there are several things that can be done to significantly reduce the risk of shock. One typical safety measure is to wear rubber gloves and rubber sleeves to prevent electrical contact with the body. A rubber sleeve covers a worker's arms and shoulders. Use of rubber mats and rubber boots, specifically designed to serve as insulation, can also prevent a path to a ground. Also rubber blankets and covers can be used to temporarily cover energized equipment. Rubber line hose can be used as a temporary insulator to cover exposed power lines. Hot sticks (long, insulated sticks) can be used to handle energized wires and other equipment (see Figure 15.3).

Insulated tools are also available for use with energized equipment. Figure 15.3 shows some of this safety equipment. These are a few of the safety products available that are designed to reduce the risk of shock from working on live equipment.

15.3 ELECTROMAGNETIC HAZARDS

Radio frequency (RF) hazards are due to electromagnetic waves or radio waves. Electromagnetic waves are invisible (with the exception of visible light) and are used for all wireless communication applications, including radio, TV, cellular phones, wi-fi, and RADAR. RF hazards result from the emission of radio frequency radiation. A few common devices that emit radio frequency radiation include cell phones, microwave ovens, wi-fi, and two-way radios. Radio frequency energy can cause thermal effects such as heating and burns, and, at certain frequencies, the radiation can even cause cancer. There are two primary types of RF hazards—low-frequency and high-frequency RF hazards. High-frequency RF energy is much more dangerous than low-frequency RF energy.

15.3.1 High-Frequency Hazards

High-frequency RF radiation can cause cancer. One example of a source of high-frequency hazard is a radar transmitter. RADAR stands for RAdio Detection And Ranging. The RADAR acronym has become so common it has now simply become its own commonly recognized word, *radar*. These systems usually consist of a side-by-side transmitter and receiver. A radar transmitter transmits a high-power burst of RF energy. The reflection of this signal from targets is processed by the receiver for detection and localization purposes. Extended exposure to radar signals has been linked to a number of health problems, including leukemia and brain tumors. A 1953 study at Hughes Aircraft Corporation found "excessive amounts of internal bleeding, leukemia, cataracts, headaches, brain tumors, heart conditions, and jaundice in those employees working with radar [1]."

Exposure to high-frequency RF signals, above 10^{17} Hz, is another cause of health problems. These frequencies include gamma, beta, and alpha radiation, as well as some X-ray frequencies. Ionization causes electrons to be stripped from atoms. This can produce molecular changes, potentially leading to DNA and biological tissue damage. Ionizing radiation is well known to cause cancer, but exposure to this radiation is not likely in most workplaces.

RF energy can also cause thermal effects, the heating of tissue, and even burns. Exposure to high levels of RF energy can be harmful because it can heat biological tissue rapidly. Power density levels of 1 to 10 mW/cm^2 can result in measurable heating of tissue. Very high power levels, 100 mW/cm^2 or higher, can easily cause heating of tissue and increasing of body temperature. This is especially of concern with the eyes and testes, as these areas have less blood flow, and thus less ability to dissipate the extra heat.

There has been much debate over the health effects of using cell phones. Cell phones use frequencies of 850, 900, or 1850 MHz, which are in the microwave (RF) range for wireless communication. Cellular systems consist of two main components: (1) base stations (cell towers), which communicate with a number of mobile users that are located in their coverage area simultaneously and (2) mobile stations (users). The distance of base stations and their output power (which is typically less than 100 W) avoids any considerable effect on users. However, there is a debate on the relationship of cell phones and cancer.

Is there a link between cell phones and cancer? The transmitting power of a cell phone is often in the order of 1 W. However, when you hold the cell phone close to your head, it can affect your body. Some studies show that extensive use can cause change in a user's DNA, a precursor to cancer [2]. The Austrian Chamber of Physicians has claimed that long-term use can cause cancer and recommends that children should never use it [3]. Studies in 1999 and 2001 showed that long-term use could disrupt brain wave patterns [4] and even cause eye cancer [5]. There are several class-action lawsuits against the cell-phone industry, claiming that the devices can cause brain tumors.

However, other studies conclude that there is no cell phone–cancer link. The American Cancer Society denies any link. "The link between cell-phone use and cancer, especially brain cancer is a myth" says Ted Gansler, Director of Medical Content, American Cancer Society [6]. Anthony Swerdlow of the Institute of Cancer Research concludes that there is no substantial risk of cancer after the first decade of cell-phone use [7]. The general consensus between both sides is that more research is needed for a definite answer.

Cell phones give off radiation at all times when turned on, even when not in use. However, more radiation is emitted when talking or text messaging. Also, when far from a base station, more power is required for communication even when not using the phone. To determine how much radiation your phone emits, check its specific absorption rate (SAR). The SAR for each specific phone model can be found at http://www.fcc.gov/cgb/sar/.

Research has also suggested a link between carrying a cell phone in one's pocket and male infertility. Ashok Agarwal, Ph.D., Director of the Center for Reproductive Medicine at the Cleveland Clinic, recently completed a study in which cell phones were set down for 1 h in talk mode, next to sperm samples in test tubes. He found that the sperm's motility and viability were significantly reduced, and levels of harmful free radicals increased after exposure [8].

Radio and TV transmitters pose similar risks to cell phones. These use frequencies from 500 kHz to 1 GHz.

EXAMPLE 15.3 Power Density and High-Frequency RF Hazard

For a radio transmitter, the power transmitted is 100 kW and the antenna gain is 50. (a) If an employee is standing 2 m from the antenna, what is the power density of the radio frequency radiation passing through his body? (b) What if the employee stands 20 m from the antenna?

The power density is given by:

$$D = \frac{P_t \cdot G_t}{4\pi d^2}$$

where D is the power density (W/m^2); P_t is the transmitted power; G_t is the antenna gain; d is the distance (m).

SOLUTION

a. Considering a 2-m separation, the power density corresponds to

$$D = \frac{100{,}000 \times 50}{4\pi 2^2} = 99.472 \text{ kW/m}^2$$

b. Considering a 20-m separation, the power density is:

$$D = \frac{100{,}000 \times 50}{4\pi 20^2} = 994.72 \text{ W/m}^2$$

Notice that the extra distance in this scenario greatly reduces the power density. If the distance becomes 10 times greater, the power density would reduce by 100 times.

EXAMPLE 15.4 Power Density and RF Hazard

A person holds a cell phone 5 mm away from his head. If the phone is transmitting 3 W, and the antenna gain is 10, what is the power density at the surface of his head? Should this power density level be of concern?

SOLUTION

Using the equation introduced in Example 15.3:

$$D = \frac{P_t \cdot G_t}{4\pi d^2} = \frac{3 \times 10}{4\pi (0.005)^2} = 95.49 \text{ kW/m}^2$$

A power density of 100 mW/cm^2 is considered very high. This value must first be converted into mW/cm^2.

$$95{,}490 \text{ W/m}^2 \times \frac{1 \text{ m}^2}{100^2 \text{ cm}^2} \times \frac{1000 \text{ mW}}{1 \text{ W}} = 9549 \text{ mW/cm}^2$$

9549 mW/cm^2 is much greater than 100 mW/cm^2. So this radiation should be of great concern to the user. Holding a cell phone about 5 cm away from head could reduce this density to 95.4 mW/cm^2.

15.3.2 Low-Frequency Hazards

High-tension power lines emit RF radiation at a frequency of 60 Hz. Is this radiation dangerous? A 2001 California health department review found that living near high-voltage power lines resulted in "Added risk of miscarriage, childhood leukemia, brain cancer, and greater incidence of suicide [9]." But, ultimately, they added that risk was a "very small increased lifetime risk."

How close can people safely live to power lines? It depends on the voltage and current through the power lines. How can the voltage of a power line be determined? The "High Voltage" signs near power lines do not typically give the voltage. One way of determining the voltage is by observing the size and other characteristics of the transmission towers carrying the lines.

The values in Table 15.2 indicate that, for higher voltages, the towers are generally taller; however, this is not always the case. Note the overlapping as evidenced by the minimum and

TABLE 15.2 Typical Transmission Tower Specifications [10]

	Conductor Spacing (m)			Transmission Tower Height (m)		
Voltage (kV)	***Median***	***Min***	***Max***	***Median***	***Min***	***Max***
345	7	5	10	32	18	43
500	9	6	13	39	30	52
765–800	13	12	16	42	35	53

maximum heights. The conductors (wires) for higher voltages are usually spaced farther apart. The conductor spacing is the minimum distance between any two conductors on a tower. Therefore, the voltage level of a power line can be roughly determined by (1) the distance between the conductors and (2) the height of the tower. Higher voltage levels utilize larger transmission towers and wider conductor spacing. Figure 15.4 shows transmission towers bearing different voltages, and Figure 15.5 shows close-up views of the power lines.

As can be seen from Figure 15.5, some conductors are combined in multiple conductors. Figure 15.5(b) shows two-conductor pairs, and Figure 15.5(c) and (d) shows four conductors. Paired conductor lines carry more power as compared to single conductor lines, for example, four conductors can carry four times more current. As discussed in Chapter 9, current flows on the surface of the conductor. To maximize current, it is desired to maximize conductor surface area while minimizing the conductor size. Thus, more current can flow through four individual conductors than can flow through a single large one with a diameter equivalent to the combination of four individual conductors.

(a) (b) (c) (d)

FIGURE 15.4 (a) 12.47-kV line (Courtesy of Shah & Associates, Inc., www.shahpe.com.); (b) 345-kV line (Courtesy of Argonne National Laboratory.); (c) 500-kV line (Courtesy of © TebNad/Fotolia.); (d) 765-kV line (Courtesy of American Electrical Power.).

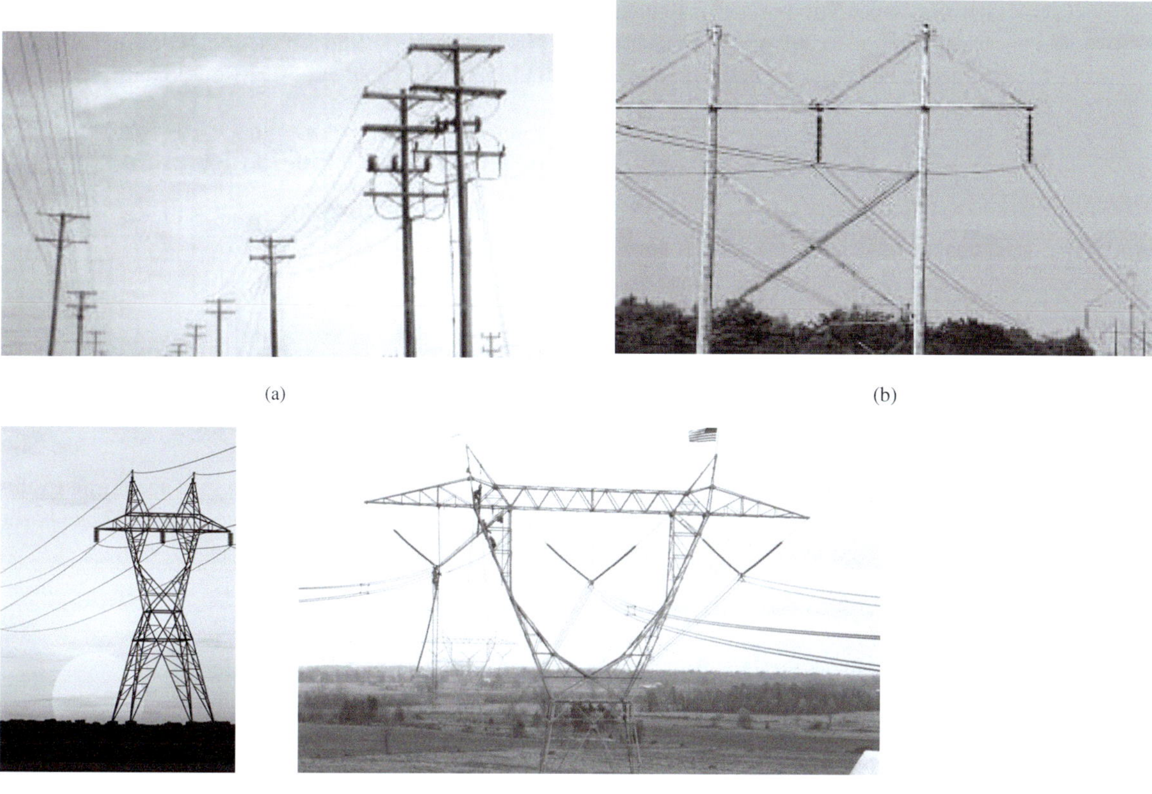

FIGURE 15.5 Close-up views of (a) 12.47-kV line (Courtesy of Shah & Associates, Inc., www.shahpe.com.); (b) 345-kV line (Courtesy of Argonne National Laboratory.); (c) 500-kV line (Courtesy of © TebNad/Fotolia.); (d) 765-kV line (Courtesy of American Electrical Power.).

The electromagnetic radiation is directly proportional to the current level through the power lines. Table 15.3 represents typical current ratings for different voltage power lines. Please note that these are the rated currents; the current at a given time more or less depends on the load on the wire at that time.

The magnitude of the magnetic field produced by a (straight) power line is given by:

$$B = \frac{\mu_0 \times i}{2\pi r} \tag{15.1}$$

where B is the magnitude of the magnetic field, in tesla (T); μ_0 is the magnetic permeability of free space; i is the current (A); r is the distance from the conductor (m).

The strength of the magnetic field is measured by a device called a magnetometer (also called a guassmeter). A typical magnetometer is shown in Figure 15.6.

How much radiation is safe? In the United States, no national limit has been set. The US Federal government does not acknowledge electromagnetic radiation as dangerous and, therefore,

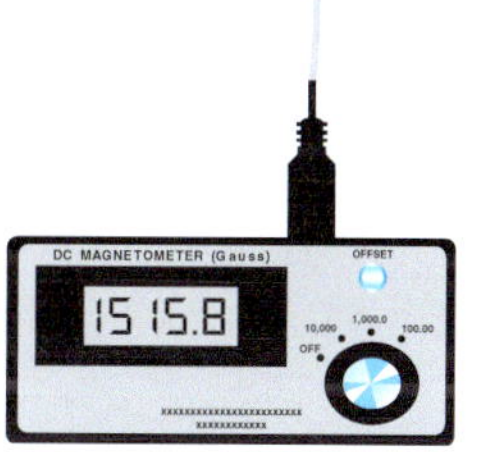

FIGURE 15.6 Typical magnetometer. (Photo courtesy AlphaLab, Inc.)

TABLE 15.3 Current Levels Through Different Power Lines

Voltage (kV)	Current (A)	Safe Distance (m)
12.47	100	~5
345	1000–2500	25
500	2000–3000	30
745	2500–4000	40

has not issued any regulations concerning building near power lines. However, some US states limit radiation to 200 mG. One tesla equals 10,000 Gauss. The government of Sweden has stricter limits on health issues and has set the maximum magnetic field exposure limit to only 2.5 mG.

EXAMPLE 15.5 Power Line Hazard; Magnetic Field Calculation

If a house is placed 50 m from a 500-kV power line bearing 3000 A, what is the strength of the magnetic field? Is this field safe (according to the US state standards)?

SOLUTION

Using Equation (15.1):

$$B = \frac{4\pi \times 10^{-7} \times 3000}{2 \times \pi \times 50} \times 10{,}000 = 120 \text{ mG}$$

The maximum allowable in some states is 200 mG. This value is safe according to this specification. However, the Swedish maximum is 2.5 mG. This value greatly exceeds the Swedish maximum.

EXAMPLE 15.6 Power Line Hazard; Magnetic Field Calculation

Find the maximum magnetic field magnitude directly under a 745-kV line.

SOLUTION

From Table 15.2, the minimum height of a 745-kV transmission tower is 35 m. From Table 15.3, the maximum current through this line is 4 kA. Therefore,

$$B = \frac{4\pi \times 10^{-7} \times 4000}{2 \times \pi \times 35} \times 10{,}000 = 228.6 \text{ mG}$$

EXAMPLE 15.7 Power Line Hazard; Magnetic Field Calculation

According to the Swedish standard, what is the minimum distance homes should be placed from a 12.47-kV line rated at 500 A? As 12.47-kV lines are often located in or near neighborhoods, is this distance practical?

SOLUTION

Using Equation (15.1) and the Swedish standard of 2.5 mG:

$$B = \frac{4\pi \times 10^{-7} \times 500}{2\pi r} \times 10{,}000 = 2.5 \text{ mG}$$

The safe distance, r, can then be calculated from this equation:

$$r = 400 \text{ m}$$

The safe distance, 400 m, is not practical because city lots are generally less than 100 m in length. Even if a house is located in the center of the lot and the power lines run along the property line at the rear of the lot, there is not enough room to implement this standard. Accordingly, the lines are placed under the ground. The ground reduces the RF intensity.

EXERCISE 15.1

It is notable that in some cities, it is required that the power lines are located under the ground. This improves the sight by removing the towers; it may save installation costs and reduce the radiation effects. You are encouraged to investigate how installing the cables under the ground would impact their radiation. Typically, considering the lines are located 1 m under the ground, calculate the magnetic field at the distance of 10 to 50 m.

15.3.3 Avoiding Radio Frequency Hazards

Here are some tips to avoid RF hazards:

- Use a "hands-free" device instead of holding a cell phone directly to your head.
- Whenever possible, store your cell phone away from your body to minimize exposure.
- Avoid standing near radio antennas and transmitters.
- Do not live in close proximity to high-tension power lines.

15.4 ARCS AND EXPLOSIONS

Explosions in any context can be a serious safety concern. Safety precautions can help to reduce or eliminate the risk of explosions. Because this chapter covers electrical safety, it will focus mostly on electrical causes of explosions and how to prevent them. Explosions are caused by a mixture of explosive materials and an ignition source. The most common electrical cause of explosion ignition is arcs. To reduce explosion hazards, both the amount of explosive materials in the atmosphere and the risk of arcs can be reduced.

Factors contributing to an explosive atmosphere include flammable gasses and organic compounds. These materials, when released into the air (above certain concentrations) may pose an explosion hazard if exposed to an ignition source. These materials include (but are not limited to) acetylene, ethyl ether, gasoline, alcohols, acetone, methane, propane, and hydrogen. It is also important to note here that high concentrations of oxygen increase the risk of fire and explosion. While oxygen itself cannot explode, it is required for any combustion to occur. Increased oxygen levels result in increased potential for combustion and explosions.

15.4.1 Arcs

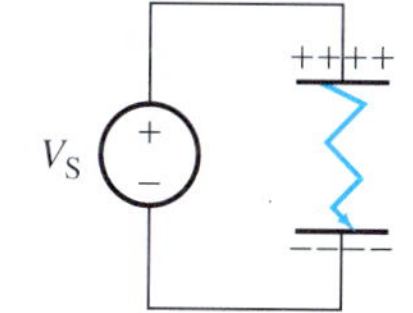

FIGURE 15.7 Electrical causes of arcs.

An explosive atmosphere does not result in an explosion unless an ignition source is available. Arcs and sparks are common ignition sources. A spark is basically a short-lived arc. An arc occurs when a substantial amount of current flows through superheated air. The energy flowing through the air causes the air to break down and forms plasma, which is an ionized gas, that is, an electrical conductor. The current then flows through this plasma mixture consisting of a vaporized electrical conductor and ionized air particles.

An electric arc occurs when two conductors having a large potential difference are located in close proximity, as shown in Figure 15.7.

High voltage causes the gas between the conductors to ionize, allowing the current to arc through it, resulting in a spark. Lightning (Figure 15.8) is an example of a large arc.

FIGURE 15.8 Lighting is a large arc. (Used with permission from Jhaz Photography/Shutterstock.com.)

Arcs release extremely high levels of thermal energy. An arc can generate temperatures as high as 50,000 K at its terminals. These high temperatures can create fatal burns at distances of more than 8 ft from the arc. The amount of energy (and heat) in an arc is proportional to the maximum short-circuit power at that point of the circuit. Low-voltage systems are just as susceptible to arcing as high-voltage systems.

When ionized, oxygen (O_2) is converted to ozone (O_3). The unique smell caused by sparks is the ozone, that is, the result of the spark. While ozone is toxic, minimal amounts are produced by most sparks, and it quickly decays back to oxygen (O_2). So, ozone generated by sparks does not pose any health hazard. Ionization is also caused by the electric field around power lines. This effect is called *corona discharge*. The crackling or buzzing sound typical of power lines is caused by corona discharge. While this effect does not present a health hazard (corona discharge is used to generate ozone by air purifiers), it is undesirable in power lines because it results in energy loss.

EXAMPLE 15.8 Arc Power

Suppose an arc occurs across the circuit breaker between points A and B in Figure 15.9. (Assume the breaker is open.) (a) What is the maximum power available to the arc? (b) What is the maximum power of an arc between A and the ground?

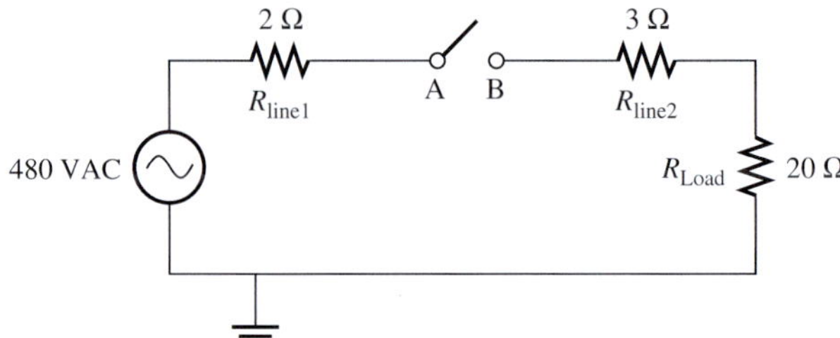

FIGURE 15.9 Circuit for Example 15.8.

SOLUTION

a. An arc across points A–B acts as a short circuit across these points. The maximum current through this short circuit is simply the current through the series circuit.

$$I_{SC} = \frac{480}{2 + 3 + 20} = 19.2 \text{ A}$$

The maximum voltage occurs when the switch is open, that is, 480 V. As a result, the maximum power is:

$$P_{max} = 480 \times 19.2 = 9.22 \text{ kW}$$

b. Since an arc between point A and the ground acts as a short circuit between these two points, the current only flows through the 2 Ω line resistance.

$$I_{SC} = \frac{480}{2} = 240 \text{ A}$$

The maximum voltage is 480 V. Therefore,

$$P_{max} = 480 \times 240 = 115.2 \text{ kW}$$

A very common electrical cause of arcs is transients in high-voltage circuits. A transient can occur during the changing of the state of a circuit, such as by flipping a switch or adding a resistor. A simple switch and light bulb circuit (such as a ceiling light or a flashlight) has small inductances contained in both the connecting wires and the bulbs (see Figure 15.10). Recall from Chapter 4 that an inductor resists changes in current. If the current changes, the inductor tries to maintain the current at the previous level.

When a switch is turned off, the inductances try to maintain the previous current level. However, because the circuit is open, this is not possible. Ohm's law states that $V = I \times R$. Opening the switch changes the circuit resistance to infinity ($R = \infty$). Because the inductor

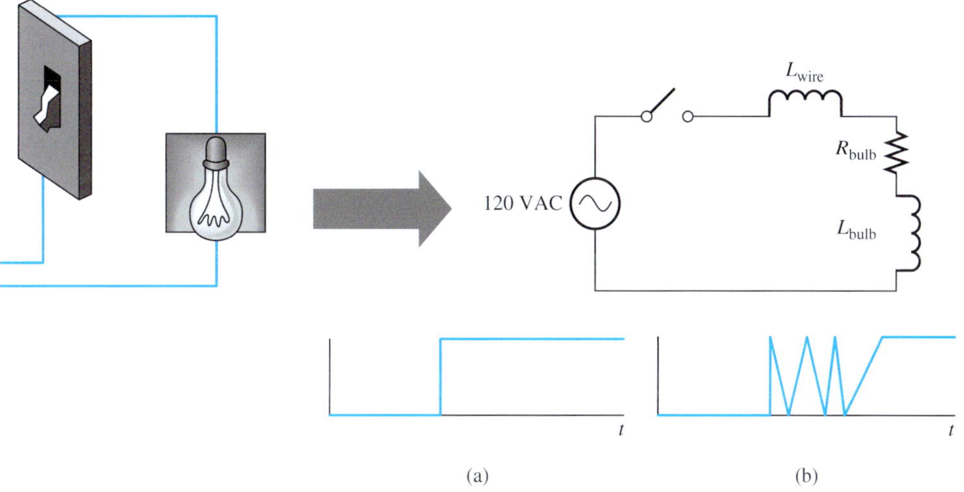

FIGURE 15.10 Equivalent model for a simple circuit shown to the left.

FIGURE 15.11 (a) Ideal switch response; (b) switch exhibiting switch bounce.

tries to maintain the current at the previous level, it produces a brief voltage across the circuit of $I \times R = I \times \infty$. This voltage appears across the break in the circuit, that is, across the switch. Therefore, the inductances in the circuit cause a brief spike in the voltage across the switch contacts as they open, which can generate sparks.

A switch may also create arcs when it is closed. This is due to the switch bounce effect. Because a switch has mechanical contacts, it doesn't turn on immediately, but fluctuates between off and on for a brief period of time (on the order of microseconds). For more details on switch bounce, see Examples 5.2, 10.18, and 14.10. This rapid fluctuation, in combination with the closeness of the contacts, can create sparks. Figure 15.11 compares the ideal switch response with what is observed in switch bounce.

A switch can also create a spark when it closes. However, the sparks when it opens are usually larger than those occurring when it closes due to the high voltage involved. In high-voltage systems, the narrowing gap between the switch contacts may allow current to arc across the gap, due to the high voltage. For this reason, when power is reconnected to a large city, it is not connected to the entire city all at once. Rather, it is reconnected sequentially to one small subset of the city at a time. This reduces the current load handled by each reconnect switch and, therefore, the maximum arcing power.

Switches are not the only cause of electric sparks. Sparks can occur at any moving contact and even in fixed contacts. Fixed contacts include motor contacts, screw terminals, wire connections, and contacts in outlets. Welding equipment by its very nature is a great cause of sparks and should not be used anywhere near explosive atmospheres. Another cause of sparks is friction between moving surfaces.

15.4.2 Blasts

Arcs can cause *blasts* or electrical explosions. An arc superheats air instantaneously. This causes a rapid expansion of air, resulting in air pressures of 100 to 200 lb/ft^2. Note that an explosive atmosphere is not required for a blast. A blast can occur without any outside fuel. However, an atmosphere containing explosive elements will make the blast more severe. Also note that not all arcs are accompanied by a blast.

While sparks are a common trigger of explosions, explosions can also be ignited by heat and fire sources, such as cigarettes, matches, candles, and pilot lights.

15.4.3 Explosion Prevention

To prevent explosions, an explosive atmosphere must be avoided. Proper ventilation can be used to remove explosive material from the air. Also, it is important to keep flammable materials

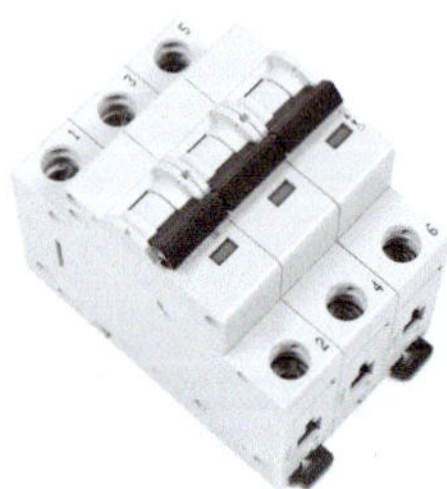

FIGURE 15.12 Safety switches (© tarczas/ Alamy).

tightly sealed to prevent the release of fumes resulting in an explosion hazard. This reduces or eliminates the fuel required for an explosion. To reduce the explosion risk, it is also necessary to prevent sparks or keep sparks that can ignite an explosion and the explosive environment separate. Because it is impossible to prevent the sparks from occurring across switch contacts, the sparks must be isolated from the surrounding environment. To do this, safety switches are used to isolate the switch contacts from any nearby explosive environment.

A *safety switch* is a special spark-resistant switch that must be used where explosion hazards exist (see Figure 15.12). The switch has completely enclosed contacts, which isolate any sparks from the outside environment. This prevents transient sparks from igniting explosions.

In conclusion, colocated explosive materials and ignition sources of explosions should be avoided. The ignition source is usually a spark. To reduce this explosion hazard, both the explosive material and the spark hazard must be reduced.

15.5 THE NATIONAL ELECTRIC CODE

The National Electric Code (NEC) provides standards and requirements regarding electrical wiring and safety in the United States. It is published by the National Fire Prevention Association (NFPA). The NEC is not a law, but its use is commonly mandated by state and local codes. It is intended to make use of electricity safer by reducing or eliminating most electrical safety hazards. Electricians are required to follow the NEC when wiring a building. When they are finished, a building inspector ensures that the wiring is "up to code." If it is not, it must be fixed before the building is occupied.

The NEC was established in 1897. The NEC focuses on electric utilization systems and does not address generation, transmission, or distribution systems. The NEC also does not apply to ships, trains, mines, or communications equipment that are under the control of a communication utility.

The NEC covers all industrial, commercial, and residential electric systems. The NEC covers four basic installation types: electrical conductors within or on public or private buildings or structures, including recreational vehicles and mobile homes, conductors and equipment that connect to electric supplies, any other outside conductors and equipment, and fiber optic cable [12].

15.5.1 Shock Prevention

The NEC offers guidelines aimed at preventing shocks by requiring proper polarization, grounding, and requiring special outlets called *ground fault circuit interrupters* (GFCIs) or *arc fault circuit interrupters* (AFCI) (discussed in Section 15.5.2.2) in certain areas.

15.5.1.1 POLARIZATION

Figure 15.13 shows a typical outlet used in the United States. You can see that the neutral slot is longer than the hot slot. This is called *polarization*. It is intended to allow a plug to be inserted into the outlet only in one way. By its nature, AC is not polarized, because it is constantly reversing direction. With this in mind, what is the purpose of polarization? Some electrical devices, if they do not have a grounding prong, use the neutral or return as the ground. The neutral wire is grounded at the circuit breaker panel. Polarization ensures that the internal ground is not connected to the hot conductor to better avoid any shock hazard.

FIGURE 15.13 Slots on a standard outlet in the United States. (Used with permission from © B.A.E. Inc./Alamy.)

APPLICATION EXAMPLE 15.9 Polarization

Suppose a man touches a metal toaster and is shocked. The toaster, having a polarized plug, is inspected, and nothing is found amiss. The inspector then looks at the outlet. He measures 120 V between the long slot and ground, and 0 V between the short slot and ground. What is the problem?

SOLUTION

The long slot is the neutral connection. This slot should be connected to ground at the circuit breaker panel, thus having a voltage of 0 V with respect to ground. The short slot is the hot connection and should have a voltage of 120 V with respect to ground. The inspection has revealed that the measurements are the opposite of what they should be. Therefore, the outlet is wired backward, with the hot and neutral wires swapped.

15.5.1.2 EQUIPMENT GROUNDING

Equipment grounding connects the circuit of a piece of equipment electrically to the ground, or the earth. The NEC details the proper methods to achieve this ground connection. Some of these ways include putting a rod into the ground or clamping a wire to a cold water pipe, which ultimately makes its way back to the ground. The NEC must be followed to ensure that the ground connection works properly. The outlets must be wired properly so that their ground slot is actually connected to a ground. Figure 15.13 shows the grounding slot on a standard outlet. This slot is used to connect metal cases and chassis of electrical devices such as computers, refrigerators, and microwaves to the ground.

The electrical wiring is grounded at one point, typically at the main circuit breaker panel. This serves to prevent ground loops (see Section 11.4). The return or neutral conductor is also grounded along with the ground conductor. It is typical to select the return path conductor as the reference node (ground) in electronic circuits (see Section 3.5).

Grounding prevents electric shock by redirecting loose current (e.g., from a broken wire resting on a metal case) to the ground. Because the resistance through the ground connection is much lower than the connection through your body to ground, the current will take the path of least resistance and travel through the ground wire to ground, and not through you.

APPLICATION EXAMPLE 15.10 Grounding

Figure 15.14 presents an example of how grounding works. The black wire (which is the incoming or hot wire) has broken loose and is touching the metal case. When proper grounding is used, this stray current is redirected to the ground (as shown by the dashed arrows) and the shock hazard is eliminated. When there is no grounding or grounding is not properly implemented, the stray current is not redirected, the housing is energized, and an unseen shock hazard is created. An example of this is using a grounding adapter to plug into a two-slot outlet and not connecting the grounding tab to the screw (see Example 15.12).

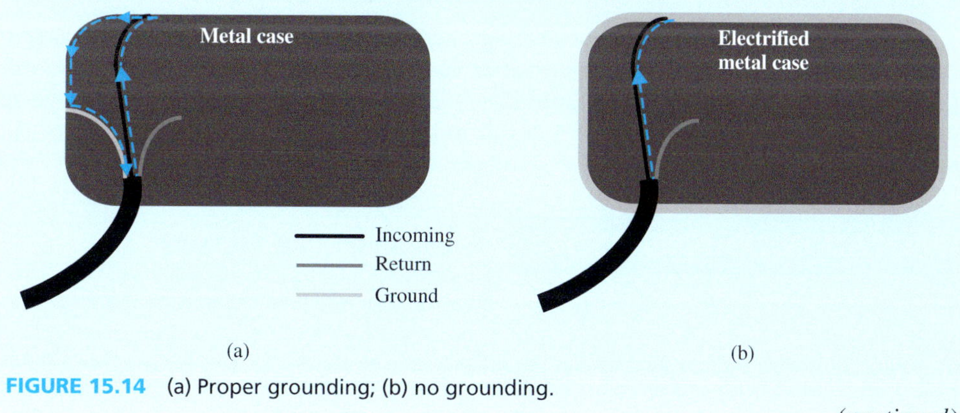

FIGURE 15.14 (a) Proper grounding; (b) no grounding.

(*continued*)

APPLICATION EXAMPLE 15.10 Continued

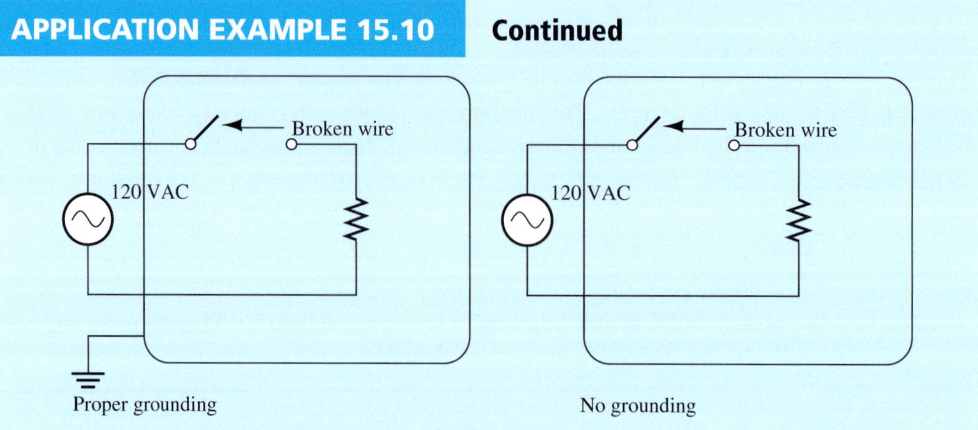

FIGURE 15.15 Schematic diagram of proper grounding.

Figure 15.15 shows the equivalent circuit for Figure 15.14. The stray current from the broken wire is redirected to the ground. Since the ground is at a zero potential (0 V), the loose current flows through it. Normally, this loose current is powerful enough to blow a fuse or trip a circuit breaker. This interrupts the current and alerts the user to the problem. However, when there is no ground, the power stays connected, leaving the operator unaware of a potentially fatal shock hazard. Proper grounding is used to eliminate this situation.

APPLICATION EXAMPLE 15.11 Checking for a Proper Ground

A voltmeter can be used to verify a proper ground. First, measure the line voltage by inserting the probes in the two vertical slots [see Figure 15.16(a)]. This serves two purposes. First, power to the outlet is verified and, second, the voltage can be observed for reference purposes. Now, insert a voltmeter probe in the incoming slot (the shorter slot), and in the ground slot [Figure 15.16(b)]. If the ground is working properly, the measured voltage should be very close to the line voltage.

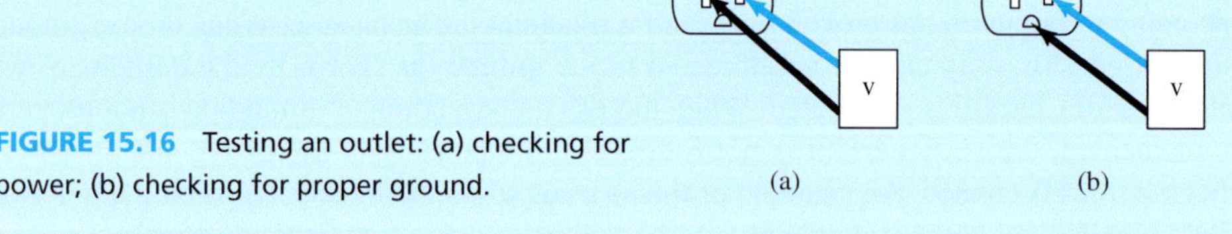

FIGURE 15.16 Testing an outlet: (a) checking for power; (b) checking for proper ground.

In older buildings, outlets may not have a ground slot. However, these outlets may have an internal ground connection. To test for this connection, measure the voltage between the incoming slot and the faceplate screw (see Figure 15.17). If this voltage matches the line voltage, an internal ground is available. Example 15.12 explains how to safely connect a grounded plug to a two-slot outlet.

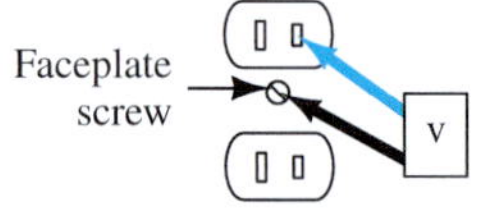

FIGURE 15.17 Checking for ground on older outlets.

APPLICATION EXAMPLE 15.12 Grounding Older Systems

If it is necessary to plug a three-prong, grounded plug into a two-slot outlet, a device called a *grounding adapter* can be used. This adapter has a three-slotted outlet on one side, and a two-prong plug with a grounding tab on the other side (see Figure 15.18). Before using this adapter, it is necessary to verify the existence of an internal ground (see Example 15.11).

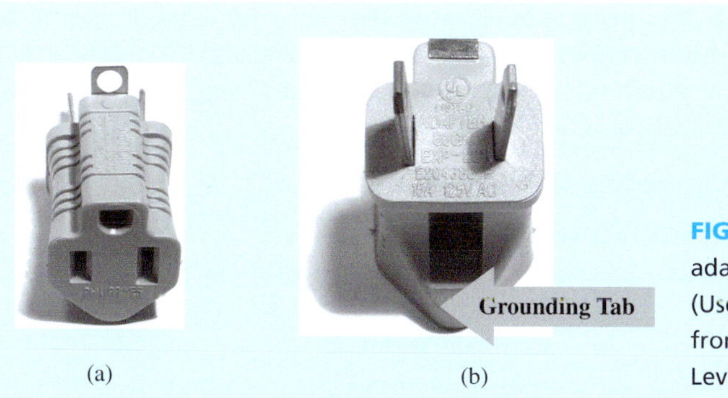

FIGURE 15.18 Grounding adapter: (a) front; (b) rear. (Used with permission from Alexander Remy Levine/Shutterstock.com.)

To use the adapter, first remove the faceplate screw from the outlet. Plug in the adapter and replace the screw, using it to connect the grounding tab to the outlet. The outlet must have an internal ground, and the tab must be connected to the screw to ensure proper grounding. If either condition is not fulfilled, the grounding has been defeated, and a shock hazard can be present. If the building wiring lacks an internal ground, it is highly recommended that the wiring be completely replaced.

Note that some devices such as CMOS (see Chapter 8) need careful grounding protection. They are very sensitive to the static electricity of the body. Therefore, when working with these devices, it is highly suggested to connect the body (e.g., using a wristband) to the ground.

Note that similar electronic devices are put and sold in conductive bags in order to prevent electric shock from body. A conductive bag keeps the electric charges in the outside layers of the bag and avoids harming the electric device inside. Similarly, in some workshops, conductive mats are used to conduct the electric charges of the body to pass through to the ground and avoid hazards for the devices.

EXERCISE 15.2

Airplanes are subject to high-voltage electric shocks due to lightening or passing through clouds that carry electric charges [11]. Explain why passengers are safe inside of the airplane.

15.5.1.3 THE GROUND FAULT CIRCUIT INTERRUPTER

A *ground fault circuit interrupter* (GFCI) is a special kind of circuit breaker that the NEC requires in certain locales to prevent shocks. This device is most commonly available as an electrical outlet, which is readily distinguished by its "Test" and "Reset" buttons. Figure 15.19 shows a standard GFCI outlet with its "Test" and "Reset" buttons.

FIGURE 15.19 GFCI outlet. (Used with permission from © Ted Foxx/Alamy.)

A GFCI compares the current levels in both the incoming and return conductors. If there is even a minor difference between these levels, the GFCI shuts off the current immediately. It can shut off current in a fraction of a second. This can increase the safety of the outlet. For example, if you come into contact with electricity, some of the current flows through your body. The current tries to find some other path to make the potential equal to zero rather than going through the return wire in the circuit. Kirchhoff's current law says that the current flowing into and out of a node must sum to zero. If the incoming and return currents do not sum to zero, the electricity is taking an additional path (i.e., through your body). Thus, a properly functioning GFCI will shut off the current fast enough to prevent tissue damage from the shock. Example 15.13 shows how a GFCI works.

Even though it is a valuable safety device, the GFCI is not foolproof. While it prevents shocks between the incoming and ground connections, it doesn't prevent shocks between the incoming and returning conductors (unless there is a path to ground). For example, if you touch the incoming conductor with one hand and the returning with the other, while wearing rubber boots, the GFCI will not trip and you will be shocked (see Example 15.14).

The NEC requires GFCI protection in areas where electricity is used close to water or dampness. Water on skin reduces its resistance, increasing the conductivity and the potential for fatal shocks. Some of these locations include outdoors, bathrooms, garages, kitchens, unfinished basements, crawlspaces, and near pools and hot tubs.

APPLICATION EXAMPLE 15.13 Ground Fault Circuit Interrupter

A GFCI monitors the currents I_{in} and I_{out} measured by ammeters A_1 and A_2, respectively. Under normal circumstances, $I_{in} = I_{out}$. If the current is redirected through a body to a ground, the current in will not be equal to the current out. When this happens, the GFCI will "trip," and shut off current to the circuit. For the circuit shown in Figure 15.20, will the GFCI trip?

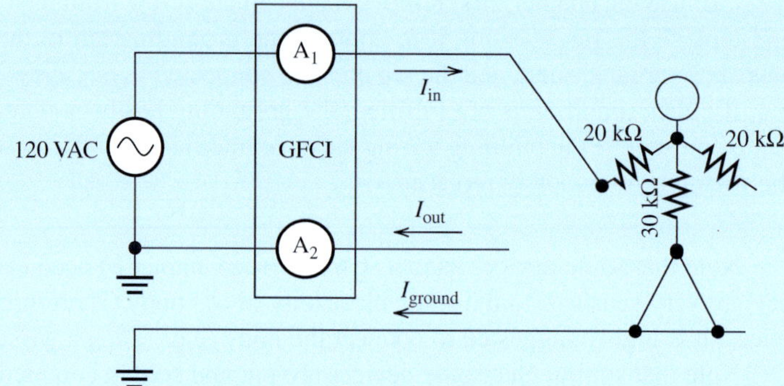

FIGURE 15.20 GFCI for Example 15.13.

SOLUTION

By inspection, it is observed that the entire current, I_{in}, passes through the person's body, and none of it passes through the return wire. A_1 measures a current, while A_2 measures no current (0 A). Therefore, the GFCI will trip.

Note:

If the person in Figure 15.20 does not touch the wire, there would not be any current to the ground, and both I_{in} and I_{out} will be equal to 0 A. Therefore, in this case, the GFCI would not trip.

APPLICATION EXAMPLE 15.14 GFCI

An individual comes into contact with the incoming and return conductors of a GFCI protected outlet, as shown in Figure 15.21. He is wearing rubber boots, which isolate his feet from the ground. (a) Find the currents I_{in}, I_{out}, and I_{ground}. (b) Does the GFCI trip?

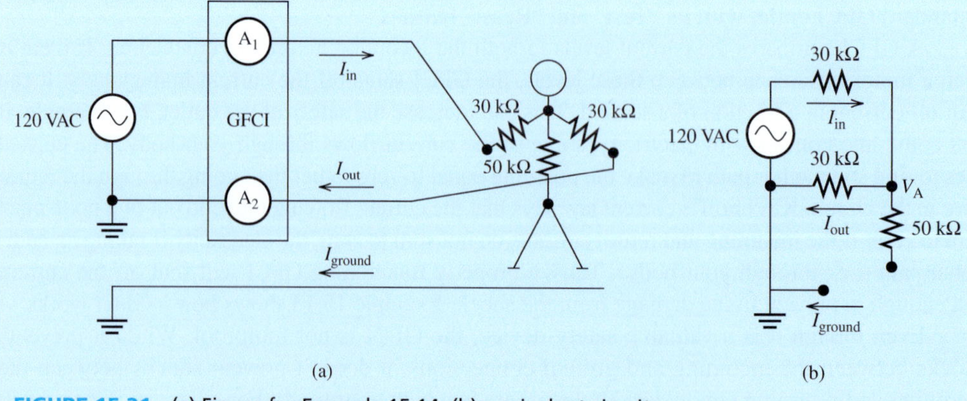

FIGURE 15.21 (a) Figure for Example 15.14; (b) equivalent circuit.

SOLUTION

a. The circuit can be modeled as shown in Figure 15.21(b). In this situation, because the person is wearing rubber shoes, and the resistance of those shoes is almost infinity, no current flows through the 50-kΩ resistor. Applying KVL:

$$\frac{V_A - 120}{30{,}000} + \frac{V_A}{30{,}000} = 0$$

Using this equation:

$$V_A = 60 \text{ V}$$

Now:

$$I_{in} = I_{out} = \frac{V_A}{30{,}000} = 2 \text{ mA}$$

b. Because I_{out} equals I_{in}, the GFCI does NOT trip; thus, the person in this example gets shocked as the current flows through his body.

EXERCISE 15.3

Repeat Example 15.14 assuming the resistance of the shoes is not infinity, rather assume it is about 100 kΩ.

15.5.2 Fire Prevention

The NEC also seeks to reduce electrical fire hazards [13]. This is mainly achieved by preventing over-current and arcs. These topics are discussed in the following subsections.

15.5.2.1 OVER-CURRENT PROTECTION

Excessive current causes conductors to dissipate more power. This leads to extreme heat in the conductors, which can ignite surrounding material. The NEC regulations, therefore, attempt to reduce or eliminate the risk of this over-current. To protect from over-current, the NEC stipulates the proper wire gauge that must be used in specific situations. The wires must have a diameter capable of handling the current and avoiding overheating and, therefore, fire. Common wire gauges and their current capacities are shown in Table 15.4. Note that in the American Wire Gauge system, a smaller number means a larger wire diameter.

Extension cords can overheat when the wrong gage of wire is used. Recall that:

$$R = \rho \times \frac{l}{A} \qquad \textbf{(15.2)}$$

When the ratio of length to cross-sectional area is unfavorable, R increases. The dissipated power corresponds to $P = I^2 \times R$. Thus, an increase in R is equivalent to higher voltage drop and higher dissipated power in the extension cord.

TABLE 15.4 Common Wire Gauges

Wire Gauge	Uses	Current Capacity
12	In-wall wiring	20 A
14	In-wall wiring	15 A
16	Extension cords, power strips	10 A
18	Most power cords	7 A

EXAMPLE 15.15 Over-Current

Three devices rated at 1000 W, 5 A, and 200 W, respectively, are connected to a 120-V, 14-gauge line on a 15-A circuit breaker. (a) Does this make the circuit breaker trip? (b) Does the wire overheat?

SOLUTION

It is clear that devices are connected in parallel in buildings. Note that if devices were connected in series, a disconnection in one device would lead to the disconnection of the circuit and loss of power to all devices. In addition, connecting in series would make it impossible to serve all loads with the same voltage.

a. Due to the parallel connection of devices, the total current is a sum of all currents. The currents through the 1000-W and 200-W devices respectively correspond to:

$$\frac{1000}{120} = 8.33 \text{ A}$$

and

$$\frac{200}{120} = 1.67 \text{ A}$$

Due to the parallel connection of devices, the total current is:

$$i_{\text{total}} = 8.33 + 1.67 + 5 = 15 \text{ A}$$

By comparing 15 A with the circuit breaker and wire ratings, it can be determined that the circuit breaker will trip.

b. Because the circuit breaker trips, the wire does not overheat, and no fire hazard is present.

To minimize over-current, the NEC mandates specific requirements on the number of outlets in a room. More specifically, it defines a minimum number of outlets per wall. If a wall is more than 4 ft wide, it must have an outlet. Outlets must be spaced a maximum of 10 ft from each other along each wall. This is intended to eliminate the need for extension cords, because most electrical devices have 6-ft cords and—with this spacing requirement—can reach an outlet regardless of their position along the wall. Extension cords can overheat, get frayed and damaged, and ignite electrical fires.

The NEC also specifies a maximum number of outlets per circuit. A *circuit* includes everything connected to a certain circuit breaker in the circuit breaker box. Too many outlets on one circuit can encourage the user to overload the circuit with too many electrical devices. The NEC restricts the number of outlets to around eight on a standard circuit.

Circuit breakers and fuses are over-current prevention devices. They disconnect current when it exceeds the maximum rated value. A fuse is a strip of metal enclosed in a glass case that melts or *blows* when the maximum current level is exceeded. A circuit breaker is a special kind of switch that *trips* when high currents flow through it. Fuses are destroyed when their maximum level is exceeded and, therefore, must be replaced. On the other hand, circuit breakers trip (move to an "off" position) and are simply reset. There is no need to replace a circuit breaker when it trips.

A *short circuit* occurs when the line conductor makes contact with the return conductor. Because the resistance of the wires is very low (nearly 0 Ω), the current may spike to dangerously high levels. An overload occurs when too many devices are plugged into one circuit, drawing more current than the circuit can handle. Either of these situations can generate a fire. Therefore, when the current exceeds the maximum level of the over-current device, it shuts off, preventing overheating and fire.

Important Note: Contrary to popular belief, circuit breakers and fuses do not prevent electric shock. This is because they are rated in the ampere range (commonly 15 or 20 A). Recall that only 0.1 A is required to be fatal.

15.5.2.2 THE ARC FAULT CIRCUIT INTERRUPTER

The arc fault circuit interrupter (AFCI) is a special circuit breaker that detects arcs in electrical circuits. Recall from Section 15.4.1 that an arc occurs when a substantial amount of current flows through superheated air. The 2005 NEC requires AFCI's in circuits feeding bedroom outlets. The 2008 NEC code requires AFCIs in circuits feeding outlets in "family rooms, dining rooms, living rooms, parlors, libraries, dens, bedrooms, sunrooms, recreation rooms, closets, hallways, or similar rooms or areas…". They are not required in areas where GFCI protection is required.

15.6 WHAT DID YOU LEARN?

- Electric shock is caused by the contact of two points with different electric potential.
- The level of electrical *current* passing through the human body is the main cause of fatality in the case of shock.
- High-frequency RF radiation can cause cancer. High RF energy can burn biological tissue.
- There is currently a debate over the potential level of radiation hazard presented by cell phones.
- High-tension power lines produce a magnetic field in their vicinity. Exposure to strong magnetic fields (>200 mG) is not safe.
- Explosive atmospheres are caused by flammable substances in the air.
- Electric sparks can cause explosions. Sparks need to be isolated from dangerous atmospheres.
- A *safety switch* is a special spark-resistant switch that can prevent explosion.
- The NEC is a set of regulations that intends to make the use of electricity safer.
- An outlet with *polarization* has a neutral slot that is longer than the hot slot. By doing this, the internal ground (with a longer lead) is not connected to the hot wire and a shock is avoided.
- All equipment should be electrically connected to a ground.
- A ground fault circuit interrupter (GFCI) is required in areas where electricity is used close to water or dampness.
- Fuses and arc fault circuit interrupters (AFCIs) are used to prevent over-current, which results in fire hazards.

References

1. Maisch, D. 2001. Mobile phone use: it's time to take precautions. *Journal of the Australasian College of Nutritional and Environmental Medicine* 20(1): 4.
2. Wired News. 20 Dec 2004. Lab tests show mobile-phone risk. http://www.wired.com/gadgets/wireless/news/2004/12/66097. Accessed 8 July 2008.
3. IOL. 30 Aug 2005. Physicians warn of cellphone cancer risk. http://www.int.iol.co.za/index.php?set_id=1&click_id=31&art_id=qw112540806283B223. Accessed 8 July 2008.
4. van Grinsven, L. 2001. Cellphones may harm by speeding up the brain, *IOL*. 21 Sept 2001. http://www.int.iol.co.za/index.php?set_id=1&click_id=31&art_id=qw1001086502683B243. Accessed 8 July 2008.
5. Stang, Andreas, A. et al. 2001. The possible role of radiofrequency radiation in the development of uveal melanoma. *Epidemiology* 12(1). http://www.epidem.com/pt/re/epidemiology/pdfhandler.00001648-200101000-00003.pdf;jsessionid=DDKXjIV2hVRM74W17Nr71fnL7FkpEbC9POtSIBM23dGj9A2EaL6l!-1660146838!-949856145!9001!-1. Accessed 8 July 2008.
6. Brown, K. 2005. Residents worry, but companies say cellphone antennas are safe. *Queens Chronicle* 6 Oct 2005. http://www.zwire.com/site/index.cfm?newsid=15340673&BRD=2731&PAG=461&dept_id=574995&rfi=8. Accessed 8 July 2008.
7. Rediff News. 1 Aug 2005. Cell phones don't cause cancer: study. http://in.rediff.com/news/2005/aug/31cellphone.htm. Accessed 8 July 2008.
8. Evans, J.A. nd. The cell tolls for thee. MSN Health and Wellness. http://health.msn.com/health-topics/cancer/articlepage.aspx?cp-documentid=100211877&page=1. Accessed 5 Aug 2008.
9. ENS. 16 July 2001. Power lines, wiring pose health risks. http://www.feb.se/EMFguru/Elf/calif-power.html. Accessed 8 July 2008.
10. Electric Power Research Institute. 1982. *Transmission Line Reference Book 345 kV and Above*, 2 edn, pp. 45–61. Palo Alto: Electric Power Research Institute.
11. DetectorTechnologies.com. nd. DC magnetometer (Gauss). http://www.detectortechnologies.com/store/detail.aspx?ID=7. Accessed 5 Aug 2008.
12. Cadick, J. 1994. *Electrical Safety Handbook*, 4.15. New York: McGraw-Hill.
13. National Fire Prevention Association. 2007. *National Electric Code*, 2008 edn, Quincy: National Fire Prevention Association, Section 210.12.

Problems

*B refers to basic, A refers to average, H refers to hard, and * refers to problems with answers.*

SECTION 15.2 ELECTRIC SHOCK

15.1 (B)* If a person comes into contact with 120 V, what body resistance can potentially result in a fatal shock?

15.2 (A) An electrical power cord falls into a swimming pool, as shown in Figure P15.2. If the pool water has a resistivity of 175 kΩ m, find the current, I, that flows through the water. Assume that the pool bottom acts as a ground (return conductor), and the pool walls act as insulators.

15.3 (B)* A current is going through your body and you do not feel it, what is the possible value of the current?

(a) 100 mA
(b) 10 mA
(c) 5 mA
(d) 0.1 mA

15.4 (H) When shoes are scuffed on a rug on a dry winter day, our bodies can be easily charged up to a potential

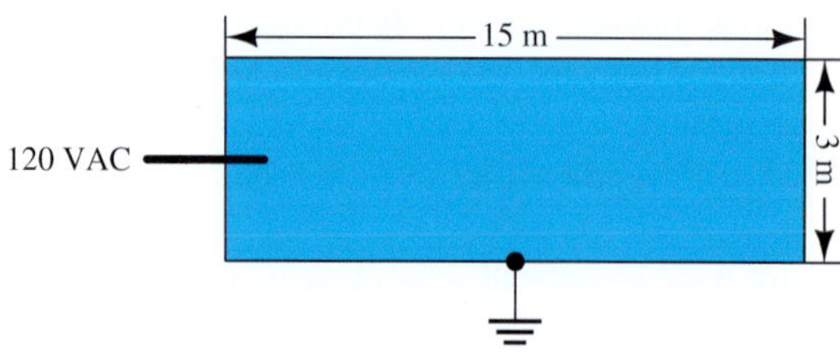

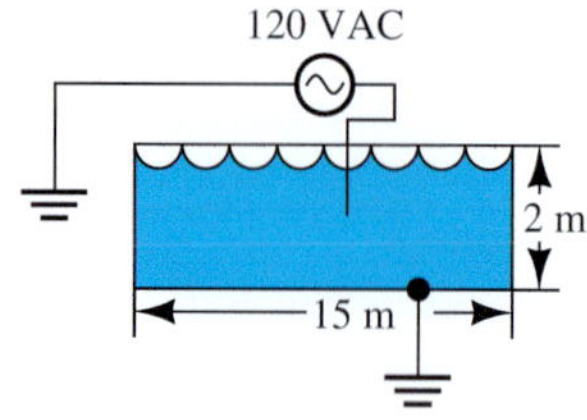

FIGURE P15.2 Swimming pool diagram for Problem 15.2.

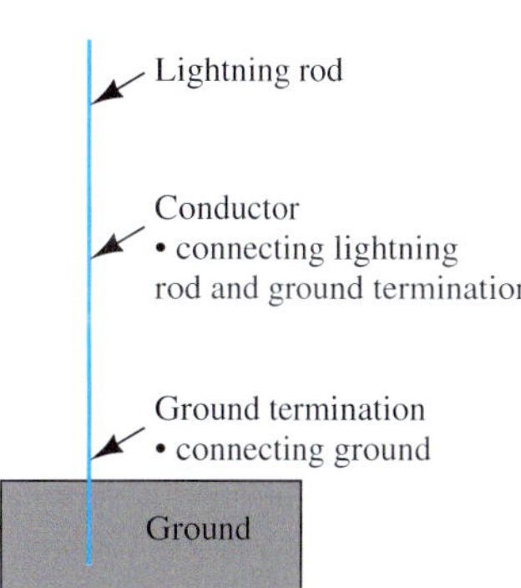

FIGURE P15.12 Building with a lightning rod and the basic structure of it. (Used with permission from © Michael Hawkridge/Alamy.)

of several thousand volts (about 1000 to 3000 V) static electricity with respect to the ground. A person shocked by this high-voltage static will not die. However, if the same person were shocked by a power supply with 200 V, the person would die. Explain the reason.

15.5 (B) In China and Europe, the voltage of the power supply for houses and apartments is 50 Hz, 220 to 240 V, while in the United States the power supply voltage is 60 Hz, 100 to 127 V. Assume the resistance of a body is 500 to 1000 Ω; compare the two power supply voltages' impact if someone is shocked.

15.6 (A)* The actual resistance of the body varies depending on the points of contact and the skin condition (moist or dry). Usually, the resistance from hand to foot is close to 500 Ω. Assume a person stands on the ground with bare feet and one hand touches a 120-V power supply cable. Is it fatal?

15.7 (B) Which of the following cannot help prevent electric shock?

(a) Rubber gloves
(b) Wet wood chair
(c) Metal workbench
(d) Dry wood workbench

15.8 (B) You need to replace a broken fuse in an instrument. To avoid shock, what will you do?

15.9 (H) Explain the reason that a high-power transmitter design lab is always equipped with wooden bar stools, and why the bar stools should be kept dry.

15.10 (H)* Assume that the density of a flash of lightning hitting the ground is 4 lightning strikes per km^2 per year, on average. Also assume that the lightning-attractive area of one house is 1250 m^2 (which means if lightning hits a spot within 1250 m^2 of a house, it can be attracted by the house and hit it). Calculate the probability of a lightning strike on the house. On average, how many years will it take for the house to be hit once?

15.11 (A)* What should you not do during a lightning storm?

(a) Go underneath trees or rain/sun shelters
(b) Touch metal objects or water
(c) Use the telephone
(d) Go into fully enclosed all-metal vehicles or substantial buildings

15.12 (B) Why can a lightning rod installed on the top of a building protect the building from lightning strikes (see Figure P15.12)?

15.13 (A) Describe the method used to protect an airplane in a storm from lightning.

15.14 (H) A safety fuse has been installed in a surge protector to prevent large currents from destroying electronic equipment. The surge protector is a safety fuse that opens the circuit automatically if the current within the circuit increases beyond a threshold. Sketch the circuit for the case when a lightning strike occurs and induces large current into a circuit that has radio equipment connected in series with a surge protector. Then, explain how the safety fuse works in this case.

SECTION 15.3 ELECTROMAGNETIC HAZARDS

15.15 (A)* For a radio transmitter, the power transmitted is 50 kW, and the antenna gain is 20.

(a) If an employee is standing 10 m from the antenna, what is the power density of the radio frequency radiation passing through his body?
(b) At what distance is the power density of the radiation 1 kW/m^2?

15.16 (A) A cell phone, held 5 mm from the head, transmits 0.3 W at an antenna gain of 1.5. What is the power density on the skin? Is this radiation level of concern?

15.17 (A)* If a house is placed 30 m from a 745-kV power line bearing 4000 A, what is the strength of the magnetic field? Is this field safe (according to the US state standards)?

15.18 (B) Find the maximum magnetic field magnitude directly under a 12.47-kV line. The line is 5 m above the ground and is rated at 800 A.

15.19 (B) If the maximum acceptable magnetic field magnitude is 200 mG, how far should houses be built from a 345-kV line rated at 2000 A based on the US state standards?

15.20 (B) List four kinds of radiations that can cause cancer.

15.21 (H) A cell phone's transmitting power is 1 W and its antenna power gain is 1.5. You can either use the cell phone close to your ear (0.5 in. to your head) by hand holding it or use a hands-free headset that allows the phone to be 3 ft away from your body. Compare the power density and select the method from the two that will best minimize the high-frequency impact on your body.

15.22 (A)* A hot pot consumes 2200 W and the power supply voltage is 120 V. Assume you are sitting close to the hot pot cable at a distance of 1 ft. Is this safe, according to US state standards?

15.23 (H) Why are overhead high-voltage electric power transmission lines more commonly used rather than underground high-voltage electric power transmission lines? *Hint:* compare them in terms of cost, heat dissipation, and environmental impacts.

15.24 (A) If someone's bone is broken, the doctor will use an X-ray machine to check the condition of the bone. Explain why a person should not be exposed to the X-ray for an extended period of time.

SECTION 15.4 ARCS AND EXPLOSIONS

15.25 (A)* What is the maximum power available for an arc across the load (between point C and the ground) in Figure P15.25? Also, find the maximum power for an arc across the circuit breaker between points A and B.

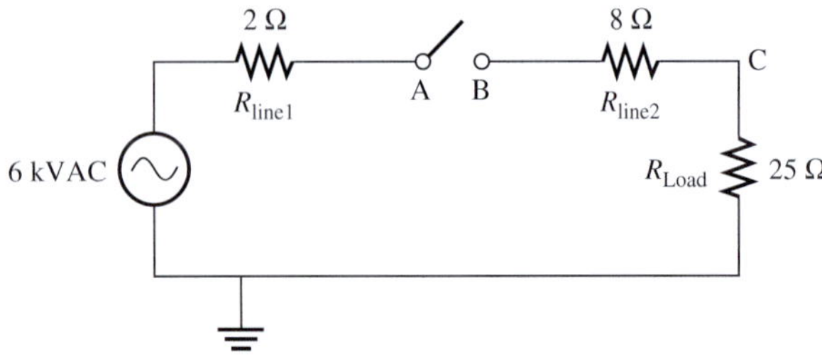

FIGURE P15.25 Circuit for Problem 15.25.

15.26 (B) Figure P15.26 shows a basic flashlight circuit. When the switch is opened, a spark occurs across its contacts. What is the maximum power in this spark?

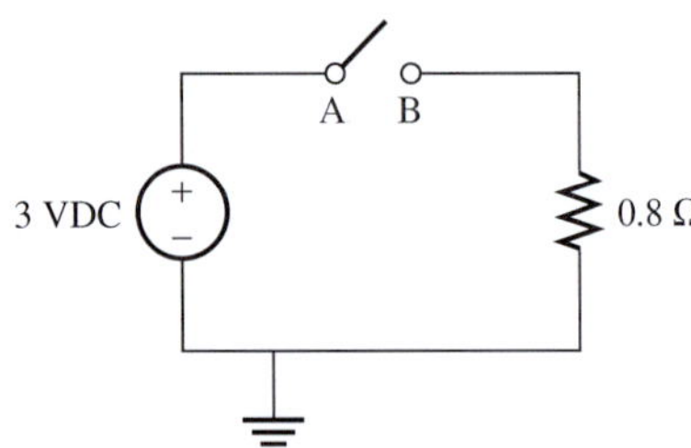

FIGURE P15.26 Circuit for Problem 15.26.

15.27 (H) Why is a spark-resistance switch needed in a milling plant?

15.28 (H) Why on a dry day in the winter is it easy to be shocked by an arc when you touch a metal door handle after you walk on a carpet? How can you avoid this kind of shock?

15.29 (A) Explain the reason why there are no walls around gas stations.

15.30 (A)* Calculate the energy dissipated in an arc generated on the switch in the lamp circuit shown in Figure P15.30. Assume the arc period is 1 ms.

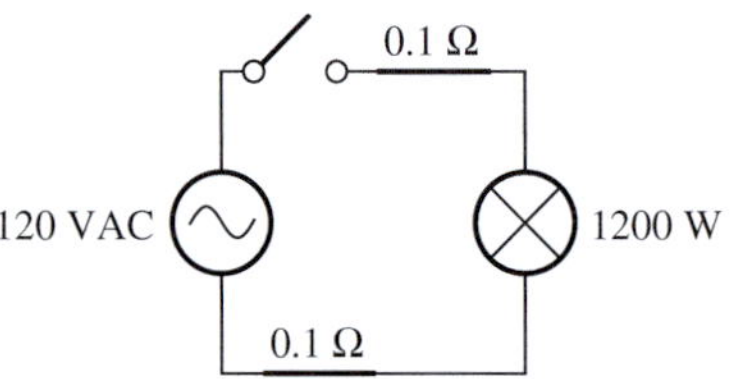

FIGURE P15.30 Lamp circuit.

15.31 (H) Explain the basics of explosion prevention.

SECTION 15.5 THE NATIONAL ELECTRIC CODE

15.32 (B) Based on the measurements in Figure P15.32, does this outlet have a proper ground?

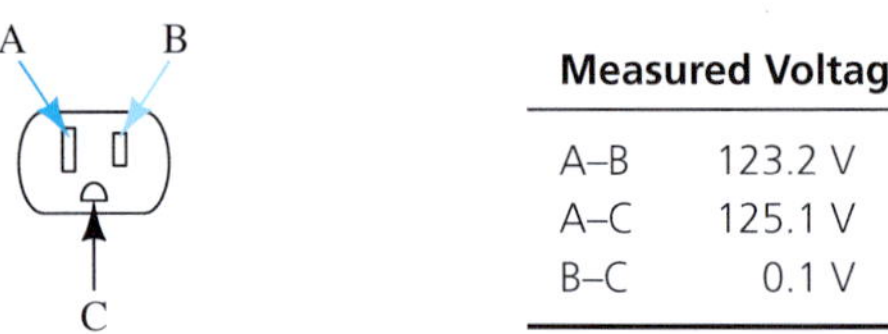

Measured Voltage	
A–B	123.2 V
A–C	125.1 V
B–C	0.1 V

FIGURE P15.32 Outlet for Problem 15.32.

15.33 (B) Can a grounding adapter be safely used with this older outlet?

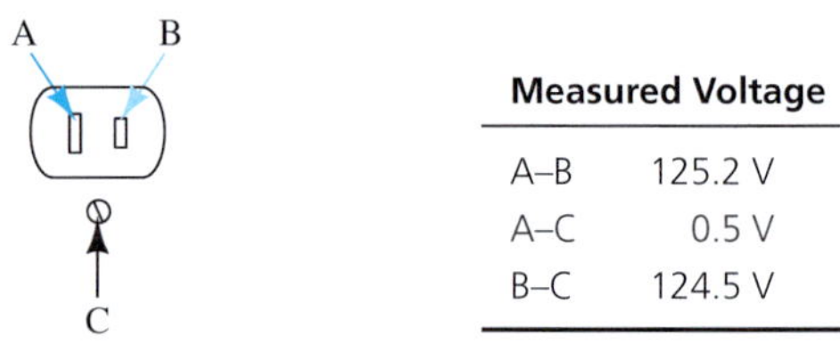

Measured Voltage	
A–B	125.2 V
A–C	0.5 V
B–C	124.5 V

FIGURE P15.33 Outlet for Problem 15.33.

15.34 (A) For each of the scenarios described below, state when current will flow through the person's body when he is in contact with incoming and return conductors.
(a) With a rubber glove and a rubber boot
(b) With a rubber glove
(c) With a rubber boot

15.35 (A)* An individual comes into contact with the incoming and return conductors of a GFCI-protected circuit, as shown in Figure P15.35.
(a) Find the currents I_{in}, I_{out}, and I_{ground}.
(b) Does the GFCI trip? If not, how much current flows through his body?

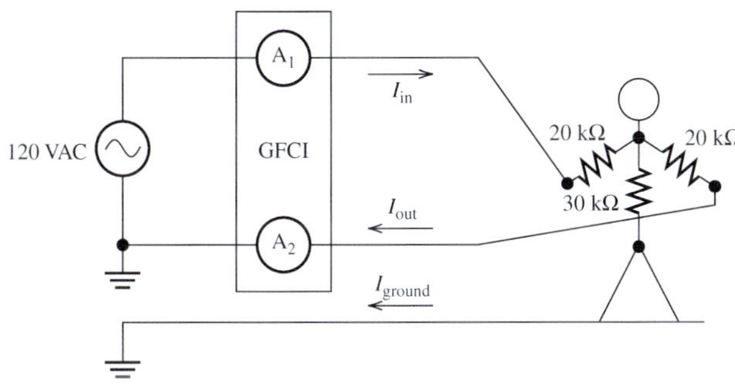

FIGURE P15.35 Circuit diagram for Problem 15.35.

15.36 (A) An individual comes into contact with the incoming and return conductors of a GFCI-protected circuit, as shown in Figure P15.36. He is wearing rubber boots, which isolate his feet from the ground.

(a) Find the currents I_{in}, I_{out}, and I_{ground}.

(b) Does the GFCI trip? If not, how much current flows through his body?

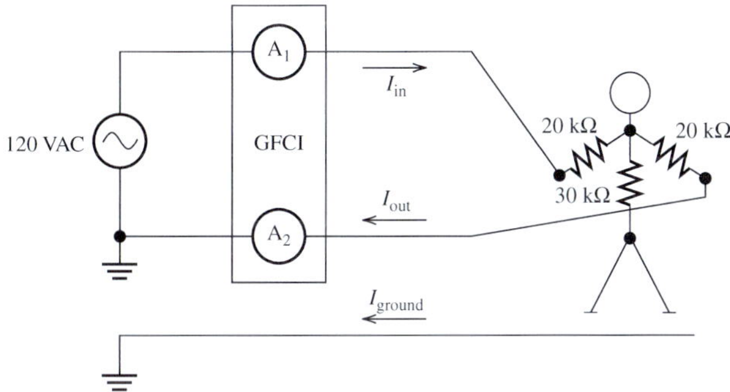

FIGURE P15.36 Circuit diagram for Problem 15.36.

15.37 (A) A 120-V circuit has devices rated at 200 W, 1500 W, 2 A, and 300 W, respectively. The 12-gauge line is connected to a 20-A circuit breaker.

(a) Does the circuit breaker trip?

(b) Is a fire hazard present?

15.38 (A)* A 120-V circuit on a 20-A circuit breaker has devices rated at 1, 200 W, 3 A, 60 W, 100 W, and 300 W, respectively. The line gauge is 14.

(a) Does the circuit breaker trip?

(b) Is a fire hazard present?

15.39 (H) The following devices, rated at 440, 520, 360, 25, and 110 W, respectively, are connected to a socket with a 120-V voltage source. The line gauge is 16.

(a) Is a fire hazard present?

(b) If there is a fire hazard, what kind of advice can be given to prevent it?

15.40 (B) What is the purpose of NEC?

15.41 (A)* There is old outlet shown in Figure P15.41. Which of the following equipment cannot be connected to the old outlet?

(a) Cell-phone charger transformer

(b) Hot-water pot with plastic case

(c) Personal computer

(d) Electric fan

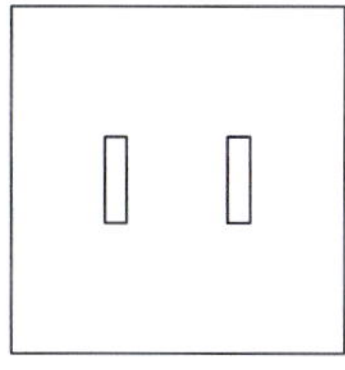

FIGURE P15.41 Old outlet.

15.42 (A) Compare the advantages and disadvantages of fuses and circuit breakers.

15.43 (A) As an instrument ages, the connection between the ground and the instrument's metal case may decay and generate a resistance, for example $R_1 = 100\ \Omega$ in Figure P15.43. Assume that the incoming line voltage is 120 V and that there is a broken point that connects to the instrument's metal case, and $R_2 = 100\ \Omega$. Please use the figure to explain the importance of dependable grounding in the building's wiring system, assuming a body's resistance to be $R_3 = 500\ \Omega$.

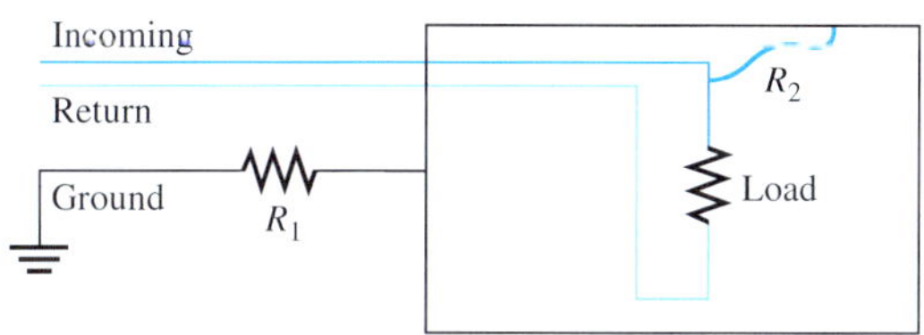

FIGURE P15.43 Example of a fault ground.

15.44 (B) What will happen if a line is carrying a current greater than the rated value? For example, what happens when a 10-A rated line carries 15-A current?

15.45 (A) Compare the functions of AFCI and GFCI protected circuits.

APPENDIX A

Solving Linear Equations

A set of linear equations will have a unique solution if the determinant of the system matrix is nonzero. The solution to the linear system $A\,x = b$ is $x = A^{-1}b$. Finding the inverse of a matrix is difficult, especially in cases where the matrix is large. Cramer's rule provides a method that makes it easier to determine the solution to the linear system.

CRAMER'S RULE

Suppose A is an $n \times n$ invertible matrix. The solution to the system $A\,x = b$ is given by:

$$x_1 = \frac{\det(A_1)}{\det(A)},\ x_2 = \frac{\det(A_2)}{\det(A)},\ \ldots,\ x_n = \frac{\det(A_n)}{\det(A)}$$

where A_i is the matrix found by replacing the ith column of A with b.

EXAMPLE A.1 Cramer's Rule

Use Cramer's rule to determine the solution to the following linear equations:

$$4x_1 + 5x_2 = 3$$
$$2x_1 - x_2 = 1$$

SOLUTION

From the problem, develop matrix:

$$A = \begin{bmatrix} 4 & 5 \\ 2 & -1 \end{bmatrix}$$

and:

$$b = \begin{bmatrix} 3 \\ 1 \end{bmatrix}$$

Calculating the determinant of matrix A results in $\det(A) = -14 \neq 0$. Thus, this method results in a unique solution for this linear system.

The matrices A_1 and A_2 are found by respectively replacing the first and the second columns of A with b. Therefore:

$$A_1 = \begin{bmatrix} 3 & 5 \\ 1 & -1 \end{bmatrix}$$

$$A_2 = \begin{bmatrix} 4 & 3 \\ 2 & 1 \end{bmatrix}$$

(continued)

EXAMPLE A.1 Continued

Now, according to Cramer's rule:

$$x_1 = \frac{\det(A_1)}{\det(A)} = \frac{-8}{-14} = \frac{4}{7}$$

$$x_2 = \frac{\det(A_2)}{\det(A)} = \frac{-2}{-14} = \frac{1}{7}$$

MATLAB software can also be used to determine the solution to a linear system. The MATLAB function "inv" determines the inverse of a matrix. Using this function, the command $x = \text{inv}(A)*b$ can be used in order to find the system solution.

APPENDIX B

Laplace Transform

The Laplace transform or simply *L-transform* is defined as the transformation from the time domain to the frequency domain (here, called the s-domain). It can be expressed mathematically as:

$$L(x(t)) = X(s) = \int_0^{\infty} x(t)e^{-st}\mathrm{d}t \qquad \textbf{(B.1)}$$

where $x(t)$ is the t-domain signal and $X(s)$ is the s-domain signal. The complex angular frequency, s, is given by:

$$s = \sigma + j\omega \qquad \textbf{(B.2)}$$

Notice that the time domain signal is expressed in lowercase (e.g., $x(t)$); however, uppercase notation (e.g., $X(s)$) is used for s-domain signals.

EXAMPLE B.1 Laplace Transform

Find the Laplace transform for the step voltage defined by:

$$v(t) = \begin{cases} E & t \geq 0 \\ 0 & \text{else} \end{cases}$$

SOLUTION

$$\begin{aligned} V(s) &= \int_0^{\infty} v(t)e^{-st}\mathrm{d}t \\ &= \int_0^{\infty} Ee^{-st}\mathrm{d}t \\ &= E\int_0^{\infty} e^{-st}\mathrm{d}t \\ &= E\frac{-1}{s}e^{-st}\Big|_0^{\infty} \\ &= \frac{E}{s} \end{aligned}$$

Note: The last equality is true when the real part of s, $R(s) > 0$ so that $e^{-st} \to 0$ as $t \to \infty$.

EXAMPLE B.2 Laplace Transform

Find the Laplace transform for the decaying exponential function defined by:

$$x(t) = \begin{cases} e^{-at} & t \geq 0 \\ 0 & \text{else} \end{cases}$$

SOLUTION

$$\begin{aligned} X(s) &= \int_0^{\infty} x(t)e^{-st}\mathrm{d}t \\ &= \int_0^{\infty} e^{-at}e^{-st}\mathrm{d}t \end{aligned}$$

(continued)

EXAMPLE B.2 Continued

$$= \int_0^\infty e^{-(s+a)t} dt$$

$$= \frac{-1}{s+a} e^{-(s+a)t} \Big|_0^\infty$$

$$= \frac{1}{s+a}$$

Note: The last equality is true when $R(s) > -a$ so that $e^{-(s+a)t} \rightarrow 0$ as $t \rightarrow \infty$.

EXAMPLE B.3 Laplace Transform

Find the Laplace transform of:

$$x(t) = 5 + e^{-2t}$$

SOLUTION

From the previous examples, it is known that L (5) = $5/s$, and L (e^{-2t}) = $1/(s+2)$. Because the Laplace transform is an integration process it possesses the linearity property, therefore:

$$X(s) = \text{L}(5) + \text{L}(e^{-2t}) = \frac{5}{s} + \frac{1}{s+2} = \frac{6s+10}{s(s+2)}$$

Table B.1 represents the Laplace transforms of some important functions. This table can be used to find the Laplace transform as well as the inverse Laplace transform. The following two examples help to explain how to find the inverse Laplace transform and how to solve a differential equation.

TABLE B.1 The Laplace Transform of Basic Functions and Some of its Properties

x(t)	*X(s)*
Basic Functions	
$\delta(t)$	1
1	$\frac{1}{s}$
t	$\frac{1}{s^2}$
t^n	$\frac{n!}{s^{n+1}}$
e^{at}	$\frac{1}{s-a}$
$\cos(\omega t)$	$\frac{s}{s^2+\omega^2}$
$\sin(\omega t)$	$\frac{\omega}{s^2+\omega^2}$
s-domain shift	
$x(t)e^{at}$	$X(s-a)$
te^{at}	$\frac{1}{(s-a)^2}$
$\cos(\omega t)e^{at}$	$\frac{s-a}{(s-a)^2+\omega^2}$

(continued)

TABLE B.1 Continued

x(t)	*X(s)*
$\sin(\omega t)e^{at}$	$\dfrac{\omega}{(s-a)^2+\omega^2}$
t-domain shift	
$x(t-a)$	$X(s)e^{-as}$
Differentiation	
$x'(t)$	$sX(s)-x(0)$
$x''(t)$	$s^2X(s)-sx(0)-x'(0)$
$x^{(n)}(t)$	$s^nX(s)-s^{n-1}x(0)-s^{n-2}x'(0)-\cdots-x^{(n-1)}(0)$
Integration	
$\int_0^t x(\tau)d\tau$	$\dfrac{1}{s}X(s)$
Convolution	
$x_1(t) * x_2(t)$	$X_1(s) \cdot X_2(s)$
$x_1(t) \cdot x_2(t)$	$X_1(s) * X_2(s)$

EXAMPLE B.4 Inverse Laplace Transform

Find the inverse Laplace transform of $Y(s)$ given by:

$$Y(s) = \frac{8}{s^3 + 4s^2 + 3s}$$

SOLUTION

To solve this problem, use the partial fraction procedure:

$$Y(s) = \frac{8}{s(s^2 + 4s + 3)} = \frac{8}{s(s+1)(s+3)} = \frac{k_0}{s} + \frac{k_1}{s+1} + \frac{k_2}{s+3}$$

where k_0, k_1, and k_2 are constants (called the residues) of the poles 0, 1, and 3, respectively. Note that the poles of $Y(s)$ are the zeros of its denominator. These residues can be calculated as follows:

$$k_0 = sY(s)\big|_{s=0} = \frac{8}{(0+1)(0+3)} = \frac{8}{3} = 2.67$$

$$k_1 = (s+1)Y(s)\big|_{s=-1} = \frac{8}{-1\times(-1+3)} = \frac{8}{-2} = -4$$

$$k_1 = (s+3)Y(s)\big|_{s=-3} = \frac{8}{-3\times(-3+1)} = \frac{8}{+6} = +1.33$$

and; therefore:

$$Y(s) - \frac{2.67}{s} - \frac{4}{s+1} + \frac{1.33}{s+3}$$

Using Table B.1 and the linearity property of Laplace transform, $y(t)$ corresponds to:

$$y(t) = 2.67 - 4e^{-t} + 1.33e^{-3t}, \qquad t > 0$$

EXAMPLE B.5 Solving Differnetial Equations Using Laplace Transform

Find the solution of the first-order differential equation:

$$x'(t) + 3x(t) - 24 = 0$$

a. Assuming zero initial conditions
b. $x(0) = 2$

SOLUTION

Applying Laplace transform:

$$sX(s) - x(0) + 3X(s) - \frac{24}{s} = 0$$

a. $x(0) = 0$, thus:

$$(s + 3)X(s) = 0 + \frac{24}{s} = \frac{24}{s}$$

or:

$$X(s) = \frac{24}{s(s + 3)} = \frac{8}{s} - \frac{8}{s + 3} = 8\left(\frac{1}{s} - \frac{1}{s + 3}\right)$$

Its inverse Laplace transform corresponds to:

$$x(t) = 8(1 - e^{-3t})$$

b. $x(0) = 2$, thus:

$$(s + 3)X(s) = 2 + \frac{24}{s} = \frac{2s + 24}{s}$$

or:

$$X(s) = \frac{2s + 24}{s(s + 3)} = \frac{8}{s} - \frac{6}{s + 3}$$

Here, the second equality is based on partial fraction procedure explained in Example B.4. Its inverse Laplace transform corresponds to:

$$x(t) = 8 - 6e^{-3t}, \qquad t > 0$$

EXAMPLE B.6 Solving Differnetial Equations Using Laplace Transform

Find the solution of the second-order differential equation, assuming all initial conditions are zeros:

$$x''(t) + 5x'(t) + 6x(t) - 50 = 0$$

SOLUTION

Applying Laplace transform:

$$s^2X(s) - sx(0) - x'(0) + 5sX(s) - 5x(0) + 6X(s) - \frac{50}{s} = 0$$

Setting $x(0) = x'(0) = 0$:

$$(s^2 + 5s + 6)X(s) = \frac{50}{s}$$

$$X(s) = \frac{50}{s(s^2 + 5s + 6)} = \frac{50}{s(s + 2)(s + 3)} = \frac{8.33}{s} - \frac{25}{s + 2} + \frac{16.67}{s + 3}$$

Using Table B.1 to find inverse Laplace transform results in:

$$x(t) = 8.33 - 25e^{-2t} + 16.67e^{-3t}, \qquad t > 0$$

APPENDIX C

Complex Numbers

Any sinusoidal signal, like the one shown in Figure C.1, can be represented by its amplitude and its phase. These signals can be expressed using complex numbers.

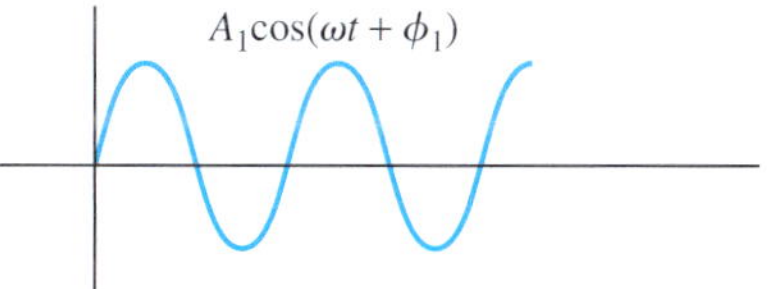

FIGURE C.1 A sinusoidal signal.

The signal shown in Figure C.1 can also be expressed in exponential form, that is, $A_1 \cos(\omega t + \phi_1) = \text{Re}\{A_1 e^{j\omega t} e^{j\phi_1}\}$. Here, $A_1 e^{j\phi_1}$ is the complex number in exponential form.

From the 13th century to the early 17th century, numbers were only used to represent real things, for example, ONE piece of cake or TWO cars, as seen below.

At that time, a mathematician solving $X^2 = 4$ would have arrived at a single solution, $X = 2$.

In the middle of the 17th century, negative numbers were added to the number system to represent things like debt, height below sea level, for example, −2$, −30 m, and so on.

Once negative numbers appeared, solving $X^2 = 4$ results in the solutions $X_1 = 2$ and $X_2 = -2$.

By the end of 18th century, the notion of complex numbers had been developed in the mind of scientists. However, the scientists could not intuitively explain the concept using real identities.

Now that complex numbers were available, solving $X^2 = -4$ resulted in the solution $X = \pm j2$, that is, $(j2)^2 = -4$, $(-j2)^2 = -4$, where $j = \sqrt{-1}$.

DEFINITION OF A COMPLEX NUMBER

First, $j^2 = -1$; correspondingly $j = \sqrt{-1}$. Here, j represents the square root of −1. Sometimes, symbol i is used instead of j.

Second, an *imaginary number* is defined as the product of a real number and the imaginary operator j; for example, $j5$ is an imaginary number, or, a pure imaginary number.

A *complex number* is defined as the sum of a real number and an imaginary number, that is, $a + jb$, where a and b are real numbers. $2 + j3$ and $1.1 + j5.3$ are examples of complex numbers. A complex number, for example, $Z = a + jb$ is said to have a real part a and an imaginary part b.

Actually, a *real number* is simply a complex number with an imaginary part equal to zero, that is, $a + j0$. For example, $1.1 + j0 = 1.1$, $2.4 + j0 = 2.4$ . . . . A pure imaginary number is a complex number with a real part equal to zero.

A *complex number* can be expressed in rectangular form $a + bj$, in exponential form $re^{j\theta}$, or in polar form $r\angle\theta$.

The addition, subtraction, multiplication, and division of complex numbers in rectangular form can be accomplished in a way similar to algebraic arithmetic.

OPERATIONS OF COMPLEX NUMBERS IN RECTANGULAR FORM

Assume: $Z_1 = a + jb$ and $Z_2 = c + jd$.

- The conjugate of the complex number, Z_1, is (changing the sign of the imaginary part):

$$Z_1{}^* = a - jb, \text{ where } * \text{ represents the conjugate.}$$

- If $Z_1 = Z_2$, then $a = c$ and $b = d$.

If two complex numbers are equal, then each number's real part and imaginary part must be equal to the real part and imaginary part of the other number, respectively.

- Addition: $Z_1 + Z_2 = (a + c) + j(b + d)$.

To add two complex numbers, add the real and the imaginary parts of them, respectively.

- Subtraction: $Z_1 - Z_2 = (a - c) + j\,(b - d)$.

To subtract two complex numbers, subtract the real part of Z_2 from the real part of Z_1, and subtract the imaginary part of Z_2 from the imaginary part of Z_1.

- Multiplication: $Z_1 \times Z_2 = (ac - bd) + j(bc + ad)$.

To multiply two complex numbers, use the same rules as in the algebraic arithmetic.

$$Z_1 \times Z_2 = (a + jb)(c + jd) = ac + jad + jbc - bd = (ac - bd) + j(bc + ad).$$

- Division:

$$\frac{Z_1}{Z_2} = \frac{a + jb}{c + jd} = \frac{(a + jb)(c - jd)}{(c + jd)(c - jd)} = \frac{(ac + bd) + j(bc - ad)}{c^2 + d^2}$$

To divide two complex numbers, multiply the numerator and the denominator by the complex conjugate of the denominator.

EXAMPLE C.1 Operations of Complex Numbers

For the given complex numbers: $Z_1 = 3 + j4$; $Z_2 = 9 + j12$.

- Addition: $Z_1 + Z_2 = (3 + 9) + j(4 + 12) = 12 + j16$
- Subtraction: $Z_1 - Z_2 = (3 - 9) + j(4 - 12) = -6 - j8$
- Multiplication: $Z_1 \times Z_2 = 27 + j36 + j36 - 48 = -21 + j72$
- Division: $\frac{Z_1}{Z_2} = \frac{(3 + j4)(9 - j12)}{(9 + j12)(9 - j12)} = \frac{75}{81 + 144} = \frac{75}{225} = \frac{1}{3}$
- Conjugate of Z_1: $Z_1^* = 3 - j4$

COMPLEX PLANE

A complex number can be represented in a *complex plane*, like the one shown in Figure C.2 (x-axis: real part; y-axis: imaginary part).

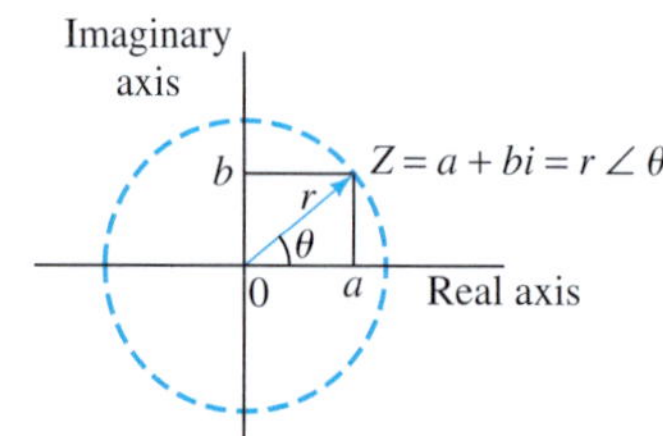

FIGURE C.2 A Complex Plane.

A complex number has three forms:

Rectangular Form: $Z_1 = a + jb$

Exponential Form: $Z_1 = re^{j\theta}$

$$\begin{cases} r = \sqrt{a^2 + b^2} \\ \theta = \tan^{-1}\dfrac{b}{a} \end{cases} \qquad \begin{cases} a = r \cdot \cos\theta \\ b = r \cdot \sin\theta \end{cases} \tag{C.1}$$

Polar Form: $Z_1 = r\angle\theta$

In the complex plane, the length, r, of the arrow represents the magnitude of the complex number, and θ is the angle between the arrow and the positive real axis.

Using Equation (C.1), a complex number can be converted from rectangular form to exponential form, or vice versa. It is easy to convert the exponential form to polar form, because both forms use magnitude and angle to express the complex number.

EXAMPLE C.2 Rectangular Form to Exponential and Polar Form Conversion

Convert the complex number $Z_1 = 4 + j3$ from rectangular form to exponential and polar forms.

SOLUTION

Using Equation (C.1):

$$r = \sqrt{3^2 + 4^2} = 5, \quad \text{and}$$

$$\tan\theta = \frac{3}{4} = 0.75, \quad \theta = \tan^{-1}0.75 = 36.87°$$

Therefore:

$$Z_1 = 5e^{j36.87°}$$

The corresponding polar form is:

$$Z_1 = 5\angle 36.87°$$

EXAMPLE C.3 Exponential to Rectangular and Polar Form Conversion

Convert the complex number $Z_1 = 2.828e^{j45°}$ from exponential form to rectangular form and polar forms.

SOLUTION

The polar form for $Z_1 = 2.828e^{j45°}$ is:

$$Z_1 = 2.828\angle 45°$$

Using Equation (C.1):

$$a = r\cos\theta = 2.828 \times \cos(45°) = 2$$
$$b = r\sin\theta = 2.828 \times \sin(45°) = 2$$

Therefore, the rectangular form is:

$$Z_1 = a + jb = 2 + j2$$

EXAMPLE C.4 Polar to Exponential and Rectangular Form Conversion

Convert the complex number $Z_1 = 10\angle 60°$ from polar form to exponential and rectangular forms.

SOLUTION

The exponential form for $Z_1 = 10\angle 60°$ is:

$$Z_1 = 10e^{j60°}$$

Using Equation (C.1):

$$a = r\cos\theta = 10 \times \cos(60°) = 5$$
$$b = r\sin\theta = 10 \times \sin(60°) = 8.66$$

Therefore, the rectangular form is:

$$Z_1 = a + jb = 5 + j8.66$$

OPERATIONS IN EXPONENTIAL AND POLAR FORMS

Assume that complex numbers Z_1 and Z_2 are expressed in exponential form: $Z_1 = r_1e^{j\theta}$, $Z_2 = r_2e^{j\phi}$.

- Complex conjugates are:

$$Z_1^* = r_1e^{-j\theta} = r_1\angle(-\theta) \quad \text{and} \quad Z_2^* = r_2e^{-j\phi} = r_2\angle(-\phi)$$

- Addition and subtraction:

If the complex numbers are expressed in exponential or polar form, they need to be converted into rectangular form for addition and subtraction.

- Multiplication:

$$Z_1 \times Z_2 = r_1e^{j\theta} \cdot r_2e^{j\phi} = r_1r_2e^{j(\theta+\phi)} = r_1r_2\angle(\theta + \phi)$$

To multiply two complex numbers in exponential or polar form, multiply the magnitude of them and add the angles.

- Division:

$$\frac{Z_1}{Z_2} = \frac{r_1e^{j\theta}}{r_2e^{j\phi}} = \frac{r_1\angle\theta}{r_2\angle\phi} = \frac{r_1}{r_2}\angle(\theta - \phi)$$

To divide two complex numbers in exponential or polar form, divide the magnitude of them and subtract the angle of the divisor from the angle of the dividend.

- Power, n, of a complex number:

$$Z_1^n = (r_1e^{j\theta})^n = r_1^ne^{jn\theta} = r_1^n\angle n\theta,$$

The magnitude is powered to n and the n of the angles are added together.

- n root of a complex number:

$$Z_1^{\frac{1}{n}} = (r_1e^{j\theta})^{\frac{1}{n}} = r_1^{\frac{1}{n}}e^{j\theta\cdot\frac{1}{n}} = r_1^{\frac{1}{n}}\angle\left(\frac{\theta + 2k\pi}{n}\right), \quad k = 0, \pm1, \pm2, \ldots$$

EXAMPLE C.5 Multiplicaton and Division in Polar Form

Assume $Z_1 = 2.828\angle 45°$ and $Z_2 = 3\angle 30°$, calculate Z_1Z_2 and Z_1/Z_2.

SOLUTION

$$Z_1 \times Z_2 = (2.828 \times 3)\angle(45^\circ + 30^\circ) = 8.484\angle 75^\circ$$

$$\frac{Z_1}{Z_2} = \frac{2.828}{3}\angle(45^\circ - 30^\circ) = \frac{2.828}{3}\angle 15^\circ = 0.9427\angle 15^\circ$$

EXAMPLE C.6 Complex Operations

Assume $Z_1 = 3\angle 45^\circ$, $Z_2 = 4\angle 45^\circ$, calculate $Z_1 + Z_2$, Z_1Z_2, and Z_1/Z_2.

SOLUTION

First, convert Z_1 and Z_2 to rectangular form, then add them.

Using Equation (B.1):

The real part a of the Z_1 is:

$$a = 3 \times \cos(45^\circ) = 2.1213$$

The imaginary part b of the Z_1 is:

$$b = 3 \times \sin(45^\circ) = 2.1213$$

The real part c of the Z_2 is:

$$c = 4 \times \cos(45^\circ) = 2.8284$$

The imaginary part d of the Z_2 is:

$$d = 4 \times \sin(45^\circ) = 2.8284$$

Therefore:

$$Z_1 + Z_2 = 2.1213 + j2.1213 + 2.8284 + j2.8284 = 4.9497 + j4.9497$$

Convert the result to polar form:

$$Z_1 + Z_2 = 7\angle 45^\circ$$

$$Z_1 \times Z_2 = (3 \times 4)\angle(45^\circ + 45^\circ) = 12\angle 90^\circ$$

$$\frac{Z_1}{Z_2} = \frac{3}{4}\angle(45^\circ - 45^\circ) = 0.75\angle 0^\circ$$

EULER'S IDENTITY

Euler's identities state that:

$$e^{j\theta} = \cos\theta + j\sin\theta \quad \text{(C.2)}$$

$$e^{-j\theta} = \cos\theta - j\sin\theta \quad \text{(C.3)}$$

$$\cos\theta = \frac{e^{j\theta} + e^{-j\theta}}{2} \quad \text{(C.4)}$$

$$\sin\theta = \frac{e^{j\theta} - e^{-j\theta}}{j2} \quad \text{(C.5)}$$

Using the Euler's identities, the cosine and sinusoidal functions can be expressed as complex numbers, and the calculation can be simplified.

EXAMPLE C.7 Application of Euler's Identity

Calculate e^{1-j1}.

SOLUTION

Method 1: Angle in radians:

$$e^{1-j1} = e^1 \cdot e^{-j1} = e^1(\cos 1 - j\sin 1) = 2.71828(0.5403 - j0.8415) = 1.469 - j2.29$$

Method 2: Angle in degrees:

$$e^{1-j1} = e^1 \cdot e^{-j1} = e^1(\cos 1 - j\sin 1) = e^1(\cos 57.3^\circ - j\sin 57.3^\circ) = 2.71828(0.5403 - j0.8415)$$
$$= 1.469 - j2.29$$

Here, "1" is the angle in radians. To convert radians into degrees or vice versa, use the following equation:

$$\frac{\theta_{\text{rad}}}{\theta_{\text{deg}}} = \frac{\pi}{180}$$

where, $\pi = 3.1415926$.

SUMMARY

A complex number can be expressed as:

$$Z_1 = a + jb = \text{Re}[Z_1] + j\text{Im}[Z_1] = re^{j\theta} = \sqrt{a^2 + b^2}e^{j\tan^{-1}(b/a)} = \sqrt{a^2 + b^2}\angle\tan^{-1}(b/a).$$

SELECTED SOLUTIONS FOR PROBLEMS

Chapter 2

2.9: $5t$	**2.10:** 25 C	**2.11:** −3 A	**2.14:** 15 J	**2.16:** 2.122 J
2.22: −1 A, −2 A, 8 A	**2.23:** 13 V	**2.26:** 25 mA	**2.28:** $7R/5$	**2.30:** 0.2 A
2.36: (a) 0.25 A, (b) $R_1/R_2 = 1/2$	**2.37:** $7R/5$	**2.41:** 70 V	**2.45:** 120 mA, 133.3 kΩ	**2.48:** 0.001284, 2.364×10^{-4}, 9.304×10^{-8}
2.52: 500 W	**2.58:** (a) 30 W, (b) −2 A	**2.62:** 2.5 V	**2.66:** 6.002 V, 4.8 W	

Chapter 3

3.6: 60 Ω	**3.8:** $10R/3$	**3.11:** (a) 0 (b) $1.760R$	**3.13:** 5 Ω
3.15: $15R/8$	**3.19:** 8 A, 4 A, 4 A, 2 A	**3.22:** 10 mA, 7 mA	**3.28:** 26.32 V
3.30: A	**3.35:** 25 V, 75 V	**3.38:** −22.72 V, −6.22 V, 3.33 V	**3.44:** 64.84 V, 44.55 V, 38.55 V
3.48: 0	**3.53:** 11 V, 16.1 mA, 683 Ω	**3.57:** 1 A	**3.64:** 1.4 A
3.69: 5 Ω	**3.71:** 683.3 Ω	**3.74:** 989 μA, 1.8 mA, 6 mA	**3.75:** 12.51 mV

Chapter 4

4.1: (d)	**4.5:** 60 μC	**4.7:** 42.4 nm	**4.14:** (a) 33.3 V (b) 667 μJ	**4.19:** 4 V
4.22: 2 A	**4.24:** 1.36×10^{-4} J	**4.28:** 41.69 μF	**4.32:** (b) 5 Gal	**4.35:** 27.8 V
4.43: 30.2 mH	**4.45:** 55.49 mH	**4.48:** 89.54 mH		

Chapter 5

5.3: $10 - 10e^{-100t}$	**5.5:** $5\,e^{-t/1.5\times10^{-3}}, t \geq 0$	**5.8:** 8.63×10^{-3} s	**5.14:** $T = RC(V_f/V_i)$
5.15: R	**5.16:** R and V_f $RV_f =$ constant	**5.22:** $6.667e^{-10t} - 666.67e^{-100t}$	**5.25:** 0 A; 0.05 A
5.28: $3750e^{-3125t}$ V	**5.32:** $\frac{40}{3} - \frac{40}{3}e^{-37.5}$ V	**5.35:** 0.0594 A; 0 A; 0.286 A	**5.38:** 1 A; 5 V

5.43: Doubled; three multiple; halved

5.46: Over-damped

5.49: $0.396e^{-1000t} - 0.396\cos(100t) + 3.96\sin(100t)\ A$

5.56: $v_c(t) = 10\cos(50t) + 10\sin(50t) - 5e^{-50t}$

Chapter 6

6.1: $-7.32 + j1.04$
6.3: 50 V
6.8: 110.31 V
6.11: $\sqrt{22}$ V
6.13: $4\cos(4\omega t + 60°) + 2.5$ V
6.17: $340.3\cos(120\pi t + 51.6°)$ V
6.20: $177.58 + j68.97\Omega$
6.24: $50 - j1970\ \Omega$
6.28: $450 + j8.75\Omega$
6.30: $R_1C_1 = R_2C_2$
6.33: $35.8\cos(200t + 56.57°)$ mA
6.36: $7.032\angle 124.86°$
6.39: $1.961\cos(200t - 11.31)$ A; $0.392\angle 78.69$
6.43: \$0.37
6.46: $Z_N = 6.24 + j39\Omega$; $I_N = 4\cos(200t + 30°)$.
6.50: $93.84 - 2.25j\Omega$
6.53: $\dfrac{j3RZ_L + 2R^2}{2R + jZ_L}$
6.58: 127.85 W; 0; 127.85 W
6.62: $0.657\angle 53.2°$
6.64: (1) 0.9629; (2) 1 W.
6.67: 91.75 μF
6.69: $47.14\angle 0°A$; $166.7\angle 66.42°A$
6.71: 0.2; 16.5 μF

Chapter 7

7.1: $\dfrac{1}{1 + j(f/20 \times 10^6)}$
7.4: $\dfrac{1}{1 + j(f/34)}$
7.6: $\dfrac{3}{2\pi}$ Hz, $\dfrac{2}{\pi}$ Hz
7.9: $\dfrac{1}{1000\sqrt{\pi}}$ F
7.13: 500 Hz, 0.1
7.14: $\dfrac{1}{1 + j(f/400)}$
7.16: $3\cos(2\times 10^5\pi\cdot t - 120°)$
7.20: $8\cos(2 \times 10^5\pi\cdot t + 150°)$
7.24: $\dfrac{j\frac{f}{4}}{1 + j\frac{f}{4}}$
7.28: 200 Hz, $\dfrac{j(f/200)}{1 + j(f/200)}$
7.32: 17.68 to 19.89 nF
7.35: 20
7.38: A series resonance band pass filter; 45 Hz
7.40: 9.9 to 86.9 nF
7.46: $\dfrac{1}{1 + j(f/500)}$

Chapter 8

8.2: (c)
8.6: Two
8.11: (a) 3.65 mA; (b) −1.71 mA
8.14: 0.0229 A
8.18: (a) 3.3 V, 5.3 V; (b) −0.2 A
8.20: If $v_S < 21$ V, $v_O = 0$; else $(93.75v_S(t) - 328.125)$ mV.
8.21: 41.67 mA; 41.67 mA; 0
8.27: 6.4 V
8.29: $R < 25.5\ \Omega$
8.31: 516.67 μF; 5.24
8.35: (c)
8.37: 99.6
8.40: 27.4 kΩ
8.44: 1 V
8.48: −0.857; −18.57; 75 kΩ; 3 kΩ
8.53: 2.93 V
8.64: NAND gate
8.65: AND gate
8.70: (a) 4; (b) 16 V
8.71: 8 V
8.73: $\dfrac{V_O}{V_S} = -\left[\dfrac{R_2}{R_1} + \dfrac{R_3}{R_1} + \dfrac{R_3}{R_1} \times \dfrac{R_2}{R_4}\right]$
8.75: 328.8 mA
8.78: $V_{out}(t) = -\dfrac{1}{RC}\displaystyle\int_0^t V_{in}(\tau)d\tau$

Chapter 9

9.3: 165.4 μF; 55.14 μF	**9.5:** 9.68 kW; 29.04 kW	**9.9:** 8 μF	**9.12:** 960 W
9.14: $I_A = 12$ A, $I_B = 24$ A, $I_C = 22.28$ A	**9.18:** 30 Ω	**9.20:** 14 mH	**9.25:** (a) 15248.11 V; (b) add a capacitive load to make phase current 50.2 A
9.27: $P_2 = P_1$, resistive; $P_2 > P_1$, inductive; $P_1 > P_2$, capacitive	**9.30:** $325.5 \times 10^6 \Omega$.m; 1618.25 Ω	**9.34:** 152410 $\angle -49.41°$ V	
9.38: 9.55 Ω	**9.42:** 3.37 m	**9.45:** 1.13 Ω	**9.48:** 11.33 pF/m; 10.89 pF/m
9.52: 15			

Chapter 10

10.3: 139	**10.6:** 13.8125	**10.13:** 28D.D3	**10.19:** 8130
10.23: E9.D	**10.24:** 1333.0302	**10.32:** 1	**10.35:** $\overline{A} + (\overline{B} \cdot C)$
10.38: 1	**10.41:** A	**10.42:** $\overline{A} \cdot B$	**10.45:** (d)
10.48: $A \cdot B + C$	**10.51:** $(A + B)(\overline{CD})$	**10.54:** $\overline{A} \cdot \overline{B} + \overline{C \cdot D} + C \cdot D \cdot B \cdot C + C \cdot D \cdot \overline{C} \cdot \overline{B}$	
10.58: $D = C + A.B$	**10.60:** 4	**10.69:** 1	**10.70:** 0
10.71: (c)			

Chapter 11 Partial Answers

11.3: (b)	**11.6:** (a)	**11.9:** (c)	**11.11:** 1,500
11.13: $k = 1.6$; $b = 0.2$	**11.15:** 18 μF	**11.20:** 22.5 m/s	
11.24: $v_o(t) = 1.252 \cos(1000\pi \cdot t + 264.61°)$		**11.28:** from 1,000 to 2,000	
11.30: (d)	**11.35:** $N \geq 6$	**11.38:** $N = 9$	**11.42:** $N >= 12$
11.46: 15.97 V	**11.48:** 0.1 V		

Chapter 12

12.3: 1.39×10^{-12} N	**12.5:** 2.5 kA	**12.7:** 30°	**12.11:** 0.1875 sin (200t) mWb
12.17: 2,100 A·t; 15.834 mWb	**12.19:** 0.7927 A	**12.21:** 3.519 mWb	**12.24:** 39.68 mH; 6.35 mH
12.28: 1,000	**12.31:** (a) 315 W; (b) 97%	**12.33:** 15.178	

Chapter 13

13.4: 42.44 Nm; 133.3 V

13.7: 50; 1.25

13.9: 1,030 rad/min

13.11: 141 V

13.13: $\omega = \dfrac{0.02k_T V_S + 900R}{(0.02k_a k_T + R)}$.

13.15: 6 Ω; 1 Nm

13.17: 948.33 rpm

13.20: 437.5 Nm

13.24: 0.275 A

13.27: 40 Ω; 0.5 Ω

13.30: 23.08%

13.33: 375.5 V

13.36: 182.18 A

13.40: 0.05

13.46: 0.528; 28.83 Hz

13.50: 4,197 Nm; 23.08%

13.56: $\phi = \dfrac{5}{2\pi}$ Wb

Chapter 14

14.1: Series: 5,450– 6,550 Ω; Parallel: 784–882 Ω.

14.5: Min: 1.72; max: 2.27

14.7: 0.71 V; 0.5 V

14.10: 0.16 Ω

14.13: 6.25%

14.16: Point 1

14.19: 1 V

14.23: 4 V; 2 V.

14.25: 400 mV; 75 mV

14.30: 0.03 W

14.34: No

14.37: $v(t) = 0.5 + 0.7 \sin(2\pi \times 10000 \times t + \theta)$

14.40: 15 mV_{pp} ~ 15 V_{pp}

Chapter 15

15.1: 1.2 kΩ

15.3: (d)

15.6: It is fatal

15.10: 1/200 years

15.11: A(a), B, C(b), (c)

15.15: 795.77 W/m^2; 8.92 m

15.17: 266.7 mG; not safe

15.22: Safe

15.25: 3.6 MW; 102.9 kW

15.30: 1.18 J

15.35: 3.75 mA, 2.25 mA, 1.5 mA; it does trip

15.38: No; yes

15.41: (c)

INDEX

M

N

O

P